CARPENTRY AND
BUILDING CONSTRUCTION
THIRD EDITION

Carpentry
and Building Construction

THIRD EDITION

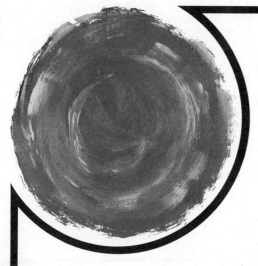

JOHN L. FEIRER
and
GILBERT R. HUTCHINGS

Bennett & McKnight
a division of
Glencoe Publishing Company

Glencoe Publishing Company
17337 Ventura Boulevard
Encino, CA 91316

10 9 8 7 6 5 4

ISBN 0-02-667390-8

Fourth Printing, 1989

Library of Congress Catalog Number 84-63103

Printed in the United States of America

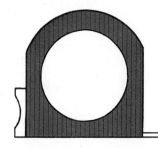

PREFACE

CARPENTRY AND BUILDING CONSTRUCTION has been designed as a basic text for students in industrial education classes at the high school, vocational school, and the community junior college levels. It has been written to prepare the learner to do quality work on the job in building construction. This book will also be helpful in apprenticeship training and for do-it-yourselfers in building and remodeling.

Wood is a remarkable space-age material. It possesses the same fundamental composite structure as some of the best man-made materials developed for space flights, yet it is a renewable material, one that this country need never deplete. Wood can be easily worked, with relatively simple equipment. With such advantages, it is natural that wood remains the basic material for seven of every ten homes built in the United States.

Craftspersons in the building trades represent the largest single group of skilled workers in our nation's labor force, comprising about three of every ten skilled workers. Carpenters alone number nearly one-third of all construction trade workers. Carpenters are actually in short supply and will continue to be in the years ahead. Therefore, it is important for schools at secondary level and above to consider adding a program in carpentry and building construction.

In preparing this text, the authors have reviewed courses of study from schools, trade unions, and public agencies to determine what should be included. They have also carefully reviewed technical materials from companies and other organizations involved in building construction. Based on this material, they have composed a text that is current and technically correct. Each area contains concise descriptions of materials and techniques. The book is generously illustrated with drawings and photographs to give the reader a thorough grasp of materials and techniques, technical know-how and related information. The text contains much current information that is not found in similar books, including a unit on manufactured housing and recreational vehicles.

This book can be used in many ways. It is an ideal text for anyone who is building a home on site. It can also serve as a course of instruction for students building sectional homes or modules in a shop or laboratory. Where physical facilities are limited, the text can be used to build a model home to scale.

Students who follow the instructions in this text should achieve the following behavioral objectives:

• To know the career opportunities in the building trades.

• To understand the basic materials used in residential home construction.

• To be able to read prints and technical material necessary to build a home.

• To demonstrate competency in the use of hand and machine tools used in house construction.

• To do acceptable workmanship in building construction, based on recognized building codes.

The many companies and individuals who have contributed to this publication are recognized in the list of acknowledgments.

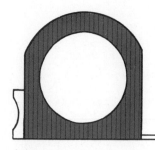

ACKNOWLEDGMENTS

PERSONAL

Susan M. Hutchings
Daniel W. Irvin
John Polderman
William K. Purdy
Walter C. Schwersinske

**COMMERCIAL, PROFESSIONAL,
AND GOVERNMENTAL**

The Abitibi Corporation
Acorn Products Company
Allied Chemical, Barrett Division
Alsco, Incorporated
Aluminum Company of America
American Concrete Institute
American Forest Products Industries,
    Incorporated
American Hardboard Association
American Ladder Institute
American Olean
American Optical Corporation
American Plywood Association
American Society of Heating,
    Refrigerating, and Air-Conditioning
    Engineers
Amerigo
Andersen Corporation
Armstrong Cork Company
Atlas
Automated Building Components,
    Incorporated
Baker-Roos, Incorporated
Benjamin Moore and Company
Berger Instruments
Berry Industries, Incorporated
Bethlehem Steel
Billings
Bird and Son, Incorporated
The Black & Decker Manufacturing
    Company

Boice-Crane Company
Bostitch, Incorporated
Boston Chamber of Commerce
Bowater Board Company
British Columbia Lumber Manufacturers
    Association
Bruce Oak Flooring
Buck Brothers, Incorporated
California Redwood Association
Canadian Wood Council
Caterpillar Tractor
The Celotex Corporation
Clark Equipment
Clausing Corporation
Condon-King Company, Incorporated
Consumers Power Company
Continental Homes
Copper Development Association
Council of the Forest Industries of
    British Columbia
Crawford Door Company
Curtis Companies, Incorporated
E. I. du Pont de Nemours and Company,
    Incorporated
Delta
Desa Industries, Incorporated
De Walt
Dexter Lock Company
Disston Saw Company
The Donley Brothers Company
Douglas Fir Plywood Association
Dow Chemical Company
Feather-Lite Manufacturing Company
Federal Pocket Door Frame Company
The Flintkote Company
Formica Corporation
General Electric Company
Georgia-Pacific Corporation
Grand Rapids Sash and Door Company
Greenberg-May Prod. Inc.

Greenlee Brothers and Company
Harr, Hedrich-Blessing
H C Products Company
Heart Truss and Engineering
Independent Nail and Packing Company
Inland Steel Company
International Conference of Building
 Officials
International Harvester Company
Irwin
The I-XL Furniture Company,
 Incorporated
Johns-Manville
Johnson-Howard Lumber Company
Kemper Brothers, Incorporated
Koppers Company, Incorporated
Lennox Industries, Incorporated
Los Angeles Building and Construction
 Trades Council
Macklanburg-Duncan Co.
Manor House Cupolas
Manpower Magazine
Maple Flooring Manufacturers
 Association
Marvin Windows
Masonite Corporation
McKee Door Company
Memphis Moldings, Incorporated
Metric Monitor
Miller Lumber Company
Millers Falls Company
Millwork Supply Company
Minnesota & Ontario Paper Company
Mobile Home Manufacturers
 Association
Modernfold Industries
Morgan Company
Mueller Brass Company
Mutschler Brothers Company
National Association of Home Builders
 Research Foundation, Incorporated
National Forest Products Association
National Gypsum Company
National Hardwood Lumber Association
National Lumber Manufacturers
 Association
National Manufacturing Company
National Oak Flooring Manufacturers
 Association

National Research Council
National Woodwork Manufacturers
 Association
Nautilus Industries, Incorporated
Nord
Nordahl Sliding Door Pockets
Norton Company Construction Products
 Division
Nu-Wood Sheathing
Oliver Machinery Company
Overhead Door Corporation
Owens Corning Fiberglas Corp.
Paslode Company
Patent Scaffolding Company
Frank Paxton Lumber Company
Pease Company
Permabilt Manufactured Homes,
 Incorporated
Pittsburgh Plate Glass Company
Porta-Table Corporation
Porter-Cable Machine Company
Portland Cement Association
Powermatic, Incorporated
Raynor Manufacturing Company
Red Cedar Shingle & Handsplit Shake
 Bureau
Redman Industries, Incorporated
Remington Arms Company,
 Incorporated
Republic Steel Corporation
Research Products Corporation
Reynolds Aluminum Company
Rock Island Millwork
Rockwell Manufacturing Company
Rolscreen Company
Russell Jennings
Safway Steel Products, Inc.
St. Regis Paper Company, Panelyte
 Division
Sanford Roof Trusses
H. J. Scheirich Company
Scholz Homes, Incorporated
Scovill, Caradco Division
Sears Roebuck and Company
Senco Products, Incorporated
Shakertown Corporation
The Sherwin-Williams Company
Shopmate
Simonds Saw and Steel Company

Simplex Products Group
Simpson Timber Company
Southern Pine Association
Spotnail, Incorporated
Stanley Iron Works, Incorporated
Stanley Tools
Sterling Homes, Incorporated
Structural Clay Products Institute
Textron Company
Timber Engineering Company
United States Department of Agriculture
United States Department of
Commerce, National Bureau of
Standards
United States Forest Service, Forest
Products Laboratory
United States Gypsum Company
United States Plywood Corporation
United States Savings and Loan
League

United States Steel Corporation
The Upson Company
Van Mark Products Corporation
Vega Industries, Incorporated
Weiser Company
Western Electric Company
Western Red Cedar Lumber
Association
Western Wood Moulding and Millwork
Producers
Western Wood Products Association
Westinghouse
Weyerhaeuser Company
Woodbridge Ornamental Iron Company
Wood Conversion Company
Wood-Mode Kitchens
Wood-Mosaic Corporation

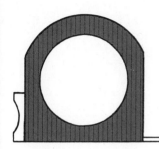

TABLE OF CONTENTS

SECTION I. Introduction   16

SECTION II. Materials   62

Contents

SECTION IV. Foundations   **202**

SECTION V. Framing   **290**

Contents

SECTION VII. Completing the Interior   **708**

INTRODUCTION

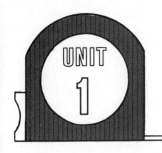

Careers in Construction

THE CONSTRUCTION INDUSTRY

The construction industry directly or indirectly employs about 15% of our working population. This means that the chances are one in six or seven that you will work in some area of building construction, which is one of America's largest industries. Fig. 1-1. Your career may be that of a carpenter, architect, drafter, a lumberyard salesperson or any of the hundreds of other building construction opportunities. Fig. 1-2.

Building construction is divided into two major divisions: *light* and *heavy* construction. Light construction includes homes, farm structures and small churches, schools, and industrial buildings. Heavy construction deals with such projects as roads, major government installations, and large shopping centers, churches, schools, factories, and office buildings.

Being a part of the construction industry is a challenging career. Most of these jobs give you a chance to work actively on construction projects. Variety, usefulness, vitality—these are the rewards, along with good pay and good working conditions, for the individuals who make a career in construction.

Every construction project is a team effort requiring people of varied educational backgrounds and skills, including architects, contractors, and workers in the crafts, and business people who supply the materials.

A great deal of light building construction is now done in the factories. There is increased use of prefabricated components such as trusses, prehung doors, and many other items that are built in the factory and installed as complete units at the job site. Even more important, many complete homes (modular form) are now produced in the factory. Such homes have plumbing, heating, and electrical work already installed and are ready for moving to the site.

CAREER LEVELS

In construction there are three career levels that require special training and education: craft, technical, and professional.

1-1. *You should consider a career in building construction since it offers such a wide variety of opportunities. There are many ways to learn more about these occupations, including interviewing a person in the field. This student (left) is talking to a successful interior designer.*

1-2. *Many business and sales personnel also understand building construction.*

CRAFTS

Workers in the building crafts represent the largest group of skilled workers in the United States labor force. Altogether, more than 3.3 million were employed in 1976. This amounts to about 3 of every 10 skilled workers in the total labor force. Fig. 1-3.

There are more than two dozen skilled building trades. Several major trades—carpenter, painter, plumber, pipefitter, bricklayer, operating engineer (construction machinery operator), and construction electrician—each had more than 100,000 workers in a recent year. Carpenters alone numbered over 1,000,000, nearly one-third of all workers in the building crafts. By contrast, only a few thousand workers were employed in each of several trades such as marble setter, terrazzo worker, glazier, and stonemason.

Building trades workers are employed mainly in the construction, maintenance, repair, and alteration of homes and other buildings, highways, airports, and similar structures. Because construction materials

1-3 *This chart shows employment opportunities in the major building craft areas.*

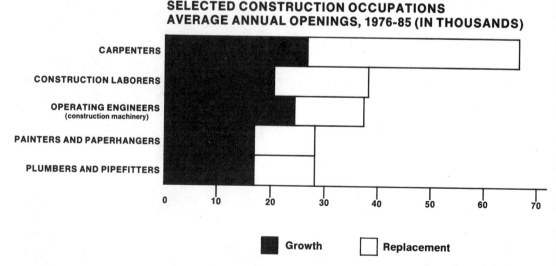

SELECTED CONSTRUCTION OCCUPATIONS
AVERAGE ANNUAL OPENINGS, 1976-85 (IN THOUSANDS)

Source: Bureau of Labor Statistics

and skills are so varied, specialization of work operations has occurred. Thus, building trades workers who use essentially the same materials or skills have tended to become identified with distinct trades. For example, bricklayers and stonemasons both work with masonry materials. Fig. 1-4. Although operating engineers do not work with particular materials, they have related skills that enable them to operate equipment for excavating, grading, and hoisting.

The building trades consist primarily of workers who generally have a high level of skill and a sound knowledge of assembly and construction operations. Most of these workers are union members. In unions, a worker in the crafts is often called a *journeyman*. These workers often are assisted by apprentices, tenders, and laborers.

The work of journeymen may be grouped into three broad classifications: structural, finishing, and mechanical. However, some—carpenters, for example—may do finishing as well as structural work. Generally, each building trade is classified in one of these three categories, as follows:

Occupations mainly concerned with *structural* work—carpenter, operating engineer (construction machinery operator), bricklayer, structural ironworker, ornamental ironworker, cement mason, reinforcing ironworker (who puts reinforcing steel in concrete forms), rigger and machine mover, stonemason, and boilermaker.

Occupations mainly concerned with *finishing* work—lather, plasterer, marble setter, tile setter, terrazzo worker, painter, paperhanger, glazier, roofer, floor covering installer, and asbestos worker.

Occupations mainly concerned with *mechanical* work—plumber, pipefitter, construction electrician, sheet-metal worker, elevator constructor, and millwright.

Since most of the work in this course is directly related to carpentry, a more detailed description of this occupation follows. Carpenters are employed in almost every type of construction activity. They erect the wood framework of the build-

1–4. *Bricklayers must work with various ceramic materials.*

ing, including subflooring, sheathing, partitions, floor joists, studding, and rafters. When the building is ready for trimming, they install molding, wood paneling, cabinets, window sash, doorframes, doors and hardware, as well as build stairs and lay floors. Carpenters, when doing finish work, must concern themselves with appearance as well as structural accuracy of the work.

Carpenters also install heavy timbers used to build docks, railroad trestles, and similar structures. They build the forms needed to pour concrete decks, columns,

1–5. In addition to on-the-job training, carpentry apprentices may spend one day a week in the classroom.

piers, and retaining walls used in the construction of bridges, buildings, and other structures. They erect scaffolding and temporary buildings at the construction site. Carpenters also may install linoleum, asphalt tile, and similar soft floor coverings.

Carpenters saw, fit, and assemble plywood, wallboard, and other materials. They use nails, bolts, wood screws, or glue to fasten materials. Carpenters use hand tools such as hammers, saws, chisels, and planes, and power tools such as portable power saws, drills, and rivet guns.

As mentioned earlier, because of the wide scope of the work performed in the trade, some carpenters specialize in a particular type of carpentry. For example, some carpenters specialize in installing acoustic panels on ceilings and walls; others specialize in the installation of millwork and finish hardware (trimming), laying hardwood floors, or building stairs. Specialization is more common in the large cities; in small communities carpenters often do many types of carpentry. In rural areas, carpenters may do the work of

some other crafts, particularly painting, glazing, or roofing. However, as a general rule carpenters stay in a particular field such as home, bridge, or highway construction, or in industrial maintenance.

For individuals who want to enter construction through trades, apprenticeship training programs have been established to prepare for journeyman work. Apprentices in the building trade are generally required to be between 18 and 25 years of age and in good physical condition. These programs, which are three to four years in length, combine on-the-job training with basic instruction in tools and procedures of the trade. In many cases, the apprentice will actually go to school one day each week for practical classroom and shop instruction. Fig. 1-5. In every instance, whether on-the-job or in the class, the apprentice receives a regular hourly wage which is a percentage of the journeyman's rate. A good way to get started in an apprenticeship is to have a good background in carpentry and building construction in your high school classes or in a vocational school.

Technical Careers

There are many careers in building construction that require additional preparation in a technical institute or community junior college. A technician is one who works on a team with engineers, architects, contractors, and people in the skilled trades. Technical specialists in construction go where the action is. In the field, at construction sites or in the architect's office, the technician has the knowledge and

1-6. *A two-year associate of arts degree in architectural drafting is an excellent way to get into the building construction industry.*

1-7. *The architect must be a creative person. He or she must also understand construction practices and be a good business manager to be really successful. There are an estimated 33,000 registered (licensed) architects in the United States.*

technical skill to do work that lies in an occupational area between the craftsperson and the architect or engineer. Some of the common two-year technician programs are in drafting and general building construction. Fig. 1-6. These programs provide a good background in design, construction practices, and procedures. They may include such courses as surveying, architectural drafting, plan reading, basic estimating, business administration, and building techniques.

Professional Careers

If you are planning to go on to college, there are many opportunities for professional careers related to construction. Some of the most important include architect, interior designer, industrial education teacher, and landscape engineer. Fig. 1-7. All of these professional opportunities require a knowledge of building construction.

YOUR FUTURE IN CONSTRUCTION

There are countless opportunities for promotion in the construction industry. The individual who begins as a carpenter with a journeyman's license can continue to positions such as craft foreman, general foreman, or superintendent. The technical school or community college graduate may start as a drafter and become an assistant engineer, job superintendent, or project manager. College graduates may start as assistant architect, or assistant engineer, and work up to full professional status, eventually becoming a partner or owner of a business.

If you are interested in a career in carpentry or related work, here are some sources you can contact for information:

● Local union of the United Brotherhood of Carpenters and Joiners of America.

- Local carpentry contractors or general contractors.
- Local joint union-management apprenticeship committee.

- Local office of state employment service or state apprenticeship agency.
- Bureau of Apprenticeship, U.S. Department of Labor.

QUESTIONS

1. Approximately what percent of the U.S. working force is involved directly or indirectly in the construction industry?

2. List several typical construction projects and tell whether they are light or heavy construction.

3. Describe the three levels of careers in construction.

4. Approximately how many million people were employed in 1970 in building crafts?

5. What is the largest group of workers in the building trades? How many are employed in this craft?

6. What are the three broad classifications of journeyman work in the building trade?

7. Explain how you can become a carpenter through an apprenticeship program.

8. What are some of the opportunities as a technician in the building industry?

9. How many architects are there in the United States?

UNIT 2

Planning to Build a Home

One of every six United States families looks for a new place to live every year. A family looking for housing has three main choices: to *rent;* to *buy* an existing home (including mobile homes and condominiums); or to *build.* If the choice is either of the last two, then an understanding of building construction is very important. Fig. 2-1. If the family decides to build a home, a great deal of planning must precede the actual construction. Fig. 2-2. Some of the major decisions are as follows.

DETERMINING WHAT TO SPEND

The amount to spend for housing depends on the family's requirements and financial situation. A general rule is that the cost of housing should not exceed 2½ times the average annual income of the family. Not more than 25% of monthly income should be spent for all housing expenses, including mortgage payments, utilities, and repairs. However, some families want to spend more on housing, expecting that their income will rise, and also that it would cost more to buy a home in the future, because of inflation. Others find that due to other heavy expenses it is prudent to spend less. Also, banks are often unwilling to lend as much money as the borrower needs to buy the house he or she wants.

Of the total available for building, approximately 12 to 15% should be budgeted for the lot and the remainder for the house and landscaping. Fig. 2-3.

Selecting and Purchasing a Lot

There are many factors involved in selecting and purchasing a lot that will be suitable for a particular house. Usually it is necessary to look for a lot in a new development or an individual lot in a built-up neighborhood. Fig. 2-4, page 26. The following are points to consider:

• Will the location be suitable for job and recreational needs?

• What are the community facilities and services, including schools, public safety, garbage and trash disposal, water and sewage, hospital and medical services, shopping, and recreational activities? If water and gas lines, sewers, curbs, and similar items are not yet installed, find out if there will be an assessment for them later.

• Lot shape and contour. Is the lot wide enough and deep enough for the house? Is it contoured up or down so that it will require a specially designed home, unlike the average home that can be built on a flat lot?

• Future prospects of the neighborhood. Is it likely to remain relatively stable, or will the nature of the area change as the city grows? Since the home is the largest single investment for most families, it is important to protect it through careful planning.

• Local zoning restrictions. Every city or area is divided into various zones, generally in the following order: single-family dwellings, multiple-family dwellings, apartments (cooperatives or condominiums), light commercial, heavy commercial, light industrial, and heavy industrial. Most families want a lot in a neighborhood that is restricted to single-family dwellings.

• Deed restrictions. Within any zoning area there may be individual deed restrictions on the lot. These may specify the minimum size house that can be built on the lot, the setback allowance from the

2-1. *A knowledge of good construction is valuable to many people—those in the building industry, those in real estate, and anyone who buys a new or existing home.*

street to the first solid wall of the home, the distance from either lot line in which a house cannot be built, and similar points. These are all limiting factors that affect the size and shape of the house that can be placed on a particular lot.

• Local building codes. Before even selecting house plans, it is important to learn the local building codes since this will greatly affect the cost of the house. Building codes indicate the minimum standards of quality and construction demanded. For example, the building code may call for rigid conduit for electrical installation rather than for flexible wiring; this would increase the cost of this part of the home substantially. Also, mortgage and loan associations and the Federal Housing Administration establish certain requirements for home construction which must be met in order to get a mortgage on the home.

The terms *real property* and *real estate* refer to land and the buildings on it. To buy such property it is necessary to be concerned with at least three legal documents.

23

2–2. *Here is a list of products and services that may be used for a new home. Some require decisions before construction begins. Others can be chosen during construction and still others don't need to be taken care of until the house is finished. With so many things to decide, it is wise to reduce confusion by scheduling decisions in an orderly fashion.*

Items	Are we going to have this?	Amount budgeted	Approximate date to order	Arrangements made
Lot Cost				
Attorney Fees or Legal Expenses				
Survey				
Building Permit				
Insurance				
Excavating—Skin, Backfill and Grading				
Basement Footing, Wall and Post Footing, Windows and Taring				
Footing Drain or Tile				
Step and Porch Footing				
Garage Footing and Wall				
Garage Floor				
Basement Floor				
Cement Porches and Walks				
Gravel or Concrete Drive				
Brick Veneer				
Planters—Exterior				
Chimney, Fireplace Flashing				
Subfloor				
Beam, Post, Stairs				
Basic House				
Interior Trim				
Insulation				
Flooring and Underlayment				
Floor Finish				
Kitchen Cabinets				
Kitchen Countertops				
Interior Planters, Dividers and Bookshelves				
Paneling				
Complete Carpenter Labor				
Built-Ins				
Dry Wall or Plaster				
Plumbing with Water Heater				
Heating—Complete				
Electric Wiring and Fixtures				
Linoleum and Tile Floors				
Slate and Ceramic Tile Floors				
Wall Tile				
Sump Pump				
Well or Water Hook-Up				
Septic System or Sewer Hook-Up				
Gas Hook-Up				

(Continued on page 25)

(Continued from page 24)

Items	Are we going to have this?	Amount budgeted	Approximate date to order	Arrangements made
Gutter and Drains				
Painting—Interior				
Painting—Exterior				
Storm Windows and Doors—Installed				
Vanities—with Top				
Bath Accessories				
Window Wells				
Landscaping—Trees, Seed, etc.				
Clean-Up—Building, Yard, Windows				
Heat				
Power				
Interest or Loan Expense				
Closing Costs				
Miscellaneous Extras				

2–3. *Families usually know what they want their house to look like, even before they search for a lot.*

• The *survey*, which shows the boundaries of the property.

• The *deed,* which is evidence of ownership, and by which ownership is transferred.

• The *abstract of title,* a history of the deeds and other papers affecting the ownership of the property, usually for the most recent 60 years.

Before buying a piece of real property, the purchaser should have a survey made to make certain that the property meets building needs. An abstract is usually prepared by a company that specializes in tracing the ownership of property through legal documents. Real estate agents often help the purchaser to obtain the necessary legal papers. Also, buyers and sellers often hire lawyers to examine the abstract and other legal papers connected with the transfer of real property.

When the property is finally purchased, the buyer should retain a copy of the official survey, the abstract, and the deed. All of these are needed to secure financing and building permits.

SELECTING HOUSE PLANS

There are several methods of obtaining house plans for the lot. One way is to get a catalog of stock house plans from a lumberyard. Other good sources are magazines and other publications devoted to such plans. The purchaser looks through the designs until he or she finds one that is suitable. The purchaser can then buy several copies of the complete set of working drawings, materials lists, and specifications. This is the least expensive way to find a house plan.

House plans can also be obtained from local building dealers who have limited architectural service. If the purchaser agrees to buy all the necessary materials from the dealer, he will gladly help the purchaser in selecting and doing some simple redesigning of the house plan. However, if the house is to be completely and individually designed, an architect must be employed. The fee usually ranges from 5 to 10% of the total building cost, depending on whether the architect only designs the house or also supervises its construction.

Still another way to choose house plans is through a company that specializes in precut material. The company will then supply all the precut materials and a contractor to build the house. It is also possible to purchase an entire prefabricated home or a modular home which can be moved to the site.

FINAL ARRANGEMENTS

After deciding on the house and the plans, the next step is to ask for contractor's bids on the cost of construction. To get a rough idea of how much the house should cost, find out the average cost per square foot of residential building in your area. Then multiply this figure by the number of square feet in the plans. Fig. 2-5. However, this cost will vary greatly according to such features as fireplaces, built-ins, and other refinements that may be included in the home. Fig. 2-6. The only way to

2–4. *This new development is well planned with roads and lots of varying sizes. Note also the open space for parks and recreation.*

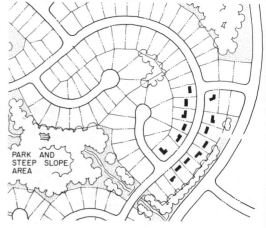

PARK AND
STEEP SLOPE
AREA

get an accurate price estimate is to ask contractors to bid on the cost of construction. It is usually desirable to get two or three bids and to talk to contractors in detail about their bids before deciding who is to build the home. Often the contractor and the architect work in close cooperation, or the contractor and the building and supply company may work together.

After a contractor has been chosen, it is usually necessary to obtain financing from a bank or savings and loan company. If the arrangements meet with the approval of the lending agency, financing up to a certain percent of the total cost can be obtained. This percentage varies, but 80% is not an unusual amount. Fig. 2-7. The borrower must have the down payment from savings or an equivalent value in a house lot. The bank or savings and loan company will arrange a mortgage loan which is really a *lien* on the property: this means that the lender is given some security for the repayment of the money. Usually

the lender can claim the property itself if the borrower does not make payments. Typically a mortgage may run from 12 to 40 years, at the prevailing interest rate. This rate varies but it has been considered high—10% or more—much of the time since the late 1970s. Often the interest cost alone on a mortgage over the years will equal or exceed the actual building cost of the house. The mortgage will often include money for *closing costs*—charges for paper work and similar items needed to make the loan transaction. Mortgages are often arranged so that the monthly payment includes not only amounts for principal and interest on the loan but also amounts for insurance and taxes on the property. The lender holds these funds *in escrow,* for payment when the tax and insurance bills are due.

Once financing has been arranged, contracts must be signed for the construction. From this stage on, it is the responsibility of the contractor and/or architect to make

2-5. *A reasonable estimate of the cost can be found by checking the number of square feet in the layout and multiplying by the average cost of building (per sq. ft.) in your locality.*

2-6. *Such items as this built-in wall add substantially to the cost of a home.*

2-7. *When making final plans for the building, it is a good idea to hire a lawyer. He or she can take care of the legal aspects of contract signing, closing procedures, and taking possession. The lawyer can also work with the bank officials to discuss mortgage arrangements.*

sure that the building goes as planned. The contractor will usually obtain a building permit from the city or county, have the lot cleared of trees and the grading done, and install a temporary electrical hook-up which can be used when building.

Contractors are usually paid a certain portion of the construction cost before the home is started. They are paid again at certain stages of construction, such as when the rough framing is done and the roof is on, and when the exterior is completed. They get their final payment when the house is completed.

QUESTIONS

1. How many families each year look for a new place to live?

2. What are the three common ways of finding permanent housing?

3. In terms of annual income, what is the maximum that should be spent on housing?

4. List several points that should be checked before purchasing a lot.

5. Approximately how much of one's total building budget should be spent on a lot?

6. Are all local building codes uniform? Explain.

7. Describe what a deed is.

8. Why is it important to have a survey made?

9. What is an abstract of title?

10. Name three ways of obtaining house plans.

11. What is the range in cost for architectural services? Explain why these costs vary.

Reading Prints

The ability to read and understand drawings, prints, and plans is basic to all construction. Sketches, drawings, and prints are a kind of language. They tell you everything you need to know to build something, including the materials needed. Fig. 3-1. By means of *lines, symbols,* and *dimensions,* the ideas of the designer or architect are conveyed to you. To be a good builder you must be able to interpret correctly the sketch, drawing, or print so you can visualize the size and shape of the product to be built. Fig. 3-2. It would not be possible to convey this information in any other way.

A *sketch* or *drawing* of something to be built is the original idea put on paper. A *print* is an exact copy of the drawing. In the building industry most prints are called *blueprints;* the paper has a blue back-

3-1. *This floor plan is just one of many drawings you would have to understand if you were to build a home.*

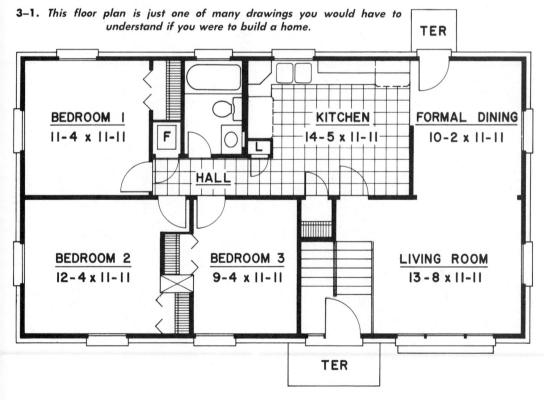

TER

BEDROOM I
11-4 x 11-11

F

KITCHEN
14-5 x 11-11

FORMAL DINING
10-2 x 11-11

L

HALL

BEDROOM 2
12-4 x 11-11

BEDROOM 3
9-4 x 11-11

LIVING ROOM
13-8 x 11-11

TER

ground with white lines. Blueprints are commonly used in house construction and other building trades because they do not fade when exposed to sunlight. They are made on chemically treated paper which shows the drawing in white against a blue background.

An architect, designer, or drafter usually has the responsibility for making the original drawing. Many carpenters also have the ability to make a good sketch. They must often take measurements "on the job" and then make sketches which are sometimes used to do the building or remodeling. At other times the sketches are reviewed, refined, and then made into a set of drawings and prints. Pictorial sketches are sometimes used.

METHODS OF MEASUREMENT*

Two common systems of measurement are used worldwide. The United States currently uses the customary (English) system of measurement while all of the other industrial nations in the world use the metric system. The United States is moving toward the metric system; so you should be acquainted with it. The three common units of measure used are those for *length, liquid measure,* and *weight.* Fig. 3-3a.

In the customary (English) system, lengths are given in inches, feet, yards, and miles. In the metric system, lengths

*A metric edition of this book is available from the publisher.

3–2. *Sometimes isometric sketches are made so that the consumer can get a better picture of what it will look like.*

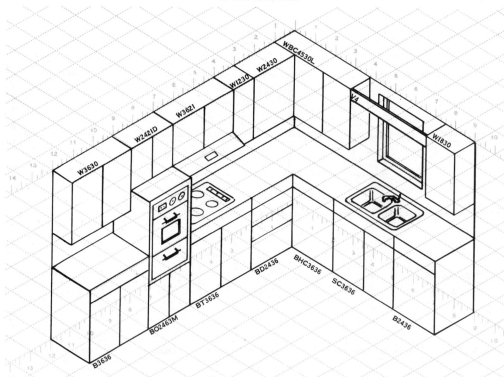

are given in millimetres, centimetres, metres, and kilometres. Fig. 3-3b. A metre, which is the basic unit of length, is slightly longer than a yard (39.37"). Since the entire metric system is based on units of ten, the millimetre is equal to 1/1000 of a metre and a centimetre is 1/100 of a metre. A kilometre is 1000 times a metre. Actually 1" is equal to 25.4 millimetres. The two common length measures used in the customary system for buildings are the inch and the foot, while the only two length measurements used in the metric system are the millimetre and the metre. It is easy to convert from one to the other as you can see in Fig. 3-4. In building construction all millimetre measurements are rounded off to the closest full measurement. For example, 1" is equal to 25 millimetres.

The *liquid* measure in the customary system is in quarts and gallons, while in the metric system it is in litres. A litre is

about 5% more than a quart. Liquid measure for finishing materials including paints is normally given in litres, half-litres and quarter-litres. *Weight* measure in the customary system is given in pounds, while in the metric system it is in kilograms. A kilogram is approximately 2.2 pounds.

In metric countries, particularly Britain, a standard module is 300 millimetres, which is very close to one foot. In architectural drawings, all building measurements are given in millimetres and all site measurements in metres and fractions of a metre. Fig. 3-6a & b, pages 34 and 35.

To get better acquainted with both systems of measurement, it is a good idea to use measuring tools marked both in inches and feet and in millimetres, centimetres, and metres. Some rules are numbered in centimetres, with ten small divisions (millimetres) between each one. However, most rules show measurements in milli-

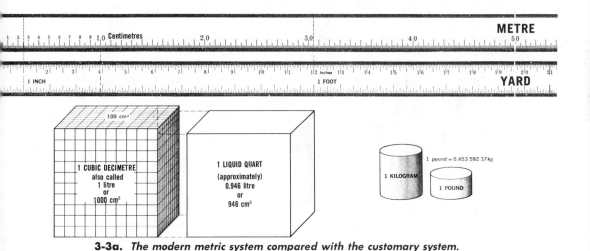

3-3a. *The modern metric system compared with the customary system.*

3-3b *Note that one inch is approximately 25 mm.*

3-4. *A conversion table can be used in woodworking.*

Conversion Table for Woodwork

Customary (English)	Metric				
	Actual	Accurate Woodworkers' Language	Tool Sizes	Lumber Sizes	
				Thickness	Width
1/32 in	0.8 mm	1 mm bare			
1/16 in	1.6 mm	1.5 mm			
1/8 in	3.2 mm	3 mm full	3 mm		
3/16 in	4.8 mm	5 mm bare	5 mm		
1/4 in	6.4 mm	6.5 mm	6 mm		
5/16 in	7.9 mm	8 mm bare	8 mm		
3/8 in	9.5 mm	9.5 mm	10 mm		
7/16 in	11.1 mm	11 mm full	11 mm		
1/2 in	12.7 mm	12.5 mm full	13 mm	12 mm	
9/16	14.3 mm	14.5 mm bare	14 mm		
5/8 in	15.9 mm	16 mm bare	16 mm	16 mm	
11/16 in	17.5 mm	17.5 mm	17 mm		
3/4 in	19.1 mm	19 mm full	19 mm	19 mm	
13/16 in	20.6 mm	20.5 mm	21 mm		
7/8 in	22.2 mm	22 mm full	22 mm	22 mm	
15/16 in	23.8 mm	24 mm bare	24 mm		
1 in	25.4 mm	25.5 mm	25 mm	25 mm	
1¼ in	31.8 mm	32 mm bare	32 mm	32 mm	
1⅜ in	34.9 mm	35 mm bare	36 mm	36 mm	
1½ in	38.1 mm	38 mm full	38 mm	38 mm(or 40 mm)	
1¾ in	44.5 mm	44.5 mm	44 mm	44 mm	
2 in	50.8 mm	51 mm bare	50 mm	50 mm	
2½ in	63.5 mm	63.5 mm	64 mm	64 mm	
3 in	76.2 mm	76 mm full		75 mm	75 mm
4 in	101.6 mm	101.5 mm		100 mm	100 mm
5 in	127.0 mm	127 mm			125 mm
6 in	152.4 mm	152.5 mm			150 mm
7 in	177.8 mm	178 mm bare			
8 in	203.2 mm	203 mm full			200 mm
9 in	228.6 mm	228.5 mm			
10 in	254.0 mm	254 mm			250 mm
11 in	279.4 mm	279.5 mm			
12 in	304.8 mm	305 mm bare			300 mm
18 in	457.2 mm	457 mm full	460 mm		
24 in	609.6 mm	609.5 mm			
36 in	914.4 mm	914.5 mm		**Panel Stock Sizes**	
48 in—4'	1219.2 mm	1220 mm or 1.22 m		1220 mm or 1.22 m width	
96 in—8'	2438.4 mm	2440 mm or 2.44 m		2440 mm or 2.44 m width	

metres only. Whenever possible, use a rule that is numbered in millimetres since these are easier to read. There will be less chance of error.

In architectural layouts, the general practice is to show all residential drawings in millimetres. Site plans are dimensioned in metres.

Building materials come in standard sizes, and most homes are built according to certain dimensional design standards. When the building industry converts to metrics, new standards will replace the customary ones. The design module for building materials will be 100 mm, which is slightly less than four inches.

Softwoods will not change in size, except that dimensions will be given in millimetres. For example, a standard 2 × 4, which measures 1½″ × 3½″, will be listed as 38 × 89 mm. This is a soft conversion. But, all panel stock will undergo a hard

conversion, and the standard size will be 1200 × 2400 mm.

READING AN ENGLISH (INCH) RULE

Rules used in carpentry are one foot (twelve inches), or multiples of one foot up to one hundred feet, in length. (See pages 111-112.) Measurements are usually given in feet, inches, and parts of an inch. You should not find it hard to measure feet in exact inches. You already know that there are twelve inches in a foot and three feet, or thirty-six inches, in a yard.

Let's take a look at the chart shown in Fig. 3-5a. It is in the English system. The distance between 0 and 1 represents one inch. At line A you see that the inch is divided into two equal parts. Each half represents ½″. On a rule this ½″ division line is the longest line between the inch marks. At line B the inch is divided into four equal parts. The first line is ¼″; the second line is ²⁄₄″ or ½″; the third line is ¾″. At line C you will notice that the inch is divided into eight equal parts so that each small division is ⅛″. Two of these divisions make ²⁄₈″ or ¼″ (as shown on line B). Four of these divisions make ⁴⁄₈″ or ²⁄₄″ or ½″. At line D the inch is divided into sixteen parts. This is usually the smallest division on rules used in drawing. Notice again that ⁴⁄₁₆″ is equal to ²⁄₈″ or ¼″. One line past ¼″ is equal to ⁵⁄₁₆″. You will see on your rule that between the inch marks, the ½″ mark is the longest one. The ¼″ mark is the next longest, the ⅛″ mark the next, and the ¹⁄₁₆″ mark is the shortest.

To read a part or fraction of an inch, count the number of small divisions beyond the last inch mark. For example,

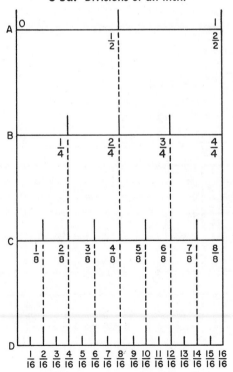

3-5a. *Divisions of an inch.*

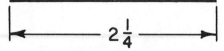

3-5b. *Use a rule to measure this line.*

when measuring the line in Fig. 3-5b, you will find that it is 2" plus four small divisions. This is 2⁴⁄₁₆", which is the same as 2²⁄₈" or 2¼". One small division past ½" would be ⁹⁄₁₆" (⁸⁄₁₆" + ¹⁄₁₆").

SCALE

Drawings must often be reduced from actual size so they will fit on a piece of paper. Care is taken to make such drawings according to *scale;* that is, exactly in proportion to full size. Fig. 3-7. For example, an architect can represent any size of building on a single piece of paper by drawing it to a certain scale. The scale is not a unit of measurement but represents the ratio between the size of the object as drawn and its actual size. If the drawing is exactly the same size as the object itself, it is called a full-size or full-scale drawing. If it is reduced, as most scale drawings are, it will probably be drawn to one of the following common scales.

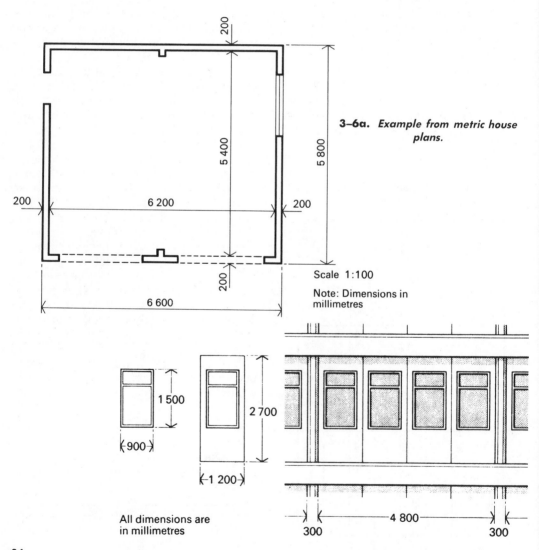

3–6a. *Example from metric house plans.*

Scale 1:100

Note: Dimensions in millimetres

All dimensions are in millimetres

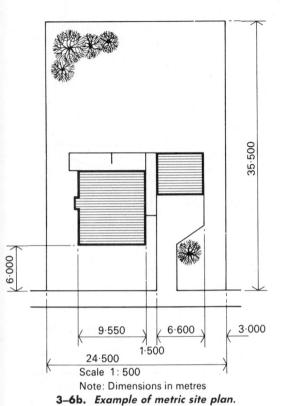

35·500

6·000

9·550   6·600   3·000

1·500

24·500

Scale 1 : 500

Note: Dimensions in metres

3–6b. *Example of metric site plan.*

Customary Scales

6″   equals 1′ (read "six inches equals one foot"): half size.
3″   equals 1′: one-fourth size.
1½″   equals 1′: one-eighth size.
1″   equals 1′: one-twelfth size.
¾″   equals 1′: one-sixteenth size.
½″   equals 1′: one twenty-fourth size.
⅜″   equals 1′: one thirty-second size.
¼″   equals 1′: one forty-eighth size.
3/16″ equals 1′: one sixty-fourth size.
⅛″   equals 1′: one ninety-sixth size.

A scale of ¼″ equals 1′ is often used for drawing buildings and rooms. Detail drawings, which show how parts of a product are made, are prepared to scales of ⅜″, ½″, ¾″, or 1¼″ equal 1′.

Metric Scales

The preferred metric scales are as follows:

1 equals 1: full size   1 equals 1250
1 equals 2: half size   1 equals 2500
1 equals 5
1 equals 10
1 equals 20
1 equals 50

3–7. *Two types of tools for making scale drawings—the architects' scale (for customary measurements) and the metric scale.*

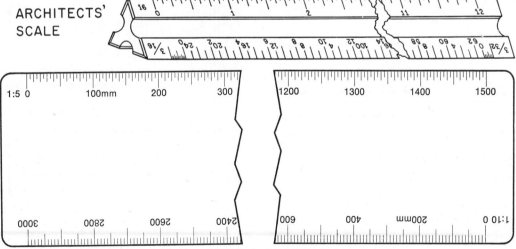

ARCHITECTS' SCALE

METRIC SCALE

ELEMENTS OF DRAWING

A drawing consists of lines, dimensions, symbols, and notes. *Lines* show the shape of a product and include many details of construction. Fig. 3-8. *Dimensions* are numbers that tell the sizes of each part as well as overall sizes. The craftsman must follow these dimensions in making the materials list and the layout. *Symbols* are used to represent things that would be impractical to show by drawing, such as doors, windows, electrical circuits, and plumbing and heating equipment. Fig. 3-9. Some drawings also contain *notes* or written information to explain something not otherwise shown. Frequently in these notes *abbreviations* are given for common words.

3-9a. *Symbols for building materials.*

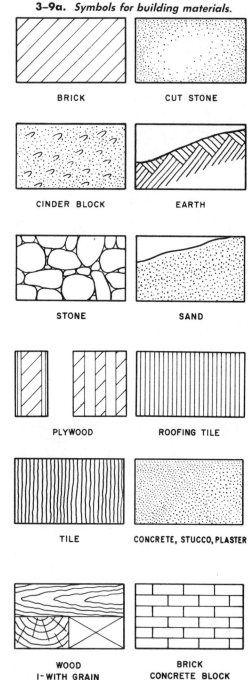

3-8. *Alphabet of lines.*

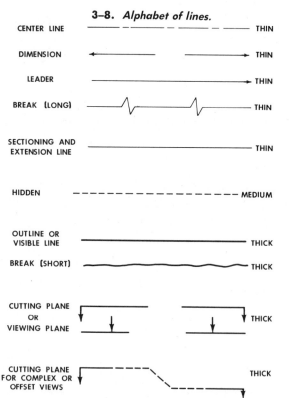

Lines

The lines described below are used for all drawings.

Centerlines. These are composed of long and short dashes, alternately and evenly spaced with a long dash at each end; at intersections the short dashes cross. Very short centerlines may be broken if there is no confusion with other lines.

Dimension Lines. Dimension lines terminate in arrowheads at each end. On construction drawings they are unbroken. On production drawings they are broken only where space is required for the dimension.

Leader Lines. These lines are used to indicate a part or portion to which a note or other reference applies. They terminate in an arrowhead or a dot. Arrowheads should always terminate at a line; dots should be within the outline of an object. Leaders should terminate at any suitable portion of the note, reference, or dimension.

Break Lines. Short breaks are indicated by solid, freehand lines. Full, ruled lines with freehand zigzags are used for long breaks.

Sectioning Lines. Sectioning lines indicate the exposed surfaces of an object in a sectional view. They are generally full, thin lines, but they may vary with the kind of material shown.

Extension Lines. Extension lines indicate the extent of a dimension and should not touch the outline.

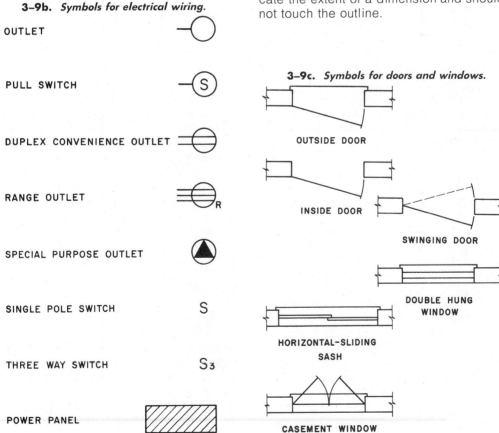

3–9b. *Symbols for electrical wiring.*

OUTLET

PULL SWITCH

DUPLEX CONVENIENCE OUTLET

RANGE OUTLET

SPECIAL PURPOSE OUTLET

SINGLE POLE SWITCH

THREE WAY SWITCH

POWER PANEL

3–9c. *Symbols for doors and windows.*

OUTSIDE DOOR

INSIDE DOOR

SWINGING DOOR

DOUBLE HUNG WINDOW

HORIZONTAL-SLIDING SASH

CASEMENT WINDOW

Hidden Lines. Hidden lines consist of short dashes evenly spaced and are used to show the hidden features of a part. They always begin with a dash in contact with the line from which they start, except when such a dash would form the continuation of a full line. Dashes touch at corners. Arcs start with dashes at the *tangent* points (where they touch each other).

Outline or Visible Lines. The outline or visible line represents those lines of the object which can actually be seen.

Cutting Plane Lines. These lines show where a section has been taken from the building drawings for detail representation.

ARCHITECTURAL WORKING DRAWINGS

Architectural drawings, Fig. 3-10, are prepared as *presentation drawings* or as *working drawings.* Presentation drawings require techniques of pictorial drawing, such as perspective (showing depth) and shading. A construction draftsman is not concerned with presentation drawings. He prepares architectural working drawings consisting of plans, elevations (which show heights), sections and details

3-9d. *Symbols for plumbing fixtures.*

TUB

BATH

STOOL

LAV.

(close-up views), and isometric views. (Isometrics are constructed around three basic lines that form 120° angles. See Fig. 3-2.)

Structural Members

In working with architectural drawings, you will find the following structural members referred to often. Therefore you will need to be familiar with them. Here they are classified according to use.

Footings. Footings rest on soil material and transmit their received load onto the soil. The natural material on which a footing rests is called the *foundation bed.* Footings support columns, piers, pilasters, walls, and similar loads. Usually, footings are made of concrete, although wood or timber may be used.

Vertical Members. Vertical members are in compression; that is, they support loads acting downward at the top. Columns, posts, studs, and piers are those most often encountered.

● *Columns.* Columns may be steel, timber, or concrete. They rest on footings and are the principal load-carrying vertical members.

● *Piers.* Piers are of concrete, timber, or masonry construction. They rest on footings and support horizontal or vertical members. In bridge construction, a pier is an intermediate support for the adjacent ends of two bridge spans.

● *Studs and posts.* Studs are vertical members used in wood-frame construction, spaced close together in walls. Posts are heavier vertical members used in wood-frame construction, usually at corners.

Horizontal Members. Those most frequently encountered are:

● *Joists.* These are lightweight beams spaced four feet or less from the center of one to the center of the next. Joists take the load directly.

● *Beams.* Beams, like joists, take the load of the floor directly, but they are spaced wider than four feet on center.

• *Girders*. Girders take the load of either joists or beams and are generally the heaviest horizontal members in a structure.

• *Lintels*. Lintels are beams which span door or window openings and carry the structure above those openings.

Roof Members. Those most frequently encountered are:

• *Common rafters*. Those members that run square with top plate and extend to the ridge board.

• *Hip rafters*. Those that extend from the outside angle of the plates toward the apex of the roof.

• *Jack rafters*. Those that are square with the top plate and intersect a hip rafter.

• *Valley rafters*. Those that extend from an inside angle of the plates toward the ridge.

• *Cripple rafters*. Those that cut between valley and hip rafters.

• *Purlin*. A timber that supports several rafters at one or more points, or one that supports the roof sheathing directly.

• *Trusses*. Structural members that connect together to span the space between the walls of a building. They support the roof load or floors.

Flooring

Subflooring is laid atop joists or trusses. Building paper is put between subflooring and finished flooring where required.

Sheathing, Siding, and Roofing

Structural members are covered with suitable materials to form the outside walls and the roof. Insulation is placed between sheathing and interior materials.

Utilities

Heating, air conditioning, wiring, and plumbing are the utilities or mechanical systems of a building. They are represented by drawings.

Finishing and Painting

Glazing, plastering, finish trim, and painting complete the building.

PRINCIPLES OF CONSTRUCTION DRAWING

Construction drawings are based on the same general principles as are all other technical drawings. The shape of a structure is described in *orthographic* (multiview) drawings, made to scale. Its size is described by *figured dimensions,* whose extent is indicated by dimension lines, arrowheads, and extension lines. Overall relationships are shown in *general drawings* similar to assembly drawings. Important specific features are shown in *detail drawings* usually drawn to a larger scale than the general drawings. Additional information about size and material is furnished in the specific and general *notes.* If you are familiar with other types of working drawings, you will find obvious similarities in construction drawings. However, there are certain terms and uses of drawings that are found only in the construction field. Chiefly these are related to the materials and methods of construction and the conventional practices of construction drawing.

Views in Construction Drawings

The views of a structure are presented in general and detail drawings. General drawings consist of plans and elevations; detail drawings are made up of sectional and specific detail views.

PLANS

A plan is a top view—a projection on a horizontal plane. Several types of plan views are used for specific purposes, such as site plans, foundation plans, and floor plans.

Site Plan. A site plan shows the building site with boundaries, contours, existing roads, utilities, and other physical details such as trees and buildings. Site plans are drawn from notes and sketches based upon a survey. The layout of the structure is superimposed on the contour drawings, and corners of the structure are located by reference to established natural objects or other buildings.

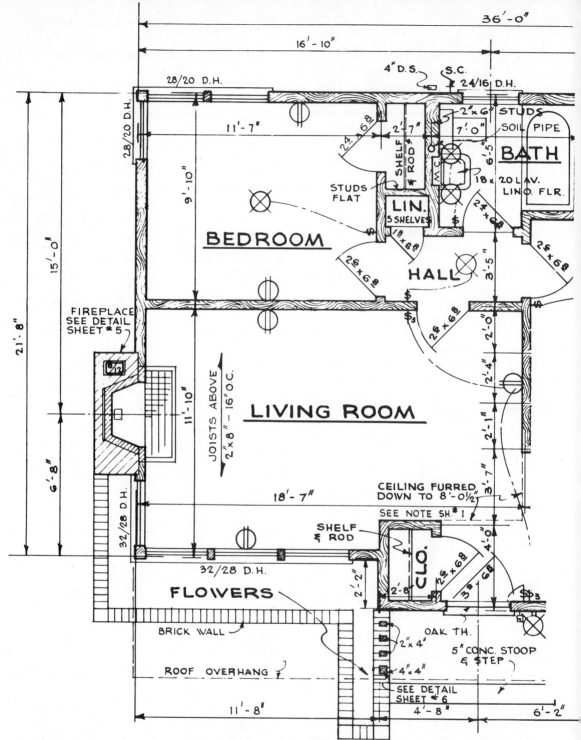

3-10a. *A complete set of working drawings for building a home.*

FLOOR

SCALE:

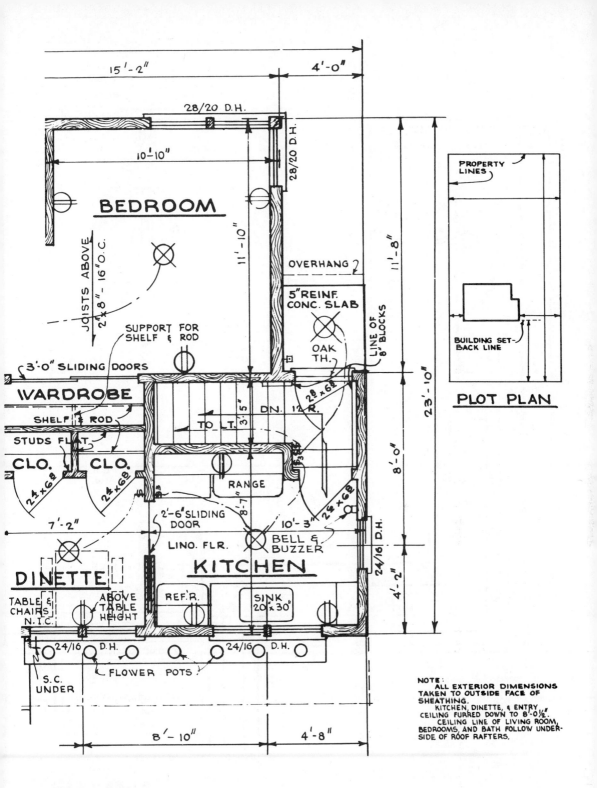

PLOT PLAN

PROPERTY LINES

BUILDING SET-BACK LINE

PLAN

¼" = 1'-0"

15'-2" 4'-0"

28/20 D.H.

10'-10"

28/20 D.H.

BEDROOM

11'-10"

11'-8"

JOISTS ABOVE 2"x8" - 16" O.C.

SUPPORT FOR SHELF & ROD

OVERHANG

5" REINF. CONC. SLAB

LINE OF 8" BLOCKS

OAK TH.

23'-10"

3'-0" SLIDING DOORS

WARDROBE

SHELF & ROD

STUDS FLAT

3'-5"

DN. 12 R.

TO LT.

28 x 68

8'-0"

CLO. 24x68 **CLO.** 24x68

7'-2"

RANGE

8'-7"

24 x 68

10'-3"

2'-6" SLIDING DOOR

LINO. FLR.

BELL & BUZZER

24/16 D.H.

4'-2"

DINETTE

TABLE & CHAIRS N.I.C.

ABOVE TABLE HEIGHT

REF. R.

KITCHEN

SINK 20 x 30

24/16 D.H.

24/16 D.H.

FLOWER POTS

S.C. UNDER

8'-10" 4'-8"

NOTE:
ALL EXTERIOR DIMENSIONS TAKEN TO OUTSIDE FACE OF SHEATHING.
KITCHEN, DINETTE, & ENTRY CEILING FURRED DOWN TO 8'-0½".
CEILING LINE OF LIVING ROOM, BEDROOMS, AND BATH FOLLOW UNDERSIDE OF ROOF RAFTERS.

41

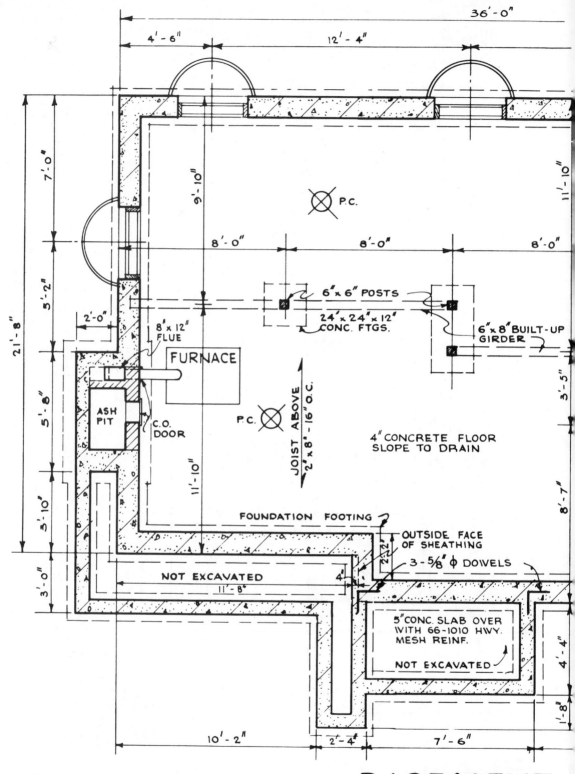

36'-0"

4'-6"   12'-4"

7'-0"

9'-10"

11'-10"

P.C.

5'-2"

8'-0"   8'-0"   8'-0"

21'-8"

2'-0"

6"x 6" POSTS

8"x 12"
FLUE

24"x 24"x 12"
CONC. FTGS.

6"x 8" BUILT-UP
GIRDER

FURNACE

3'-5"

5'-8"

ASH
PIT

C.O.
DOOR

P.C.

JOIST ABOVE
2"x 8"- 16" O.C.

4" CONCRETE FLOOR
SLOPE TO DRAIN

11'-10"

8'-7"

3'-10"

FOUNDATION FOOTING

OUTSIDE FACE
OF SHEATHING

3'-0"

NOT EXCAVATED

11'-8"

2'-2"

4"

3-5/8" φ DOWELS

5"CONC. SLAB OVER
WITH 66-1010 HWY.
MESH REINF.

NOT EXCAVATED

4'-4"

1'-8"

10'-2"   2'-4"   7'-6"

BASEMENT

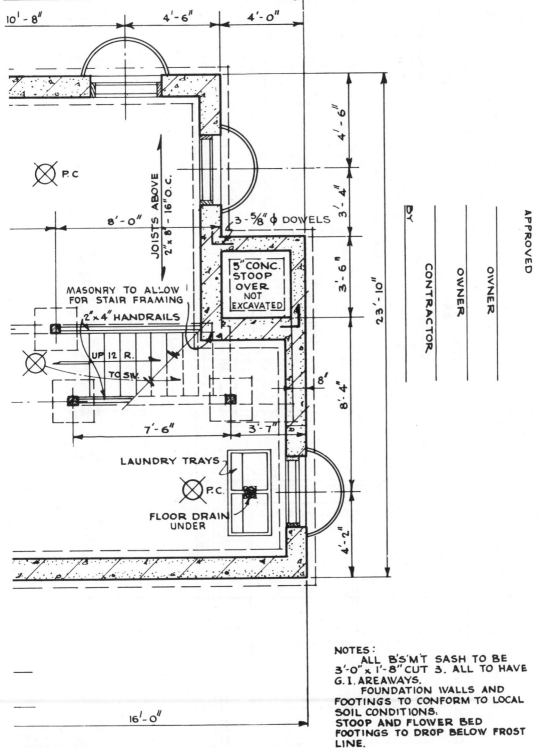

10'-8"   4'-6"   4'-0"

P.C.

8'-0"

JOISTS ABOVE
2" x 8" - 16" O.C.

4'-6"

3'-4"

3-5/8" φ DOWELS

MASONRY TO ALLOW
FOR STAIR FRAMING

2" x 4" HANDRAILS

5" CONC.
STOOP
OVER
NOT
EXCAVATED

3'-6"

23'-10"

UP 12 R.

TO SW.

7'-6"   3'-7"

8'

8'-4"

LAUNDRY TRAYS

P.C.

FLOOR DRAIN
UNDER

4'-2"

16'-0"

BY

CONTRACTOR

OWNER

OWNER

APPROVED

NOTES:
   ALL B'S'M'T SASH TO BE
3'-0" x 1'-8" CUT 3. ALL TO HAVE
G.I. AREAWAYS.
   FOUNDATION WALLS AND
FOOTINGS TO CONFORM TO LOCAL
SOIL CONDITIONS.
STOOP AND FLOWER BED
FOOTINGS TO DROP BELOW FROST
LINE.

PLAN

43

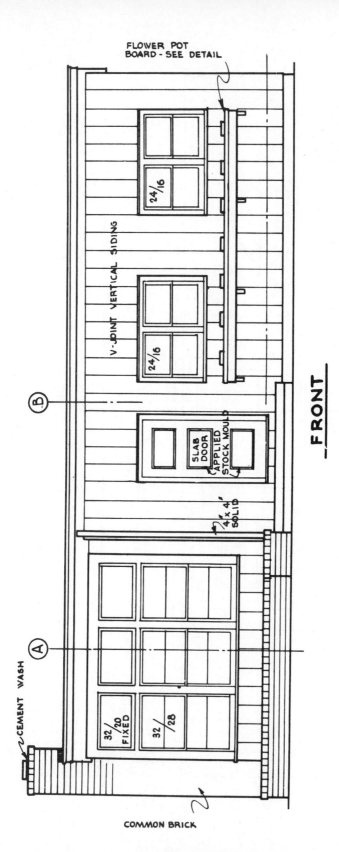

FLOWER POT
BOARD - SEE DETAIL

24/16

24/16

V-JOINT VERTICAL SIDING

B

SLAB DOOR

APPLIED STOCK MOULD

4"x 4" SOLID

A

32/20 FIXED

32/28

CEMENT WASH

COMMON BRICK

FRONT

ELEVATIONS

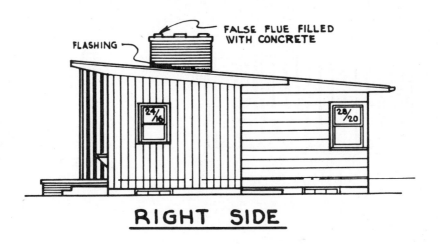

FALSE FLUE FILLED WITH CONCRETE

FLASHING

24/16

28/20

RIGHT SIDE

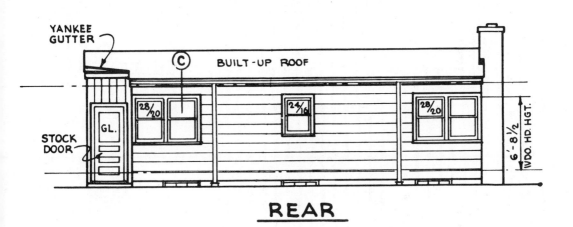

YANKEE GUTTER

© BUILT-UP ROOF

STOCK DOOR

GL.

28/20

24/16

28/20

6'- 8 1/2 WDO. HD. HGT.

REAR

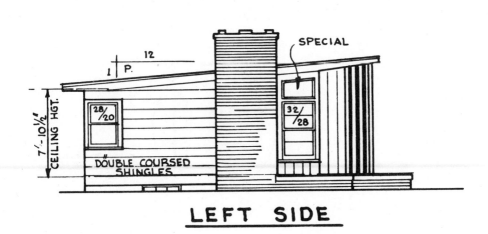

12

1 P.

SPECIAL

7'- 10 1/2" CEILING HGT.

28/20

32/28

DOUBLE COURSED SHINGLES

LEFT SIDE

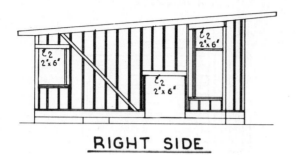

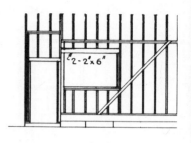

RIGHT SIDE

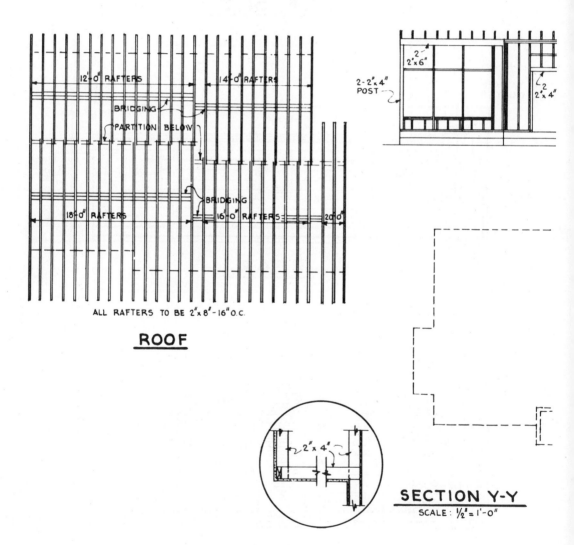

ROOF

SECTION Y-Y

SCALE: $\frac{1}{2}'' = 1'-0''$

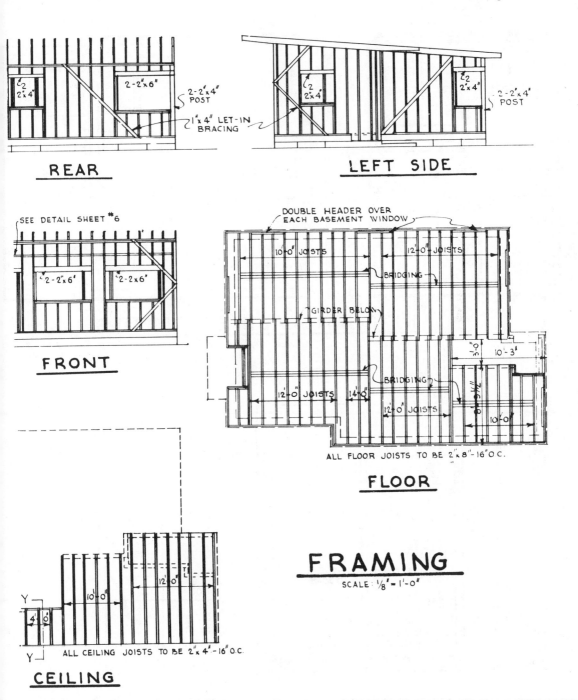

REAR

LEFT SIDE

FRONT

FLOOR

CEILING

FRAMING

SCALE: 1/8" = 1'-0"

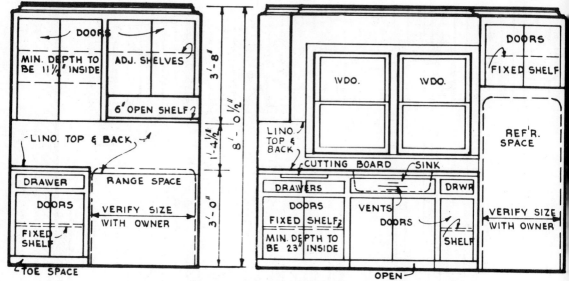

KITCHEN CABINET DETAILS

SCALE: ⅜" = 1'-0"

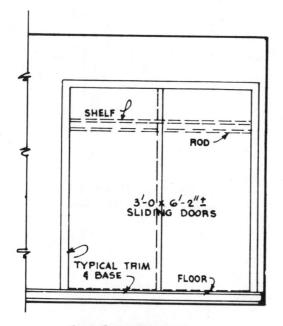

WARDROBE

SCALE: ⅜" = 1'-0"

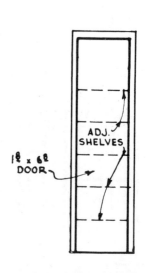

LINEN CLOSET

SCALE: ⅜" = 1'-0"

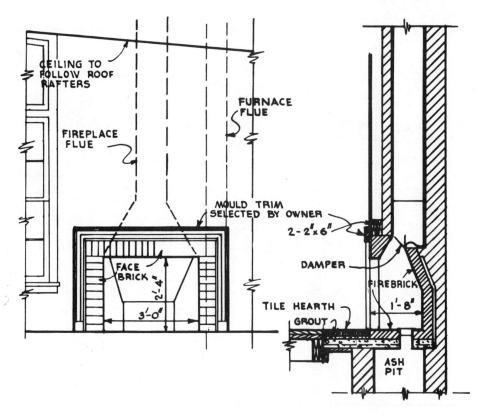

CEILING TO FOLLOW ROOF RAFTERS

FIREPLACE FLUE

FURNACE FLUE

MOULD TRIM SELECTED BY OWNER

FACE BRICK

2'-4"

3'-0"

2 - 2"x 6"

DAMPER

FIREBRICK

TILE HEARTH

GROUT

1'- 8"

ASH PIT

FIREPLACE DETAILS
SCALE : 3/8" = 1'-0"

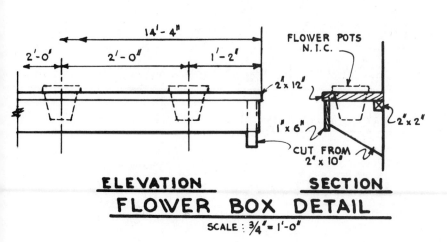

14'- 4"

2'-0" 2'-0" 1'-2"

FLOWER POTS N.I.C.

2"x 12"

1"x 6"

2"x 2"

CUT FROM 2" x 10"

ELEVATION

SECTION

FLOWER BOX DETAIL
SCALE : 3/4" = 1'-0"

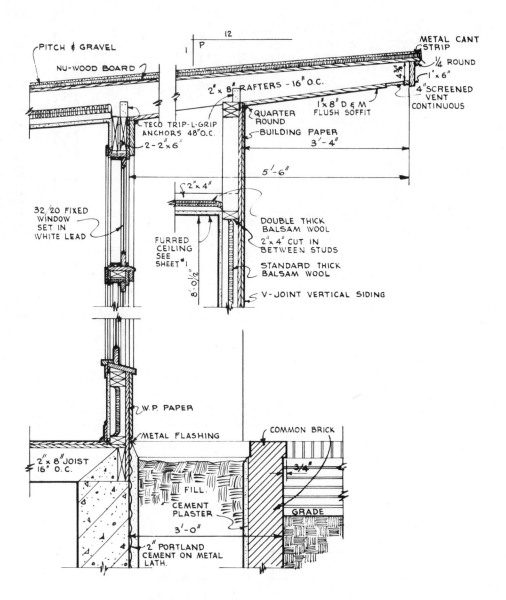

PITCH & GRAVEL

NU-WOOD BOARD

12

P

1

METAL CANT STRIP

¼ ROUND

1" x 6"

4" SCREENED VENT CONTINUOUS

2" x 8" RAFTERS - 16" O.C.

4¾"

1" x 8" D & M FLUSH SOFFIT

QUARTER ROUND

TECO TRIP-L-GRIP ANCHORS 48"O.C.

2 - 2"x 6"

BUILDING PAPER

3'-4"

5'-6"

2" x 4"

32/20 FIXED WINDOW SET IN WHITE LEAD

FURRED CEILING SEE SHEET #1

8'-0½"

DOUBLE THICK BALSAM WOOL

2" x 4" CUT IN BETWEEN STUDS

STANDARD THICK BALSAM WOOL

V-JOINT VERTICAL SIDING

W.P. PAPER

METAL FLASHING

COMMON BRICK

2" x 8" JOIST 16" O.C.

FILL

CEMENT PLASTER

3'-0"

2" PORTLAND CEMENT ON METAL LATH.

GRADE

¾"

SECTIONS A & B

DETAILS

SCALE: ¾" = 1'-0"

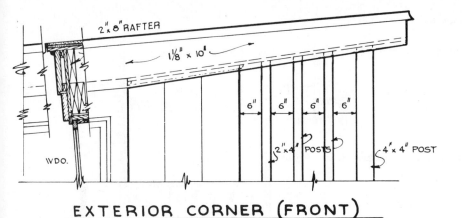

EXTERIOR CORNER (FRONT)

2"x 8" RAFTER

1⅛" x 10"

6" 6" 6" 6"

2"x 4" POSTS

4" x 4" POST

WDO.

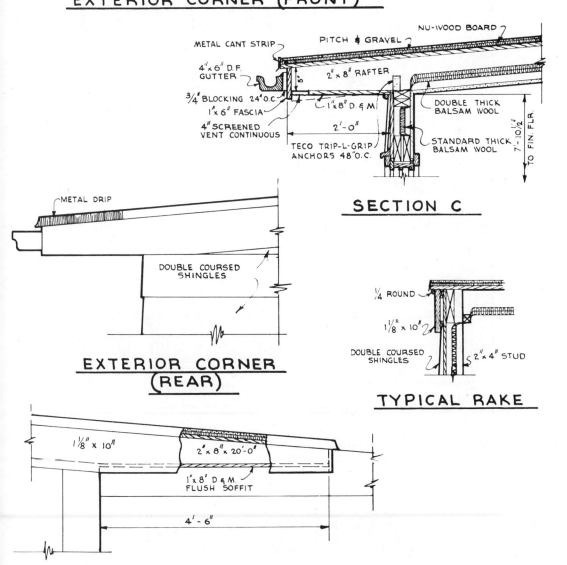

METAL CANT STRIP

PITCH & GRAVEL

NU-WOOD BOARD

4" x 6" D.F. GUTTER

2" x 8" RAFTER

¾" BLOCKING 24" O.C.

1" x 6" FASCIA

DOUBLE THICK BALSAM WOOL

4" SCREENED VENT CONTINUOUS

1" x 8" D. & M.

2'- 0"

TECO TRIP-L-GRIP ANCHORS 48" O.C.

STANDARD THICK BALSAM WOOL

7'- 10½" TO FIN. FLR.

SECTION C

METAL DRIP

DOUBLE COURSED SHINGLES

EXTERIOR CORNER (REAR)

¼ ROUND

1⅛" x 10"

DOUBLE COURSED SHINGLES

2" x 4" STUD

TYPICAL RAKE

1⅛" x 10"

2" x 8" x 20'-0"

1" x 8' D & M. FLUSH SOFFIT

4'- 6"

REAR ENTRY OVERHANG

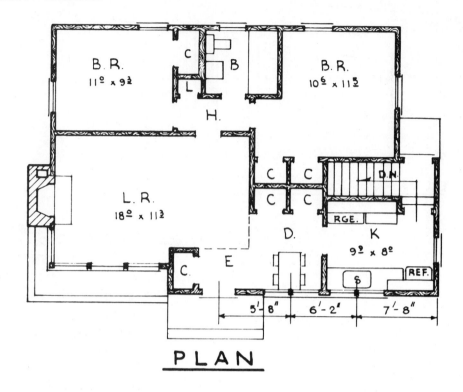

PLAN

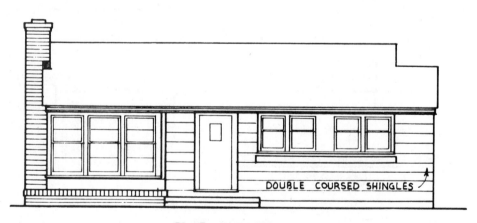

FRONT

SCALE: $\frac{1}{8}'' = 1'-0''$

POSSIBILITY NO. 1

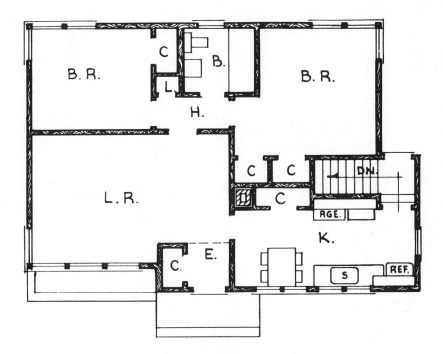

PLAN WITH NO FIREPLACE

THIS PLAN MAY BE USED WITH EITHER EXTERIOR

POSSIBILITY NO. 2

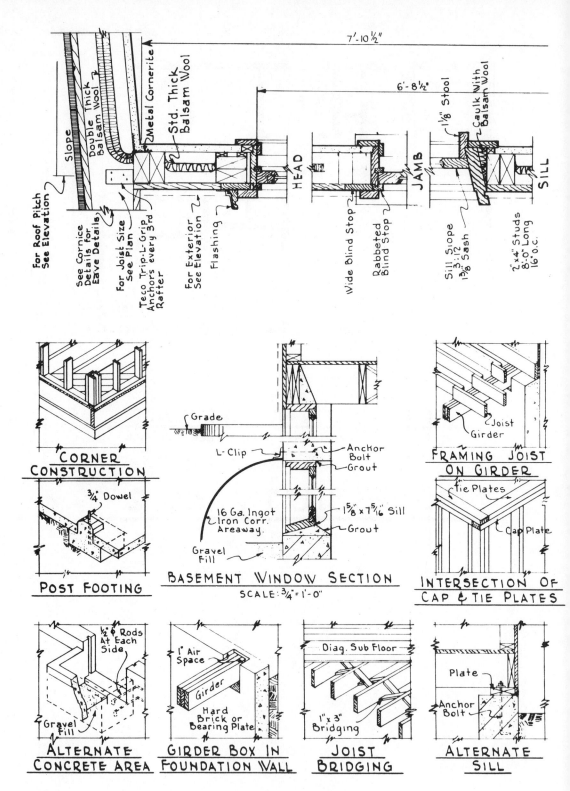

7'-10½"

6'-8½"

For Roof Pitch See Elevation

See Cornice Details for Eave Details

For Joist Size See Plan

Teco Trip-L-Grip Anchors every 3rd Rafter

For Exterior See Elevation

Slope

Double Thick Balsam Wool

2 Metal Cornerite

Std. Thick Balsam Wool

Flashing

HEAD

Wide Blind Stop

Rabbeted Blind Stop

JAMB

1⅛" Stool

Caulk With Balsam Wool

Sill Slope 3":12

1⅜" Sash

SILL

2"x4" Studs 8'-0" Long 16" o.c.

CORNER CONSTRUCTION

POST FOOTING

¾" Dowel

Grade

L-Clip

Anchor Bolt

Grout

16 Ga. Ingot Iron Corr. Areaway.

1⅝" x 7⁵⁄₁₆" Sill

Grout

Gravel Fill

BASEMENT WINDOW SECTION
SCALE: ¾" = 1'-0"

FRAMING JOIST ON GIRDER

Joist

Girder

INTERSECTION OF CAP & TIE PLATES

Tie Plates

Cap Plate

ALTERNATE CONCRETE AREA

½ ⌀ Rods At Each Side

Gravel Fill

GIRDER BOX IN FOUNDATION WALL

1" Air Space

Girder

Hard Brick or Bearing Plate

JOIST BRIDGING

Diag. Sub Floor

1" x 3" Bridging

ALTERNATE SILL

Plate

Anchor Bolt

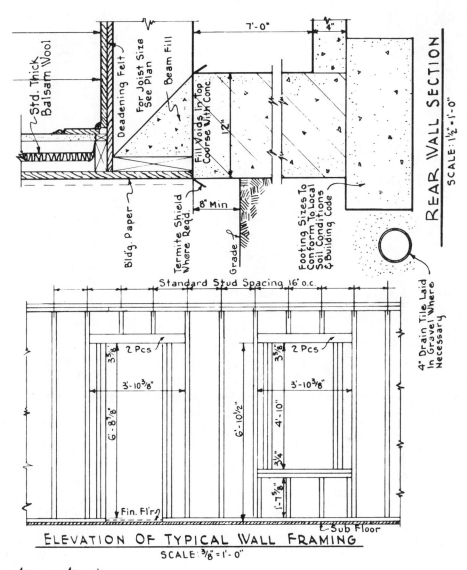

Std. Thick
Balsam Wool

Deadening Felt

For Joist Size
See Plan

Beam Fill

Fill Voids In Top
Course With Conc.

7'-0"

4"

12"

Bldg. Paper

Termite Shield
Where Req'd.

8" Min

Grade

Footing Sizes To
Conform To Local
Soil Conditions
& Building Code

4" Drain Tile Laid
In Gravel Where
Necessary

REAR WALL SECTION
SCALE: 1½"=1'-0"

Standard Stud Spacing 16" o.c.

3⅜"

2 Pcs

3'-10⅜"

6'-8⅛"

3⅜"

2 Pcs

3'-10⅜"

6'-10½"

4'-10"

3¼"

1'-7⅝"

Fin. Fl'r

Sub Floor

ELEVATION OF TYPICAL WALL FRAMING
SCALE: ⅜"=1'-0"

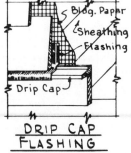

Bldg. Paper

Sheathing

Flashing

Drip Cap

DRIP CAP
FLASHING

3–10b. *This is the finished house shown in the preceding drawings.*

Foundation Plans. A foundation plan is a top view of the footing or foundation walls, showing their area and location by (1) distances between centerlines and (2) distances from reference lines or boundary lines. Foundation walls are located by dimensions from the corner of the building to the wall itself. All openings in foundation walls are shown.

Floor Plan. Floor plans, commonly referred to as plan views, are cross-section views of a building. The horizontal cutting plane is placed so that it includes all doors and window openings. A floor plan shows the outside shape of the home; the arrangement, size, and shape of rooms;

types of materials; thickness of walls and partitions; and the types, sizes, and locations of doors and windows for each story. A plan may also include details of framework and structure, although these features are usually shown on separate drawings called framing plans.

ELEVATIONS

Elevations are external view of a structure; they may be drawn to show views of the front, rear, and right or left side. They correspond to front, rear, and side views in orthographic projections on vertical planes. An elevation is a picturelike view of a building that shows exterior materials and the height of windows, doors, and rooms. It may also show the ground level surrounding the structure, called the *grade*.

FRAMING PLANS

Framing plans show the size, number, and location of the structural members constituting the building framework. Separate framing plans may be drawn for the floors, walls, and roof. The floor framing plan must specify the sizes and spacing of joists, girders, and columns used to support the floor. Detail drawings are added, if necessary, to show the methods of anchoring joists and girders to the columns and foundation walls or footings. Wall framing plans show the location and method of framing openings and ceiling heights so that studs and posts can be cut. Roof framing plans show the construction of the rafters used to span the building and support the roof. Size, spacing, roof slope, and all necessary details are shown.

Floor Framing. Framing plans for floors are basically plan views of the girders and joists. The unbroken double-line symbol is used to indicate joists which are drawn in the positions they will occupy in the completed building. Double framing around openings and beneath bathroom fixtures is shown where used.

Wall Framing. Wall framing plans are *detail drawings* showing the locations of studs, plates, sills, and bracing. They show one wall at a time. Usually they are elevation views.

Roof Framing. Framing plans for roofs are drawn in the same manner as floor framing plans. A draftsman should imagine that he is looking down on the roof before any of the roofing material (sheathing) has been added. Rafters are shown in the same manner as joists.

SECTIONAL VIEWS

Sectional views, or sections, provide important information as to height, materials, fastening and support systems, and concealed features. They show how a structure looks when cut vertically by a plane. The cutting plane is not necessarily continuous but, as with the horizontal cutting plane in building plans, may be staggered to include as much construction information as possible. Like elevations, sectional views are vertical projections. Being detail drawings, they are drawn to large scale. This facilitates reading and provides information that cannot be given on elevation or plan views. Sections may be classified as *typical* and *specific*.

Typical Sections. Typical sections are used to show construction features that are repeated many times throughout a structure.

Specific Sections. When a particular construction feature occurs only once and is not shown clearly in the general drawing, a cutting plane is passed through that portion. The cutting plane is indicated by lines on the general drawing. These lines, which sometimes have arrowheads, have a letter at each end. The same letter is used at each end of the line. These letters then become part of the title of the section drawing—for example, Section Y-Y, on pages 46 and 47, Fig. 3-10a. Thus the cutting-plane lines show the relationship between the general drawing and the section drawing. They show the location of that portion of the general drawing which is represented in the section drawing.

DETAILS

Details are large-scale drawings showing the builders of a structure how its various parts are to be connected and placed. Details do not use a cutting-plane indication, but they are closely related to section drawing because sections are often used as parts of detail drawings. The construction at doors, windows, and eaves is customarily shown in detail drawings. Such drawings are also used whenever the information provided in elevations, plans, and sections is not clear enough for the mechanics on the job. They are usually grouped so that references may be made easily from general drawings.

Dimensioning Construction Drawings

Plan views are dimensioned both outside and inside the building lines. Outside dimensions describe changes and openings in the exterior wall in addition to overall dimension. Inside measurements locate partitions relative to each other and to exterior walls. All horizontal dimensions are shown in a plan view.

Notes in Construction Drawings

Notes in a set of construction drawings are clear, direct statements regarding such matters as materials, construction, and finish. They are included wherever necessary to provide information not clearly indicated by the dimensions. There are two kinds of notes—specific and general.

All notes on the drawings themselves are *specific*. Such notes may add to the dimensioning information or they may explain a procedure or material standard. When more than one line of explanatory notes is placed on a drawing, lower case lettering is used. Titles and subtitles are always in upper case letters. Many terms frequently used on construction drawings are abbreviated to save space.

General notes are usually grouped according to material of construction in a tabular form called a *schedule.*

The notes with a set of construction drawings are so extensive that they cannot all be placed on the drawings themselves. All of the general notes and many specific notes are made into a separate list called the *specifications.* These notes tell the manner in which work will be performed, designate what materials and finishes are to be used, and establish the responsibility of the unit performing the work.

Although it is not a drafter's or a carpenter's job to prepare specifications, these workers should be familiar with such notes. This is because the specifications give detailed instructions regarding materials and methods of work and are therefore an important source of information related to the drawings.

A drafter experienced in preparing construction drawings can be of assistance to the specifications writer. Specifications should be written clearly and briefly. Fig. 3-11.

Bill of Materials in Construction Drawings

A bill of materials is a table of information that tells the requirements for a given project. It shows the item number, name, description, quantity, kind of material, stock size, and sometimes the weight of each piece. Fig. 3-12.

QUESTIONS

1. Why is it important to learn to read prints?
2. What is a blueprint?
3. Define scale.
4. Describe the basic elements of drawing.
5. Name some of the common lines.
6. Name the common views needed in a construction drawing.
7. What is an elevation?
8. What are framing plans?
9. Why are sectional views shown?
10. What are detail drawings and what is their purpose?
11. What are specifications?
12. Define a bill of materials.

3–11. *Specifications for the house shown in Fig. 3–10. This is the first of four pages. Other main topics covered are Miscellaneous Iron, Framing and Carpentry, Sheet Metal, Lath and Plaster, Painting and Finishing, Electric Wiring, Plumbing, and Heating.*

SPECIFICATIONS

The house is to be built for_____Owner,

residing at (Number)_____(Street)_____

_____
               (City or Town)   (County)   (State)

and is to be built upon the Owner's property located as described below:

_____

_____

_____

LOCATION OF HOUSE ON LOT — The location of the house shall be as shown and dimensioned on the Plot Plan included in the Working Drawings.

GENERAL CONDITIONS OF THE SPECIFICATIONS

GENERAL DESCRIPTION OF THE WORK—The Contractor shall supply all labor, material, transportation, temporary heat, fuel, light, equipment, scaffolding, tools and services required for the complete and proper shaping of the work in strict conformity with the Drawings and Specifications. All work of all trades included in the Specifications shall be performed in a neat and workmanlike manner equal to the best in current shop and field practice.

BIDS—In receiving bids for the work specified herein, the Owner incurs no obligations to any bidder and reserves the right to reject any and all bids.

CONTRACT DOCUMENTS—The Contract Documents consist of the Drawings, Specifications, Plot Plan and the Agreement. The Contract Documents are complementary and what is called for by one shall be as binding as if called for by all. The intent and purpose of the Contract Documents is to include all material, equipment, transportation and handling necessary for the complete and proper execution of the work.

PERMITS AND INSPECTIONS—The Contractor shall give all notices, secure and pay for all permits and inspections and shall comply with all laws, ordinances and regulations governing construction, fire prevention, health and sanitation bearing on the conduct of the work.

PROTECTION—The Contractor shall fully and continuously protect all parts of the work from damage, and shall protect the Owner against all loss or injury arising in connection with the execution of the Contract. He shall protect adjacent property as required by law, and shall provide and maintain all passageways, guard fences, lights and other facilities for protection as required by public authority or local conditions. The Contractor shall protect all trees, shrubs, walks and curbs from damage during building operations. The Owner shall provide adequate fire and tornado insurance during construction.

CONTRACTOR'S LIABILITY INSURANCE—The Contractor shall insure himself against claims under Workmen's Compensation Acts and from all other claims for damage for personal injury, including death, which may arise from operations under this Contract, whether such operations be by himself or by any Sub-Contractor or by anyone directly or indirectly employed by either of them. Certificate of such insurance shall be furnished and shall be subject to the Owner's approval for adequacy of protection.

CASH ALLOWANCES—All cash allowances specified shall be included in the Contract sum. If the Owner's selections total more or less than the allowances specified, the Contract sum shall be adjusted accordingly.

EXTRA WORK—Work shall not be started on any item not included in the Plans, Specifications and Contract until the Owner and Contractor agree in writing to the specific quantity and quality intended and to the cost of the extra work. Owner and Contractor shall operate in strict conformity with this requirement for their mutual protection.

CLEANING—The Contractor shall at all times keep the premises free from accumulations of waste materials and rubbish, and at the completion of the work all rooms and spaces shall be left broom clean.

WORK NOT INCLUDED—The following items of work are excluded from the Contract, however, may be included if noted under "Special Items Included."

Blasting	Furniture and Furnishings
Sub-soil Drain	Venetian Blinds
Waterproofing	Window Shades
Driveways and Walks	Refrigerator
Finished Grading, Planting	Cooking Range
and Landscaping	Bathroom Accessories
Fences	Weatherstripping

EXCAVATION AND GRADING

The General Conditions of the Specifications apply to this Section.

WORK INCLUDED—The work under this Section shall consist of furnishing all equipment and performing all necessary labor to do all excavating and rough grading work shown or specified. Excavate to dimensions one foot greater in size than the outside dimensions of the masonry and to the depth required or to solid formation suitable for the foundation. The top soil removed from the excavation shall be stored on the site. Sufficient excavated materials shall be retained to bring the grade up to the necessary level to receive the top soil. If additional earth is required for the rough grading, the Contractor shall furnish it as specified under Special Items Included. Excavation shall be kept free from standing water at all times.

BACK FILLING—The Contractor shall back fill against all walls to the grade line with clean earth well tamped and wetted.

MASONRY

The General Conditions of the Specifications apply to this Section.

WORK INCLUDED—The work under this Section shall consist of furnishing and installing all material and equipment and performing all necessary labor to do all masonry work shown or specified.

FOOTINGS—Footings shall be of concrete mixed in the proportion of 1 part Portland cement, 3 parts of clean, coarse, sharp sand free from loam or vegetable matter, and 5 parts of ¾" gravel. Concrete shall be machine mixed with clean water to the proper consistency and shall be placed immediately after mixing and thoroughly puddled into the forms. Contractor shall check bearing power of soil in all cases and construct footings of sufficient size to conform to local soil requirements and building code.

3–12. *Part of the materials list for the house shown in Fig. 3–10.*

Material List

Lumber							
Item	**Grade & Species**	**Pcs.**	**Size**	**Length**	**F.B.M.**	**Price**	**Amt.**
Framing							
Posts	#1 D.F.	3	6x6	14'0"	126		
	"	3	2x4	10'0"	20		
	"	1	4x4	10'0"	14		
Girders	"	6	2x8	16'0"	128		
	"	3	2x8	8'0"	32		
1st Floor Joists	"	4	2x8	14'0"	75		
	"	30	2x8	12'0"	480		
	"	22	2x8	10'0"	294		
1st Floor Joist Headers	"	2	2x8	16'0"	43		
	"	3	2x8	12'0"	48		
	"	2	2x8	10'0"	27		
	"	3	2x8	8'0"	32		
Flower Pot Support	"	1	1x6	16'0"	8		
	"	1	2x12	14'0"	28		
	"	1	2x2	14'0"	5		
	"	1	2x10	4'0"	7		
Ceiling Joists	"	8	2x4	12'0"	64		
	"	6	2x4	10'0"	40		
	"	3	2x4	4'0"	8		
Rafters	"	3	2x8	20'0"	80		
	"	15	2x8	18'0"	360		
	"	10	2x8	16'0"	214		
	"	10	2x8	14'0"	187		
	"	15	2x8	12'0"	240		
Studs—Exterior	#2 D.F.	94	2x4	10'0"	627		
	"	33	2x4	8'0"	176		
Studs—Partition	"	5	2x6	10'0"	50		
	"	134	2x4	10'0"	893		
Plates	"	1	2x6	14'0"	14		
	"	9	2x4	16'0"	96		
	"	9	2x4	14'0"	84		
	"	21	2x4	12'0"	168		
	"	12	2x4	10'0"	80		
	"	9	2x4	8'0"	48		
Lintels (Wdo. & Door)	#1 D.F.	6	2x6	14'0"	84		
	"	2	2x6	12'0"	24		
	"	5	2x4	8'0"	27		
Diagonal Braces	#3 P.P.	4	1x4	12'0"	16		
	"	2	1x4	14'0"	9		
Stair Horses	#1 D.F.	3	2x10	12'0"	60		
Bsmt. Stair Treads	"	3	2x10	12'0"	60		
Handrail	"	2	2x4	10'0"	13		

(Continued on page 61)

(Continued from page 60)

Material List

Lumber							
Item	**Grade & Species**	**Pcs.**	**Size**	**Length**	**F.B.M.**	**Price**	**Amt.**
Framing							
Grounds	#3 P.P.	1220L′	1x1	Random	102		
Bridging	"	380L′	1x3	"	95		
Sheathing							
Sub-Flooring	#2 D.F. Shiplap		1x8	Random	940		
Wall Sheathing	#3 P.P. Shiplap		1x8	"	1456		
Roof Sheathing	"		1x8	"	1267		

Section II

MATERIALS

Imagine just climbing up & up & round & round. Who knows—you may never stop.

fences make vistas

Sculpture court.

Planting boxes.
2" western red cedar.

Plenty of levels

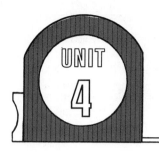

Wood as a Building Material

For all-around utility wood has few if any equals as a building material. Fig. 4-1. Wood's advantages are numerous:

- Wood is light in weight. It is easily shipped and handled.
- Wood is easily worked into various shapes.
- Wood is easily fastened with nails, staples, bolts, connectors, screws, or glue.
- Wooden buildings are easily altered or repaired. Openings can be cut and additions made without difficulty.
- Wood makes a smooth, sanitary interior, holds a decorative coating, and lends itself well to preservative treatments. (See Appendix for information on finishing wood.)
- Wood is strong. Pound for pound, certain common framing woods are actually as strong and stiff as fairly good steel, and stronger than cast iron.
- Wood has low heat conductivity, which helps keep buildings warm.
- Wood resists rusts, acids, salt water, and other corrosive agents better than many other structural materials.
- Wood has high salvage value compared with original cost.
- Wood combines well with other materials.

TREE GROWTH

A tree is a woody plant having one main, self-supporting stem or trunk, at least 10 feet tall at maturity, crowned by leafy boughs. Like human beings trees require food. They need water. They cannot grow without the sun's rays. Fig. 4-2.

The growing, "working" parts of a tree are the *tips of its roots*, the *terminal buds*,

the *leaves,* and a layer of cells just inside the bark, called the *cambium.* It is the cambium layer which produces new wood. Inside the cambium is the sapwood which carries sap from the roots to the leaves. At the center is the heartwood which is not living, but which gives strength and rigidity to the tree. Water from the soil enters a tree through its roots. The water travels

4–1. *Wood is the basic building material used in over 70 percent of the homes built in the United States.*

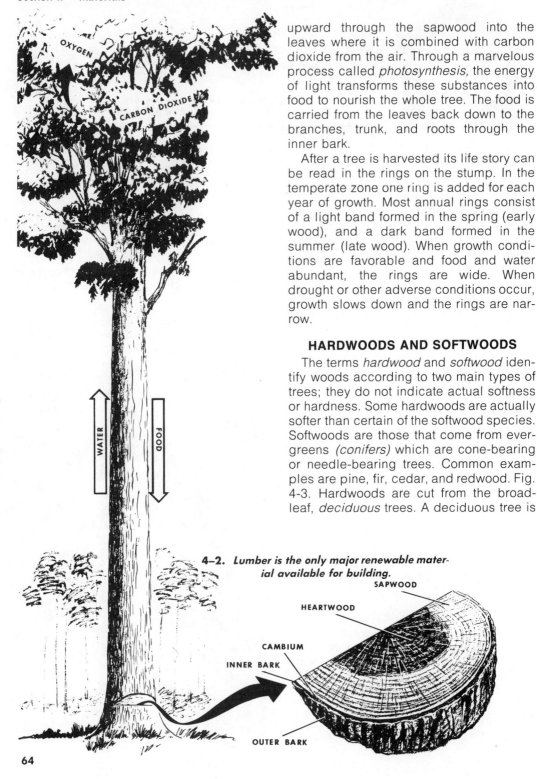

OXYGEN

CARBON DIOXIDE

WATER

FOOD

4–2. *Lumber is the only major renewable material available for building.*

SAPWOOD

HEARTWOOD

CAMBIUM

INNER BARK

OUTER BARK

upward through the sapwood into the leaves where it is combined with carbon dioxide from the air. Through a marvelous process called *photosynthesis,* the energy of light transforms these substances into food to nourish the whole tree. The food is carried from the leaves back down to the branches, trunk, and roots through the inner bark.

After a tree is harvested its life story can be read in the rings on the stump. In the temperate zone one ring is added for each year of growth. Most annual rings consist of a light band formed in the spring (early wood), and a dark band formed in the summer (late wood). When growth conditions are favorable and food and water abundant, the rings are wide. When drought or other adverse conditions occur, growth slows down and the rings are narrow.

HARDWOODS AND SOFTWOODS

The terms *hardwood* and *softwood* identify woods according to two main types of trees; they do not indicate actual softness or hardness. Some hardwoods are actually softer than certain of the softwood species. Softwoods are those that come from evergreens *(conifers)* which are cone-bearing or needle-bearing trees. Common examples are pine, fir, cedar, and redwood. Fig. 4-3. Hardwoods are cut from the broadleaf, *deciduous* trees. A deciduous tree is

4–3. *This list relates the commercial names for lumber to the botanical names of the species from which the lumber may be manufactured. In some instances more than one species is associated with a single commercial name.*

Commercial Names of the Principal Softwoods

Commercial Names for Lumber	Official Common Tree Names	Botanical Names
Cedar:	Alaska-cedar	Chamaecyparis nootkatensis
Alaska Cedar		
Incense Cedar	incense-cedar	Libocedrus decurrens
Port Orford Cedar	Port-Orford-cedar	Chamaecyparis lawsoniana
Eastern Red Cedar	eastern redcedar	Juniperus virginiana
	southern redcedar	J. silicicola
Western Red Cedar	western redcedar	Thuja plicata
Northern White Cedar	northern white-cedar	Thuja occidentalis
Southern White Cedar	Atlantic white-cedar	Chamaecyparis thyoides
Cypress:	baldcypress	Taxodium distichum
	pondcypress	T. distichum var. nutans
Fir:		
Balsam Fir	balsam fir	Abies balsamea
	Fraser fir	A. fraseri
Douglas Fir	Douglas-fir	Pseudotsuga menziesii
Noble Fir	noble fir	Abies procera
White Fir	subalpine fir	Abies lasiocarpa
	California red fir	A. magnifica
	grand fir	A. grandis
	noble fir	A. procera
	Pacific silver fir	A. amabilis
	white fir	A. concolor
Hemlock:		
Eastern Hemlock	Carolina hemlock	Tsuga caroliniana
	eastern hemlock	T. canadensis
Mountain Hemlock	mountain hemlock	Tsuga mertensiana
West Coast Hemlock	western hemlock	Tsuga heterophylla
Juniper:		
Western Juniper	Alligator juniper	Juniperus deppeana
	Rocky Mountain juniper	J. scopulorum
	Utah juniper	J. osteosperma
	western juniper	J. occidentalis
Larch:		
Western Larch	western larch	larix occidentalis
Pine:		
Jack Pine	jack pine	Pinus banksiana

(Continued on page 66)

(Continued from page 65)
Commercial Names of the Principal Softwoods

Commercial Names for Lumber	Official Common Tree Names	Botanical Names
Lodgepole Pine	lodgepole pine	P. contorta
Norway Pine	red pine	P. resinosa
Ponderosa Pine	ponderosa pine	P. ponderosa
Sugar Pine	sugar pine	P. lambertiana
Idaho White Pine	western white pine	P. monticola
Northern White Pine	eastern white pine	Pinus strobus
Longleaf Yellow Pine	longleaf pine	Pinus palustris
	slash pine	P. eliottii
Southern Yellow Pine	loblolly pine	Pinus taeda
	longleaf pine	P. palustris
	pitch pine	P. rigida
	shortleaf pine	P. echinata
	slash pine	P. eliottii
	Virginia pine	P. virginiana
Redwood:	redwood	Sequoia sempervirens
Spruce:		
Eastern Spruce	black spruce	Picea mariana
	red spruce	P. rubens
	white spruce	P. glauca
Engelmann Spruce	blue spruce	Picea pungens
	Engelmann spruce	P. engelmannii
Sitka Spruce	Sitka spruce	Picea sitchensis
Tamarack:	Tamarack	Larix laricina
Yew:		
Pacific Yew	Pacific yew	Taxus brevifolia

one that sheds its leaves annually. Some common hardwoods are walnut, mahogany, maple, birch, cherry, and oak.

METHODS OF CUTTING BOARDS FROM LOGS

There are two common ways of cutting boards. Fig. 4-4. The first is called *plain sawed* (when it is hardwood) or *flat grained* (when softwood). The log is squared and sawed lengthwise (tangent to the annual growth rings). The second method is called *quarter sawed* (when hardwood) or *edge grained* (when softwood). This lumber is not cut parallel to the grain, but sawed so that its rings form angles of 45° to 90° with the surface.

Plain-sawed lumber is usually cheaper, and its defects extend through fewer boards. It is also easier to kiln dry and produces greater widths. However, it has a high tendency to shrink and warp.

Quarter-sawed lumber has low tendency to warp, shrink, and swell, provides a more durable surface, does not tend to twist or cup, and holds paints and finishes better.

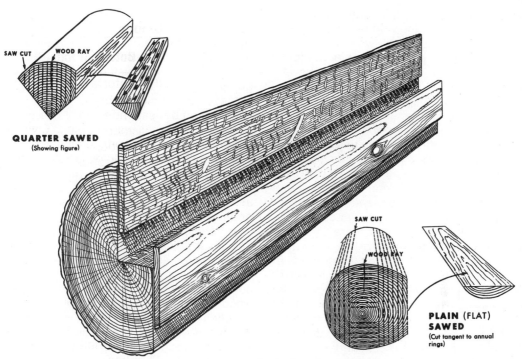

SAW CUT   WOOD RAY

QUARTER SAWED
(Showing figure)

SAW CUT

WOOD RAY

PLAIN (FLAT)
SAWED
(Cut tangent to annual
rings)

4-4. *Common methods of cutting lumber.*

SEASONING

When a tree is cut down, the wood may contain from 30 to 300% more moisture than it will after drying. For example, a cubic foot of oak may contain as much as 14 quarts of water. Fig. 4-5.

There are two methods of seasoning or drying wood: *air drying* and *kiln drying.* Air drying, or seasoning, is done out of doors. The rough lumber is stacked either on edge at an angle or in layers separated by crosspieces called *stickers.* The wood is allowed to remain stacked usually from one to three months and sometimes longer. After correct air drying, the wood should have an average moisture content of 19% or less.

In kiln drying, the lumber is stacked in piles also with stickers between the boards. It is then placed in a *kiln*—an oven in which moisture, air, and temperature are carefully controlled. Fig. 4-6. Steam is applied to the wood at low heat; then the steam is reduced and the heat increased. As the heat increases, moisture is taken out of the wood. Properly kiln-dried lumber has less than 10% moisture content. Green, 1″ lumber can be dried to 6 to 12% moisture content in three or four days in a modern kiln. Moisture content can be checked during the drying process with a moisture meter.

In both kinds of drying, seasoning causes some defects that downgrade the quality of the lumber. Among the most common are checks, honeycombs, warps, loosening of knots, and cracks causes by unequal shrinkage.

MOISTURE CONTENT

Moisture content (M.C.) is a measure of the amount of water contained in wood. M.

C. is expressed as a percentage of the ovendry weight of the wood—the lower the percentage, the drier the wood.

In the drying process lumber tends to shrink, both in width and length. However, shrinkage in length is normally so small that in almost all species it is not considered a problem.

Water exists in green wood in two conditions: *free* in the cell cavities, and *absorbed* in the cell walls. When the cell walls have absorbed all the water they can hold, but there is no water in the cavities, the wood is at the *fiber-saturation point.* Water in excess of this amount cannot be absorbed by the cell walls; therefore it fills the cell cavities. Removal of this free water has no apparent effect upon the properties of wood except to reduce its weight. However, as soon as any water in the cell walls is removed, wood begins to shrink. Since the free water is the first to be removed, shrinkage does not begin until after the fiber-saturation point is reached.

The fiber-saturation point varies from about 23 to 30% moisture content but, for practical purposes, it can be taken as approximately 28% for most woods. Therefore reductions in moisture from natural or green condition down to roughly 28% do not result in shrinkage. Fig. 4-7.

After the fiber-saturation point has been passed and the cell walls begin to give up their moisture, they shrink in all directions and not uniformly. Shrinkage is actually the contraction of cell walls. This process causes a certain amount of shrinkage across the face of wood that is edge grained (vertical) or quarter sawed and about twice as much in plain-sawed lumber. Even though lumber is dried to a certain percentage it will continue to pick up or give off water, depending on the condition of the air. If the air is damp, dry wood will swell, causing many problems. If the air is very dry, the wood will shrink. When the moisture content is in balance with the humidity of the surrounding air

4–5. *Compare the amounts of water in green and dried lumber.*

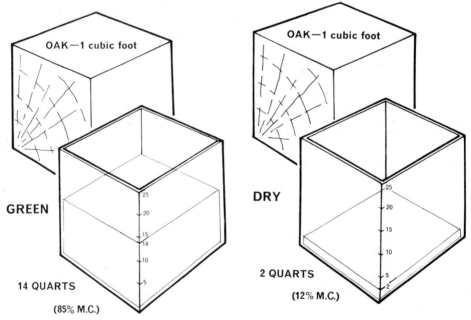

OAK—1 cubic foot

OAK—1 cubic foot

GREEN

DRY

14 QUARTS

(85% M.C.)

2 QUARTS

(12% M.C.)

and neither gains or loses moisture, the wood is said to have *equilibrium moisture content.*

Checking Moisture Content

There are two common ways of determining moisture content. One is by oven-drying methods, and the other is by using a moisture meter.

Use of a moisture meter is the more common method. Fig. 4-8. While the meter readings do not show the same accuracy as those obtained with an oven sample, they are good enough for building construction. These meters provide the only means of checking moisture content rapidly. This is often necessary when lumber is being processed through a manufacturing plant, as in the production of mass-produced housing or in on-site home building.

There are two types of moisture meters. One has needles which pierce the lumber and measure the electrical resistance of current flow through the wood. The other is a capacity-type that measures the relation between moisture content and a constant setting. This second type can be recognized by the plates or shoes which are applied to the lumber surface.

The accuracy of most meters is within the range of plus or minus 1% of the true figure—assuming that moisture content is uniform throughout the thickness of the board. However, moisture is not always uniform. For example, when lumber is

4–6. *Kiln drying is the quickest method of seasoning lumber properly.*

4–7. *Woods should be dried to different levels for various uses.*

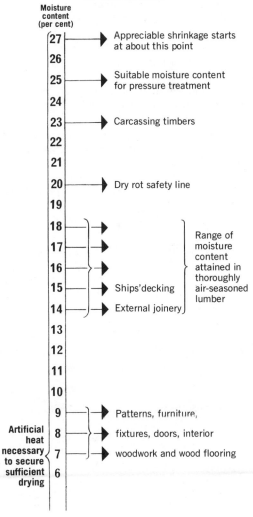

dried rapidly in a kiln, higher moisture is found in the center of the piece than on the outside. Therefore a more accurate reading can be obtained by cutting off the end of a board and inserting the needles in the end grain near the center of the piece. Meters should never be used on foggy days or when the air is excessively humid. Top boards or outside pieces that have been exposed to the weather for some time may be above or below the average of the lumber pile.

In house construction, such things as the prevention of plaster cracks and the fitting of drawers and doors depend on the use of lumber with proper moisture content.

GRADES OF LUMBER

A tree grows much like an onion, adding a new layer each year. When the tree is small, the limbs are low to the ground. As it grows larger, the lower limbs drop off and additional material fills out the tree's diameter. In the growing process, many defects develop in wood that lower its quality. For example, each limb that drops causes a knot to form, but as a tree increases in diameter, the growth covers these knots. Because of this, the best quality lumber is found in the lower part of the trunk near the outside.

So that the purchaser can know the quality of the wood he buys, lumber is classified in grades, according to the defects in it. In the grading process, a lumber piece must meet the lowest requirements for its grade. Wide differences in quality are found in the same grade because some lumber is much better than the minimum for its grade, but not quite good enough for a higher classification. In the lumberyard, the sawer (or *sawyer*) attempts to cut each log into pieces of the best possible grade. The actual grading is done by specialists either manually or electronically (using computers).

Defects in Standard Grades

Defects in lumber are the faults which detract from the quality of the piece, either in appearance or utility. About 25 characteristics and conditions which occur in softwood lumber are used when grading. These terms and their definitions are found in any set of grading rules. Some of the more common terms and definitions used by the woodworker are:

Check. Lengthwise grain separation, usually occurring through the growth rings as a result of seasoning.

Decay. Disintegration of wood substance due to action of wood-destroying fungi.

Knot. Branch or limb embedded in the tree and cut through in the process of lumber manufacture; classified according to size, quality, and occurrence. To determine the size of a knot, average the maximum length and maximum width unless otherwise specified.

Pitch. Accumulation of resin in the wood cells in a more or less irregular patch.

Pitch-pocket. An opening between growth rings which usually contains or has contained resin, bark, or both.

Shake. A lengthwise grain separation between or through the growth rings. May be further classified as ring shake or pitch shake.

Split. Lengthwise separation of the wood extending from one surface through the piece to the opposite or an adjoining surface.

Stain. Discoloration on or in lumber other than its natural color.

4–8. *Electric moisture content meters give direct reading of M.C. in wood.*

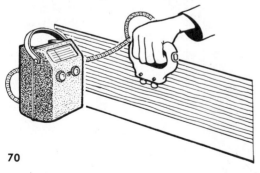

Summerwood. Denser outer portion of each annual ring, usually without easily visible pores, formed late in the growing period, not necessarily in summer.

Torn grain. Part of the wood torn out in dressing.

Wane. This is bark or lack of wood from any cause on the edge or corner of a piece.

Warp. Any variation from a true or plane surface; includes bow, crook, cup, or any combination thereof.

Bow. Deviation flatwise from a straight line from end to end of a piece, measured at the point of greatest distance from the straight line.

Crook. Deviation edgewise from a straight line from end to end of a piece, measured at the point of greatest distance from the straight line; classified as slight, small, medium, and large. Based on a piece 4″ wide and 16′ long, the distance for each degree of crook shall be: *slight crook* = 1″; *small crook* = 1-1 1/2″; *medium crook* = 3″; and *large crook* = over 3″. For wider pieces it shall be 1/8″ less for each additional 2″ of width. Shorter or longer pieces may have the same curvature.

Cup. Deviation flatwise from a straight line across the width of a piece, measured at the point of greatest distance from the line; classified as slight, medium, and deep. Based on a piece 12″ wide, the distance from each degree of cup shall be: *slight cup,* 1/4″; *medium cup,* 3/8″; *deep cup,* 1/2″. Narrower or wider pieces may have the same curvature.

The poorest quality piece permitted in each grade is specified by the grading rules. As assurance to the purchaser, many manufacturers of softwood lumber stamp the grade on the end, edge, or face of the piece.

Hardwood Lumber Grading

Hardwoods are available in three common grades, *firsts* and *seconds* (FAS), *select,* and *No. 1 common.* Generally, firsts and seconds are used for built-ins and paneling.

Grading rules for hardwood lumber have been established by the National Hardwood Lumber Association. Each different kind of hardwood lumber has a slightly different grading standard.

Relatively little hardwood is used in building construction. For more detailed information on hardwood grading consult the text CABINETMAKING AND MILLWORK, also by Dr. John L. Feirer.

Softwood Lumber Grading

Grading rules for softwood lumber are outlined in the bulletin "Voluntary Product Standard PS20-70." This standard was developed by the National Bureau of Standards (NBS) in cooperation with producers, distributors, and users of softwoods. Major lumber organizations and many governmental agencies, including the Forest Products Laboratory of Madison, Wisconsin, contributed to these standards.

According to this voluntary standard, lumber is divided into two basic groups: *green* or *unseasoned* lumber (with moisture content in excess of 19%), and *dry* or *seasoned* lumber (19% or less M. C.) Fig. 4-9. Based on this standard, each major lumber association has developed a complete set of grading rules. Since these grading rules are extensive and vary somewhat for each association, it is best for the builder who uses primary lumber from one section of the country to secure a complete set of grading rules from the organization in that area. For example, if most of the building lumber comes from the western states, then the grading rules published by the Western Wood Products Association should be used. These are available in a reference entitled *Western Woods Use Book.* Similar grading books are available from such associations as the *Southern Pine Inspection Bureau, West Coast Lumber Inspection Bureau, Redwood Inspection Service, Northeastern Lumber Manufacturers Association,* and

Northern Hardwood and Pine Manufacturers Association.

BASIC SELECTION FACTORS

Attention to the following points, and familiarity with common lumber abbreviations, will simplify the selection and specifications of softwood lumber. Fig. 4-10.

Product Classification. Identify product names for clarity. Examples: paneling, structural decking, joists, rafters, studding, beams, and siding. Fig. 4-11.

Species. Include all suitable species. With more species to choose from, you may be able to lower your costs. Check with your local supplier.

When wood color, grain, durability or other special characteristics are important, select and specify the species accordingly.

Grade. Specify standard grades as described in the official grading rules. Consider all grades suitable for the intended use. For economy, it is recommended that the lowest suitable grade be specified.

Stress Rating. When strength is a factor, specify the stress rating requirements without reference to grades. There are two methods of assigning stress values—

visually and by machine. (More detail is given later in this unit, under *Stress-Grade Lumber.*)

Grade Stamps. Specify grade-stamped framing lumber, sheathing, and other construction items. Finish lumber and decking may also be grade-stamped on ends or backs where the stamp will not be visible in use and may be so specified if desired.

Size. For standard products such as boards and framing, specify the nominal size by thickness and width in full inches. Example: 1 × 6, 1 × 8, 2 × 4, 2 × 6. Fig. 4-12. (Nominal sizes are larger than actual or dressed sizes. For example, a nominal 2 by 4 is 1½ by 3½ dressed.)

Surface Texture. Indicate whether lumber is to be smooth surface (surfaced) or rough surface (rough).

Seasoning. Specify "seasoned" lumber to assure long-range product quality, stability, increased nail-holding power, improved paintability, and workability. "Seasoned" covers both major methods of drying lumber—air dried or kiln dried.

Stress-Grade Lumber

Stress-grade lumber is structural lumber that has been scientifically graded with

4-9.
Standard Thicknesses and Widths of Common Softwood Lumber

Nominal			Minimum Dressed					
				Dry			Green	
	Exact	Rounded		Exact	Rounded		Exact	Rounded
inches	mm	mm	inches	mm	mm	inches	mm	mm
1	25.4	25	¾	19.1	19	25/32	19.8	20
1¼	31.8	32	1	25.4	25	1 1/32	26.2	26
1½	38.1	38	1¼	31.8	32	1 9/32	32.51	33
2	50.8	50	1½	38.1	38	1 9/16	39.7	40
2½	63.5	63	2	50.8	50	2 1/16	52.4	52
3	76.2	75	2½	63.5	63	2 9/16	65.1	65
3½	88.9	90	3	76.2	75	3 1/16	77.8	78
4	101.6	100	3½	88.9	89	3 9/16	90.5	90
5	127.0	125	4½	114.3	114	4 5/8	117.5	117
6	152.4	150	5½	139.7	140	5 5/8	142.9	143
7	177.8	175	6½	165.1	165	6 5/8	168.3	168
8	203.2	200	7¼	184.1	184	7 ½	190.5	190

4–10.
Lumber Abbreviations

AD—Air-dried	**EE**—Eased edged
ADF—After deduction freight	**LGTH**—Length
ALS—American Lumber Standards	**LIN**—Lineal
AVG—Average	**LNG**—Lining
AW&L—All widths and lengths	**LP**—Lodgepole pine
BD—Board	**M**—Thousand
BD FT—Board feet	**M. BM**—Thousand (ft.) board measure
BDL—Bundle	**MC**—Moisture content
BEV—Bevel	**MG**—Mixed grain
BH—Boxed heart	**MLDG**—Molding
B/L, BL—Bill of lading	**MOE**—Modulus of elasticity or "E"
BM—Board Measure	**MOR**—Modulus of Rupture
B&S—Beams and Stringers	**MSR**—Machine Stress Rated
BSND—Bright Sapwood no defect	**NBM**—Net board measure
BTR—Better	**N1E**—Nose one edge
CB—Center beaded	**PAD**—Partly Air Dried
CF—Cost and freight	**PARA**—Paragraph
CIF—Cost, insurance and freight	**PART**—Partition
CIFE—Cost, insurance, freight, exchange	**PAT**—Pattern
C/L—Carload	**PET**—Precision end trimmed
CLG—Ceiling	**PP**—Ponderosa pine
CLR—Clear	**P&T**—Posts and Timbers
CM—Center matched	**RC**—Red Cedar
CS—Caulking seam	**RDM**—Random
CSG—Casing	**REG**—Regular
CV—Center V	**RGH**—Rough
DET—Double end trimmed	**R/L, RL**—Random lengths
DF—Douglas Fir	**R/S**—Resawn
DF-L—Douglas Fir-Larch	**R/W, or RW**—Random widths
DIM—Dimension	**R/W, R/L**—Random width, Random length
DKG—Decking	**SB1S**—Single bead one side
D/S, DS—Drop siding	**SDG**—Siding
D&M—Dressed and matched	**SEL**—Select
E—Edge or modulus of elasticity	**SG**—Slash or flat grain
EB1S—Edge bead one side	**S/L or SL**—Shiplap
EB2S—Edge bead two sides	**STD. M**—Standard matched
E&CB2S—Edge & center bead two sides	**SM**—Surface measure
EV1S—Edge vee one side	**SP**—Sugar pine
EV2S—Edge vee two sides	**SQ**—Square
E&CV1S—Edge & center vee one side	**STK**—Stock
E&CV2S—Edge & center vee two sides	**STPG**—Stepping
	STR—Structural

(Continued on page 74)

(Continued from page 73)
Lumber Abbreviations

S&E—Side and edge	**LFVC**—Loaded full visible capacity
S1E—Surfaced one edge	**LGR**—Longer
EG—Edge (vertical) grain	**S2E**—Surfaced two edges
EM—End matched	**S1S**—Surfaced one side
ES—Englemann spruce	**S2S**—Surfaced two sides
f—Allowable fiber stress in bending (also Fb)	**S4S**—Surfaced four sides
	S1S&CM—Surfaced one side and center matched
FAS—Free alongside (vessel)	**S2S&CM**—Surfaced two sides and center matched
FG—Flat or slash grain	
FLG—Flooring	**S4S&CS**—Surfaced four sides and caulking seam
FOB—Free on board (Named point)	
FOHC—Free of heart center	**S1S1E**—Surfaced one side, one edge
FRT—Freight	**S1S2E**—Surfaced one side, two edges
Ft—Foot	**S2S1E**—Surfaced two sides, one edge
FT.BM—Feet board measure (also FBM)	**TBR**—Timber
	T&G—Tongued and grooved
FT.SM—Feet surface measure	**VG**—Vertical (edge) grain
H.B.—Hollow back	**WDR**—Wider
HEM—Hemlock	**WF**—White fir
H&M—Hit and miss	**WT**—Weight
H or M—Hit or miss	**WTH**—Width
IC—Incense cedar	**WRC**—Western Red Cedar
IN—Inch or inches	**WWPA**—Western Wood Products Association
IND—Industrial	
IWP—Idaho white pine	
J&P—Joists and Planks	**SYMBOLS**
JTD—Jointed	**″**—Inch or inches
KD—Kiln-dried	**′**—Foot or feet
L—Larch	**x**—By, as 4 x 4
LBR—Lumber	$^4/_4$, $^5/_4$, $^6/_4$, etc.—nominal thickness expressed in fractions
LCL—Less than carload	
LF—Light Framing	

electronic devices and stamped with information to indicate the specific load it will support.

To understand stress grades it is necessary to know something about loads and strength properties of wood. The home and each of its parts are subject to a variety of external forces or loads. These loads can be classified as either dead or alive.

Dead load is the weight of the material used in the structure. It is its own weight plus its share of the total weight of the remainder of the structure. For example, a dead roof load includes the rafters, or trusses, and all of the roofing material.

Live load, on the other hand, refers to the weight or forces *applied* to the house. Live loads may be static, repetitive, or impact.

4–11. *Product classifications of western lumber.*

Product Classification

	Thickness in.	Width in.		Thickness In.	Width In.
Board Lumber	1 "	2 " or more	**Posts & Timbers**	5 " x 5 " and larger	not more than 2 " greater than thickness
Light Framing	2 " to 4 "	2 " to 4 "	**Decking**	2 " to 4 "	4 " to 12 " wide
Joists & Planks	2 " to 4 "	6 " and wider	**Siding**	thickness expressed by dimension of butt edge	
Beams & Stringers	5 " and thicker	more than 2 " greater than thickness	**Moldings**	size at thickest and widest points	

Standard lengths of lumber generally are 6 feet and longer in multiples of 1'

4–12. *Common materials used in home construction.*

Dimensional Data/Nominal, Dressed

	Product Description	Nominal Size		Dressed Dimensions		
				Thicknesses and Widths In.		Length Ft.
		Thickness In.	Width In.	Surfaced Dry	Surfaced Unseasoned	
Framing	S1S1E S1E or S4S	2	2	1½	1⁹/₁₆	6 ft. and longer in multiples of 1'
		3	3	2½	2⁹/₁₆	
		4	4	3½	3⁹/₁₆	
		6	6	5½	5⅝	
		8	8	7¼	7½	
		10	10	9¼	9½	
		12	12	11¼	11½	
		Over 12	Over 12	Off ¾	Off ½	

		Nominal Size		Dressed Dimensions		
Timbers		5 and Larger		Thickness In. ½ Off Nominal	Width In.	Same

	Product Description	Nominal Size		Dressed Dimensions Surfaced Dry		
		Thickness In.	Width In.	Thickness In.	Width In.	Lengths Ft.
Decking	S2S, CM, EV1S, D&M	2	4	1½	3	
		3	6	2½	5	
		4	8	3½	6¾	Same
			10		8¾	
			12		10¾	
Selects And Commons S-Dry	S1S, S2S, S4S, S1S1E, S1S2E	4/4	2	¾	1½	
		5/4	3	1⁵/₃₂	2½	
		6/4	4	1¹³/₃₂	3½	
		7/4	5	1¹⁹/₃₂	4½	
		8/4	6	1¹³/₁₆	5½	Same
		9/4	7	2³/₃₂	6½	
		10/4	8 and wider	2⅜	¾ Off nominal	
		11/4		2⁹/₁₆		
		12/4		2¾		
		16/4		3¾		
Flooring	(D & M), (S2S & CM)	⅜	2	⁵/₁₆	1⅛	
		½	3	⁷/₁₆	2⅛	
		⅝	4	⁹/₁₆	3⅛	Same
		1	5	¾	4⅛	
		1¼	6	1	5⅛	
		1½		1¼		

(Continued on page 76)

(Continued from page 75)
Dimensional Data/Nominal, Dressed

		Nominal Size		Dressed Dimensions		
		Thickness In.	Width In.	Thickness In.	Width In.	Lengths Ft.
Rustic And Drop Siding	(D & M) If ⅜" or ½" T & G specified, same over-all widths apply.	⅝ 1	4 5 6 8 10	9/16 23/32	3⅛ 4⅛ 5⅛ 6⅞ 8⅞	Same
	(Shiplapped, ⅜-in. lap)	⅝ 1	4 5 6	9/16 23/32	3 4 5	Same
	(Shiplapped, ⅜-in. lap)	⅝ 1	4 5 6 8 10 12	9/16 23/32	2⅞ 3⅞ 4⅞ 6⅝ 8⅝ 10⅝	Same
Ceiling And Partition	(S2S & CM)	⅜ ½ ⅝ ¾	3 4 5 6	5/16 7/16 9/16 11/16	2⅛ 3⅛ 4⅛ 5⅛	Same
Bevel Siding Grades	Bevel Siding Western Red Cedar Bevel Siding available in ½", ⅝", ¾" nominal thickness. Corresponding thick edge is 15/32", 9/16" and ¾".	½ 9/16 ⅝ ¾ 1	4 5 6 8 10 12	7/16 butt, 3/16 tip 15/32 butt, 3/16 tip 9/16 butt, 3/16 tip 11/16 butt, 3/16 tip ¾ butt, 3/16 tip	3½ 4½ 5½ 7¼ 9¼ 11¼	Same
	Wide Bevel Siding (Colonial or Bungalow)	¾	8 10 12	11/16 butt, 3/16 tip	7¼ 9¼ 11¼	
Finish And Boards S-Dry	S1S, S2S, S1S2E	⅜ ½ ⅝ ¾ 1 1¼ 1½ 1¾ 2 2½ 3 3½ 4	2 3 4 5 6 7 8 and wider nominal	5/16 7/16 9/16 ⅝ ¾ 1 1¼ 1⅜ 1½ 2 2½ 3 3½	1½ 2½ 3½ 4½ 5½ 6½ ¾ off	3' and longer. In Superior grade, 3% of 3' and 4' and 7% of 5' and 6' are permitted. In Prime grade, 20% of 3' to 6' is permitted.
Factory And Shop Lumber	S2S*	1 (4/4) 1¼ (5/4) 1½ (6/4) 1¾ (7/4) 2 (8/4) 2½ (10/4) 3 (12/4) 4 (16/4)	5 and wider (4" and wider in 4/4 No. 1 Shop and 4/4 No. 2 Shop)	25/32 (4/4) 1 5/32 (5/4) 1 13/32 (6/4) 1 19/32 (7/4) 1 13/16 (8/4) 2⅜ (10/4) 2¾ (12/4) 3¾ (16/4)	(See Rough Sizes Below)	6 ft. and longer in multiples of 1'

*These thicknesses also apply to Tongue &
Groove (T&G).
See coverage estimator chart above for
T&G widths.

Minimum Rough Sizes   Thicknesses and Widths Dry or Unseasoned All Lumber (S1E, S2E, S1S, S2S)
80% of the pieces in a shipment shall be at least ⅛" thicker than the standard surfaced size, the remaining 20% at least 3/32" thicker than the surfaced size. Widths shall be at least ⅛" wider than standard surfaced widths.
When specified to be full sawn, lumber may not be manufactured to a size less than the size specified.

Abbreviations
Abbreviated descriptions appearing in
the size table are explained below.
S1S—Surfaced one side.
S2S—Surfaced two sides.

S4S—Surfaced four sides.
S1S1E—Surfaced one side, one edge.
S1S2E—Surfaced one side, two edges.
CM—Center matched.

D & M—Dressed and matched.
T & G—Tongue and grooved.
EV1S—Edge vee on one side.
S1E—Surfaced one edge.

Static loads are those which are applied slowly and remain constant, or are repeated relatively few times. The furniture in a house is a static load.

Loads that are applied a large number of times are called *repetitive* or *fatigue* loads. People walking across the floor are illustrations of this type.

Impact loads are sudden or instantaneous forces. A high wind or a tree falling against the roof is an impact load.

To handle these loads, wood and other materials used in the home must have certain strength properties. Fig. 4-13. Loads cause stress in a piece of wood. *Stress* refers to a force acting on a piece of material and tending to change the shape of the piece. The most common stresses are tensile, compressive, and shear.

Tensile (or *tension*) stress occurs when forces tend to elongate the piece. *Compressive* (or *compression*) *stress* is a result of squeezing or crushing. *Shear stress* is caused by opposite forces that meet head on, like the blades of a pair of shears. For example, when a floor joist is under a load it develops three kinds of stress. Compression develops along the upper edge of the joist, tension along the lower edge, and horizontal shear through the middle. Fig. 4-14. There is also a vertical shear where the joists rest on the foundation.

A load or a weight on the ends of joists, beams, and similar pieces of lumber causes stress that tends to compress the fibers. A load that causes the joist to bend produces tension in the extreme fibers on the face farthest from the applied load, and it also causes compression in the extreme fibers along the face nearest to the applied load.

The *modulus of elasticity* is a measure of the stiffness of material or a measure of resistance to deflection (bending). A piece may deflect slightly or a great deal depending on the size, span, load, and modulus of elasticity for the particular species. Each species of wood and each grade of lumber within that species has its own strength properties. For example, Douglas

4–13. *Live and dead loads.*
Diagram Showing Method Of Figuring Loads For House Framing

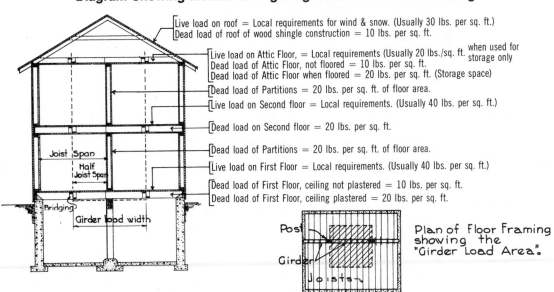

Live load on roof = Local requirements for wind & snow. (Usually 30 lbs. per sq. ft.)
Dead load of roof of wood shingle construction = 10 lbs. per sq. ft.

Live load on Attic Floor, = Local requirements (Usually 20 lbs./sq. ft. when used for storage only
Dead load of Attic Floor, not floored = 10 lbs. per sq. ft.
Dead load of Attic Floor when floored = 20 lbs. per sq. ft. (Storage space)

Dead load of Partitions = 20 lbs. per sq. ft. of floor area.

Live load on Second floor = Local requirements. (Usually 40 lbs. per sq. ft.)

Dead load on Second floor = 20 lbs. per sq. ft.

Dead load of Partitions = 20 lbs. per sq. ft. of floor area.

Live load on First Floor = Local requirements. (Usually 40 lbs. per sq. ft.)

Dead load of First Floor, ceiling not plastered = 10 lbs. per sq. ft.
Dead load of First Floor, ceiling plastered = 20 lbs. per sq. ft.

Joist Span
Half Joist Span
Bridging
Girder load width

Post
Girder
Joists

Plan of Floor Framing showing the "Girder Load Area".

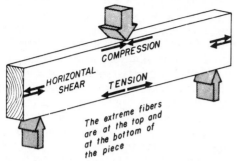

4-14. Note how the various forces act on a floor joist.

4-15. It is a good idea to purchase lumber that has been properly stored at the lumberyard.

fir has a higher modulus of elasticity (stiffness) than western cedars. Southern yellow pine has greater compressive strength endwise than ponderosa pine. The National Lumber Manufacturers Association and other trade associations such as the Western Wood Products Association compile tables that indicate the strength properties of all the common species and grades.

4-16. Recommended average moisture content for lumber to be used for interior woodwork in buildings in various areas of the U.S. In Canada the recommended moisture contents are as follows: Vancouver, 11%; Saskatoon, 7%; Ottawa, 8%; Halifax, 9%. (These cities represent four major geographical areas.)

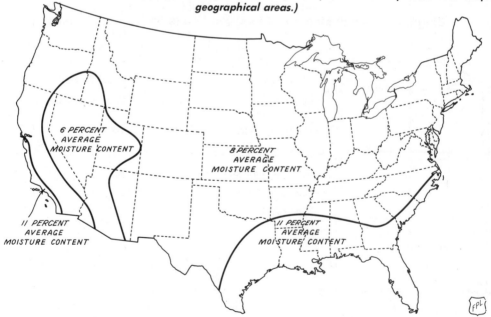

Saw mills also produce some lumber that is stress-rated and marked with a grade stamp that gives exact information about the strength properties of that particular piece of wood. The rating may be done on the basis of appearance (visual stress-rated lumber) or it may be done mechanically (machine stress-rated lumber). Most retail lumberyards do not handle stress-rated lumber. The typical retail yard stocks only *yard lumber,* which is adequate for average house construction. A house built of the proper-size yard lumber for joists, beams, and rafters, will be of adequate strength. However, for certain items, when strength properties are critical, the architect may specify stress-rated lumber of a certain species and grade. For example, manufacturers of roof trusses will often use stress-grade lumber since these parts are so important to the strength of the home.

STORING LUMBER

In Lumberyards. Lumber stored in yards should be indoors under covered sheds. If stored outdoors it should be protected by covering with plastic or waterproof paper. Fig. 4-15. If the sides, top, and ends of a piece of lumber are properly covered, it will not absorb too much moisture outdoors even in damp, rainy, or humid weather. In indoor storage, moisture content can be controlled by regulating the temperature in the storage shed.

On the Job. Lumber received on the job for house construction should be stored carefully to prevent shrinking and swelling. It should be off the ground and piled carefully with stickers between each layer. It should then be covered with waterproof material until used.

Lumber kept indoors will pick up or lose moisture until it reaches a balance with the moisture of air in the room. Fig. 4-16 is a map of the United States which shows the proper moisture content for wood to be used in the interior parts of heated buildings. If lumber, when delivered, has a lower moisture content than was ordered, it will pick up additional moisture until it balances with that in the air. Fig. 4-17. Even paint does not keep wood from absorbing moisture, since it does not completely seal the wood.

FIGURING BOARD FEET

Sizes of softwood or building construction lumber have been standardized for convenience in ordering and handling. Building materials sizes run 6, 8, 10, 12, 14, 16, 18, and 20 feet in length, 2, 4, 6, 8, 10, and 12 inches in width, and 1, 2, and 4 inches in thickness. As mentioned briefly earlier, the actual width and thickness of dressed lumber are considerably less than the standard, or nominal, width and thickness. For the relative differences between standard and actual sizes of construction lumber, see Fig. 4-13a. Hardwoods which have no standard length or widths run $1/4$, $1/2$, 1, $1 1/4$, $1 1/2$, 2, $2 1/2$, 3, and 4 inches in thickness. Plywoods run from 4 feet in width to 8 feet in length, and vary in thickness from $1/8$ to 1 inch. Stock panels are usually available in 48-inch widths and in lengths varying in multiples of 16 inches up to 8 feet. Panel lengths run in 16-inch multiples because the accepted spacing for studs and joists is 16 inches. The amount of lumber required is measured in board feet. A *board foot* is a unit of measure representing a piece of lumber having a flat surface area of 1 square foot and a thickness of 1 inch nominal size. The number of board feet in a piece of lumber can be computed by the arithmetic method or by the tabular method.

Arithmetic Method

In order to determine the number of board feet in one or more pieces of lumber, the following formula is used:

$$\frac{\text{Pieces} \times \text{Thickness (in.)} \times \text{Width (in.)} \times \text{Length (ft.)}}{12}$$

Example 1: Find the number of board feet

in a piece of lumber 2" thick, 10" wide and 6' long. Fig. 4-18.

$$\frac{2 \times 10 \times 6}{12} = 10 \text{ bd. ft.}$$

Example 2: Find the number of board feet in 10 pieces of lumber 2" thick, 10" wide and 6' long.

$$\frac{10 \times 2 \times 10 \times 6}{12} = 100 \text{ bd. ft.}$$

If all three dimensions are expressed in inches, the same formula applies except the divisor is changed to 144.

Example: Find the number of board feet in one piece of lumber 2" thick, 10" wide, and 18" long.

$$\frac{1 \times 2 \times 10 \times 18}{144} = 2\frac{1}{2} \text{ bd. ft.}$$

Tabular Method

The standard Essex board measure table found on the back of the blade of the framing square is a quick and convenient aid in computing board feet. Fig. 4-19. In

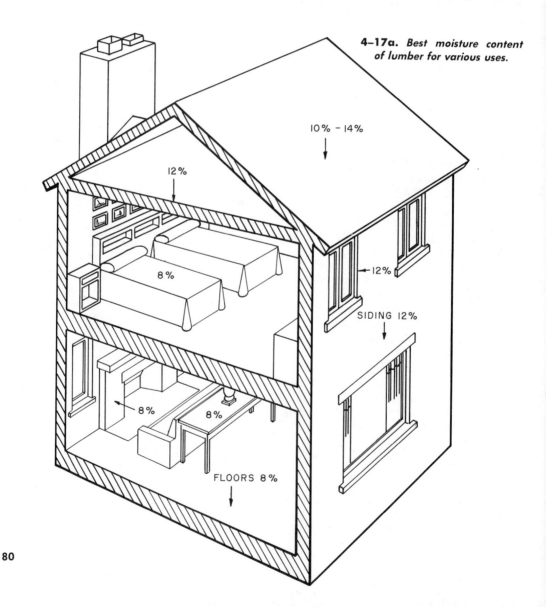

4–17a. *Best moisture content of lumber for various uses.*

10% – 14%

12%

8%

12%

SIDING 12%

8%

8%

FLOORS 8%

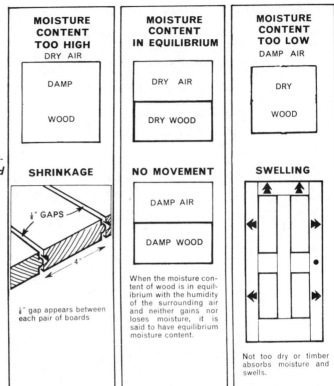

MOISTURE CONTENT TOO HIGH	MOISTURE CONTENT IN EQUILIBRIUM	MOISTURE CONTENT TOO LOW
DRY AIR		DAMP AIR
DAMP WOOD	DRY AIR / DRY WOOD	DRY WOOD
SHRINKAGE	**NO MOVEMENT**	**SWELLING**

4–17b. *Wood at the correct moisture content reduces swelling and shrinkage to a minimum.*

⅛″ GAPS

4″

⅛″ gap appears between each pair of boards

DAMP AIR

DAMP WOOD

When the moisture content of wood is in equilibrium with the humidity of the surrounding air and neither gains nor loses moisture, it is said to have equilibrium moisture content.

Not too dry or timber absorbs moisture and swells.

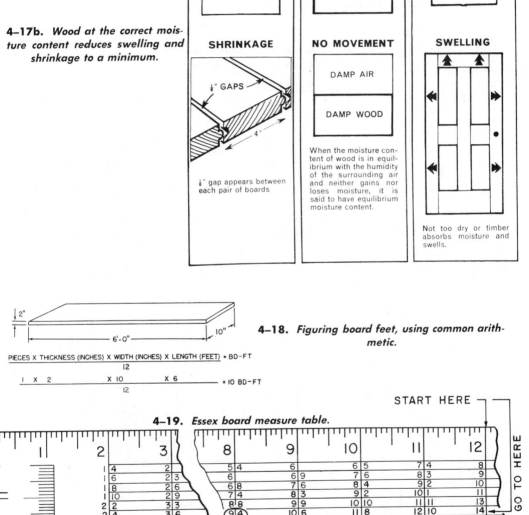

2″

6'-0″

10″

4–18. *Figuring board feet, using common arithmetic.*

$$\frac{\text{PIECES} \times \text{THICKNESS (INCHES)} \times \text{WIDTH (INCHES)} \times \text{LENGTH (FEET)}}{12} = \text{BD-FT}$$

$$\frac{1 \times 2 \times 10 \times 6}{12} = 10 \text{ BD-FT}$$

4–19. *Essex board measure table.*

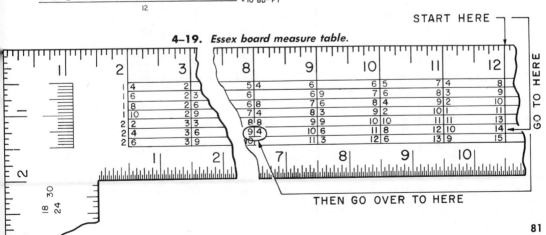

START HERE

GO TO HERE

THEN GO OVER TO HERE

4–20. *A rapid way to estimate board feet.*

Estimating Board Feet

Width	Thickness	Board feet
3 "	1 " or less	¼ of the length
4 "	1 " or less	⅓ of the length
6 "	1 " or less	½ of the length
9 "	1 " or less	¾ of the length
12"	1 " or less	Same as the length
15 "	1 " or less	1¼ of the length

using the table, all computations are made on the basis of 1" thickness. The inch markings along the outer edge of the blade represent the width of a board 1" thick. The third dimension, length, is provided in the vertical column of figures under the "12" on the outer edge representing a 1" thick board that is 12" wide.

To compute the number of board feet in

4–21. *A table of board-foot measure.*

Board-Foot Measure

Nominal size (in.)	Actual length in feet								
	8	10	12	14	16	18	20	22	24
1 x 2	....	1⅔	2	2⅓	2⅔	3	3⅓	3⅔	4
1 x 3	....	2½	3	3½	4	4½	5	5½	6
1 x 4	2⅔	3⅓	4	4⅔	5⅓	6	6⅔	7⅓	8
1 x 5	....	4⅙	5	5⅚	6⅔	7½	8⅓	9⅙	10
1 x 6	4	5	6	7	8	9	10	11	12
1 x 7	....	5⅚	7	8⅙	9⅓	10½	11⅔	12⅚	14
1 x 8	5⅓	6⅔	8	9⅓	10⅔	12	13⅓	14⅔	16
1 x 10	6⅔	8⅓	10	11⅔	13⅓	15	16⅔	18⅓	20
1 x 12	8	10	12	14	16	18	20	22	24
1¼ x 4	....	4⅙	5	5⅚	6⅔	7½	8⅓	9⅙	10
1¼ x 6	....	6¼	7½	8¾	10	11¼	12½	13¾	15
1¼ x 8	....	8⅓	10	11⅔	13⅓	15	16⅔	18⅓	20
1¼ x 10	....	10⁵/₁₂	12½	14⁷/₁₂	16⅔	18¾	20⅚	22¹¹/₁₂	25
1¼ x 12	....	12½	15	17½	20	22½	25	27½	30
1½ x 4	4	5	6	7	8	9	10	11	12
1½ x 6	6	7½	9	10½	12	13½	15	16½	18
1½ x 8	8	10	12	14	16	18	20	22	24
1½ x 10	10	12½	15	17½	20	22½	25	27½	30
1½ x 12	12	15	18	21	24	27	30	33	36
2 x 4	5⅓	6⅔	8	9⅓	10⅓	12	13⅓	14⅔	16
2 x 6	8	10	12	14	16	18	20	22	24
2 x 8	10⅔	13⅓	16	18⅔	21⅓	24	26⅔	29⅓	32
2 x 10	13⅓	16⅔	20	23⅓	26⅔	30	33⅓	36⅔	40
2 x 12	16	20	24	28	32	36	40	44	48
3 x 6	12	15	18	21	24	27	30	33	36
3 x 8	16	20	24	28	32	36	40	44	48
3 x 10	20	25	30	35	40	45	50	55	60
3 x 12	24	30	36	42	48	54	60	66	72
4 x 4	10⅔	13⅓	16	18⅔	21⅓	24	26⅔	29⅓	32
4 x 6	16	20	24	28	32	36	40	44	48
4 x 8	21⅓	26⅔	32	37⅓	42⅔	48	53⅓	58⅔	64
4 x 10	26⅔	33⅓	40	46⅔	53⅓	60	66⅔	73⅓	80
4 x 12	32	40	48	56	64	72	80	88	96

a piece of lumber 4" thick, 8" wide, and 14' long, find number 14 in the vertical column under the "12" mark on the outer edge. Follow the guideline under figure 14 laterally to the left across the blade until it reaches the figure on that line directly under the inch mark corresponding to the width of the piece. Under the 8" mark on the guideline indicated by the 14, the figures 9 and 4 appear. The figure to the left represents feet and that on the right represents inches. In this case, these figures mean that there are 9⁴/₁₂ or 9¹/₃ board feet in a piece of lumber 14' long, 8" wide, and 1" thick. Remember, however, that we were checking a board 4" thick. Therefore we must still multiply by 4, for a final answer of 37¹/₃ board ft.

Rapid Estimation

Rapid estimation of board feet can be made by the use of the tables in Figs. 4-20 and 4-21.

METRIC LUMBER MEASURES

Most major wood-producing countries cut lumber into the customary (English) sizes with thicknesses and widths in inches and lengths in feet. However, lumber to be shipped to most foreign countries is marked with metric sizes. Millimetres (mm) are used to indicate thickness and width, and metres (m) for length. The common sizes of lumber are almost identical in either system of measurement. For example, the basic thickness is 1" or 25 mm, and nominal width is 100 mm or 4". The metric lengths range from 1.8 m, which is approximately 6', and increase in steps of 300 mm (or 0.3 m) to 6.3 metres. Note that 300 mm is close to but slightly shorter than a foot. Fig. 4-22.

The metric size of panel stock such as plywood is 1220 mm (or 1.22 m) by 2440 mm (or 2.44 m), which is, for all practical purposes, the same as a 4' × 8' sheet. In countries "going metric" a 1200 × 2400 mm size is used, based on a 100 mm building module. As in customary lumber measurement the S2S (surfaced on two sides, or planed) lumber will always measure less than the rough sawn (nominal) size. An allowance should be made for this when stating required sizes.

4-22. Metric lumber sizes. Note that the thickness and width are given in millimetres (mm) and length in metres (m). Most of the metric sizes in cross section are just slightly smaller than standard sizes in inches.

Standard Range of Sizes for Sawn Softwood

mm	75	100	125	150	175	200	225	250	300
16	x	x	x	x					
19	x	x	x	x					
22	x	x	x	x					
25	x	x	x	x	x	x	x	x	x
32	x	x	x	x	x	x	x	x	x
36	x	x	x	x					
38	x	x	x	x	x	x	x		
40	x	x	x	x	x	x	x		
44	x	x	x	x	x	x	x	x	x
50	x	x	x	x	x	x	x	x	x
63		x	x	x	x	x	x		
75		x	x	x	x	x	x	x	x
100		x		x		x		x	x
150				x		x			x
200						x			
250								x	
300									x

Lengths: 1.8 to 6.3 m x 0.3 m   Allow: 4 mm for planing

QUESTIONS

1. List five advantages of wood as a building material.

2. Describe the difference between earlywood and latewood.

3. Name the two common methods of cutting boards from a log.

4. Tell how much moisture a typical tree may contain when it is first cut down,

as compared with the moisture in dried lumber.

5. Describe what is meant by *fiber saturation point.*

6. Describe the two methods of drying lumber.

7. Define moisture content.

8. Name two ways of checking the moisture content of lumber.

9. List several defects in standard grades of lumber.

10. What are the three types of warp?

11. What organization handles hardwood lumber grades?

12. How does a new standard of softwood lumber differ from the old standard?

13. Why is it necessary to secure a complete set of grading rules for detailed help on selecting lumber?

14. When is *stress-grade lumber* used?

15. Tell how lumber should be stored for on-the-job building.

16. Define the term "board foot."

17. How many board feet are in a piece of lumber 2″ × 8″ × 10′?

18. What unit is used for the thickness and width of lumber in the metric system?

19. What is the minimum length of lumber in metres?

20. What is the size of a standard plywood sheet in the metric system?

Plywood

Plywood is a most versatile building material. Fig. 5-1. It can be used in nearly every step of home building, including:
• Roof decking, where plywood adds strength and rigidity.
• Gable ends and soffits, where plywood provides a smooth, clean, easy-to-paint surface. (For information on finishing and painting see Appendix.)
• Subfloors and underlayment. Plywood makes a firm, smooth base for flooring.
• Interior walls. Plywood has warmth of real wood.
• Base for tile, cork, and decorative walls. Plywood is smooth, flat, and durable.
• Built-ins. These can be planned for beauty, convenience, and need.

• Sheathing, to produce walls that are warm, strong, and rigid.
• Fences, windbreaks, and patio screens, to provide privacy and comfort.
• Exteriors. Smart and modern plywood exteriors can be arranged to fit any home styling.
• Concrete forms. Plywood used this way can be reused for sheathing.

WHAT PLYWOOD IS

Plywood consists of glued wood panels made from layers and/or plies of veneer or veneer and wood. The grain of each layer is at an angle, usually a right angle, to the grain of the nearest layer on each side. The grain of the outer plies always runs in the

5-1a. *Common uses of plywood in building construction.*

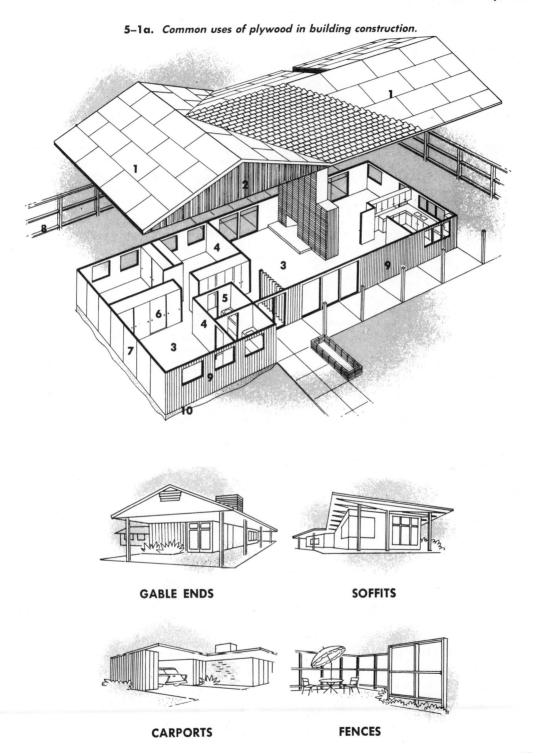

GABLE ENDS

SOFFITS

CARPORTS

FENCES

5-1b. *Plywood is being used for sheathing the exterior of this home.*

same direction. The outer plies are called *faces,* or *face* and *back.* The center ply is called the *core.* The ply or plies between the core and the faces are called *crossbands.*

Veneer is a very thin sheet of wood that is sawed, peeled, or sliced from a log. Construction and industrial plywoods (formerly called softwood plywoods) are made entirely of veneer of three, four, five, six, or seven plies. Hardwood plywoods may be made entirely of veneer (in which case they are called *veneer-core* plywoods), or of veneer bonded to a core of glued-up lumber, (called *lumber-core* plywood). Sometimes hardwood plywood has a particle-board core. Fig. 5-2.

MANUFACTURE OF CONSTRUCTION AND INDUSTRIAL PLYWOOD

This chapter deals primarily with construction and industrial plywood because that material, rather than hardwood plywood, is used most in construction. An examination of how this material is made

will help you to understand the properties which make it so important in the building construction industry. Fig. 5-3.

Only selected logs qualify as plywood "peelers," that is, logs from which veneers for plywood are cut. Such veneers are generally cut from the lower, larger portion of the trunk where the wood is mature and strong, with the clearest grain. The log comes from the forest and goes first into the storage pond. Next it is cut to lathe size, and then it moves into the plywood factory and through the debarker. Now it's a peeler "block"—pick of the forest crop—ready to become plywood.

Huge tongs from an overhead crane lift it into the lathe. Metal gripper chucks clamp both ends and the block begins to spin, revolving swiftly against a long, keen, razor-sharp blade of steel.

The blade bites in. A continuous strip of thin wood is peeled off—unwound from the log like paper from a roll.

This is *veneer,* not plywood. The lathe cuts it to an exact and specified thickness—from $1/16''$ to $5/16''$—and sends it streaming over conveyors to the clipping machine.

Giant knives, controlled by a skilled operator, cut the veneer to proper width for full utilization. It passes on into long mechanical dryers, where the moisture content of the wood is scientifically reduced to provide for greatest panel stability and best glue bond.

Next comes patching. Wood is never perfect; so the natural defects are cut out, to be replaced with solid wood or synthetic patching materials, carefully glued into place.

Next, some of the sheets go through the glue spreader, where large rollers cover both sides with a uniform thickness of adhesive. These glue-covered pieces are stacked alternately with dry veneers, to make up panels of the desired thickness. This is called the *lay-up* process.

Always there is an *odd* number of layers (3, 5, or 7). And, as mentioned earlier, there is always a cross-graining of layers. The grain of each layer is placed at *right angles* to that of adjacent layers. (Some plywood has two plies glued together with the grain parallel to form one layer.) The result is a balanced panel of cross-laminated wood veneers—strong, rigid, split proof, able to resist great impact.

Several important steps remain before the panel is finished. After lay-up, panels go into presses where the wood and glue are bonded into one homogeneous material. Two important variations are made here.

If the panel is to be *exterior-type* plywood, for permanent outdoor or marine use, the glue must be completely waterproof. Also, heat as well as pressure is required on the press. The result is a highly durable bond.

If the panel is to be *interior-type* plywood—for ordinary indoor and construction use—the adhesives are not waterproof, and may be cured with or without heat. Even these glues, however, are highly resistant to moisture. The interior panel can be produced more economically than the exterior type.

From the press, panels move through machines which trim them to exact,

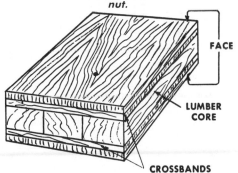

5–2a. *Veneer core. The core is made of thick wood veneer.*

5–2b. *Lumber core. The core consists of strips of lumber bonded together. Good woods for lumber-core plywood are basswood and chestnut.*

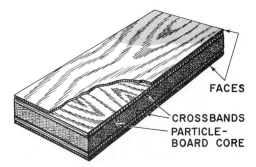

5-PLY WITH PARTICLE-BOARD CORE

5–2c. *Particle-board core. The core is made of particle board, a wood composition material sometimes referred to as flakeboard or chipboard. This kind of plywood is commonly used for cabinet doors because it is very stable.*

5–3. *Cutaway view of a typical piece of plywood.*

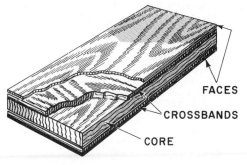

5-PLY WITH VENEER CORE

precision-squared sizes and sand them down to proper thickness. Then comes inspection, repair of blemishes in face plies, and final grading.

CONSTRUCTION AND INDUSTRIAL PLYWOOD

Construction and industrial plywood is manufactured under U.S. Product Standard PS 1-74 replacing the previous standard PS 1-66 for *softwood plywoods*. The standards are different from old regulations in such areas as the use of hardwoods. Also, common softwood panels are now constructed of an even number of plies, and a grade trademark change has been made from DFPA to APA. These standards affect the *use of plywood* in building construction and the *way in which materials are ordered*.

Types. Plywood is manufactured in two major types—*exterior,* with 100 percent waterproof glue, and *interior,* with highly moisture-resistant glue. A third type is designated as *structural* panels. Fig. 5-4. The structural panels are those designed for special engineering use, which meet such properties as tension and compression. These property standards are in addition to all of the other standards that apply to the interior and exterior types.

Veneer Quality. Top quality is N. The

5–4a.
Exterior Type Grades (a)

Panel Grade Designations	Minimum Veneer Quality			
	Face	Back	Inner Plys	Surface
Marine, A-A, A-B, B-B, HDO, MDO[b]				Sanded 2 sides
Special Exterior, A-A, A-B, B-B, HDO, MDO[c]				Sanded 2 sides
A-A	A	A	C	Sanded 2 sides
A-B	A	B	C	Sanded 2 sides
A-C	A	C	C	Sanded 2 sides
B-B (concrete form)[d]				
B-B	B	B	C	Sanded 2 sides
B-C	B	C	C	Sanded 2 sides
C-C Plugged	C Plugged	C	C	Touch-sanded
C-C	C	C	C	Unsanded[e]
A-A High Density Overlay	A	A	C Plugged	—
B-B High Density Overlay	B	B	C Plugged	—
B-B High Density Concrete Form Overlay	B	B	C Plugged	—
B-B Medium Density Overlay	B	B	C	—
Special Overlays	C	C	C	—

(a) Available also in Structural I and Structural II classifications

(b) Marine grades shall meet the requirements of Exterior type and shall be of one of the following grades: A-A, A-B, B-B, High Density Overlay, or Medium Density Overlay

(c) Special Exterior—An Exterior type panel that may be produced of any species covered by this Standard. Except in regard to species, it shall meet all of the requirements for Marine panels (see b) and be produced in one of the following grades: A-A, A-B, B-B, High Density Overlay, or Medium Density Overlay.

(d) B-B concrete form panels—Face veneers shall be not less than B grade and shall always be from the same species group. Inner plys shall be not less than "C" grade. This grade of plywood is produced in two classes and panels of each class shall be identified accordingly. Panels shall be sanded two sides and mill-oiled unless otherwise agreed upon

(e) Except for decorative grades, panels shall not be sanded, touch-sanded, surface textured, or thickness sized by any mechanical means.

quality of the others ranges from A to D. D is the lowest quality. Fig. 5-5.

Species. Plywood is manufactured from seventy different species of wood including native and imported hardwoods. However, most plywood will continue to be produced primarily of softwoods such as Douglas Fir. On the basis of stiffness and other factors, these species are divided into five groups. The strongest woods are found in Group 1. Fig. 5-6.

Plywood Construction. Plywood is constructed of three, four, five, six, or seven plies. A *ply* is a sheet of veneer cut to various thicknesses ranging from as thin as 1/16" to as thick as 5/16". Note that an even number of plies can be used to make

up the thickness. However, the number of *layers* used in panel construction is always odd, such as three, five, or seven. Fig. 5-7 (Pages 92, 93). This is possible because two of the inner plies may be glued together with the grain parallel to form one inner layer. The grain of all layers must be perpendicular to the grain of the adjacent layer. For example, plywood that is 1/2" thick can be constructed of four plies. The two inner plies are glued together with the grain parallel to form the inner layer. A face and a back veneer are glued to the center layer at right angles. The result is plywood of four plies made up of three layers.

The same 1/2" plywood could be manufactured of three plies with three layers. In

5–4b.

Interior Type Grades

Panel Grade Designations	Minimum Veneer Quality			Surface
	Face	Back	Inner Plies	
N-N	N	N	C	Sanded 2 sides
N-A	N	A	C	Sanded 2 sides
N-B	N	B	C	Sanded 2 sides
N-D	N	D	D	Sanded 2 sides
A-A	A	A	D	Sanded 2 sides
A-B	A	B	D	Sanded 2 sides
A-D	A	D	D	Sanded 2 sides
B-B	B	B	D	Sanded 2 sides
B-D	B	D	D	Sanded 2 sides
Underlayment	C Plugged	D	C & D	Touch-sanded
C-D Plugged	C Plugged	D	D	Touch-sanded
Structural I C-D[a]				Unsanded[b]
Structural I C-D Plugged, Underlayment[a]				Touch-sanded
Structural II C-D[a]				Unsanded[b]
Structural II C-D Plugged, Underlayment[a]				Touch-sanded
C-D	C	D	D	Unsanded[b]
C-D with exterior glue	C	D	D	Unsanded[b]
(See para. 3.6.6)				

(a) Structural panels—These panels are especially designed for engineered applications such as structural components where design properties, including tension, compression, shear, cross-panel flexural properties and nail bearing may be of significant importance.

(b) Except for decorative grades, panels shall not be sanded, touch-sanded, surface textured, or thickness sized by any mechanical means.

this construction the center ply would be much thicker than in the four-ply construction.

Thickness and Size. Thickness of standard plywood ranges from $1/4''$ to $1 1/4''$ and greater in $1/8''$ increments. The standard thickness of unsanded panels ranges from $5/16''$ to $1 1/4''$, in increments of $1/8''$ for thicknesses over $3/8''$.

Plywood is available in panel widths of 36", 48", and 60" and in lengths ranging from 60" to 144" in 12" increments. Plywood 48" wide by 96" long (a typical 4' × 8') and 48" wide by 120" long (4' × 10') are most commonly available.

Grade Markings. Grade marks are stamped on the face and edge of all ply-wood that meets the American Plywood Association's standards. Typical grade trademarks are shown in Fig. 5-8 (Page 94) together with notations of what each element means.

Sheathing Panels. Certain grades of plywood panels are used for subflooring, interior and exterior wall sheathing, and roof decking. These include grade trademarks of C-C, C-D, structural C-C, and structural C-D.

When these panels are used for sheathing purposes, they must include an identification as shown in Fig. 5-9. A careful study should be made of this information since it is important in selecting the correct sheathing for certain kinds of construction.

Note that the sheathing is identified by the group of wood species used, the type as indicated by interior, exterior and struc-

5–4c.
Structural Panels

Grade	Glue Bond	Species
Structural I C-D(a) C-D Plugged(a) Underlayment (a)	Exterior	Face, back and all innerplies limited to Group 1 species
Structural II C-D(a) C-D Plugged(a) Underlayment(a)	Exterior	Face, back and all inner plies may be of any Group 1, 2, or 3 species
Structural I All Exterior grades (see table 3)	Exterior	Face, back and all inner plies limited to Group 1 species
Structural II All Exterior grades (see table 3)	Exterior	Face, back and all inner plies may be of any Group 1, 2, or 3 species

(a) Special limitations applying to Structural (C-D, C-D Plugged, Underlayment) grade panels are:

— In D grade veneers white pocket in any area larger than the size of the largest knothole, pitchpocket or split specifically permitted in D grade shall not be permitted in any ply.

— Sound tight knots in D grade shall not exceed 2-1/2 inches measured across the grain.

— Plugs, including multiple repairs, shall not exceed 4 inches in width.

5–5.
Veneer Quality

N	Intended for natural finish. Selected all heartwood or all sapwood. Free of open defects. Allows some repairs.
A	Smooth and paintable. Neatly made repairs permissible. Also used for natural finish in less demanding applications.
B	Solid surface veneer. Repair plugs and tight knots permitted. Can be painted.
C	Sanding defects permitted that will not impair the strength or serviceability of the panel. Knotholes to 1½" and splits to ½" permitted under certain conditions.
C plugged	Improved C veneer with closer limits on knotholes and splits. C plugged face veneers are fully sanded.
D	Used only in interior type for inner plies and backs. Permits knots and knotholes to 2½" in maximum dimension and ½" larger under certain specified limits. Limited splits permitted.

tural, and the nominal thickness of the panel. The index numbers, as you will note, are particularly useful when panels are used for subflooring and roof sheathing to describe the recommended maximum span in inches under nominal conditions. Fig. 5-10 (Page 96). The left-hand number refers to the spacing of the roof framing and the right-hand number refers to the spacing of floor framing.

For example, suppose you wish to select a 3/8″ thickness of plywood sheathing to use on a roof in which the joists are spaced 24″. Under these conditions any one of several different grades in Group 1 and 2 could be used. Fig. 5-9. For example, you could select a panel made of Group 1 wood species with an exterior grade of C-C.

Again, suppose you need to select a sheathing panel for subflooring in which the joists are 16″ on center. You could select any sheathing grade in 1/2″ thickness in the first column, 5/8″ thickness in the second column and 3/4″ thickness from the third column. Fig. 5-9. As you can see, a thinner panel made of wood in the stronger wood species can be used in place of a thicker panel in the weaker wood species. For example, for this use you may select a 1/2″ thick Group 1 *interior* C-D type or a 3/4″ thick Group 4 *interior* C-C type to do the same job.

Ordering. To order plywood you should designate the species, number of pieces, width, length, number of plies, type, grade, and finished thickness, in that order. A typical order might be as follows:

5–6. *Species of wood from which plywood is manufactured.*
Classification of Species

Group 1		Group 2	Group 3	Group 4	Group 5
Apitong(a) (b)	Cedar, Port Orford	Maple, Black	Alder, Red	Aspen	Basswood
Beech, American	Cypress	Mengkulang(a)	Birch, Paper	Bigtooth	Fir, Balsam
Birch	Douglas Fir 2(c)	Meranti, Red(a) (d)	Cedar, Alaska	Quaking	Poplar, Balsam
Sweet	Fir	Mersawa(a)	Fir, Subalpine	Cativo	
Yellow	California Red	Pine	Hemlock, Eastern	Cedar	
Douglas Fir 1(c)	Grand	Pond	Maple, Bigleaf	Incense	
Kapur(a)	Noble	Red	Pine	Western Red	
Keruing(a) (b)	Pacific Silver	Virginia	Jack •	Cottonwood	
Larch, Western	White	Western White	Lodgepole	Eastern	
Maple, Sugar	Hemlock, Western	Spruce	Ponderosa	Black (Western Poplar)	
Pine	Lauan	Red	Spruce	Pine	
Caribbean	Almon	Sitka	Redwood	Eastern White	
Ocote	Bagtikan	Sweetgum	Spruce	Sugar	
Pine, Southern	Mayapis	Tamarack	Black		
Loblolly	Red Lauan	Yellow Poplar	Engelmann		
Longleaf	Tangile		White		
Shortleaf	White Lauan				
Slash					
Tanoak					

(a) Each of these names represents a trade group of woods consisting of a number of closely related species.

(b) Species from the genus Dipterocarpus are marketed collectively: Apitong if originating in the Philippines; Keruing if originating in Malaysia or Indonesia.

(c) Douglas fir from trees grown in the states of Washington, Oregon, California, Idaho,

Montanna, Wyoming, and the Canadian Provinces of Alberta and British Columbia shall be classed as Douglas fir No. 1. Douglas fir from trees grown in the states of Nevada, Utah, Colorado, Arizona and New Mexico shall be classed as Douglas fir No. 2.

(d) Red Meranti shall be limited to species having a specific gravity of 0.41 or more based on green volume and oven dry weight.

5–7a.

Panel Constructions

Panel Grades	Finished Panel Nominal Thickness Range (inch)	Minimum Number of Plies	Minimum Number of Layers

Exterior
  Marine
  Special Exterior (See para. 3.6.7)
  B-B concrete form
  High Density Overlay
  High Density concrete form overlay
	Through ⅜
	Over ⅜, through ¾
	Over ¾
		3
		5
		7
			3
			5
			7

Interior
  N-N, N-A, N-B, N-D, A-A, A-B, A-D
  B-B, B-D
  Structural I (C-D, C-D Plugged and
    Underlayment)
  Structural II (C-D, C-D Plugged and
    Underlayment)
 	
Exterior
  A-A, A-B, A-C, B-B, B-C
  Structural I and Structural II
  Medium Density and special overlays
	Through ⅜
	Over ⅜, through ½
	Over ½, through ⅞
	Over ⅞
		3
		4
		5
		6
			3
			3
			5
			5

Interior
  (including grades with exterior glue)
    Underlayment
 	
Exterior
  C-C Plugged
	Through ½
	Over ½, through ¾
	Over ¾
		3
		4
		5
			3
			3
			5

Interior
  (including grades with exterior glue)
  C-D
  C-D Plugged
 	
Exterior
  C-C
	Through ⅝
	Over ⅝, through ¾
	Over ¾
		3
		4
		5
			3
			3
			5

Note: The proportion of wood based on nominal finished panel thickness and dry veneer thickness before layup, as used, with grain running perpendicular to the panel face grain shall fall within the range of 33 percent to 70 percent.

The combined thickness of all inner layers shall be not less than ½ of panel thickness based on nominal finished panel thickness and dry veneer thickness before layup, as used, for panels with 4 or more plies.

5–7b. *Types of panel construction.*

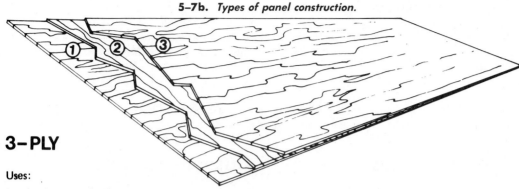

3–PLY

Uses:

Roof sheathing, subflooring, wall sheathing in residential and commercial construction, and in many industrial and agricultural applications.

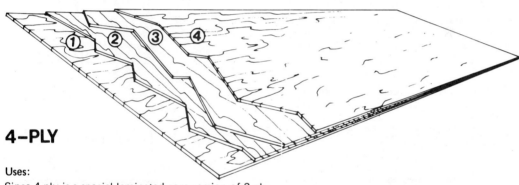

4–PLY

Uses:

Since 4-ply is a special laminated-core version of 3-ply plywood, it may be used in the same applications as 3-ply, including roof sheathing, subflooring, and wall sheathing.

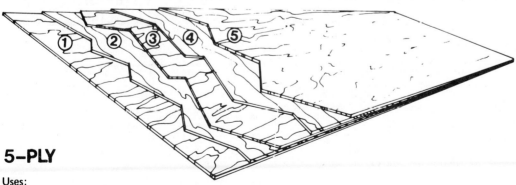

5–PLY

Uses:

Normal residential, commercial, industrial applications.

Group 2 plywood: 100 pcs. 48 inch by 96 inch, 3-ply interior type, A-D grade, sanded two sides to ¼ inch thickness.

The above order means that you want 100 pieces of 4′ × 8′ plywood made up of three plies with a face of the second highest veneer quality and a back of the lowest veneer quality, sanded on both sides to a finished thickness of ¼″. The material would be suited for all interior work.

5–8. *Typical grade markings. The top four show how these appear as back stamps while the lower one shows an edge marking.*

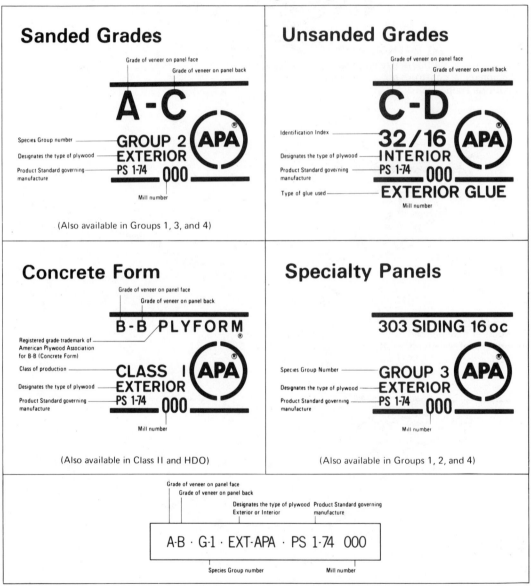

Another typical order might be: Group 3 plywood: 100 pcs. 48 inch by 96 inch, 3-ply exterior type, A-A grade, sanded two sides to ³/₈ inch thickness (add further special requirements).

For more detailed information concerning all plywood, obtain a copy of U.S. Product Standard PS 1-74 for Construction and Industrial Plywood with typical APA Grade-Trademarks from the American Plywood Association, Tacoma, Washington.

WORKING WITH PLYWOOD

Storage to Avoid Warpage. The best method of storing plywood is to lay the sheets flat. If this is not possible, they should be stored on edge with the sheets supported in a vertical position. Never lay plywood at an angle, especially the thinner panels, as it will warp.

Cutting to Avoid Splintered Edges. When hand sawing, always place plywood with the good face up and use a saw that has at least 10 to 15 points to the inch. Make sure that the panel is supported firmly so it will not sag. Hold the saw at a low angle when cutting and, if possible, place a piece of scrap stock underneath. When using a circular saw, install a special plywood blade or a sharp combination blade. The blade should be adjusted so that the teeth just clear the top of the stock. When cutting plywood on a table saw, always place the good side of the plywood up. When cutting with a portable, power handsaw, place the good face down.

Using Nails and Screws. Nails or screws do not hold well in the edges of plywood. It is important to remember this, especially when attaching hinges. Whenever possible, hinges for plywood doors should be the kind that attach to the face rather than to the edge.

When nailing plywood, always choose nail size in terms of panel thickness. Nails should be selected as follows: for ³/₄" plywood, 6d casing nails or 6d finishing nails, for ⁵/₈", 6d or 8d finishing nails; for ¹/₂", 4d or 6d; for ³/₈", 3d or 4d; and for ¹/₄"

5-9.
Identification Index (a)
Table for Sheathing Panels

Species of face and back	Grade			
Group 1	C-C Str. I C-C, C-D Str. II C-C, C-D(c) C-D ⎤ —(b)			
Group 2	C-C Str.II C-C, C-D C-D ⎤ (d)	C-C Str. II C-C, C-D C-D ⎤ —(b)		—(b)
Group 3		C-C Str. II C-C, C-D C-D ⎤ —(b)		
Group 4			C-C ⎤ —(d) C-D	C-C ⎤ —(b) C-D

Nominal Thickness				
5/16	20/0	16/0	12/0
3/8	24/0	20/0	16/0
1/2	32/16	24/0	24/0
5/8	42/20	32/16	30/12
3/4	48/24	42/20	36/16
7/8		48/24	42/20
(e)			

(a) Identification Index refers to the numbers in the lower portion of the table which are used in the marking of sheathing grades of plywood. The numbers are related to the species of panel face and back veneers and panel thickness in a manner to describe the bending properties of a panel. They are particularly applicable where panels are used for subflooring and roof sheathing to describe recommended maximum spans in inches under normal use conditions and to correspond with commonly accepted criteria. The left-hand number refers to spacing of roof framing with the right-hand number relating to spacing of floor framing. Actual maximum spans are established by local building codes.

(b) Panels of standard nominal thickness and construction.

(c) Panels manufactured with Group 1 faces but classified as Structural II by reason of Group 2 or Group 3 inner plies.

(d) Panels conforming to the special thickness and panel construction provisions.

(e) Panels thicker than ⅞ inch shall be identified by group number.

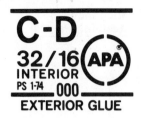  

5–10. *Typical grade markings on sheathing panels.*

use ³/₄″ or 1″ brads. For very careful installations, predrill to keep the nails from splitting out at the edge. The drill should be slightly smaller in diameter than the nail. Space nails about 6″ apart for most work. Closer spacing may be necessary only when nailing thin plywood to avoid buckling between the joints. Nails and glue together produce a strong joint. Flathead wood screws are needed when nails will not provide adequate holding power. Glue should also be used whenever possible.

The following gives plywood thicknesses, diameter, and the length of the smallest screws recommended:

³/₄″ plywood — No. 8—1¹/₂″
⁵/₈″ plywood — No. 8—1¹/₄″
¹/₂″ plywood — No. 6—1¹/₄″
³/₈″ plywood — No. 6—1″
¹/₄″ plywood — No. 4—³/₄″

(Use longer screws when the work permits.)

Screws or nails should be countersunk and the holes filled with wood dough, putty, or plugs. Apply filler until it is slightly higher than the plywood surface; then sand it level after it is dry.

Drilling. If the back side of plywood is going to show, chipped edges can be avoided by placing a wood block under the back when drilling.

HARDWOOD GRADE STANDARDS

Hardwood plywoods are used primarily for interior paneling, built-ins, cabinets and similar finish carpentry applications. A complete description of this material can be found in the textbook, *Cabinetmaking and Millwork,* also written by John L. Feirer and published by Bennett & McKnight Publishing Company

QUESTIONS

1. Describe the three common types of plywood construction.

2. Explain the steps in the manufacture of plywood.

3. List six uses for plywood in building construction.

4. Explain how exterior-type plywood differs from interior plywood.

5. How does grade "A" differ from grade "B" in plywood?

6. Name three species of wood from the strongest group for plywood construction.

7. Explain some of the precautions that must be taken in working with plywood.

Building Board—Hardboard, Particle Board, and Others

Many basic building materials, including wood, cement, and brick, are not uniform in quality, weight, flexibility, and insulating value. Individual pieces also vary widely in size and strength. To eliminate these disadvantages, many types of *building board products* are manufactured. Fig. 6-1. All of these board products are made of wood or other materials that have been reduced to small components and then reassembled into panel form. They are all flat, relatively thin, and come in standard widths and lengths (usually 4' × 8'). Fig. 6-2. Plywood is a good example of a building board (see Unit 5). Other common building boards include hardboard, particle board, structural insulating board, gypsum board, and plastic foam board.

HARDBOARD

Hardboard is an all-wood panel manufactured from wood fibers. Logs are cut

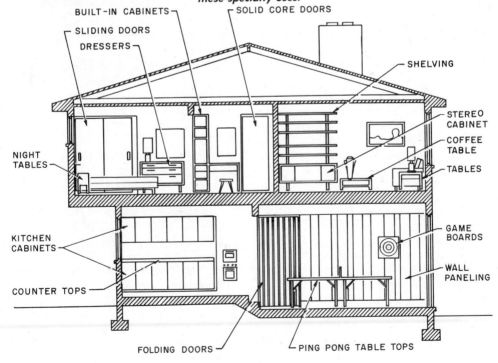

6–1. *Hardboard and particle board can be used for many purposes in home construction, including these specialty uses.*

BUILT-IN CABINETS

SOLID CORE DOORS

SLIDING DOORS

DRESSERS

SHELVING

NIGHT TABLES

STEREO CABINET

COFFEE TABLE

TABLES

KITCHEN CABINETS

GAME BOARDS

COUNTER TOPS

WALL PANELING

FOLDING DOORS

PING PONG TABLE TOPS

6-2. *A major advantage of using panel stock—uniformity of size and shape—is illustrated in these stacks of standard size hardboard.*

6-3. *Decorative exterior hardboard such as this V-grooved panel simplifies construction.*

into small wood chips which are reduced to fibers by steam or mechanical processes. These fibers are refined, then compressed under heat and pressure in giant presses to produce a sturdy, quality building material.

Some of the advantages of hardboard are:

- It has exceptional strength.
- It has superior wear resistance.
- It does not split, crack, or splinter.
- It has high abrasive resistance.
- It has permanent resistance to moisture.
- It is easy to work with ordinary tools.
- It is easy to fasten and bend.

Hardboard has many uses for interiors and exteriors of new and remodeled homes. About half of all hardboard is used in building construction for siding, interior paneling, underlayment, kitchen cabinets, and similar purposes. Fig. 6-3.

Types and Sizes of Hardboard

Hardboard is made in three basic types: *standard, tempered,* and *service.* Standard hardboard is given no additional treatment after manufacture. This board has high strength and good water resistance. It is commonly used in cabinetwork because it has a very smooth surface and finishes well. Tempered hardboard is standard board to which chemical and heat-treating

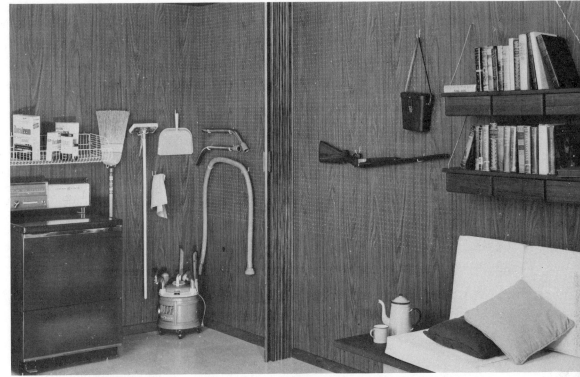

6-4. *Perforated hardboard can be useful in many rooms of a house.*

processes have been applied to improve stiffness, hardness, and finishing properties. Service hardboard has somewhat less strength than standard. It is used where low weight is an advantage. It does not have quite as smooth a surface as standard.

Hardboard is manufactured with one or both sides smooth. One side smooth is known as S1S and two sides smooth is S2S. Hardboard is available in thicknesses from $1/8''$ to $3/8''$, the common thicknesses being $1/8''$, $3/16''$, and $1/4''$. The standard panel size is 4' × 8', but widths up to 6' and lengths to 16' are also available.

SPECIALTY HARDBOARDS

Because hardboard is manufactured, it can be made in shapes, sizes, and surfaces to meet varying needs of the building industry. The following paragraphs list a few of the many kinds available.

Perforated hardboard has very closely spaced holes punched or drilled in the surface. The openings may be round, square, or diagonal and can be fitted with metal hooks, holders, supports, or similar fittings. Fig. 6-4. Such hardboard is in common use not only in homes, but in stores for display and storage.

Embossed patterns are available in simulated leather, wood grain, and basket weave.

Acoustical hardboard has perforations which improve its properties for controlling sound. It makes an excellent covering for ceilings and walls.

Wood-grain hardboard is printed with grain to match the color and texture of oak, walnut, mahogany, and many other woods.

6–5. *Both hardboard and particle board, shown here, make excellent underlay material for finish flooring.*

It is popular for interior paneling. Many other types of hardboard panels are designed for building construction.

Hardboard exterior siding is available in both horizontal lap and vertical panel style, fully prefinished or factory-primed. Lap siding, which is made in widths of 6" to 12" and up to 16' in length, is applied horizontally as clapboard. The vertical panel siding is available in 4' widths and in lengths up to 16'. Some lap and panel sidings are embossed with wood grain or other textures. Panel sidings come with V-grooved surfaces or in plain panels for board-and-batten style application.

Filigree hardboard panels are decora-

tive items. They are made of tempered hardboard, smooth on both sides, and die-cut to provide the filigree patterns. They are normally 1/8" thick and available in sizes up to 4' × 8'. They are used in cabinet doors, room dividers, folding screens, sliding doors, and as accent wall panels, among others.

Hardboard underlayment is service-grade hardboard, planed to a uniform thickness. It comes in 3' × 4' and 4' × 4' sizes. These sheets are nailed (with ringed underlayment nails) or stapled over sub-flooring or old finish to serve as a uniform base for floor tile, linoleum, and carpeting. Fig. 6-5. Panels should be fitted in staggered fashion, spaced about 1/32" apart. Joints should not coincide with those on the subfloor. Use 4" spacing between fasteners across the entire area of the panel. Hardboard manufacturers have their own instructions for applying underlayment, which should be followed.

Other special-purpose hardboard products include ceiling blocks, attractive ceiling beams, and barn siding, among others. There is also a wide range of accessories to simplify the use of hardboards. Examples of accessories are metal, wood, vinyl, or hardboard moldings, color-matched to the prefinished paneling; color-matched nails; and special adhesives for nail-free applications.

PARTICLE BOARD

Particle board is made by combining wood flakes or particles with resin binders, and hot-pressing them into panels. The panels range from 1/4" to 1 1/2" thick, from 3' to 8' wide, and up to 24' in length. Almost any dimension may be obtained by cutting or gluing segments together. In the manufacturing process, the composition and properties of each panel are carefully controlled.

There are two basic types of particle board—*extruded* and *mat-formed*. Panels made by the extrusion process are generally used by the manufacturer that produced them. Mat-formed wood particle board is sold for a wide variety of uses in the building and woodworking trades.

Uses in Architecture and Home Building

Particle board is an important material for providing flat, true surfaces which are needed in most homes. Uses for this material in home building include nearly all types of counters; cabinets; drawers; shelving; vanities; sliding, folding, solid-core, or accordion doors; room dividers, and many other kinds of built-ins.

One of the largest uses of particle board in the home is in the construction of kitchen counters, sink tops, and cabinets. Fig. 6-6. Components made with particle board may be purchased from cabinet and counter top manufacturers or built on the site. Many cabinet and sink top manufacturers use particle board because of its smooth, grain-free surface, which is a good base for high-pressure laminates. Also, doors made of particle board are warp-free, requiring a minimum of on-site adjustment.

Counter tops and cabinets of particle board can be built on the construction site. Often this is done with panels veneered with popular kitchen cabinet woods. These are commonly available through lumber and building materials dealers. Particle board also makes good shelving, which can be painted easily.

The largest single use of particle board in the construction industry is for floor underlayment. It provides a smooth, stable base for resilient tile or carpeting.

Selecting and Specifying Particle Board

A wide choice of particle board products is available. This is partly because there are many manufacturers and they use different raw materials and processes. Another reason for the variety is that since particle board is a man-made product, it can be

6–6. *Applying plastic laminate to a particle-board core for counter tops.*

produced in different forms to meet specific needs. Fig. 6-7.

To select the right particle board product for a specific job, be sure to read the manufacturers' descriptions of their materials. Such descriptions will tell you the specific characteristics of each product. Also you will need to know the requirements for the job you are doing.

For example, one type of particle board may have very high stiffness; it would be best suited for use as a core material for very large table tops, large folding partitions or wall panels, or for use where intermediate supports are limited. However, another type with lower stiffness might be just as good if adequately supported or overlaid.

Besides stiffness, there are other properties—machinability, edge-finishing characteristics, hardness, surface smoothness—which are important in selecting a specific panel. It is important to know how these and other properties vary from one type of panel to another.

6–7.

Types and Uses of Particle Board

1. Corestock	Products of flakes or particles, bonded with urea-formaldehyde or phenolic resins with various densities and related properties.	For furniture, casework, architectural paneling, doors and laminated components.
2. Wood Veneered Particle Board	Corestock overlaid at the mill with various wood veneers.	For furniture, panels, wainscots, dividers, cabinets, etc.
3. Overlaid Particle Board	Particle board faced with impregnated fiber sheets, hardboard or decorative plastic sheets.	For applications such as furniture doors, wall paneling, sink tops, cabinetry and store fixtures.
4. Embossed Particle Board	Surfaces are heavily textured in various decorative patterns by branding with heated roller.	For doors, architectural paneling, wainscots, display units and cabinet panels.
5. Filled Particle Board	Particle board surface-filled and sanded ready for painting.	For painted end-products requiring firm, flat, true surfaces.
6. Exterior Particle Board	Made with phenolic resins for resistance to weathering.	For use as an exterior covering material.
7. Toxic-Treated Particle Board	Particle board treated with chemicals to resist insects, mold and decay producing fungi.	For tropical or other applications where wood products require protection against insect attack or decay.
8. Primed or Undercoated	Factory painted base coat on either filled or regular board—exterior or interior.	For any painted products.
9. Floor Underlayment	Panels specifically engineered for floor underlayment.	Underlay for carpets or resilient floor coverings.
10. Fire-Retardant Particle Board	Particles are treated with fire retardants.	For use where building codes require low flame spread material, as in some schools, office buildings, etc.

6–8.

Typical Working Characteristics Of Manufactured Board

	Thick Panels—¼"—1½"		Thin Panels—⅛"—⅜"		
	Flake Board 42# cu. ft.	Particle Board 40# cu. ft.	Hardboard		
			Standard	Tempered	Specialties
Bending	Fair	Fair	Good	Excellent	Good
Drilling	Excellent	Good	Good	Excellent	Excellent
Hardness	High	Medium	Medium	High	High
Laminating	Excellent	Good	Good	Excellent	Excellent
Nailing	Good	Good	Fair	Good	Good
Painting	Unfilled—Good Filled—Excel.	Unfilled—Fair Filled—Good	Fair	Excellent	Excellent
Punching	Fair	Fair	Fair	Excellent	Good
Routing	Excellent	Good	Fair	Excellent	Good
Sanding	Excellent	Good	Fair	Excellent	Good
Sawing	Excellent	Good	Fair	Excellent	Good
Screw Holding	Excellent	Good	Fair	Good	Good
Shaping	Excellent	Good	Fair	Excellent	Good
Water Resistance	Interior or Exterior	Interior or Exterior	Interior	Exterior	Interior or Exterior

There are certain variables which can be controlled during the manufacturing process to produce a certain type of panel. These variables include density, flake or particle shape, amount and type of resin, and moisture content.

DENSITY

Stock panels range in density from 24 to 62 pounds per cubic foot. Ordinarily, a panel with high density will also have greater strength and a smoother, tighter edge than a low-density panel.

PARTICLES USED

Two strong influences on the properties of board products are:
- The size and shape of individual flakes and particles.
- The ratio of resin to particles.

Particle shape and resin content can be controlled to create a given set of physical properties. The size, type, and position of the particles also will influence a panel's surface smoothness.

RESIN

Two types of resin, urea-formaldehyde and phenol-formaldehyde, are used in the manufacture of particle board. Urea-formaldehyde is the most common and is suitable for interior use. Phenol-formaldehyde is used where the panel is subjected to extreme heat or humidity or for exterior applications.

MOISTURE CONTENT

It is especially important to control the moisture content of those particle board panels which are to be overlaid. This is because the core material and the overlay material must have nearly equal moisture content. Panels normally are shipped from the mill at a moisture content of 7 to 9% unless otherwise specified.

SPECIAL PROPERTIES

Panels can be given special finishes or treatments at the mill. They may be filled or primed for easy painting, or embossed for decorative, textured surfaces. The edges of panels may be banded with lumber, cut to size, or given special sanding or overlays. Laminating and/or edge-gluing is done to make panels of unusual sizes.

Working with Particle Board and Hardboard

These materials are free from the cracks and other imperfections commonly found in wood. They present none of the problems related to grain in wood. Fig. 6-8.

Generally, particle board and hardboard are worked with standard woodworking tools. These materials can be sawed, routed, rabbeted, shaped, and drilled cleanly, with good edges and corners. Fig. 6-9. Since these boards are made to exact thicknesses and finish-sanded at the mill, there is little need for further surface preparation. (For information on finishing these materials, see Appendix.)

All types of joints usually employed in casework or used with architectural assemblies are readily made with particle board. Architectural panels may be butted or V-grooved, splined, or the joint emphasized with a batten or decorative molding. In cabinetry, joint types include ordinary miter, splined or lockmiter, dowel, mortise and tenon, dovetail, and tongue and groove.

The absence of voids gives these boards a full, uniform contact surface for glued joints. This assures strong glued butt joints, permitting short lengths to be glued into longer sections for a minimum of waste.

STRUCTURAL INSULATING BOARD

Most structural insulating board is made from wood fibers. It comes in two grades—*sheathing* and *insulation*.

There are two types of sheathing-grade insulation board. In one type all the surfaces and edges are covered with asphalt. In the other type the fibers are impregnated with asphalt during manufacture. These

6-9. *There is no chipping or splintering when cutting hardboard or particle board.*

boards usually come in 4' × 8' sheets, 25/32" thick. They also come in 2' × 8' sheets. Sheathing grade is used for insulation and sound control as well as for structural sheathing.

Insulation grade is made in decorative panels, decorative ceiling tile, V-notched plaster base, and roof insulation. The standard thicknesses of this type of board are ½", ⅝", ¾", or 1", although thicker boards up to 2" to 3" for roof insulation are also made. Sometimes thicker board is made by using an insulating board in the middle and a ¼" hardboard on both surfaces. Ceiling tiles are made in a wide variety of sizes with tongue-and-groove edges, and with a choice of finishes. They are also made with a series of small holes to improve the sound control. Ceiling tile can be cemented, clipped, stapled, nailed, or interlocked in place. Acoustical tile absorbs up to seventy percent of the noise in a room.

GYPSUM

Gypsum wallboard is a fire-resistant sheathing for home interiors. It is used in dry-wall construction. It is made of a core of gypsum (plaster of paris) covered on both sides with heavy kraft paper. The paper on the exposed surface is ivory in color while the other side is gray. The sheets are made in 2' and 4' widths and in common lengths from 6' to 16'. Although the standard thickness is ½", it is also

available in ¼", ⅜", ⅝", and 1" thicknesses. The board is available with a variety of edge joints—square, tapered, round, beveled, and tongue-and-groove. The material is also available with aluminum backing. Another kind of gypsum wallboard has a base of gypsum mixed with fiber glass for improved fire resistance. For interior walls, this board is also available with vinyl exposed surface.

Gypsum lath is the same kind of material made in sizes of 16" × 48" or 96". While the most common thickness is ⅜", it is also available in ½". Laths ⅜" × 16" × 48" are usually packaged six to a bundle. This lath is used as a base for plastering. For improved fire protection, a perforated lath is used which allows a heavier layer of plaster to be applied to it.

PLASTIC FOAM BOARD

Plastic foam sheets of polystyrene and polyurethane are excellent building materials for certain structural and insulating purposes. For example, these materials can be used as insulation around a concrete slab and for wall and deck insulation. These sheets are made in ½" to 3" thicknesses, 12" to 24" widths, and 4' to 8' lengths.

OTHER TYPES OF BOARDS

Other types of building board are strawboard, corkboard, mineral fiberboard wa-

6-10. *Waferboard. Note that the wafers are not aligned.*

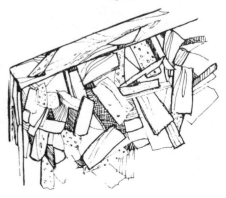

6-11. *Waferboard is excellent for wall and roof sheathing.*

ferboard, and oriented strandboard. Strawboard is made of compressed wheat straw covered with kraft paper. Corkboard is made of a mixture of ground cork and synthetic resin. Mineral fiberboard is made from glass or rock wool covered by stiff paper.

Waferboard is an exterior bonded structural panel. It is made of wafers cut from roundwood bonded with a waterproof phenolic resin. Fig. 6-10. Waferboard is different from particleboard, which is intended for interior and furniture use. Because waferboard has great strength, rigidity, and moisture resistance, it is widely used for subflooring and exterior sheathing. It is easy to saw, drill, nail, plane, file, or sand since it contains nothing but natural wood, waterproof resin, and a small amount of wax. Waferboard can be worked with regular carpenter tools. Its nail-holding properties make it ideal for sheathing and subflooring. Fig. 6-11. The panel can be finished with any good quality paint. It can be produced in large sizes. For example, a single sheet of 8′ × 28′ can replace seven 4′ × 8′s as a subfloor of a trailer.

Oriented strand board (OSB) is similar to waferboard since it is manufactured from reconstituted wood strands bonded with resins under heat and pressure. It differs from waferboard in that the strands are mechanically aligned. This improves the properties in the aligned direction but reduces properties perpendicular to alignment. Oriented strand materials may be produced as the center layer of composite panels, or may be cross-laminated in layered, nonveneered panels. Fig. 6-12.

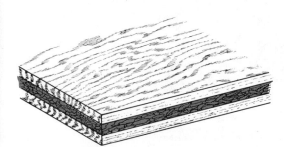

6-12a. *Oriented strand board as a core between layers of veneer.*

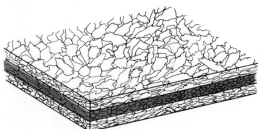

6-12b. *OSB cross-laminated upon itself. Note that the strands of OSB are oriented in one direction.*

QUESTIONS

1. What are the advantages of manufactured building board products?

2. List five manufactured building boards.

3. List five specialty hardboards and several uses of each.

4. How is particle board made?

5. What are the two basic types of particle board?

6. What is one of the largest uses of particle board in the home?

Section III

TOOLS AND MACHINES

UNIT 7

Safety

In building construction, safety is regarded as an important part of the total operation for at least two reasons. One is the natural concern for people's welfare. Another is financial. Building contractors know that injuries to employees are costly since they reduce the efficiency of the working force and may result in expensive medical bills and law suits. Therefore many contractors have safety programs which are intended to protect employees.

Safety is of prime importance in the operation of the power tools you will be using in building construction. Therefore you should not just read but *really learn* the safety rules for any job you do, and put the rules into practice. Do this for each piece of equipment you use. Learn to make safety a habit as you develop your skills.

This unit contains information on how to work safely in the building construction trades.

The important thing to remember is that serious accidents will not occur if workers are safety conscious and follow recommended precautions. The following information should be valuable in aiding you to develop the correct attitudes toward safety and acquaint you with certain safety precautions.

DRESS

● Wear safety glasses whenever your work involves a threat to your eyes. (Sometimes this is required by law.) Fig. 7-1.

● Clothing should suit your individual needs and be suitable for the prevailing weather conditions. Avoid loose fitting clothing that would restrict freedom of movement.

● In a shop or on a job site, you will probably work harder and safer in clothes you are not afraid to get dirty. If at all possible, as you enter the shop each day you should exchange your regular clothing for coveralls or other accepted working attire.

● Avoid wearing pants or overalls that are too long. Cuffs made by turning up the legs of pants tend to catch heels, causing falls.

● To avoid catching on nails, keep the sleeves of shirts or jackets buttoned.

7-1. Always wear adequate eye protection for the job to be performed. Sometimes it is necessary also to wear a mask for operations such as spray painting.

7-2. Supervisory and sales personnel should never enter the job site without wearing a hard hat and eye protection.

- Keep hair cut short or keep long hair in place with a visor or hair net.
- To protect feet from protruding nails, wear shoes with thick, sturdy soles.
- To protect feet from falling objects, wear safety shoes or boots with steel toe caps.
- Wear a hard hat when exposed to overhead work or whenever there is danger from falling objects. Fig. 7-2.
- Remove neckties; also rings, wristwatches, neck chains, and other jewelry.

MATERIAL HANDLING

- Long pieces of material should be carried by two persons. Fig. 7-3.
- Use the muscles in your legs and knees to lift heavy objects. Do not lift with your back muscles. To do so could result in painful back injuries.
- Observe caution when carrying planks or other objects across frozen, wet, or otherwise slippery footing.

AVOID FALLS

- Learn to watch your footing; avoid objects that could trip you.

- Check scaffolding and temporary walkways before walking on them. Be sure the supports are strong and secure.
- Use only ladders which are in good condition and set up properly.

PRACTICE GOOD HOUSEKEEPING

- Materials and equipment should be stacked straight and neat.
- Keep aisles and walkways clear of tools, materials, and debris.
- To prevent fires and reduce hazards which cause accidents, dispose of scraps and rubbish daily.
- Whenever you see protruding nails, remove them or bend them down immediately.
- When working above other people, place tools and materials where they will not fall and cause injuries.

GENERAL SAFETY

The following is a list of general safety rules to be used in the shop or on the job site as you work on and around machines and construction. These rules will help to protect you and others who are near you while you work.

- Always walk—do not run.
- Never talk to or interrupt anyone who is working on a machine.
- Remove power plug or turn off power supply to a machine when changing cutters or blades.
- Never leave tools or pieces of stock lying on the table surface of a machine being used.
- When finished with a machine, turn off the power and wait until the blade or cutter has come to a complete stop before leaving.
- Always carefully check stock for knots, splits, metal objects, and other defects before machining.
- Do not use a machine until you understand it thoroughly. Any tool with a sharp cutting edge can cause serious injury if mishandled.

• Use guards on power equipment. It should be understood that using guards does not necessarily prevent accidents. Guards must be used correctly if they are to provide fullest protection. Also, it is impossible to do some operations, especially on the circular saw, with the regular guard in place. Therefore there are times when special guards should be used.

• Always keep your fingers away from the moving cutting edges. The most common accident is caused by trying to run too small a piece through a machine.

• Keep the floor around the machine clean. The danger from falling or slipping is always great.

• Make all adjustments with the power off and the machine at a dead stop.

• Always use a brush to clean the table surface.

• Always keep your eyes focused on where the cutting action is taking place.

• Always use sharp tools.

• When using tools for set-up work on a machine: (1) Select the right tool for the job. (2) Keep it in safe condition. (3) Keep it in a safe place.

• Report strange noises or faulty operation of machines to the instructor.

• Follow the suggestions for each machine given in this book.

The Congress of the United States in April of 1971 made the Federal Occupational Safety and Health Act (OSHA) an official part of the national labor law. The purpose of this law is ". . . to assure so far as possible every working man and woman in the nation safe and healthful working conditions and to preserve our human resources." This law affects all employees who are working in the building trades where one or more workers are employed. As an individual employed in the building trades it is just as important to develop safe work attitudes and habits as outlined by this law as it is to develop the skills of your trade. Building trades employers will be looking for men and women with these

traits for their benefit and welfare as well as yours as a skilled tradesman. Thus it is important to know and follow safety rules.

GENERAL SAFETY RULES
FOR
PORTABLE POWER TOOLS

• Never use portable power tools in contact with water, including rain, or if any part of your body is in contact with moisture. Be sure the power plug is removed before making any adjustments.

• Portable power tools should be properly grounded with a three-prong grounded plug. If a grounded receptacle is not available, use a three-to-two prong adapter plug which has been properly grounded.

• Always wear approved eye protection.

• Always disconnect the power plug when the work is completed.

• Be sure the switch is in the "off" position before connecting the power plug.

• Always use the recommended extension cord size. Fig. 7-4.

NOTE: In the following units on tools and machines, many illustrations show dangerous operations being performed on machines without guards. The guards have been removed so that the photographs will show the operations more clearly. Whenever a drawing of a guard appears with an illustration, **a guard must be used** for the operation that is shown. Fig. 7-5.

7–3. Heavy loads are often moved with a fork lift.

7–4.
Recommended Extension Cord Sizes For Use With Portable Electric Tools

Name-plate Amperes	Cord Length In Feet																			
	25	50	75	100	125	150	175	200	225	250	275	300	325	350	375	400	425	450	475	500
1	16	16	16	16	16	16	16	16	16	16	16	16	16	16	16	16	16	16	16	14
2	16	16	16	16	16	16	16	16	16	16	14	14	14	14	14	12	12	12	12	12
3	16	16	16	16	16	16	14	14	14	14	12	12	12	12	12	12	10	10	10	10
4	16	16	16	16	16	14	14	12	12	12	12	12	12	12	10	10	10	10	10	10
5	16	16	16	16	14	14	12	12	12	12	10	10	10	10	10	8	8	8	8	8
6	16	16	16	14	14	12	12	12	10	10	10	10	10	8	8	8	8	8	8	8
7	16	16	14	14	12	12	12	10	10	10	10	8	8	8	8	8	8	8	8	8
8	14	14	14	14	12	12	10	10	10	10	8	8	8	8	8	8	8	8		
9	14	14	14	12	12	10	10	10	8	8	8	8	8	8	8	8	8			
10	14	14	14	12	12	10	10	10	8	8	8	8	8	8	8	8				
11	12	12	12	12	10	10	10	8	8	8	8	8	8	8						
12	12	12	12	12	10	10	8	8	8	8	8	8	8							
13	12	12	12	12	10	10	8	8	8	8	8	8								
14	10	10	10	10	10	10	8	8	8	8	8									
15	10	10	10	10	10	8	8	8	8	8										
16	10	10	10	10	10	8	8	8	8	8										
17	10	10	10	10	10	8	8	8	8											
18	8	8	8	8	8	8	8	8	8											
19	8	8	8	8	8	8	8	8												
20	8	8	8	8	8	8	8	8												

Notes: Wire sizes are for 3-CDR Cords, one CDR of which is used to provide a continuous grounding circuit from tool housing to receptacle.
Wire sizes shown are A. W. G. (American Wire Gauge).
Based on 115V power supply; Ambient Temp. of 30°C, 86°F.

7–5. *Whenever this drawing of a guard appears with an illustration, a guard must be used for the operation shown.*

QUESTIONS

1. Why is safety so important to the building contractor?

2. Why should students wear some form of work clothes for building construction?

3. What features should you look for when buying shoes to be used for working on a job site?

4. Do guards prevent accidents on power machines? Explain.

5. Why should portable power tools be grounded?

6. Explain why it is important to wear proper eye protection at all times, even when you are not operating equipment.

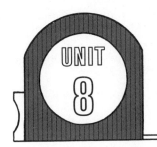

Hand Tools

LAYOUT, MEASURING, AND CHECKING DEVICES

Tool	Description	Uses
Bench Rule *Fig. 8-1.*	A 12-inch or one foot (or 300 mm) rule. One side is divided into eighths, the other into sixteenths. A metric rule is divided into centimetres or millimetres.	1. To make simple measurements. 2. To adjust dividers. *Caution.* Never use as a straight-edge.
Zig-Zag Rule *Fig. 8-2.*	A folding rule of six- or eight-foot (or 2 m) length.	1. To measure distances greater than 2′ (600 mm), place the rule flat on the stock. 2. To measure less than 2′ (600 mm) it is better to use the rule on edge. (This instrument is good for inside measurement, since the reading on the brass extension can be added to the length of the rule itself.)

8–1a.

8–1b.

Tool	Description	Uses
Flexible Tape Rules *Fig. 8-3.*	A flexible tape that slides into a metal case. Comes in lengths of 6′, 8′, 10′, 12′, 50′, and 100′ (2 m to 50 m). The steel tape has a hook on the end that adjusts to true zero.	1. To measure irregular as well as regular shapes. 2. To make accurate inside measurements. (Measurement is read by adding 2″ (50 mm) to the reading on the blade.)

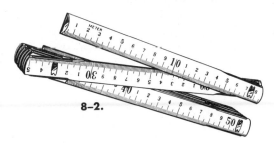

8–2.

8–3a.

Try Square
Fig. 8-4.

A squaring, measuring, and testing tool with a metal blade and a wood or metal handle.

1. To test a surface for levelness.
2. To check adjacent surfaces for squareness.
3. To make lines across the face or edge of stock.

Combination Square
Fig. 8-5.

Consists of a blade and handle. The blade slides along in the handle or head. There is a level and a scriber in the handle.

1. To test a level or plumb surface.
2. To check squareness—either inside or outside.
3. To mark and test a 45-degree miter.
4. To gauge-mark a line with a pencil.

Sliding T Bevel
Fig. 8-6.

A blade that can be set at any angle to the handle. Set with a framing square or protractor.

1. To measure or transfer an angle between 0 and 180 degrees.
2. To check or test a miter cut.

Dividers
Fig. 8-7.

A tool with two metal legs. One metal leg can be removed and replaced with a pencil. To set the dividers, hold both points on the measuring lines of the rule.

1. To lay out an arc or circle.
2. To step off measurements.
3. To divide distances along a straight line.

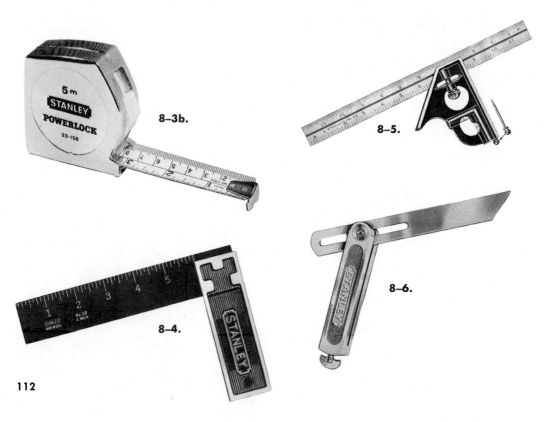

8–3b.

8–5.

8–4.

8–6.

Framing or Rafter Square *Fig. 8-8.*	A large steel square consisting of a blade, or body, and a tongue.	1. To check for squareness. 2. To mark a line across a board. 3. To lay out rafters and stairs.
Carpenter's Level *Fig. 8-9.*	A rectangular metal or wood frame with several level glasses.	To check whether a surface is level or plumb.
Marking Gage *Fig. 8-10.*	A wood or metal tool consisting of a beam, head, and point.	To mark a line parallel to the grain of wood.
Scratch Awl *Fig. 8-11.*	A pointed metal tool with handle.	1. To locate a point of measurement. 2. To scribe a line accurately.
Trammel Points *Fig. 8-12.*	Two metal pointers that can be fastened to a long bar of wood or metal.	1. To lay out distances between two points. 2. To scribe arcs and circles, larger than those made with dividers.

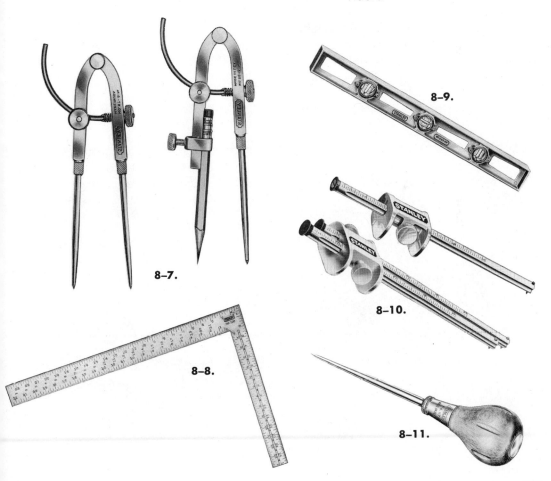

8–7.

8–8.

8–9.

8–10.

8–11.

| Plumb Bob and Line *Fig. 8-13.* | A metal weight with a pointed end. The opposite end has a hole for attaching the cord. | 1. To determine the corners of buildings.
2. To establish a vertical line. |

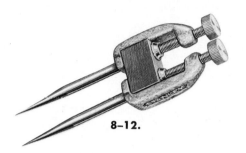

8–12.

8–13.

SAWING TOOLS

Tool	Description	Uses
Back Saw *Fig. 8-14.*	A fine-tooth crosscut saw with a heavy metal band across the back to strengthen the thin blade.	1. To make fine cuts for joinery. 2. To use in a miter box.
Crosscut Saw *Fig. 8-15.*	A hand saw in lengths from 20″ to 26″ with from 4 to 12 points per inch. A 22″, 10-point saw is a good one for general purpose work.	1. To cut across grain. 2. Can be used to cut with the grain. *Caution:* Never cut into nails or screws. Never twist off strips of waste stock.
Rip Saw *Fig. 8-16.*	A hand saw in lengths from 20″ to 28″. A 26″, 5¹/₂-point saw is good for general use.	To cut with the grain. *Caution:* Support the waste stock. Never allow end of saw to strike the floor.

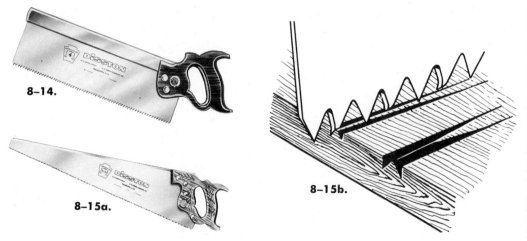

8–14.

8–15a.

8–15b.

Compass Saw *Fig. 8-17.*	A 12″ or 14″ taper blade saw.	1. To cut gentle curves. 2. To cut inside curves.
Keyhole Saw *Fig. 8-18.*	A 10″ or 12″ narrow taper saw with fine teeth.	To cut small openings and fine work.
Miter Box Saw *Fig. 8-19.*	A longer back saw (24″ to 28″).	Used in a homemade or commercial miter box for cutting miters or square ends.
Coping Saw *Fig. 8-20.*	A U-shaped saw frame permitting $4^5/_8$″ or $6^1/_2$″ deep cuts. Uses standard $6^1/_2$″ pin-end blades.	1. To cut curves. 2. To shape the ends of molding for joints. 3. For scroll work.
Dovetail Saw *Fig. 8-21.*	An extremely thin blade with very fine teeth.	For smoothest possible joint cuts.

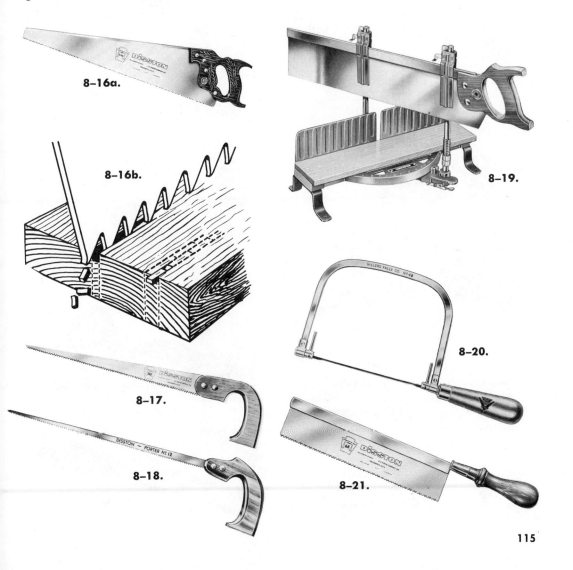

8–16a.

8–16b.

8–19.

8–17.

8–18.

8–20.

8–21.

EDGE-CUTTING TOOLS

Tool	Description	Uses
Smooth Plane *Fig. 8-22.*	A 7″ to 9″ plane.	1. For general use. 2. For smaller work.
Jack Plane *Fig. 8-23.*	A 14″ or 15″ plane.	1. Ideal for rough surface where chip should be coarse. 2. Also used to obtain a smooth, flat surface.
Fore Plane *Fig. 8-24.*	An 18″ plane.	For fine flat finish on longer surfaces and edges.
Jointer Plane *Fig. 8-25.*	A 22″ or 24″ plane.	1. To smooth and flatten edges for making a close-fitting joint. 2. For planing long boards such as the edges of doors.
Router Plane *Fig. 8-26.*	A cutting tool with several cutters.	To surface the bottom of grooves and dadoes.
Block Plane *Fig. 8-27.*	A small plane with a single, low-angle cutter with the bevel up.	1. To plane end grain. 2. For small pieces. 3. For planing the ends of molding, trim, and siding.

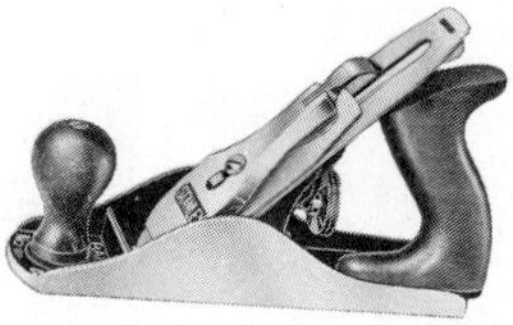

8–22.

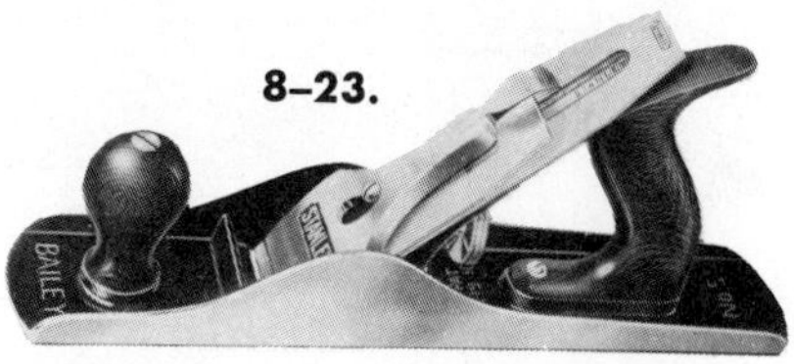

8–23.

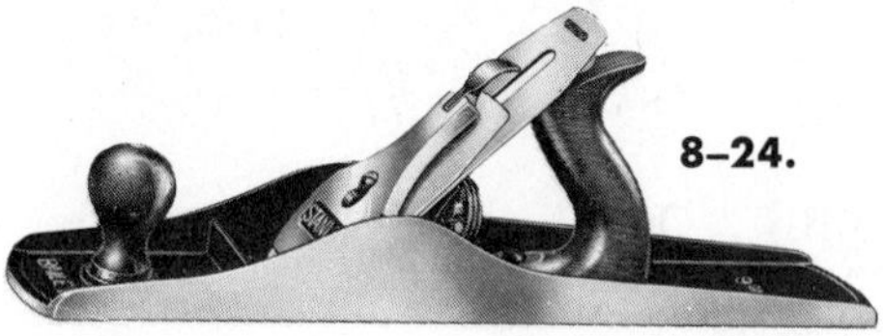

8–26.

8–24.

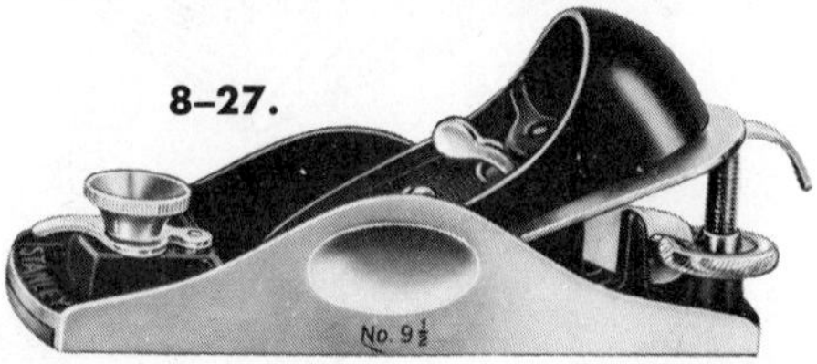

8–27.

8–25.

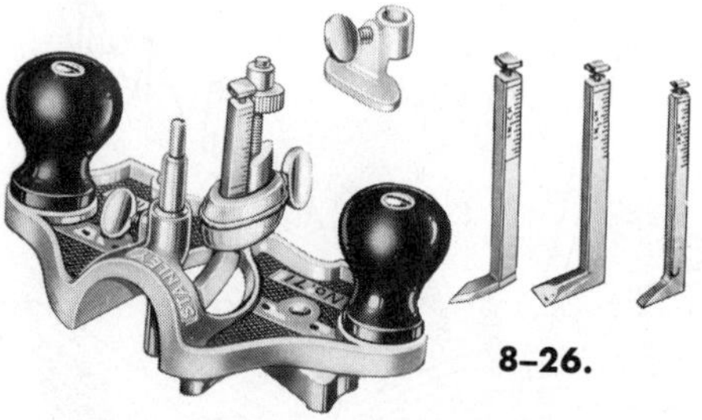

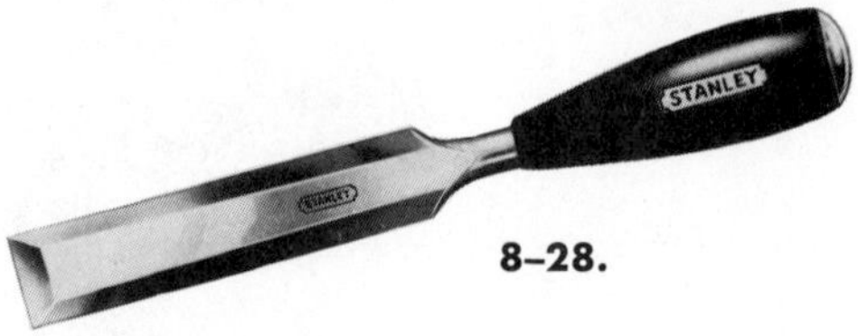

8–28.

Tool	Description	Uses
Chisels *Fig. 8-28.*	A set usually includes blade widths from ⅛″ to 2″.	To trim and shape wood.
Surform Tool® *Fig. 8-29.*	Available in plane file type. Also round, or block-plane types. A blade with 45-degree cutting teeth.	For all types of cutting and trimming.
Hatchet *Fig. 8-30.*	A cutting tool with a curved edge on one side and a hammer head on the other. Has hammer-length handle.	To trim pieces to fit in building construction. For nailing flooring.
Utility Knife *Fig. 8-31.*	An all-purpose knife with retractable blade.	1. To cut and trim wood, veneer, hardboard, and particle board. 2. To make accurate layouts.

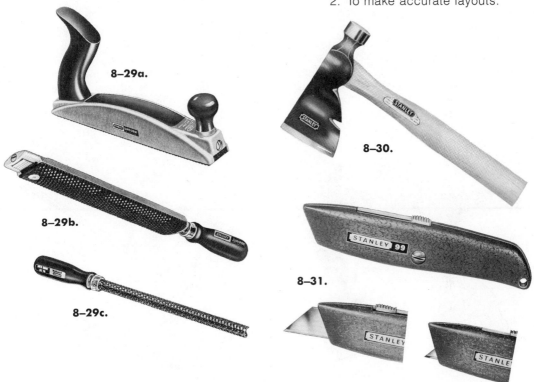

8–29a.

8–29b.

8–30.

8–31.

8–29c.

FASTENING AND ASSEMBLY TOOLS

Tool	Description	Uses
Vise *Fig. 8-32.*	Metal vise has replaceable wooden faces to protect wood. Larger vises attach permanently to work bench. Others clamp to table or counter.	Holds work for sawing, planing sanding, and many other jobs.

Claw Hammer *Fig. 8-33.*	Has curved claw. Heads weigh from 5 to 20 ounces. Face may be flat, bell, or checkered. Handle may be of wood, steel, or fiberglass.	For driving or removing nails. Use 16-ounce head with flat or bell face for general construction. Steel or fiberglass handles are better than wood.
Rip Hammer *Fig. 8-34.*	Has wedge-shaped claw.	For prying apart pieces that have been nailed together.
Mallet *Fig. 8-35.*	Heads with two striking surfaces, often made of wood, rubber, or plastic.	Used for striking blows where steel hammers would mar or damage the surface.
Ripping bars *Fig. 8-36.*	Available in lengths up to 8'. Three-foot bar is suited for general use.	For pulling large nails and for removing old materials during renovation.
Nail Set *Fig. 8-37.*	Concave tip will not slip from nail head. Tip diameter from $\frac{1}{32}''$ to $\frac{5}{32}''$.	For driving nails below the surface on interior trim. Nail holes can then be filled.
Screw Drivers *Fig. 8-38.*	For slotted and Phillips head screws. Standard slotted head widens from tip to shank. Cabinet slotted head is a uniform width to reach recessed screws.	Select head width that most closely fits screw slot. Use No. 1 Phillips for screw gauges 0-4, No. 2 for gauges 5-9, No. 3 for gauges 10-16, and No. 4 for gauges 17 and larger.

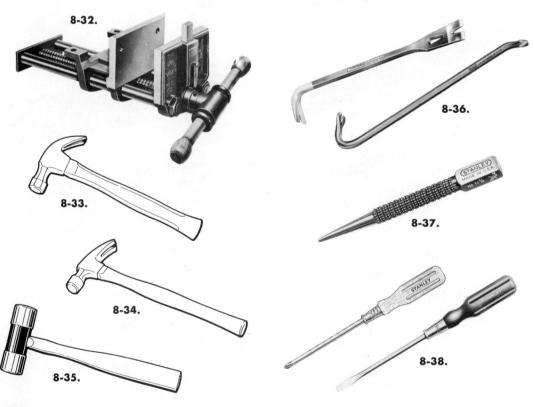

8-32.

8-33.

8-34.

8-35.

8-36.

8-37.

8-38.

| Stapler Gun *Fig. 8-39.* | Heavy-duty models drive up to 9/16″ staples with spring-driven plungers. Some require use of hammer or mallet. | For attaching ceiling tile, insulation, screen, and other soft or thin materials. |
| Pneumatic Nailer-Stapler *Fig. 8-40.* | Drives nails, pins, staples, and fasteners. Runs on compressed air and requires an air compressor. | For framing, roofing, siding, and other assembly work. Drives nails up to 16d size (3½″). |

8-39.

8-40.

DRILLING AND BORING TOOLS

Tool	**Description**	**Uses**
Auger Bit *Fig. 8-41.*	May be either single-twist or double-twist bit. Comes in sizes from No. 4 (¼″) to No. 16 (1″).	1. To bore holes ¼″ or larger. 2. Single twist bit is better for boring deep holes.
Dowel Bit *Fig. 8-42.*	A shorter bit with a sharper twist.	To bore holes for making dowel joints.
Expansion Bit *Fig. 8-43.*	A bit that holds cutters of different sizes. Sometimes this tool is called an expansive bit.	1. To bore a hole larger than 1″. 2. One cutter will bore holes in the 1″ to 2″ range. 3. A second cutter will bore holes in the 2″ to 3″ range.

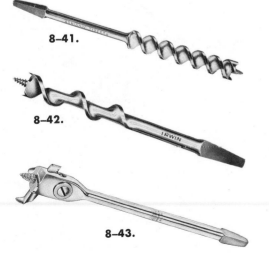

8–41.

8–42.

8–43.

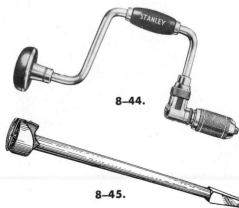

8–44.

8–45.

Brace Fig. 8-44.	Two common types—the plain for a full swing, and the ratchet for close corners.	To hold and operate bits.
Foerstner Bit Fig. 8-45.	A bit with a flat cutting surface on the end.	1. To bore a shallow hole with a flat bottom. 2. To bore a hole in thin stock. 3. To bore a hole in end grain. 4. To enlarge an existing hole.
Bit or Depth Gauges Fig. 8-46.	Two types—one is a solid clamp, the other a spring type.	To limit the depth of a hole.
Twist Drill (a) or Bit Stock Drill (b) Fig. 8-47.	A fractional-sized set from $1/64''$ to $1/2''$ is best.	To drill small holes for nails, screws, etc.
Hand Drill Fig. 8-48.	A tool with a 3-jaw chuck.	To hold twist-drills for drilling small holes.
Automatic Drill Fig. 8-49.	A tool with drill points and handle. Drill point sizes: #1 = $1/16''$; #2 = $5/64''$; #3 = $3/32''$; #4 = $7/64''$; #5 = $1/8''$; #6 = $9/64''$; #7 = $5/32''$; #8 = $11/64''$.	To drill many small holes.

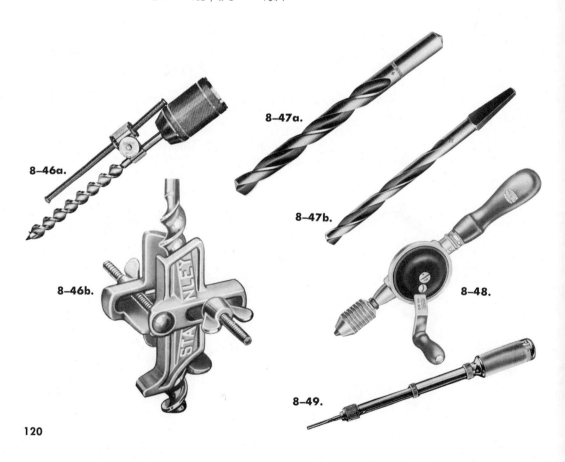

8-47a.

8-46a.

8-47b.

8-46b.

8-48.

8-49.

METALWORKING TOOLS*

Tool	Description	Uses
Hacksaw *Fig. 8-50.*	A U-shaped frame with handle. Uses replaceable metal-cutting blades.	To cut all types of metal fasteners, hardware, and metal parts.
Cold Chisel *Fig. 8-51.*	A tool-steel chisel with cutting edge especially hardened and tempered for cutting metal. Angle between bevel surface is about 60 degrees.	1. To cut off a rivet or nail. 2. To get a tight or rusted nut started.
Adjustable Wrench *Fig. 8-52.*	An extra-strong, lightweight, thin-jawed tool with one adjustable jaw. Wrench develops greatest strength when hand pressure is applied to the side that has the fixed jaw.	1. To make adjustments on machines, when there is plenty of clearance. 2. To install and replace knives and blades.
Open-end Wrench *Fig. 8-53.*	A non-adjustable wrench with accurately machined openings on either end. Sizes of openings are stamped on the tool. For variety of work, a complete set is needed.	1. To make adjustments on machines where there is plenty of clearance. 2. To install and replace knives and blades.
Box Wrench *Fig. 8-54.*	A metal wrench with two enclosed ends. Heads are offset from 15 to 45 degrees.	To make adjustments where there is limited space for movement.

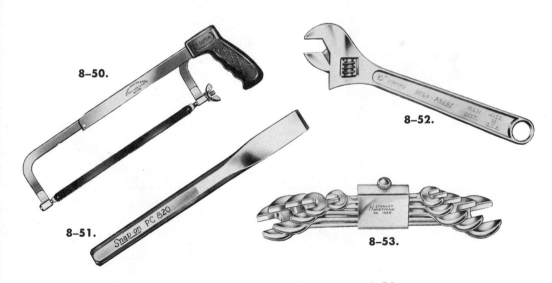

8–50.

8–51.

8–52.

8–53.

8–54.

*In building construction, many metalworking tools are needed to set up and adjust machinery and to work with metal hardware and fasteners.

Socket Wrench Set *Fig. 8-55.*	A series of sockets using a variety of handles.	To assemble and disassemble machinery. Fits many sizes of bolts and nuts.
Vise-grip Wrench *Fig. 8-56.*	An all-purpose tool with double-lever action that locks the jaws on the work.	Used as a substitute for a vise, clamp, pipe wrench, fixed wrench, or adjustable wrench.
Pipe Wrench *Fig. 8-57.*	A tool with hardened, cut teeth on the jaws.	Used on pipes and rods, never on nuts or bolts.
Allen Wrenches *Fig. 8-58.*	Hexagonal steel bars with bent ends.	To tighten and loosen set screws that are often used to hold jointer and planer knives in cutterhead.
Combination Pliers *Fig. 8-59.*	An all-purpose, slip-joint adjustable pliers.	To hold and turn round pieces. Never used on heads of nuts or bolts.

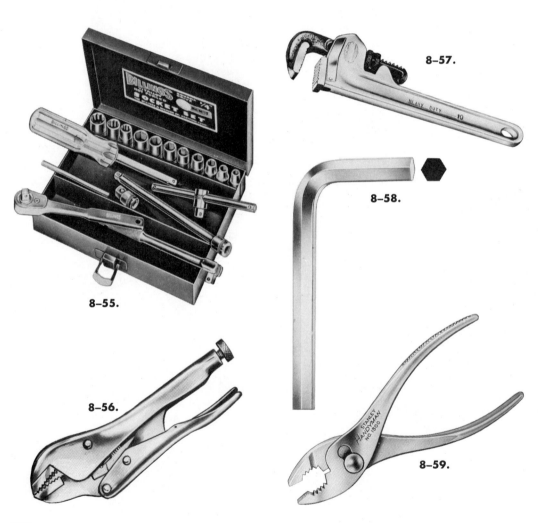

8–57.

8–58.

8–55.

8–56.

8–59.

Long, Flat-nose Pliers *Fig. 8-60.*	Pliers with long, thin, flat nose.	To hold and bend thin wire and metal fittings.
Box-joint Utility Pliers *Fig. 8-61.*	A larger pliers with a slip joint at four positions.	To hold and turn large, round parts.
Tin Snips *Fig. 8-62.*	Available with jaws from 2" to 4" in length.	For cutting sheets of metal, plastic, asphalt, and other construction materials.

8-60. 8-61. 8-62.

QUESTIONS

1. Name several kinds of common rules.

2. What are four uses for the combination square?

3. Why are flexible tapes useful measuring tools?

4. What is the difference between a crosscut saw and a ripsaw? Describe their teeth.

5. Name the hand saws that are used for cutting irregular curves.

6. What is the difference between a back saw and a dovetail saw?

7. Why are planes made in different lengths?

8. Arrange the following planes in order of length: (a) fore, (b) jack, (c) jointer, (d) smooth.

9. What kind of plane is used to clean out the bottoms of grooves and dadoes?

10. Name two types of metal hammers and explain how their uses differ.

11. When would you use a mallet instead of a claw hammer?

12. How would you determine what size Phillips screw driver to use?

13. What force powers the pneumatic nailer-stapler?

14. How does a bit differ from a drill?

15. Name the tool used for operating bits. For operating drills?

16. Describe several uses for the Foerstner bit and the two devices that are used to limit the depth of a hole.

17. Why are metalworking tools needed in a wood shop?

18. Name four kinds of wrenches that are of specific size to fit certain bolts or nuts.

19. Should a pipe wrench be used on bolts and nuts? Explain.

20. Name several common kinds of pliers.

Radial-Arm Saw

SAFETY

- Always keep the safety guard and the anti-kickback device in position.
- Make sure the clamps and locking handles are tight.
- When crosscutting, adjust the anti-kickback device (sometimes called "fingers") to clear the top of the work by about 1/8". This acts as a guard to prevent your fingers from coming near the revolving saw.
- Make sure the stock to be cut is held tightly against the fence.
- For crosscutting, dadoing, and similar operations, pull the saw into the work.
- Return the saw to the rear of the table after each cut.
- For ripping, make sure that the blade is rotating upwards toward you. Use the anti-kickback device to hold work firmly against the table. Feed the stock from the end opposite the anti-kickback device.
- Keep your hands away from the danger area—that is, the path of the saw blade.
- Be sure the power is off and the saw is *not* rotating before making any adjustments.
- Always use a sharp saw or cutter.
- Allow the saw to reach full speed before making a cut.
- Hold the saw to prevent it from coming forward, before turning on the power.
- This saw tends to feed itself into the work. Therefore it is necessary to regulate the rate of cutting by holding back the saw. Otherwise it will feed faster than it can cut, causing the motor to stall.
- Use a brush or stick to keep the table clear of all scraps and sawdust.

The radial-arm saw, Fig. 9-1, is a very versatile machine. It can be used for ripping, dadoing, grooving, and various combinations of these cuts. Many of these operations can be performed more easily on the radial-arm saw than on any other machine. For instance, a long board can be cut into shorter lengths easily because the board remains stationary on the table while the saw is pulled through the stock. Fig. 9-2 (Page 128). Another advantage is that the saw blade is on top of the work so that when dadoes, grooves, and stop cuts are made, the cut is always in sight.

124

INSTALLING THE SAW BLADE

Remove the guard by removing the wing nut on top of the motor housing. Fig. 9-3. Raise the blade so it will clear the table top when it is removed.

To remove the arbor nut, hold the arbor with one wrench and turn the nut clockwise with a second wrench. Fig. 9-4. *Do not attempt to hold the blade with a block of wood while you loosen the nut.* If you do, the saw will climb onto the block and will thus be forced out of alignment.

Place the blade on the arbor. Make certain the teeth at the bottom are pointing away from you and toward the column. Replace the collar, recessed side against the blade. Replace and securely tighten the nut. Then replace the guard. Fig. 9-5.

CROSSCUTTING

1. Mount a crosscutting or combination saw blade on the arbor.

2. Adjust the radial arm to zero (at right angles to the guide fence) and set the

9–1a. *Study the names of the parts and controls. You must know them to follow directions for making adjustments and cuts. See Fig. 9–1b.*

RADIAL ARM CONTROLS
RIGHT SIDE

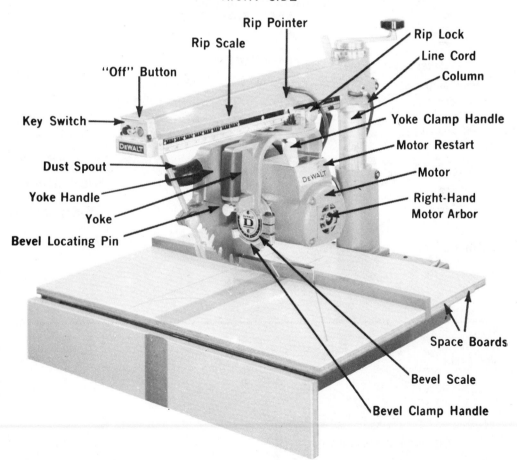

motor so that the blade will be at right angles to the table top. Lock the radial arm with the miter clamp handle.

3. Turn the elevating handle down until the teeth are about $1/16''$ below the surface of the wood table. (The blade should follow the saw kerf already cut in the table.)

4. Adjust the anti-kickback fingers about $1/8''$ above the work surface.

5. With one hand hold the work on the table firmly against the guide fence. The layout line should be in line with the path of the saw.

6. Turn on the power and allow the saw to come to full speed. Grasp the motor yoke handle and pull the saw firmly but slowly through the work. Fig. 9-6.

7. When the cut is completed, return the saw behind the guide fence. Then turn off the power.

MITER CUTS

Flat miter. For miter cuts, loosen the miter clamp handle and lift the miter latch. Swing the arm to the desired angle. Reclamp and make the cut as described under crosscutting. Fig. 9-7. The flat miter can also be cut by clamping or nailing a

9–1b.
RADIAL ARM CONTROLS
LEFT SIDE

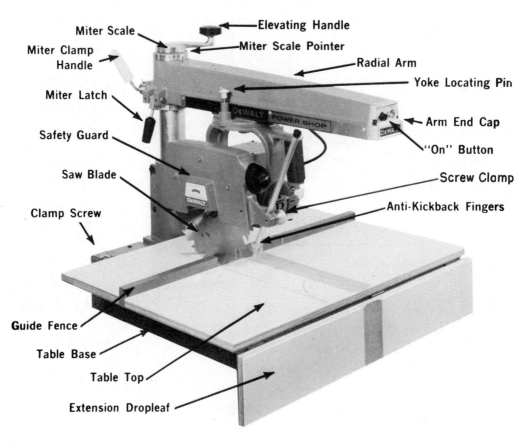

Miter Scale
Elevating Handle
Miter Clamp Handle
Miter Scale Pointer
Radial Arm
Miter Latch
Yoke Locating Pin
Safety Guard
Arm End Cap
Saw Blade
"On" Button
Clamp Screw
Screw Clamp
Anti-Kickback Fingers
Guide Fence
Table Base
Table Top
Extension Dropleaf

piece of stock on the table top at the required angle. Fig. 9-8.

Edge miter (Bevel). To make an edge miter, loosen the bevel clamp handle and pull out the bevel locating pin. Tilt the motor to the desired angle and reclamp. The saw will have to be elevated so the blade will clear the table top when the motor is tilted. Proceed with the cut as described under crosscutting. Fig. 9-9.

Compound miter (Double bevel). To cut the compound miter, the arm is set as described under flat miter cuts and the motor is tilted as for edge miters. The correct settings for the arm and motor can be determined by referring to Fig. 9-10a (Page 130). Many rafter cuts are compound miters. These cuts require special saw settings which are shown in Fig. 9-10b. The cut is then made as described under crosscutting. Fig. 9-11 (Page 132).

Ripping

1. Mount a combination or ripping blade. Pull the entire motor carriage to the front of the arm. Pull up on the locating pin above the yoke. Rotate the yoke 90 degrees *clockwise* until the blade is parallel to the guide fence. The motor should be "outboard" (that is, away from the column) so it will not obstruct the cutting. Fig. 9-12. When ripping wide panels it is necessary to rotate the yoke counterclockwise so the motor will be "inboard" (that is, toward the column). This will increase the ripping capacity. Fig. 9-13.

9–1c. *This machine has the double arm with the arm track pivoting from the upper arm directly over the work area. This places the saw cut nearer the center of the table on both the left- and right-hand miters.*

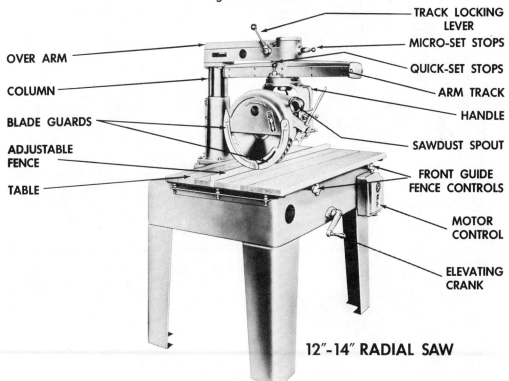

TRACK LOCKING LEVER
MICRO-SET STOPS
QUICK-SET STOPS
ARM TRACK
HANDLE
SAWDUST SPOUT
FRONT GUIDE FENCE CONTROLS
MOTOR CONTROL
ELEVATING CRANK

OVER ARM
COLUMN
BLADE GUARDS
ADJUSTABLE FENCE
TABLE

12"-14" RADIAL SAW

9-2. *Rafters which are awkward to handle are easily cut on the radial-arm saw. Shown here is a special cutter mounted on the saw arbor for cutting the "bird's-mouth" in the rafter. By clamping the rafters to the fence, seven rafters can be cut accurately at one time.*

9-3. *The guard is held firmly on the motor housing by a wing nut.*

9-4. *The arbor is held in place with a hex wrench. On some machines the arbor is held by using an open-end wrench on a flat area between the blade and motor housing. Use the correct size wrench to turn the arbor nut.*

9-5. *The teeth on the bottom of the saw blade point away from the operator and toward the column. Not all blades will have an arrow to show the direction of rotation.*

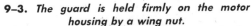

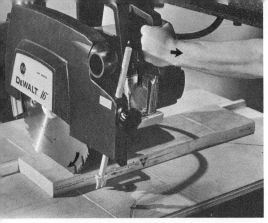

 9–6a. *When cutting stock to length, place the workpiece against the fence and slowly pull the saw into the stock. (Note the guard printed over this caption. This means that a guard must be used for this operation.)*

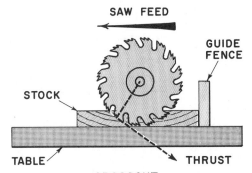

SAW FEED

GUIDE FENCE

STOCK

TABLE

THRUST

CROSSCUT

9–6b. *In crosscutting, the saw's thrust is downward and to the rear, thus holding the stock firmly against the guide fence.*

 9–7. *Cutting a flat miter. The workpiece is held firmly against the fence and the saw is pulled slowly into the stock as in crosscutting.*

 9–8. *When cutting a flat miter on a piece of molding, it is necessary to make a right-hand cut on one end, and a left-hand cut on the other end. These cuts can be made on some radial-arm saws, such as the one shown here, without swinging the arm to a different side for each of the cuts.*

9–9. *To make an edge miter (bevel), hold the workpiece with the left hand and pull the saw into the work with the right hand.*

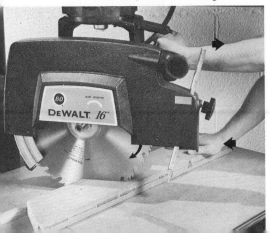

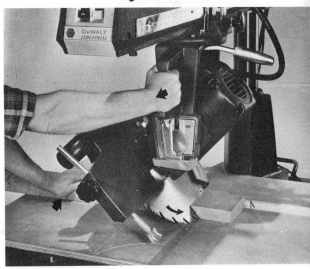

2. Move the motor assembly along the radial arm until the correct width is shown on the rip scale. Tighten the *rip clamp* (on opposite side of radial arm from locating pin). Lower the saw until the blade just touches the wood table.

3. Adjust the guard so that the infeed end clears the work slightly (about ⅛"). Adjust the anti-kickback device so that the points are ⅛" below the surface of the workpiece. Fig. 9-14a and b.

4. Turn on the power. Make sure the saw is rotating upwards toward you. Hold the work against the guide fence and feed it into the blade as shown in Fig. 9-14c. Never feed the work from the anti-kickback end. Use a push stick to complete the cut. Fig. 9-15.

9–10a. *Table of compound angles for use on the radial-arm saw. Note that angles are given for both butt and miter joints. NOTE: The track arm angle given in the chart is the number of degrees off 90°. A four-sided box with mitered corners and with the sides tilted 25° requires a track arm setting of 23° off 90° (90° − 23° = 67°). The track arm is set at 67°.*

Compound Angles

The figures in the table below are degrees to nearest quarter-degree, and are for direct setting of track-arm and blade tilt. Taper per inch given in second column applies only to front elevation and only to a 4-side figure.

Tilt of Work	Equivalent Taper per Inch	4-Side Butt		4-Side Miter		6-Side Miter		8-Side Miter	
		Blade Tilt	Track-Arm	Blade Tilt	Track-Arm	Blade Tilt	Track-Arm	Blade Tilt	Track-Arm
5°	.087	½	5°	44¾	5°	29¾	2½	22¼	2
10°	.176	1½	9¾	44¼	9¾	29½	5½	22	4
15°	.268	3¾	14½	43¼	14½	29	8¼	21½	6
20°	.364	6¼	18¾	41¾	18¾	28¼	11	21	8
25°	.466	10	23	40	23	27¼	13½	20¼	10
30°	.577	14½	26½	37¾	26½	26	16	19½	11¾
35°	.700	19½	29¾	35¼	29¾	24½	18¼	18¼	13¼
40°	.839	24½	32¾	32½	32¾	22¾	20¼	17	15
45°	1.000	30	35¼	30	35¼	21	22¼	15¾	16¼
50°	1.19	36	37½	27	37½	19	23¾	14¼	17½
55°	1.43	42	39¼	24	39¼	16¾	25¼	12½	18¾
60°	1.73	48	41	21	41	14½	26½	11	19¾

9-10b. Rafters can be cut accurately with the radial-arm saw by setting the saw according to this table. To use the table, locate rise in inches per foot of run (first column) or fractional pitch of roof (second column). Read across to the column for the rafter to be cut and read angle for saw setting. With saw blade set at 90° to table surface, swing the arm to the right and set it at angle indicated. To complete "bird's-mouth" and make plumb and tail cuts, swing saw to the left 90° from first setting.

Conversion Table—Fractional Pitches into Degrees and Minutes

Rise in Inches per Foot of Run	Fractional Pitch	Pitch in Degrees & Minutes	Commons		Hips and Valleys		Jacks		
			Seat & Plumb Cut on Framing Square	Seat Cut in Degrees[1]	Side Cut on Framing Square	Side Cut in Degrees[2]	Side Cut on Framing Square	Side Cut in Degrees[2]	Seat Cut for Hip or Valley[1]
2	1/12	9° - 30'	12" & 2"	9° - 30'	12" 11 15/16	44° - 45'	12" 11 13/16	44° - 35'	6° - 45'
3	1/8	14° - 5'	12" & 3"	14° - 5'	12" 11 13/16	44° - 30'	12" 11 5/8	44° - 5'	10° - 0'
4	1/6	18° - 25'	12" & 4"	18° - 25'	12" 11 11/16	44° - 15'	12" 11 3/8	43° - 20'	13° - 15'
5	5/24	22° - 40'	12" & 5"	22° - 40'	12" 11 1/2	43° - 50'	12" 11 1/16	42° - 40'	16° - 25'
6	1/4	26° - 35'	12" & 6"	26° - 35'	12" 11 5/16	43° - 20'	12" 10 3/4	41° - 45'	19° - 25'
7	7/24	30° - 15'	12" & 7"	30° - 15'	12" 11 1/16	42° - 40'	12" 10 3/8	40° - 45'	22° - 25'
8	1/3	33° - 40'	12" & 8"	33° - 40'	12" 10 7/8	42° - 10'	12" 10	39° - 45'	25° - 10'
9	3/8	36° - 55'	12" & 9"	36° - 55'	12" 10 5/8	41° - 25'	12" 9 5/8	38° - 45'	27° - 55'
10	5/12	39° - 50'	12" & 10"	39° - 50'	12" 10 3/8	40° - 50'	12" 9 1/4	37° - 40'	30° - 30'
11	11/24	42° - 30'	12" & 11"	42° - 30'	12" 10 1/8	40° - 5'	12" 8 7/8	36° - 30'	32° - 55'
12	1/2	45°	12" & 12"	45°	12" 9 7/8	39° - 25'	12" 8 1/2	35° - 15'	35° - 15'
13	13/24	47° - 15'	12" & 13"	47° - 15'	12" 9 5/8	38° - 45'	12" 8 1/8	34° - 10'	37° - 25'
14	7/12	49° - 25'	12" & 14"	49° - 25'	12" 9 3/8	38°	12" 7 13/16	33°	39° - 30'
15	5/8	51° - 20'	12" & 15"	51° - 20'	12" 9 1/16	37°	12" 7 1/2	32°	41° - 25'
16	2/3	53° - 10'	12" & 16"	53° - 10'	12" 8 3/4	36° - 10'	12" 7 3/16	30° - 55'	43° - 15'
17	17/24	54° - 45'	12" & 17"	54° - 45'	12" 8 1/2	35° - 20'	12" 6 15/16	30° - 5'	45° - 0'
18	3/4	56° - 20'	12" & 18"	56° - 20'	12" 8 1/4	34° - 30'	12" 6 11/16	29° - 15'	46° - 40'

[1]/Bevel scale setting for cutting "bird's-mouth" with the setup shown in Fig. 9-2. To set up for the plumb or tail cut on a common rafter, set the angle of the arm at the difference between 90° and the angle shown in this column.

[2]/Angle of arm with saw set at 45° on the bevel scale.

9-11. *When making a compound miter cut, the arm or track and motor unit must be carefully set to the correct angles according to the chart in Fig. 9-10a.*

9-12. *When ripping, adjust the "fingers" on the anti-kickback device to project about ⅛" below the surface of the workpiece. As the work is fed, the "fingers" will ride up on the surface of the work. Always use a push stick to feed the workpiece past the saw blade as the cut is finished.*

9-13. *The ripping capacity can be increased by rotating the saw counterclockwise and ripping with the blade "outboard" (away from the column).*

Ripping Angles

The motor is positioned as described in ripping. Then the saw is elevated and the motor is tilted in the yoke to any desired angle from the horizontal to the vertical position. Lower the saw until the teeth are ¹⁄₁₆" below the wood table. Make the cut as described under ripping. Several cuts of this type are shown in Fig. 9-16.

USING A DADO HEAD

The same types of dado heads which are used on the circular saw can be used on the radial-arm saw. Be sure the arbor hole is the correct size. Mount the dado head as described on page 161. Make certain the saw teeth next to the table top are pointed back toward the column, the same as when mounting a saw blade. Fig. 9-17. Replace the guard and rotate the dado head by hand to make sure it turns freely.

Plain dado. A dado is cut across the grain of the wood. Mount the correct combinations of cutters for the desired width of cut. Take a piece of scrap stock of the same thickness as your finished stock, and lay it on the table top. Lower the blade until it just touches the surface of the scrap

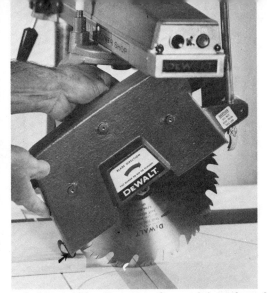

9–14a and b. *For your safety and the safety of others working around you, always set the guards properly for maximum protection.*

stock. Remove the scrap piece. On most radial arm saws one revolution of the elevation crank lowers the blade $1/8''$. If a $1/4''$-deep cut is desired, turn the elevating crank two complete turns; then proceed with the cut as you would for crosscutting. Fig. 9-18.

Blind dado. Place a clamp stop on the radial arm to limit the travel of the saw and insure that all dadoes will be the same length. Should you want the dado to be blind on both ends, raise the saw by turning the elevating handle. Clamp the stock in place. Locate the saw over the point

where the cut is to begin and turn on the machine. Lower the blade by turning the elevating handle to the desired depth. Then pull the saw until the carriage hits the front clamp stop. Turn the power off, then raise the saw and push it back against the rear clamp. Fig. 9-19.

9–15. *Use a push stick to complete the cut. Push the stock about 2'' beyond the saw, then pull the stick directly back.*

9–14c. *When ripping stock, always feed it into the rotation of the blade as shown here.*

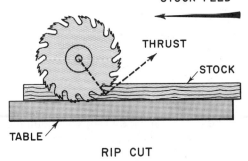

STOCK FEED

THRUST

STOCK

TABLE

RIP CUT

133

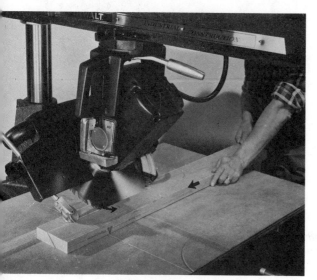

9–16a. *Ripping a bevel.*

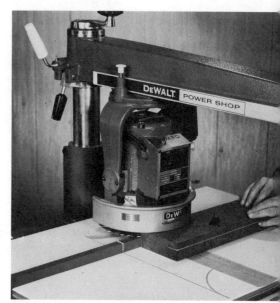

9–16c. *A groove can be cut on the edge of a piece of stock by rotating the motor unit to a vertical position. Notice the use of the special guard.*

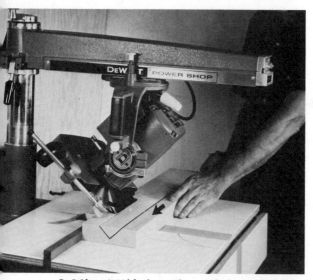

9–16b. *A V-block can be made by making two bevel rip cuts to a set depth.*

Grooves

Install the dado head as described on page 161 and rotate the motor unit counterclockwise as for ripping. Clamp the rip lock and set the depth of cut. Proceed as in ripping. Fig. 9-20. *Be sure that the direction of feed is as shown in Fig. 9-14c.*

To cut a groove on the edge of a piece of stock, loosen the yoke clamp handle, pull the yoke locating pin, turn the yoke 90 degrees *counterclockwise,* reset the yoke pin, and tighten the yoke clamp handle. Raise the radial arm about 2″ by turning the elevating handle. Then loosen the bevel clamp handle, pull the bevel locating pin, and pivot the motor to the vertical position. Lower the radial arm so that the saw is the correct height off the table top. Place the yoke in the correct position and tighten the rip lock. Be sure the blade is properly guarded. Again, refer to Fig. 9-14c to see the direction of feed. Fig. 9-21.

Rabbet

Install the dado head, set up the saw, and make the cut the same as for cutting a groove. Fig. 9-22. The end rabbet can be made with a combination saw blade. Raise

9–16d. *A rabbeted edge can be made on an angle by making the first cut as shown in Fig. 9–16c and then setting the saw as shown here to complete the cut.*

9–18. *Cutting a plain dado.*

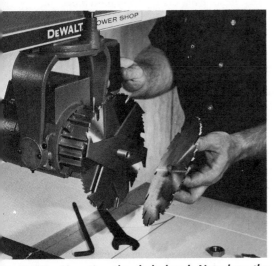

9–17. *Mounting the dado head. Note how the chippers are placed an equal distance apart.*

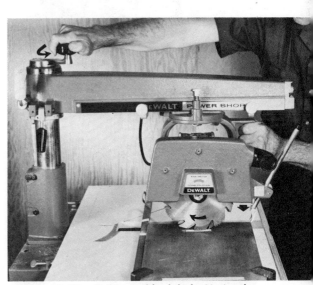

9–19. *Cutting a blind dado. Notice the use of two clamps on the arm to limit the travel of the saw.*

the radial arm about 2″, loosen the bevel clamp, pull the bevel pin, and turn the motor unit to the vertical position. Push the pin in, tighten the clamp, and lower the saw to the desired height. The material to be cut should be placed on a wooden auxiliary table, so the guard will clear the table top. Figs. 9–23 and 9–24. The shoulder cut is made with the saw in the regular crosscutting position. Adjust the saw blade to the correct depth of cut. Fig. 9–25.

9-20. *Making a groove with a dado head. This cut is sometimes called ploughing.*

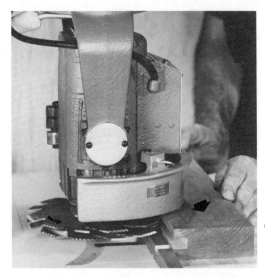

 9-21. *Cutting a groove with a dado head. Note the plywood placed over the table so the groove just clears the guide fence.*

 9-22. *Cutting a rabbet.*

9-23. *Auxiliary wood table for horizontal cutting. This table is installed in place of the standard guide fence. To do this, release the clamp screws, lift out the guide fence, slide in the auxiliary table, then retighten the clamp screws.*

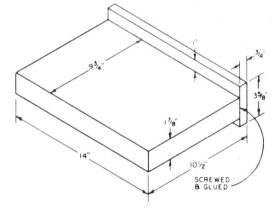

9–24. *To make an end rabbet, place the stock flat against the auxiliary table and fence (Fig. 9–23) and then make the first cut with the saw in horizontal position.*

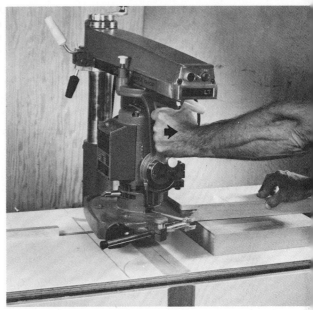

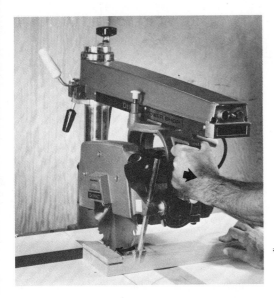

9–25. *The second cut for an end rabbet is a simple crosscut with the blade set for correct depth.*

QUESTIONS

1. Describe the main safety precautions to follow in operating a radial-arm saw.

2. Tell how to do crosscutting on a radial-arm saw.

3. In ripping stock on a radial-arm saw, does the work or the saw move? Explain the action.

4. What are the fundamental operations of the radial-arm saw?

5. Explain why it is easier to cut a blind dado on the radial-arm saw than on the circular saw.

6. Describe the procedure for cutting a rabbet on the radial-arm saw.

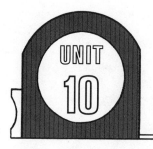

Jointer

SAFETY

• *Always keep the knives of the jointer sharp.* Dull knives tend to cause kickback and also result in poor planing.

• *The fence should be tight.* Never adjust the fence while the jointer is running.

• Adjust the depth of cut before the jointer is turned on.

• *See that the guard is in place and operating easily.* If the regular guard is removed, a special guard must be provided.

• *Always allow the machine to come to full speed before using it.*

• Check the stock for knots, splits, metal particles, and other imperfections before jointing. Defective stock may break up or be thrown from the jointer.

• *Keep the left hand back from the front end of the board when feeding.*

• *Stand to the side of the jointer, never directly behind it.* In case of kickback you will be out of the way.

• *Cut with the grain, never against it.*

• *Always use a push stick or push block.*

• *Do not try to take too heavy a cut.*

• *Use common sense about when stock is too thin or too short to joint safely.*

• Never apply pressure to the board with your hand directly over the cutterhead.

• Use a brush to clean shavings off the table. Never use your hand.

Although the jointer is not used for a great variety of operations, it is one of the most frequently used machines in a typical shop. Common uses of the jointer are for surfacing a board and for planing an edge or an end. It can also be used for cutting a rabbet, bevel, chamfer, or taper.

The jointer has a circular cutterhead which usually has three or four blades (or knives). The blades rotate, shearing off small chips of wood, thus producing a smooth surface on the workpiece.

SIZE

The size of a jointer is indicated by the length of the knives. Since most jointing operations are performed on the edge of stock, a 6″ or 8″ jointer is most common. Fig. 10-1.

The length of the bed also affects the usefulness of the jointer, since a longer bed provides better support for jointing longer pieces.

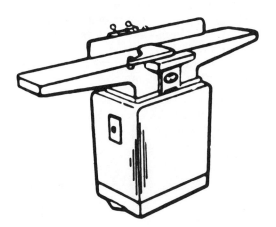

PARTS

The *frame or base* of the jointer has two tables—the *front or infeed table* and the *rear or outfeed table.*

Fig. 10-2. On most machines, both of these tables are adjustable, although there are some on which only the infeed table can be raised or lowered.

The *cutterhead* is the heart of the jointer. As mentioned, it consists of the head itself and three or more knives. This assembly usually operates on two roller bearings. Fig. 10-3.

The fence provides support for the work while it is fed, on edge or on end, through the machine. The fence can be adjusted to various angles, usually up to 45 degrees both ways from the vertical position.

The guard is a protective device covering the cutterhead. It either swings out of the way or lifts up. Most operations, except rabbeting on some jointers, and certain tapering, should be done with the guard in place.

The jointer usually operates at about 4000 rpm.

ADJUSTMENTS
Aligning and Adjusting the Outfeed Table

The top of the outfeed table must be at exactly the same height as the knife blades at their highest point of revolution. If the table is too low, the board will drop down onto the knives as it leaves the infeed table. This will cause a recess to be cut at the end of the board. If the table is too high, the board will be slightly tapered. Fig. 10-4 shows correct and incorrect cuts.

To align the knives with the table as just mentioned, turn the cutterhead until one blade is at its highest point. Release the table locking screw on the side of the jointer. Lower the outfeed table until it is below the blade; then place a straightedge on the outfeed table with one end projecting over the blade. Fig. 10-5. Turn the table up slowly until it is in line with the knife at the highest point. Turn the cutterhead over

slowly by hand until there is very light contact between the knives and the bottom of the straightedge. Tighten the lock nut. Once the outfeed table is set, it does not require changing except for certain cuts such as stop chamfers and bevels, and recess cuts. If the outfeed table is the fixed kind, raise or lower the cutterhead until the knives are even with the outfeed table.

Adjusting the Infeed Table

The distance the infeed table is below the knives determines the depth of cut. The

10-1. *This 6″ jointer (top) and 8″ jointer (bottom) are popular sizes for edge jointing and specialty cuts. Note that the rear (outfeed) table is adjustable on both of these machines, making them even more versatile.*

depth of cut to be taken will depend on:
- The width of the surface being jointed.
- The kind of wood and grain pattern.
- Whether you are making a rough or finish cut.

Loosen the lock on the side of the infeed table, then turn the handle beneath the table to raise or lower it. There is a pointer and scale, indicating the depth of cut, which must be checked periodically for accuracy. Fig. 10-6.

Adjusting the Position of the Fence

For most operations it is desirable to have the fence at an exact right angle to the table. To adjust the fence, loosen the knob or lever that holds it in position; then set the fence at a 90-degree angle to the table. To check that the angle is correct, hold a square against the table and fence.

Fig. 10-7. The fence can be moved in or out. When cutting, never expose any more of the blade than necessary.

The fence can also be *tilted* 45 degrees to right or left. This can be set on the tilt scale and checked with a protractor head of a combination square set or a sliding T bevel. There is a pointer and scale to indicate the tilt. Fig. 10-8.

BASIC PROCEDURES

1. Check the fence for squareness and the infeed table for depth of cut before turning on the machine. If the jointer has been used for some other operation, make a trial cut after resetting it.

2. Adjust the *depth of cut* with these things in mind:
- The amount of stock to be removed. Take a light cut for such operations as face

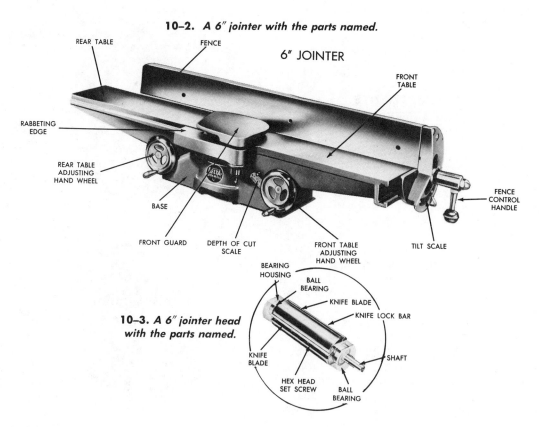

10–2. *A 6″ jointer with the parts named.*

6″ JOINTER

REAR TABLE   FENCE   FRONT TABLE

RABBETING EDGE

REAR TABLE ADJUSTING HAND WHEEL

BASE

FRONT GUARD   DEPTH OF CUT SCALE   FRONT TABLE ADJUSTING HAND WHEEL   TILT SCALE

FENCE CONTROL HANDLE

10–3. *A 6″ jointer head with the parts named.*

BEARING HOUSING   BALL BEARING   KNIFE BLADE   KNIFE LOCK BAR   SHAFT

KNIFE BLADE   HEX HEAD SET SCREW   BALL BEARING

planing or end planing and a slightly heavier cut for edge planing.

• The kind of wood. A light or heavy cut may be made on soft woods; a light cut is best on hard woods.

10–4. *The jointer must be adjusted so the outfeed table is at exactly the same height as the cutterhead knife at its highest point. Otherwise a taper or a recess will be cut. (A recess is sometimes called a "snipe.")*

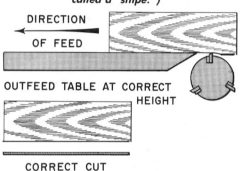

DIRECTION OF FEED

OUTFEED TABLE AT CORRECT HEIGHT

CORRECT CUT

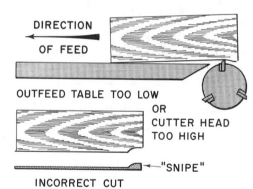

DIRECTION OF FEED

OUTFEED TABLE TOO LOW OR CUTTER HEAD TOO HIGH

←—"SNIPE"

INCORRECT CUT

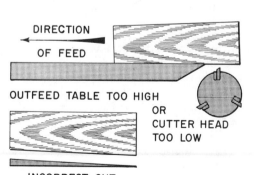

DIRECTION OF FEED

OUTFEED TABLE TOO HIGH OR CUTTER HEAD TOO LOW

INCORRECT CUT

• The kind of planed surface. Take a heavier cut for removing stock and a lighter cut for finishing.

3. Change the position of the fence periodically to distribute the wear on the jointer knives.

4. When duplicate parts are needed, do the jointing operations first; then cut the stock into the desired smaller pieces.

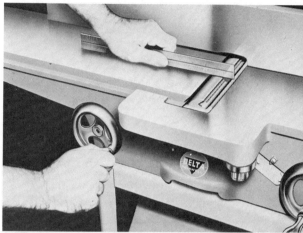

10–5. *Adjusting the outfeed table. Raise the table slowly until the straightedge rests evenly on the table and the knife. Always replace the guard after making this adjustment.*

10–6. *Always check the depth of cut before making a cut on the jointer.*

10-7. Use a try square to make sure the fence is set at right angles to the table.

10-8. Loosen the fence control handle to adjust for the angle of cut. The fence can also be moved in and out to distribute the wear on the knives, especially for edge jointing.

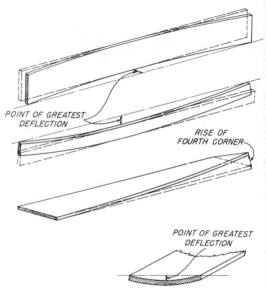

POINT OF GREATEST DEFLECTION

RISE OF FOURTH CORNER

POINT OF GREATEST DEFLECTION

10-9. Common kinds of warp that can be removed on the jointer.

5. If you are right-handed, stand to the left of the jointer with your left foot forward and right foot back and beneath the infeed table. Move your body along as you do the planing operation.

6. Always check a board for warp and wind first. Fig. 10-9. Place a concave surface down for the first cuts. If the board has twist, balance it on the high corners to take the first cuts.

Planing a Surface

1. Check the board for warp and for direction of grain. Be certain the jointer is correctly adjusted.

2. Hold the board firmly on the infeed table with your left hand toward the front of the board and your right hand on the push block. Fig. 10-10a. The push block is hooked on the end of the board over the infeed table. Fig. 10-10b. Apply equal pressure with both hands. Fig. 10-11.

3. Turn on the machine and allow it to come to full speed.

4. Move the stock forward, keeping your left hand back of the cutterhead. When

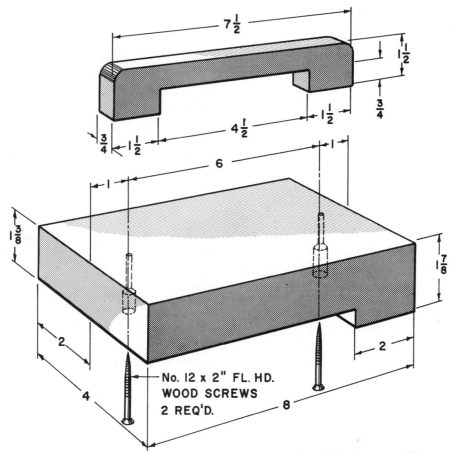

10–10a. *A drawing of a one-handed push block.*

10–10b. *Using the one-handed push block.*

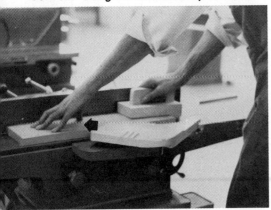

10–11a. *Using a push block to do facing on short stock. Note the use of the push block or hold-down. The knob is held in the left hand and the handle in the right.*

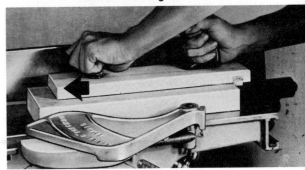

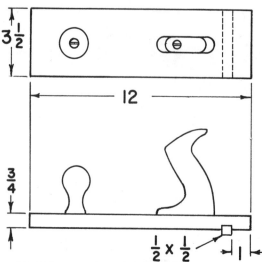

10–11b. Drawing of a push block. This one has a knob and handle from a hand plane.

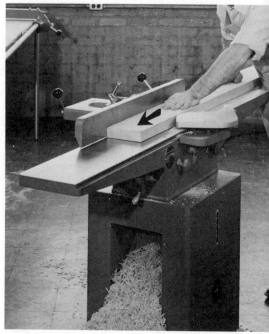

10–12. Face planing or surfacing on an 8″ jointer. Note how the left hand is kept back from the front edge of the board.

about half to two-thirds of the board has passed the cutterhead, move the left hand to the board over the outfeed table. Fig. 10-12.

5. After most of the board has passed over the cutter, move the right hand to the portion of the board over the outfeed table to finish the cut. *Never place your hand directly over the cutterhead.* Fig. 10-13.

Planing an Edge

The most common use for the jointer is planing or jointing an edge. Fig. 10-14a. An edge is said to be jointed when it is at right angles to the face of the board and is true along its entire length. Fig. 10-14b.

1. Check the fence for squareness. Generally, for safest operation, it is best to set the fence as close as possible to the left side of the machine.

2. Select the best edge and determine the grain direction.

3. Adjust for proper depth of cut. To insure parallel edges on the stock, rip to width and allow just enough extra stock to joint off the sawn edges.

4. Hold the stock firmly against the infeed table and the fence. The jointed or planed surface of the board should be against the fence.

5. For the right-handed person, the left hand is a guide and the right hand pushes the stock across the cutterhead. Move the left hand along with the board until the major portion of the board is over the outfeed table; then move the right hand to the other side of the cutterhead, to the stock over the outfeed table. Fig. 10-15. Do not push the board too fast, as this will make a rippled edge.

Planing End Grain

This operation is very dangerous, especially with stock less than 10″ wide. This is because the cutters must shave off the ends of the wood fibers, which are tough.

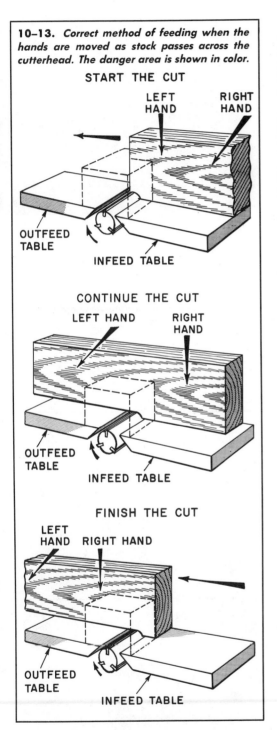

10-13. *Correct method of feeding when the hands are moved as stock passes across the cutterhead. The danger area is shown in color.*

START THE CUT

LEFT HAND   RIGHT HAND

OUTFEED TABLE

INFEED TABLE

CONTINUE THE CUT

LEFT HAND   RIGHT HAND

OUTFEED TABLE

INFEED TABLE

FINISH THE CUT

LEFT HAND   RIGHT HAND

OUTFEED TABLE

INFEED TABLE

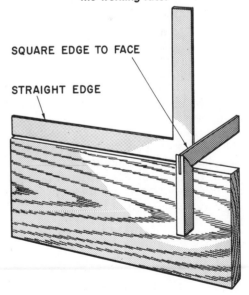

10-14a. *Jointing the edge of a board.*

10-14b. *A properly jointed edge is straight along its entire length and forms a 90° angle with the working face.*

SQUARE EDGE TO FACE

STRAIGHT EDGE

10–15. *Three steps in jointing an edge or surface.*

Always set the machine for a very light cut for this operation.

If both edges of the board are surfaced, proceed as follows:

1. Take a light cut about 1″ in length along the end grain.

2. Reverse the board and finish the cut.

Make sure you hold the board firmly over the outfeed table as the end of the cut is made. Fig. 10-16. As in hand planing, running the board completely across would split out the edge.

 10–16. *Planing end grain.*

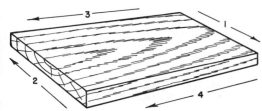

10–17. *Steps in jointing the edges and ends of a board.*

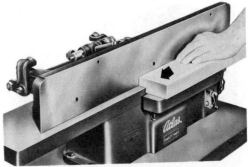

10–18. *Cutting a rabbet. On some machines it is not necessary to remove the guard, as must be done here.*

If only one edge of the board is surfaced, the jointing can be done from the finished edge all the way across, since a splinter can be removed when the rough edge is jointed later. Fig. 10-17.

Cutting a Rabbet

One of the best ways to cut a rabbet with the grain is on the jointer.

1. Adjust the fence so that the distance between the end of the knives and the fence is equal to the width of the rabbet.

2. Lower the infeed table an amount equal to the depth of the rabbet. If the rabbet is quite deep, it may be necessary to cut it in two passes. In that event the table is lowered an amount equal to about half the depth of the rabbet for the first pass, then lowered again to the desired depth to complete the cut.

3. Remove the guard, if necessary. Hold the stock firmly on the infeed table and move it along slowly. Fig. 10-18.

Many types of simple moldings can be rabbeted in this way.

QUESTIONS

1. What function do rotary cutters on a jointer perform?

2. How is the size of a jointer indicated.

3. How is the depth of cut adjusted on a jointer?

4. What happens if the outfeed table of a jointer is too low?

5. What happens if the outfeed table of a jointer is too high?

6. List five safety rules for operating a jointer.

7. What three factors should be considered in adjusting for depth of cut?

8. When should a push block be used?

9. Indicate the most common use for a jointer.

10. When planing end grain, what should be the narrowest width of board?

11. Describe the proper procedure for cutting a deep rabbet on the jointer.

Circular Saw

SAFETY

- Make all adjustments when the power is off and the blade has stopped revolving.
- Always adjust the saw blade so it protrudes just enough above the stock to cut completely through.
- Never reach over a revolving saw; instead, bring the cut piece back around the side of the machine.
- Keep your fingers away from the saw blade at all times.
- Always keep the guard and splitter in place unless this is impossible for the kind of cut you are making.
- If the cut you are making doesn't permit use of the regular guard, use a feather board or a special guard.
- When crosscutting with the miter gauge, never use the fence for a stop unless a clearance block is used.
- Always push the stock through with a push stick when ripping stock that cannot be fed safely by hand.

- Never stand directly behind the blade.
- Always use a sharp blade.
- When ripping, place the jointed edge against the fence.
- Keep the saw table clean. Remove all scraps with a brush or push stick, *never with your fingers*.
- Remove rings, watches and other items that might catch in the saw. Wear garments with short or tight sleeves.
- Use the proper saw blade for the operation being performed.
- Always hold the stock firmly against the miter gauge when crosscutting and against the ripping fence when ripping.
- Be certain the fence is clamped securely.
- When a helper assists you, he should not *pull* the stock. He only *supports* the stock.
- Do not saw warped material on the circular saw.
- If stock must be lowered onto the revolving blade for certain cuts, use stops and guards. Never have your hands in line with the blade.

Many fundamental woodworking operations can be done with the circular saw. It can be used not only for cutting stock to size but also for cutting many joints. Fig. 11-1.

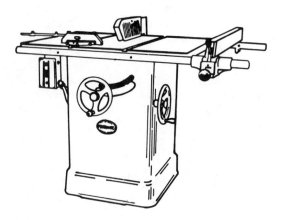

SIZE

The size of the circular saw is indicated by the diameter of blade recommended for its use. Typical sizes are the 8" or 10".

11-1. *The circular saw is essential to any well-equipped woodworking shop. The operator should always closely follow the recommended safety procedures in its use.*

11-2. *Cutting a bevel on a circular saw with a blade that tilts to the left.*

11-3. *Cutting a chamfer with a blade that tilts to the right.*

These saws are made in either bench or floor type and are called *variety saws*. On most models the blade tilts to the right while on some it tilts to the left. The table remains permanently fastened in the horizontal position. Figs. 11-2 and 11-3.

BLADES

There are six basic kinds of saw blades. In selecting a blade, make sure that you secure one with the correct diameter arbor hole size. Never attempt to install a blade that has too large a hole.

The six common kinds of blades are:

• The *cutoff* or *crosscut* blade. This has teeth similar to the hand crosscut and is used primarily for trimming stock to length and squaring. Fig. 11-4a.

• The *hollow-ground* (or *planer*) blade. This is used for fine cabinetwork. Fig. 11-4b. The teeth of this blade are not set (bent). The necessary side clearance is ground in the body of the blade as the cross section drawing shows. (Most saws have some teeth that are bent to the right and others to the left. The teeth are bent this way to make the kerf, or saw cut, slightly wider than the blade. This provides clearance so the blade will not stick in the kerf.)

• The *ripsaw* blade. This has chisel-like teeth and is used for ripping operations. Fig. 11-4c.

• The *combination saw* blade. This has a combination of ripping and crosscut teeth

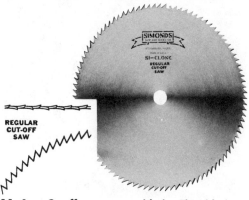

11–4a. *Cutoff or crosscut blade. This blade is designed for cutting only across the grain of the lumber.*

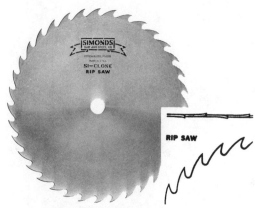

11–4c. *Rip blade. This blade is designed for sawing only with the grain of the lumber.*

11–4b. *Hollow-ground or planer blade. It should be used only where a smooth finish cut is needed.*

and is used for a great variety of cutting. Several styles of teeth are available, and each has a particular application. Fig. 11-4d.

• The *easy-cut* blade. This has only a few large teeth and is considered to be the safest blade since it practically eliminates kickback. It does, however, make a rather wide kerf and does not cut so smoothly as the cutoff, hollow-ground, and combination saws. Fig. 11-4e.

• The *plywood saw* blade, as the name indicates, is a special blade for cutting

plywood. It cuts with a minimum of chipping and leaves an extremely smooth edge. The steel is specially tempered to give the teeth a longer life for cutting through the many glue lines in plywood. Fig. 11-4f.

Dado Head

In addition to standard equipment, a *dado head* can be purchased that will cut all common widths of grooves or dadoes. Fig. 11-4g. This will be described later.

PARTS

Study the parts of the circular saw, as shown in Fig. 11-5. Notice that this is a tilting arbor saw. The table top has two grooves cut in it into which the miter gauge fits. These are parallel to the saw blade. The miter gauge comes equipped with a stop rod which can be adjusted in length for cutting duplicate parts. A fence clamps to the table for all ripping operations. Also available are table extensions which can be fastened to the sides of the table top and are especially convenient when cutting long or large stock such as a sheet of plywood. An opening in the center of the table is covered by a throat plate. A guard, which drops over the blade, is always fastened to the back or side of the table.

150

11–4d. Combination blades. These blades are designed to be used for either ripping or crosscutting, and are convenient when it is impractical to change blades frequently. Style S teeth are recommended for bench saws which require a fine cut. Style U teeth are recommended for radial-arm saws. Style V teeth are the fastest cutting of the various combination saws and are recommended for use on all types of machines.

This should be kept in place whenever possible. (There are some operations for which regular guards cannot be used; a special guard or a feather board should then be used.) There is also a splitter which is usually a part of the back of the guard. This fits directly back of the saw blade and is slightly thicker than the blade. It keeps the saw kerf open as the cutting is done.

ADJUSTMENTS

Installing or Removing a Saw Blade

1. Remove the throat plate. This usually snaps in and out of position.

2. Select a wrench to fit the arbor nut. On some saws the arbor has a left-hand thread and must be turned clockwise to loosen. However, some saws have a right-hand thread. If so, you must turn it counterclockwise to remove. Always check the thread before loosening. A good rule to remember is that the nut always loosens by turning it in the direction the teeth are pointing (direction of blade rotation). If the nut doesn't come off easily, hold a piece of

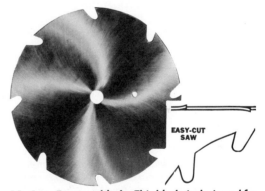

11–4e. Easy-cut blade. This blade is designed for ripping or cutoff and rough combination work. The high back of the tooth prevents overfeeding. Kickback tendency is minimized and relatively little power is needed. For use on all types of machines.

scrap wood against the blade to keep the arbor from turning. Fig. 11-6 (Page 154).

3. Remove the nut and the collar, and take off the old blade.

4. Replace the blade in the correct position, with the saw teeth pointing in the

direction of blade rotation. Replace the collar and nut. Tighten it firmly, but not too tight. The nut tightens against the rotation and will not come off. Replace the throat plate.

Raising the Saw Blade

There is a wheel or lever on the front of the machine to raise or lower the blade. Often, in addition, there is a lock that must be loosened when making this adjustment. To raise the blade to the proper position, hold the workpiece against the side of the blade and carefully turn the blade until the top tooth is at the correct height. For most cutting, the top of the blade should extend no more than 1/8″ above the stock. On many joint cuts, however, the blade must be set for the exact depth of cut.

Tilting the Saw Blade

A lever or handle on the side of the machine tilts the blade. A pointer or scale on the front indicates the degree of tilt. There is usually a lock to hold the blade in position when it is tilted.

Adjusting the Fence

A ripping fence is fastened to the table for all ripping operations and for many other cutting jobs. It is usually placed to the right of the blade. To adjust the fence to the correct position, first move it to an approximate location. Hold a rule or try square at right angles to the fence and carefully measure the distance from the fence to one tooth bent toward the fence. Fig. 11-7.

On some machines there is a pointer on the fence and a scale on the front of the table to indicate the width of cut. This should be checked frequently to make sure it is accurate. Each time the blade is changed, this scale will have to be checked, because the amount of set in the saw blade will affect the distance between the fence and the saw kerf. It is a good practice to use this scale for rough setups only. Accurate setups should be made by making a test cut and checking it with a rule or by superimposing (mock assembly of parts).

Adjusting the Miter Gauge

The miter gauge, which is used for crosscutting operations, can be used in either groove of the table but usually is placed in the groove to the left of the blade. There is a pointer and scale on the miter gauge for setting it to any degree right or left. Most gauges have automatic stop positions at 30, 45, 60, and 90 degrees.

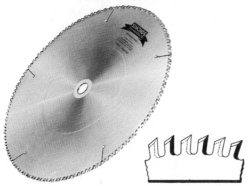

11–4f. *Plywood blade. This blade is designed for cutting across the grain and can be used for either hard or soft woods, or plywood. When used for plywoods, the finer tooth saws are recommended.*

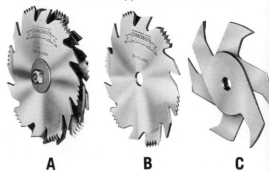

11–4g. *A. An assembled dado head. B. An outside cutter. C. Chippers or inside cutters.*

A   **B**   **C**

RIPPING

Install a ripping, combination, or easy-cut blade for these operations:

Cutting Wide Stock to Width

1. When the width of the board to be cut is 6 inches or more, this is considered a wide cut. Adjust the fence and blade accordingly.

2. Turn on the machine. Place the board over the table. Apply pressure against the fence with the left hand, and push the board forward with the right. If the board is longer than 6 or 8 feet, have a helper stand behind the saw to hold the piece up after it passes the blade. If a helper is not available, use a roller stand as shown in Fig. 11-8.

3. Feed the stock at an even speed into the blade about as fast as it will cut. Be careful not to overload the saw. Hold your right hand close to the fence as you push the end of the board through the saw. Fig. 11-9a.

4. If extremely thick or hard wood is being cut, it is often necessary to cut partway through the board, then turn the board over and complete the cut.

11–5. *Parts of a 10" tilting-arbor circular saw.*

UNIGUARD WITH "DISAPPEARING" SPLITTER

MITER GAGE

TABLE

SINGLE LOCK FENCE

SWITCH

SAW TILT SCALE

SAW RAISING HANDWHEEL

SAW TILT HANDWHEEL

LOCK KNOB

SAWDUST CLEAN-OUT

CABINET

CABINET BASE

FRONT GRADUATED GUIDE BAR

FENCE CLAMP HANDLE

FENCE MICRO-SET KNOB

"T" SLOT FOR MITER GAGE

MOTOR COVER

POWER CORD

10" TILTING ARBOR UNISAW®

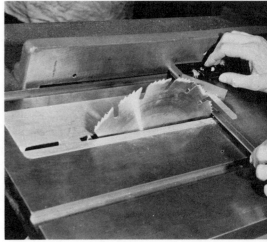

11–6. *Hold the blade with a piece of scrap wood. The nut will loosen if turned in the direction the teeth are pointing.*

11–7. *Adjusting the fence for the correct width of cut. Make a small test cut on the workpiece and measure it to double-check the setup before making the complete cut.*

Cutting Narrower Stock to Width

1. When cutting stock narrower than 6 inches, observe the same general practices as in starting the cut on wide stock.

2. As the end of the board reaches the front of the table, use a push stick to do the work you began with your right hand, guiding the board between the blade and the fence. Fig. 11-9b. *Never under any circumstances cut narrow stock without a push stick.* It is good practice to hang the push stick conveniently at the side of the saw so that you don't take a chance and cut without it.

3. If very narrow stock is being cut, it may be a good idea to cut half the length of the stock, pull it back out, reverse it, and complete the cut from the other end. Fig. 11-10.

CROSSCUTTING

Install a crosscutting, hollow-ground, or combination blade. Use the miter gauge for all crosscutting operations. For added support of the workpiece some operators like to fasten permanently a long support board to the miter gauge. Always remove the ripping fence for crosscutting. Carefully mark a line on the edge of the stock nearest the blade. (You must be able to see the mark easily, so you can begin the cut accurately.) Be sure the miter gauge is set to cut the correct angle. It is also a good practice to use the stop rod as an aid to prevent the stock from moving while the cut is made.

Cutting Short Boards

Place the gauge in the groove in the side toward the longest portion of the board. Fig. 11-11. Hold the stock firmly against the gauge and advance it slowly into the blade. Never drag the cut edge back across the blade.

Cutting Long Pieces

If the board is longer than 6', have a helper support the other end.

Cutting Plywood

Because of its construction and often because of its size, plywood presents spe-

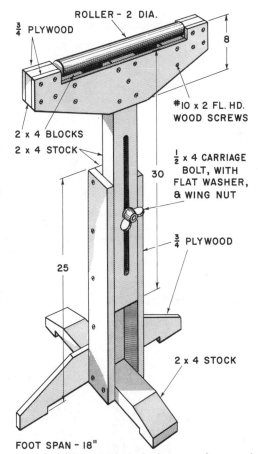

ROLLER - 2 DIA.

¾ PLYWOOD

8

#10 x 2 FL. HD. WOOD SCREWS

2 x 4 BLOCKS

2 x 4 STOCK

30

½ x 4 CARRIAGE BOLT, WITH FLAT WASHER, & WING NUT

¾ PLYWOOD

25

2 x 4 STOCK

FOOT SPAN - 18"

11–8. A roller stand used to support long stock when ripping.

11–9a. Ripping on the circular saw.

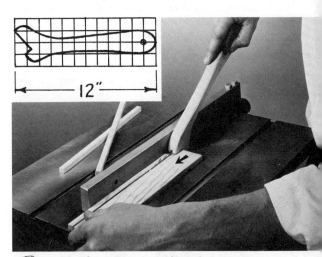

12"

11–9b. Using a push stick in ripping narrow stock.

cial cutting problems. Since grain directions of alternating plies are at right angles to each other, there is a tendency to split out the ends of cross-grain layers, no matter what the direction of the cut. The glue lines are also a problem in that they dull the blade. Finally, since plywood is glued up in large sheets, the workpiece is often too large to fit conveniently on the table of a circular saw.

To reduce these problems to a minimum, adjust the blade so it will barely clear the top of the plywood, and place the stock with the good side up. Fig. 11-12. Then use one of the three following methods to guide the stock:

• The miter gauge can be reversed in the groove when the cut is started, to guide the stock for as long a cut as possible. Fig. 11-13. Then the gauge can be removed and slipped into its regular position to complete the cut.

• Another suggestion for sawing plywood is to clamp a straightedged board on the underside of the plywood. This will act as a

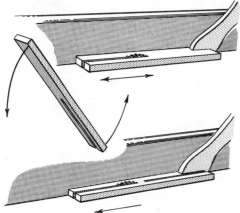

11-10. *Another method of ripping narrow stock. Saw halfway through and then move the stock back out of the saw. Turn the stock end for end and complete the cut.*

11-12. *Plywood should be cut with the good face up. Use a combination, cross-cut, or plywood blade.*

11-11. *Crosscutting narrow pieces.*

11-13. *Starting a cut on a piece of plywood with the miter gauge reversed.*

guide against the edge of the table. Fig. 11-14.

• The ripping fence can be used as a guide in cutting plywood to size.

Cutting Identical Pieces to Length

There are many ways of cutting identical pieces to length:

• For cutting many short pieces, clamp a stop block to the ripping fence in front of the cutting edge of the blade. Adjust the fence to cut the proper length of stock. By placing the end of the board against this stop, you can cut the correct length and there will be plenty of clearance between the fence and the finished pieces to prevent kickback. Fig. 11-15.

• A second method is to adjust the stop rod on the gauge for the correct length of the cut. Fig. 11-16.

11–14. *Clamp a piece of scrap stock, with a straight edge, to the underside of a piece of plywood to act as a guide when cutting.*

• A third method is to clamp a stop block to the auxiliary board fastened to the miter gauge. Fig. 11-17.
• A fourth method is to clamp a stop block to the table. Fig. 11-18.

11–15. *This method is recommended for cutting several short pieces of the same length from a long piece of stock.*

SPECIAL CUTS
Cutting a Bevel or Chamfer with the Grain

Tilt the blade to the correct angle for the chamfer or bevel. Place the fence on the table so the blade tilts away from the fence. Adjust the height of the blade to clear the top of the board slightly. Hold the work firmly against the fence as the cut is made. Fig. 11-19.

11–16. *Using the stop rod on the miter gauge is a quick, accurate way of cutting several pieces to the same length. Remember to square one end of each piece first.*

Cutting a Bevel or Chamfer across Grain

Adjust the blade to the correct angle and place the miter gauge in the groove on the side toward which the blade tilts. Hold the stock firmly against the gauge to make the cut. Sometimes it's a good idea to clamp the stock to the gauge for making this kind of cut. Fig. 11-20.

JOINT CUTS

All cuts for making joints should be done with a crosscut, hollow-ground, or combi-

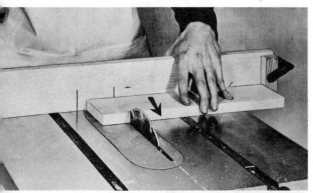

11-17. *Using a stop block on an auxiliary board fastened to a miter gauge for cutting pieces to identical length.*

11-20. *Cutting a chamfer across grain by tilting the saw blade to 45 degrees.*

11-18. *Using a stop block clamped to the table.*

11-19. *Cutting a bevel with the grain.*

nation blade, since it is important to have a very smooth cut.

Cutting a Rabbet

There are two common methods of cutting a rabbet on the circular saw:

METHOD A. MAKING TWO CUTS WITH A SAW BLADE.

Lay out the width and depth of the rabbet on the end or edge of stock so that the lines can be easily seen during the cutting. Adjust the saw blade to a height equal to the depth of the rabbet. If the rabbet is cut with the grain, place the stock face down on the table, with the edge against the fence, and make the first cut. Fig. 11-21. If the rabbet is cut across grain, place the stock face down and hold against the miter gauge with the end firmly against the fence. Fig. 11-22. For the second cut, adjust the blade to a height equal to the width of the rabbet and adjust the fence with the saw blade just inside the waste stock. Fig. 11-23. Hold the surface away from the rabbet firmly against the fence and carefully make the second cut. If the surface or edge of the board that was on the table for the first cut is held against the

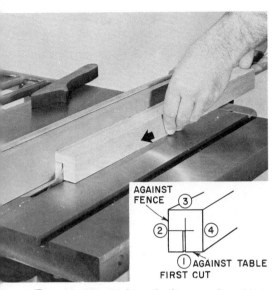

AGAINST
FENCE ③

② ④

① AGAINST TABLE
FIRST CUT

11-21. Making the first cut of a rabbet with the grain. After the cut is started, hold the work against the fence with your left hand and push it along with a push stick. Notice that side No. 1 is against the table and side No. 2 is against the fence.

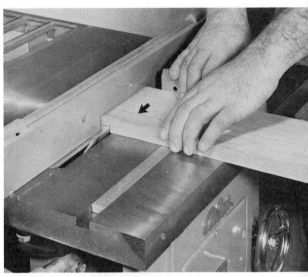

11-22. Making the first cut of a rabbet across the grain.

fence when the second cut is made, the strip of wood which is cut out will kick back with considerable force.

METHOD B. USING A DADO HEAD.

Set up the dado head as described on page 261. Take a piece of wood about the size of the ripping fence and clamp it to the fence. Keep the clamps up off of the table so they will not interfere with the stock during the cut. Set the fence for the desired width of cut, and adjust the height of the dado head for depth of cut. Take care to hold the stock firmly on the table to avoid an uneven cut. Fig. 11-24.

Cutting a Miter

Miter cuts are usually made at a 45-degree angle, although they can be made at any angle. If any shape other than a rectangle is to be cut, you must find the

correct angle for the miter cuts. To do this, divide 180 by the number of sides; then subtract that answer from 90. The result will be the number of degrees for each miter cut. For example, to make cuts for a five-sided figure:

$$180 \div 5 = 36$$
$$90 - 36 = 54$$

Make the cuts at a 54-degree angle.

Making a flat miter cut: Adjust the miter gauge as required, usually to 45 degrees. If two miter gauges are available, the cutting will be simplified. Adjust both gauges to turn inward toward the saw blade. Place a stop rod on the left miter gauge to equal the exact length of the sides to be cut. Hold the stock firmly against the right miter gauge and cut the first miter from the inside edge toward the corner. Fig. 11-25. To make the second cut, hold the mitered end against the stop rod and cut as before.

Making a miter on edge: Tilt the saw blade to an angle of 45 degrees and set the miter gauge at 90 degrees. Place the miter gauge in the groove so the blade tilts

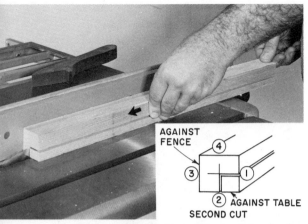

AGAINST FENCE
4
3
1
2 AGAINST TABLE
SECOND CUT

11-23. *Making the second cut of a rabbet with the grain. With this method the waste stock falls away from the blade without binding or kickback. Notice that side No. 2 is against the table and side No. 3 is against the fence.*

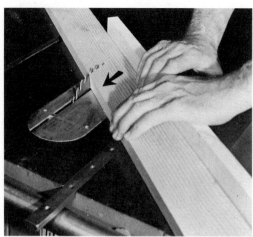

11-25. *Making a flat miter cut. With wood screws a piece of scrap stock is fastened to the miter gauge. This prevents "tear out" as the blade cuts through the edge of the molding.*

11-24. *Cutting a rabbet with a dado head. Hold the workpiece firmly on the table to insure an even cut.*

11-26. *Making an edge miter.*

away from it. Adjust it to the correct height, and make the miter cut as shown in Fig. 11-26.

Cutting a Groove

A groove is a rectangular opening cut with the grain of wood. There are three simple ways of doing this operation.

METHOD A. MAKING TWO OR MORE CUTS WITH A STANDARD BLADE.

Adjust the blade to a height equal to the depth of the groove. Adjust the fence so the blade will remove stock to form one side of the groove. Make the first cut. Turn the stock around so the second face is against the fence and make the second

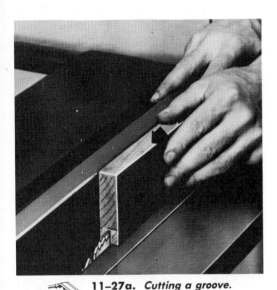

11-27a. *Cutting a groove.*

STOCK TO BE GROOVED

FENCE PROTECTIVE
STRIP

A→ ←B   B→ ←A

FENCE

SAW TABLE

SAW BLADE

11-27b. *Here are the steps for cutting a groove with a single saw blade.*

cut. This will insure that the groove will be in the center of stock. When the two outside cuts have been made on all the pieces, move the fence over, if necessary, and make cuts as needed to clean out the remaining stock. Fig. 11-27.

METHOD B. USING WOBBLE WASHERS.

Wobble washers replace the regular washers that hold the blade on the arbor. These can be set so the blade wobbles and cuts a groove of a specific width. There are marks on the washers for setting the width of cut. Adjust the blade to the correct height. Set the fence and make the cut. *This is somewhat dangerous because the blade does not run smoothly. Fig. 11-28.*

METHOD C. USING A DADO HEAD.

This is the safest and fastest method of cutting grooves. As mentioned, a dado head consists of two outside cutters with chippers placed between them. Fig. 11-29. Remove the throat plate and the saw blade. Place one of the dado head blades on the arbor and then put on the correct number of chippers for the desired width of

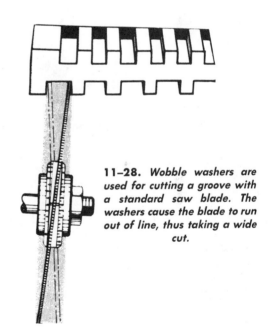

11-28. *Wobble washers are used for cutting a groove with a standard saw blade. The washers cause the blade to run out of line, thus taking a wide cut.*

groove. Finally, add the second blade. (Usually the blades and cutters are 1/16", 1/8" and 1/4" wide, making it possible to cut a groove of any standard width.)

Turn the chippers until the points are evenly spaced and the swaged (enlarged)

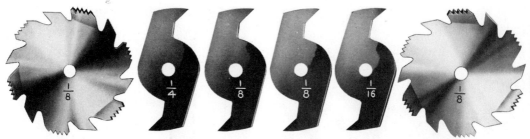

11–29. *With this dado head set you can cut grooves from ⅛″ to ¹³/₁₆″ in intervals of ¹/₁₆″.*

SWAGED CHIPPER

11–30. *The dado head mounted on the saw arbor. The swaged ends on the chippers must be set in the gullet of the outside cutters, because of the additional thickness. This thickness will clean the bottom of the cut when paper or cardboard shims are used if the dado head must be set to an interval of less than ¹/₁₆″.*

cutting edge of the chipper is in the gullet of the outside cutter. Fig. 11-30. For example, if three cutters are used, they should be set 120 degrees apart. This makes the dado head operate smoothly. Install a

11–31. *A special throat plate is needed for installing a dado head.*

throat plate of the type made specially for a dado head. Fig. 11-31. Adjust the dado head and the fence as required, and proceed with the cut. Fig. 11-32.

Adjustable Dado Head

The adjustable dado head is easy to use and will give a clean cut. Fig. 11-33. The width can be set by loosening the arbor nut and rotating the center section of the head until the width mark on this part is opposite the desired dimension. Fig. 11-34a. The cut can be easily and accurately varied to any width from ¼″ to ¹³/₁₆″. Fig. 11-34b.

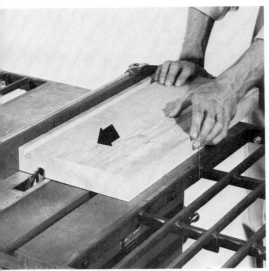

11-32. *Cutting a groove, using a dado head.*

11-33. *Using the adjustable dado.*

Cutting a Stopped Groove

A stopped groove is not cut along the entire length of the stock. Such grooves are usually toward the center of the stock. Fasten hand screws to the fence to control the length of the cut. Turn on the machine. Hold one end of the stock against the first clamp, and lower it into the saw. Push the stock along until it strikes the second clamp. Carefully raise the stock. Fig. 11-35.

Cutting Dadoes

Plain dado. A dado is a groove cut across grain. It can be done in any of the ways described for cutting a groove. The fence can serve as a stop block while the work is held against the miter gauge. When cutting a regular dado, pass the stock completely across the cutter, then remove the stock. Fig. 11-36. Do not draw the board back across the dado head. This is a very important precaution when using the dado head.

Blind dado. A blind dado or gain is cut only partly across the board. Clamp a stop

11-34a. *An adjustable dado head.*

block to the fence to control the length of the dado. Cut the dado as before until the board hits the stop block. Fig. 11-37. Then slowly raise the board or turn off the machine and remove the board.

163

← ¼" to 13/16" WIDE →

UP to ¾" DEEP

11–34b. *Drawing of an adjustable dado head. Note that the depth as well as the width of cut is adjustable.*

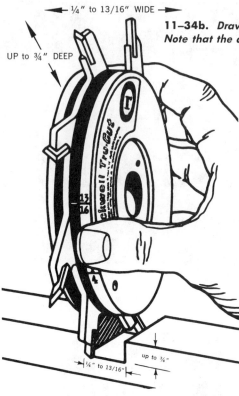

up to ¾"

¼" to 13/16"

11–35. *Cutting a stopped groove. Notice the use of hand screws as stop blocks to control the length of cut.*

11–36. *Cutting a plain dado.*

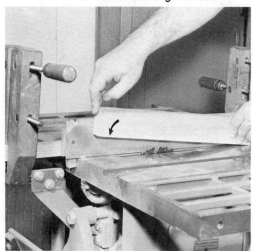

11–37. *Cutting a blind or stop dado. A stop block clamped to the table controls the length of cut.*

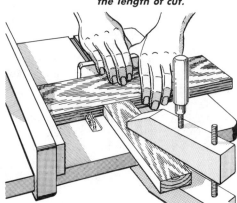

QUESTIONS

1. Name five kinds of blades that can be used on a circular saw.

2. Name the two devices that are used to guide stock when cutting on a circular saw.

3. Tell how to remove a saw blade.

4. Explain two ways of supporting long stock for ripping.

5. In ripping narrow stock, what safety device should be used?

6. What is meant by resawing?

7. When cutting plywood, should the good side be up or down?

8. Briefly tell when and why a stop block should be used.

9. How many cuts are necessary to make a rabbet on a circular saw with a single saw blade?

10. Tell how to cut a groove with a single saw blade.

11. Describe how to assemble a dado head for cutting a $^{13}/_{16}''$ groove.

12. What is a stop groove?

13. What is another name for a blind dado?

Portable Power Saw

SAFETY

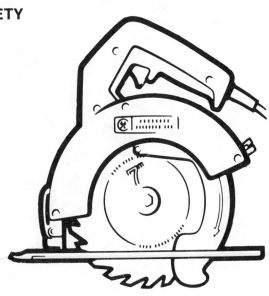

● Make sure the teeth of the blade are sharp and set correctly.

● Never make an adjustment on a saw when it is running.

● Don't stand directly in line with the saw blade. If the blade binds, it may kick the saw back out of the cut. If this happens, turn off the switch immediately.

● Always keep the guard in place and the blade adjusted for the correct depth of cut.

● Use the correct blade for the work to be done.

● Disconnect the power source to change a blade. Make sure that the teeth are pointing in the direction of rotation and that the arbor nut is tight.

● Allow the saw to reach full speed before starting a cut.

- Always keep your hands clear of the cutting line.
- Never stand in line with the cut.
- When finished with a cut, release the switch and wait until the blade comes to a dead stop before setting the saw down.

Portable power saws are sometimes called cutoff electric circular hand saws. (Also, these saws are often known by their trade names.) These saws are primarily used for straight cutting on lumber and plywood. They are made in a wide variety of styles and types. Fig. 12-1. One type has a reversible motor with two separate bases so that cuts can be made from either end of a 4 × 8 plywood sheet. Fig. 12-2. With the proper kind of blade, plastic laminate and nonferrous metals can also be cut. These saws are excellent tools for cabinetmakers, carpenters, and plastic laminate fabricators. Fig. 12-3.

The size of a portable power saw is determined by the blade diameter, which ranges from 6″ to 10″. A common size is a 1/2-horsepower motor with an 8 1/2-inch blade. Because the saw cuts from the bottom of the material, it leaves a smoother cut at the bottom than at the top. Fig. 12-4. Therefore, plywood should always be cut with the good side down. The blade is on the right side of the motor which makes it convenient for the right-handed person.

PARTS

The saw consists of a motor, a handle, a baseplate or shoe, a fixed and a movable guard, a blade, and a switch. Fig. 12-1a. Blades used are the same type as for the circular saw. Be certain that the blade is of the correct diameter and that the arbor hole in the blade is of the right size and shape. Fig. 12-5.

STRAIGHT CUTS

1. Mark the cutoff line on the right end of the board whenever possible. This will give better support as the cut is made.

12–1a. *Parts of a portable power saw.*

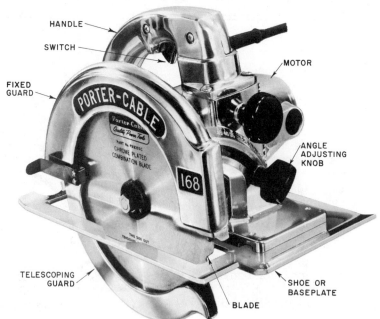

Place the work over the saw horses or support it securely in some other way so that the cutoff line is clear.

2. Loosen the nut or clamp to adjust the depth of cut. Only about ⅛" of the blade should show below the stock. Fig. 12-6. Place the baseplate, or shoe, on the work with the blade in line with the layout line. Turn on the power and allow it to come up to full speed. Guide the saw across the board firmly but without too much pressure, following the layout line. Fig. 12-7. A

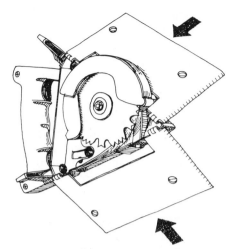

12–2b. *The saw has two separate bases so angle cuts can be made from either direction.*

12–1b. *This saw has a worm gear drive. Notice the saw has two handles, with a switch on each handle.*

12–2a. *Ripping with the reversible saw from right to left.*

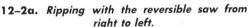

12–3a. *The reversible saw turned over and in position to rip from left to right.*

12–3b. *Making a pocket cut from right to left.*

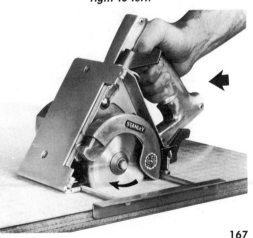

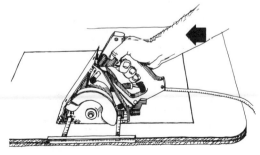

guide like the one in Fig. 12-8 will make crosscutting much more accurate. A long ripping cut can be made freehand following the layout line. It is much better, however, to use a ripping fence as shown in Fig. 12-9.

MITER CUTS

Angle or miter cuts can be made freehand except that it is more difficult to start the cut on the layout line. A protractor attachment is ideal to use for making miter cuts. This device is marked in degrees and can be set to cut any angle by moving the projecting arm to the correct degree. To use this attachment, the shoe is lined up

with the protractor straightedge. Then saw the same as for any cut.

BEVEL CUTS

On most saws the shoe can be adjusted between 45 and 90 degrees. Loosen the wing nut or handle and tilt the shoe to the desired angle. Then retighten the wing nut or handle. Adjust the saw for the correct depth of cut. Make the bevel cut freehand or use a jig to guide the saw. Fig. 12-10.

COMPOUND ANGLE CUTS

A compound angle cut can be made by tilting the saw blade and using a protractor guide. Fig. 12-11.

POCKET CUTS

Internal or pocket cuts can be made in a panel. An example of this operation is cutting the opening in a counter top for a sink. Swing the saw guard out of the way and keep it there. Place the front edge of

12–3c. *This type of blade is specially designed to cut when rotating in either direction, for use on saws with reversible motors.*

12–4. *The cutting action of the portable power saw is different from that of the circular saw. The portable saw cuts from the bottom up.*

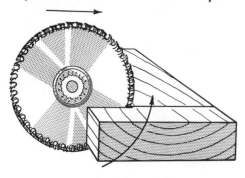

12–5. *Arbor hole bore sizes and shapes for portable power saw blades.*

D-T
SKIL SAW, THOR

½" & 13/16"
SQUARE
MALL, WAPPAT,
CUMMINS, PET

⅝" ROUND
PORTER-CABLE,
SIOUX, SYNTRON,
MILWAUKEE,
CRAFTSMAN,
STANLEY,
MONTGOMERY
WARD

1⅛" & 1⅜"
ROUND
BLACK & DECKER
(also ⅝" round)

the saw base on the work. Start the saw and let it come to full speed. Then, using the front edge of the saw base as a pivot, slowly lower the blade into the work at the guide line. Fig. 12-12.

CUTOFF TABLE AND MITER BOX

To provide a firm, convenient surface for making crosscuts, angle cuts, and bevel cuts, a portable metal table is available for use with the portable power saw. Fig.

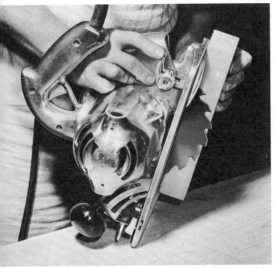

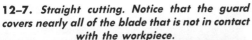

12–6. *Adjusting the saw for the depth of cut.*

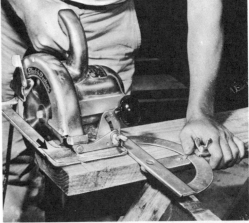

12–8. *Using a protractor guide for cutting. This can be adjusted to any angle to make miter cuts.*

12–7. *Straight cutting. Notice that the guard covers nearly all of the blade that is not in contact with the workpiece.*

12–9. *Ripping with a fence. When ripping a long board, either walk slowly with the saw or stop the saw and pull it back in the kerf a little way, taking a new position to finish the cutting.*

12-13. This table can be helpful to carpenters; to mechanics working with aluminum, vinyl, or composition siding; and to cabinetmakers either inside or outside the house.

The table also has a stopping block which clamps to the fence or to the cutoff extension, providing consistently accurate repeat cuts. Fig. 12-14. Miter stops pop up from the table surface to hold the stock in position. This makes a flat miter cut possible with minimum setup time. Fig. 12-15. Edge miters are made by tilting the saw to the proper angle and proceeding as in crosscutting. Fig. 12-16.

THE POWER MITER BOX

This tool—a power saw mounted in a miter box—is an important labor-saving device, especially for the carpenter. Fig. 12-17. It is highly useful in cutting miters for such operations as fitting and installing moldings. This machine cuts a clean, accurate joint and can be used for general cutoff work.

The machine may be set up to cut 45° miters either right or left. The table contains positive (locked) stops at 90° and 45°. A variety of materials can be cut on the saw, including compositions, plastic, wood, and soft, lightweight aluminum extrusions. Fig. 12-18.

12–10. *Making a bevel cut.*

12–11. *Making a compound angle cut.*

12–12. *Starting a pocket cut. Notice that the guard must be held out of the way. Be sure to release the switch and let the blade come to rest before lifting the saw out. Clean out the corners with a hand saw.*

FIVE FT. EXTENSION ADJUSTABLE BUTTSTOP MITER BOX, POP-UP JIGS FOLDING STEEL LEGS

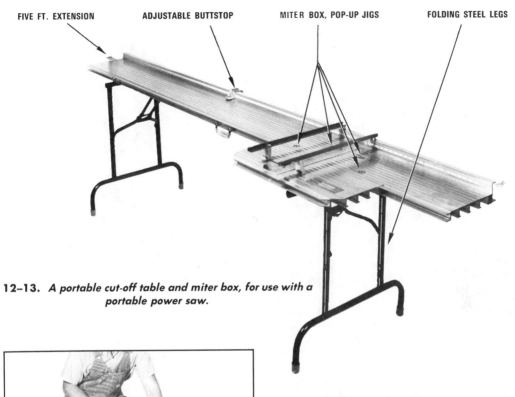

12–13. *A portable cut-off table and miter box, for use with a portable power saw.*

12–14a. *Cutting a piece of siding to length. Notice the use of the stop block clamped to the fence.*

12–14b. *The cut-off extension is self-storing. When used, it is pulled out and clamped into position. The piece to be cut off butts against the stop at the end of the extension.*

12-17. *A power miter box.*

12-15. *Cutting flat miters, using the pop-up miter stops as a guide.*

12-16. *Edge miters are cut by tilting the portable power saw to the desired angle.*

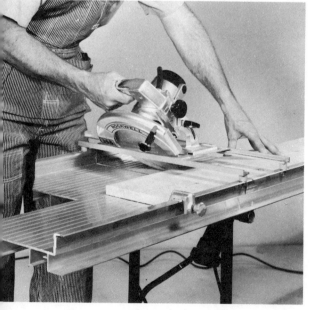

Removing the Blade

To remove the saw blade from the power miter box, first disconnect from the power source. Then place the hex wrench in the hex slot in the end of the arbor, to hold the arbor stationary. Fig. 12-19. Remove the arbor nut with an open-end wrench by applying pressure on the wrench in the direction of the blade rotation. Remove the nut, the collar, and the blade. To install a blade, place it on the arbor. Make certain the teeth at the bottom are pointing away from you and toward the fence. Replace the collar, with the recessed side against the blade. Replace and securely tighten the nut.

Operating the Power Miter Box

To turn on the saw, pull the switch trigger as shown at arrow #1. Fig. 12-20. Make the cut by pivoting the saw down into the wood. As soon as the cut is completed, release the switch trigger, return the cutter-head to the up position, and press down on the brake button as shown at arrow #2.

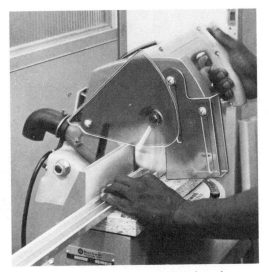

12–18a. *Cutting stock to length.*

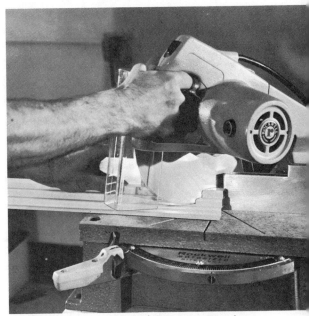

12–18b. *Cutting a 2″ × 4″ at a 45° angle.*

12–18c. *Cutting soft aluminum extrusions. When cutting aluminum, a stick wax should be applied to the side of the blade.*

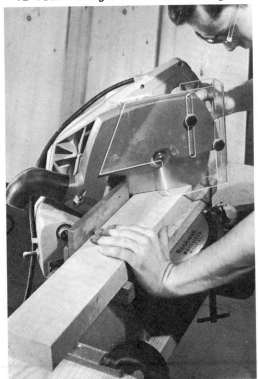

12–19. *Removing a saw blade.*

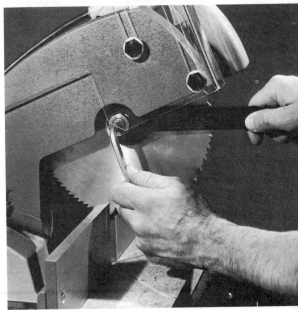

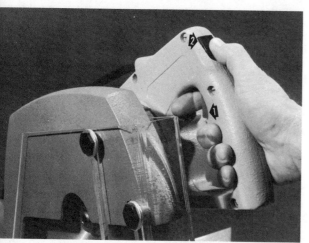

12-20. *To stop the machine, release the trigger (arrow #1) and press down on the brake button (arrow #2).*

12-23a. *To change the angle of the saw cut, release the cam lock (arrow #1): move the indicator to the proper angle and reclamp.When setting the saw to cut at 90° or 45°, release the cam lock and depress the positive lock (arrow #2) until it engages the desired stop on the machine; then re-engage the cam lock.*

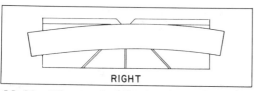

12-21. *When cutting flat bowed pieces, make sure the material is positioned on the table as shown.*

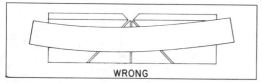

12-22. *Material positioned as shown here will pinch the blade.*

Crosscutting with the Power Miter Box

When cutting flat pieces, first check to see if the material is bowed. If it is, make sure the material is positioned on the table as shown in Fig. 12-21. If the material is positioned as shown in Fig. 12-22, the material will pinch the blade near the completion of the cut.

To make angle cuts, release the friction clamp or the cam lock, as shown at arrow #1. Fig. 12-23. Move the indicator to the angle to be cut and reclamp the cam lock. When making 45° or 90° angle cuts, release the cam lock (#1). Depress the positive lock as shown at arrow #2 in Fig. 12-23. Move the handle over until the positive lock (#2) makes contact with the stop on the machine; then re-engage the cam lock (#1).

As an aid in cutting crown moldings (trim where walls meet ceiling), construct a filler block as shown in Fig. 12-24. Fasten the filler block to the fence by drilling two holes in each side of the fence and securing the block to the fence with roundhead wood screws from the rear. When the filler block is installed in this way, the crown molding will be on the table of the miter box in the same position as it would be when nailed between the ceiling and the wall.

12–23b. *Cutting a piece of base molding at a 45° angle.*

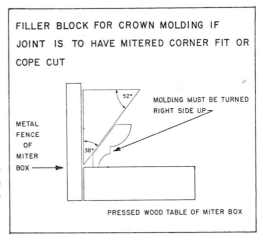

FILLER BLOCK FOR CROWN MOLDING IF JOINT IS TO HAVE MITERED CORNER FIT OR COPE CUT

52°

MOLDING MUST BE TURNED RIGHT SIDE UP

METAL FENCE OF MITER BOX

38°

PRESSED WOOD TABLE OF MITER BOX

12–24. *Using a filler block to hold the crown molding in the correct position for sawing miters for outside corners, or for sawing to provide an outline for coping an inside corner.*

QUESTIONS

1. List three safety rules for the portable power saw.

2. What are portable power saws normally used for?

3. How does the cutting action of the portable power saw differ from that of the circular saw?

4. When making straight cuts with a portable power saw, how deep should the saw blade be set?

5. Describe two methods of making miter cuts.

6. Describe how a bevel can be cut.

7. What is a pocket cut?

8. List several reasons why a cut-off table and miter box would be convenient for use with the portable power saw.

9. How does the power miter box differ from the cut-off table and miter box combination accessory?

Router and Electric Plane

SAFETY

- Review general safety rules for use of portable electric tools.
- Make certain the fence or guide is securely clamped.
- When using the power tool, keep both hands on the handles.
- Feed in the correct direction.
- Always lay the power tool down with the cutter pointing away from you and be alert to the coasting cutter.
- Always hold onto the power tool when it is turned on.
- Make certain the workpiece is securely clamped.
- Make adjustments only when the cutter is at a dead stop.

- Be certain the power switch is off before connecting the power plug.
- When installing or removing cutters, be sure the power plug is disconnected.

The portable router is used for shaping the surfaces and edges of stock and for jointery. With accessories the router can also be used for plastic laminate trimming, hinge butt routing, or as an electric hand plane.

The portable router consists of a motor with a chuck attached to the spindle. Fig. 13-1. This motor screws into a base to which two handles are attached. A guide for straight or curved routing also can be secured. With an attachment the portable router can be held in an inverted position and used as a small shaper.

ROUTER BITS

These bits come in many shapes for doing grooved or decorative work in the surface or edge of stock. Some of the common ones are straight, rounding-over, beading, cove, and chamfer bits. Fig. 13-2. In addition, a dovetail bit is needed for cutting a dovetail joint. All router bits cut on their sides rather than on end. Fig. 13-3.

Installing Cutters

1. Disconnect the power plug.
2. Lock the shaft or hold it with a wrench, depending on the kind and size of router.

176

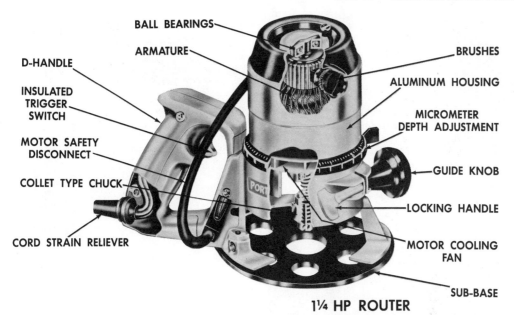

BALL BEARINGS

ARMATURE

D-HANDLE

INSULATED
TRIGGER
SWITCH

MOTOR SAFETY
DISCONNECT

COLLET TYPE CHUCK

CORD STRAIN RELIEVER

BRUSHES

ALUMINUM HOUSING

MICROMETER
DEPTH ADJUSTMENT

GUIDE KNOB

LOCKING HANDLE

MOTOR COOLING
FAN

SUB-BASE

1¼ HP ROUTER

13–1. *Portable router with parts named.*

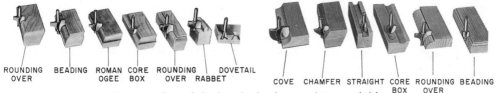

ROUNDING
OVER BEADING ROMAN
OGEE CORE
BOX ROUNDING
OVER RABBET DOVETAIL

COVE CHAMFER STRAIGHT CORE
BOX ROUNDING
OVER BEADING

13–2. *A few of the hundreds of router bits available.*

3. Insert the shank of the bit into the chuck at least ½".

4. Tighten the chuck with a wrench.

5. Unlock the shaft.

6. Adjust the depth of cut by moving the motor unit up and down in the base.

7. Make a test cut in scrap stock. *Hold onto the router when turning on the power, to overcome the starting torque of the motor.*

13–3a. *The amount the router can move sideways can be controlled in five ways: A straight-edge can be clamped to the stock and the router base held in contact with it.*

13–3b. *A straight or circular guide can be attached to the base to control the lateral movement. A guide is used here to make a cut near the edge of a round table top.*

13–3e. *You can operate the router freehand, as in making this sign.*

13–3c. *A template or pattern can be used. A sleeve or guide is attached to the bottom of the base and this rides against the template.*

13–3d. *Many cutters have pilot edges to control the amount of cut.*

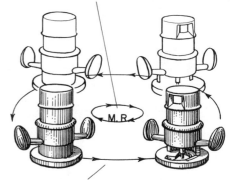

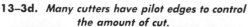

DIRECTION OF MOTOR ROTATION (M.R.)

M.R.

DIRECTION OF FEED

13–4. *The router bit revolves clockwise. Therefore, when cutting straight edges, move the router from left to right. When making circular cuts, move the router counterclockwise.*

Feeding the Hand Router

Always feed against the direction of motor rotation. When making a cut on a straight edge, feed from left to right. When cutting on circular stock, feed in a counterclockwise direction. Fig. 13-4.

Your judgment of how fast to feed will have to be developed by practice. The speed at which the best cut is made will

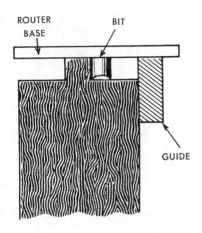

ROUTER BASE · BIT · GUIDE

13–5a. *Cutting a groove in the edge of a board. Notice the extra piece of wood attached to the guide. This will give the router added support when riding on the narrow edge of the workpiece. (In the photograph, a previous cut removed much of the stock.)*

13–5b. *Cutting a dado.*

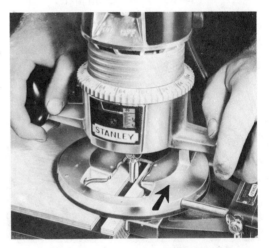

depend on the size cut to be made and on the hardness of the wood. Most hand routers run at speeds of about 21,000 rpm. This is the speed without a load; the motor will slow down when under a work load. If the router is fed too fast, the motor will slow down too much, causing a poor cut. If the router is fed too slowly, the bit will get hot, possibly drawing the temper from the cutting edge or burning the wood. Don't force the cut; allow the bit to cut freely. Listen to the motor for an indication of whether it is working at its most efficient speed.

USING A PORTABLE ROUTER
Cutting Grooves, Dadoes, Gains, or Mortises

To make these cuts, fasten a straight router bit in the chuck. Screw the motor into the base until the router bit extends the desired depth beyond the base. Attach a guide to the base to control the cut. Lay out the cut and locate the guide. Start at one side or end and move the router along to make the cut. Fig. 13-5.

13–5c. *Cutting a mortise.*

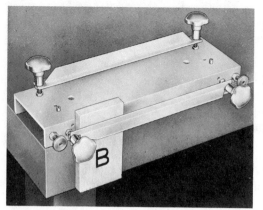

13–6a. *Clamp side in place.*

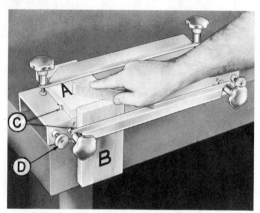

13–6b. *Clamp front or back in place.*

13–6c. *Place template over piece.*

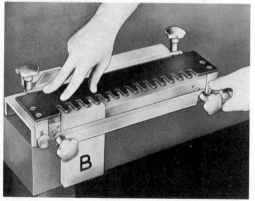

13–6d. *Cutting the dovetail.*

Cutting Decorative Edges

As mentioned, cutters of many shapes are used for cutting decorative edges. Many of these have a pilot tip which does not cut but merely rides against the uncut edge of the wood. Attach the bit in the machine and adjust for depth. Secure a piece of scrap the same thickness as the finished stock and clamp it to the top of a bench. Hold the router firmly against the top of the work and move the cutter into the stock. Check to see that the desired shape is being cut. The shape can be changed by moving the motor up and down in the base. Fig. 13-3d.

Freehand Routing

Signs and decorations are sometimes made freehand with the router. Carefully lay out the areas to be removed. Secure a cutter bit of the desired diameter and fasten it into the machine. Clamp the work to the table top, then lower the router into the design. Move the router along to follow the outline. Fig. 13-3e.

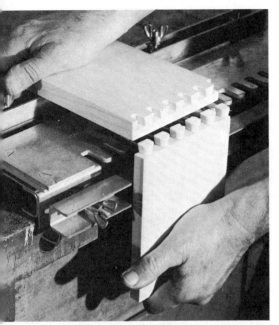

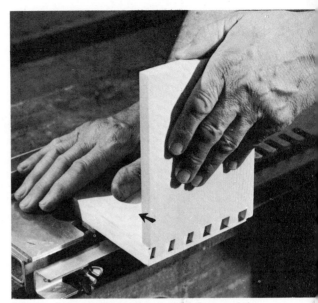

13–6f. *Back of drawer side is tipped up and the fit of the joint is checked.*

13–6e. *Drawer front and side are held on the outside of the fixture, in the same position in which they were cut.*

Dovetail Joint

The best joint for most drawer construction is the dovetail. This joint is sometimes used for other wooden "box" construction as well. It is a difficult joint to make by hand but simple with a router and dovetail attachment.

1. Clamp the dovetail attachment to a bench or table. Square up the stock to be used for the front and sides of the drawer or box. Fasten a template guide to the base of the router and install a dovetail bit. Adjust the dovetail bit to extend below the base the desired amount. This amount can usually be determined by making a trial cut and readjusting the router to the exact depth.

2. Select a board which will be one side of the box or drawer. Clamp it face side in against the front of the base and protruding ½" or more above the top surface of the base. This is shown as board B, Fig. 13-6a.

3. Clamp board A, which will be the front or back of the drawer or box, inner side up. Make sure that it is in full contact with board B. It should be set flush with the top end of B. Both boards must be in contact with the locating pins. See arrow "C" on Fig. 13-6b.

4. Place the dovetail template over the two pieces of stock and clamp in place. Fig. 13-6c.

5. Make a trial cut, being sure that the template guide follows the template. Fig. 13-6d. If the trial joint is too *loose,* adjust to make a *deeper cut.* If the trial joint is too *tight,* adjust for a *shallower cut.* Fig. 13-6e and f.

6. Fig. 13-7a shows a drawer side which does not fit far enough into the front. When this happens, turn *in* the template adjusting nut (D in Fig. 13-6b). This allows a deeper cut into the drawer front. If the drawer side fits in too deeply, Fig. 13-7b, turn the template adjusting nut *out.* Be sure these adjusting nuts on both ends of the fixture are set the same.

181

7. The completed dovetail should appear as in Fig. 13-7c. The left end of the fixture is used for cutting the right front of the drawer and the left rear corner. The left

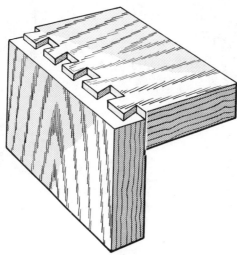

13–7a. *If the drawer side does not fit far enough into the front, turn the template adjusting nut in to allow a deeper cut in the drawer front.*

13–7b. *If the drawer side fits too deeply into the cuts in the front, turn the template adjusting nut out. (Note, however, that sometimes the side will be set low intentionally. This allows for clearance between the drawer side and the cabinet. If this is done, complete the joint by cutting a rabbet on the inside of the drawer front equal to the depth of the projection.)*

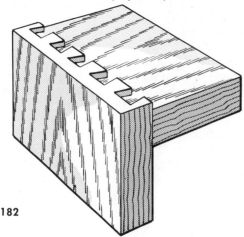

front and right rear corner are cut on the right end of the dovetail fixture.

Dovetail Dado

The dovetail dado is a good joint to use where extra strength is needed, because the joint will pull the two pieces together and hold tightly. It is an ideal joint for fastening a drawer side to a front when the front "lips over" the cabinet sides. Fig. 13-8.

Install the dovetail bit in the router and attach the guide. Adjust the depth of cut and set the guide for the first cut. Fig. 13-9a. If the slot is to be cut in a drawer front, it will not be necessary to clamp a piece of wood on either side of the work to support the router. When the first cut has been completed, set the second piece in the vise with a piece of scrap stock on each side. The scrap pieces should be at least 3/4" thick and both the same thickness. Leave the depth of cut set the same as for the slot, and readjust the guide so one cut is made on each side to form the tenon. The width of the tenon is cut to fit the slot by adjusting the guide. Fig. 13-9b. It is recommended that a trial cut be made in

13–7c. *The dovetail joint is found in the finest drawer construction.*

scrap stock of the same thickness to insure a good fitting joint.

Hinge Butt Routing

A special template is available for hinge butt routing. Because the bits leave a slight curve at the corner of the cut, it is necessary to chisel the corner square for the hinges. Fig. 13-10 shows a metal template in position on a door. This guides the router so that the hinge mortises are cut easily and quickly, in the proper location, to uniform size and depth.

PORTABLE PLANES
Electric Hand Plane

Sometimes called the *portable electric plane*, this tool greatly reduces the time

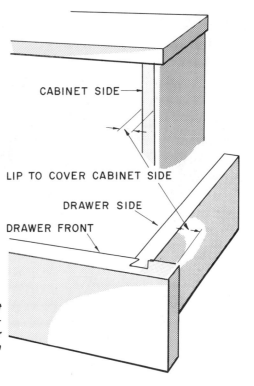

CABINET SIDE

LIP TO COVER CABINET SIDE

DRAWER SIDE

DRAWER FRONT

13-8. *The dovetail dado joint is an excellent joint to use for drawer construction on Contemporary furniture. This joint is also recommended for attaching the sides of a cabinet to the top, when the cabinet top overhangs the side panels.*

13-9a. *Cutting the dovetail mortise or slot. Note that in the picture it is cut in the edge of the workpiece. For a drawer front or cabinet top the cut would be made on a surface.*

13-9b. *Cutting the dovetail tenon or stub. A cut is made on each side, leaving the tenon in the center.*

13–10. *Cutting a hinge mortise on a door. If a square corner hinge is used, the corners of the cut must be squared with a chisel. However, hinges are available with round corners so they fit directly into the opening. After gains or mortises are cut on the door, the template guide can be transferred to the doorframe for cutting hinge mortises on the jamb.*

and labor involved in planing by hand. Fig. 13-11. Such a plane can be purchased, or it can be assembled by using an attachment with the router. It is particularly useful for installing large doors and paneling, because it will make a smooth, accurate cut. The actual cutting tool is either a spindle-and-plane-cutter combination or a one-piece plane cutter with a threaded shank. For most jobs the one-piece plane cutter is used.

The cutter must be set at zero before setting for depth of cut. Move the depth adjustment lever (located at the front of the plane) to the zero position and turn the plane over. To the left of the handle and directly behind the motor bracket is a cutter-adjusting lever. Turn this lever toward the rear of the plane and lay a straightedge across the cutter so that it rests on both the front and rear shoes. Turn the cutter by hand until it lifts the straightedge. Then adjust the lever until the tip of

13–11. *Parts of a 16″ portable electric plane.*

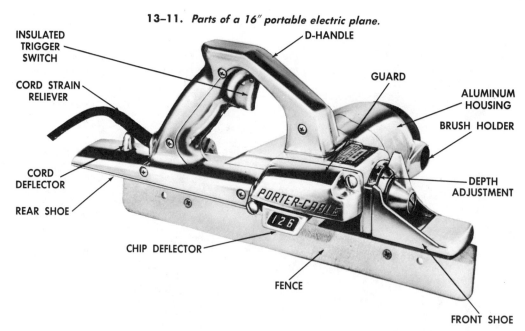

INSULATED TRIGGER SWITCH

D-HANDLE

GUARD

ALUMINUM HOUSING

CORD STRAIN RELIEVER

BRUSH HOLDER

CORD DEFLECTOR

DEPTH ADJUSTMENT

REAR SHOE

CHIP DEFLECTOR

FENCE

FRONT SHOE

16″ PORTABLE PLANE

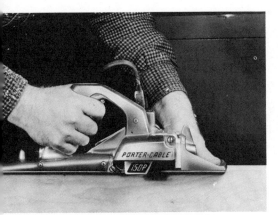

13-12a. *Using a portable electric plane to surface the edge of a door.*

13-12b. *Planing a bevel on the edge of stock.*

the cutting edge just touches the straight-edge when it (the straightedge) rests evenly on both shoes. Now the desired depth of cut can be set by rotating the depth-adjustment lever.

To use, place the plane on the work, with the front shoe and fence held firmly against it. Fig. 13-12. Turn on the power. Then apply steady, even pressure as you do the planing. Never overload or push the plane too hard. When working on thick plywood or hardwoods, do not attempt to cut as fast as for softer woods. The plane can also be set for outside bevel cuts from zero to 45 degrees. In planing the edges of plywood, there is danger of breaking out the cross-grain plies at a corner. The best way to prevent this is to clamp a piece of scrap wood at the end of the plywood before the cut is made. As you near the edge, move the plane very slowly.

Power Block Plane

The power block plane is a small tool with a spiral cutter similar to that of the electric hand plane. Fig. 13-13. It can be used for many procedures such as surface planing, edge planing, making and cleaning up rabbet cuts, planing cupboard doors to size, and beveling the edges of

plastic laminates that have been bonded to a wood base. Fig. 13-14. Just under the handle is a knob for adjusting depth of cut. A small fence clamps to the bottom of the plane but can be removed for face planing. This lightweight tool can be controlled with one hand.

13-13. *Parts of the power block plane.*

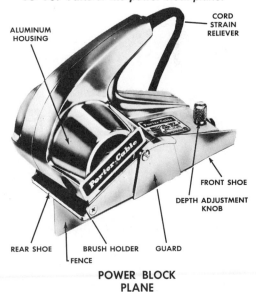

CORD STRAIN RELIEVER

ALUMINUM HOUSING

FRONT SHOE

DEPTH ADJUSTMENT KNOB

REAR SHOE   BRUSH HOLDER   GUARD

FENCE

POWER BLOCK PLANE

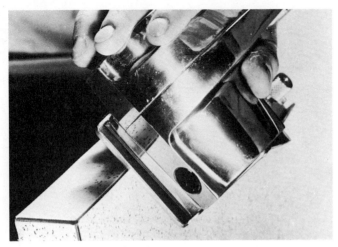

13-14. *Beveling the edge of plastic laminate.*

QUESTIONS

1. Besides using the router for shaping surfaces and edges, briefly describe what other operations can be performed by using accessories.

2. Name six common kinds of router bits.

3. Can a portable router be used for freehand routing? Explain.

4. When making a cut along the outside edge of a piece of stock, in which direction should the router be fed?

5. What accessory is necessary to cut a dovetail joint with a portable router?

6. List several reasons why an electric hand plane would be a valuable tool for a carpenter or cabinetmaker.

Portable

Electric Drill

SAFETY

● Review general safety rules for use of portable electric tools.

● Disconnect the power plug before installing or removing drills.

● Make certain the drill is clamped securely in the chuck.

● Be sure the key has been removed.

● Do not force the drill—use an even steady pressure.

● Never use a bit with a square tang or a lead screw.

● When laying the drill down, always have the point away from you, even when it is "coasting" to a stop.

- Never drill through cloth.
- Always clamp small pieces; do not hold them with your fingers when drilling.

The portable electric drill is an excellent tool for drilling, boring holes, and for many other uses. The tool consists of a housing with a handle, a motor, and a chuck. Most of these drills have a key-type chuck. Fig. 14-1. These tools are made in a variety of shapes. The most common shapes are the pistol-grip drill which usually will hold drill bits up to ¼", and the spade handle drill which usually has a chuck capacity up to ½". Fig. 14-2.

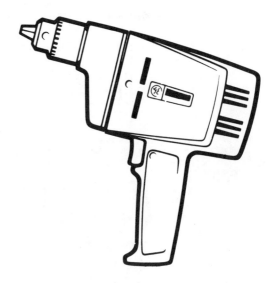

14-1. *Parts of a ¼" portable drill.*

FIELD WINDING   ARMATURE   REDUCTION GEARS

BRUSHES

ALUMINUM HOUSING

BALL BEARING

PISTOL GRIP HANDLE

INSULATED TRIGGER SWITCH

GEARED KEY CHUCK

BALL BEARINGS

MOTOR COOLING FAN

CORD STRAIN RELIEVER

HEAVY DUTY PORTABLE DRILL

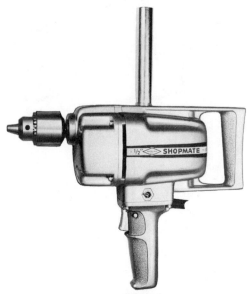

14–2. *The ¹/₂″ capacity electric hand drill with spade handle.*

14–3a. *The spade bit is often used to bore large holes with a ¹/₄″ capacity hand drill.*

14–3b. *Combination drills are available in most of the common woodscrew sizes. For example, if a 1″ #8 woodscrew is used, a 1″ #8 combination drill should be used.*

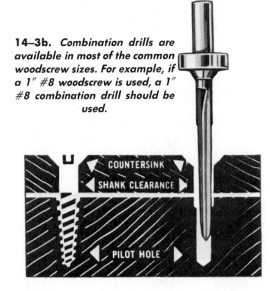

COUNTERSINK

SHANK CLEARANCE

PILOT HOLE

Some drill housings are made of plastic in order to reduce the danger from shock. Others have a built-in variable speed unit, with the speed varying directly with the amount of pressure exerted on the trigger switch. The cutting tools can be twist drills, auger bits (if they have straight shanks), or spade (speed) bits. Fig. 14-3a shows a spade bit.

The combination drill and countersink is a convenient tool for inserting woodscrews. It is always necessary to drill the correct size holes for woodscrews to prevent splitting the workpiece and to get the maximum holding power. This is usually a three-step procedure, if the screw is to be countersunk. The combination drill, however, will drill the pilot hole, shank hole, and countersink, all in one operation. Fig. 14-3b.

INSTALLING A BIT

Before inserting a bit, make sure that it has a straight shank. Turn the drill chuck by hand until the jaws are open wide

14–4. *Always tighten the chuck securely to prevent the bit from slipping.*

enough to take the desired size bit. Insert the bit shank in the chuck as far as possible, then close the jaws by hand. Next, tighten the chuck by inserting the key wrench in each of the key holes in succession. Use all three holes to avoid slippage as much as possible. Fig. 14-4. To release the bit, only one hole needs to be used. Remember, always unplug the drill when changing bits.

DRILLING HOLES

Always use the correct bit or accessory. Fig. 14-5. Make sure the tool (preferably of high-speed steel) is sharp. Apply just enough pressure to the drill to keep it cutting. Too little pressure will make the drill dull; too much pressure may cause it to stall or break. To prevent break-through splintering, clamp a piece of scrap wood behind the piece being drilled. Always clamp the wood in a vise, or hold it securely with a clamp. Hold the tool at right angles to the work when drilling a straight hole. Fig. 14-6. This can be checked by using a try square to align the tool. If the drill is equipped with a variable speed unit, it can be used as a powerful screwdriver by inserting a screwdriver bit. Fig. 14-7. Start the screw slowly, increase the

speed as the screw moves into the stock, and finish by slowing to a stop. On a standard-speed drill, a speed reduction ratchet attachment must be used for installing screws.

DRILLING HOLES FOR WOOD SCREWS

Select the correct size drill for the wood screws. Fig. 14-8. The pilot hole is for the threaded portion of the screw. In hardwood, it is good practice to bore the pilot hole the same size as the root diameter. (Consult Fig. 14-8 under heading *Root Diameter*.) In softwood, drill the pilot hole about 15 percent smaller.

Drill the pilot hole through the first piece and into the second piece to the desired depth. Drill the shank hole through the first piece. For flathead screws, use an 82-degree, rose-type countersink to enlarge the end of the shank hole. If the screw is to be covered with a plug, bore the hole for this first, then the shank hole, and finally the root diameter hole.

A second method of drilling holes for screws is to drill the shank hole first. Then hold the first piece over the second and mark the location for the pilot hole with a scratch awl.

14–5.
Suggested Drilling Speeds

Material Or Job	Bit or Accessory	Suggested Speed	
		¼" Drill	⅜" Drill
Wood	Twist Drill	High	High
Wood	Spade Bits	High	High
Wood	Auger Bits	Med. to Low	High to Med.
Wood	Hole or Dial Saw	Med. to Low	Medium
Heavy Metal	Twist Drill*	Medium	High
Light Metal	Twist Drill*	High	High
Plastics	Twist Drill	Medium	High
Driving Screws	Driver Bit	Low	Low

*Only high speed steel twist drills should be used for drilling in metal.

14–6. *Drilling a hole in flat stock. Be sure to hold the tool at right angles to the workpiece.*

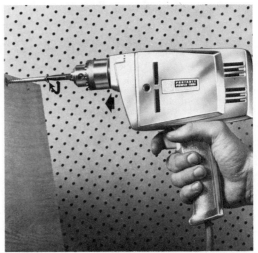

14–7. *Using the variable-speed hand drill as a power screwdriver.*

QUESTIONS

l. List three safety rules that should be observed when using the portable electric drill.

2. Why are some portable electric drill housings made of plastic?

3. Why is it necessary to drill the correct size hole for wood screws?

4. When drilling holes for wood screws, what is the advantage of a combination drill?

5. How far should the drill bit be inserted in the chuck?

6. When drilling a hole, what should be done to prevent splintering the workpiece as the hole is completed?

7. What is the purpose of a pilot hole?

8. The shank hole should have the same diameter as what part of the wood screw?

Boring Chart for Wood Screws

NO. OF SCREW	MAXIMUM HEAD DIAMETER	SHANK DIAMETER			ROOT DIAMETER		THREADS PER INCH	NO. OF SCREW
		BASIC DEC. SIZE	NEAREST FRACTIONAL EQUIVALENT		AVERAGE DEC. SIZE	NEAREST FRACTIONAL EQUIVALENT		
0	.119	.060	1/16	OVERSIZE .002	.040	3/64 OVERSIZE .007	32	0
1	.146	.073	5/64	OVERSIZE .005	.046	3/64 BASIC SIZE	28	1
2	.172	.086	3/32	OVERSIZE .007	.054	1/16 OVERSIZE .008	26	2
3	.199	.099	7/64	OVERSIZE .010	.065	1/16 UNDERSIZE .002	24	3
4	.225	.112	7/64	UNDERSIZE .003	.075	5/64 OVERSIZE .003	22	4
5	.252	.125	1/8	BASIC SIZE	.085	5/64 UNDERSIZE .007	20	5
6	.279	.138	9/64	OVERSIZE .002	.094	3/32 BASIC SIZE	18	6
7	.305	.151	5/32	OVERSIZE .005	.102	7/64 OVERSIZE .007	16	7
8	.332	.164	5/32	UNDERSIZE .007	.112	7/64 UNDERSIZE .003	15	8
9	.358	.177	11/64	UNDERSIZE .005	.122	1/8 OVERSIZE .003	14	9
10	.385	.190	3/16	UNDERSIZE .002	.130	1/8 UNDERSIZE .005	13	10
11	.411	.203	13/64	BASIC SIZE	.139	9/64 OVERSIZE .001	12	11
12	.438	.216	7/32	OVERSIZE .003	.148	9/64 UNDERSIZE .007	11	12
14	.491	.242	1/4	OVERSIZE .008	.165	5/32 UNDERSIZE .009	10	14
16	.544	.268	17/64	UNDERSIZE .002	.184	3/16 OVERSIZE .003	9	16
18	.597	.294	19/64	OVERSIZE .003	.204	13/64 UNDERSIZE .001	8	18
20	.650	.320	5/16	UNDERSIZE .007	.223	7/32 UNDERSIZE .004	8	20
24	.756	.372	3/8	OVERSIZE .003	.260	1/4 UNDERSIZE .010	7	24

No. 0
No. 1
No. 2
No. 3
No. 4
No. 5
No. 6
No. 7
No. 8
No. 9
No. 10
No. 11
No. 12
No. 14
No. 16
No. 18
No. 20
No. 24

14–8. *Drill Size Selection Chart. Root diameters are average dimensions measured at the middle of the threaded portion. Shank diameters are shown as a decimal in thousandths of an inch. To use this chart, read down either of the columns headed Number of Screw to the screw size used. For example, if a No. 5 woodscrew is used, the shank diameter of the screw will require a 1/8" hole. The pilot hole will need to be 5/64". Note that the pilot hole is 0.007" undersized. This is about right for softwood. In hardwood the next larger drill size, 3/32", should be used. The column at right shows actual woodscrew shank sizes. To determine the size of a screw visually, lay the screw shank on the silhouette.*

UNIT
15

Saber Saw

SAFETY

- Review general safety rules for use of portable electric tools.
- Select the correct blade for the work and properly secure it in the chuck.
- Be certain the material to be sawed is properly clamped.
- Keep the cutting pressure constant. Do not force the cut.
- When finished, turn off the power switch and allow the saw to come to a dead stop before setting the saw down.
- Hold the base down securely on the work when cutting.

The *portable jig* (also called *saber* or *bayonet*) *saw* is the best choice for an on-the-job cutting tool for straight or irregular cutting. This tool can do the same cutting as a floor-type jig or band saw, with the added convenience of a hand tool. A larger jig saw can cut through material 2" thick. It also can cut through a 2" × 4" piece in less than 15 seconds. It will cut metal, wood, plastic, and many other materials. Most hand jig saws use orbital action (cutting the material on the up stroke and moving away from it on the down stroke). Because of this, cutting speed is greatly increased and the saw cuts with a cleaner edge.

PARTS

The design of this tool varies with the manufacturer. However, all of these tools consist of a motor, a handle, a mechanism to change rotary action into up-and-down action, and a baseplate or shoe. Fig. 15-1. Select the correct blade. Fig. 15-2. At least three teeth must be on the cutting surface at all times. To install the blade, loosen the set screws or clamp, and slip the blade into the slot under the chuck cover until you are sure it is tightly seated. Then tighten the set screw or clamp.

OPERATIONS
Straight and Irregular Cutting

Mount the work so it is held rigid. Fig. 15-3. Make a layout line that can be followed. Set the shoe of the tool on the work. Start the motor and allow it to come up to full speed. Then move the saw along slowly. Fig. 15-4. Don't force the cutting. Use

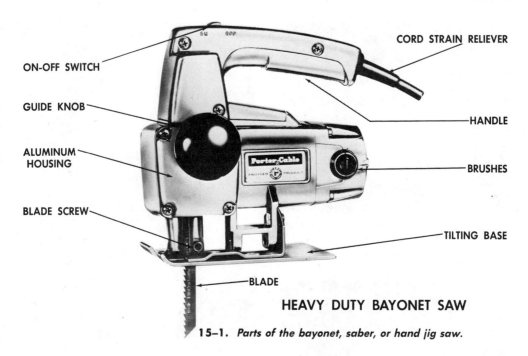

ON-OFF SWITCH

GUIDE KNOB

ALUMINUM HOUSING

BLADE SCREW

CORD STRAIN RELIEVER

HANDLE

BRUSHES

TILTING BASE

BLADE

HEAVY DUTY BAYONET SAW

15–1. *Parts of the bayonet, saber, or hand jig saw.*

15–2. *Guide for selecting the correct blade.*

Blade Selection

Heavy cuts 2″ × 4″ at 45°	6 teeth per inch
General cutting	7, 10
Smooth cuts	12
Plywood	12
Hardboard	12
Cardboard	Knife
Leather	Knife

only enough pressure to keep the saw cutting at all times. The tool is always held in one hand. The other hand can hold the work or steady the saw. For more accurate straight cutting, a ripping fence can be installed. Fig. 15-5. The ripping fence can also be used for cutting circles. A nail or peg must be driven into the center of the circle.

Plunge or Internal Cutting

The portable jig saw can be used to cut out an internal area without first drilling a hole. This is called plunge cutting. With a pencil, mark out the area to be cut. Choose a convenient starting place inside the waste stock. Tip the tool forward with the shoe resting on the surface of the material and the top of the blade clear of the work surface. Fig. 15-6a. Turn on the power. When the blade reaches full speed, slowly lower the back of the machine until the blade cuts through the material to the full depth. Then cut out the opening. Fig. 15-6b.

Bevel Cutting

The shoes of some saws can be adjusted from 0 to 45 degrees for bevel cutting. Fig. 15-7. Such cutting can be done free-hand, as shown in Fig. 15-8, or with a guide.

Circle Cuts

For circle cuts, remove the guide and turn it over. Set the guide into position for the radius desired and tighten it. Make a pocket cut or drill a pilot hole on the

15-3. *Cutting a curve. Notice how the work is clamped to the table.*

15-4a. *Straight cutting.*

15-4b. *Two methods of cutting an exterior corner. One way is to make a slightly curved cut at the corner; then trim this off with a second cut. Another method is to make a complete circle in the waste stock.*

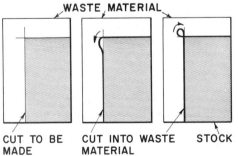

WASTE MATERIAL

CUT TO BE MADE   CUT INTO WASTE MATERIAL   STOCK

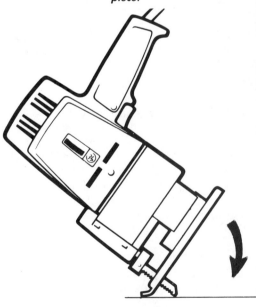

15-5. *Using a fence to do ripping.*

15-6a. *Tip the saw up on the front of the baseplate. Turn it on and allow it to come to full speed before lowering the blade into the work piece.*

15–6b. *Making an internal cut.*

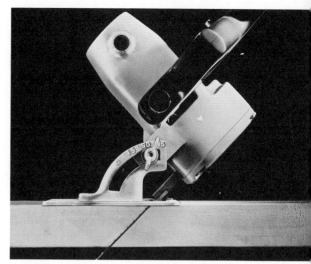

15–8. *Cutting a bevel with the saber saw.*

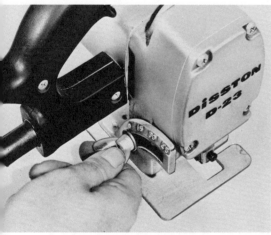

15–7. *Adjusting the shoe, or baseplate, for bevel cutting.*

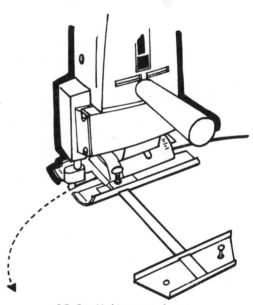

15–9. *Making a circle cut.*

circumference of the desired circle and insert the saw blade. You will find it very difficult to make perfect circles starting from the edge of the board. With the blade on the circumference of the circle, locate the center of the circle and drive a small nail through the hole in the guide at that point. Now begin the cutting, but do not force the saw. Let it do the cutting and you will have a perfect circle. Fig. 15-9.

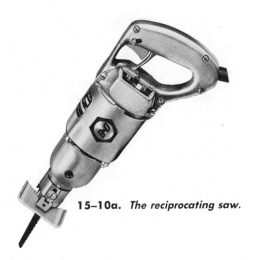

Reciprocating Saw

Another type of all-purpose saw operates with a back-and-forth movement, like a hacksaw without a frame. This saw is most commonly used for remodeling or cabinetwork, since it can cut wood, plastic, metal, ceramics, and many other materials. Fig. 15-10.

15-10a. *The reciprocating saw.*

15-10b. *Making a plunge cut with a reciprocating saw.*

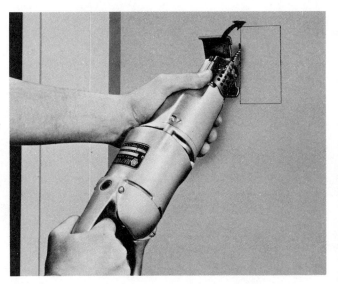

QUESTIONS

1. List three safety rules that should be observed when using a saber saw.

2. By what other names is the saber saw sometimes known?

3. What materials can be cut with a portable electric jig saw?

4. When using the saber saw, at least how many teeth should be on the cutting surface at all times?

5. How much pressure should be used when cutting with the saber saw?

6. What is meant by plunge cutting?

7. Describe how circles can be cut with the saber saw.

Portable Sander

SAFETY

- See general safety rules for operation of portable electric tools.
- Be sure the abrasive is in good condition and that its grit is correct for the work to be done.
- Be sure the abrasive belt is installed with the correct tension and is tracking properly.
- Keep your hands away from abrasive surfaces.
- Never touch the edge of a belt or disc.
- Be sure there are no nicks or tears in the edge of a disc or belt.
- Disconnect the power plug when changing abrasives.
- Make certain the switch is in the "off" position when plugging in the power cord.

Portable electric sanders are excellent for finish-sanding projects after assembly.

PORTABLE BELT SANDER

The portable belt sander operates in a manner similar to the floor-model belt-sanding machine, except that the revolving belt is placed on the work instead of the work against the belt. Fig. 16-1. The size of the machine is determined by length and width of belt.

Installing the Belt

Install the belt so that the arrow on the inside of the belt points in the same direction as the arrow on the side of the sander. When the belt has been placed on the

pulleys, it can be made to run straight by adjusting the tracking screw. Fig. 16-2a. The belt should not be allowed to rub against the left side of the machine.

Using a Portable Belt Sander

The proper technique is to put the cord over your right shoulder, hold the machine with both hands, turn the machine on, and lower the back of it slowly onto the wood. Then do the sanding by moving the machine back and forth, and at the same time moving it slowly from one side to the other. Fig. 16-2b.

FINISHING (OR PAD) SANDER

There are many kinds of finishing sanders. Parts of a typical one are shown in Fig. 16-3a. All of these sanders operate on one of the three basic principles shown in Fig. 16-3b. Those with straight-line action are the least likely to leave cross-grain

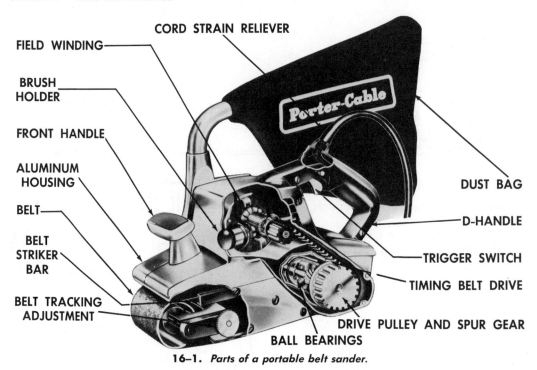

FIELD WINDING

BRUSH HOLDER

FRONT HANDLE

ALUMINUM HOUSING

BELT

BELT STRIKER BAR

BELT TRACKING ADJUSTMENT

CORD STRAIN RELIEVER

Porter-Cable

DUST BAG

D-HANDLE

TRIGGER SWITCH

TIMING BELT DRIVE

DRIVE PULLEY AND SPUR GEAR

BALL BEARINGS

16–1. *Parts of a portable belt sander.*

16–2a. *A two-speed portable belt sander. Note the round knob near the bottom (indicated by arrow) used for tracking the belt. The lever just above this knob is used to release belt tension when replacing a belt.*

scratches. The size of this machine is determined by the size of the abrasive sheet used. Finishing sanders are primarily used for fine finish-sanding after the project is assembled. To replace a sheet of abrasive, first cut a sheet of paper to the required size. The paper is held on the pad at either end and pressure keys are used to lock the paper in place. Release the pressure keys at either end. Fasten the paper in one end and lock the key at that end. Pull the paper tightly over the pad, slip the loose end under the other clamp, and tighten this lock. The exact method of fastening the paper to the pad will vary with different sanders.

Using a Finishing Sander

The finishing sander should rest evenly on the stock. Apply a moderate amount of pressure and move the sander back and forth, working from one side to the other. Fig. 16-4.

HEEL TOUCHES FIRST

DROP DOWN

SHORT STRAIGHT OVERLAPPING STROKES FOR UNIFORM SURFACE

16–2b. *Lower the sander slowly to the surface. Move the machine in the pattern shown.*

16–3a. *Parts of a half-pad finishing sander.*

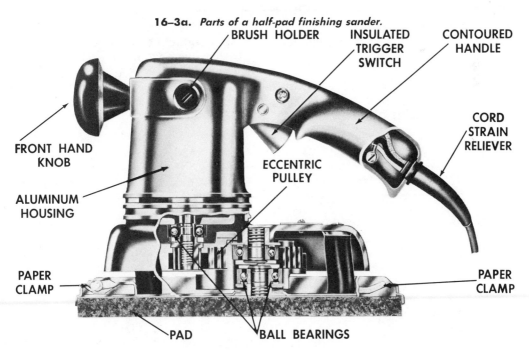

BRUSH HOLDER

INSULATED TRIGGER SWITCH

CONTOURED HANDLE

CORD STRAIN RELIEVER

FRONT HAND KNOB

ALUMINUM HOUSING

ECCENTRIC PULLEY

PAPER CLAMP

PAPER CLAMP

PAD

BALL BEARINGS

199

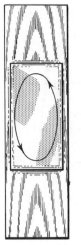
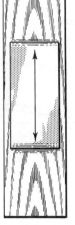

ORBITAL   STRAIGHT   MULTI
ACTION   LINE   MOTION
          ACTION   ACTION

16–3b. *Three kinds of action in finishing sanders. Orbital and multi-motion action may cause some cross-grain scratches. Straight-line action is like hand sanding and results in the best surface.*

16–4. *Using a finishing sander.*

QUESTIONS

1. List three safety rules that should be observed when using the portable sander.

2. How is the size of a portable belt sander determined?

3. When installing a belt on a portable electric sander, in which direction should the arrow on the inside of the belt point?

4. Describe the proper technique for using a portable belt sander.

5. Finishing sanders have three kinds of actions. Name them.

6. Describe the proper technique for using a finishing sander.

FOUNDATIONS

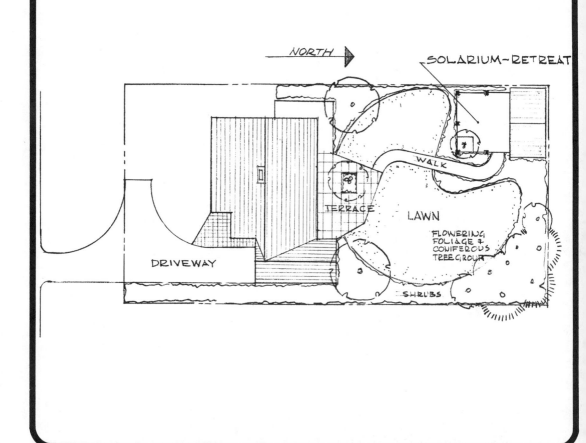

Locating the House on the Building Site

Most city building regulations require that a plot plan be a part of the house plans, to show the location of the building on the lot. Before the exact location of the house is determined, check local codes for minimum setback and sideyard requirements; the location of the house is usually affected by such codes. Sometimes the setback is established in accordance with existing houses on neighboring property.

After the site is cleared, the location of the outer walls of the house is marked out. Usually the surveyor will mark the corners of the lot after making a survey of the plot of land. Sometimes the surveyor will also be asked to do a rough marking of the corners of the building.

There are two basic ways of accurately determining the location of the proposed building on the property:
- By measuring from an established reference line.
- By using an instrument such as an optical level or a transit level.

LAYING OUT FROM A REFERENCE LINE

Sometimes a building or excavation may be planned to parallel an identifiable line, such as a street or property line. Such a line can then be used as a guide or reference point. This makes it possible to stake out the site without using a transit level or an optical level. When working in this way, it is best to make a drawing of the property before attempting to stake out the site itself. Fig. 17-1 shows such a drawing. In that illustration, rectangle ABCD represents the property lines, and boundary AB is the identifiable line. Refer to the drawing

as you study the following paragraphs. The letters in the following paragraphs refer to points and lines on the drawing. They will help you to understand the actual operation of staking out the site. To stake out, proceed as follows:

1. Check the plot plan to find the setback distance. Along boundaries AC and BD measure this distance back from front line AB. The setback is shown by segments AO and BO.

2. Stretch a line tightly between the points marked O. This line shows where the front of the building will be.

3. On line 00, locate the front corners of the building. There are two ways to do this. You can obtain the measurement from the plot plan to see how far the corners will be from the side boundaries. Then along line

17-1. Staking out a rectangular building without the use of an optical or transit level.

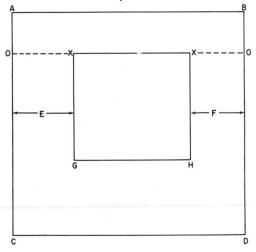

00 measure in the indicated distances from AC and BD. Or, if the building is to be centered between the side boundaries, you need not refer to the plot plan. Instead, subtract the length of the building from the length of 00, then measure in half this distance from each end of 00. (An X represents each front corner of the building on Fig. 17-1). Measure the distance between the two points marked X and check this distance with the plans. The distance XX represents the length of the building and it must be accurate.

4. Check the plans to learn the depth of the project (how far back it will extend from the front corners). Mark off the depth by extending lines back from the two points marked X. If the boundary lines of the lot form a 90° angle at the corners, these lines should be parallel to AC and BD. Note that E is the same as OX, and F the same as XO. Thus E and F show the distance between sides of the building and the side boundary lines of the lot. Points G and H are used to represent the rear corners of the building.

If the boundary lines of the lot are not at right angles to each other, make certain that when the building is located on the site the minimum front and sideyard requirements are established at the building's closest point to the boundary line. Under these circumstances the building lines will probably not be parallel to the boundary lines of the lot. Therefore it will be necessary to establish the corner of the building which will be closest to the boundary line and then lay out the building from this point by using the method described on page 207, "Laying Out a Right Angle." Also see "Laying Out a Simple Rectangle" on the same page.

5. Establish a line to indicate the rear of the building. (This is shown by GH on the drawing.)

6. If the building is not rectangular, divide it into smaller rectangles. Then follow the steps just given to find the front, back, and sides of each smaller rectangle. In other words, more lines such as 00 will have to be established to indicate the front of each rectangular area of the building. You can get the necessary information from the plans. Then Steps 3, 4, and 5 are carried out for each rectangle. The result will be a group of adjoining rectangles which will show the total outline of the building.

LAYING OUT WITH A TRANSIT OR OPTICAL LEVEL

The Instruments

Two instruments commonly used by builders and contractors are:

- The *optical level,* sometimes called the *dumpy level* or *builder's level.* Fig. 17-2a and b.
- The *transit level,* often called simply a *transit.* Fig. 17-2c and d.

The basic difference between these two instruments is that the optical level is fixed in a horizontal position. It can be used only for measuring horizontal angles because it cannot be tilted up and down. The telescope of the transit level can be moved up and down as well as sideways. Because of its vertical movement, the transit level can be used to:

- Determine if a wall is plumb.
- Measure vertical angles.
- Run straight lines.

The layout procedures described in the following pages can be carried out with either a transit or an optical level, unless otherwise stated.

Reference Points

To lay out a building by means of a transit or an optical level you must have a basic starting point. This point is usually called the *bench mark.* It is a reference point from which measurements can be made.

In built-up areas the surveyors will often have provided a bench mark. It may appear as a mark, or point, on the foundation of a nearby building. More often it is a stone marker in the ground at a designated location. Sometimes the level of a nearby sidewalk, street, or curbing is used as the bench mark.

The bench mark may appear on the architect's drawings. If so, the plans will usually be oriented to that point.

Another key reference point is the *grade line,* which will be discussed later.

Setting Up a Level

Fig. 17-3a shows a typical location for setting up a transit or optical level in relation to a building site. The point over which the level is directly centered (point A in the drawing) is called the *station mark.* This is the point from which the layout is to be sighted (or shot). It may be a bench mark or a corner of the lot. In Fig. 17-3a point B is the bench mark.

Set up the level where the area of the plot can be conveniently sighted. (If the bench mark and the station mark are not the same, you should also be able to sight the bench mark conveniently.) Under the head of the level there is a hook. A plumb bob suspended from the hook is used to center the level directly over the station mark. Adjust the tripod so that it rests firmly on the ground, with the sighting tube at eye level.

17-2b. *Using an optical level.*

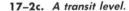

17-2c. *A transit level.*

17-2a. *An optical level, sometimes called a dumpy level.*

17-2d. *Using a transit level.*

To level the head of the instrument, loosen the horizontal clamp screw (Fig. 17-3b, arrow 1) and turn the telescope until the bubble (arrow 2) is in line with two opposite leveling screws (arrows 3). Grip the screws with the thumb and forefinger of each hand. Loosen one screw as you tight-

en the other the same amount. This is done by moving your thumbs toward or away from each other. Keep the screws snug on the foot plate. Do not overtighten them. Continue to adjust the screws until the bubble is centered. Then rotate the telescope 90° so that it is over the other two leveling screws and repeat the previously described leveling procedures. Return the telescope to the first position, check the bubble, and readjust if necessary. Recheck the second position. Continue to alternately check the bubble between the two positions until the bubble is within one graduation on either side of center in the bubble tube.

Once the level is properly set up, *be careful not to move or jar the tripod.*

Establishing Points on a Line

For this operation you must use a transit level. With a plumb bob, level the instrument and center it accurately over a point on the line. Sight the telescope on the most distant visible known point of that line. Lock the horizontal motion clamp screw to keep the telescope on line. Then adjust the tangent screw to place the vertical cross hair exactly on the distant point. Now, by tipping the telescope in the vertical plane (up or down), you can determine the exact location of any number of stakes on that same line. Fig. 17-4.

17-3a. *Laying out a building site with a transit level or optical level.*

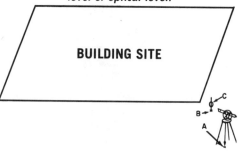

BUILDING SITE

17-3b. *An optical level.*

Laying Out a Right Angle

Using a plumb bob, set up an optical level or a transit level directly over the line at the point where the right angle is to be. This is shown as point A in Fig. 17-5. Sight a reference point on that line and set the 360° scale at zero. (B is the reference point in Fig. 17-5.) Turn the telescope until the scale indicates that an arc of 90° has been completed. Establish a leveling rod (Fig. 17-6) in position along this line of sight at the desired distance. (D in Fig. 17-5 indicates this distance.) A line from the rod to the point from which the sighting was taken will be perpendicular to the base line. Thus a right angle will be formed where the lines intersect. (This would be at point A of Fig. 17-5.)

Staking Out the Building

When the location and alignment of a building have been determined, a rectangle comprising the exterior dimensions of the structure is staked out. If the building is to be a simple rectangular structure, the staked-out area will follow the exact exterior of the foundation line. (Layout of an irregular building is discussed later.)

Laying Out a Simple Rectangle

To perform this operation you work from an established line such as a road or street line, a property line, or other reference line. This line is shown as AB in Fig. 17-7. Locate the point which represents the lateral (side) limit for a front corner of the project (C in Fig. 17-7).

Still referring to Fig. 17-7, set up the optical or transit level at point C. Sight point E, turn the telescope 90°, and estab-

lish point D, a front corner of the project. Then move the instrument and set it up at point E. This should be a greater distance along line AB from point C than the intended length of the project. Set a stake at F by sighting C and turning the telescope 90°. Point F should be the same distance from AB as D. The distance between points C and D will be equal to the distance between points E and F.

The front line of the building is established by marking off the length of the project (DG) along line DF. The two front corners of the building are located at D and G. (This is assuming that the building is to be parallel to the established line, such as a road or property line. If it is not, check the architect's plans or the plot plan for the correct setback and establish the two front corners of the building accordingly.)

Now return the instrument to point C and set it up. Sight point E, then swing the instrument 90° and sight along this position to establish H, a rear corner of the building.

Now move the instrument and set up at G; sight D and swing the sight tube 90° and identify I, the other rear corner of the building.

To prove the work (that is, check its accuracy) measure distance IH. If IH is equal to DG, the work is correct; if not, the work must be repeated.

Laying Out an Irregularly Shaped Building

Where the outline of the building is other than a rectangle, the procedure in establishing each point is the same as de-

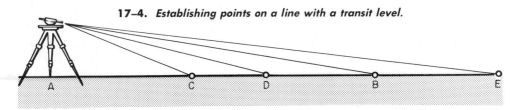

17–4. *Establishing points on a line with a transit level.*

scribed for laying out a simple rectangle. However, more points have to be located, and the final proving of the work is more likely to reveal a small error. When the building is not to be regular in shape, it is usually best first to lay out a large rectangle which will comprise the entire building or the greater part of it. This is shown in Fig. 17-8 as HOPQ. Having once established this accurately, the remaining portion of the layout will consist of small rectangles, each of which can be laid out and proved separately. These rectangles are shown as LMNP, ABCQ, DEFG, and IJKO in Fig. 17-8.

Batter Boards

The next step, after the corners of the house have been established, is to determine lines and grades as aids in keeping the work level and true. *Batter boards* are horizontal boards fastened to small posts and placed near where the corners of the building will be located. Use of these boards is one method of locating and marking the outline of the building. Fig.

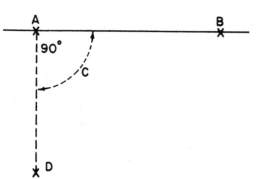

17-5. *Laying out a right angle with a transit level or an optical level.*

17-9. The height of the boards is sometimes established to conform to the height of the foundation wall.

To set up batter boards, the first step is to locate the corners of the building precisely by one of the methods previously discussed in this chapter. Nails are driven into the tops of the stakes to indicate the outside line of the foundation walls. To be certain that the corners are square, measure the diagonals of the completed layout to see if they are the same length. The corners can also be squared by using the 3-4-5 system. This is done by measuring along one side a distance in 3' units such as 6, 9, 12, and along the adjoining side the same number of 4' units (8, 12, and 16). The diagonals will then measure the equal of 5' units (10, 15, and 20) when the unit is square. Thus a 9' distance on one side and a 12' distance on the other should result in a 15' diagonal measurement for a true 90° corner. Fig. 17-10.

The batter boards are set up after the corners have been located. Three 2" × 4" or larger stakes of suitable length are driven at each location 4' (minimum) beyond the lines of the foundation; then 1" × 6" or 1" × 8" boards are nailed horizontally so the tops are all level and are all the same distance from the grade line. Next, twine or stout string (carpenter chalkline) is held across the tops of opposite boards at two corners and adjusted so it will be exactly over the nails in the corner stakes at either end; a plumb bob is handy for setting the lines. Saw kerfs at the outside edge are cut where the lines touch the boards so that they may be replaced if broken or disturbed. After similar cuts are located in all eight batter boards, the lines of the house will be established. Check the

17–6a. *A leveling rod with target. Note that the rod is divided into feet, inches, and eighths of an inch.*

diagonals again to make sure the corners are square. The area for an L-shaped building, for example, can be divided into rectangles, treating each separately or as an extension of one or more sides.

Grade Line

Another important reference point from which measurements are made is the grade line. This is found on the architect's plan and refers to the level of the ground where it will touch the foundation of the completed building. The grade line must be established accurately because it is used for making important measurements. From the grade line you can find the depth of the excavation and also establish certain elevations such as floor and foundation levels.

Sometimes the bench mark is used as a reference point for establishing the grade line. At other times this line may be located in relation to the level of an existing street, sidewalk, or curbing. The grade line is indicated on a stake driven into the ground outside the excavation area.

Establishing Elevations

Check the architect's plan to see which elevations are to be determined. Set up and level the instrument as previously described. Be certain that the locations of the elevations can be seen through the telescope. Place a leveling rod (sometimes called a measuring rod) upright on any point to be checked. Then sight through the telescope at the leveling rod and take a reading by means of the horizontal cross hair in the telescope. Then move the rod to the second point to be established. Raise

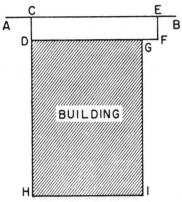

17–7. *Laying out a regular square or rectangular building with an optical level or a transit level.*

17–6b. *The target on the leveling rod can be moved up or down by releasing the clamp on the back of the target. Shown here is an engineer's leveling rod. It is divided into whole feet, tenths, and hundredths of a foot.*

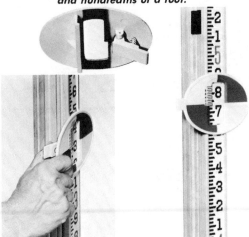

17–8. *Laying out an irregular building made up of a series of squares and rectangles with an optical level or a transit level.*

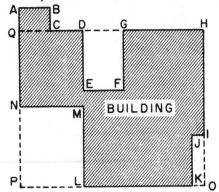

or lower the rod until the reading is the same as for the first point. The bottom of the rod is then at the same elevation as the original point.

In accurate work a spirit level (carpenter's level) may be attached to the leveling rod to check if the rod is being held plumb. The rod can also be kept plumb by aligning it with the vertical cross hair in the telescope. The person at the telescope can tell the rod holder which way to move the top of the rod. An assistant should hold the leveling rod, and should move the target on the rod up or down until the crossline on the target comes in line with the cross hair sights in the sighting tube.

Measuring Difference in Elevation

To learn the difference in elevation between two points, such as A and B in Fig. 17-11, set up and level either a transit level or an optical level at an intermediate point. With the measuring rod held on point A, note the reading where the horizontal cross hair in the telescope crosses the graduation marks on the rod. Then with the rod held on point B, sight on the rod and note where the horizontal cross hair cuts the graduations on the rod. The difference between the reading at A (5') and the reading at B (5'6") is the difference in elevation between A and B. Thus the ground at point B is 6" *lower* than the ground at point A.

Sometimes it is not possible to sight two points from a single point between them. A high mound can cause this difficulty. To solve this problem, one or more additional intermediate points (such as C and D) must be used for setting up the instrument, as shown in Fig. 17-12.

HEIGHT OF FOUNDATION WALLS

The proposed height of the foundation walls above grade determines the depth of the excavation. That is, the excavation is dug to a depth which will give the foundation walls the correct height above grade. To determine this depth it is common practice to use the highest elevation on the perimeter of the excavation as the reference point. Fig. 17-13. This is true for

17-9. *Using batter boards to establish the outline and height of the foundation wall. The top edge of the batter board represents the height of the foundation wall.*

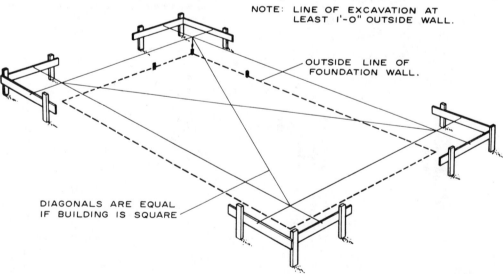

NOTE: LINE OF EXCAVATION AT LEAST 1'-0" OUTSIDE WALL.

OUTSIDE LINE OF FOUNDATION WALL.

DIAGONALS ARE EQUAL IF BUILDING IS SQUARE

graded and ungraded sites. This method will insure good drainage if sufficient foundation height is allowed for the sloping of the finish grade. Fig. 17-14. (When the grading is completed, the level of the ground is referred to as *finish grade*.) Foundation walls at least 7'4" high are desirable for full basements, and 8' walls are commonly used.

Foundation walls should be extended above the finish grade around the outside of the house. This is done so that the wood finish and framing members will be adequately protected from soil moisture and will be well above the grass line. Thus, in termite-infested areas, there will be an opportunity to observe signs of termites between the soil and the wood. Protective measures can then be taken before damage develops.

The top of the foundation wall should usually be at least 8" above the finish grade at the wall line. The finish grade at the building line (perimeter of the building)

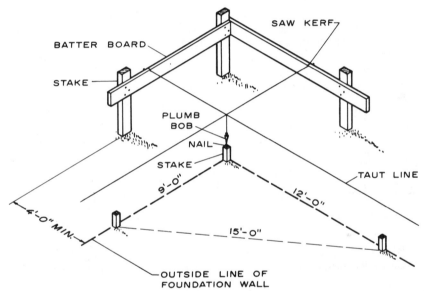

17–10. *Checking the corners of the house layout to make certain they are square, using the 3-4-5 system. In this drawing, a multiple of 3, 4, 5 was used—9, 12, and 15—which is three units of each:*
$$3 \times 3 = 9, \ 3 \times 4 = 12, \ and \ 3 \times 5 = 15.$$

17–11. *Obtaining the difference in elevation between two points that are visible from an intermediate point.*

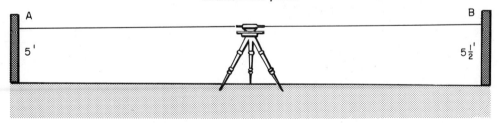

will often be 4″ to 12″ above the original ground level. In lots sloping upward from front to rear, this increase may amount to more than 12″. In very steeply sloped lots, permanent retaining wall at the rear of the wall line is often necessary.

Enough height should be provided in crawl spaces to permit inspections for termites, and also to install soil covers. Such covers reduce the effect of ground moisture on framing members. Ordinarily there should be at least 18″ between the undersides of the joists and the highest point of ground enclosed by the foundation walls.

If the interior ground level is excavated or is otherwise lower than the outside finish grade, measures must be taken to assure good drainage. Fig. 17-14 shows one method of dealing with this problem. In-

stalling a drainage system would be another possibility. This is discussed in Chapter 19.

EXCAVATION

Before excavating for a new home, determine the subsoil conditions by test borings or by checking existing houses constructed near the site. A rock ledge may be encountered, requiring costly removal; or a high water table may require design changes from a full basement to crawl space or concrete slab construction. If there has been a landfill on the building site, the footings should always extend through to undisturbed soil. Any variation from standard construction practices will increase the cost of the foundation and footings. Thus it is good practice to examine the types of foundations used in neigh-

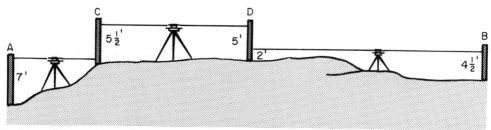

17–12. *Obtaining the difference in elevation between two points not visible from a single intermediate point.*

17–13. *To establish the depth of the excavation, use the highest elevation of the excavation as the control point.*

17–14. *If necessary, add fill to bring the finish grade above the original grade and insure drainage away from the house.*

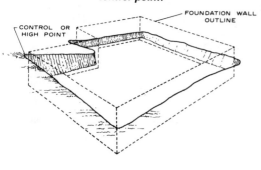

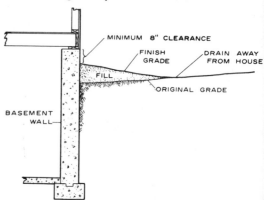

boring houses. This might influence the design of the house you are building.

Excavation for basements may be accomplished with one of several types of earth-moving equipment. Topsoil is often stockpiled by bulldozer or front-end loader for future use. Fig. 17-15. Excavation of the basement area may be done with a front-end loader, power shovel, or similar equipment.

Power trenchers are often used in excavating for the walls of houses built on a slab or with a crawl space, if soil is stable enough to prevent caving. This eliminates the need for forming below grade when footings are not required.

It is best to excavate only to the top of the footings or the bottom of the basement floor, because some soil becomes soft upon exposure to air or water. Thus it is advisable not to make the final excavation for footings until nearly time to pour the concrete unless form boards are to be used.

Excavation must be wide enough to provide space to work. This can involve not only constructing the walls, but possibly also waterproofing them and laying drain tile, if these operations are necessary, as in poor drainage areas. Fig. 17-16. The steepness of the back slope of the excavation is determined by the subsoil encountered. With clay or other stable soil, the back slope can be nearly vertical. When sand is encountered, an inclined slope is required to prevent caving.

When excavating for basements, some contractors roughstake only the perimeter of the building for the removal of the earth. When the proper floor elevation has been reached, the footing layout is made and the earth is removed. After the concrete is poured and set, the building wall outline is established on the footings and marked for the formwork or concrete block wall.

Estimating Excavation

Excavation costs are based on the total cubic yards of earth to be removed. Using

17-15. *A grader being used to strip the topsoil from a building site in preparation for excavating the basement.*

the house plan in Fig. 17-17, figure the earth to be removed for the excavation for the basement area. To determine the volume of material to be removed, multiply the length times the width times the depth. This is usually done in feet and decimals of a foot rather than in feet and inches. Fig. 17-18. When the length, width, and depth of the excavation are in feet, the volume will read in cubic feet.

In the example, Fig. 17-17, multiply 7′ (depth of excavation) × 28′ (26′ width of house + 2′ clearance between the excavation and the outside of the foundation wall) × 42′ (40′ length of house + 2′ clearance). The answer is 8,232 cubic feet.

To convert this to cubic yards, divide by 27 (because there are 27 cubic feet in 1 cubic yard):

$$\frac{8,232}{27} = 304.8$$

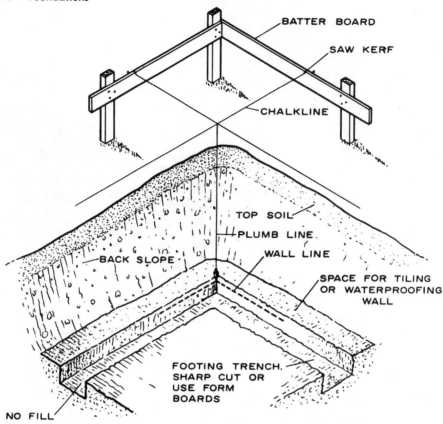

17-16. *The excavation of the basement area is back-sloped to eliminate cave-ins. Note the use of the batter board and chalk line to drop the plumb line for accurately locating the foundation wall line.*

17-17. *With a 7' deep excavation, the basement for this house will require the removal of 305 cubic yards of material.*

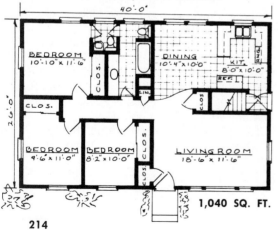

17-18. *A conversion table of inches to decimal fractions of a foot.*

Inches	Feet
1	0.083
2	0.167
3	0.250
4	0.333
5	0.417
6	0.500
7	0.583
8	0.667
9	0.750
10	0.833
11	0.916
12	1.000

17–19. *To use this table, use as an example an excavation 24′ × 30′ and 6′ deep. 24′ × 30′ = 720 square feet. In the table, the 6′ depth shows 0.222 cubic yards for each square foot. The total amount of the excavation is figured: 720 square feet (total area of excavation) × 0.222 (cubic yards per square foot for a 6′ depth) = 159.84 cubic yards of material to be excavated. (Another example is given in the text.)*

Excavation Factors

Depth	Cubic Yards per Square Foot
2″	0.006
4″	0.012
6″	0.018
8″	0.025
10″	0.031
1′0″	0.037
1′6″	0.056
2′0″	0.074
2′6″	0.093
3′0″	0.111
3′6″	0.130
4′0″	0.148
4′6″	0.167
5′0″	0.185
5′6″	0.204
6′0″	0.222
6′6″	0.241
7′0″	0.259
7′6″	0.278
8′0″	0.296
8′6″	0.314
9′0″	0.332
9′6″	0.350
10′0″	0.369

Thus, in even numbers, approximately 305 cubic yards of material will have to be excavated.

The table in Fig. 17-19 provides another way of determining the cubic yards of material to be removed. It can be used if the depth of the excavation is one of the standard ones shown on the table. Refer again to Fig. 17-17. The excavation shown there is 28′ wide and 42′ long, and is to be dug 7′ deep. Multiplying the width by the length you find that the area of the excavation is 1,176 square feet. On the table you see that for an excavation 7′ deep, 0.259 cubic yards of material are removed for each square foot. Thus by multiplying 1,176 by 0.259 you find that approximately 305 (actually 304.5) cubic yards of material will be removed.

Trench excavations, such as those that might be dug for utilities, can be figured by using the table shown in Fig. 17-20. For example, if a trench is to be 42″ deep and

17–20. *Trench excavations—cubic yard content per 100 lineal feet.*

Trench Excavations
Cu. Yd. Content Per 100 Lineal Ft.

Depth in Inches	Trench Width in Inches						
	12	18	24	30	36	42	48
6	1.9	2.8	3.7	4.6	5.6	6.6	7.4
12	3.7	5.6	7.4	9.3	11.1	13.0	14.8
18	5.6	8.3	11.1	13.9	16.7	19.4	22.3
24	7.4	11.1	14.8	18.5	22.2	26.0	29.6
30	9.3	13.8	18.5	23.2	27.8	32.4	37.0
36	11.1	16.6	22.2	27.8	33.3	38.9	44.5
42	13.0	19.4	25.9	32.4	38.9	45.4	52.0
48	14.8	22.2	29.6	37.0	44.5	52.0	59.2
54	16.7	25.0	33.3	41.6	50.0	58.4	66.7
60	18.6	27.8	37.0	46.3	55.5	64.9	74.1

18" wide, the table shows that 19.4 cubic yards of material will be removed for every 100 lineal feet. Such a trench might be dug for a house with a 30' setback. To deter- mine how much material would be re- moved, divide 19.4 by 0.30 (since 30' is 0.30 of 100'). The answer is that 5.82 cubic yards of material would be removed.

QUESTIONS

1. Explain briefly how it is possible to locate a house on a building site without the use of either an optical level or a transit level.

2. What is the difference between an optical level and a transit level?

3. Explain briefly how to learn the difference in elevation between two points.

4. What is the procedure for laying out an irregularly shaped building?

5. Why are foundation walls extended above the finish grade around the outside of the house?

6. Why are test borings taken of the subsoil conditions before excavating for a new home?

7. Why doesn't the excavator remove the soil for the footings at the same time the soil is removed for the basement?

8. Why is the excavation for a basement somewhat larger than the outside dimensions of the foundation wall?

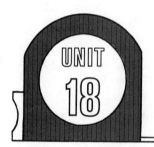

UNIT 18

Scaffolds and Ladders

SCAFFOLDS

A scaffold is a temporary or movable platform to stand on when working at a height above the floor or ground. The scaffold must also support the weight of the worker's tools and materials. Scaffolds make it possible to work safely, in a comfortable and convenient position, with both hands free.

Scaffolding is of two general types:
• Wood scaffolding, constructed on the job.
• Manufactured scaffolding.

Wood Scaffolding

When constructing wood scaffolding, select clear straight lumber for maximum strength. Fig. 18-1. Use adequate bracing and nail securely using a duplex head nail. Fig. 18-2. These nails can be driven in tightly and still be easily pulled when dismantling.

Manufactured Scaffolding

Manufactured scaffolding is designed to be readily assembled or dismantled. Fig. 18-3. The scaffold planks may be set at various heights for comfort and safety. For interior use, casters are installed for easy movement. The end frames may be assembled in a staggered position, making it possible to work off a stairway. Fig. 18-4. Where additional height is necessary, the units may be stacked. Fig. 18-5.

Scaffolding Safety

• All scaffolding should be plumb and level. Use adjusting screws, not blocks, to adjust to uneven grading conditions.

• Adequate support should be provided. Use base plates, making sure that they rest firmly on the ground.

• All braces should be fastened securely.

• Cross braces should never be climbed. Access to scaffolds should be by stairs or fixed ladders only.

• Wall scaffolds should be securely anchored.

• Free-standing scaffold towers must be secured by guying (attaching guy ropes or wires) or other means.

• Proper guard rails should be provided. Add toe boards when required on planked or staged areas.

• Ladders or makeshift equipment should never be used on top of the scaffold.

• A scaffold should never be overloaded. Inspect the scaffolding assembly regularly.

• Lumber used for scaffold planks must be properly inspected and graded for that

18–1. *A scaffold must be safe. Use adequate supports and bracing. Note the blocks at the large arrows. These blocks are attached to the building. They are notched to receive the 2″ × 6″ boards.*

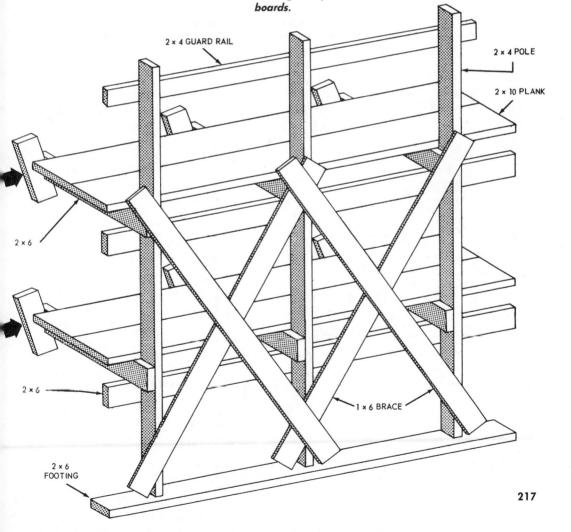

2 x 4 GUARD RAIL

2 x 4 POLE

2 x 10 PLANK

2 x 6

2 x 6

1 x 6 BRACE

2 x 6
FOOTING

18–2. *A duplex head nail.*

purpose. Both ends of planks must be cleated to prevent planks from sliding off supports. If planking is to be continuous, it should have at least a 12″ overlap, and should extend at least 6″ beyond the center of the support. Also do not extend the

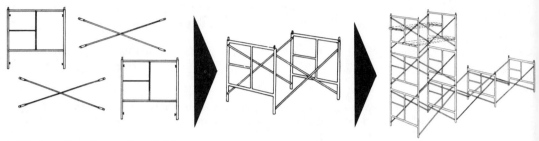

18–3a. *Manufactured scaffolding components may be assembled in a variety of sizes and shapes, depending on the job requirements.*

18–3b. *Assembling manufactured scaffolding. Note that the ladder is a part of the unit.*

18–4. *Manufactured scaffolding is versatile equipment.*

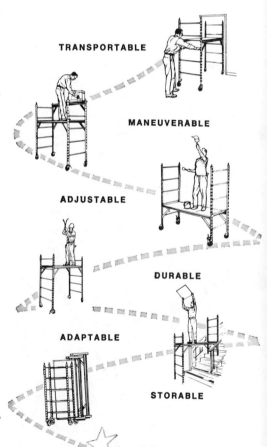

TRANSPORTABLE

MANEUVERABLE

ADJUSTABLE

DURABLE

ADAPTABLE

STORABLE

★ **SAFE**

plank too far beyond the supports because such planks tend to be unstable.

● Whenever necessary for stability, planks should be nailed or clamped to the scaffold.

BRACKETS, JACKS, AND TRESTLES
Brackets

Scaffold planks may be supported by special brackets which are available for sidewall and roof installations. There are various styles of *sidewall* and *corner brackets*. Some are nailed to the studs while others are bolted or hooked directly around the studding. Fig. 18-6. The nail-attached wall and corner brackets are secured to the wall with 20d nails driven at an angle into the wall stud through the tapered holes in the bracket. The brackets may be easily removed without pulling the nails. Any nails remaining after brackets are removed are driven flush.

Many styles of *roofing brackets* are available for various applications. Roofing brackets are attached with nails through the roof sheathing and into the rafters. They can be removed without pulling the nails. When they are removed, the nails are driven flush with the roof, and the shingles cover the nails that had held the bracket in place. One style holds a 2″ × 4″ or 2″ × 6″ flat against the roof. Others will hold the 2″ × 4″ or 2″ × 6″ on edge at right angles to the roof. A third type positions a 2″ × 6″ so that it provides a level walkway on a roof. Fig. 18-7.

Ladder Jacks

A ladder jack is a device for hanging a scaffold plank from a ladder. A jack can be used over or under any ladder that has rungs rather than steps. It has two hooks at the top and two at the bottom which fit close to the ladder siderails, preventing excessive loads on the ladder rungs. The ladder jack adjusts to the pitch of the ladder and can be hooked over the rungs with one hand. Fig. 18-8 (Page 222). The scaffold plank is then placed onto the horizontal projection to provide a conve-

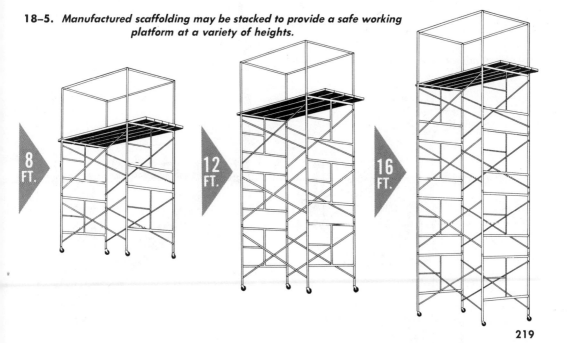

18–5. *Manufactured scaffolding may be stacked to provide a safe working platform at a variety of heights.*

8 FT.

12 FT.

16 FT.

nient work platform so you don't have to move the ladder frequently. Fig. 18-9.

Trestles

A trestle consists of two jacks as shown in Fig. 18-10. Trestles are available in a wide range of sizes, from 16″ to 12′, each with height adjustments at approximately 3″ intervals. Timber is used as a *ledger* (a stationary support) between the jacks to support the scaffold. This type of scaffold is sometimes used by plasterers and gyp-

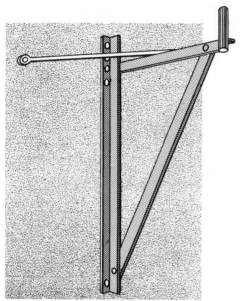

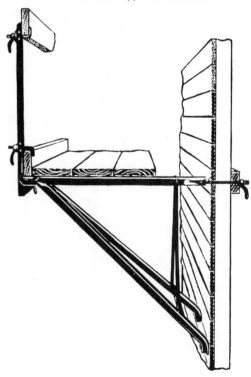

18–6c. *A studding bracket. The scaffold plank support on this bracket hooks around a stud (see the insert).*

18–6d. *Bolt-attached brackets. Note the guard rail used with this bracket. This rail can also be used with other types of brackets.*

18–6a. *A nail-attached wall scaffold bracket.*

18–6b. *A nail-attached corner scaffold bracket.*

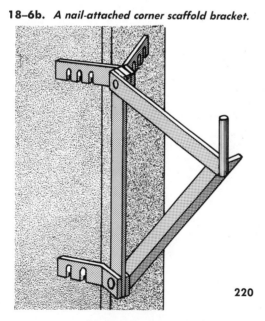

sum dry-wall applicators for working on ceilings. The scaffolding is set up over the entire floor area, on trestles about 18" to 20" high. The workers then can work continuously without stopping to take the scaffolding apart, move it, and set it up again.

LADDERS

Carpenters often must use ladders for high work. Ladders are usually made of wood or aluminum, and they come in many sizes. Commonly they are made in lengths from 3' to 50', with special three-section

18–7a. *Roofing brackets. The folding roof bracket adjusts to various roof pitches, from 90° to level walkways. It is ideal for use on steep roofs and will handle planks as large as 2" × 10". A. This roofing bracket holds a 2" × 4" or 2" × 6" flat against the roof. B. This bracket supports a 2" × 6" at a right angle to the roof. C. This bracket positions a 2" × 6" so that it provides a level walkway. D. This bracket supports a 2" × 4" at a right angle to the roof.*

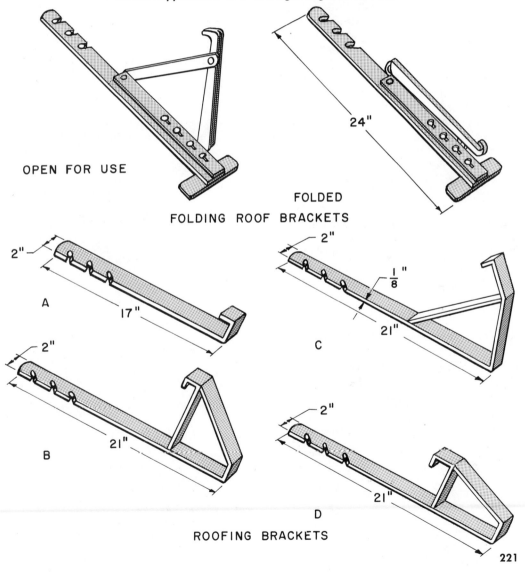

OPEN FOR USE

24"

FOLDED

FOLDING ROOF BRACKETS

2"
A
17"

2"
C
$\frac{1}{8}$"
21"

2"
B
21"

2"
D
21"

ROOFING BRACKETS

ladders available for reaching greater distances. Fig. 18-11.

There are three basic types of ladders (Fig. 18-12):
- Folding (stepladder).
- Straight.
- Extension.

To set up a straight or extension ladder, place the lower end against a base so it cannot slide. Then grasp a rung at the upper end with both hands. Raise the top end and walk forward under the ladder, moving the hands to grasp other rungs as you proceed. Fig. 18-13. When the ladder is erect, lean it forward to the desired position. Check the angle, height, and stability at top and bottom. Fig. 18-14.

When using a stepladder, always be certain that the four legs are firmly supported and that the spreaders are straight and level. Never stand on the top step of the ladder. The ledge on the back of the ladder is for holding tools and materials. Do not use it as a step.

When working from a ladder, set it where the work can be reached with ease. Never lean out far to one side. Relocate the ladders as necessary so the work area can be reached without much leaning.

When going up or down, grip the ladder firmly and place your feet squarely on the rungs. Make certain your shoes and the

rungs are free of mud and grease. While working, it is recommended that one leg be wrapped around a rung. When using a ladder for access to a roof, the ladder should extend above the edge of the roof by at least 3'. Fig. 18-15.

Ladder Safety

- Inspect ladders carefully. Keep nuts, bolts, and other fastenings tight. Oil moving metal parts frequently. Do not allow makeshift repairs. Never straighten or use a bent metal ladder.

18–8. *Ladder jacks may be used over or under any ladder with rungs.*

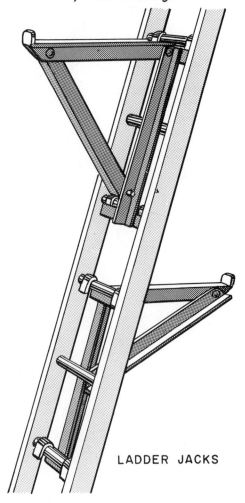

LADDER JACKS

18–7b. *A ladder hook is used to support a ladder for use on a very steep roof.*

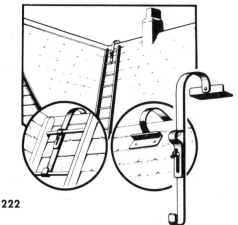

• Ladders must stand on a firm, level surface. Always use safety feet with non-slip bases.

• Face the ladder when climbing up or down.

• Always place the ladder close enough to the work to avoid dangerous overreaching.

• Keep your weight centered between both side rails.

• Keep steps and rungs free of oil, grease, paint, or other slippery substances.

• Be sure that stepladders are fully open and the spreader straight.

• Never stand or climb on the top, pail rest, or rear rungs of a stepladder.

• Never place ladders in front of doors or openings unless appropriate precautions are taken.

• Always insure that the working length of the ladder will reach the support height required. It should be lashed or otherwise secured at the top to prevent slipping and should extend at least 3' above a roof or other elevated platform. Never stand on the top three rungs.

• Position the ladder so the horizontal distance of the ladder foot from the top

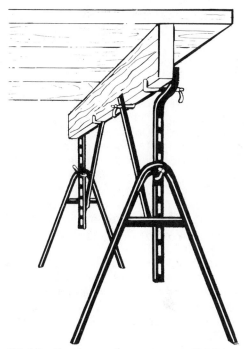

18–10. *Trestles used to support scaffolding.*

18–11. *Use this table to determine the correct ladder length for safe working conditions.*

If Vertical Height	Minimum Working Length
12 ft.	16 ft.
14 ft.	18 ft.
16 ft.	20 ft.
20 ft.	24 ft.
24 ft.	28 ft.
28 ft.	32 ft.
32 ft.	36 ft.
34 ft.	40 ft.
38 ft.	44 ft.
44 ft.	50 ft.

For heights of 50' and over, three-section ladders are available. Chart allows for lap required in two-section ladders.

18–9. *A ladder jack mounted under the ladder with a scaffold plank in place.*

support is one-fourth the working length of the ladder (75° angle). (*Working length* is the distance from the ground to the top support.) Always make sure that both side rails are fully supported top and bottom.

• Overlap extension-ladder sections by the following amounts: 3' for total extended lengths up to 36'; 4' for total lengths of 36' to 48'; and 5' for total lengths of 48' to 60'.

• When using a two-section extension ladder, place it so the upper section is outermost.

• Be sure all locks on extension ladders are securely hooked over rungs before climbing. Adjust the height of an extension ladder only when standing at the base of the ladder.

• Metal and water conduct electricity. *Do not use metal, metal reinforced, or wet ladders where direct contact with a live power source is possible.* Provide for temporary insulation of any exposed electrical conductors near the place of work.

• A ladder is intended to carry only one person at a time. Do not overload.

• Never use ladders in a horizontal position.

• Store ladders in dry, cool, ventilated places. Fig. 18-16.

• Never use ladders after prolonged immersion in water or exposure to fire, chemicals, or fumes which could affect their strength.

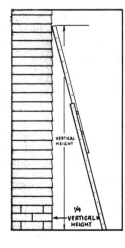

18-14. *For safety on a ladder, the pitch or angle should be such that the horizontal distance at the bottom is one-fourth the working length of the ladder.*

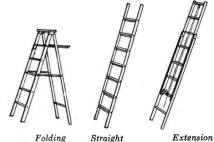

Folding   Straight   Extension

18-12. *The three types of ladders: folding, straight, and extension.*

18-15. *The top of the ladder should extend above the edge of the roof at least three feet.*

18-13. *To raise an extension ladder, walk forward under the ladder as shown, moving the hands to grasp other rungs as you proceed.*

18-16. *A ladder should be stored horizontally on supports to prevent sagging. Never store ladders exposed to weather or near heat.*

STORAGE

QUESTIONS

1. What is a scaffold?
2. What are the two types of scaffolds?
3. What is a duplex head nail?
4. List several kinds of scaffold brackets.
5. What is a ladder jack?
6. List the three basic types of ladders.

7. Describe the method of setting up an extension ladder.

8. At what angle should an extension ladder be set against a building?

9. How much should the sections of an extension ladder overlap when working at a height of 39'?

UNIT 19

Concrete and Footings

CONCRETE

One of the most important construction materials is concrete. It is a *synthetic* material—that is, it is made from other materials. Specifically, it is made by mixing together cement, fine aggregate (usually sand), coarse aggregate (usually gravel or crushed stone), and water in the proper proportions. Fig. 19-1. The product is not concrete unless all four of these ingredients are present. Without coarse aggregate, a mixture of cement, sand, and water is mortar or grout. Fig. 19-2.

The fine and coarse aggregate in a concrete mix are called the *inert ingredients,* while cement and water are the *active ingredients.* The inert ingredients and the cement are thoroughly mixed together first. As soon as the water is added, a chemical reaction between the water and the cement begins. It is this reaction (called *hydration*) that causes the concrete to harden.

Notice the difference between hydration and dehydration. In dehydration a drying out takes place. This is not what happens when concrete hardens. In fact, concrete will harden just as well under water as in air. This shows that hydration is truly a reaction between the water and the cement, not just a drying out of the concrete. Actually, rather than drying out, concrete must be kept as moist as possible in the early stages of hydration. Drying out would cause the water content to drop below the amount needed for satisfactory hydration.

Concrete Ingredients

Concrete used for residential construction consists of portland cement, water, and fine and coarse aggregate. It is important to use good-quality ingredients.

Cement is measured by the sack. One sack of portland cement weights 94 lbs. and is equal to 1 cu. ft. by volume. Cement should be stored in a dry place to prevent lumps from forming. Lumps that cannot be pulverized by squeezing in the hand should not be used.

The water used to mix concrete must be clean and free from oil, alkali, or acid. A good rule to follow is that it must be suitable to drink.

Fine aggregate consists of sand or other suitable materials up to ¼" in diameter.

19–1. Study the picture of this large construction project and see how many different phases of concrete work you can identify.

Coarse aggregate consists of gravel, crushed stone, or other suitable material larger than 1/4". All aggregates should be clear and free of loam, clay, or vegetable matter which would reduce concrete strength.

The size of the aggregate should vary, depending on the kind of work for which the concrete is to be used. In walls, the largest pieces of aggregate should not be more than one-fifth the thickness of the finished wall section. For slabs, the aggregate should not be more than approximately one-third the thickness of the slab. The largest piece of aggregate should never be larger than three-fourths the width of the narrowest space through which the concrete will be required to pass during pouring.

Proportions

The proportions of fine and coarse aggregate, amount of cement, and water content should follow the recommendations of the American Concrete Institute. Strength, durability, watertightness, and wear resistance of the concrete are controlled by the amount of water per sack of cement.

A large amount of concrete is supplied by ready-mix plants, even in rural areas. Concrete in this form is normally ordered by the number of bags of cement per cubic yard of concrete, in addition to aggregate size and water-content requirements. Five-bag mix (that is, five bags per cubic yard) is considered minimum for most work; where high strength is needed or where steel reinforcement is used, six-bag mix is commonly specified.

When concrete is mixed on the job site, the quantities of cement and aggregate must be figured separately for each cubic yard of concrete. Fig. 19-3a shows the number of bags of portland cement and the cubic feet of aggregates required to produce 1 cu. yd. (27 cu. ft.) of mixed concrete for several suggested trial mixes. Fig. 19-3b shows the amount of water to use in trial mixes with sand of varying moisture content.

19-2. *Using mortar to lay up a brick wall.*

MEASURING MATERIALS

For accurate proportioning of materials, a bottomless measuring box may be used. This device is a frame made of 1" or 1 1/2" material with a capacity of not less than 1 cu. ft. If larger, it should be of 2, 3, or 4 cu. ft. capacity. These frames should be marked on the inside to show levels at which the volume will equal certain amounts, such as 1 cu. ft., 2 cu. ft., or smaller batches. Handles on the side of the box make it easier to lift after the material has been measured. Fig. 19-4.

To measure the materials, the box is placed on the mixing platform and filled with the required amount of material. The box is then lifted and the material remains on the platform. Pails are often used for proportioning materials. Fig. 19-5. For example, a batch of concrete could be measured by using one pail of portland cement, two pails of sand, and three pails of gravel or crushed stone. (This would be called a *1:2:3 batch.*) Measuring can also be done with shovels or wheelbarrows, depending on the amount of material required. Suggested trial mixes are shown in Figs. 19-3 & 19-6.

All concrete should be thoroughly mixed until it is uniform in appearance and all materials are uniformly distributed in the mixture.

Pouring Concrete

Concrete should be poured continuously wherever possible and kept practically

19–3a. *This table suggests the mixture for trial batches and the materials necessary for a cubic yard of concrete with various maximum sizes of aggregate.*

Proportions for Various Trial Mixes of Concrete		Cement bags**	Aggregates	
			Fine cu. ft.	Coarse cu. ft.
With ¾" maximum size aggregate	Mixture for 1 bag trial batch*	1	2	2¼
	Materials per cu. yd. of concrete	7¾	17 (1550 lbs.)	19.5 (1950 lbs.)
With 1" maximum size aggregate	Mixture for 1 bag trial batch	1	2¼	3
	Materials per cu. yd. of concrete	6¼	15.5 (1400 lbs.)	21 (2100 lbs.)
With 1½" maximum size aggregate (preferred mix)	Mixture for 1 bag trial batch	1	2½	3½
	Materials per cu. yd. of concrete	6	16.5 (1500 lbs.)	23 (2300 lbs.)
With 1½" maximum size aggregate (alternate mix)	Mixture for 1 bag trial batch	1	3	4
	Materials per cu. yd. of concrete	5	16.5 (1500 lbs.)	22 (2200 lbs.)

*Mix proportions will vary slightly depending on gradation of aggregates. A 10 percent allowance for normal wastage has been included in the above figures for fine and coarse aggregate.

* *One bag of cement equals 1 cu. ft.

19–3b. *Trial mix proportions for sand of various moisture contents. This table also specifies the amount of water to be added to the mix depending on the climate exposure.*

Trial Mix Proportions	Gallons of water added to 1-bag batch if sand is:				Suggested mixture for 1-bag trial batches[4]		
	Dry	Damp[1]	Wet[2]	Very wet[3]	Cement bags (Cu. ft.)	Aggregates (Cu. ft.) Fine	Coarse
For mild exposure 1½" max. size aggregate	7	6¼	5½	4¾	1	3	4
For normal exposure 1" max. size aggregate	6	5½	5	4¼	1	2¼	3
For severe exposure 1" max. size aggregate	5	4½	4	3½	1	2	2¼

[1]"Damp" describes sand that will fall apart after being squeezed in the palm of the hand.
[2]"Wet" describes sand that will ball in the hand when squeezed but leaves no moisture on the palm.
[3]"Very wet" describes sand that has been subjected to a recent rain or recently pumped.
[4]Mix proportions will vary slightly depending on gradation of aggregates.

level throughout the area being poured. All vertical joints should be keyed to hold the parts together. *Spade* or *vibrate* the concrete to remove air pockets and force the concrete into all parts of the forms. Fig. 19-7.

In hot weather, protect concrete from rapid drying. It should be kept moist for several days after pouring. Rapid drying lowers its strength and may damage the exposed surfaces of sidewalks and drives.

In very cold weather, keep the temperature of the concrete above freezing until it has set. The rate at which concrete sets is affected by temperature, being much slower at 40° F and below than at higher temperatures. In cold weather, the use of heated water and aggregate during mixing is good practice. In severe weather, insulation or heat is necessary until the concrete has set.

19-5. *Using a pail for proportioning concrete materials.*

FOOTINGS

Foundation walls are enlarged at their base in order to furnish a larger bearing surface against the soil beneath. These enlarged bases are called *footings.*

Footings are an important part of the foundation, so particular attention should be paid to their size and shape. Footing details will generally be specified on the architectural plans. The FHA Minimum Property Standards specify footing sizes based on soil of average bearing value. Approximately 2,000 lbs. per sq. ft. or better is used as a guide. These minimum footing sizes are shown in Fig. 19-8.

Local codes will specify the type and size of footings suitable for the soil condition. In cold climates the footings should

19-6. *Water proportions for mixing small batches of concrete.*

Pints Of Water To Add To Mixer For Batches Using 1/2, 1/4, 1/5 And 1/10 Sacks Of Cement

Size of Batch	Pints of mixing water to add			
	Very wet sand	Wet sand	Damp sand	Dry sand
5 Gal. Water Per Sack Of Cement				
½ sack	14	16	18	20
¼ sack	7	8	9	10
⅕ sack (18.8 lb.) . .	5⅗	6⅖	7⅕	8
¹⁄₁₀ sack (9.4 lb.) . .	2⅘	3⅕	3⅗	4
6 Gal. Water Per Sack Of Cement				
½ sack	17	20	22	24
¼ sack	8½	10	11	12
⅕ sack	6⅘	8	8⅘	9⅗
¹⁄₁₀ sack	3⅖	4	4⅘	4⅘

19-4. *A 4-cubic-foot measuring box used for accurate proportioning of concrete.*

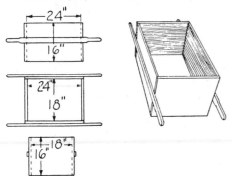

be far enough below ground level to be protected from frost. Local codes usually establish this depth, which is often 4' or more in northern sections of the United States.

Poured concrete footings are more dependable than those of other materials. Thus they are recommended for use in house foundations. Where fill has been used, the foundations should extend below the fill to undisturbed earth. In areas where irregular settlement of the foundation and the building it supports are likely to occur, local practices that have been successful should be followed.

Wall Footings

Well-designed wall footings are important in preventing settling or cracks in the wall. One rule for determining the size of wall footings is based on the proposed wall thickness. This two-part rule, which is often followed for normal soils, states:

• The footing depth should be equal to the wall thickness. Fig. 19-9.

• The footing should project beyond each side of the wall one-half the thickness of the wall. Thus the total width of the footing would be twice the wall thickness.

This is a general rule which may have to be varied. For example, if the soil is of low load-bearing capacity, wider reinforced footings may be required. Also, local regulations often set certain requirements for

19–7b. Spading tool in use.

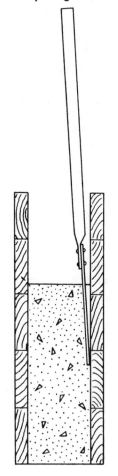

19–7a. Spading tool.

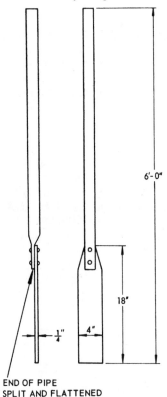

6'-0"

18"

1/4"   4"

END OF PIPE
SPLIT AND FLATTENED

wall footings as well as column and fireplace footings.

A few rules that apply to footing design and construction are:

• Footings must be at least 6″ thick (8″ or more is preferable).

• If footing excavation is too deep, fill with concrete—never replace with dirt.

• Use formboards for footings where soil conditions prevent sharply cut trenches.

• Place footings *below* the frostline.

• Reinforce footings with steel rods where they cross pipe trenches.

• Use key slot for better resistance to water entry at wall location. Figs. 19-9 and 19-14.

• In freezing weather, cover with straw or supply heat.

FOOTING FORMS

A common type of wall footing form is shown in Fig. 19-10. The sides of the footing are molded by boards (called *haunch boards*). The bottom is natural soil. Set the footing form boards (haunch boards) level and nail the spreaders in place with duplex head nails. Fig. 19-11. For many footings, artificial form work is not used at all; the footing is cast with bottom and sides against the natural earth.

19–7c. *Using a vibrator to insure that the concrete will be consolidated.*

19–8. *FHA minimum footing sizes as specified in the minimum property standards for one and two living units.*

Footing Sizes

Number of stories	Frame		Masonry or masonry veneer	
	Min. thickness (inches)	Projection each side of wall (inches)	Min. thickness (inches)	Projection each side of wall (inches)
One story:				
No basement	6	2	6	3
Basement	6	3	6	4
Two story:				
No basement	6	3	6	4
Basement	6	4	8	5

Pier, Post, and Column Footings

A footing for a *pier,* a *post,* or a *column* should be square. Also it should have a *pedestal*—a raised area on which the member will bear. To anchor a wood post, a protruding steel pin is ordinarily set in the pedestal. Fig. 19-12. Bolts for the bottom plate of steel posts are usually set when the pedestal is poured. At other times, steel posts are set directly on the footing and the concrete floor poured around them.

Footings vary in size, depending on the weight the soil will support and the spacing of the piers, posts, or columns. Common sizes are 24″ × 24″ × 12″ and 30″ × 30″ × 12″. The pedestal is sometimes poured after the footing. The minimum height should be about 3″ above the finish basement floor and 12″ above finish grade in crawl-space areas.

Footings for fireplaces, furnaces, and chimneys should ordinarily be poured at the same time as other footings.

Stepped Footings

Stepped footings are often used where the lot slopes to the front or rear and the garage or living areas are at basement level. The vertical part should be poured at the same time as the rest of the footing. The bottom of the footing is always placed on undisturbed soil below the frostline. Each run of the footing should be level.

The vertical step should be at least 6″ thick and the same width as the rest of the footing. Fig. 19-13. The height of the step should not be more than three-fourths of the adjacent horizontal footing. On steep slopes, more than one step may be required. It is good practice, when possible, to limit the vertical step to 2′. In very steep slopes, special footings—perhaps larger and/or reinforced—may be required.

FOOTING DRAINS

Foundation or footing drains must often be used around foundations which enclose basements or rooms which people will

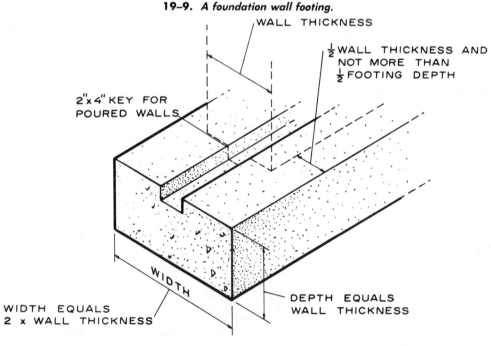

19–9. A foundation wall footing.

WALL THICKNESS

$\frac{1}{2}$ WALL THICKNESS AND NOT MORE THAN $\frac{1}{2}$ FOOTING DEPTH

2″ x 4″ KEY FOR POURED WALLS

WIDTH

WIDTH EQUALS 2 x WALL THICKNESS

DEPTH EQUALS WALL THICKNESS

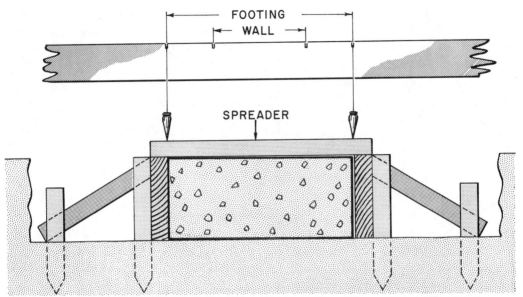

19–10a. *Footing form detail. Note the notches in the batter board, from which the plumb bobs are hung, to designate the footing width and wall thickness.*

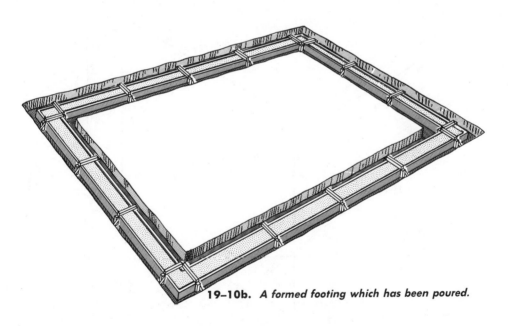

19–10b. *A formed footing which has been poured.*

19–11. *A duplex head nail is used when fabricating forms. This enables the forms to be securely nailed but allows the nail to be easily pulled when the forms are to be removed.*

233

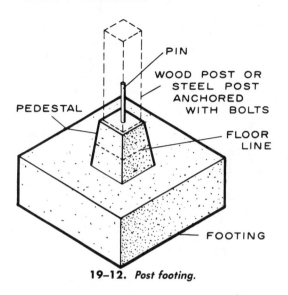

19–12. *Post footing.*

occupy below the outside finish grade. Fig. 19-14. These may especially be needed in sloping or low areas or any location where it is necessary to drain away subsurface water. This precaution will help prevent damp basements and wet floors. The use of drains is often necessary where basement rooms are to be occupied or where houses are located near the bottom of a long slope subjected to heavy runoff.

Drains are installed at or below the area to be protected. They should drain toward a ditch or into a sump where the water can be pumped to a storm sewer. Clay or concrete drain tile, 4" in diameter and 12" long, is ordinarily placed at the bottom of the footing level on top of a 2" gravel bed. Fig. 19-14. Tile are placed end to end and

19–13. *Stepped footing details.*

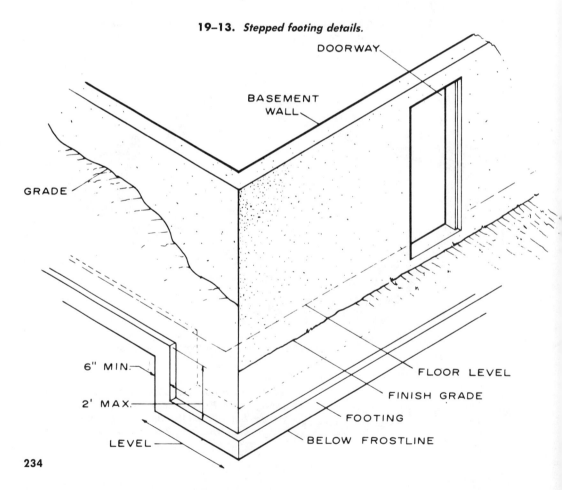

spaced about ¹/₈″ apart. The tops of the joints between the tile are covered with a strip of asphalt-saturated felt or a building paper. Then 6″ to 8″ of gravel is placed over the tile. Dry wells for drainage water are used only when the soil conditions are favorable. Local building regulations vary somewhat and should be consulted before starting to construct a drainage system.

ESTIMATING
Concrete

As mentioned earlier, concrete is measured and sold by the cubic yard (27 cubic feet). To calculate the amount of concrete required for a job, figure the volume of the forms in cubic feet (thickness × width × length) and divide by 27 to obtain the total number of cubic yards. For example, to pour a 4″ thick driveway which is 20′ wide and 40′ long, calculate the cubic footage by multiplying the thickness, 4″ (or ¹/₃ of a foot) times 20′ (width) times 40′ (length). This multiplies out to 266²/₃ or 267 cubic

feet. Converting this to cubic yards, ²⁶⁷/₂₇ = 9²⁴/₂₇ or 10 cubic yards of concrete to complete the driveway.

Fig. 19-15 provides a shortcut which can be used in estimating amounts of concrete. For the 4″ × 20′ × 40′ driveway, read down the left column of the table to the depth of the concrete in inches, which is 4″. Reading across you find that one cubic yard at that depth will fill 81 sq. ft. The area of the driveway is 800 sq. ft. (which you learn by multiplying the width times the length—20′ × 40′). Next divide the total sq. ft. (the area) by the number of sq. ft. covered by one cubic yard of cement:

$$\frac{800}{81} = 9.88$$

Thus you can see that roughly 10 cubic yards of concrete will be needed for the driveway.

This method of calculating concrete requirements will provide the necessary in-

19–14. *Drain tile. Footing drains are frequently used around foundations enclosing a living area below an outside finish grade. Notice that the bottom of the wall is keyed to the footing. This will tie the wall and footing together and help keep out the moisture.*

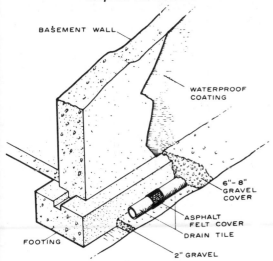

19–15. *This table indicates the area in square feet that 1 cubic yard of concrete will fill for a variety of thicknesses. For example, 1 cubic yard of concrete will fill a form area of 40 square feet for a wall 8″ thick.*

Concrete Estimating
1 Cu. Yd. of Concrete Will Fill

Thickness in.	Sq. ft.	Thickness in.	Sq. ft.	Thickness in.	Sq. ft.
1	324	4¾	68	8½	38
1¼	259	5	65	8¾	37
1½	216	5¼	62	9	36
1¾	185	5½	59	9¼	35
2	162	5¾	56	9½	34
2¼	144	6	54	9¾	33
2½	130	6¼	52	10	32.5
2¾	118	6½	50	10¼	31.5
3	108	6¾	48	10½	31
3¼	100	7	46	10¾	30
3½	93	7¼	45	11	29.5
3¾	86	7½	43	11¼	29
4	81	7¾	42	11½	28
4¼	76	8	40	11¾	27.5
4½	72	8¼	39	12	27

formation for ordering ready-mix concrete or for estimating a job. However, if the concrete is to be mixed on the job site, the amount of aggregate and cement will have to be figured. Do this by using the information provided in Fig. 19-3a. If a 6-bag mix is desired, read across the line of the chart under "Cement bags." Note that a 6-bag mix requires 1500 lbs. of fine aggregate and 2300 lbs. of coarse aggregate for each cubic yard of concrete. Applying this to the driveway in the example, multiply these figures by 10 (yards required for the driveway to obtain the total requirements of cement and aggregate.

Footings

MATERIALS

To determine the amounts of materials required to pour the footings in a home, figure the volume of the forms as described under "Estimating Concrete" (page 235). The table shown in Fig. 19-16 may be used as an aid.

To see how this works, calculate the footing requirements for the house plan in Fig. 19-17. Note that the foundation measures 42' by 24'. Thus the perimeter—the total of the four sides—is 132 lineal feet. Assuming that the footing size will be 8" × 12", consult Fig. 19-16. This shows that for 8" × 12" footings, 2.5 cubic yards of concrete are needed for 100 lineal feet. However, the footings in the example are longer than 100', so additional calculations are necessary. The problem now is: If 2.5 cubic yards of concrete will fill 100', how many will be needed for 132'? Divide 132 by 100. The answer is 1.32. This indicates that the house footings are 1.32 times larger than 100. Therefore by multiplying 1.32 × 2.5 you can find out the total amount of concrete needed. The final answer is 3.3 cubic yards.

Labor for Excavation

The labor for excavation of the footings can also be determined by referring to the

19–16. A table for estimating the material and labor needed for footings.

FOOTINGS	Material			Labor	
Size	Cubic Feet Concrete Per Linear Foot	Cubic Feet Concrete Per 100 Lin. Feet	Cubic Yards Concrete Per 100 Lin. Feet	Excavation Hours per 100 Linear Feet	Placement Hours per Cubic Yard
6 × 12	0.50	50.00	1.9	3.8	2.3
8 × 12	0.67	66.67	2.5	5.0	2.3
8 × 16	0.89	88.89	3.3	6.4	2.3
8 × 18	1.00	100.00	3.7	7.2	2.3
10 × 12	0.83	83.33	3.1	6.1	2.0
10 × 16	1.11	111.11	4.1	8.1	2.0
10 × 18	1.25	125.00	4.6	9.1	2.0
12 × 12	1.00	100.00	3.7	7.2	2.0
12 × 16	1.33	133.33	4.9	9.8	2.0
12 × 20	1.67	166.67	6.1	12.1	1.8
12 × 24	2.00	200.00	7.4	15.8	1.8

Note: Excavation—Reduce hours by ¼ for sand or loam.
Soil—Increase hours by ¼ for heavy clay soil.

Placement Labor based on ready-mixed concrete.

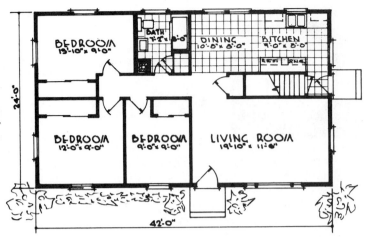

19–17. *House plan for use in calculating materials and labor for footings. (See examples in text.)*

table in Fig. 19-16. For the footing size in the previous example (8″ × 12″), the table shows it will take 5 hours to excavate 100 lineal feet. Because the perimeter is 132 lineal feet, allowance must be made for the excess over 100. Thus again you divide 132 by 100 and get 1.32. Multiplying 1.32 × 5, you arrive at a final answer of 6.6 hours of labor. In other words if it takes 5 hours to excavate 100 lineal feet, it will take 6.6 hours for the excavation of 132 lineal feet.

Labor for Placing Concrete

The table in Fig. 19-16 also provides information for figuring the labor of placing the concrete in the forms, based on the use of a ready-mixed concrete. Using the house in Fig. 19-17, which required 3.3 cubic yards of concrete, calculate the labor for placing the concrete. The table shows that for an 8″ × 12″ footing, it will take 2.3 hours to place one cubic yard of concrete in the form. To figure the total time, multiply 2.3 (placement hours per cubic yard) × 3.3 (total amount of concrete to be placed) = 7.59 hours or 7.6 hours of labor. This also includes the time for forming the footings. These items may have to be corrected depending on the nature of the soil, as noted at the bottom of the table shown in Fig. 19-16.

QUESTIONS

1. What are the active ingredients in concrete?

2. Explain hydration.

3. To mix 1 cubic yard of 6¼-bag-mix concrete, how many pounds of fine aggregate are required?

4. When proportioning materials for mixing concrete, what does 1:2:3 represent?

5. What is the purpose of a footing?

6. What are the common sizes of footings for piers, posts, and columns?

7. How is concrete measured and sold?

Poured Concrete Foundation Walls

Foundation walls form an enclosure for basements or crawl spaces. Also they carry wall, floor, roof, and other building loads. The two types of walls most commonly used are *poured concrete* and *concrete block* (which is discussed in Unit 21 of this book).

The job of forming and pouring concrete foundation walls and footings is usually *subcontracted*. This means that the building contractor hires another contractor— one who specializes in this type of work. These specialists have the necessary tools, materials, and equipment. Therefore the purpose of this unit is mainly to make the reader familiar with the general procedure and requirements for poured-concrete foundation walls, as well as the terms commonly used.

POURED CONCRETE FOUNDATION WALLS

Wall thickness and types of construction are ordinarily controlled by local building regulations. Thickness of poured-concrete basement walls may vary from 8″ to 10″.

Clear wall height should be no less than 7′ from the top of the concrete basement floor to the bottom of the joists; greater clearance is usually desirable to provide headroom under girders, pipes, and ducts. Many contractors pour walls that are 8′ high above the footings. This provides a clearance of 7′ 8″ from the top of the finished concrete floor to the bottom of the joists.

The forms for poured-concrete walls must be tight. They must also be braced and tied to withstand the forces of the pouring operation and the fluid concrete. Fig. 20-1.

Poured-concrete walls should be *double-formed* (formwork constructed for each wall face). Reusable forms are used for most poured walls. Panels may consist of wood framing with plywood facings. Such panels are fastened together with clips or other ties. Fig. 20-2. Whether reusable or not, formwork should be plumb, straight, and braced sufficiently to withstand the pouring operations. Fig. 20-3.

Frames for cellar windows, doors, sleeves for utilities, and other openings are set in place as the forms are erected. This is also true of forms for the beam pockets which are located to support the ends of the floor beam. Fig. 20-4.

As mentioned above, forms must have sufficient blocking and bracing to keep them in place during pouring operations. For reusable forms the use of horizontal bracing members is usually sufficient. Forms constructed with vertical studs and waterproof plywood or lumber sheathing require horizontal whalers and bracing. (Whalers are lumber used to stiffen concrete forms.)

Level marks of some type, such as nails along the form, should be used to assure a level foundation top. This will provide a good level sill plate and floor framing.

Concrete should be poured continuously, without interruption. It should be constantly worked to remove air pockets and to get the material under window frames and other blocking. Such working is called *puddling*. Fig. 20-5. If wood spacer blocks are used, they should be removed and not

20-1. *Wood forms for a large commercial building project. Notice the extensive use of reinforcing rod.*

20-2. *Form for a poured-concrete foundation wall.*

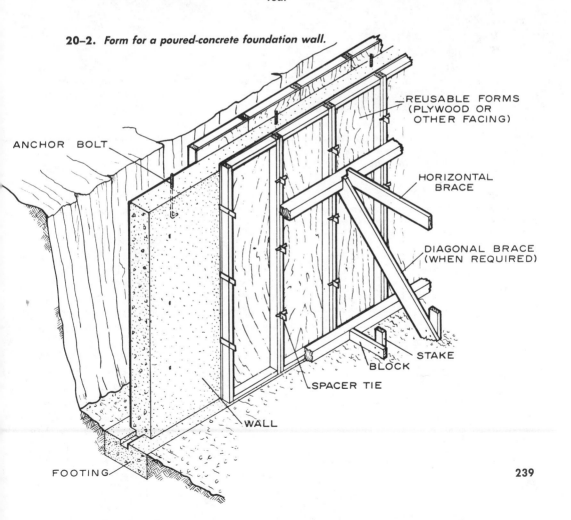

REUSABLE FORMS (PLYWOOD OR OTHER FACING)

ANCHOR BOLT

HORIZONTAL BRACE

DIAGONAL BRACE (WHEN REQUIRED)

STAKE

BLOCK

SPACER TIE

WALL

FOOTING

20-5. *Using a lightweight vibrator to remove air pockets and settle the concrete in the form.*

20-3. *Pouring concrete. The forms are made of plywood. Notice the worker spading the concrete next to the forms to eliminate voids or honeycombing at the sides.*

20-4. *Insert a tin or wood sleeve to provide openings through the basement wall for utilities.*

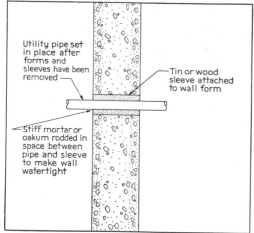

Utility pipe set in place after forms and sleeves have been removed

Tin or wood sleeve attached to wall form

Stiff mortar or oakum rodded in space between pipe and sleeve to make wall watertight

permitted to become buried in the concrete. Anchor bolts for the sill plate should be installed very soon after the concrete is placed.

Freshly placed concrete should be protected and possibly heated when temperatures are below freezing.

Forms should not be removed until the concrete has hardened and acquired sufficient strength to support loads imposed during early construction. At least two days (and preferably longer) are required when temperatures are well above freezing, and perhaps a week when outside temperatures are below freezing.

Poured concrete walls can be *dampproofed* with one heavy cold or hot coat of tar or asphalt. The coat should be applied to the outside from the footings to the finish grade line. Such coatings are usually sufficient to make a wall watertight against ordinary seepage (such as may occur after a rainstorm), but should not be applied until the surface of the concrete has dried enough to assure good adhesion. In poorly drained soils, a waterproof membrane of roofing felt or similar material can be applied, with shingle-style laps of 4″ to 6″ over the tar or asphalt coating. A hot coating of tar or asphalt is then commonly applied over the membrane. This covering will prevent leaks if minor cracks develop.

Masonry Construction for Crawl Spaces

In some areas of the country, the crawl-space house is often preferred to those constructed over a basement or on a concrete slab. Three important points in such construction are:

• A good soil cover must be used under the house.

• The crawl space must be ventilated. (A small amount of ventilation is usually sufficient.)

• There must be adequate insulation to reduce heat loss through the floor.

More will be said about these points in later units.

One of the primary advantages of the crawl-space house as compared with the full-basement house is the reduced cost. Little or no excavation or grading is required except for the footings and walls. In mild climates, the footings are located only slightly below the finish grade. However, in the northern states where frost penetrates deeply, the footing is often located four or more feet below the finish grade. This, of course, requires more masonry work and increases the cost. The footings should be poured over undisturbed soil and never over fill unless special piers and grade beams are used.

The construction of a masonry wall for a crawl space is much the same as that required for a full basement except that no excavation is required within the walls. Waterproofing and drain tile are normally *not* required for this type of construction. The masonry pier replaces the wood or steel posts used to support the center

beam of a full-basement house. Footing size and wall thickness vary somewhat by location and soil conditions. Common minimum thicknesses for walls in single-story frame houses are 8″ for hollow concrete block and 6″ for poured concrete.

For crawl-space houses, the minimum footing thickness (depth) is 6″; minimum widths are 12″ for concrete block and 10″ for the poured foundation wall. However, well-constructed houses usually have 8″ walls and 16″ × 8″ footings.

Poured-concrete or concrete-block piers are often used to support floor beams in crawl-space houses. They should extend at least 12″ above the ground line. The minimum size for a concrete block pier should be 8″ × 16″ with a 16″ × 24″ × 8″ footing. Solid blocks are used for the top course. Poured concrete piers should be at least 10″ × 10″ with a 20″ × 20″ × 8″ footing.

Unreinforced concrete piers should be no greater in height than 10 times their least dimension. Concrete block piers should be no higher than four times the least dimension.

When exterior wall beams and interior girders are set at right angles to the floor joists, the spacing of piers should not exceed 8′ on center. Under exterior wall beams set parallel to the floor joists, the spacing of piers should not exceed 12′ on center. Exterior wall piers should not extend above grade more than four times their least dimension unless supported laterally (on the sides) by masonry or concrete walls. As for wall footing sizes, the size of the pier footings should be based on the load they must bear and the capacity of the soil. Review "Footings," page 229.

Sill Plate Anchors

In wood-frame construction, the *sill plate* should be anchored to the foundation wall with ½″ bolts hooked and spaced about 8′ apart. Fig. 20-6. In some areas, sill plates are fastened with masonry nails, but such nails do not have the uplift resistance of bolts. In high-wind and storm areas, well-anchored plates are very important. A *sill sealer* is often placed under the sill plate on poured walls to smooth any irregularities which might have occurred during curing of the concrete. Anchor bolts should be embedded 8" or more in poured concrete walls and 16" or more in walls made of blocks with concrete-filled cores. A large plate washer should be used at the head end of the bolt for the block wall. If termite shields are used, they should be installed under the plate and sill sealer.

Reinforcing Poured Walls

Poured concrete walls normally do not need to be reinforced with steel except over window or door openings located below the top of the wall. This type of construction requires that a properly designed steel or reinforced-concrete *lintel* be built over the door or window opening. Fig. 20-7. In poured walls, the rods are laid in place while the concrete is being poured so that they are about 1½" above the opening. Frames should be prime painted or treated before installation to minimize the absorption of moisture from the concrete. For concrete block walls, a similar lintel of reinforced poured concrete or a precast lintel is commonly used.

Where concrete work includes a connecting porch or garage wall not poured with the main basement wall, it is necessary to provide reinforcing-rod ties. Fig. 20-8. These rods are placed during pouring of the main wall. Depending on the size and depth, at least three ½" deformed rods should be placed at the intersection of each wall. The deformation provides maximum holding power. Keyways may also be used to resist lateral movement by forming an interlock between the walls. Such connecting walls should extend below normal frostline and be supported by undisturbed ground.

Masonry Veneer over Frame Walls

If *masonry veneer* is used for the outside finish over wood-frame walls, the foundation must include a supporting ledge or offset about 5" wide. Fig. 20-9. This results

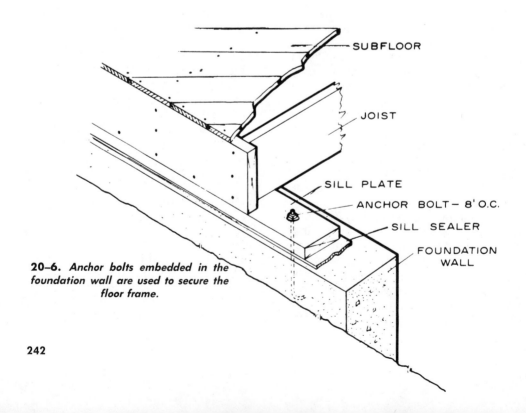

SUBFLOOR

JOIST

SILL PLATE

ANCHOR BOLT – 8' O.C.

SILL SEALER

FOUNDATION WALL

20–6. Anchor bolts embedded in the foundation wall are used to secure the floor frame.

in a space of about 1" between the masonry and the sheathing for ease in laying the brick. A base flashing is used at the brick course below the bottom of the sheathing

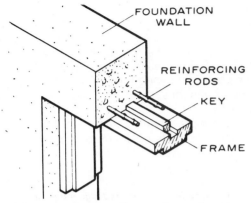

20-7. *Foundation walls must be reinforced over window and door frames.*

and framing. Fig. 20-9. The flashing should be lapped with sheathing paper. *Weep holes* (to provide drainage) are also located at this course and are formed by eliminating the mortar in a vertical joint. Corrosion-resistant metal ties—spaced about 32" apart horizontally and 16" vertically—should be used to bond the brick veneer to the framework. Where other than wood sheathing is used, secure the ties to the studs.

Brick and stone should be laid in a full bed of mortar; avoid dropping mortar into the space between the veneer and sheathing. Joints that will be exposed to moisture should be tooled to a smooth finish to get the maximum resistance to water penetration. (Tooling is discussed in detail on page 258.)

Masonry laid during cold weather should be protected from freezing until after the mortar has set.

20-8. *Connecting walls are tied to the main foundation wall by reinforcing rods.*

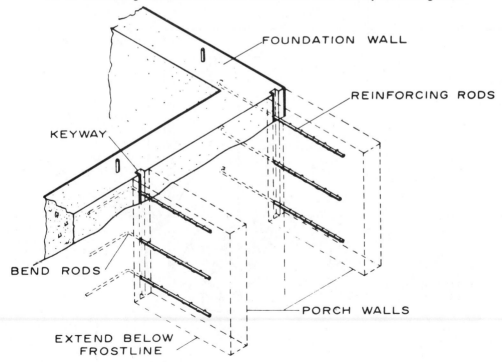

Notch for Wood Beams

A wall notch or pocket is needed for wooden basement beams or girders. The notch should be large enough to allow at least 1/2" of clearance at sides and ends of the beam for ventilation. Fig. 20-10. Unless the wood is treated, there is a decay hazard if beams and girders are so tightly set in wall notches that moisture cannot readily escape. A waterproof membrane, such as roll roofing, is commonly applied under the end of the beam to minimize moisture absorption.

PROTECTION AGAINST TERMITES

Certain areas of the country, particularly the Atlantic Coast, Gulf States, Mississippi and Ohio Valleys, and southern California, are infested with wood-destroying termites. In such areas, wood construction over a masonry foundation should be protected by one or more of the following methods:

- Poured concrete foundation walls.
- Masonry foundation walls capped with reinforced concrete.
- Metal shields made of rust-resistant material. (Such shields are effective only if they extend beyond the masonry walls and are continuous, with no gaps or loose joints.)
- Wood-preservative treatment. (But, this method protects only the members treated.)

20-9. *A foundation wall under a wood-frame house must include a supporting ledge if a masonry veener is to be applied.*

STUDS

SHEATHING PAPER

METAL TIES.
FASTEN TO STUDS

SHEATHING

BASE FLASHING
EXTEND BEHIND
SHEATHING PAPER

SILL

WEEP HOLES
(4' O.C.)

5"

FOUNDATION

MASONRY VENEER

• Treatment of soil with poison. (This is one of the most common and effective methods.)

See Unit 52 for further details on protection against termites.

ESTIMATING

Forming

The foundation wall forms are set on the footings after the footings have cured adequately. Using the house plan in Fig. 20-11, which measures 40′ × 26′, determine the total foundation wall area, assuming the wall to be 8′ high. Multiply 8′ (height of foundation wall) × 132′ (the perimeter of the building). The answer is 1056 sq. ft. Assuming a wall thickness of 8″, use the chart in Fig. 20-12 and read down under the column headed "Wall Thickness" to 8″. Then read across to the column headed "Forming." Remember, the wall is to be 8′ high. The chart shows that the wall will require 7.75 hours per 100 square feet of wall area.

Next calculate the total time for installing the forms, as follows:

Since you know it will take 7.75 hours for each 100 sq. ft., divide the total square feet by 100 and multiply by 7.75.

$$\frac{1056 \text{ (total sq. ft.)}}{100 \text{ (sq. ft. installed in 7.75 hours)}} = 10.56$$

10.56 × 7.75 = 81.8 hours to install forms for the wall in the example.

Next figure the time needed to remove the forms. Consulting Fig. 20-12 you see that 3 hours will be needed to remove forms for 100 sq. ft. of a high wall. Thus it can be readily seen that something over 30 hours will be needed to remove forms for 1056 sq. ft. The precise calculation is as follows:

$$\frac{1056 \text{ (total sq. ft.)}}{100 \text{ (sq. ft. removed in 3 hours)}} = 10.56$$

10.56 × 3 = 31.68 (or approx. 31²/₃ hours total labor time for removing the forms.)

Material

The amount of material needed for a concrete foundation wall can be figured using Fig. 20-12. As an example, take the same wall used in figuring for the forms in the preceding paragraphs. That wall, you remember, was 8″ thick and had 1056 sq. ft. of total area.

Find the 8″ thickness in the column at left. Reading across, under "Material" you find that 2.47 cubic yards of concrete will be needed for every 100 sq. ft. of wall. Therefore you must divide the total area by 100 (to see how many *hundreds* of square

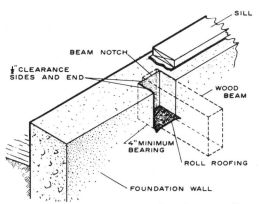

20–10. A beam notch in a foundation wall.

20–11. A foundation wall 8″ thick and 8′ high for this home will require 26¹/₄ cubic yards of concrete.

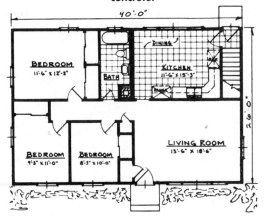

feet there are in the area of the wall) and then multiply by 2.47.

$$\frac{1056 \text{ (total sq. ft.}}{100} = \frac{10.56 \text{ (wall area}}{\text{expressed in}} $$

1056 (total sq. ft. of wall area) / 100 = 10.56 (wall area expressed in hundreds of sq. ft.)

10.56 × 2.47 (cubic yards per 100 sq. ft.) = 26.08 (cu. yds. of concrete for total wall)

For estimating, round off to the next larger 1/4 cu. yd; so a total of 26¹/4 cu. yds. of concrete would be needed.

Labor

To place concrete in the forms for the wall, Fig. 20-12 shows that it takes an average of 3.25 hours for 1 cubic yard. Multiply the total cubic yards of concrete by the time required to pour 1 cubic yard. This will tell you the total time required for placing the concrete.

In the preceding example, 26¹/4 cubic yards of concrete were required for the foundation walls. Therefore: 26¹/4 (cu. yds. of concrete) × 3.25 (hours to place 1 cu. yd.) = 85.31 or, rounded off, 85¹/3 hours labor.

20–12. *Estimating table for concrete foundation walls.*

WALLS	Material			Forming			Concrete Placement
	Per 100 Square Feet Wall			Hours per 100 Square Feet			Hours Per Cubic Yard
Wall Thickness	Cubic Feet Required	Cubic Yards Required		Place		Remove	
				0' 4'	4' 8'		
4"	33.3	1.24		4.7	7.13	2.0	Average 3.25 Hours
6"	50.0	1.85		4.7	7.75	Varies	
8"	66.7	2.47		5.0	7.75	as to	
10"	83.3	3.09		5.0	7.90	Height	
12"	100.0	3.70		5.0	7.90	3.0	

QUESTIONS

1. What are the purposes of foundation walls?

2. What are the common thicknesses and heights of a basement foundation wall?

3. What is meant by double-formed?

4. What is meant by dampproofing?

5. What is the maximum height recommended for an 8" × 16" block tier?

6. What is the purpose of a sill plate anchor?

7. How is masonry veneer secured to a wood-frame wall?

8. Why is ventilation required at the end of a wood beam or girder when it rests on a concrete wall?

9. What areas of the country are most troubled with termites?

10. For the home in Fig. 20-11, a foundation wall 8" thick and 8' high will require 26¹/4 cubic yards of concrete. For a front exterior of brick veneer, the thickness of the foundation wall would have to be increased to 10". Calculate the additional amount of concrete required.

UNIT 21

Concrete Block Foundation Walls

Concrete block construction is one of the most popular methods for building a foundation. One major reason for this popularity is that the work can be done without expensive, heavy equipment. Fig. 21-1a.

GENERAL CONSIDERATIONS

Any hollow masonry unit (or block) is called concrete block, but the most common type is made with sand and gravel or crushed stone. Concrete blocks are available in many shapes and sizes for a large variety of applications. Those that are most widely used are 8″, 10″, and 12″ wide. Fig. 21-1b. Modular (standard-size) blocks allow for the thickness and width of the mortar joint; thus they are usually about $7^5/_8″$ high by $15^5/_8″$ long. This results in blocks which measure 8″ high and 16″ long from centerline to centerline of the mortar joints.

The thicknesses of concrete block walls vary from 8″ to 12″ depending on story heights and the lengths of unsupported walls. Such walls are customarily constructed of 11 courses above the footings, with a 4″ solid cap-block. This leaves about 7′ 4″ between the joists and the basement floor.

Concrete block walls require no formwork. Block courses start at the footing and are laid up with about $^3/_8″$ mortar joints, usually in a *common bond*. This is the overlapping arrangement shown in Fig. 21-1c. Joints should be tooled smooth to resist water seepage. Mortar should be spread fully on all contact surfaces of the block. (Such spreading is called *full bedding* of mortar.)

Pilasters are columnlike projections which may be used to strengthen a wall, as by supporting a beam or girder. Some building codes require them. Pilasters are placed on the interior side of the wall and are constructed as high as the bottom of the beam or girder they support. Basement door and window frames should be set with keys for rigidity and to prevent air leakage. Fig. 21-1c.

When exposed block foundation is used as a finished wall for basement rooms, the *stack bond pattern* (Fig. 21-1c) may be employed for a pleasing effect. This is done by placing blocks directly above one another, resulting in continuous vertical mortar joints. However, when this system is used, it is necessary to add some type of joint reinforcement at every second

21-1a. *Building a concrete block wall does not require heavy equipment, but it does require considerable skill.*

247

course. This usually consists of small-diameter steel rods arranged in a grid pattern. The common bond does not normally require this reinforcing, but when additional strength is desired, it is good practice to use this bonding system.

Freshly laid block walls should be protected in temperatures below freezing. Freezing of the mortar before it has set will often result in low adhesion, low strength, and joint failure.

To provide a tight, waterproof joint between footing and wall, an elastic calking compound is often used. The wall is waterproofed by applying a coating of cement-mortar over the block and forming a cove where the wall joins with the footing. Fig. 21-1c. When the mortar is dry, a coating of asphalt or other waterproofing will normally assure a dry basement.

Sometimes added protection is needed, as when wet soil conditions may be encountered. A waterproof membrane of roofing felt or similar material, with shingle-type laps of 4" to 6", can be applied over the coating. Hot tar or hot asphalt is commonly used over the membrane. This covering will prevent leaks from minor cracks that develop in the blocks or joints between the blocks.

MIXING MORTAR AND TRUCKING THE BLOCK

To build a strong, solid wall you need good mortar. The strength of the mortar bond depends on such items as:
- The type and quantity of cementing material.
- The workability or plasticity of the mortar.
- The surface texture of the mortar-bedding areas.
- The rate at which the masonry units absorb moisture from the mortar.
- The water retention of the mortar.
- The quality of workmanship in laying up the units.

Masonry walls subject to severe frost or stress require mortars that are stronger and more durable than walls exposed to ordinary conditions. Mortar mixes shown in Fig. 21-2 are recommended for the type of

21–1b. *Typical shapes and sizes of concrete masonry units. Dimensions shown are actual unit sizes. A 7⅝" × 7 ⅝" × 15⅝" unit is commonly known as an 8" × 8" × 16" concrete block. Half-length units are usually available for most of the units shown. Check local suppliers to learn what is available in your area.*

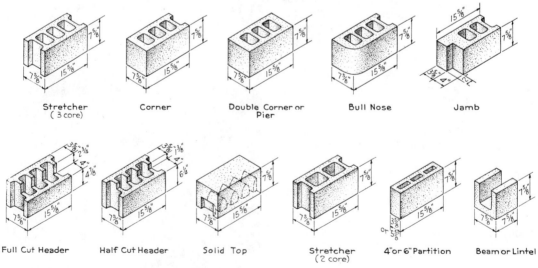

Stretcher (3 core)   Corner   Double Corner or Pier   Bull Nose   Jamb

Full Cut Header   Half Cut Header   Solid Top   Stretcher (2 core)   4" or 6" Partition   Beam or Lintel

service indicated. Mortar should be mixed in power mixers except for very small jobs where it may be mixed by hand.

Mortar will stiffen on the mortar board because of evaporation or hydration. When evaporation occurs, the mortar can be re-tempered to restore its workability by thorough remixing and by the addition of water as required. Fig. 21-3. Mortar stiffened by hydration (setting) should be discarded. When unused mortar stiffens, it is not easy to tell whether evaporation or hydration is the cause. However, a judgment can usually be made on the basis of time elapsed after initial mixing. Mortar should be used within 2 1/2 hours after original mixing when air temperature is 80° F or higher, and within 3 1/2 hours when air temperature is below 80° F. Mortar not used within these time limits should not be used.

Mortar must be sticky so that it will cling to the concrete block when it is laid into the

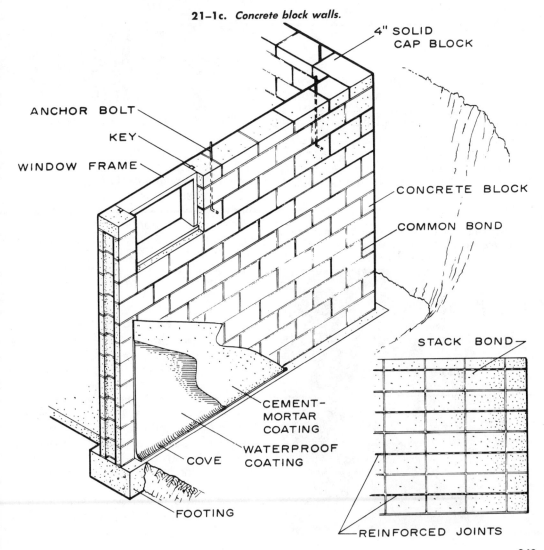

21-1c. *Concrete block walls.*

4" SOLID CAP BLOCK

ANCHOR BOLT

KEY

WINDOW FRAME

CONCRETE BLOCK

COMMON BOND

CEMENT-MORTAR COATING

WATERPROOF COATING

COVE

FOOTING

STACK BOND

REINFORCED JOINTS

21-2. Recommended mortar mixes—proportions by volume.

Type of service	Cement	Hydrated lime	Mortar sand in damp, loose condition
For ordinary service	1—masonry cement* or 1—portland cement	— 1 to 1¼	2 to 3 4 to 6
Subject to extremely heavy loads, violent winds, earthquakes or severe frost action. Isolated piers.	1—masonry cement* plus 1—portland cement or 1—portland cement	— 0 to ¼	4 to 6 2 to 3

*ASTM Specification C91, Type II

wall. When taking a trowel full of mortar from the mortar board, shake the trowel with a quick vertical snap of the wrist to make the mortar stick to the trowel and shake off the excess. Fig. 21-4. This also keeps the mortar from falling off the trowel when it is applied to the edges of the block. Block and mortar should be placed

21-4. Picking up mortar with the trowel.

on the scaffold near final position to minimize your movements. Fig. 21-5.

Care must be taken to keep blocks dry on the job. They should be stockpiled on planks or other supports, free from contact with the ground, and covered for protection against wetting. Fig. 21-6. *Concrete block must not be allowed to get wet just before or during laying in the wall.*

First Course

The hollow cores in concrete blocks are larger at one end than at the other. Blocks should be laid so that the surfaces with the smaller holes are up. This provides a larger mortar-bedding area.

After locating the corners, place the blocks for the first course in position *without mortar* in order to check the layout. Fig. 21-7. A chalked snap-line can be used to mark the footing and help to align the block accurately. A full mortar bed is then spread with a trowel. Make sure there is plenty of mortar along the footing to rest the bottom edges of the blocks. Fig. 21-8.

21-3. Add water and thoroughly remix to restore the workability of mortar which has begun to dry out.

21-5. Distribute the blocks and the mortar to the areas in which they will be used.

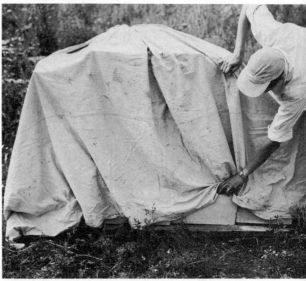

21-6. Care should be taken to keep concrete blocks off the ground and covered for dryness.

The corner block should be laid first and carefully positioned. Fig. 21-9.

For vertical joints, only one end of the block is covered with mortar. By placing several blocks on end, you can apply mortar to the vertical face shells of three or four blocks in one operation. Fig. 21-10. Each block is then brought over its final position and pushed downward into the mortar bed and against the previously laid block. Fig. 21-11.

After three or four blocks have been laid, use the mason's level as a straightedge to assure correct alignment of the block. Fig. 21-12. Blocks are then carefully checked with the level and brought to proper grade. Fig. 21-13. They are made plumb by tapping with the trowel handle. Fig. 21-14. The first course of concrete masonry should be laid with great care to make sure it is properly aligned, leveled, and plumbed. This will assist you in laying other courses and in building a straight, true wall.

After the first course is laid, apply mortar to the horizontal faceshells of the block. This is called *faceshell mortar bedding.*

21-7. Check the layout by setting the blocks in position on the footings, allowing space for the mortar.

21-8. *Be certain to get a full bed of mortar on the footing when setting the first course.*

21-11. *Push the block downward into the mortar bed and against the previously laid block.*

21-9. *Here the corner block was carefully laid first. Correct positioning of other blocks depends on this.*

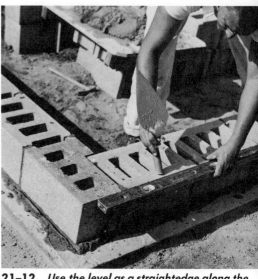

21-12. *Use the level as a straightedge along the faces of the block to check the alignment.*

21-10. *Apply mortar to the vertical face shells of several blocks in one operation.*

Fig. 21-15. Mortar for the vertical joints can be applied to the vertical face shells of the next block or to the vertical edges of the block previously laid. Some masons apply mortar to the vertical face shells of both blocks to insure well-fitted joints.

The corners of the wall are built first, usually four or five courses higher than the center of the wall. As each course is laid at the corner, check it with a level for alignment, Fig. 21-16, for levelness, Fig. 21-17, and for plumbness, Fig. 21-18. Check each block carefully with a level or straight-

21-13. *The level is used to check the top edge for levelness.*

21-15. *Mortar bedding the faceshell in preparation for laying up additional courses.*

21-14. *The level is used to be certain the outside surface is plumb.*

21-16. *Check the alignment of the blocks frequently.*

edge to make certain that the faces of the block are all in the same plane. Fig. 21-19. This is necessary to insure true, straight walls.

A *story pole* or *course pole* is simply a board with markings 8" apart. The use of such a pole provides an accurate method of positioning the top of the masonry for each course. Fig. 21-20. Mortar joints for concrete masonry should be ³/₈" thick.

Each course in building the corners is stepped back a half block. Check the horizontal spacing of the block by placing

21-17. *Build the corners up first. Be sure to check the blocks for levelness as the work progresses.*

21–18. *After the corners have been built up, be sure to check the corner for plumb before continuing.*

21–20. *Using a story pole or course pole.*

21–19. *Use the level as a straightedge to check the faces of the block, making sure they are all on the same plane.*

21–21. *If the blocks have been stepped back correctly, the alignment can be checked by holding a level or straightedge diagonally across the corners of the block.*

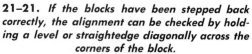

21-22. *After the corners have been built up, stretch a mason's line from corner to corner for each course. Between the corners, set the blocks so their top edges align with the mason's line.*

21-23. *Set the block down carefully, making sure to align it with the block previously laid.*

the level diagonally across the corners of the block. Fig. 21-21.

Between Corners

When filling in the wall between the corners, a mason's line is stretched from corner to corner for each course; the top, outside edge of each block is laid to this line.

The way the block is handled or gripped is important and is learned with practice. Fig. 21-22. Tipping the block slightly toward yourself, you can see the upper edge of the course below; thus you can place the lower edge of the block directly over the course below. Fig. 21-23.

By rolling the block slightly to a vertical position and shoving it against the adjacent block, it can be laid to the mason's line with minimum adjustment. *All adjustments to final position must be made while the mortar is soft and plastic. Any adjustments made after the mortar has stiffened will break the mortar bond.* By tapping lightly with the trowel handle, each block is leveled and aligned to the mason's line. Fig. 21-24. The use of the mason's level

21-24. *The block can be tapped into position with the handle of the trowel.*

21-25. *Cut off the excess mortar with the trowel.*

21-28. *These blocks have a full mortar bed—that is, mortar has been applied to the cross webs as well as to the faceshells.*

21-26. *Sometimes the mortar that has been cut from the joints can be applied to the vertical face shells of the block just laid.*

21-27. *Well-filled joints will result if mortar has been applied to the vertical joints of the block already in the wall and also to the block being set.*

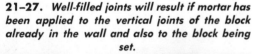

between corners is limited to checking the face of each block to keep it lined up with the face of the wall.

To assure good bond, mortar should not be spread too far ahead of actual laying of the block or it will stiffen and lose its plasticity. As each block is laid, excess mortar at the joints is cut off with the trowel. Fig. 21-25. If the work is going well, the excess mortar can be applied directly to the vertical faceshells of the block just laid. Fig. 21-26. Should there be a delay long enough for the mortar to stiffen on the block, the mortar should be put back on the mortar board and reworked. The application of mortar to the vertical joints of the block already in the wall and to the block being set insures well-filled joints. Fig. 21-27. "Dead" mortar that has been picked up from the scaffold or from the floor should not be used.

In some localities, a full mortar bed may be specified on all concrete block construction. Fig. 21-28.

Closure Block

The block which fills the final gap in a course between corners is called the *closure block.* To install this block, spread mortar on all edges of the opening and all four vertical edges of the block itself. Fig. 21-29. The closure block should be carefully lowered into place. Fig. 21-30. If some

256

21-29. *When installing the closure block, mortar is applied to all four vertical edges of the opening as well as to the closure block.*

21-30. *The closure block is carefully placed in position.*

of the mortar falls out leaving an open joint, the closure block should be removed, fresh mortar applied, and the operation repeated.

Intersecting Bearing Walls

Bearing walls built of intersecting concrete blocks should not be tied together in a masonry bond, except at the corners. Instead, one wall should terminate at the face of the other wall, with a *control joint* at that point. (This is a joint which controls movement caused by stress in the wall. For more details see page 260.) For lateral support, bearing walls are tied together with a metal tiebar 1/4" thick, 1/4" wide, and 28" long, with 2" right angle bends on each end. Fig. 21-31. These tiebars are spaced not over 4' apart vertically. The bends at the ends of the tiebars are embedded in cores filled with mortar or concrete. Fig. 21-32. Pieces of metal lath placed under the cores support the concrete or mortar filling. Fig. 21-33.

If the control joint at the intersection of the two bearing walls is to be exposed to view or the weather, it should be constructed and sealed with a calking compound as described on page 260.

Intersecting Nonbearing Walls

For tying nonbearing block walls to other walls, strips of metal lath or 1/4" mesh

21-31. *A metal tiebar is placed in the mortar joint to tie an intersecting wall to the main wall.*

21-32. *Be sure to embed the ends of the tiebar in cores filled with mortar or concrete.*

257

21-33. *Notice in Fig. 21-32 that a piece of metal lath has been placed under the tiebar. When the core is filled with mortar or concrete as in this picture, the lath will support the filling.*

21-35. *If the main wall is built first, be sure to place metal lath in alternate courses. Thus it can be tied in as the intersecting wall is laid up.*

21-34. *Metal lath placed across the joint is used to tie a nonbearing intersecting wall to the main wall.*

galvanized hardware cloth are placed across the joint between the two walls. Fig. 21-34. The metal strips are placed in alternate courses in the wall. When one wall is constructed first, the metal strips are built into the wall and later tied into the mortar joint of the second wall. Fig. 21-35.

Where the two walls meet, the vertical mortar joint is raked out to a depth of 3/4″ if it is exposed to view in the finished build-

ing. Calking compound is then packed into this recess, as described on page 261.

Tooling

Weathertight joints and neat appearance of concrete block walls depend on proper tooling. After a section of the wall has been laid and the mortar has become "thumbprint hard," (so that the thumb makes no indentation) the mortar joints should be tooled. The tooling operation compacts the mortar and forces it tightly against the masonry on each side of the joint. Proper tooling also produces joints of uniform appearance, with sharp, clean lines. Unless otherwise specified, all joints should be tooled either concave or V-shaped.

The jointer for tooling horizontal joints should be at least 22″ long, preferably longer, and upturned on one end to prevent gouging the mortar. A suitable handle should be located approximately in the center for handling ease. For concave joints, a tool made from a 5/8″ round bar is satisfactory. Fig. 21-36. For V-shaped joints, a tool made from a 1/2″ square bar is generally used. Fig. 21-37. Tooling of the horizontal joints should be done first, followed by striking the vertical joints with a small S-shaped jointer. Fig. 21-38. After

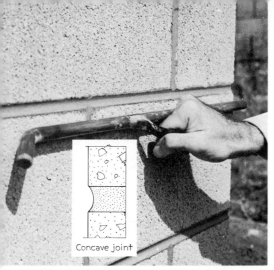

Concave joint

21–36. *A jointer for concave joints can be made from a piece of ⅝″ round bar.*

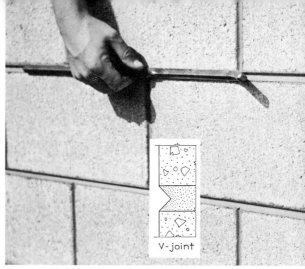

V-joint

21–37. *A jointer for V-shaped joints can be made from a ½″ square bar.*

the joints have been tooled, a trowel is used to trim off mortar burrs flush with the face of the wall. Fig. 21-39. Burrs can also be removed by rubbing with a burlap bag.

CAUTION: Do not move or straighten the block in any manner once the mortar has stiffened, or even partly stiffened. Final positioning of the block must be done while the mortar is soft and plastic. Any attempt to move or shift the block after the mortar has stiffened will break the mortar bond and allow the penetration of water. Fig. 21-40 shows such a break.

Finishing Foundation Walls

Foundation walls of hollow concrete block must be capped with a course of solid masonry to help distribute the loads from floor beams and to act as a termite barrier. Solid-top blocks, in which the top 4″ is of solid concrete, are available in some areas. Fig. 21-41. When stretcher blocks are used, a strip of metal lath wide enough to cover the core spaces is placed in the mortar joints under the top course. Fig. 21-42. The cores are then filled with concrete or mortar and troweled smooth. Fig. 21-43. Sometimes 4″ solid units are used to cap concrete block foundation walls. Fig. 21-44. All vertical joints must be completely filled, and slushing of the joints

21–38. *A small S-shaped jointer is used for striking the vertical joints.*

should not be permitted. That is, mortar should not be thrown into the joint with the trowel. The mortar must be placed on the end of the block before it is set in place.

Installing Anchor Bolts

Wood plates on which the house framing bears are fastened to tops of concrete block walls. This is done by means of anchor bolts ½″ in diameter and 18″ long,

259

21-39. *Remove any mortar burrs with a trowel.*

21-42. *When stretcher blocks are used for the top course, place a strip of metal lath in the mortar joint below.*

21-40. *Never attempt to move or shift the block after the mortar has stiffened. This will break the mortar bond and permit water seepage.*

21-41. *A solid-top block is often used for the top course.*

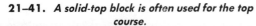

spaced not more than 4' apart. These anchor bolts are placed in cores of the top two courses of block, with the cores filled with concrete or mortar. Pieces of metal lath are placed in the second horizontal mortar joint from the top of the wall and under the cores to be filled. Fig. 21-45. The lath will support the concrete or mortar filling. The threaded end of the bolt should extend above the top of the wall. Fig. 21-46. When the filling has hardened, the wood plate can be securely fastened to the wall.

Control Joints

Stress in masonry walls causes movement which must be controlled. Control joints are one method of solving this problem. These joints are used to control cracking resulting from unusual stresses. The joints are built into the wall in a way that permits slight wall movement without cracking the masonry. They are continuous from the top of the wall to the bottom. To keep control joints from being too noticeable, care must be taken to make them plumb and of the same thickness as the other mortar joints. When a control joint cannot be concealed or when it will be exposed to weather, it should be sealed with calking compound, following the manufacturer's directions.

21–43. *With metal lath in place, the cores of the top-course blocks can be filled and troweled smooth.*

21–44. *Sometimes 4″ solid blocks are used to cap a wall.*

Unless a control joint is to be calked, it should be laid the same as other vertical joints. For a joint to be calked, make a recess for the calking material by raking out mortar to a depth of about ³/₄″. Do this after the mortar is quite stiff. Fig. 21-47. Use a thin, flat, calking trowel to force the compound into the joint. Fig. 21-48.

The spacing and location of control joints will depend on a number of factors:
- The length of the wall.
- Architectural details.
- Local experience about such joints.

Control joints should be placed at the junctions of bearing as well as nonbearing walls, at places where walls join columns and pilasters, and in walls weakened by openings. In long walls, control joints are ordinarily spaced at approximately 20′ intervals, again depending on local experience.

21–45. *Place a piece of metal lath under the core to be filled. This will support the concrete or mortar.*

21–46. *Fill the core with concrete or mortar and insert the bolt so that the threads extend above the top of the wall.*

Lintels

Precast concrete lintels are often used over door and window openings. Fig. 21-49. For modular window and door openings, precast concrete lintels are designed with an offset (or recessed area) on the underside. Fig. 21-50. Steel angles (pieces of steel bent at right angles) are also used for lintels to support block over

21–47. *Mortar is raked out of a vertical joint to provide a recess for calking material.*

21–48. *Calking compound is forced into a control joint.*

openings. To fit modular openings the steel lintel angles must be installed with an offset on the underside. Fig. 21-51.

RELATED OPERATIONS
Cutting Block

Concrete masonry units usually are available in half-length as well as full-length units. However, to fit special job conditions it is sometimes necessary to cut a block with a brick hammer and chisel. The block is scored on both sides to make a clean break. Fig. 21-52. For fast, neat cutting, masonry saws are often used. Fig.

21-53. Block should be cut dry when masonry saws are used so as not to increase the moisture content of the block.

Patching and Cleaning Block Walls

Any patching of the mortar joints or filling of defects in the blocks should be done with fresh mortar. Particular care should be taken to prevent smearing mortar onto the surface of the block. Mortar smears will mar the appearance of the finished wall. Once hardened they cannot be removed, and paint will not always hide them. Acid wash should not be used to

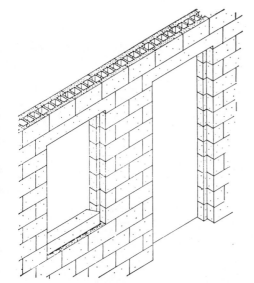

21–49. *Precast lintels set in place over openings.*

21–50. *A cross section of a precast concrete lintel.*

remove smears or mortar droppings from concrete block walls. Therefore care should be taken to keep the wall clean during construction. Any mortar droppings that stick to the block wall should be allowed to dry before removal with a trowel. Fig. 21-54. The mortar may smear if removed while soft. When dry and hard, most of the remaining mortar can be removed by rubbing with a small piece of block. Fig. 21-55. Brushing the rubbed spots removes practically all of the mortar. Fig. 21-56.

Protection

Boards, building paper, or tarpaulins are used to cover the tops of unfinished block walls at the end of the day's work. This will prevent rain or snow from entering the cores. Fig. 21-57.

ESTIMATING
Number of Blocks

The number of blocks necessary for a building can be determined by the area of the wall to be built. Nine 8″ × 8″ × 16″ blocks will make eight sq. ft. of vertical wall area. Therefore take the total number of square feet in the wall and divide it by

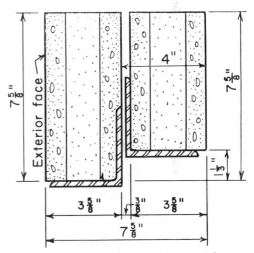

21–51. *Steel angles are sometimes used to support block over openings.*

eight. Multiply the result by nine and you will have a good estimate of the number of blocks necessary for the wall.

$$\frac{\text{sq. ft. wall area}}{8} \times 9 = \text{No. of blocks}$$

For example, consider a house with a 25′ × 40′ foundation and a 7′-high finished basement. The simplest way to find

21-52. *For a clean break, score the blocks along both sides with a chisel.*

21-54. *Remove mortar droppings with the trowel after they have dried.*

21-53. *A masonry saw used for cutting block is fast and accurate.*

21-55. *Rubbing with a small piece of block will remove dry mortar from the surface of the wall.*

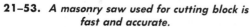

the total sq. ft. of wall area is to multiply the perimeter times the depth. The perimeter is the length of the four walls added together. In this instance the perimeter is 130′ (25′ + 25′ + 40′ + 40′ = 130′). Multiply this times 7′ (the depth of the basement) and you find that the total area of the four basement walls is 910 sq. ft.

Now apply the formula given above:

$$910 \div 8 = 113.75$$
$$113.75 \times 9 = 1,023.75$$

Rounded off, your answer would be 1,024. This, however, is not the number of blocks needed. One further step is necessary. Because the courses overlap or interlock at the corners, subtract ½ block for each corner of each course. The wall in the example would be 11 blocks high; therefore subtract 5½ blocks for each corner or 22 blocks. This shows that a total of 1,002 blocks would be needed. This number would be reduced somewhat more to allow for windows or other openings.

The number of concrete blocks necessary for a wall can also be determined by referring to the chart in Fig. 21-58. In the left column, find the size of the block used. If you select an 8″ × 8″ × 16″ block, the chart indicates 110 concrete blocks for each 100 sq. ft. of wall. The walls in the example, you will recall, had an area of 910 sq. ft. Divide this by 100 to find the number of sq. ft. expressed in hundreds.

$$\frac{910 \text{ (total sq. ft.)}}{100} = 9.1 \text{ (hundreds of sq. ft.)}$$

The chart shows that 110 blocks are needed for each 100 sq. ft. Therefore by multiplying 9.1 times 100 you find the total number of blocks needed.

9.1 × 110 = 1,001 total blocks needed

Note that some adjustment may still be necessary if there are openings in the wall. However, the chart allows for the overlapping of blocks at the corners, so it is not necessary to subtract for this as in the previous example.

21–56. *After the wall has been cleaned with the trowel or a piece of block, brush it for a final cleaning.*

21–57. *At the end of each day, cover the walls to prevent water from getting on surfaces where blocks will be laid.*

Note also that the answer is not precisely the same as when calculated by the method given previously. However, the estimates are very close, which shows that both methods are reliable.

Amount of Mortar

The number of cubic feet of mortar needed for a block wall can also be determined from Fig. 21-58. For the walls in the preceding example, the chart shows that 3.25 cu. ft. of mortar would be needed for every 100 sq. ft. of wall area. You will recall that there were 9.1 hundreds of sq. ft. in the walls. Thus by multiplying 9.1 times 3.25 you find the total amount of mortar needed to lay up the walls.

$$9.1 \times 3.25 = 29.5 \text{ cu. ft. of mortar}$$

Labor Costs

To determine labor costs for laying up the walls, again consult Fig. 21-58. You will see that 8″ × 8″ × 16″ blocks are laid at a rate of 18 per hour. Taking the figure 1,001 for the total number of blocks, you can divide that number by 18 to learn the number of hours needed for laying the block.

$$\frac{1,001}{18} = 55.6 \text{ hours}$$

Multiply the hours needed by the hourly rate of pay for laying block. This will give you the total labor cost.

SURFACE BONDING OF CONCRETE BLOCK

A new material that eliminates the need for conventional mortar above the first course in concrete block walls is now available to builders. This material is a mixture of specially formulated alkali-resistant glass fibers $\frac{1}{2}$″ long, hydrated lime, portland cement, and a water-resistant agent. The glass fibers provide strength by acting like steel mesh does in reinforced concrete. There are no epoxies or resins in this surface bonding mix. This fiberglass reinforced mortar is applied to the wall surface after the blocks have been stacked dry (with the exception of the base course).

The material is designed to surface bond concrete masonry blocks to form above-grade (not exceeding two floors or 25 ft.) and below-grade walls, load bearing and nonbearing, in residential and light commercial construction.

In tests conducted by the National Concrete Masonry Association and in actual field installations, surface bonding provided greater flexural strength than conventionally mortared walls.

21–58. Estimating chart.

MASONRY BLOCKS		Material 100 Square Feet Of Wall					Labor
	Wall Thickness	Concrete Block		Lightweight Block			Blocks Per Hour
Size		No. of Units	Mortar Cubic Feet	No. of Units	Mortar Cubic Feet		
8 × 4 × 12	4″			146	4.0		24
8 × 4 × 16	4″			110	3.25		22
12 × 4 × 12	4″			100	3.25		30
8 × 6 × 16	6″			110	3.25		21
8 × 8 × 16	8″	110	3.25				18
8 × 10 × 16	10″	110	3.25				16
8 × 12 × 16	12″	110	3.25				13

Note: Motar quantities based on ⅜″ mortar joints, plus 25% waste. For ½″ joints add 25%.

21–59. *Lay the first course in a full mortar bed on the footings. The head joints should be butted tightly without mortar.*

21–60. *Level and plumb every third or fourth course.*

Since individual block joints are not mortared, users of this material realize important labor savings. Other significant benefits include:
- Moisture protection without additional treatment.
- Finishing flexibility; walls can be painted, faced, or left as is.
- Do-it-yourself homeowners can easily layup block walls.

Layout

The layout of the running bond for the first course is very important because it helps determine whether it will be necessary to cut an odd-sized block for the closure.

The corner units should be laid first, leveled, and aligned. Then the first course should be laid with great care because it will assist you in laying succeeding courses and in building a straight, plumb wall. Fig. 21-59.

Once three or four units have been laid, use a level to check them for alignment, grade, and plumbness. Fig. 21-60.

In laying the first course, place a full mortar bed on the foundation, the full thickness of the wall. It is not necessary to apply mortar to the head joints.

Stacking

Once the first course is set in mortar, stacking of block may begin. The corners are usually laid up about four courses high, stepping back each course by one-half a unit length.

The wall is then filled in between these stepped corners. Stretch a mason's line tightly from corner to corner at each third course to assist you in laying the top outside edge of each block to proper line and grade. Fig. 21-61. Blocks should be tightly butted together.

Smoothing the top of the block by scraping the surface of two blocks together eliminates any excess material or burrs. This will aid in the stacking procedure.

It is recommended that the reinforced mortar be applied at each scaffold height. Block may, however, be stacked to a height not exceeding 10 ft. before applying the reinforced mortar.

21-61. *Use a mason's line to assure proper line and grade every third course.*

21-63. *The metal strap anchor in place. The block has been notched (see arrow) to accept the anchor.*

21-64. *Control joint locations.*

Control Joints

Wall Location	Length of Wall		
	0'-25'	26'-59'	Over 60'
*Below Grade	No Control Joints Required		Control Joint Every 50'
Above Grade	No Control Joints Required	Use Length to Height Ratio of 3, with a Maximum Length of 50' between Control Joints, regardless of Height.	

*Must be backfilled within 96 hours.

ground block will eliminate any variability found in standard block and allow stacking without shimming.

Anchoring Intersecting Walls

Intersecting masonry walls should be anchored with metal straps as in conventional construction. However, since there is no mortar joint, the webs of the block must be notched to accept the anchor. Fig. 21-63.

Control Joints

Build control joints into the wall as you would in a conventional mortared block wall. See "Control Joints," page 260. The control joints must extend through the reinforced mortar surface coating and, on an exterior wall, must be watertight. Fig. 21-64.

21-62. *Joints are shimmed where necessary to insure a plumb and level wall.*

Shimming

Since there are variations in block from different producers and different areas, keeping the walls level and plumb may require shimming. If shimming is necessary, use sand, sheet metal, or mortar to level and align. Fig. 21-62. The use of

1.

Determine rough opening dimension required.

2.

Blocks are laid dry up to the bottom of the window sill. Note: If proper height is not attained in natural coursing, a solid masonry unit can be placed under the sill. Also, the sill unit can be laid in mortar or reinforced mortar if additional height is required. (See Step 5)

3.

Mark the window and door width dimensions on the bottom sill course previously laid up.

4.

Lay up the leads for the openings at these marks.

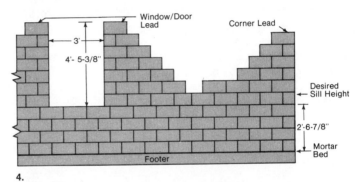

5.

Fill in each course with block as the walls require. A cut block may be necessary to complete each course. If necessary, place solid masonry units under sill to obtain proper height. If a metal masonry window or door is used, install the metal frame, then place lintel on top of opening. If wood casement windows are specified, follow Step 5 with the exception of installing window casement.

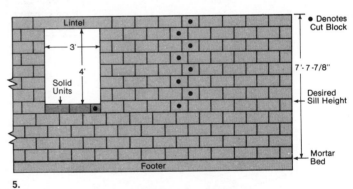

21–65. *Compensating for rough opening dimension differences.*

Window and Door Placement

Doors and windows are placed in walls in the same manner as with conventional mortared block walls. Because surface bonding eliminates the need for placing mortar in the joint between the blocks, the rough opening of windows and doors will be altered. To compensate for this dimen- sion difference when installing windows, use steps 1–5 in Fig. 21-65. When install- ing doors, omit Step 2.

The Top Course

The top course of concrete block walls must be constructed to insure adequate distribution of vertical loads. The top course construction for surface bonding is the same as with conventional mortared walls. If solid top masonry units and anchor bolts are specified, notch these units to allow clearance for the bolts. Fig. 21-66.

Changes in Wall Thickness

When changing wall thicknesses, cap the lower portion in the same manner as a top course. The first course of the new wall thickness must then be started in a mortar bed. Fig. 21-67.

Mixing Fiberglass Reinforced Mortar

Mix according to the manufacturer's instructions. Use a clean mortar mixer and make sure the water is clean. Blend the mortar for one to two minutes or until all materials are thoroughly wetted. Additional water may be added to achieve a consistency similar to mortar.

Excessive mixing can cause lumps to form and should be avoided. Total mixing time should not exceed five minutes.

Tools

Standard masonry tools are used with the fiberglass reinforced mortar. To speed application and reduce lines, it is recommended that the corners of the mason's trowel be rounded to a $3/8''$ radius. Fig. 21-68.

A sheet metal device called a joint guide is used when the application is interrupted for more than an hour. Place the joint guide over the last course of laid block and apply the reinforced mortar up to the straight edge. Fig. 21-68.

Application of Reinforced Mortar

The wall must be free of dirt, oil, paint or other foreign matter. It must be thoroughly wetted down prior to fiberglass mortar mix application. Fig. 21-69.

The mix should be troweled smoothly on both sides of the wall to a minimum thickness of $1/8''$. Press firmly to ensure a good bond. Fig. 21-70.

Fiberglass reinforced mortar that has begun to "set," or which is not used within $1\frac{1}{2}$ hours after initial mixing, should be

21–66. *Notch the solid blocks for the anchor bolts.*

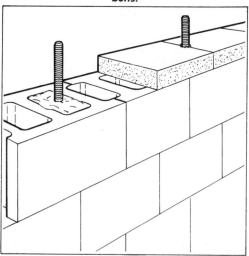

21–67. *Changing the thickness of a wall. Note that a conventional mortar bed is indicated at the arrow.*

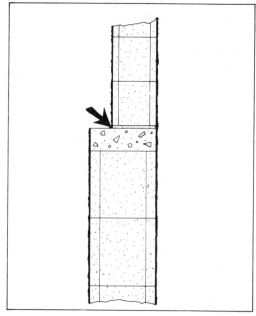

discarded. Fiberglass reinforced mortar that has stiffened due to evaporation within a 1½ hour period can be retempered to restore its workability.

When it is necessary to discontinue application of the surface bonding for more than one hour, the joint in the reinforced fiberglass mortar must not coincide with the joints in the block. In addition, for the best appearance, the stoppage or discontinuance at vertical joints should occur at a corner, pilaster, or control joint. In this way one day's work can be blended into the next.

Finishing

Various conventional finishes can be applied to fiberglass reinforced mortared walls. Walls that will receive a second treatment such as stucco can be left with a rough finish that will serve as the base coat. The finish coat may be either hand- or spray-applied. Furring strips for supplemental surfaces may be attached to the wall by any conventional method, including nailing and adhesives.

Walls that will be left exposed can be given a smooth finish by taking extra time and care in troweling. These walls can be painted with any standard masonry paint.

Curing

Within 24 hours after application of the reinforced mortar, the wall should be dampened with a water mist when temperature is above 40° F. This prevents premature drying. Depending on the temperature and humidity, surface bonding applications can be wetted down earlier than 24 hours.

In the event of heavy rain, freshly applied reinforced mortar should be protected with a waterproof covering until it has cured for at least eight hours. Fig. 21-71.

Backfilling

Avoid backfilling the foundation walls before the first floor is in place, or brace the walls by some means until the first floor

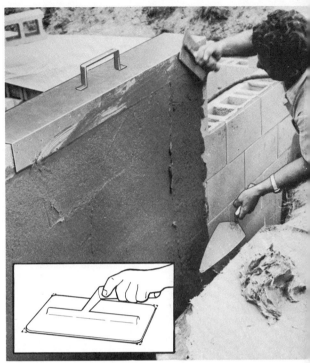

21-68. *Use a joint guide when application is to be stopped for more than one hour. Insert: Round the corners of a conventional trowel for use when applying the reinforced mortar.*

21-69. *Wet down the wall prior to the application of the surface bonding material.*

271

21-70. Trowel the reinforced mortar on both sides of the wall to a thickness of ⅛".

21-72. Remove a 2" wide section of reinforced mortar around a damaged area.

is installed. A completed wall must cure 48 hours before backfilling and 24 hours before structural work is begun. Avoid standing on the walls during this period. If high compaction is required or mechanical compacting equipment is to be used, the surface bonding must be allowed to cure

at least two weeks prior to beginning such work.

The finished grade should be sloped away from the foundation walls for good surface drainage.

Settlement of backfill material should be anticipated. Operating heavy equipment

21-71. Freshly applied reinforced mortar should be covered in the event of heavy rain.

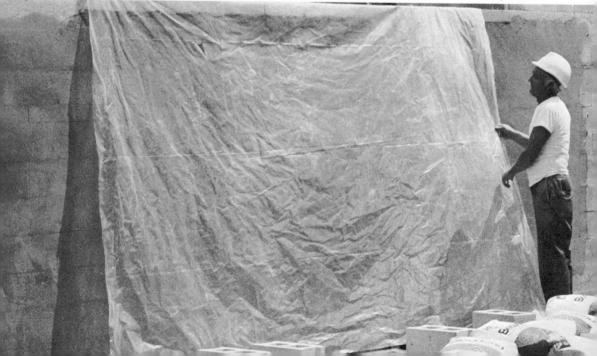

21-72b. *Reapply the reinforced mortar to the damaged area.*

any closer to a wall than a distance equal to the height of this fill should be avoided.

Repairs

If a block is knocked out of alignment after application of the fiberglass reinforced mortar, or if cracks appear as a result of foundation settlement, repairs can be made by removing a l" strip on either side paralleling the damage. If the reinforced mortar is still workable, scrape it off. If it has set, chip the area clean and reapply as indicated in the application section. Fig. 21-72.

Other Construction Details

The following construction techniques are the same as with conventional mortared block walls.
• Bracing for above- and below-grade walls.
• Vertical and horizontal reinforcement.
• Wall footings and drainage construction.
• Wall thicknesses.
• Lintel and sill installation (apply reinforced mortar over lintel faces).
• Electrical and plumbing installation.
• Pilasters.
• Beam pockets.

21-73. *Estimating chart.*

Table I

Actual Height of Units, Inches	No. of Units per 100 sq.ft.	Bags per 100 sq.ft. of Wall*
7-5⁄8 (Modular)	121	2
3-5⁄8 (Modular)	255	2
8 (Nonmodular)	115	2

*Includes application to both sides 100 sq. ft. wall.

Table II
(Standard 8″ Block)

No. Block	Length of Wall	Height of Wall
1	1' 3⅝"	8"
2	2' 7¼"	1' 3⅝"
3	3' 10⅞"	1' 11¼"
4	5' 2½"	2' 6⅞"
5	6' 6⅛"	3' 2½"
6	7' 9¾"	3' 10⅛"
7	9' 1⅜"	4' 5¾"
8	10' 5"	5' 1⅜"
9	11' 3⅝"	5' 9"
10	13' 0¼"	6' 4⅝"
11	14' 3⅞"	7' 0¼"
12	15' 7½"	7' 7⅞"
13	16' 11⅛"	8' 3½"
14	18' 2¾"	8' 11⅛"
15	19' 6⅜"	9' 6¾"

ESTIMATING

Because mortar joints are eliminated more block is needed per square foot of wall area.

The number of concrete masonry units and amount of surface bonding required can easily be calculated for each 100 square feet of wall by using Table I in Fig. 21-73.

Wall dimensions, door openings, windows and wall heights should be checked to compensate for the elimination of all mortar joint thickness except the bed layer (see Table II, Fig. 21-73). Using full-sized block (full 8" × 8" × 16" dimensions) where available will eliminate this situation.

QUESTIONS

1. What is the actual size of most concrete blocks?

2. What is the standard height of a concrete block basement wall? How many courses is this?

3. Describe a stack bond pattern.

4. Name three factors that determine the strength of mortar.

5. When laying a concrete block foundation wall, which block should be laid first?

6. What is a good method to insure that the top of each masonry course is exactly 8" high?

7. When laying concrete blocks, which end is laid face up? Why?

8. What is a closure block?

9. What is a good test to determine when the joints should be tooled?

10. What is a control joint?

11. Why should mortar droppings that stick to the wall be allowed to dry before they are removed?

UNIT 22

Slab and Flatwork

Concrete flatwork consists of flat areas of poured concrete, usually 5" or less in thickness. Examples are concrete floor construction for no-basement houses, basement floors, driveways, and walks.

CONCRETE FLOOR SLABS ON GROUND

The number of new one-story houses with full basements has declined steadily, particularly in the warmer areas of the United States. This is due in part to lower construction costs of houses without basements and an apparent decrease in the need for the basement space. For residences without basements or for unexcavated portions under houses, concrete floors on the ground are an excellent type of construction. Recent studies have shown that when properly designed and constructed, concrete slabs give better results than other types of floors for no-

basement construction. Houses without basements, regardless of the type of floor construction, should not be erected in low-lying areas that are damp or in danger of flooding from surface water.

Two types of concrete floor construction are the *combined* (or *unified*) *slab and foundation* and the *independent slab and foundation*.

Combined Slab and Foundation

Sometimes referred to as the thickened-edge slab, this construction is useful in warm climates where frost penetration is not a problem and where soil conditions are especially favorable. It consists of a footing poured over a vapor barrier with the slab thicker and reinforced at the edges. Fig. 22-1. The bottom of the footing should be at least 1' below the natural grade line and supported on solid, unfilled, well-drained ground.

Independent Concrete Slab and Foundation Walls

In areas where ground freezes fairly deep during winter, the walls of the house must be supported by foundations or piers which extend below the frostline to solid bearing on unfilled soil. In such construction, the concrete slab and foundation walls are usually separate. Three typical systems are suitable for such conditions. Figs. 22-2 through 22-4.

CONCRETE FLOOR SLAB REQUIREMENTS

To provide a satisfactory concrete floor slab, the following basic construction requirements should be met:
● The finish floor level should be high enough above the natural ground level so

that finish grade around the house can be sloped away for good drainage. Top of slab should be no less than 8″ above the ground and the siding no less than 6″.
● Topsoil should be removed. Sewer and water lines are installed, then covered with 4″ to 6″ of gravel or crushed rock well tamped in place.
● A vapor barrier consisting of a heavy plastic film such as polyethylene, asphalt-laminated sheet, or heavy roofing should be put under the concrete slab. Joints should be lapped at least 4″ and sealed. The barrier should be strong enough to resist puncturing during placing of the concrete.
● A permanent, waterproof, nonabsorbent type of rigid insulation should be installed around the perimeter of the wall. Insulation

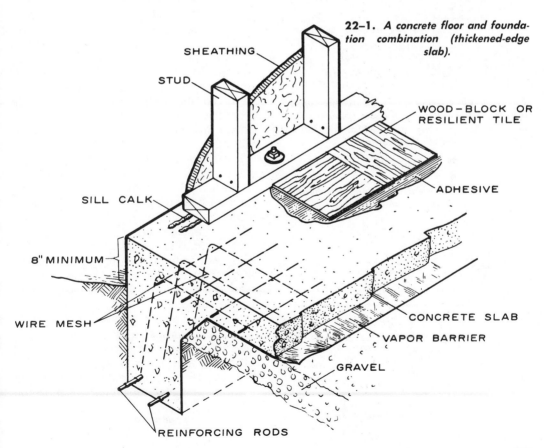

22-1. *A concrete floor and foundation combination (thickened-edge slab).*

SHEATHING
STUD
WOOD-BLOCK OR RESILIENT TILE
SILL CALK
ADHESIVE
8″ MINIMUM
WIRE MESH
CONCRETE SLAB
VAPOR BARRIER
GRAVEL
REINFORCING RODS

may extend down on the inside of the wall vertically or under the slab edge horizontally.

• The slab should be reinforced with 6" × 6" No. 10 wire mesh or other effective reinforcing. The concrete slab should be at

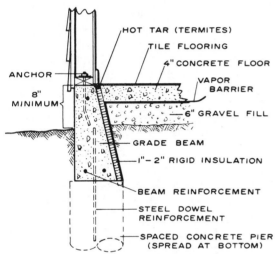

22–2. *A reinforced grade beam used to support a concrete floor. The grade beam spans the area between the concrete piers which are located below the frostline.*

22–3. *Installation details for perimeter heating in a concrete floor.*

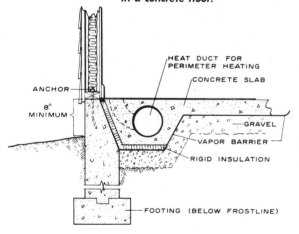

least 4" thick. A thickened-edge slab is preferred in termite areas. Fig. 22-1.

• After leveling (as by screeding) the surface should be smoothed with wood or metal floats while it is still plastic. (Screeding and floating equipment and operations are discussed on pages 281-3). If a smooth, dense surface is needed for the installation of wood or resilient tile with adhesives, the surface should be given a final smoothing with a steel trowel.

CONSTRUCTING A CONCRETE FLOOR SLAB*
Subgrade

It is important that the subgrade (the earth below the slab) be well and uniformly compacted to prevent any unequal settlement of the floor slab.

All organic matter such as sod and roots should first be removed and the ground leveled off. Any holes or other irregularities in the subgrade and any trenches for utilities should be filled in layers not exceeding 6" deep and thoroughly tamped. Material for fill should be of uniform character and should not contain large lumps, stones, frozen chunks, or material which will rot.

The entire subgrade should be rough-graded to an elevation slightly above the finished grade and then thoroughly compacted by tamping or rolling. The finished subgrade should be carefully checked for elevation and profile.

Soil cannot be properly compacted if it is too wet or too dry. A rough idea of the proper moisture content of ordinary soils, except very sandy ones, may be obtained by squeezing some in the hand. With proper moisture content the soil will cling together but will not be plastic or muddy. If the soil is too dry, it should be sprinkled with water and mixed before compacting. If the soil is too wet, it must be allowed to dry. If there is any question that sandy soil

---

* Adapted from material provided by courtesy of the Portland Cement Association.

may be too dry, water should be added, since an excess of water in such soil does not prevent compacting.

Granular Fill

A coarse fill should be placed over the finished subgrade. The fill should be brought to the desired grade and then thoroughly compacted. Fig. 22-5. This granular (grainy) fill should consist of coarse slag, gravel, or crushed stone, preferably ranging from 1/2" to 1" in diameter. This fill is intended both to insulate and to protect against moisture from the ground.

The fill material particles should be of uniform size to insure a maximum volume of air space in the fill. If necessary, the material should be screened to remove any *fines* (finely crushed or powdered material). The large volume of air space will add to the insulating qualities and keep subsoil moisture from being absorbed.

A line of drain tile should be placed around the outside edge of the exterior wall footings. The tiles are connected to drains to help keep ground moisture out of this granular fill. Fig. 22-6. Such moisture would reduce its insulating value. Tile is

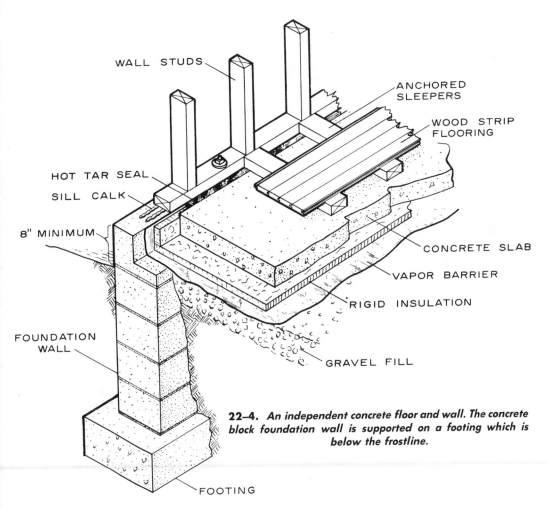

22–4. An independent concrete floor and wall. The concrete block foundation wall is supported on a footing which is below the frostline.

22–5. *Compacting coarse granular fill before the application of the grout coat.*

not necessary where the floor is to be located on relatively high ground, where subsoil is well drained, or in a dry climate.

Provisions for Mechanical Trades

Ducts for heating systems as well as supply and waste plumbing lines can be placed under the granular fill. Water service supply lines, if placed under the floor slab, should be installed in trenches of the same depth as those outside the building. This will prevent damage from freezing if the building is not occupied during cold weather. Connections to these utilities can be brought to a point above the finished concrete floor level prior to concreting. The electric supply line and all distributing lines for electricity and plumbing are carried in the walls or partitions.

Dampproofing

After the granular fill has been compacted and graded, a stiff grout coat should be placed over it. This will provide a smooth surface for installing a membrane-type material for dampproof-

ing. The stiff grout should consist of one part portland cement and three parts sand. This should be at least ½" thick and should be broomed or floated in place.

The grouted surface, when hardened and dried, should be mopped with hot asphalt. So should the tops of any bearing partition footings. As the mopping proceeds, a layer of 15-pound asphalt-saturated roofing felt should be placed on the hot asphalt with edges of felt well lapped. (This felt is the dampproofing membrane mentioned earlier.) Two layers of roofing felt are recommended, with hot asphalt mopped between layers and also on top of the second layer. Fig. 22-7. This membrane dampproofing should be continuous over the entire floor area and carried up on the inside of the foundation walls to a point 1" or more above the finished floor level. Fig. 22-6. Workers must be cautioned against puncturing the membrane when placing the concrete floor.

Concrete

Concrete for the floor slab and bearing partition footing should be made with durable, well-graded aggregate. The concrete should contain not more than 6 gallons of water for each sack of cement, including the moisture contained in the aggregate. If the sand is average-moist, do not add more than 5 gallons of water per sack of cement. The mix should consist of one part portland cement to approximately 2¼ parts of fine aggregate well graded from ¼" down, and three parts of coarse aggregate well graded in size from ¼" up to 1".

The concrete should be workable, so it can be placed without honeycombing or developing excess water on the surface. If necessary, the proportion of fine and coarse aggregate should be adjusted to obtain a mix of desired workability. After placing, the concrete should be thoroughly compacted by vibrating or by tamping and spading. Then it should be screeded to proper grade. The subsequent steps in finishing the concrete surface will depend upon the floor finish specified. (See "Finishing Concrete," page 280.)

For an even, true surface, the concrete should be worked with a wood float in a manner which will compact the surface and leave no depressions or inequalities of any kind. After the concrete has hardened sufficiently to prevent fine material from working to the top (when the sheen or shiny film of water on the surface has disappeared) it should be steel-troweled; excessive troweling should be avoided.

The concrete should be kept moist for at least two days. However, when the finished floor is to be exposed concrete, at least five days of moist curing are required. Moist burlap or canvas or a waterproof concrete curing paper may be used to cover the floor slab during this period. Curing should begin as soon as the concrete is hard enough to prevent damage. If burlap or canvas is used, it should be kept wet by sprinkling with water.

Reinforcement

Metal reinforcement weighing not less than 40 pounds per 100 sq. ft., with reinforcement in a grid pattern, should be placed in the concrete slab 1½" from the top surface.

22-7. Applying dampproofing over the grout coat in preparation for a concrete floor slab.

Insulation

A 1" thick, continuous, rigid, waterproofed insulation strip should be provided between the foundation walls and the edge of the floor slab. Recent studies by the National Bureau of Standards have indicated that this edge insulation is highly

22-6. Construction details of a concrete floor showing damp-proofing and insulation requirements.

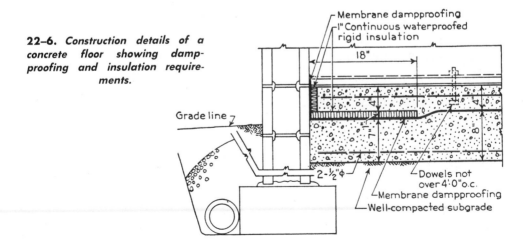

22-8. *A completed floor slab. Note the metal clips for attaching the wood sleepers.*

22-9. *The bottom plates for the partitions are attached to the floor slab. Note the plumbing installed in the partitions.*

important. The granular fill and membrane dampproofing under the floor slab act as an insulating material; usually these substances do a satisfactory job of reducing the heat loss to the subgrade.

COMPLETING THE CONSTRUCTION

Construction cannot continue until the concrete has cured. When curing is complete the *sleepers* are installed. These are heavy wooden members which lie flat on the concrete slab. They are attached by means of metal clips which were embedded earlier. Fig. 22-8. The sleepers will provide support for the wooden floor. Next the wall plates are laid out on the slab. They are attached with anchor bolts which were also embedded in the concrete earlier.

When wood flooring is to be installed over the slab, the subfloor and finish floor are installed over the sleepers. After the wallplates have been laid out and attached, plumbing is installed in the partition. Fig. 22-9. Wall framing can then proceed as will be discussed in Unit 25.

Floor Finishes

The most common coverings for concrete floors on ground are terrazzo, concrete tile, ceramic tile, asphalt tile, wood flooring, linoleum, small rugs, and wall-to-wall carpeting. When linoleum, asphalt tile, or similar resilient-type flooring material is to be applied, the concrete surface should be given a smooth, steel-troweled finish. Fig. 22-10. When other types of floor covering are contemplated, it is recommended that the manufacturer's advice be obtained as to their suitability and the methods of application.

Some concrete floors are troweled smooth in their natural color and may be sealed, then waxed and polished.

Color can be added to concrete floors by incorporating pure mineral oxide pigments as the concrete is placed. After the concrete has hardened and dried, the colored surface should be waxed and buffed. Color can also be produced by using commercial stains guaranteed by the manufacturer to produce satisfactory results. Stains can generally be obtained from local paint or hardware dealers and should be applied as the manufacturer recommends.

Finishing Concrete

A concrete surface which is not completely enclosed by a form, such as a floor, driveway, or sidewalk slab, must be finished. The finish will depend on the use. A good, non-slippery finish is best for sidewalks; a coarse, scored surface for driveways; and a smooth-troweled finish for porches and basement floors. A stiff, coarse broom is useful in giving the surface a scored finish. The broom is run crosswise to the slab. Smooth finishes are produced with a steel trowel.

SCREEDING

The first step in finishing a slab is called *screeding*. A hand-operated screed and the method of using it are shown in Fig. 22-11. The chief purpose of screeding is to level the surface of the slab by striking off (removing) the excess concrete. The concrete is struck off just after it is placed in the forms. The screed rides on the edges of the side forms or on wood or metal strips set up for the purpose. Two people move the screed along the slab. The movement should be like that of a saw.

Screeding may also be done by means of mechanical equipment. Fig. 22-12.

EDGING AND JOINTING

When all water and water sheen has left the surface and the concrete has started to stiffen, other finishing operations can be done. Edging, if necessary, can be done at this time. This operation produces a rounded edge on the slab to prevent chipping or damage. The edger should be run back and forth until a finished edge is produced. The cement mason should be careful that all coarse aggregate particles are covered and that the edger does not leave too deep an impression in the top of the slab. Otherwise the indentation may be difficult to smooth out with subsequent finishing operations.

Immediately following edging, the slab is jointed (or grooved). This means that a jointing tool is used to cut *control* or *contraction joints* about 3/4" deep in the slab. The purpose of these joints is to control cracking. Sometimes shrinkage stresses are present in the slab as a result of

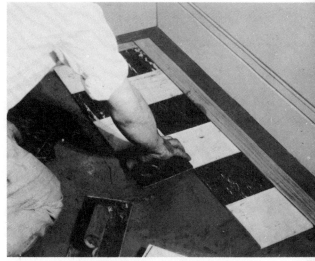

22–10. *Installing asphalt tile directly on the concrete slab.*

temperature changes or dryness. These stresses can cause the concrete to crack. However, where the joints are cut, the thickness of the slab is reduced. Thus cracks are likely to occur only at these intentionally weakened points. When the concrete shrinks, these joints open slightly, thus preventing irregular and unsightly random cracks.

In sidewalk and driveway construction, the tooled joints are usually spaced at intervals equal to the width of the slab, but not more than 20' intervals. As mentioned, for control joints the jointer should have a 3/4" bit. However, if the slab is to be grooved only for decorative purposes, jointers with shallower bits may be used.

It is good practice to use a straight 1" × 8" or 1" × 10" board as a guide when making the groove in the concrete slab. If the board is not straight it should be planed true. The tooled joints should be perpendicular to the edge of the slab. The same care must be taken in running joints as in edging, for a tooled joint can add to or detract from the appearance of the finished slab.

22-11. *Striking off a concrete slab. Notice that this slab will be quite large. Therefore the two outside sections were formed and poured, leaving the center section open to allow working from both edges. When the two outside sections have cured sufficiently, the forms can be removed. The center section is poured even with the two outside slabs which serve as a form for the center section.*

FLOATING

Screeding leaves a level surface with a fairly coarse finish. When a smoother finish is desired, screeding is followed by *floating*. The wood float is used in making an even gritty surface, as for sidewalks. Fig. 22-13. The wood float is also used to fill up the hollows and to compact the concrete.

Floating is done shortly after screeding, while the concrete is still plastic enough to allow mortar to be brought to the surface. However, the concrete should not be too plastic. Premature floating brings an excess amount of fines and moisture to the surface. This will cause fine "hair" cracks (crazing) or the appearance of a powdery material (dusting) at the surface. Only enough floating should be done to remove defects and to bring up enough mortar to produce the desired finish. If floating is to be the last step in finishing, it may be

22-12. *A power screed.*

22-13a. *A wood float.*

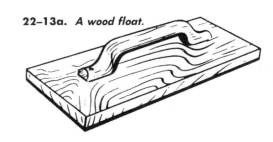

22–13b. *Floating a concrete slab.*

22–14b. *Troweling a concrete slab.*

necessary to float the surface a second time after the concrete has hardened slightly.

TROWELING

For a dense, smooth finish, floating is followed by *steel troweling*. Fig. 22-14. For large areas, a mechanical trowel is convenient. Fig. 22-15. Troweling is not begun until the concrete has hardened enough to prevent fine material and water from being worked to the surface. In fact, troweling should be delayed as long as possible. A surface which is troweled too early lacks durability, whereas one which is troweled too late is too hard to finish properly.

Troweling should leave the surface smooth, even, and free of marks and ripples. A fine-textured surface may be obtained by following the first troweling

immediately with a second one. In this second operation the trowel, held flat, is passed lightly and with a circular motion.

For a *hard steel-troweled finish*, the second troweling should be delayed until the concrete has become hard enough to

22–15. *A power trowel.*

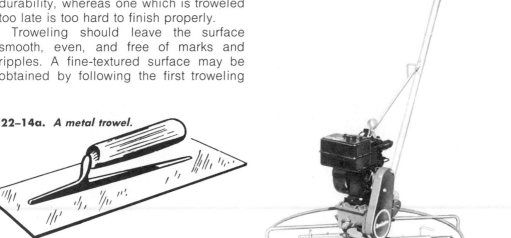

22–14a. *A metal trowel.*

make a ringing sound under the trowel. In hard steel-troweling the trowel is tilted slightly, and heavy pressure is applied to compact the surface.

Kneeboards. When troweling or floating a large surface, kneeboards (or kneeling boards) may be used. Fig. 22-16. These boards, which measure about 12″ × 24″, are placed on the concrete to support the weight of the finisher. One board supports the knees, while the second board is placed on the concrete just behind to support the feet of the finisher. As it becomes necessary to move from one area to another, the finisher stands on the rear board. He then places the kneeboard in another area requiring troweling, steps onto the kneeboard, and repositions the second board. The finisher is now in position to continue the floating and troweling operations without having stepped directly onto the fresh concrete.

DRIVEWAYS, SIDEWALKS, AND BASEMENT FLOORS

A new home is not complete until driveways and walks have been installed, ready for landscaping. Because the automobile is an important element of life, the garage is usually a prominent part of house design. This in turn establishes the location of driveways and walks.

Concrete and bituminous (blacktop) pavement is most commonly used in the construction of walks and drives, especially in areas where snow removal is important. In some areas of the country, a gravel driveway and a flagstone walk are satisfactory and reduce the cost of improvements.

Basements are normally finished with concrete floors of some type. These floors are poured after all improvements such as sewer and waterlines have been connected. Concrete slabs should not be poured on recently filled areas.

Driveways

The grade, width, and radius of curves in a driveway are important to consider when

22–17. *Recommended dimensions for a single-slab driveway turnaround.*

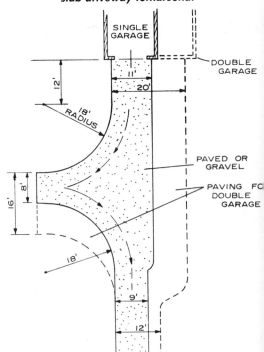

22–16. *Using kneeboards (or kneeling boards) to support the weight of the concrete finisher.*

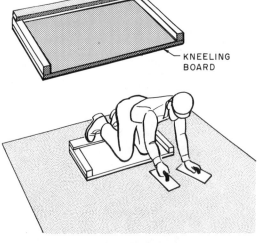

establishing a safe entry to the garage. Driveways that have a grade more than 7% (7' rise in 100') should have some type of pavement to prevent wash. Driveways that are long and require a turn-around need especially careful design. Fig. 22-17 shows a driveway and turn-around which allow the driver to go forward to the street or highway. In areas of heavy traffic, this is much safer than having to back into the roadway. A double garage should be serviced by a wider entry and turn-around.

Steep driveways should have a near-level area from 12' to 16' long in front of the garage for safety.

Two basic types of driveways are:
• The slab or full-width type, which is the more common.
• The ribbon type, with a strip of grass between two strips of concrete.

When driveways are fairly long or steep, the full-width type is the more practical. Fig. 22-18. The ribbon driveway is cheaper and perhaps less conspicuous because of the grass strip between the concrete runners. However, it is not practical if curved, steep, or long. Fig. 22-19.

The width of the single-slab drive should be 9', although 8' is often considered minimum. When the driveway is also used as a walk, it should be at least 10' wide to allow for a parked car as well as a walkway. The width should be increased by at least 1' at curves. The radius of the drive at the curb should be at least 5'. Relatively short double driveways should be at least 18' wide, and 2' wider when they also serve as a walk from the street.

The concrete strips in a ribbon driveway should be at least 2' wide and located so that they are 5' on center (that is, so their centers are 5' apart). When the ribbon is also used as a walk, the width of strips should be increased to at least 3'.

Pouring a concrete driveway over an area that has been recently filled is poor

22–18. *Single-slab driveway details. A concrete slab should drain from the center outward.*

practice unless the fill, preferably gravel, has settled and is well tamped. A gravel base is not ordinarily required on sandy, undisturbed soil but should be used on all other soils. Concrete should be about 5" thick. Side forms are often built of 2" × 6" boards. These members establish the elevation and alignment of the driveway and are used to support the strike board for striking off the concrete.

Under most conditions, the use of steel reinforcing is good practice. Steel mesh, 6" × 6", will normally prevent or minimize cracking of the concrete. Expansion joints with asphalt-saturated felt strips should be used where the driveway joins with the public walk or curb, at the garage slab, and about every 40 feet on long driveways. A 5- or 5½-bag commercial mix concrete is ordinarily used for driveways. However, a 5½- to 6-bag mix containing an air-entraining mixture should be used in areas having severe winter climates. Concrete made with these cements contains tiny, well-distributed and completely separated air bubbles.

Blacktop driveways, normally constructed by paving contractors, should also have a well-tamped gravel or crushed rock base. The top should be slightly crowned for drainage.

Sidewalks

Main sidewalks should extend from the front entry to the street, to a front walk, or to a driveway leading to the street. A 5% grade is considered maximum for sidewalks; any greater slope usually requires steps. Walks should be at least 3' wide.

Concrete sidewalks are constructed much the same as concrete driveways. They should not be poured over filled areas unless the fill is settled and very well tamped. This is especially true of areas near the house after basement excavation backfill has been completed.

Minimum thickness of concrete over normal undisturbed soil is usually 4". As described for concrete driveways, contraction joints should be used and spaced on 4' centers.

When slopes to the house are greater than a 5% grade, stairs or steps should be used. For gentler slopes, other construction may be acceptable, such as a ramp sidewalk, a flight of stairs at a terrace, or a continuing sidewalk. Fig. 22-20. Steps should have 11" treads and 7" risers when the stair is 30" or less in height. When the total rise is more than 30", the tread should be 12" and the riser 6". For a moderately uniform slope, a stepped ramp may be satisfactory. Fig. 22-21. Generally, the rise should be about 6" to 6½", with enough space between risers for two or three normal paces.

Walks can also be made of brick, flagstone, or other types of stone. Brick and stone are often placed directly over a well-tamped sand base. However, this system is not completely satisfactory where the soil freezes. For a more durable walk in cold climates, the brick or stone topping should be embedded in a freshly laid reinforced concrete base. Fig. 22-22.

As mentioned earlier for blacktops, all concrete sidewalks and curbed or uncurbed driveways should have a slight

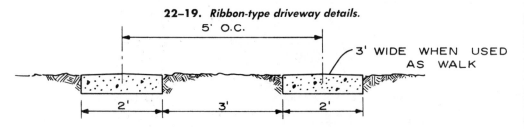

22–19. Ribbon-type driveway details.

5' O.C.

3' WIDE WHEN USED AS WALK

2'   3'   2'

crown for drainage. (That is, the center should be slightly higher than the edges.) Joints between brick or stone may be filled with a cement mortar mix or with sand.

Basement Floors

Basement floor slabs should be no less than 3¹/₂″ thick and sloped toward the floor drains. A 2″ × 4″ (3¹/₂″ wide) is often used on edge for form work. There should be at least one drain in a basement floor, and for large floors two are more satisfactory. One should be located near the laundry area.

To assure a dry basement floor, a polyethylene film or similar vapor barrier can be installed under the concrete slab. However, basement areas or multilevel floors used only for utility or storage do not usually require a vapor barrier. When finished rooms have concrete floors, the use of a vapor barrier is normally required.

PIPE-SCREED STRIPS

When concrete is poured in an enclosed area, forms are not necessary. A basement floor is an example. The foundation walls serve as forms to confine the concrete. However, such a surface still must be screeded for correct thickness and levelness. This is done by means of rail-like devices on which a screed will ride. The rails are made of sections of 1″ pipe set on stakes which are driven into the subgrade.

22–20. Sidewalks placed on a grade should have a 5% maximum slope.

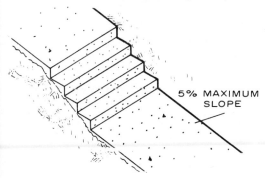

Fig. 22-23. A board with a straight edge can be used as the screed. The pipes used as rails are called *screed strips*. The stakes are driven deep enough so that when the pipes are set on them, the tops of the pipes will be at the level desired for the surface of the slab. After screeding, the pipes and stakes are removed; a float is used to pack concrete in the resulting voids.

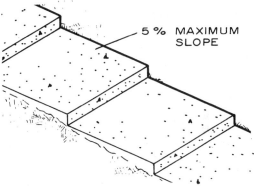

22–21. A stepped ramp is sometimes used on a moderate grade.

22–22. Masonry paved walks: A. Brick. B. Flagstone.

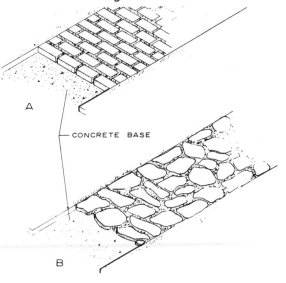

ESTIMATING

Material

To calculate the amount of material required for the basement floor for the home shown in Fig. 22-24, first figure the area of the slab. The house measures 26' × 40', so the total area of the basement slab will be 1040 sq. ft.

Assume the floor to be 4" thick. Consulting Fig. 22-25, you see that 81 sq. ft. of area 4" thick can be placed with one cubic yard of concrete. To calculate the total amount of concrete required, divide the total slab area (1040 sq. ft.) by the sq. ft. from one cubic yard (81).

$$\frac{1040}{81} = 12.84 \text{ cubic yards.}$$

Rounding this off to the next quarter of a yard, you will need an estimated 13 cubic yards of concrete.

Labor

FORMS AND SCREED STRIPS

To calculate the amount of labor involved in placing the screed strips for the basement floor in the example, refer again to Fig. 22-25. Here you see that an average of 30 linear feet of screed strips can be set in place per hour.

For the basement in the example, assume a screed strip down each of the long walls and a center screed strip. This makes a total of 3 screed strips, each 40' (the length of the building), for a total of 120'. To calculate the time required, divide the total length of screeds (120') by the

22–23. *A pipe screed used as a guide for the straightedge when leveling a basement floor. Workers should pull the screed toward them by the handles shown at the right and left ends. The arrows indicate the movement of the board. Note that it moves from side to side as well as toward the workers.*

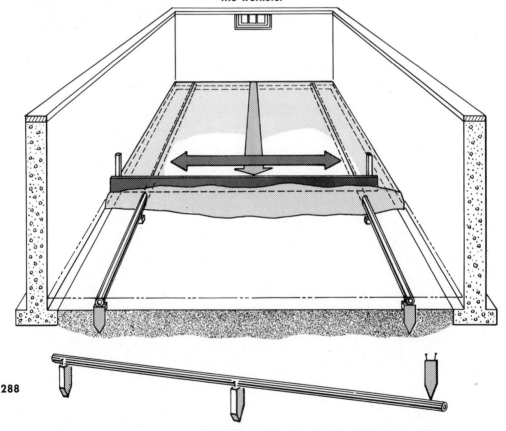

average linear feet of screed strips that can be placed in an hour (30').

$$\frac{120}{30} = 4$$

Thus you see that four hours will be needed to set the screed strips.

PLACEMENT OF CONCRETE

Next, determine the labor for the placement of the concrete for the basement floor in the example. Referring to Fig. 22-25, under the heading "Labor" you see that 100 sq. ft. of surface can be placed on the average in 3.6 hours. Divide the total slab area (1040 sq. ft.) by 100, to get the area as expressed in hundreds of sq. ft.

$$\frac{1040}{100} = 10.40$$

Next multiply by the time required to place 100 sq. ft. of surface (3.6 hours).

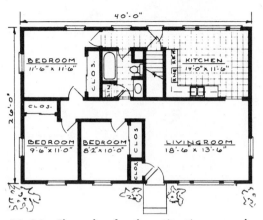

22–24. *Floor plan for the estimating examples given in the text.*

$$10.40 \times 3.6 = 37.44 \text{ hours}$$

Rounding off, you will have to allow for approximately 37½ hours to place and finish the basement floor in the example.

22–25. *Estimating table for concrete slabs.*

SLABS	Material		Labor	
Thickness	Per Square Foot Cubic Feet of Concrete	Square Feet from One Cubic Yard	Forms and Screeds 100 Linear Feet	Placement 100 Sq. Ft. of Surface
2"	0.167	162		
3"	0.25	108	Average	Average
4"	0.333	81	30 Linear Feet	3.6
5"	0.417	65	Per Hour	Hours
6"	0.50	54		

Note: Placement includes finishing with topping. If topping omitted deduct 1.2 hours.

QUESTIONS

1. How deep should the footing be on a combined slab and foundation?
2. When constructing a concrete floor slab, what is the purpose of the granular fill over the subsoil?
3. How is concrete compacted?
4. Describe how the concrete slab is cured.
5. What types of floor coverings require a steel-troweled finish?
6. What is screeding?
7. What problems will occur if floating is done prematurely?
8. What is the purpose of troweling?
9. What is a kneeboard?
10. What is meant by a 6% grade?
11. At what intervals should expansion joints be used on driveways?
12. How thick should a basement floor slab be?

FRAMING

Framing Methods

Most homes in the United States and Canada are of wood-frame construction. Many are covered with wood siding; other common coverings include wood shingles, composition shingles or siding, brick veneer, and stucco.

Wood-frame houses have several important advantages. In general, frame construction costs less than other types. It provides more house for a given price and better insulation, thereby increasing comfort to the occupants and reducing heating and air-conditioning costs.

Wood is easily worked and is suitable for use with a wide variety of exteriors. This flexibility allows architects and builders to produce nearly any architectural style.

A well-built wood-frame home is very durable. Some of the oldest existing buildings in North America are Paul Revere's house in Boston, built in 1677, and the "House of Seven Gables" in Salem, Mass., constructed in 1668. Fig. 23-1.

TYPES OF WOOD FRAMING

Buildings framed of lumber usually belong in one of two main classes:
- *Multiple-member assemblies*, often called *conventional framing*.

23-1. *Paul Revere House. This is the oldest home in Boston, built around 1677. Revere left from here on his historic ride to Lexington. The house is still open to the public as a museum.*

• *Plank-and-beam framing*, which consists of heavier members, more widely spaced.

Each of these framing methods has its advantages, which will be discussed.

Conventional Framing

Conventional framing consists of multiple small members (joists, studs, and rafters) so joined that they act together and share the loads in supporting the structure. When assembled and sheathed, these members form complete floor, wall, and roof surfaces. Two common types of conventional framing are *platform* and *balloon*.

PLATFORM-FRAME CONSTRUCTION

As the name indicates, in this type of construction the floors are complete platforms, independent of the walls. The subfloor extends to the outside edges of the building and provides a platform upon which exterior walls and interior partitions are erected. Platform construction is generally used for one-story houses. It is also used alone or in combination with balloon construction for two-story structures. Building techniques in most parts of the United States have been developed almost entirely around the platform system. Fig. 23-2. This book will therefore concentrate on platform-frame construction.

Compared with balloon framing, platform construction is easier to erect because at each floor level it provides a flat surface on which to work. It is also easily adapted to various methods of prefabrication. Each level of a two-story house is constructed separately. With a platform-framing system, it is common practice to assemble the wall framing on the floor and then tilt the entire unit into place.

BALLOON-FRAME CONSTRUCTION

The feature which identifies this type of construction is that the studs are continuous from sill to top plate. Fig. 23-3. Studs and first-floor joists rest on the anchored

sill. Second-floor joists bear on 1" × 4" ribbon strips which have been let into the inside edges of the studs.

In this type of construction there is less cross-grain wood framing than in conventional construction. Wood expands and contracts across the grain but is relatively stable with the grain. Therefore balloon-frame construction is less likely to be affected by expansion and contraction than conventional construction. This is an advantage for certain types of buildings. Specifically, balloon framing is good for two-story buildings on which the exterior covering is of brick veneer, stone veneer, or stucco. With such buildings, movement of the wood framing under the masonry veneer can be a serious problem.

If exterior walls are of solid masonry, it is also desirable to use balloon framing for interior bearing partitions. Again, this is because there is relatively little cross-grain wood framing. This minimizes dimensional changes in the walls and reduces variations in settlement which may occur between exterior walls and interior supports.

In balloon framing, blocks are placed between the joists to serve the dual purpose of solid bridging and fire stopping. Solid bridging holds the joists' ends in line; fire stopping prevents the vertical and horizontal spaces from acting as flues in the event of fire. Fig. 23-4.

Balloon-frame construction is rarely used today. The longer framing members required for this type of construction are not readily available, and they cost more than the materials used in platform-frame construction. Therefore balloon framing construction techniques will not be discussed in detail in this book.

Plank-and-Beam Construction

In the plank-and-beam method of framing, plank subfloors or roofs, usually a 2" nominal thickness, are supported on beams spaced up to 8' apart. The ends of the beams are supported on posts or piers.

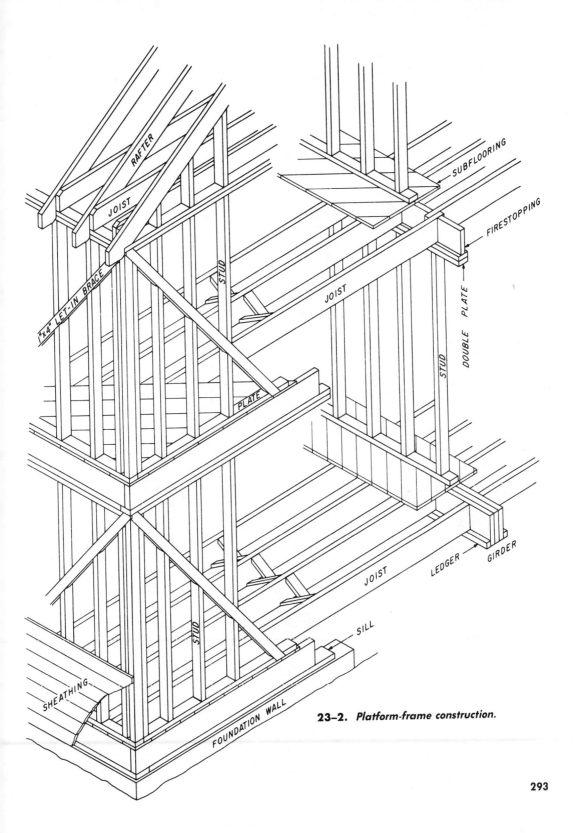

23-2. *Platform-frame construction.*

Labels in figure: RAFTER, JOIST, 1"x4" LET-IN BRACE, STUD, SUBFLOORING, FIRESTOPPING, JOIST, DOUBLE PLATE, STUD, PLATE, LEDGER, GIRDER, JOIST, STUD, SILL, SHEATHING, FOUNDATION WALL

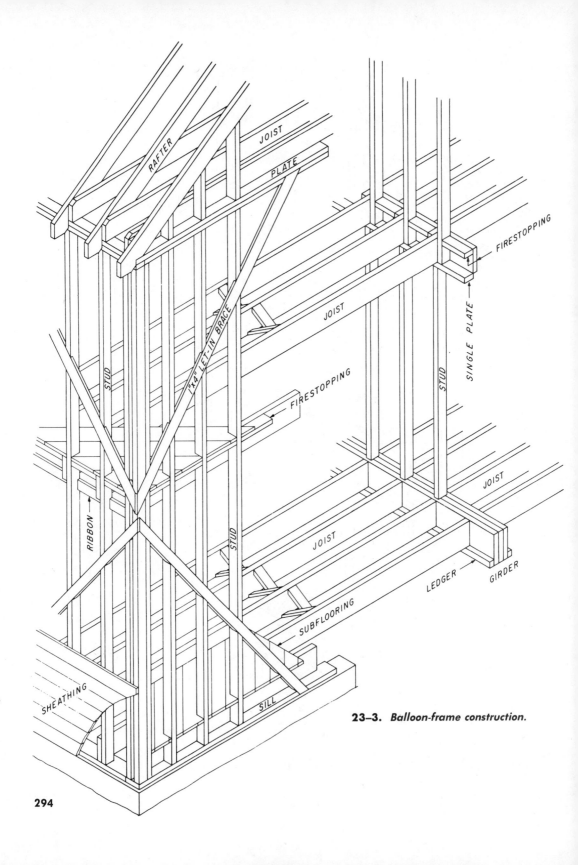

23–3. Balloon-frame construction.

Wall spaces between posts are provided with supplementary framing as needed for attachment of exterior and interior finish. This extra framing and its covering also provide lateral bracing for the building.

Conventional framing uses joists, rafters, and studs spaced 12" to 24" on centers; the plank-and-beam method calls for fewer and larger-sized pieces, spaced farther apart.

There are many advantages to be gained through the use of the plank-and-beam system of framing. One of the best points is the architectural effect provided by the exposed plank-and-beam ceiling. In this type of construction, the roof plank serves as the ceiling, which provides added height to the living area at no additional cost. Generally, the planks are selected for appearance; therefore no further ceiling treatment is required except the application of a finish.

In a well-planned plank-and-beam framed structure, there are important savings in labor. As mentioned, the pieces are larger, and there are fewer of them than in conventional framing. The cross-bridging of joists is eliminated, and larger and fewer nails are required. This results in a substantial reduction of labor on the job site.

In plank-and-beam framing, the ceiling height is measured to the underside of the plank, but in conventional construction it is measured to the underside of the joists. The difference between the thickness of the plank and the depth of the joist gives the building a smaller volume and also reduces the height of the interior walls.

Fig. 23-5a compares the height of a plank-and-beam house with that of a conventionally framed house. Fig. 23-5b compares the two framing methods.

METAL FRAMING

Metal framing members, sections, and accessories are factory-fabricated for easy job-site assembly. Structural members, such as columns, beams, joists, plates, studs, corner posts, partition tees, window

and door members, truss members, and fascia are all available for framing with this system. Fig. 23-6. This method of framing has the advantage of permitting the utiliza-

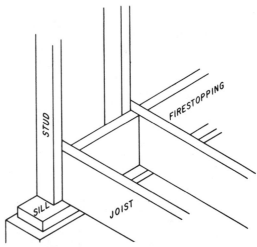

23–4. *First-floor framing at exterior wall—balloon-frame construction.*

23–5a. *Comparison of height of plank-and-beam house with conventionally framed house.*

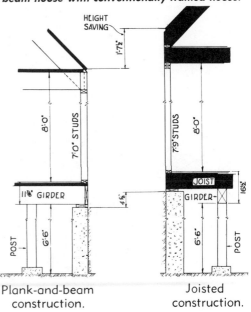

Plank-and-beam
construction.

Joisted
construction.

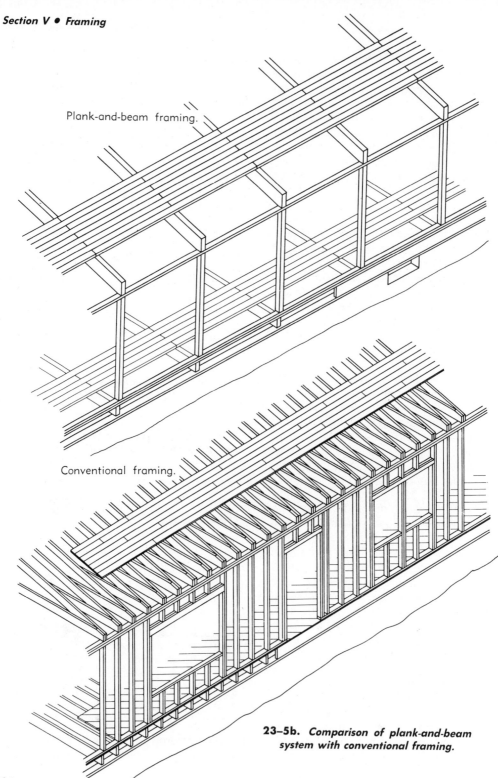

Plank-and-beam framing.

Conventional framing.

23–5b. *Comparison of plank-and-beam system with conventional framing.*

23–6. *This drawing is based on a catalog page. It shows cross sections of members available for a typical metal-frame building system.*

ALUMIFRAME BUILDING SYSTEM MEMBERS

EVERY STRUCTURAL MEMBER NEEDED FOR FRAMING A HOUSE.

FLOOR MEMBERS:

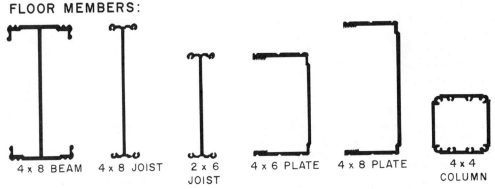

| 4 x 8 BEAM | 4 x 8 JOIST | 2 x 6 JOIST | 4 x 6 PLATE | 4 x 8 PLATE | 4 x 4 COLUMN |

2 x 4 STUDS:

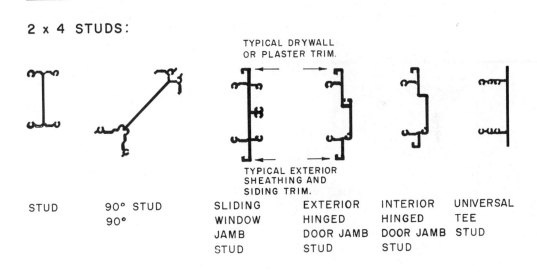

TYPICAL DRYWALL OR PLASTER TRIM.

TYPICAL EXTERIOR SHEATHING AND SIDING TRIM.

STUD   90° STUD 90°   SLIDING WINDOW JAMB STUD   EXTERIOR HINGED DOOR JAMB STUD   INTERIOR HINGED DOOR JAMB STUD   UNIVERSAL TEE STUD

2 x 4 PLATES:

TYPICAL CEILING DRYWALL SUPPORT.

PLATE   SINGLE BASEBOARD AND PLATE   SINGLE FLANGED PLATE

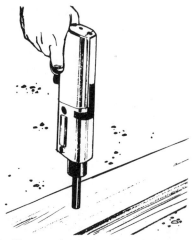

23–7. The powder-actuated stud driver is powered with a .22 caliber charge that will drive a stud (special nail) into concrete or metal. This is a quick, efficient way of fastening wood to concrete or metal.

23–8. Assembling and sheathing an exterior wall section on the job site.

tion of splice-free lengths up to 40' long for continuous beams, joists, rafters, and truss members.

The framing members are assembled with power screwdrivers using self-drilling, self-tapping screws. The floor assembly is fastened to the foundation with studs (special nails) driven through the perimeter plate by a powder-actuated stud driver. Fig. 23-7. The plywood subfloor is installed over the floor framing system with self-drilling, self-tapping screws and a structural adhesive. Wall sections can be assembled at the job site or delivered as preassembled panels from an off-site assembly plant. Fig. 23-8. Conventional sheathing is attached to the framework with screws.

Doorframes for both the interior partitions and exterior walls are integral with the system. They are prepainted and come complete with hinges, lock, rubber stops, and weather stripping as necessary. The windows are also integral parts of the system, prefabricated and painted. These units include the interior and exterior trim which is designed to accept 1/2" wallboard and 1/2" sheathing plus siding on the outside.

The plumbing is installed in the prepunched stud webs. The wiring is passed through insulated grommets which are inserted in the prepunched webs of the studs and plates. Wall and ceiling fixtures are mounted by attaching wood blocking spaced between the flanges of the wall studs or trusses. Friction-tight insulation is installed by placing the batts (bundles of insulating material) between the studs on the exterior walls. Studs are spaced 2' on center, providing adequate space for the installation of heating and air-conditioning ducts.

Roof trusses are set in place and roof sheathing is attached in the same manner as described previously for the subfloor. After the sheathing is in place, roofing, such as shingles, is then installed in a conventional manner.

QUESTIONS

1. List several advantages of wood-framed houses.

2. How does platform-frame construction differ from balloon construction?

3. How does plank-and-beam construction differ from platform-frame construction?

4. Can you think of some advantages of metal framing?

UNIT 24 Floor Framing

Floor framing consists of posts, girders, sills, joists or floor trusses, and subflooring. Fig. 24–1. All members are tied together to support the load on the floor and to give support to the exterior walls.

POSTS

A post is a wooden or steel member which supports girders. A wood post must be solid and not less than 6″ × 6″ in size. It should rest on the top of a masonry pedestal that is at least 3″ above the floor. Fig. 24–2a. Steel posts may be H-Sectional, I-Sectional, or round. They have steel bearing plates at each end. Fig. 24–2b.

Wood posts should be square at both ends and securely fastened to the girder. Fig. 24–3. When necessary, a bearing plate should be placed between the post and the girder. Fig. 24–4 (Page 302). Posts are generally spaced 8′ to 10′ on center depending on the size and strength of the girder in relation to the load it must support.

24–1b. *Metal hangers that fit over the sill are often used with floor trusses.*

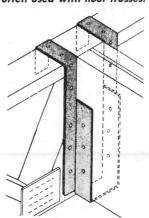

24–1a. *Floor trusses of various designs are often used instead of solid wood joists.*

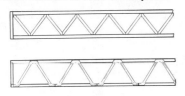

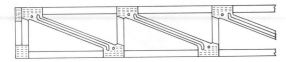

GIRDERS

Girders are large principal beams used to support the floor joists. They may be of wood or steel.

For steel girders, I-beams are commonly used. An advantage of steel is that it does not present the problem of shrinkage that wood does.

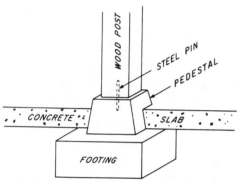

24–2a. *The foundation and pedestal for a basement or cellar post. If the post is wood, a steel pin must be set into the pedestal to secure the post.*

24–2b. *A steel post used with a steel I-beam. A wood beam may also be used with a steel post. Notice the flanges, welded onto each end of the post, for bolting the post into position.*

Common types of wood girders include *solid,* *built-up,* and *hollow.* Built-up wood girders consist of planks nailed together with two rows of 20d nails. The nails in each row should be spaced about 30″ apart with the end joints over the supports. Fig. 24-5. Glued laminated members are sometimes used. Hollow beams resemble a box made of 2 × 4s, with plywood webs. Fig. 24-6.

The ends of wood girders should bear at least 4″ on masonry walls or pilasters. A ½″ air space should be provided at each end and at each side of a wood girder framed in masonry. (When steel bearing plates are placed at ends of girders between the masonry and the girder, they should be of full bearing size.) Fig. 24-7. Tops of wood girders should be level with tops of sill plates on foundation walls unless there are ledger strips or notched joists. Fig. 24-8 (Page 304).

Installing a Girder or Beam

If the beam or girder is of metal, it may be installed in one of three locations:

● The top of the beam may be flush with

24–3a. *A wood post under a built-up girder. Notice that the joint in the girder is over the support.*

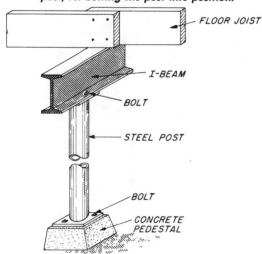

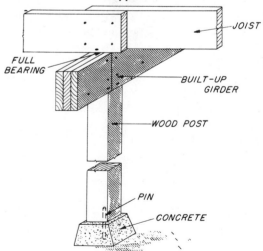

the plate to act as a bearing for the joists. Fig. 24-9.

● A wood plate may be placed on top of the metal beam to carry the joists. Fig. 24-10.

● When the top of the beam is set slightly higher than the wall sill plate, the joist is framed into the beam. Fig. 24-11.

Whether the beam is wood or metal, make sure that it is aligned from end to end and side to side. Also make sure that the length of the bearing post under the girder is correct so that it will properly support the girder.

Installing Joists Over Girders

In the simplest floor and joist framing, the joist rests on top of the girder. In Fig. 24-12, the top of the girder is aligned with the top of the sill plate. This method is used where basement height provides adequate headroom below the girder. The main disadvantage is that shrinkage is greater than when ledger strips are used. If more clearance is wanted under the girder, ledger strips are securely nailed to each side of the girder to support the joist. The joists are toenailed to the wood girders and nailed to each other where they lap

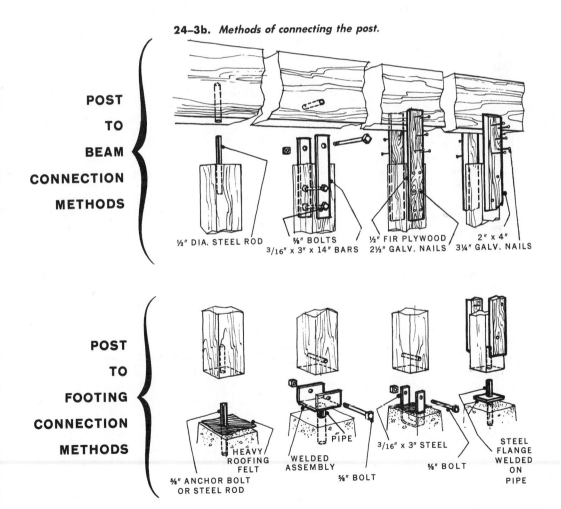

24–3b. *Methods of connecting the post.*

POST TO BEAM CONNECTION METHODS

½" DIA. STEEL ROD   ⅝" BOLTS   ½" FIR PLYWOOD   2" x 4"
3/16" x 3" x 14" BARS   2½" GALV. NAILS   3¼" GALV. NAILS

POST TO FOOTING CONNECTION METHODS

⅝" ANCHOR BOLT OR STEEL ROD   HEAVY ROOFING FELT   WELDED ASSEMBLY   ⅝" BOLT   PIPE   3/16" x 3" STEEL   ⅝" BOLT   STEEL FLANGE WELDED ON PIPE

over the girder. Care should be taken to obtain full bearing on the tops of ledger strips. Fig. 24-13 (Page 306).

SPACED GIRDERS

To provide space for heat ducts in a partition supported on the girder, a spaced girder is sometimes installed. Solid blocking is used at intervals between the two members. Fig. 24-14. A single post support for a spaced girder usually requires a

brace, preferably of metal, with a span sufficient to support the two members.

WOOD-SILL CONSTRUCTION

There are two types of wood-sill construction over foundation walls—one for platform construction and one for balloon-frame construction.

24-5b. *This table can be used for determining the amounts of lumber and nails necessary for a built-up girder. A 4' × 6' girder 20' long contains 43 board feet of lumber (20 × 2.15 = 43).*

Built-Up Girders

Size of Girder	Bd. Ft. per Lin. Ft.	Nails per 1000 Bd. Ft.
4 × 6	2.15	53
4 × 8	2.85	40
4 × 10	3.58	32
4 × 12	4.28	26
6 × 6	3.21	43
6 × 8	4.28	32
6 × 10	5.35	26
6 × 12	6.42	22
8 × 8	5.71	30
8 × 10	7.13	24
8 × 12	8.56	20

CHANNEL CAP OR CHANNEL RIVETED TO I BEAM AND BOLTED TO POST.

WOOD POST

24-4. *If a wood post is used with a steel girder, a cap should be provided for bolting the post in position.*

24-5a. *A built-up wood beam. The nails in the nailing pattern are 20d.*

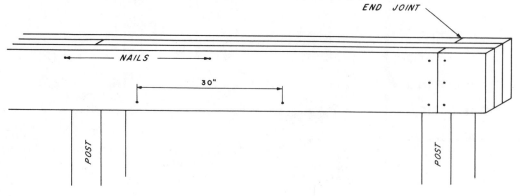

END JOINT

NAILS

30"

POST

POST

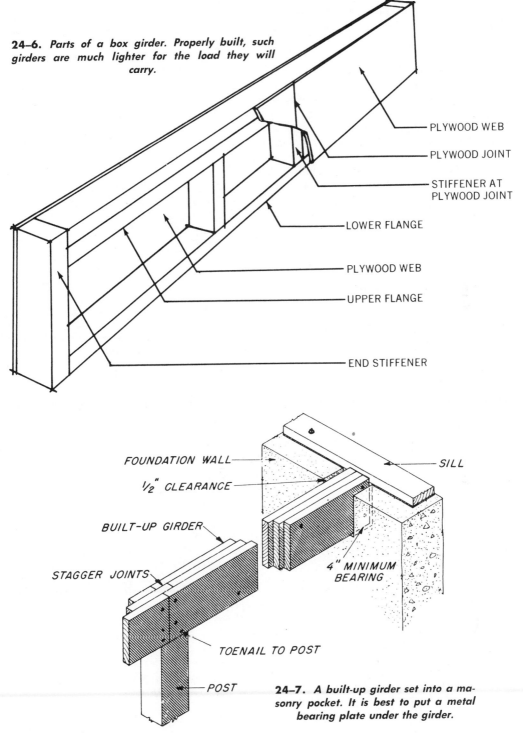

24-6. Parts of a box girder. Properly built, such girders are much lighter for the load they will carry.

PLYWOOD WEB

PLYWOOD JOINT

STIFFENER AT PLYWOOD JOINT

LOWER FLANGE

PLYWOOD WEB

UPPER FLANGE

END STIFFENER

FOUNDATION WALL

½" CLEARANCE

BUILT-UP GIRDER

STAGGER JOINTS

SILL

4" MINIMUM BEARING

TOENAIL TO POST

POST

24-7. A built-up girder set into a masonry pocket. It is best to put a metal bearing plate under the girder.

The *box sill* is usually used in platform construction. It consists of a sill or sill plate anchored to the foundation wall for supporting and fastening the joists, with a header (band) at the ends of the joists resting on the foundation wall. Fig. 24-15.

Balloon-frame construction also has a sill plate upon which the joist rests. The studs also bear on this plate and are nailed both to the floor joist and to the plate. Fig. 24-16.

The sill or sill plate is the lowest member of the frame structure that rests on the foundation. Insulation material and a metal termite shield can be placed under the sill, if desired. Fig. 24-17. The sill should consist of one or two thicknesses of 2" lumber placed on the foundation walls to provide a full and even-bearing surface. Fig. 24-18. Sills should be anchored to the foundation with 1/2" bolts spaced approximately 6' to 8' apart, with at least two bolts in each pair of sills. Fig. 24-19. The preferred method is to spread a bed of mortar on the foundation

and lay the sill upon it at once, tapping gently to secure an even bearing surface along the entire length. The nuts are then put in place over the washers and tightened gently with the fingers. After the mortar has set for a day or two, tighten the

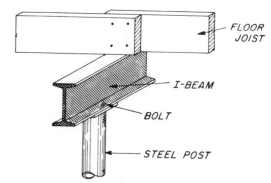

24-9. This method of running the floor joists over the girder usually requires 2" × 4" spacing blocks between the joists to prevent movement on the beam.

24-8. When the top of the girder is not set even with the top of the sill plate, it is necessary to install a ledger strip and notch the joists at the girder.

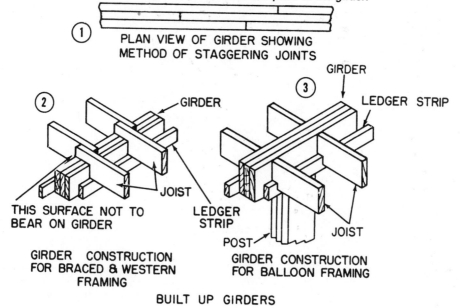

① PLAN VIEW OF GIRDER SHOWING METHOD OF STAGGERING JOINTS

② GIRDER CONSTRUCTION FOR BRACED & WESTERN FRAMING

THIS SURFACE NOT TO BEAR ON GIRDER

③ GIRDER CONSTRUCTION FOR BALLOON FRAMING

BUILT UP GIRDERS

nuts. This method provides a good bearing for the sill, and it also prevents leakage of air between the sill and the foundation wall.

Installing Sills or Plates on the Foundation

Establish the building line points at each of the corners of the foundation. Pull a chalk line very tight at these established points and snap a line for the location of the sill. Square up the ends of the sill stock. Then place the sill on edge and mark the location of the anchor bolts. With a square, extend these marks across the width of the sill. The distance "X" in Fig. 24-20a (Page 308) shows how far from the edge of the sill to bore the holes. Locate the midpoints between the lines representing the bolt locations and bore holes. If the sheathing is to rest on the foundation walls,

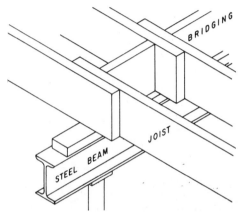

24–10. *This is the method used most frequently to install framing on a steel beam. Fasten a wood sill plate to the beam so that the joists can be nailed to the plate. Note the solid bridging which holds the joist ends vertical.*

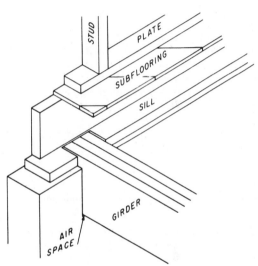

24–12a. *First-floor framing at the girder and exterior wall in platform-frame construction. Note that the girder is flush with the top of the sill.*

24–11. *A method of notching the joist into a steel girder.*

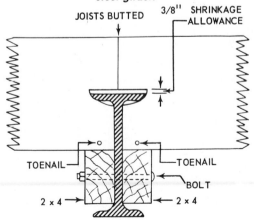

24–12b. *Another view of the construction shown in Fig. 24–12a. Note the placement of the joist.*

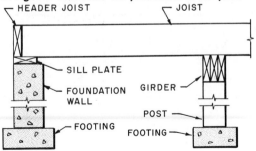

allow for this by subtracting the sheathing thickness from the distance "X". Locate and bore holes as previously explained. If insulation and termite shields are used, bore holes at the same locations in the insulation and shields. Place the insulation and shields over the anchor bolts. Fig.

24-20b. Then place the sill on top. Start at the high point of the foundation and check the sill for levelness. Shim up as necessary. Add a washer and then tighten the nut to the sill. Apply grout to any openings between sill and foundation.

In hillside construction where there is a step foundation, short sills are placed on each of the steps and then a longer sill is placed in the highest position for the entire

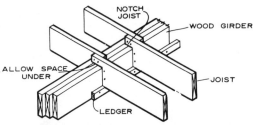

24–13a. *More headroom in the basement is gained by notching the joists and using a ledger strip for support on the girder.*

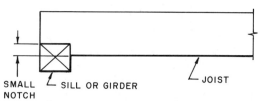

24–13b. *When using the method shown in Fig. 24–13a, do not notch the joist more than one-third of its depth.*

24–14. *Spaced girder. The space is used for running plumbing or heating pipes. A piece of wood called a scab is used to span the opening above the girder.*

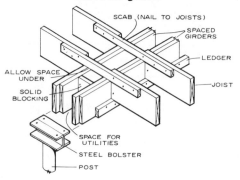

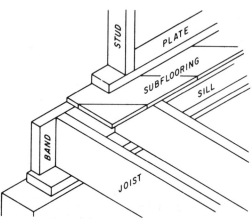

24–15. *First-floor framing at the exterior wall in platform-frame construction.*

24–16. *First-floor framing at the exterior wall in balloon-frame construction.*

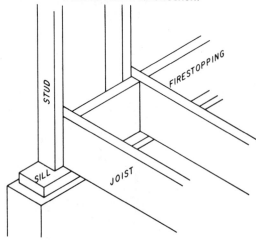

length of the building. Cripple studs are then set onto the short sills and cut to a length which will support the longer sill in a level position to carry the floor joists.

Floor Joists for Platform Construction

Check the house plans to determine the size and direction of the joists. Fig. 24-21a & b. On the sill or wall plate lay out the desired joist spacing. Fig. 24-21c (Page 310). If there is to be a double joist under the outside wall, place a ³/₄″ spacer block between the first and second joists. All other joists should be 16″ on center (unless the plan calls for 24″ on center).

Lay out a double joist under each cross partition. (A *cross partition* is one which runs parallel to the floor joists.) Fig. 24-22. If the cross partition is to be used for plumbing or heating pipes or ducts, place a solid spacer block between the double joists to allow pipes to run between them. Fig. 24-14. Transfer the joist spacing

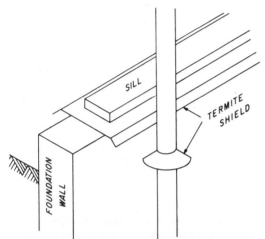

24–17. *Termite shields should be not less than 26-gauge galvanized iron, aluminum, or copper. They should be installed on top of all foundation walls and piers, and around pipes. The outer edges should be bent down slightly.*

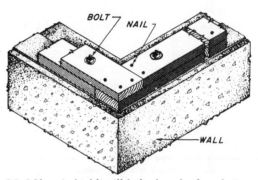

24–18b. *A double sill bolted to the foundation. The top sill is face-nailed to the bottom sill. Note the lap at the corner of the two sills.*

24–18a. *A single sill bolted to the foundation wall. Note the two pieces joined by toenailing at the corner with two 10d nails.*

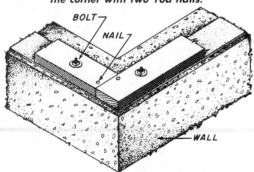

24–19. *The sill plate is anchored to the foundation wall with anchor bolts.*

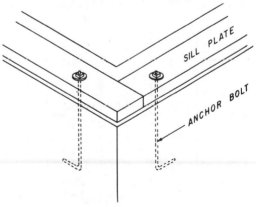

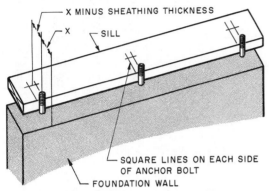

X MINUS SHEATHING THICKNESS

X

SILL

SQUARE LINES ON EACH SIDE
OF ANCHOR BOLT

FOUNDATION WALL

24-20a. *Laying out the location of the bolt holes on the sill.*

24-20b. *Roll out a strip of fiberglass sill sealer on the foundation wall just before laying the sill plate. The insulation will compress under the weight of the building. This fills irregularities and helps keep out dirt. It also keeps out drafts and reduces heat loss.*

marks from the sill onto a story pole and set it aside for later use. Place the header joist on edge, making sure it is aligned with the outside edge of the sill. Toenail the header joist in place, nailing as required by code. Now hold the outside joist in a 90-degree vertical position to the header joist along the outside edge of the sill. Spike through the header joist into the end of the outside joist to form the corner. Fig. 24-23.

Any joists having a slight bow edgewise should be so placed that the crown is on top. A crowned joist will tend to straighten out when subfloor and normal floor loads

24-21a. *The group classifications in this table refer to the species and minimum grades of nonstress-graded lumber. See Fig. 24-21b. This table was taken from the Uniform Building Code Manual. If more specific information is needed about design, loading, and deflection, refer to Vol. I of the Manual.*

Allowable Spans For Floor Joists Using Nonstress-Graded Lumber

Size Of Floor Joists (Inches)	Spacing Of Floor Joists (Inches)	Maximum Allowable Span (Feet and Inches)							
		Group I		Group II		Group III		Group IV	
		Plastered Ceiling Below	Without Plastered Ceiling Below	Plastered Ceiling Below	Without Plastered Ceiling Below	Plastered Ceiling Below	Without Plastered Ceiling Below	Plastered Ceiling Below	Without Plastered Ceiling Below
2 × 6	12	10-6	11-6	9-0	10-0	7-6	8-0	5-6	6-0
	16	9-6	10-0	8-0	8-6	6-6	7-0	5-0	5-0
	24	7-6	8-0	6-6	7-0	5-6	6-0	4-0	4-0
2 × 8	12	14-0	15-0	12-6	13-6	10-6	11-6	8-0	8-6
	16	12-6	13-6	11-0	11-6	9-0	10-0	7-0	7-6
	24	10-0	11-0	9-0	9-6	7-6	8-0	6-0	6-6
2 × 10	12	17-6	19-0	16-6	17-6	13-6	14-6	10-6	11-6
	16	15-6	16-6	14-6	15-6	12-0	13-0	9-6	10-0
	24	13-0	14-0	12-0	13-0	10-0	10-6	7-6	8-6
2 × 12	12	21-0	23-0	21-0	21-6	17-6	19-0	13-6	14-6
	16	18-0	20-0	18-0	19-6	15-6	16-6	12-0	13-0
	24	15-0	16-6	15-0	16-6	12-6	13-6	10-0	10-6

are applied. The largest edge knots should be placed on top since knots on the upper side of a joist are on the compression side of the member. Fig. 24-24. Place the joists on the sills and securely nail through the header into the ends of the joists. Now lay the story pole (made earlier) across the tops of the joists, near the center and at right angles to the joists.

After the story pole is laid parallel to the joist header, nail it to each of the joists, using the spacing marked to obtain the correct spacing of the joists. Leave this strip in place until the subfloor is laid.

After nailing through the header joist (band) into the ends of the joists, pull a line

tight along the top edge to check it for alignment. When the band has been properly aligned, nail the other ends of the joists to the girder and to the joists from the other side of the building. Lay out and frame the floor openings and install the bridging as required. The subfloor may then be laid and nailed in place.

Framing Floor Openings

When framing for large openings such as stairwells, fireplaces, and chimneys, the joists and headers framing the opening should be doubled. Fig. 24-25 (Page 312). The proper method of nailing is also shown in Fig. 24-25. Place the first trimmer joist in

24–21b.
Group Classification—Nonstress-Graded Lumber

Species	Minimum Grade	Uniform Building Code Standard Number
Group I		
Douglas Fir & Larch[1]	Construction	{25-3, 25-4}
Group II		
Bald Cypress (Tidewater Red Cypress)	No. 2	25-2
Douglas Fir (South)[1]	Construction	25-4
Fir, White	Construction	{25-3, 25-4}
Hemlock, Eastern	No. 1	25-5
Hemlock, West Coast & Western[1]	Construction	{25-3, 25-4}
Pine, Red (Norway Pine)	No. 1	25-5
Redwood, California	Select Heart	25-7
Spruce, Eastern	No. 1	25-8
Spruce, Sitka	Construction	25-3
Spruce, White and Western White	Construction	25-4[2]
Group III		
Cedar, Western	Construction West Coast Studs	25-3
Cedar, Western Red and Incense	Construction	25-4
Douglas Fir & Larch[1]	Standard West Coast Studs	{25-3, 25-4}
Douglas Fir (South)[1]	Standard	25-4
Fir, Balsam	No. 1	25-8
Fir, White	Standard West Coast Studs	{25-3, 25-4}

Species	Minimum Grade	Uniform Building Code Standard Number
Group III (Continued)		
Hemlock, Eastern	No. 2	25-5
Hemlock, West Coast & Western[1]	Standard West Coast Studs	{25-3, 25-4}
Pine, Ponderosa, Lodgepole, Sugar, Idaho White	Construction	25-4
Redwood, California	Construction	25-7
Redwood, California (studs only)	Two Star	25-7
Spruce, Engelmann	Construction	25-4
Spruce, Sitka	Standard West Coast Studs	25-3
Spruce, White and Western White	Standard	25-4[2]
Group IV [See Section 2501 (e) Uniform Bldg. Code]		
Cedar, Western	Utility	25-3
Cedar, Western Red and Incense	Utility	25-4
Douglas Fir & Larch	Utility	{25-3, 25-4}
Douglas Fir (South)	Utility	25-4
Fir, White	Utility	{25-3, 25-4}
Hemlock, West Coast & Western	Utility	{25-3, 25-4}
Pine, Ponderosa, Lodgepole, Sugar, Idaho White	Utility	25-4
Redwood, California	Merchantable	25-7
Redwood, California (studs only)	One Star	25-7
Spruce, Engelmann	Utility	25-4
Spruce, Sitka	Utility	25-3
Spruce, White and Western White	Utility	25-4[2]

[1]Two-inch by 4-inch only.
[2]Spruce (White and Western White) shall be graded under the requirements of Section 25.409 of U.B.C. Standard No. 25-4.

position. Lay out on this joist the location of the double header. Cut the first header to length and nail it in position. Lay out on this header the position of the tail beams, using regular spacing.

Place the tail beams in position and nail through the header into these joists, using three 20d nails in the end of each joist. Cut and place the second part of the double header in position, nailing it to the first header. Also nail through the first trimmer joist into the end of the second header. Now add the second trimmer joist and nail the two trimmer joists together, keeping the top edges even. Joist hangers and short sections of angle iron are often used as joist supports for the larger openings. Fig.

24-26. For further details of stairwells see "Stairs," Unit 45.

Floor Framing at Bay-Window Projections

The framing for a bay window or similar projection should be arranged so that the floor joists extend beyond the foundation wall. This allows them to carry the necessary loads. Fig. 24-27. This extension should normally not exceed 2 feet. The joists forming each side of the bay and the header for the bay should be doubled. Nailing, in general, should conform to that for floor openings. The subflooring is extended to the outer framing member and sawed flush with that member. Ceiling

24–21c. Notice that the 16″ spacing is measured from the outside edge of the first joist to the center of the second joist and then to the centers of the other joists.

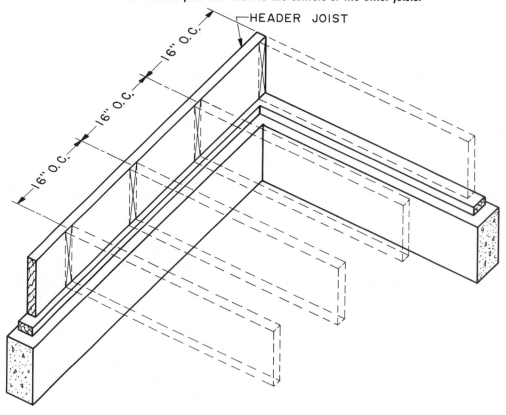

HEADER JOIST

16″ O.C.

16″ O.C.

16″ O.C.

joists should be carried by a header framed over the window opening in the projected part of the structure. Fig. 24-28 (Page 314).

Bridging

When joists are placed over a long span, they have a tendency to sway from side to side. To help solve this problem, a bracing method called *bridging* is commonly used.

Floor frames are bridged in order to:
● Stiffen the floor frames.
● Prevent unequal deflection (bending) of the joists.
● Enable an overloaded joist to receive some assistance from the joists on either side of it.

Bridging is of two kinds, horizontal (solid) bridging, Fig. 24-29, and cross (diagonal) bridging. Cross bridging is more generally used since it is very effective and requires less material. Lumber 1" × 3" or 2" × 2" is usually used for cross bridging. Fig. 24-30a. Rigid metal cross bridging with nailing flanges may also be used. Fig. 24-30b. If the joists are over 8' long, one row of bridging is installed at the center of the joist span. For joists 16' and longer, install two rows of bridging equally spaced on the joist span.

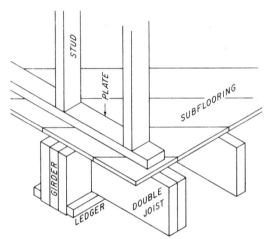

24-22. *Floor joists are doubled under nonbearing partitions.*

24-23. *Box-sill construction. The header joist is nailed to the other joists with 20d nails. Three nails are driven into the end joist and two into the others. The outside joist and header joist are toenailed to the sill on 16" centers.*

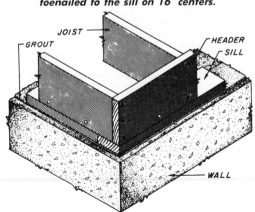

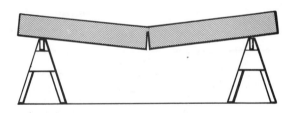

SAW CUT FROM BOTTOM OF BOARD OPENS UP.

24-24a. *With a saw cut at the bottom, the same as with a knot or other defect, the board will open up and break.*

24-24b. *A saw cut at the top of the board will close up (compress). Thus the board will retain more strength than the example in Fig. 24-24a.*

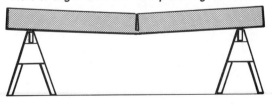

SAW CUT AT TOP OF BOARD BINDS ON THE SAW.

CUTTING AND INSTALLING DIAGONAL BRIDGING

Use the framing square to lay out diagonal bridging. The tongue of the square represents the width of the joists; the blade represents the space between the joists. Place a piece of bridging stock across the square as shown in Fig. 24-31 (Page 316), and mark the angle to be cut along the

outside edge of the tongue. Cut the piece off at the marked angle. Place this cut on the body of the square at the 14½″ mark (for 16″ O.C. joist spacing). Lay the other end of the piece of bridging across the 9¼″ mark (for a 10″ joist) and scribe a mark on the outside edge of the tongue. Cut the bridging to the finished length and try it between two joists for fit.

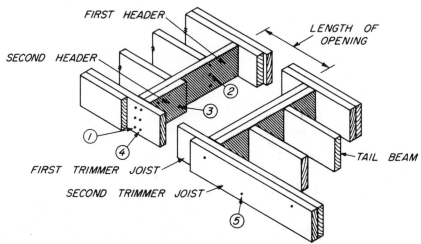

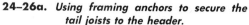

24–25. Nailing procedure for framing floor openings. 1. First trimmer is nailed to first header with three 20d nails. 2. First header is nailed to tail beams with three 20d nails. 3. Second header is nailed to first header with 16d nails spaced 6″ apart. 4. First trimmer is nailed to second header with three 20d nails. 5. Second trimmer is nailed to first trimmer with 16d nails spaced 12″ apart.

24–26a. Using framing anchors to secure the tail joists to the header.

24–26b. Using a joist hanger to secure a header joist to a trimmer joist. Double trimmers and double headers are used around floor openings.

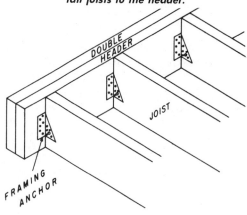

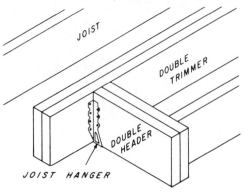

Cut the first piece for a template and use as a pattern for the other pieces. For one row of bridging, locate the center of the span and snap a chalk line across the top of the joists. Before installation, drive two 8d nails at each end of the bridging pieces. Drive the nails until the points just show through.

Start at a wall and nail one piece of bridging in position. Continue by placing one row of bridging on each side of the chalk line on the joist. Complete the nailing at the top of the bridging; however, do not nail the bottom until the subfloor has been laid. This permits the joists to adjust themselves to their final positions. The bottom ends of bridging may then be nailed, forming a continuous truss across the whole length of the floor and preventing any overloaded joist from sagging below the others.

GIRDER FLOOR FRAMING

The girder method of floor framing is widely used in warm climates where homes without basements are built. It is much faster than joist-frame construction, but requires heavier or built-up material for the girders. Fig. 24-32. The correct girder size has to be figured on the basis of the load it is to support, the span, and the type of material used. This information is found on the building plans or by checking with the local building department for maximum spans of girders. Girder spacing is usually four feet on center, with the maximum spacing of girder posts usually five feet on center. Steel girders are often used for fairly long spans.

Houses of the size and style shown in Fig. 24-33 may have girder floor framing. Such homes are built with 4" × 6" girders set on 4" × 6" posts. First, the building foundation walls and footings are poured. The locations of the piers are established, and the holes for the pier footings are dug. Fig. 24-34 (Page 318). Depending on the soil and local restrictions, these footings may require a built-up form.

The concrete is poured into the cavity provided for the footing, and the piers are set in place. Fig. 24-35. The piers should

24–27. Floor framing for a bay window. Nailing procedure: Second header is nailed to first header with 16d nails spaced 6″ apart. First header is end-nailed to each member of double-stringer joist with three 20d nails.

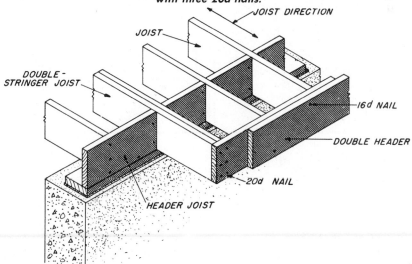

JOIST DIRECTION

JOIST

DOUBLE-STRINGER JOIST

16d NAIL

DOUBLE HEADER

20d NAIL

HEADER JOIST

be in a reasonably straight line, but the height is not critical because the posts will be cut individually to the correct length for supporting the girders. Fig. 24-36 (Page 318). The sill is then cut to size and bolted in place as described on page 305. Fig. 24-37.

The bearing posts must be cut accurately to length to provide a level floor. Pull a line tight from opposite sill plates over the piers. Make certain the line is down tight on the plates. Measure the distance from the line to the top of the pier cap or redwood block on top of the concrete pier. Record this distance, usually on each pier cap or redwood block. Repeat this operation for each line of piers until the height of each bearing post has been determined and recorded. Square one end and cut the bearing posts to length. Care should be taken with material of this size to cut it square with two of its adjacent surfaces.

Treat the end of the post for termites, fungus, and similar problems. Toenail the

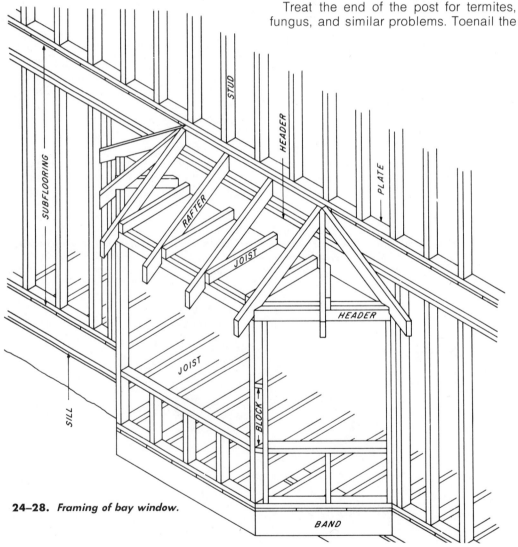

24-28. *Framing of bay window.*

treated end to the pier cap with two 8d nails on each side (a total of 8 nails). Fig. 24-38. Next square one end of the girder, cut it to length, and toenail it to the posts. Fig. 24-39. If a low house profile is desired or if the finished floor is to have a step-down area, the tops of the girders in this area are set flush with the top of the sill. When this is done, a special metal hanger must be installed to support the girder end. Fig. 24-40. Two other methods of setting the top of the girder flush with the top of the plate are shown in Fig. 24-41 (Page 320).

When there is a step (two levels) in the floor, the ends of the girders must be headed off to support the subfloor. A 2″ × 6″ is used with 4″ × 6″ girders. Fig. 24-42.

In framing for a fireplace or other openings in the floor, the tail beams and headers are of the same structural material as the girders. Fig. 24-43.

Small openings for heating or air conditioning ducts will require only 2″ × 4″ boards laid flat. Fig. 24-44. Because working space under the girders is small, the

plumbing and heating are "roughed in" before the floor is applied. Fig. 24-45. The subfloor is then cut and nailed in place. The surface is now ready for layout and erection of the sidewalls. Fig. 24-46.

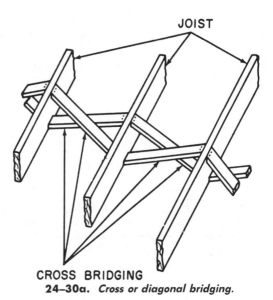

JOIST

CROSS BRIDGING

24–30a. *Cross or diagonal bridging.*

24–30b. *Metal diagonal bridging.*

24–29. *Horizontal or solid bridging. Nailing is easier if the bridging is offset 1 5/8″. Nails can then be driven directly through the joist into the end of each piece of bridging.*

JOIST

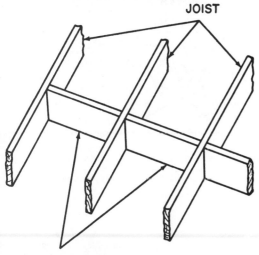

SOLID BRIDGING

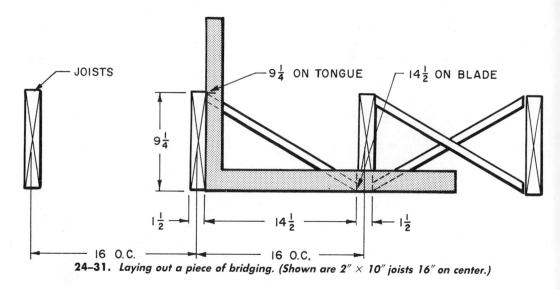

JOISTS

$9\frac{1}{4}$ ON TONGUE

$14\frac{1}{2}$ ON BLADE

$9\frac{1}{4}$

$1\frac{1}{2}$

$14\frac{1}{2}$

$1\frac{1}{2}$

16 O.C.

16 O.C.

24–31. *Laying out a piece of bridging. (Shown are 2″ × 10″ joists 16″ on center.)*

24–32. *Girder construction with box-sill framing. The 2•4•1 system is illustrated in Figs. 24–60 through 24–62. *If square-edge 2•4•1 panels are used, blocking is required at unsupported edges. **In areas of termite infestation or under conditions of adverse ground moisture, use 18″ minimum.*

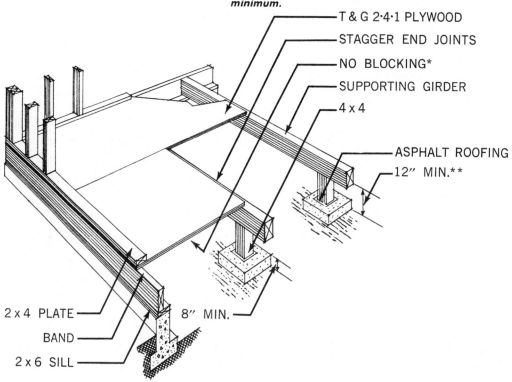

T & G 2•4•1 PLYWOOD

STAGGER END JOINTS

NO BLOCKING*

SUPPORTING GIRDER

4 x 4

ASPHALT ROOFING

12″ MIN.**

2 x 4 PLATE

BAND

2 x 6 SILL

8″ MIN.

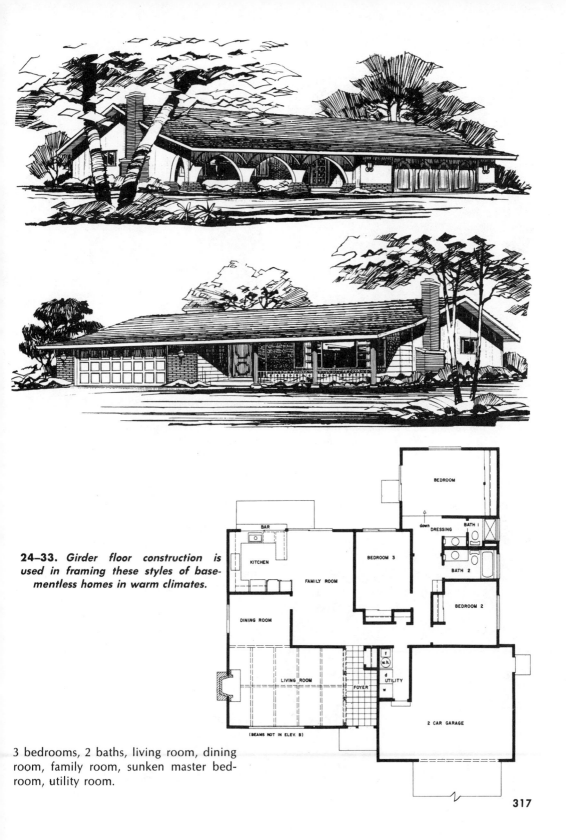

24-33. *Girder floor construction is used in framing these styles of basementless homes in warm climates.*

3 bedrooms, 2 baths, living room, dining room, family room, sunken master bedroom, utility room.

BEDROOM

down

DRESSING

BATH 1

BAR

KITCHEN

BEDROOM 3

BATH 2

FAMILY ROOM

BEDROOM 2

DINING ROOM

f
w. h.

d
UTILITY

w

LIVING ROOM

FOYER

2 CAR GARAGE

(BEAMS NOT IN ELEV. B)

24–34. *The footing hole is dug and the precast concrete pier is ready for setting onto the footing.*

24–36. *The foundation wall with the anchor bolts set for the sill. The piers are ready for the posts.*

24–35. *The piers are set in place on the footings.*

ESTIMATING FLOOR FRAMING

Determining the Number of Joists

The number of joists necessary for a building may be determined by dividing the length of the floor (in feet) by the joist spacing (in feet), and adding one for the end joist.

Conventional joist spacing is 16″ (1$\frac{1}{3}$′) on center. Dividing by 1$\frac{1}{3}$ is the same as multiplying by $\frac{3}{4}$. Therefore for joists 16″ on center, simply take $\frac{3}{4}$ the length of the building and add one.

For example, consider a building that is 40′ long. Multiply $\frac{3}{4} \times 40$. The answer is 30. Add one, for a total of 31 joists.

Note, however, that the joists may not extend from wall to wall. Let's say that the building in our example is 20′ wide and that we are using 10′ joists. The joists will

24–37. *The sill is bolted in place. The girders are laid and ready for installation.*

24–39. *The girders are nailed to the posts. The step in the girders is for a step-down area of the building.*

24–38. *The posts are cut to the correct height and toenailed to the wood pad which is set into the precast pier.*

24–40. *When the top of the girder is to be even with the top of the sill, a metal hanger is used to support the girder.*

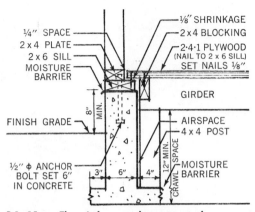

¼" SPACE
2 x 4 PLATE
2 x 6 SILL
MOISTURE BARRIER

⅛" SHRINKAGE
2 x 4 BLOCKING
2·4·1 PLYWOOD (NAIL TO 2 x 6 SILL) SET NAILS ⅛"

GIRDER

FINISH GRADE

8" MIN.

AIRSPACE
4 x 4 POST

½" φ ANCHOR BOLT SET 6" IN CONCRETE

3" 6" 4"

12" MIN.
CRAWL SPACE

MOISTURE BARRIER

24–41a. *The girder may be supported on a post set on the footing.*

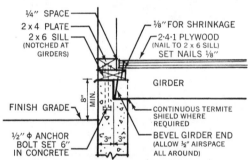

¼" SPACE
2 x 4 PLATE
2 x 6 SILL (NOTCHED AT GIRDERS)

⅛" FOR SHRINKAGE
2·4·1 PLYWOOD (NAIL TO 2 x 6 SILL) SET NAILS ⅛"

GIRDER

FINISH GRADE

8" MIN.

CONTINUOUS TERMITE SHIELD WHERE REQUIRED

½" φ ANCHOR BOLT SET 6" IN CONCRETE

3" 3"

BEVEL GIRDER END (ALLOW ½" AIRSPACE ALL AROUND)

24–41b. *The girder may be supported by a pocket in the foundation.*

24–42. *The girder ends are boxed in for the step in the floor.*

extend only from one wall to a center girder. Another 31 joists will be needed to cover the span from the girder to the opposite wall. Thus a basic total of 62 joists would be needed to span the complete floor area of the building. Also one extra joist must be added for each partition for which double joists are specified.

The table in Fig. 24-47 may also be used to determine the number of joists. In the column headed "Length of Span" find the length of the building (40' in our example). Read across to the spacing of the joists in the example (16") and you find the number of joists required (31). Again, this will have to be doubled if the joists extend only to a center girder, and extra joists must be added for partitions, as specified.

Determining the Material Cost

An accurate cost estimate can be figured by multiplying the number of joists required by the cost per joist. Sometimes, however, the builder will not make up a complete bill of materials but will want only a rough cost estimate. This can be figured without knowing the exact number

24–43. *The area in which the fireplace is to be located must be "headed off." Note the reinforcing rod which has been set into the fireplace footing.*

of pieces needed. First find the area of the floor by multiplying the length times the width of the building *for each level*. For example, a one-story building 20′ wide and 40′ long has a floor area of 800 sq. ft. (20 × 40 = 800).

The number of board feet for joists required for a building can be found by using the chart in Fig. 24-48. Again use the example of a building with 800 sq. ft. of floor area. According to the chart, if the joists are 2″ × 6″ and 16″ on center, there are 102 board feet of lumber for each 100 square feet of floor surface area. Divide the total floor area by 100, and multiply by the number of board feet which you learned from the chart. The answer to the example problem is 816 board feet (800 ÷ 100 = 8; 8 × 102 = 816).

By multiplying the cost per board foot of lumber by the number of board feet required, you can obtain a rough cost estimate.

24-44. *The heating and air conditioning ducts are installed. They are wrapped with fiberglass insulation, and the joints are taped. Notice that the 2″ × 4″ boards which support the duct will also support the flooring around the cutout for the register.*

24-45. *Plumbing is roughed in after the girders are in place.*

24-46. *Subfloor with roughed-in plumbing ready for wall layout and fabrication. An interior wall will be framed in along the line of the plumbing. All the plumbing pipes shown here will then be totally enclosed within the wall.*

24–47.

Number Of Wood Joists Required For Any Floor And Spacing

Length	Spacing of Joists									
of Span	12″	16″	20″	24″	30″	36″	42″	48″	54″	60″
6	7	6	5	4	3	3	3	3	2	2
7	8	6	5	5	4	4	3	3	3	2
8	9	7	6	5	4	4	3	3	3	3
9	10	8	6	6	5	4	4	3	3	3
10	11	9	7	6	5	4	4	4	3	3
11	12	9	8	7	5	5	4	4	3	3
12	13	10	8	7	6	5	4	4	4	3
13	14	11	9	8	6	5	5	4	4	4
14	15	12	9	8	7	6	5	5	4	4
15	16	12	10	9	7	6	5	5	4	4
16	17	13	11	9	7	6	6	5	5	4
17	18	14	11	10	8	7	6	5	5	4
18	19	15	12	10	8	7	6	6	5	4
19	20	15	12	11	9	7	6	6	5	5
20	21	16	13	11	9	8	7	6	5	5
21	22	17	14	12	9	8	7	6	6	5
22	23	18	14	12	10	8	7	7	6	5
23	24	18	15	13	10	9	8	7	6	6
24	25	19	15	13	11	9	8	7	6	6
25	26	20	16	14	11	9	8	7	7	6
26	27	21	17	14	11	10	8	8	7	6
27	28	21	17	15	12	10	9	8	7	6
28	29	22	18	15	12	10	9	8	7	7
29	30	23	18	16	13	11	9	8	7	7
30	31	24	19	16	13	11	10	9	8	7
31	32	24	20	17	13	11	10	9	8	7
32	33	25	20	17	14	12	10	9	8	7
33	34	26	21	18	14	12	10	9	8	8
34	35	27	21	18	15	12	11	10	9	8
35	36	27	22	19	15	13	11	10	9	8
36	37	28	23	19	15	13	11	10	9	8
37	38	29	23	20	16	13	12	10	9	8
38	39	30	24	20	16	14	12	11	9	9
39	40	30	24	21	17	14	12	11	10	9
40	41	31	25	21	17	14	12	11	10	9

One joist has been added to each of the above quantities to take care of extra joist required at end of span.
Add for doubling joists under all partitions.

Fig. 24-48 also has information for determining the number of nails necessary. For the floor in the example, the chart shows that 10 pounds of nails are needed for each 1000 board feet. The floor, you will recall, has only 800 board feet; so it will require about eight pounds of nails. Multiply the pounds needed by the cost per pound to find the total cost of the nails.

Determining the Labor Cost

The labor cost for framing a floor can be found by using Figs. 24-48 & 24-49. First you must know the joist size. Usually this can be learned from the building plan. In the example we have been using, the joists are 2″ × 6″ × 10′. Refer to Fig. 24-49, which indicates the number of board feet in lumber of standard sizes. The chart shows that 2″ × 6″ × 10′ boards have 10 board feet. For a building with 62 joists, as in our example, there would be a total of 620 board feet of joists (10 bd. ft. × 62 = 620 bd. ft.).

Now you can refer to Fig. 24-48 to find the labor cost. The last column shows that one worker in one hour can frame 65 board feet of 2″ × 6″ joist material. To find the total man-hours needed, divide the total board feet by the number framed in one hour. The answer is 9.5 man-hours (620 ÷ 65 = 9.5). Next multiply the man-hours by the hourly rate.

SUBFLOORING IN JOIST CONSTRUCTION

A subfloor is a wood floor laid over the floor joists, under the finished floor. Ordinarily a subfloor is nailed directly to the floor joists. In conventional joist construction, sound subflooring is virtually a "must". Most modern building codes specify subfloors. Their omission usually is poor economy, even if the finish floor is of strong, durable oak.

Purposes Served by Subflooring

Subflooring · serves several important purposes. It lends bracing strength to the building. It provides a solid base for the finish floor, making floor sag and squeaks very unlikely. By acting as a barrier to cold and dampness, subflooring helps keep the building warmer and drier in winter. In addition, it provides a safe working surface for building the house.

Selection, Nailing of Subfloor Boards

If strip flooring is to be used for the finish floor, it is generally recommended that the subflooring consist either of softwood plywood or of good quality boards about 3/4″ thick and not more than 6″ wide. Wider boards are likely to expand and contract too much.

Square-edge boards generally are preferred to tongue-and-groove boards. This is because snug joining usually is not desirable in subflooring. This is particularly true of houses in moist climates and summer homes which are not heated in the winter. Square-edge boards are more economical too. Boards should be thoroughly dry. The use of green subfloor boards frequently causes squeaks and cracks.

Subfloor boards may be applied diagonally or at right angles to the floor joists. When subflooring is at right angles to the joists, the finish floor should then be laid at right angles to the subfloor. However, it is best to lay the subflooring diagonally. This arrangement permits the finish strip flooring to be laid in any direction. For a home of two or more stories, it is best to have the subfloor boards run in opposite diagonal directions on alternate floors. Diagonal

subflooring also provides better bracing and stiffness to the building.

Laying a Diagonal Subfloor

When laying diagonal subfloor, lay a relatively long board first. The short cuttings may then be used as you approach a corner and the length decreases. Work can then proceed on both sides of this first diagonal length.

To lay out the first piece, begin at a corner of the building (B in Fig. 24-50) and measure equal distances along the header joist and the first joist (points A and C). Snap a chalk line along the top of the joists between these two points to form a 45° angle. Lay the first board along this line, cut it to fit, and nail it at each joist with two 8d nails, not more than 3/4″ from the edge.

Often two or more pieces of subfloor board will have to be used to make up one diagonal strip. Ends of these pieces are nailed so they butt together on the top of a joist, forming a joint. Do not allow two of these joints to come side-by-side on the same joist; there should be at least two boards between joints. Generally, the shortest pieces of board should be long enough to span three joists—that is, start from one joist, cross another, and be fastened to a third. Of course, shorter pieces will have to be used in the corners. Fig. 24-51. Many carpenters will use an 8d nail to space between the edges of the subfloor boards. This allows for drainage and buckling which can be caused by swelling if the floor gets wet during construction.

24–48.

Floor Joist	Material				Nails	Labor
	Board Feet Required for 100 Sq. Ft. of Surface Area				Per 1000 Bd. Ft.	Board Feet per Hr.
Size Of Joist	12″ O.C.	16″ O.C.	20″ O.C.	24″ O.C.	Pounds	
2 × 6	128	102	88	78	10	65
2 × 8	171	136	117	103	8	65
2 × 10	214	171	148	130	6	70
2 × 12	256	205	177	156	5	70

24–49.

Board Feet Content

Size of Timber in Inches	Length Of Piece In Feet							
	10	12	14	16	18	20	22	24
1 × 2	1⅔	2	2⅓	2⅔	3	3⅓	3⅔	4
1 × 3	2½	3	3½	4	4½	5	5½	6
1 × 4	3⅓	4	4⅔	5⅓	6	6⅔	7⅓	8
1 × 5	4⅙	5	5⅚	6⅔	7½	8⅓	9⅙	10
1 × 6	5	6	7	8	9	10	11	12
1 × 8	6⅔	8	9⅓	10⅔	12	13⅓	14⅔	16
1 × 10	8⅓	10	11⅔	13⅓	15	16⅔	18⅓	20
1 × 12	10	12	14	16	18	20	22	24
1 × 14	11⅔	14	16⅓	18⅔	21	23⅓	25⅔	28
1 × 16	13⅓	16	18⅔	21⅓	24	26⅔	29⅓	32
1 × 20	16⅔	20	23⅓	26⅔	30	33⅓	36⅔	40
1¼ × 4	4⅙	5	5⅚	6⅔	7½	8⅓	9⅙	10
1¼ × 6	6¼	7½	8¾	10	11¼	12½	13¾	15
1¼ × 8	8⅓	10	11⅔	13⅓	15	16⅔	18⅓	20
1¼ × 10	10⅓	12½	14½	16⅔	18⅔	20⅚	22⅚	25
1¼ × 12	12½	15	17½	20	22½	25	27½	30
1½ × 4	5	6	7	8	9	10	11	12
1½ × 6	7½	9	10½	12	13½	15	16½	18
1½ × 8	10	12	14	16	18	20	22	24
1½ × 10	12½	15	17½	20	22½	25	27½	30
1½ × 12	15	18	21	24	27	30	33	36
2 × 4	6⅔	8	9⅓	10⅔	12	13⅓	14⅔	16
2 × 6	10	12	14	16	18	20	22	24
2 × 8	13⅓	16	18⅔	21⅓	24	26⅔	29⅓	32
2 × 10	16⅔	20	23⅓	26⅔	30	33⅓	36⅔	40
2 × 12	20	24	28	32	36	40	44	48
2 × 14	23⅓	28	32⅔	37⅓	42	46⅔	51⅓	56
2 × 16	26⅔	32	37½	42⅔	48	53⅓	58⅔	64
2½ × 12	25	30	35	40	45	50	55	60
2½ × 14	29⅙	35	40⅚	46⅔	52½	58⅓	64⅙	70
2½ × 16	33⅓	40	46⅔	53⅓	60	66⅔	73⅓	80
3 × 6	15	18	21	24	27	30	33	36
3 × 8	20	24	28	32	36	40	44	48
3 × 10	25	30	35	40	45	50	55	60
3 × 12	30	36	42	48	54	60	66	72
3 × 14	35	42	49	56	63	70	77	84
3 × 16	40	48	56	64	72	80	88	96
4 × 4	13⅓	16	18⅔	21⅓	24	26⅔	29⅓	32
4 × 6	20	24	28	32	36	40	44	48
4 × 8	26⅔	32	17⅓	42⅔	48	53⅓	58⅔	64
4 × 10	33⅓	40	46⅔	53⅓	60	66⅔	73⅓	80
4 × 12	40	48	56	64	72	80	88	96
4 × 14	46⅓	56	65⅓	74⅔	84	93⅓	102½	112

Tongue-and-groove boards should have holes at intervals to allow rainwater to drain off.

24–50. *Laying out diagonal subfloor.*

SAVE
SHORT
ENDS FOR
CORNERS

A

B

C

Laying a Straight Subfloor

Check the plans to determine the way the finish floor is to be laid in relation to the direction of the floor joists. If the finish floor is to run parallel to the joists, a straight subfloor may be laid. Begin by laying the first piece of subfloor at one end or edge of the building, at right angles to the joists. Make the joints over the centers of the joists. Fig. 24-52. Laying and nailing are then done generally as for a diagonal subfloor.

Laying a 2″ Tongue-and-Groove Subfloor

Some local building codes permit the use of 2″ tongue-and-groove (T & G) subfloor over girder floor framing. This eliminates floor joists. When the girder is set into a pocket or hung with a bracket, care should be taken to insure that the top of the girder is even with the top of the sill plate. Fig. 24-53. Sometimes the girders are set on

24–51. *Floor framing nailing procedure. 1. Bridging (1″ × 3″) nailed at top and bottom with 8d nails. 2. Subfloor board nailed with two or three 8d nails (plywood subfloor also shown as alternative). 3. Header joist end-nailed to corner joists and intermediate joists with three 20d nails. 4. Header joist toenailed to sill with 10d nails 16″ on center.*

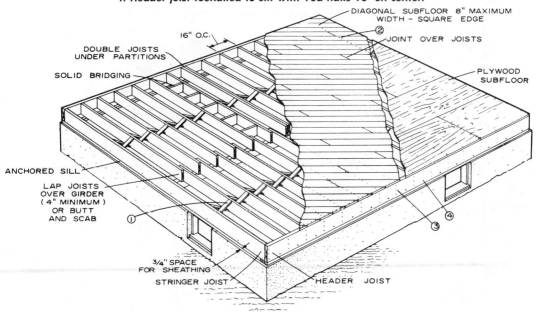

DIAGONAL SUBFLOOR 8″ MAXIMUM WIDTH – SQUARE EDGE
JOINT OVER JOISTS
16″ O.C.
DOUBLE JOISTS UNDER PARTITIONS
SOLID BRIDGING
PLYWOOD SUBFLOOR
ANCHORED SILL
LAP JOISTS OVER GIRDER (4″ MINIMUM) OR BUTT AND SCAB
3/4″ SPACE FOR SHEATHING
STRINGER JOIST
HEADER JOIST

top of the sill, giving the building a high profile; the tongue-and-groove flooring is cut even with the outside of the framing and nailed on top of the girders. Fig. 24-54. Use 16d nails to toenail at the tongue and to face-nail at a joint on all girders. This type of construction requires vents in the foundation.

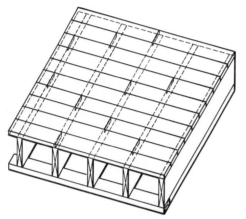

24–52. *Straight subfloor. Joints (in color) are over the joists.*

24–53. *These girders are set flush with the sill in preparation for the subfloor. Notice the ventilation openings in the foundation wall. These openings are needed in this method of framing.*

Laying a Plywood Subfloor

Plywood produces a smooth, solid, stable base for any kind of finish flooring. Big panels cover large areas fast. This reduces the laying time. The correct thickness for the various applications and the nailing requirements can be found in Fig. 24-55.

Plywood can be used as a combined subfloor and underlay for such flooring as tile, carpeting, or linoleum. Either tongue-and-groove or square-edge plywood may be chosen for this combined application. However, if square-edge plywood is used, the joints must be blocked. Fig. 24-56. When a separate underlayment is put over the subfloor, care should be taken to stagger the joints to provide adequate support. This method will not require blocking under the joints. Fig. 24-57. Plywood subflooring under strip flooring does not require blocking and permits the strip flooring to be laid in either direction. Fig. 24-58.

Begin laying the plywood subfloor by placing a full sheet even with one of the outside corners of the joist framing. The grain of the plywood should run at right angles to the joists. Drive just enough nails

24–54. *Girder construction with 2″ × 6″ tongue-and-groove flooring.*

to hold the panel in place. Place the next full panel in position at the end of the first panel. Be sure the joint is centered over the joist, and leave about $1/32''$ space between panels.

Begin the second row of panels at the end of the building, alongside the first panel laid. Cut a panel in half, lay the end flush with the outside of the building, and nail the half panel to the joists. If the joists are running at right angles to the grain direction of the plywood panel, measure and cut the panel so that the joint will be on the fourth joist. Fig. 24-58. Continue to lay and nail panels in this row. The next (third) row of panels is started with a full panel. This will stagger the joints and provide the strongest floor. Continue to lay, driving just enough nails in each panel to hold it in position until all panels are laid. Then complete the nailing as required. Fig. 24-59.

TONGUE-AND-GROOVE PLYWOOD FLOORS

As mentioned earlier, some panels are made for use as a combined subfloor and underlayment. They are made of interior-type plywood, tongue-and-groove, $1^1/8''$ thick. These panels are laid over two supporting beams placed 48" on center. Standard-size T & G panels measure 4' × 8' on the face, with additional allowance for the tongue. Fig. 24-60.

24–55. *Recommended plywood subfloor thickness and nailing requirements.*

Plywood subflooring[1][3]/For direct application of T&G wood strip and block flooring and light weight concrete.[4]

Panel Identification Index[2]	Plywood Thickness (inches)	Maximum Span[5] (inches)	Nail Size & Type	Nail Spacing (inches)	
				Panel Edges	Intermediate
$30/12$	$5/8$	12[6]	8d common	6	10
$32/16$	$1/2, 5/8$	16[7]	8d common[8]	6	10
$36/16$	$3/4$	16[7]	8d common	6	10
$42/20$	$5/8, 3/4, 7/8$	20[7]	8d common	6	10
$48/24$	$3/4, 7/8$	24	8d common	6	10
$1^1/8''$ Groups 1 & 2	$1^1/8$	48	10d common	6	6
$1^1/4''$ Groups 3 & 4	$1^1/4$	48	10d common	6	6

Notes: (1) These values apply for Structural I and II, Standard sheathing and C-C Exterior grades only.

(2) Identification Index appears on all panels except $1^1/8''$ and $1^1/4''$ panels.

(3) In some non-residential buildings, special conditions may impose heavy concentrated loads and heavy traffic requiring subfloor constructions in excess of these minimums.

(4) Edges shall be tongue and grooved or supported with blocking for square edge wood flooring, unless separate underlayment layer ($1/4''$ minimum thickness) is installed.

(5) Spans limited to values shown because of possible effect of concentrated loads. At indicated maximum spans, floor panels carrying Identification Index numbers will support uniform loads of more than 100 psf.

(6) May be 16" if $25/32''$ wood strip flooring is installed at right angles to joists.

(7) May be 24" if $25/32''$ wood strip flooring is installed at right angles to joists.

(8) 6d common nail permitted if plywood is $1/2''$.

Use plywood with these kinds of APA grade-trademarks for subfloors. See Fig. 5-9, Unit 5, for complete details.

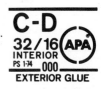

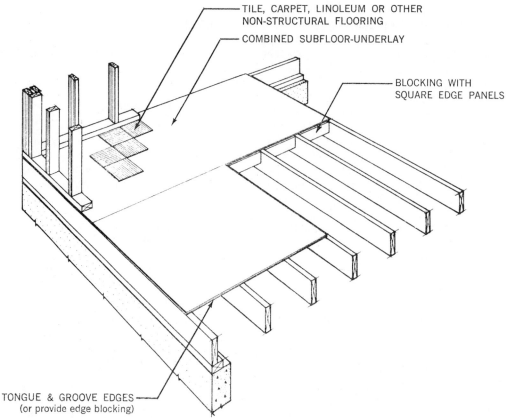

TILE, CARPET, LINOLEUM OR OTHER
NON-STRUCTURAL FLOORING

COMBINED SUBFLOOR-UNDERLAY

BLOCKING WITH
SQUARE EDGE PANELS

TONGUE & GROOVE EDGES
(or provide edge blocking)

24–56a. *A single layer of plywood may be applied to the joists as a combined subfloor underlay.*

24–56b. *Specifications for a combined subfloor underlay.*
(WSP under "Plywood Species Group" stands for Western Soft Pine.)

Plywood Species Group	Subfloor-Underlay		Nail Size (approx.) and Type (set nails $^{1}/_{16}$")	Nail Spacing (inches)	
	Min. Plywood Thickness (inches) (1)	Max. Spacing of Supports c. to c. (inches)		Panel edges	Intermediate
Douglas fir or Western larch	½	16	6d ring shank or screw type	6	10
	⅝	20	6d ring shank or screw type	6	10
	¾	24	6d ring shank or screw type	6	10
Groups 1 or 2 WSP	⅝	16	6d ring shank or screw type	6	10
	¾	24	6d ring shank or screw type	6	10

(1) In some non-residential buildings, special conditions may impose heavy concentrated loads and heavy traffic requiring subfloor-underlay constructions in excess of these minimums.

Note: For certain types of flooring such as wood block or terrazzo, sheathing grades of plywood may be used.

Typical grade-trademarks. See Fig. 5-9, Unit 5, for complete details.

UNDERLAYMENT

GROUP 2
INTERIOR
PS 1-74 **APA**
000

C-C PLUGGED

GROUP 4
EXTERIOR
PS 1-74 **APA**
000

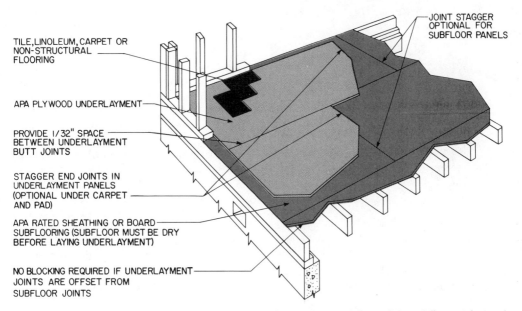

TILE, LINOLEUM, CARPET OR NON-STRUCTURAL FLOORING

APA PLYWOOD UNDERLAYMENT

PROVIDE 1/32" SPACE BETWEEN UNDERLAYMENT BUTT JOINTS

STAGGER END JOINTS IN UNDERLAYMENT PANELS (OPTIONAL UNDER CARPET AND PAD)

APA RATED SHEATHING OR BOARD SUBFLOORING (SUBFLOOR MUST BE DRY BEFORE LAYING UNDERLAYMENT)

NO BLOCKING REQUIRED IF UNDERLAYMENT JOINTS ARE OFFSET FROM SUBFLOOR JOINTS

JOINT STAGGER OPTIONAL FOR SUBFLOOR PANELS

24–57a. *A separate underlayment of plywood is placed over a subfloor. If the subfloor is plywood, the joints of the subfloor and the underlayment should be staggered.*

24–57b. *Recommended thickness, nail spacing, and related information for plywood subfloor underlay.*

Plywood Species Group	Underlayment Min. Plywood Thickness (inches)	Nail Size (approx.) and Type (set nails $^1/_{16}$")	Nail Spacing (inches)	
			Panel Edges	Intermediate
Douglas fir or Western larch; Groups 1, 2 or 3 WSP	⅜ (1) Thicker panels to match other floors	3d ring shank (also for ½" panels) (4d ring shank for ⅝" and ¾") 16 gauge staples	6 3	8 each way 6 each way

(1) FHA accepts ¼" underlayment.

These are typical grade-trademarks. See Fig. 5-9, Unit 5, for complete details.

UNDERLAYMENT
GROUP 2 **(APA)**
INTERIOR
PS 1-74 000

B-C
GROUP 3 **(APA)**
EXTERIOR
PS 1-74 000

24–57c. *Using a pneumatic stapler to attach plywood subflooring.*

Any one of several support systems may be used with this floor system if the beams are spaced up to 48″ O.C. Some suggested methods are box beams, 4″ lumber, or two 2″ joists spiked together. When the beams are set into pockets in the foundation, the plywood can bear directly on the sill. Also, conventional box sill construction may be used. Fig. 24-61. Solid lumber beams, particularly when pocketed, should be as dry as possible to minimize shrinkage.

Tongue-and-groove subfloor should be started with the tongue toward the outside of the building. Thus any pounding required to close the joints between the panels can be done on a scrap block against the groove. When the panels are nailed in place, the face grain of the panel should run across the main beams, and wherever possible, cover two openings. End joints should be staggered.

24–58a. *Plywood subflooring under 25/32″ strip flooring allows the strip flooring to be laid in either direction.*

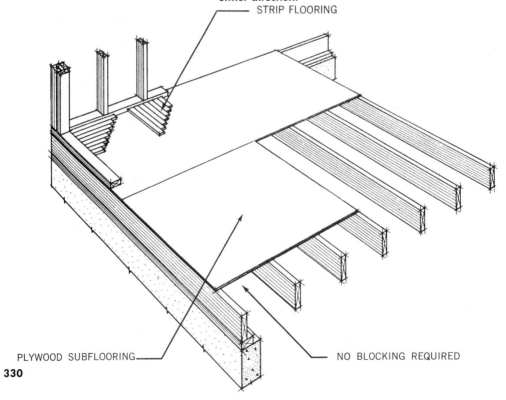

STRIP FLOORING

PLYWOOD SUBFLOORING

NO BLOCKING REQUIRED

24–58b. *Specifications for plywood subflooring used under strip flooring.*

Plywood Species Group	Subflooring		Nail Size and Type	Nail Spacing (inches)	
	Min. Plywood Thickness (inches) (3)	Max. Spacing of Supports c. to c. (inches)		Panel Edges	Intermediate
Douglas fir or Western larch; Group 1 WSP Ext. glue, and Ext. C-C only)	½ (1)	16	6d common	6	10
	⅝	20	8d common	6	10
	¾	24	8d common	6	10
Group 2 WSP	⅝ (2)	16	8d common	6	10
	¾	24	8d common	6	10

(1) If $^{25}/_{32}$" wood strips are perpendicular to supports, ½" can be used on 24" spans.
(2) If $^{25}/_{32}$" wood strips are perpendicular to supports, ⅝" can be used on 24" spans.
(3) In some non-residential buildings, special conditions may impose heavy concentrated loads and heavy traffic requiring subfloor constructions in excess of these minimums.

This is a typical grade-trademark. See Fig. 5-9, Unit 5, for complete details.

C-D PLUGGED
GROUP 2 **APA**
INTERIOR
PS 1-74 000

24–59. *The subfloor is nailed with a pneumatic nailer.*

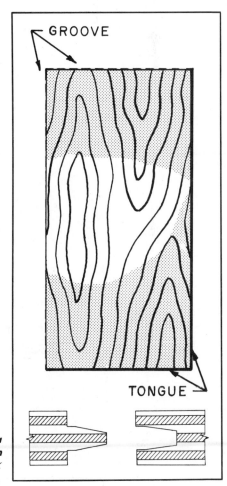

GROOVE

TONGUE

24–60. *A plywood panel 1⅛" thick with a groove on an end and an edge, and a tongue on the other end and edge. The surface is a full 4' × 8' with an additional allowance for the tongue.*

If the panels are square edged, 2 × 4 blocking is required under the edges between beams. If both the sides and ends of the panel are tongue-and-grooved, drive the side joints tight first. Then, drive end joints tight. When the floor covering is of the thin, resilient type, fill any cracks 1/16" or wider and sand the joints lightly if they are not absolutely flush.

To achieve the greatest resistance to nail popping, and the maximum in withdrawal strength, use 8d common ring-shanked or helically threaded nails. Space the nails 6" O.C. at all bearing points. However, 10d common smooth-shanked nails may be substituted if desired. Under resilient tile when the beams are not fully seasoned, set all nails 1/8" but do not fill.

24–61a. *Plywood 1 1/8" thick is often used with the girder construction system. Girders can be supported in many ways to accommodate the heavy panels.*

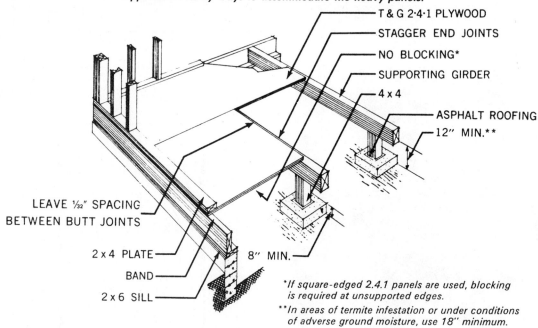

T & G 2·4·1 PLYWOOD
STAGGER END JOINTS
NO BLOCKING*
SUPPORTING GIRDER
4 x 4
ASPHALT ROOFING
12" MIN.**

LEAVE 1/32" SPACING BETWEEN BUTT JOINTS

2 x 4 PLATE
BAND
2 x 6 SILL
8" MIN.

If square-edged 2.4.1 panels are used, blocking is required at unsupported edges.

**In areas of termite infestation or under conditions of adverse ground moisture, use 18" minimum.*

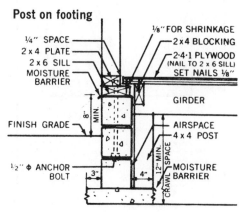

Post on footing

1/4" SPACE
2 x 4 PLATE
2 x 6 SILL
MOISTURE BARRIER

1/8" FOR SHRINKAGE
2 x 4 BLOCKING
2·4·1 PLYWOOD (NAIL TO 2 x 6 SILL) SET NAILS 1/8"

GIRDER

FINISH GRADE
8" MIN.

AIRSPACE
4 x 4 POST

1/2" Φ ANCHOR BOLT
3"
4"
12" MIN. CRAWL SPACE
MOISTURE BARRIER

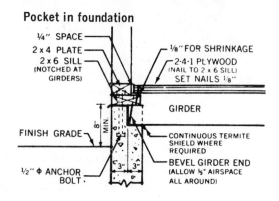

Pocket in foundation

1/4" SPACE
2 x 4 PLATE
2 x 6 SILL (NOTCHED AT GIRDERS)

1/8" FOR SHRINKAGE
2·4·1 PLYWOOD (NAIL TO 2 x 6 SILL) SET NAILS 1/8"

GIRDER

FINISH GRADE
8" MIN.

1/2" Φ ANCHOR BOLT
3" 3"

CONTINUOUS TERMITE SHIELD WHERE REQUIRED
BEVEL GIRDER END (ALLOW 1/2" AIRSPACE ALL AROUND)

(*Setting* means driving the heads below the wood surface.) Set the nails just before laying the resilient flooring. This lets you take advantage of the beam seasoning that has taken place. When the panels are tongue-and-grooved at the ends, one line of nails can be used to secure them. Drive the nails through both panels at a point near the middle of the tongue as shown in Fig. 24-62.

ESTIMATING SUBFLOORING
Material

Figure the square feet of floor area in the building. A one-story building 28 feet by 50 feet will contain 1400 square feet (28' × 50' = 1400 sq. ft.). To figure the amount of lumber required, use the chart shown in Fig. 24-63.

In the first column find the type of material to be applied, for example, S4S. Then read across to the second column for the size, for example, 1 × 8. The third column will tell you the board feet per square foot of area. In our example this is 1.15. Multiply the floor area of the building (1400) by this factor (1.15) to determine the amount of material required for the building. For the building in this example, 1610

24–61c. The 1¹/₈″ plywood may be applied over box-sill framing, as shown. It may also be laid directly on sills and girders which are supported in special brackets.

24–61b. *Nailing and related information for 1¹/₈″ subfloor-underlayment panels.*

2-4-1 Subfloor-underlayment[1] / For application of tile, carpeting, linoleum or other non-structural flooring; or hardwood flooring
(Live loads up to 65 psf—two span continuous; grain of face plys across supports)

Plywood Species Group	2-4-1			Nail Size and Type[2]	Nail Spacing (inches)	
	Plywood Thickness (inches)	Maximum Spacing or Supports c. to c. (inches)			Panel Edges	Inter-mediate
Groups 1, 2, & 3	1⅛ only	48[3]		8d ring shank recommended or 10d common smooth shank (if supports are well seasoned)	6	6

Notes: (1) For additional information, see American Plywood Association standards.
(2) Set nails ⅛″ and lightly sand subfloor at joints if resilient flooring is to be applied.
(3) In some non-residential buildings special conditions may impose heavy concentrated loads and heavy traffic, requiring support spacing less than 48″.

Typical grade-trademarks. See Fig. 5-9, Unit 5, for complete details.

2·4·1
GROUP 1 (APA)
INTERIOR
PS 1-74 000

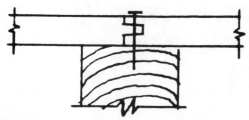

24-62. With tongue-and-groove plywood, a single row of nails at the joint is all that is necessary to nail the two panels together and to the girder.

board feet of S4S, 1" × 8" boards will be needed: 1.15 × 1400 = 1610.

To figure the number of plywood sheets necessary for the building in the example, divide the floor area of the building by the area of one sheet of plywood. A 4' × 8' sheet of plywood contains 32 sq. ft. (4' × 8' = 32 sq. ft.). Therefore the floor area in the example will require 43.75 sheets of plywood.

$$\frac{1400 \text{ sq. ft. of floor area}}{32 \text{ sq. ft.}} = 43.75$$

This is rounded off to 44 sheets.

To determine the quantity of nails required, use the chart in Fig. 24-63. In the column headed "Lbs. nails per 1000 bd. ft." find the spacing of the framing members for the type and size material to be applied. For our example, if the floor joists are spaced 16" on center, the chart shows that 23 lbs. of nails are required per 1000 board feet of lumber. Therefore 37.03 or 37 lbs. of nails will be needed (1610 ÷ 1000 = 1.61; 1.61 × 23 = 37.03).

24-63. *Use this chart to determine the amount of material needed for subflooring.*

Estimating Subflooring

Type	Size	Bd. Ft. per Sq. Ft. of Area	Lbs. Nails Per 1000 Bd. Ft.			
			Spacing of Framing Members			
			12"	16"	20"	24"
T & G	1 × 4	1.32	66	52	44	36
	1 × 6	1.23	43	33	28	23
	1 × 8	1.19	32	24	21	17
	1 × 10	1.17	37	29	24	20
Shiplap	1 × 4	1.38	69	55	46	38
	1 × 6	1.26	44	34	29	24
	1 × 8	1.21	32	25	21	17
	1 × 10	1.18	37	29	25	20
S4S	1 × 4	1.19	60	47	40	33
	1 × 6	1.15	40	31	26	22
	1 × 8	1.15	30	23	20	17
	1 × 10	1.14	36	28	24	19

24-64. *Use this chart to estimate labor time for laying subflooring.*

Subflooring Labor Time

Type of Subfloor	Bd. Ft. Installed per Hour
Straight	75
Diagonal	65
Tongue and Groove	55
Plywood	100

Labor

To figure the labor for the application of the subfloor, use the information in Fig. 24-64. For our example, the chart shows 75 board feet of subflooring can be laid each hour. Therefore it will take 21.5 hours to lay 1610 board feet of lumber (1610 ÷ 75 = 21.46, or 21.5 hours).

QUESTIONS

1. List the names of the floor framing parts.

2. What is the purpose of a post?

3. What is considered the best method of framing the joist and girder?

4. How is a sill plate attached to a foundation wall?

5. How is the header joist checked for alignment?

6. What is the purpose of bridging?

7. How does girder floor framing differ from conventional floor framing?

Wall Framing

Exterior sidewalls and, in some designs, interior walls normally support the roof load. They will also serve as a framework for attaching interior and exterior coverings. When roof trusses spanning the entire width of the house are used, the exterior sidewalls carry both the roof and ceiling loads. Fig. 25-1. Interior partitions then serve mainly as room dividers. When ceiling joists are used, interior partitions usually sustain some of the ceiling loads. Fig. 25-2.

The exterior walls of a wood-frame house normally consist of *studs, interior and exterior coverings, windows and doors,* and *insulation.* The wall-framing members used in conventional construction are generally 2″ × 4″ studs spaced 16″ on center. Fig. 25-3.

The requirements for wall-framing lumber are *stiffness, nail-holding power, freedom from warpage,* and *ease of working.* Species and grades used for all framing may, in general, follow those used for floor-framing materials—for example,

Douglas fir, the hemlocks, and southern yellow pine. Also commonly used for studs are spruce, pine, and white fir. No. 1 and No. 2 grades are most often used. Moisture content of framing members usually should not exceed 19% for on-site construction and 12% if shop fabricated.

As with floor construction, there are two general types of wall framing—*platform construction* and *balloon-frame construction.* The platform method is nearly always used by builders throughout the United States and Canada. This is primarily because platform construction is fairly simple and it lends itself to a great many architectural designs. With the platform method the floor framing should be complete, with subfloor securely fastened in place, before wall framing begins. The first step in wall framing is to lay out the wall locations on the subfloor.

LAYING OUT WALL LOCATIONS

The carpenter responsible for the layout of a building is one of the most important

25-1. *Trussed rafters are installed as complete units. They eliminate the need for bearing partitions.*

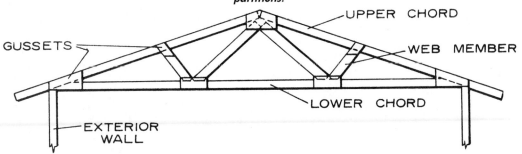

people on the job. Accuracy of layout is most important to the final overall quality of the building. So is knowledge of the structural members of the building. The carpenter must also be aware of special framing requirements for other skilled workers such as plumbers, electricians, and plasterers.

The layout is usually done by two peo-

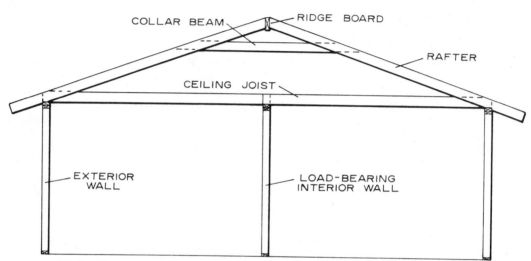

25-2. *Conventional framing requires an interior wall to support the ceiling joists.*

25-3. *Parts of a framed wall section. (Filler studs are also called jack studs.)*

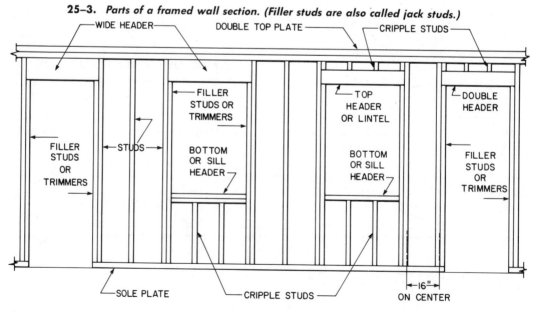

awl to hold the end of the tape or chalk line. The area to be laid out must be swept clean and all objects that might be in the way must be removed. This will help make it possible to make a clear, solid chalk line to which the plates are aligned later. Fig. 25-4.

The carpenter must study the plans for the building and have a thorough understanding of them before layout begins. During the actual layout the carpenter should not have to refer often to the plans, only to check them occasionally to "pick-off" dimensions. Fig. 25-5.

A steel tape is used to mark the location of the interior partitions and exterior walls. Fig. 25-6. A chalk line is pulled taut on these marks and snapped to indicate the exact location and alignment for the full length of an edge of the sole plate. Fig. 25-7. The carpenter thens marks an *X* on the subfloor to show on which side of the line the plate is located. After the location of the plates is laid out on the subfloor, the various wall members are cut to size in preparation for assembly of the wall sections.

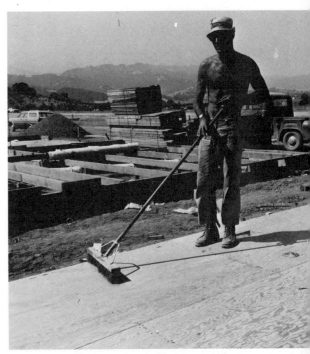

25–4. *The subfloor must be clean before beginning the wall layout.*

25–5. *Keep the floor plan readily available for reference when laying out the walls on the subfloor.*

WALL FRAMING

The following is a description of the various parts of the interior and exterior walls in platform construction:

Sole Plate

All partition walls and outside walls are finished with a piece of material corresponding to the thickness of the wall, usually a 2 × 4. Laid horizontally on the subfloor, this member carries the bottom end of the studs. Fig. 25-8. This 2 × 4 is called the "sole" or "sole plate." The sole should be nailed with two 16d nails at each joist that it crosses. If laid lengthwise on top of a girder or joist, there should be two nails every two feet.

For the bottom plate select straight material from the plate stock. Square up the ends of this material, then carefully position the end of the plate on the subfloor at

25-6. *Laying out on the subfloor. Use a steel tape to measure from the corner to the centerline of all openings and intersecting partitions.*

25-7. *This carpenter is stretching the chalk line to snap on the subfloor. The chalked line will be used to align the sole plate as it is fastened in place.*

25-8. *The box-sill assembly showing the sole plate as a link between the floor and wall units.*

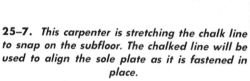

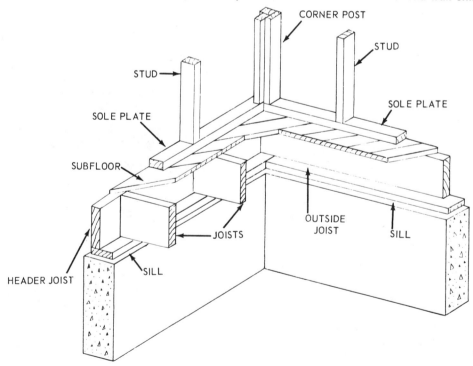

25-9. *Nailing sheathing. Note that the walls are set back so the sheathing rests on the subfloor. The outside surface of the sheathing is flush with the header joist.*

25-10. *For the sheathing-application method shown in the top drawing, the sill is set back the thickness of the sheathing. The sheathing rests on the foundation wall and covers the box-sill assembly. The lower drawing shows platform frame construction on a basementless home. The termite shield extends beyond the interior of the foundation wall.*

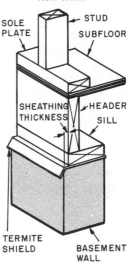

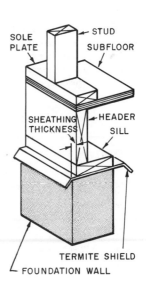

an outside corner. The plate is usually located in from the outside of the building at a distance equal to the thickness of the sheathing. Thus when the sheathing is applied it will rest on the subfloor. Fig. 25-9. Some local codes require that the sheathing cover the joist header and be set on the foundation wall; the sole then is set flush with the outside edge of the subfloor. Fig. 25-10.

Securely nail the plate at the corner. If more than one length of plate stock is necessary, butt the additional lengths up tightly against each other and toenail them together. The joints should be made at the centerline of a stud location. Square and cut off the end of the last length to the desired length and nail the plate securely at this corner.

Continue on around the outside of the building with the sole until complete. Note that the plate is continuous; the door openings are not cut out until the wall is erected and permanently braced.

Before securely nailing between the corners, straighten the sole. Usually the edge of the plate is aligned with the chalk line which was snapped on the subfloor when the walls and partitions were laid out. (See "Laying Out Wall Locations".) The sole can also be straightened by driving a nail at each corner and pulling a chalk line along the outside edge of the sole plate. Then place a 3/4" spacer block between the chalk line and the plate at each end. Measure in 3/4" from the edge of the plate to the chalk line, or use a gauge block between the chalk line and the plate. Fig. 25-11. Nail the plate securely to the subfloor as it is aligned. When more than one length of stock is used, nail at the joints first. Care must be taken to provide a straight plate line and thus a straight wall.

LAYING OUT THE SOLE PLATE

When laying out the sole plate, lay out the door and window openings, the partitions, and then the stud locations, in that order. The studs that are designated where openings occur can then be marked as "cripples." The mark usually used is an *O*.

25–11. *The chalk line is held out away from the plate with a 3/4" block at each end. The plate can then be aligned by measuring 3/4" from the plate to the chalk line or by using another 3/4" block for a gauge.*

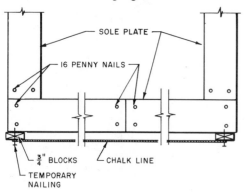

The full-length studs are usually marked with an *X*. Other special studs which will be discussed later should also be marked. Corner studs are marked with a *C*, partition studs with a *P*, and trimmer or lap studs with an *L* or *S*.

Begin the plate layout by referring to the plans to find the distance from the corner of the building to the center of the first opening. Measure off this distance and square a line across the plate at this point. Mark the line with a centerline symbol (₵) and an identification letter or number for reference when cutting other parts for this opening.

Continue around the outside wall to lay out and mark for identification all of the other openings. Now from the outside corners lay off the centers of all partitions and mark them with a P. Fig. 25-12.

Wall Openings

Studs around wall openings require special treatment. Allowances must be made for framing in doors and windows. These allowances, when added to the size of the finished openings, make up what is called a "rough opening." The rough opening is the distance between the trimmer studs. Fig. 25-13. Most window and door schedules will provide the rough opening (R.O.) size for framing. If not, the following allow-

25–12a. *Centerlines of wall openings and intersecting walls are located on the plate by measuring from an outside corner. The locations of the wall members are marked for identification so other carpenters can assemble and erect the walls later.*

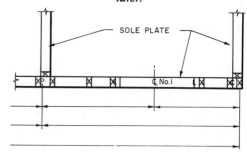

ances can be used for door and window rough opening widths and heights:
- Double-Hung Window (Single Unit):
  Rough opening width = glass width + 6".
  Rough opening height = total glass height + 10".
- Casement Window (Two Sash):
  Rough opening width = total glass width + 11¼".
  Rough opening height = total glass height + 6³/₈".
- Doors:
  Rough opening width = width of door + 2½".
  R.O. height = height of door + 3".

LAYING OUT WALL OPENINGS

After the wall openings have been located on the plate, on each side of the centerline of each opening lay off one-half the rough opening size. Fig. 25-13. For example, for a 2' 8" door, the distance between the full-length studs is 37½"— that is, 32" for the door plus 5½" for the side jambs, wedges, and *trimmer studs* (which support the header). Fig. 25-14.

After all openings and partitions have been laid out, proceed by laying out all regular and cripple studs.

Top Plate

The top plate (Fig. 25-15, page 346) has the following purposes:
- Ties the studding together at the top.
- Forms a finish for the walls.
- Furnishes a support for the lower ends of the rafters.
- Serves as a connecting link between wall and roof just as sills and girders are connecting links between floors and the walls.

The plate is made up of one or two pieces of material of the same size as the

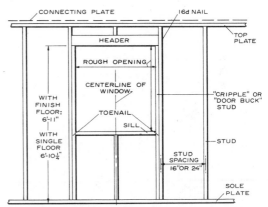

25-13. Rough framing for a window opening.

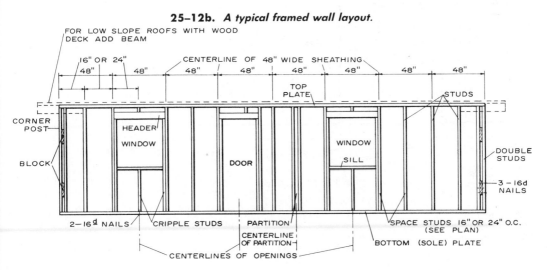

25-12b. A typical framed wall layout.

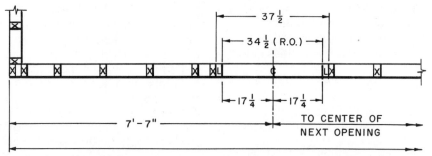

$37\frac{1}{2}$

$34\frac{1}{2}$ (R.O.)

$17\frac{1}{4}$ $17\frac{1}{4}$

7' - 7"

TO CENTER OF
NEXT OPENING

25–14a. *The centerline of a door opening is marked on the sole 7' 7" from the outside corner. This information will be found on the floor plan. Fig. 25–14c, at arrow. One-half the R. O. ($17^1/_4$") is then laid off on each side of the centerline. The thickness of the trimmer stud is laid off on each side of this. The header will rest on top of the trimmers and between the full studs marked with the X.*

25–14b. *Door framing allowance for rough opening.*

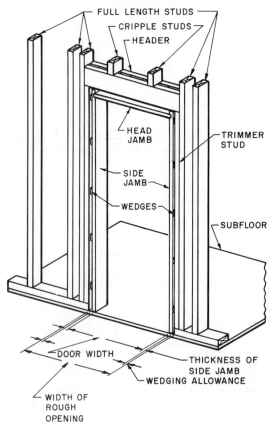

FULL LENGTH STUDS
CRIPPLE STUDS
HEADER

HEAD
JAMB

TRIMMER
STUD

SIDE
JAMB

WEDGES

SUBFLOOR

DOOR WIDTH

THICKNESS OF
SIDE JAMB
WEDGING ALLOWANCE

WIDTH OF
ROUGH
OPENING

studs. When placed on top of partition walls, the plate is sometimes called the *cap.* Where the plate is doubled, the first plate or bottom section is nailed with 16d or 20d nails to the top of the corner posts and to the studs. The connection at the corner is made as in Fig. 25-16a & c (Page 346). After the single plate is nailed securely and the corner braces are nailed into place, the top part of the plate is nailed to the bottom section with 10d nails, spaced 16" O.C. Fig. 25-17. The edges of the top section should be flush with the bottom section and the corner joints lapped as in Fig. 25-16b.

LAYING OUT TOP WALL PLATES

Select straight material for the top plates. Square off the ends and cut to the same lengths as the bottom plates. Place the top plate next to, or on top of, the bottom plate; nail it just enough to hold it in place. Then transfer all the layout marks from the bottom plate to the top plate. Fig. 25-18. There is another way of laying out these plates. Often it is better to cut top and bottom plates at the same time. The top plate can then be nailed temporarily to the sole plate and all the layout done at one time. Figs. 25-19 & 20 (Pages 347 and 348).

Studs

Studs are a series of slender wood members placed in a vertical position as supporting parts of a wall or partition. Studs should be 2 × 4s or 2 × 6s for one- and two-story buildings. As mentioned earlier, short studs that are sometimes added above or below the header of doors and windows are called *cripple studs* or *cripples*. Studs which support headers are sometimes called *trimmer studs*.

The number of full studs, trimmers, and cripples can be determined by counting the layout marks on the wall plates. The lengths of these pieces can be found on the story pole or master stud pattern, as will be explained later. Standard studs can also be purchased precut to finished length. Fig. 25-21.

The stud locations are laid out on the sole and top plate, usually 16" or 24" O.C. Begin the layout on the front wall. Measure the first 16" from the outside-corner edge of the corner post to the center of the first stud; from that point on, measure 16" O.C. for each of the studs.

When beginning the layout of the first stud on an end wall, the center of the first stud is measured from the outside of the sheathing line on the front wall. Fig. 25-22 (Page 349). The same is true for any wall parallel to an end wall, as when there is an offset in the building. The layout of the stud locations on the back wall and all parallel walls should begin from the same end of the building as the layout of the front. By using a common measuring point, the studs will bear over the floor joists, while the ceiling joists and rafters bear over the studs. This alignment creates a direct bearing of the rafter right down through to the foundation wall of the building. Fig. 25-23a (Page 351).

The studs are nailed with either two 16d nails through the plate and into the ends of the stud or they are toenailed into the plate with four 8d nails, two on each side of the stud. Fig. 25-23b (Page 352).

STORY POLE AND MASTER STUD PATTERN

A *story pole* is a fullsize layout of a cross section of a wall. A *master stud pattern* is the same for a portion of the wall. One or the other of these layout devices should be developed for any wall-framing job, as a quick reference for use by all the carpenters. The use of such a device speeds construction and helps eliminate costly errors. This is particularly true when there are many carpenters on a single job or when a building contractor is supervising several jobs.

The story pole will show floor level (or levels if the structure is a multi-story or split level), ceiling height, window and door elevations, and thicknesses of many materials used in construction. Usually, however, only a portion of the wall section is laid out fullsize. This portion would be from the top of the subfloor to the bottom of the ceiling joist; or in a two-story building it would be the distance between the finished floors. As mentioned, this drawing of a portion of the wall is called the master stud pattern. It includes information about the location and size of the window headers, sills, door headers, the heights of various openings above the subfloor, and the thicknesses of the ceiling and the finished floor. Fig. 25-24. If there are several different heights to mark off above the subfloor, it may be necessary to avoid confusion by making additional master stud patterns, labeling them for the various rooms or areas.

CORNER POST

A corner post must form an inside corner and an outside corner. The inside corner will provide two good nailing bases for inside wall covering. The outside corner will provide two good nailing bases for outside wall sheathing. Studs at the corners of the frame construction are usually built up from three or more ordinary studs to provide greater strength. Corner posts may be made in several different ways.

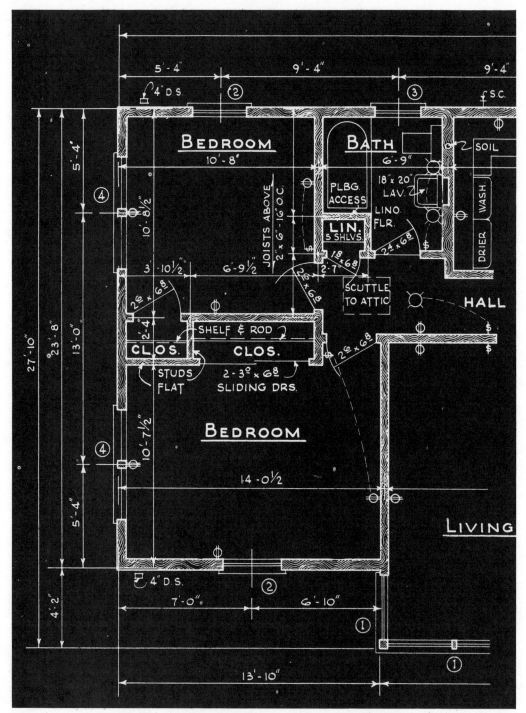

25–14c. *The floor plan from a set of working drawings.*

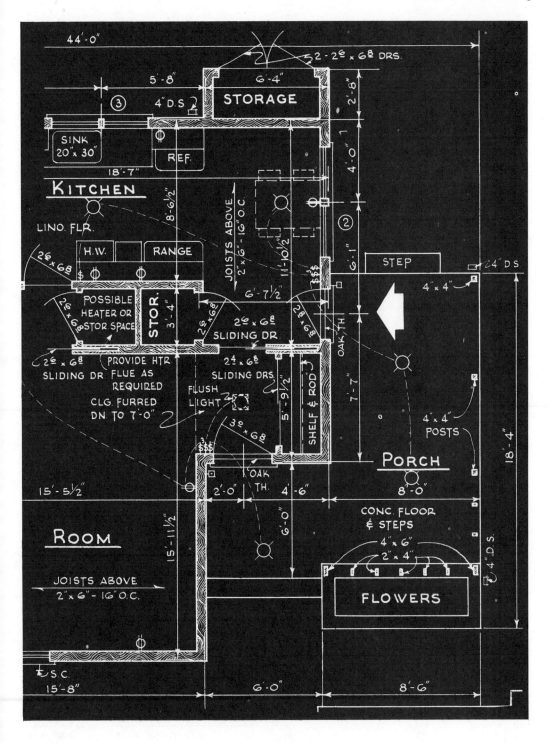

44'-0"

5'-8"   ③   4"D.S   STORAGE   6'-4"   2 - 2⁶ x 6⁸ DRS.   2'-8"

SINK
20"x 30"   REF.   4'-0"

KITCHEN   18'-7"   8'-6½"   JOISTS ABOVE
2"x 6"-16"O.C.   11'-10½"   ②   6'-1"   STEP   4" D.S

LINO. FLR.
2⁶x6⁸ $ H.W.   RANGE   6'-7½"   OAK TH.   4" x 4"

2⁶x6⁸   POSSIBLE
HEATER OR
STOR SPACE   STOR.   3'-4"   2⁶x6⁸   2⁶x6⁸
SLIDING DR.   2⁶x6⁸   7'-7"

2⁶ x 6⁸   PROVIDE HTR
SLIDING DR. FLUE AS
REQUIRED   2⁴ x 6⁸
SLIDING DRS.   5'-9½"   SHELF & ROD

CLG. FURRED
DN. TO 7'-0"   FLUSH
LIGHT   PORCH   4" x 4"
POSTS   18'-4"

3⁶x6⁸   8'-0"

15'-5½"   OAK
TH.   2'-0"   4'-6"   CONC. FLOOR
& STEPS   4"D.S

15'-11½"   6'-0"   4" x 6"
2" x 4"

ROOM

JOISTS ABOVE
2"x 6"-16"O.C.   FLOWERS

S.C.
15'-8"   6'-0"   8'-6"

25–15. *The top plate is shown at the arrow. The double or rafter plate sometimes is not fastened until the walls have been erected, plumbed, and straightened.*

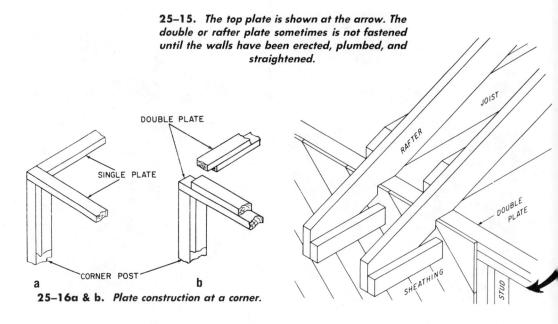

25–16a & b. *Plate construction at a corner.*

25–16c. *Metal framing anchors may also be used to attach the corner posts to the sole and top plate. Note that these brackets are made in lefts and rights and one of each is necessary.*

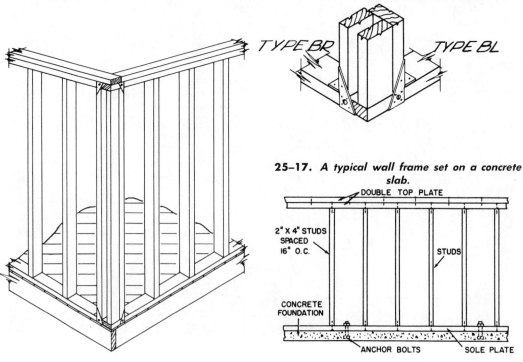

25–17. *A typical wall frame set on a concrete slab.*

25-18. *This carpenter is transferring the layout from the sole plate to the top plate. Note the use of the special template for laying out corners and partitions. Also of special interest are the two pencils in his right hand. The carpenter's pencil is used for location marks and the colored pencil is used for identification marks. The carpenter does not take time to change from one pencil to the other. He writes with either in the position they are held in his hand.*

25-19. *This drawing shows measuring to the centers of the stud locations. In actual layout a carpenter will mark only the left edge of the stud and then place an X to the right of this line. The line can then be seen along the edge of the stud for alignment.*

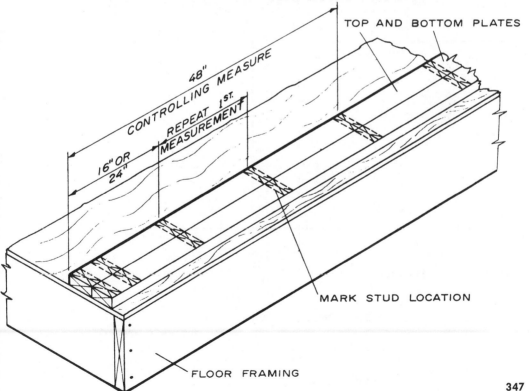

TOP AND BOTTOM PLATES

CONTROLLING MEASURE

48"

REPEAT 1ST. MEASUREMENT

16" OR 24"

MARK STUD LOCATION

FLOOR FRAMING

25–20a. *The outside wall plates have been nailed temporarily to the outside of the joist header (box or band), and the studs are being laid out with a homemade layout template. This template is 4' long,. the standard width of wall covering. The template has 4 fingers, 1¹/₂" wide and 16" O. C., representing 4 stud markings. The fingers are attached to a piece of angle iron, so that the template works like 4 try squares at once. It is made of aluminum for light weight and protection from moisture.*

25–20b. *Laying out the plates for a partition wall. Note the use, in a different position, of the template described in Fig. 25–20a.*

25–21. *A standard stud is 92⁵/₈" long. This stud length will provide a 95⁵/₈" ceiling height with a ³/₄" finished floor and a ³/₄" finished ceiling. Some studs are now being cut 93" for a 96" (8') ceiling.*

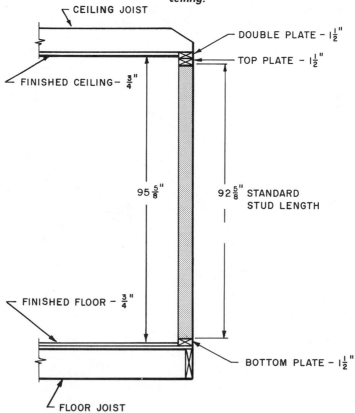

Two of the more common methods are shown in Figs. 25-25 a & b (Page 353). The number of corner posts required can be determined by counting the number of places on the plan where two walls intersect at right angles. Corner posts are nailed together with 10d and 16d nails. They are distributed on the subfloor where they will be used for assembly of the wall sections. The short pieces of 2 × 4s shown at the base of the corner post in Figs. 25-25 a & b are installed after the wall erection to provide places for nailing the corner ends of the baseboard.

PARTITION CORNER POST

Studs should be arranged at a point where a partition ties into a wall between the corners. Three common types of parti-

tion post assemblies are shown in Fig. 25-26 a, b, & c. In the type shown in Fig. 25-26a, the regular spacing of the outside wall studs is interrupted by double studs at the point where the partition ties in. The double studs are set 3″ apart. This interval allows the partition end stud (which is 3⅝″ wide) to lap the others just enough to permit nailing, while leaving most of the inner edges of the others clear to serve as nailing bases for inside wall covering. A variation of the method in Fig. 25-26a is shown in Fig. 25-26b (Page 354). This method will give more nailing surface for the inside wall covering.

In another type of partition corner assembly, the regular spacing of the outside wall studs is maintained. Fig. 25-26c. The cross blocks are made of stud stock, with

25–22a. *When laying out stud locations, allow for the thickness of the sheathing at the corners. Note that all layout dimensions begin from the left end of the building as one stands in front. This will line up the studs on the front and rear walls.*

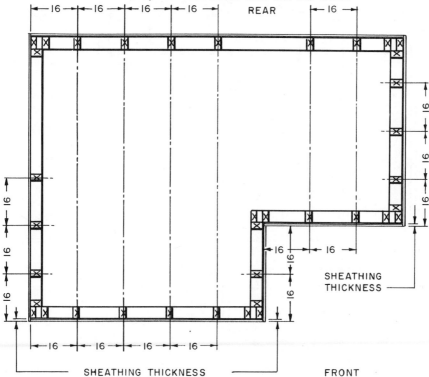

the exception of the bottom block, which is made of wider stock to provide a solid nailing base for ends of baseboard.

The number of partition corners can be counted from the plan. The partition corner posts are nailed together with 10d and 16d nails and distributed to their locations on the subfloor for wall assembly.

CRIPPLE STUDS

Cripple studs are those studs which, because of an opening in the wall, do not extend from the sole plate to the top plate. These studs are installed over window, door and fireplace headers and below window sills. Fig. 25-12b. They are located in the same place that a full stud would

25–22b. *The application of panel-siding or sheathing to a stud wall. Notice the center of the first stud on the side wall is measured from the outside of the corner post. However, the first stud on the end wall is measured from the outside surface of the wall covering material to the center of the first stud. This will permit the wall covering material to lap properly at the corner. See arrow.*

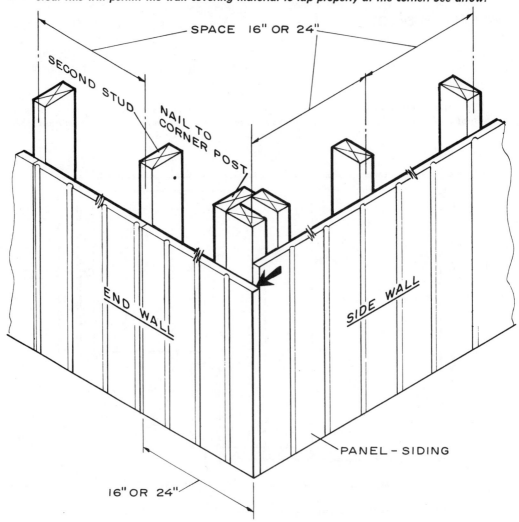

25–23a. *The layout of the stud locations coincides with the joist layout. Later the rafters can be laid out directly over the studs and joists. This creates a direct bearing from the roof right down to the footings of the building. Although this alignment is not required when a double plate is used, it is recommended.*

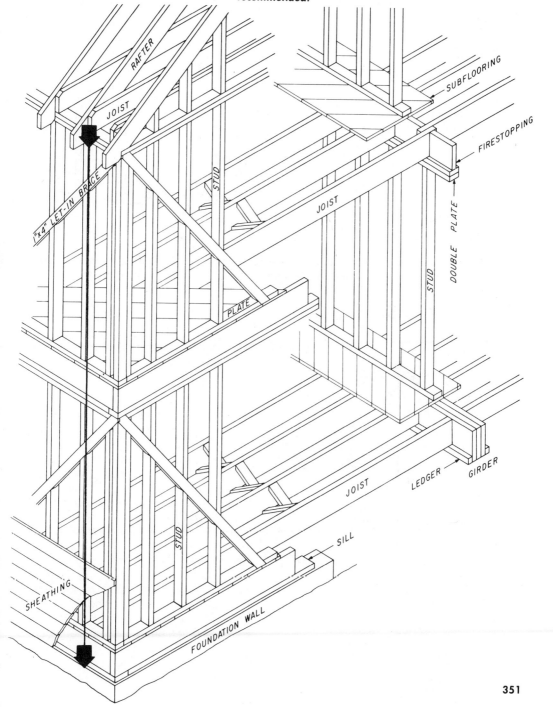

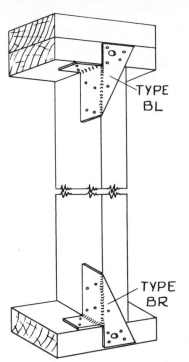

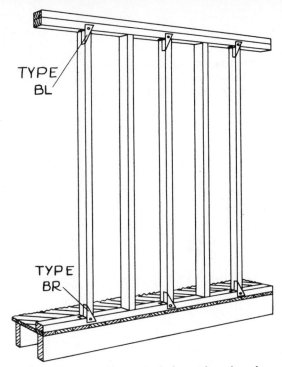

25–23b. *Metal framing anchors may be used either inside or outside to attach the studs to the sole and to the top plate. These brackets are normally used only on alternate studs.*

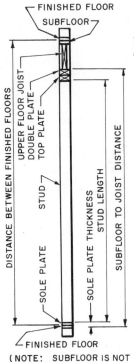

25–24a. *A master stud pattern for a two-story dwelling. Additional information may be included on this layout, as is shown in Fig. 25–24b.*

normally be placed if there were no opening. Cripples are necessary for nailing outside sheathing and inside wall covering. The lengths of the various cripples can be determined by referring to the master stud pattern. These studs are usually pre-cut to length and distributed with the door and window headers in readiness for the assembly of the wall sections.

TRIMMER STUDS

As mentioned earlier, trimmers are studs which support the header over an opening. They are shorter than standard studs, so they are sometimes called cripples. Fig. 25-27. Sometimes they are also called *double studs* or *lap studs*. These studs are nailed to a regular stud under the ends of

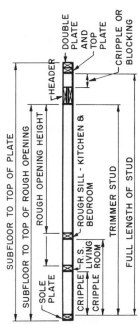

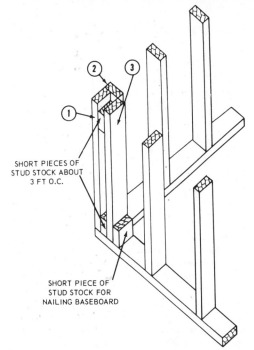

25-24b. *A master stud pattern for a one-story building. Select a clean, straight piece of 1" × 4" or 2" × 4" stock and lay it out full size. The information for this layout is taken from the building plans.*

25-25b. *This is the corner post most commonly used. The pieces numbered 1, 2, and 3 are selected straight standard studs. The short blocks are usually 10 or 12 inches long. Studs 1 and 3 are nailed to the blocks with 10d nails. Stud #2 is then nailed to the assembly of 1 and 3. Care should be taken to keep all ends and outside edges flush and even.*

25-25a. *The simplest type of corner post. The pieces numbered 1, 2, and 3, are selected straight standard studs.*

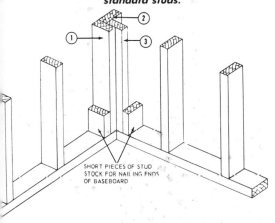

25-26a. *Double-stud partition corner assembly.*

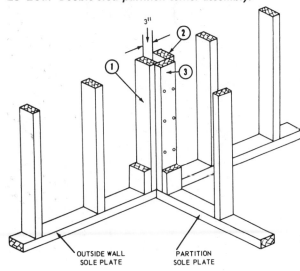

the header. To hold them in place, 10d nails are used, spaced 16″ apart and staggered as shown in Figs. 25-27 & 28. Notice that the trimmer or double stud for a door may extend from the header to the subfloor and does not rest on the sole plate. If this method is used, the sole plate must be cut away between the full studs before the trimmers can be installed. Trimmers should be cut to fit snugly under the head-

er so that they will support it properly. If a header settles, plaster cracks may develop and doors may fit improperly. The double studs in a door opening also form solid supports when the door is slammed (on the latch side) and for the weight of the door (on the hinge side).

Headers

Where windows or doors occur in outside walls or partitions, parts of some studs must be cut out. It is necessary, therefore, to install some form of header over the doorway to support the lower ends of studs that have been cut. Likewise, at the bottom of a window opening the "rough sill" supports the upper ends of studs that have been cut. Fig. 25-12b. The width of a header is determined by the length of the opening it must span. This information will be available from the building plans or local code requirements. Fig. 25-29.

Headers are sometimes built-up trusses as shown in Fig. 25-30. In some cases 4″ stock is used rather than two pieces of 2″ material nailed together. This saves work and also allows the thickness of the header

25–26b. This partition corner assembly is the one most commonly used. It will give more nailing on the inside corners than does the one shown in Fig. 25–26a. Nail the two wall studs to the spacer with 10d nails. The end stud of the partition can be nailed to these studs at this time. Or, it can be nailed up as part of the partition wall and then attached to the spacer at the time the partition is erected.

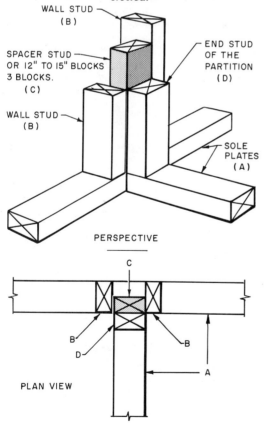

25–26c. Partition corner assembly with wall studs at regular intervals.

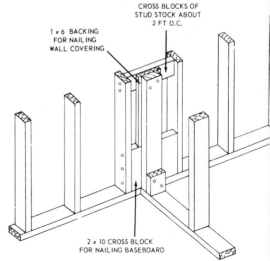

to be exactly the same as the width of a 2" × 4" stud. Fig. 25-31a. (When two 2" members are used for a header, the total thickness is only 3". This requires a ½" spacer to give the header the full 3½" width of the stud.) Fig. 25-31b and c (Page 358).

Framing wide openings such as double garage doors which require headers 16' to 18' long can be done with nailed plywood box beams. These headers can be fabricated on or off the building site. The design and construction of these headers is shown in Fig. 25-31d. The ends of the headers should be supported on studs or by framing anchors, depending on the local code requirements. Figs. 25-32 & 33 (Page 360). The header lengths are obtained by measuring the layout of the bottom wall plates. The header is measured between the full studs. Fig. 25-13. In the case of the header for the door shown in Fig. 25-14a, the header length would be 37¾".

It is best to number the openings (such as windows, doors, and fireplaces) for identification and then make a cutting schedule for all headers. One person can cut these to length and, if 2" material is used, nail them together. Use 16d nails,

25–27a. *Door opening in a wall or partition. The cripple studs (A) are nailed with four 8d nails, two on each side. The standard studs (B) are nailed to the header with four 16d nails on each side and toenailed to the sole plate with two 8d nails. Or the full stud (B) could be nailed from the bottom up through the plate, with two 16d nails if the sole is attached before the wall is erected. The trimmer is nailed with 10d nails at C and staggered 16" O.C. Two 10d nails are driven into the end of the sole at D.*

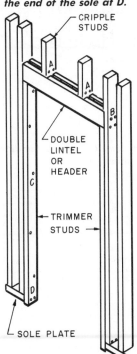

CRIPPLE STUDS

DOUBLE LINTEL OR HEADER

TRIMMER STUDS

SOLE PLATE

25–27b. *Alternate method of framing a door opening.*

SINGLE HEADER MAY BE USED FOR A NONBEARING WALL. USE A DOUBLE HEADER WITH SPACERS FOR A BEARING WALL AND OUTSIDE WALLS.

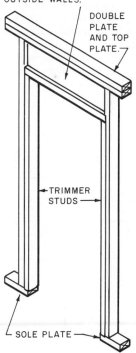

DOUBLE PLATE AND TOP PLATE.

TRIMMER STUDS

SOLE PLATE

355

two near each end, and stagger the others 16" apart along the length of the header. Don't forget to use ½" spacers between the 2" members where the nailing occurs. Headers are then distributed to their locations on the subfloor in readiness for the assembly of the wall sections.

CONTINUOUS HEADER

A *continuous header*, consisting of two 2" members set on edge, may be used instead of a double top plate. The width of the header will be the same as that required to span the largest opening. Joints in individual members should be staggered at least three stud spaces and should not occur over openings. The header is toenailed to studs and corners. At intersections with bearing partitions, the members are taped or tied with metal straps. Fig. 25-34 (Page 362).

25–28. *Window opening. The cripple studs at A are toenailed with four 8d nails, two on each side. The full stud is nailed to the header at B with four 16d nails, to the trimmer (double stud) at C with 10d nails 16" O.C., and it is toenailed to the sole at the bottom. The lower part of the double sill is nailed with two 10d nails into the ends of the cripples at D. The upper part of the sill is nailed to the lower with 10d nails 8" O.C. and staggered.*

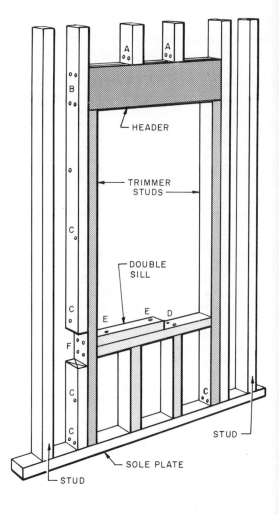

25–29. *These header widths (lintel spans) may be used for most residential buildings. Wider openings often require trussed headers.*

Maximum Spans For Lintels

Nominal depth of lintels made of two thicknesses of nominal two-inch lumber installed on edge	Interior Partitions Or Walls				Exterior Walls		
	Limited attic storage	Full attic storage, or roof load, or limited attic storage plus one floor	Full attic storage plus one floor, or roof load plus one floor, or limited attic storage plus two floors	Full attic storage plus two floors, or roof load plus two floors	Roof, with or without attic storage	Roof, with or without attic storage, plus one floor	Roof, with or without attic storage, plus two floors
4 in.	4 ft.	2 ft.	Not permitted	Not permitted	4 ft.	2 ft.	2 ft.
6 in.	6 ft.	3 ft.	2 ft. 6 in.	2 ft.	6 ft.	5 ft.	4 ft.
8 in.	8 ft.	4 ft.	3 ft.	3 ft.	8 ft.	7 ft.	6 ft.
10 in.	10 ft.	5 ft.	4 ft.	3 ft. 6 in.	10 ft.	8 ft.	7 ft.
12 in.	12 ft. 6 in.	6 ft.	5 ft.	4 ft.	12 ft.	9 ft.	8 ft.

*Supported loads include dead loads and ceiling.

EXTERIOR WALLS, ASSEMBLY AND ERECTION

Several procedures can be used when assembling wall sections. Whichever method is used, the order in which the exterior walls are assembled and erected must first be determined. Usually the front and rear walls (the longest sections) should be set up first, and the side walls then erected in between. Fig. 25-35a. When assembling wall sections, some carpenters first set and plumb the corner posts on the sole plate. Then they raise the studs and toenail them in place separately. Next, they nail on the top plate. In a very different procedure, the wall members can be assembled on the subfloor, the wall squared, sheathing applied, windows installed, and even siding applied before the wall is erected. Fig. 25-35b. However, there are two methods (with slight variation) that are used most often.

In the *first* of the common methods the studs, cripples, trimmers, headers, corners, and partition corners are all precut to length. Then they are distributed to the area on the subfloor where they will be

25–31a. *Plates and headers laid out on the subfloor in preparation for the wall assembly. Notice the use of the solid 4″ header.*

25–31b. *A header built up of two 2″ members with ¹/₂″ spacers to bring the total thickness out to 3¹/₂″. The members are nailed with 16d nails staggered on 16″ centers. Plywood makes a good spacing material.*

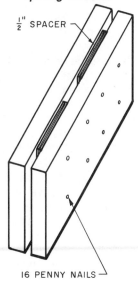

25–30. *Trussed headers.*

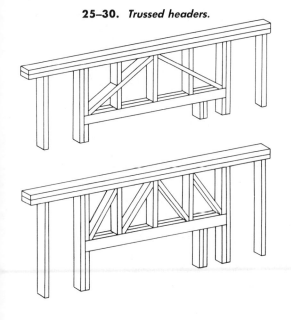

25–31c. *Nails should be driven into the header at an angle.*

In this method of assembly, the sole plate has already been aligned and nailed securely in place on the subfloor. (See "Sole Plate", page 337.) The wall is now ready to be raised. Since the studs, cripples, trimmers, corners, and partition corners are nailed only to the top plate, the bottom ends will all be loose. Push the loose ends of the assembled wall section up against the sole plate. Depending on the size of the wall section, two or more workers are needed to lift the wall to a vertical position. Fig. 25-38 (Page 364).

Nail some braces temporarily so they can be adjusted later when the wall is plumbed and straightened. Fig. 25-39. Each of the studs is toenailed down to its mark on the sole plate with 8d nails, two to each side of a stud.

In the *second* common method of assembling wall sections, the sole plate is nailed temporarily to the subfloor for layout and pulled free for assembly. In this method the nails are driven up through the bottom of the sole into the ends of the wall members. Fig. 25-40. The sheathing and siding can be fastened to the studs while the walls are still on the subfloor. Figs. 25-35b and 25-41. If preferred, the wall framing can be set up, squared, and braced, and the sheathing applied later. Figs. 25-42 and 25-43 (Page 366). When the sole plate is nailed to the other wall members before erection, a chalk line is snapped on the subfloor to show the exact location of the plate. The sole is straightened as it is nailed in position alongside the chalk line. The bottom plate is fastened to the floor framing with 16d nails spaced 16" apart and staggered when practical. The wall can now be plumbed and temporary bracing added to hold it in place in a true vertical position. Figs. 25-44 & 45.

assembled. (Some preassembly of wall members may have been done. Fig. 25-36.)

The wall sections are assembled on the subfloor. Fig. 25-37. Begin the assembly by removing the premarked top plate from its temporary nailing on the sole plate. Lay the top plate on edge on the subfloor about 8' out from the sole plate. Face the marks on the top plate toward the sole plate. Lay a stud at each mark and place the header so that the rough sill, cripples, and trimmers are in position. Also place the preassembled corners and partition corners at the marked locations. Beginning at one end of the top plate, nail these members at the locations marked. Drive two 16d nails through the top plate into each member. Be careful to keep the edges of the members flush with each other. This is necessary for a smooth application of the interior and exterior wall covering later.

Temporary Bracing

Temporary bracing may consist of 1" × 6" members nailed to one face of a stud and to a 2" × 4" block which has been

25–31d. *A nailed plywood box beam may be constructed as a header to span a large opening. First, nail the single piece 2 × 4s together flatwise with 10d nails to form the top and bottom flanges. Then 2 × 4 stiffeners are set in place at the beam ends and between the flanges behind the plywood web joints. Over this, plywood webs are then nailed in place with 8d common nails to complete the assembly.*

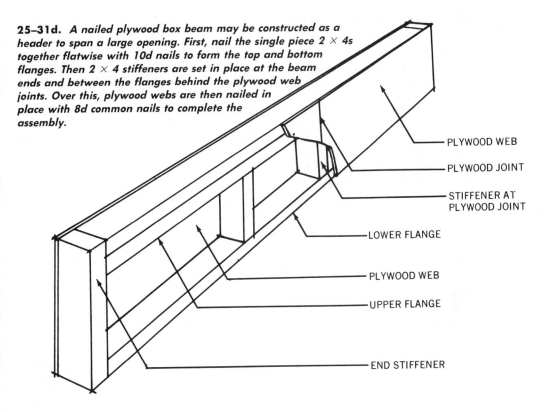

PLYWOOD WEB

PLYWOOD JOINT

STIFFENER AT PLYWOOD JOINT

LOWER FLANGE

PLYWOOD WEB

UPPER FLANGE

END STIFFENER

16′ Span Garage
Door Header 24-ft-wide building 25 psf L.L. (420 pounds lineal foot total load)

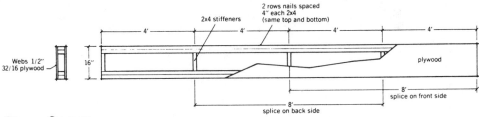

2x4 stiffeners

2 rows nails spaced 4″ each 2x4 (same top and bottom)

4′ 4′ 4′ 4′

Webs 1/2″ 32/16 plywood

16″

plywood

8′ splice on front side

8′ splice on back side

18′ Span Garage
Door Header 28-ft-wide building 40 psf L.L. (700 pounds lineal foot total load)

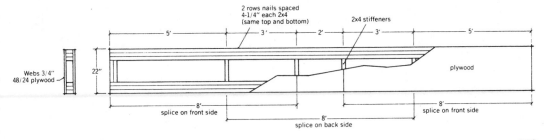

2 rows nails spaced 4-1/4″ each 2x4 (same top and bottom)

2x4 stiffeners

5′ 3′ 2′ 3′ 5′

Webs 3/4″ 48/24 plywood

22″

plywood

8′ splice on front side

8′ splice on back side

8′ splice on front side

nailed to the subfloor. Braces should be at about a 45° angle. Fig. 25-39. These braces should be nailed to the inside of the walls. This permits attaching the permanent bracing and sheathing without moving any temporary braces. Care should also be taken not to project the ends of the temporary braces above the top plate. Otherwise, the braces could interfere with ceiling and roof framing and would have to be removed. Moving these braces at this time would disturb the plumbed and straightened walls. Use enough nails to hold the wall section securely, but do not drive the nails in all the way. The nailhead should project enough to allow easy withdrawal for removal later. The temporary bracing is left in place until the ceiling and the roof framing are completed and sheathing is applied to the outside walls.

PLUMBING AND STRAIGHTENING THE WALL SECTIONS

Before the walls can be straightened, all exterior and intersecting corners must be plumbed and temporarily braced. Either a level or a plumb bob may be used to plumb the wall sections.

Using a Plumb Bob

To plumb a corner with a plumb bob, first attach to the bob a string long enough to extend to or below the bottom of the corner post. Lay a rule on top of the post so that 2" of the rule extends over the post on the side to be plumbed; then hang the bob

25-32. *Window framing with framing anchors used to support the header, sill, and studs. (R and L indicate right and left.)*

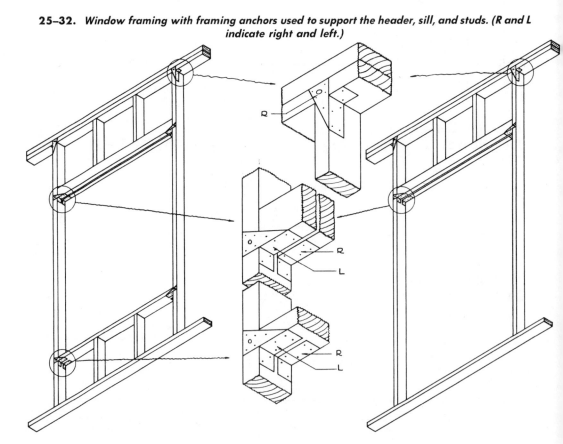

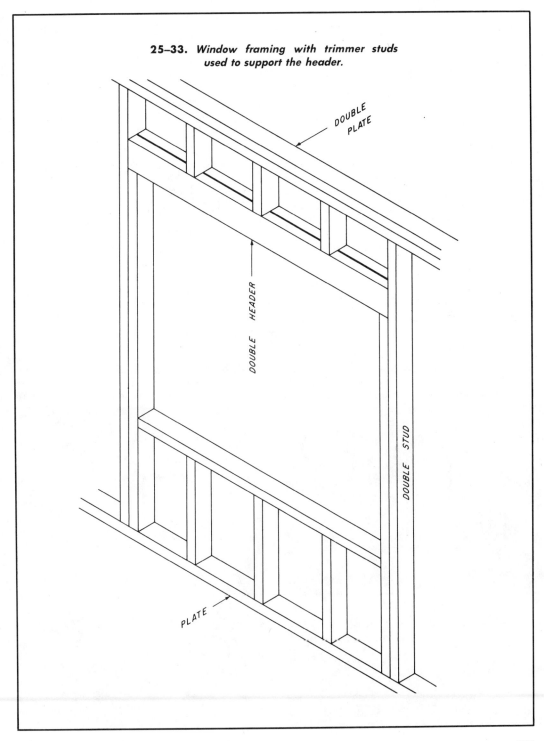

25–33. Window framing with trimmer studs used to support the header.

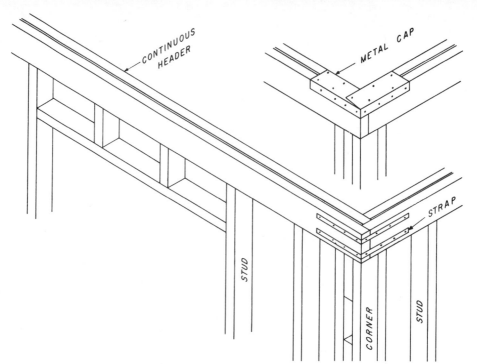

25-34. A continuous header used in framing an exterior wall. Note the alternate corner treatment.

25-35a. The front and rear walls have been set up. The materials for the end walls are being distributed in preparation for assembly.

25-35b. Nailing prefinished paneling on an exterior wall before erection.

line over the rule so that the line is 2″ from the post and extends to the bottom of it.Fig. 25-46 (1). With another rule, at the bottom of the post measure the distance from the post to the line. If the distance is not 2″, the post is not plumb. Move the post inward or outward until the distance from the post to the line is.exactly 2″. Then nail the temporary brace in place. Repeat this procedure from the other outside face of the post. The post is then plumb. This process is carried out for the remaining corner posts and partition posts.

An alternate method of plumbing a post is shown in Fig. 25-46 (2) (Page 366). Attach the plumb bob string securely to the top of the post to be plumbed. Make sure that the string is long enough to allow the plumb bob to hang near the bottom of the post. Use two blocks of wood identical in thickness as gauge blocks. Tack one block near the top of the post between the plumb bob string and the post (gauge block No.

1). Insert the second block between the plumb bob string and the bottom of the post (gauge block No. 2). If the entire face of the second block makes contact with the string, the post is plumb.

Using a Level

To plumb a corner with a level, do not place the level directly against a stud, because the face or edge of the stud is likely to be irregular in shape. Instead, make a long straightedge. Fig. 25-47 & 48. The straightedge should have a couple of lugs, for placing against the stud, and an edge on which to place the level. Check the straightedge for trueness by placing it on a level surface and checking with the builder's level. To increase accuracy when plumbing the walls, hold the level in such a position that you can look straight in at the bubble. While one worker uses the level, another should be ready to nail the brace to the block on the subfloor as soon

25–36. *The rough sill and cripples have been preassembled and laid with the header for wall assembly later.*

25–37. *Assembling the wall sections on the subfloor in preparation for erection. This man is nailing the window trimmer in place.*

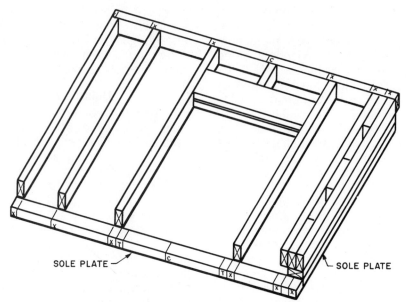

25–38a. *Studs of the assembled wall are pushed up against the sole plate opposite the marks where they are to be nailed.*

SOLE PLATE

SOLE PLATE

25–38b. *The wall is raised to the vertical position. The ends of the studs are pushed against the sole plate which has been securely nailed in place.*

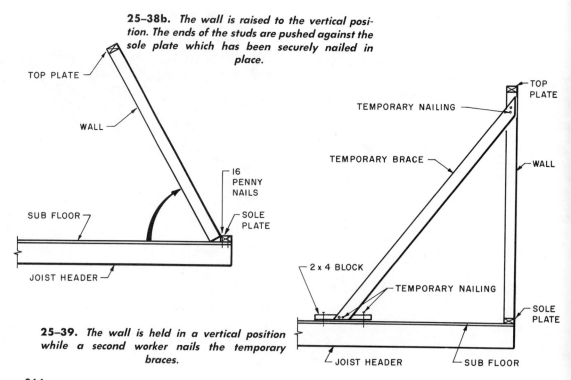

TOP PLATE

WALL

SUB FLOOR

JOIST HEADER

16 PENNY NAILS

SOLE PLATE

TOP PLATE

TEMPORARY NAILING

TEMPORARY BRACE

WALL

2 x 4 BLOCK

TEMPORARY NAILING

SOLE PLATE

JOIST HEADER

SUB FLOOR

25–39. *The wall is held in a vertical position while a second worker nails the temporary braces.*

25–40a. *This carpenter is nailing the sole plate to the ends of the studs before the wall is raised.*

25–41. *Checking the assembled wall section for squareness by measuring diagonally across the corners. The two diagonal measurements must be equal.*

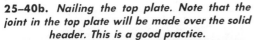

25–40b. *Nailing the top plate. Note that the joint in the top plate will be made over the solid header. This is a good practice.*

25–42. *These carpenters are using another variation of assembly. Both plates have been nailed to the studs while the wall was lying on the subfloor. After the wall is set up, the window framing will be installed either as a preassembled unit or a piece at a time.*

25-45. *This wall, including the sole plate, was nailed together on the subfloor and raised into position. Notice the pieces of 1" × 4" material nailed to the joist header. These pieces keep the assembled wall from sliding off the subfloor as it is being raised.*

25-43. *These carpenters have completed the assembly of the wall framing, including the double plate, before raising the wall section.*

25-44. *Nailing the temporary bracing.*

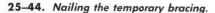

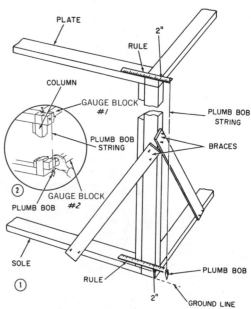

25-46. *Plumbing the post.*

as the correct position is found. This nailer works the end of the brace back and forth on signal from the worker with the level. When the nailer gets the word that the bubble is centered, he nails the end of the brace to the block.

Outside corners must be plumbed and braced both in and out and side to side.

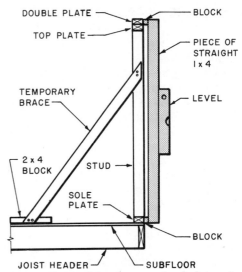

25–48. *The straightedge can simply be a piece of 1″ × 4″ stock. Make sure the edge on which the level is placed is parallel to a line drawn between the two blocks nailed on the ends.*

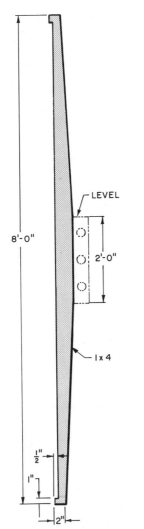

25–47. *A straightedge can be made to use with a small level for plumbing the posts.*

After all exterior corners have been plumbed and braced, the intersecting interior partition posts are plumbed and braced. The partition posts need to be plumbed in one direction only (in and out). This will plumb the exterior wall at the point of intersection. If the top and bottom plates have been correctly cut and laid out, the partition corners will be plumb from side to side after the exterior corner posts are plumbed. After all exterior and partition corner posts have been plumbed and nailed in place with temporary bracing, the wall sections between the posts should be straightened. If necessary, they should be held with additional temporary bracing.

Straightening Walls

To straighten walls, fasten a chalk line to the outside of one of the corner posts at the top. Stretch the line to the corner post at the opposite end of the building, and fasten the line to this post in the same manner as for the first post. Place a small wooden block, $3/4″$ thick, under each end of the line to give clearance. Fig. 25-49. Place additional temporary braces at intervals close enough to hold the wall straight. When the wall is far enough away from the line to permit a $3/4″$ block to slide between the line and the plate, the brace is nailed. This procedure is carried out for the entire perimeter of the building. Inside partition walls should be straightened later in the same manner.

PARTITIONS—ASSEMBLING AND ERECTING

Partition walls divide the inside space of a building. These walls in most cases are framed as part of the building. Partition walls are of two types—*bearing* and *nonbearing*. A bearing wall supports ceiling joists, while a nonbearing wall supports only itself. Partition walls are framed in the same manner as outside walls. After all the exterior walls are set up, plumbed, braced and the sole plate securely nailed, the interior walls (partitions) are assembled and erected. The top and bottom plates for the partitions are cut and laid out in the same way as described for exterior walls (pages 340 and 342). The sizes of the various partition parts (such as headers, trimmers, and cripples) can be learned from the master stud pattern and the building plans. These parts are cut to size, marked for identification, and then distri-

buted to the areas on the subfloor where they are to be assembled. Assembly of the partitions is also the same as described for outside walls. Careful planning of the order in which the partitions are assembled and erected is very important. A floor plan with a suggested sequence of assembly and erection is shown in Fig. 25-50. The next two paragraphs will explain some of the reasons for the order in which the steps are carried out.

The first operation is to raise, fasten, and temporarily brace the longest center partition. Work then proceeds from one end of the building to the other and from the center wall out to the exterior walls. To save time and effort, operations in one area are completed before moving to the next area.

Still referring to Fig. 25-50, notice that partition 2 helps support the center partition and connects it to the previously plumbed exterior wall. The third partition is at right angles to the second, and the fourth is at right angles to the third. This pattern continues to the rear of the building. This is better than erecting two parallel partitions (such as 2 and 4) and then

25-49. *Using a chalk line as a reference to straighten the walls.*

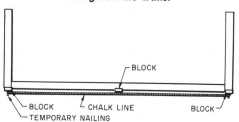

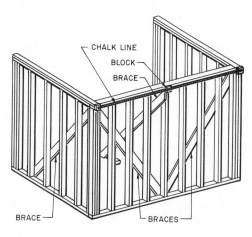

25-50. *Sequence of assembling and erecting interior partitions. After reading the text you will understand why the sequence was suggested. There are other possible sequences that would be worth considering. Can you suggest one that might have its own advantages?*

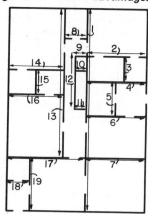

having to work in a confined area to erect the connecting partition (3).

The Double Plate or Rafter Plate

The double plate or rafter plate is applied over the top plate. Fig. 25-51. The same kind of material is used as for the top plate. The pieces must be cut accurately to length. This can be done either at the same time when the sole and top plates are cut or after the walls are erected. Note, however, that the rafter plates are not cut

25–51. *The top plate is usually fastened in place after the walls have been plumbed and straightened. Ten-penny nails are used in the nailing pattern to attach the double plate.*

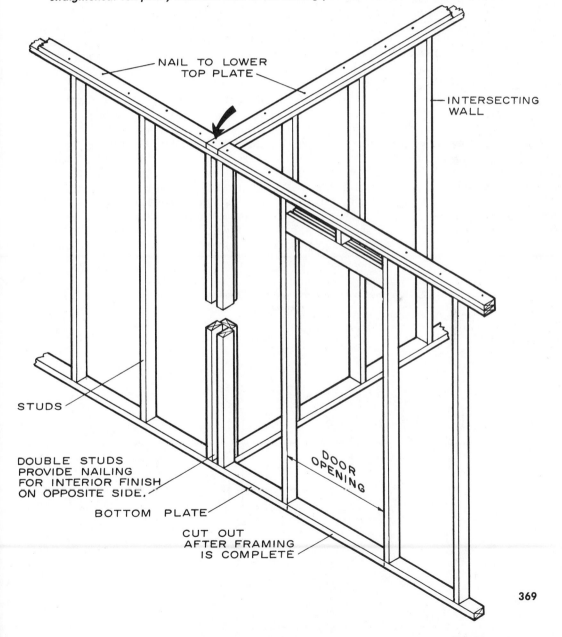

NAIL TO LOWER
TOP PLATE

INTERSECTING
WALL

STUDS

DOUBLE STUDS
PROVIDE NAILING
FOR INTERIOR FINISH
ON OPPOSITE SIDE.

BOTTOM PLATE

DOOR
OPENING

CUT OUT
AFTER FRAMING
IS COMPLETE

exactly the same length as the other two plates. Since one of the primary purposes of this plate is to tie the walls together at the top, the rafter plate laps over the joint formed at a corner of intersecting wall. Fig. 25-52. On a long wall, if joints are necessary in the rafter or double plate, they should be at least 4′ from any joint in the top plate. Fasten the double plate with 10d nails spaced 16″ O.C. End laps between adjoining plates are nailed with two 16d nails on each lap. (See arrow, Fig. 25-51.)

BRACING

Bracing stiffens framed construction and makes it rigid. Good bracing keeps corners square and plumb. It also prevents warping, sagging, and shifting. Without bracing, the shape changes in the frame would cause badly fitting doors and windows and cracked plaster.

It is important to have the frame properly nailed together. The bracing is then securely nailed to hold the framing rigid. Fig. 25-53 provides a review of correct nailing procedure and shows the bracing in place: (1) The sole plate is nailed to a joist or header joist with 16d nails spaced 16″ O.C. (2) The top plate is end-nailed to the studs with two 16d nails. (3) Studs are toenailed to the sole plate with two 8d nails on each side. (4) Doubled studs are nailed together with 10d nails 16″ O.C. (5) Top plates are spiked together with 10d nails 16″ O.C. (6) The top plates, laps, and intersections are nailed together with two 16d nails. (7) A one-inch brace is nailed to each stud and plate with two 8d nails. (8) Corner studs and multiple studs are nailed with 10d nails 12″ O.C. Other joints should be nailed to provide proportional strength.

Frame walls may be braced at the corners by diagonal members (usually 1″ × 4″) set in gains cut into the plates, studs, and corner posts. Fig. 25-53. To lay out these gains, place the bracing members in position against the framing members. Then score (mark) the outline on each stud or plate with a scratch awl. Diagonal braces must be properly nailed at each crossing stud with not less than two 8d nails and at the ends with three 8d nails. Fig. 25-54. A single diagonal extending from floor line to ceiling line should cross at least three

25–52. The double top plates are joined together with half-lap joints.

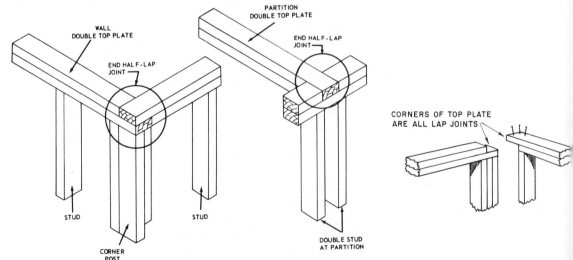

stud spaces. When a wall opening makes this impossible, a K-brace is installed. Fig. 25-55.

Sometimes diagonal bracing gains are laid out and cut into the framing before the walls are erected. Figs. 25-56 & 57. Then the braces are also nailed to the pole plate before wall erection. Fig. 25-58. When the wall sheathing provides adequate corner bracing, diagonal bracing may be omitted. Diagonal-laid wood sheathing or plywood is usually considered adequate bracing. Figs. 25-59 (Page 374) & 60. When other sheathing materials are used, a sheet of plywood is sometimes fastened to each side of each corner instead of diagonal

bracing. Also if there is not enough room to brace adequately because of an opening in the exterior wall, a sheet of plywood will provide the necessary strength for the corner. Fig. 25-61.

SPECIAL FRAMING

Special framing includes such operations as providing openings for heating vents and plumbing pipes, adding support for heavy items (such as bathtubs), and providing bases for nailing paneling or other covering materials. Such framing makes a carpenter's job easier and faster in the finishing stages and adds strength and quality to the construction.

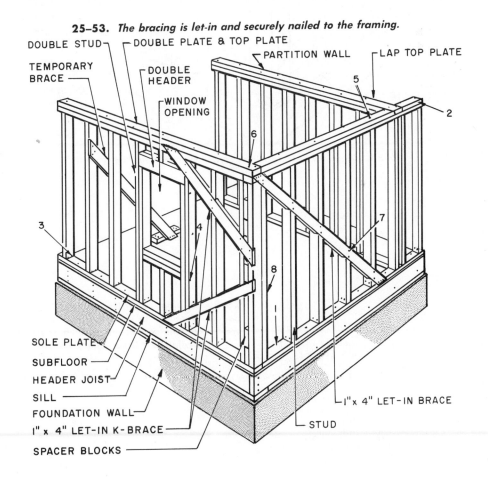

25-53. *The bracing is let-in and securely nailed to the framing.*

DOUBLE STUD ⌐DOUBLE PLATE & TOP PLATE
⌐PARTITION WALL   ⌐LAP TOP PLATE
TEMPORARY BRACE   ⌐DOUBLE HEADER
⌐WINDOW OPENING
DOUBLE STUD
SOLE PLATE
SUBFLOOR
HEADER JOIST
SILL
FOUNDATION WALL
1" x 4" LET-IN K-BRACE
SPACER BLOCKS
1" x 4" LET-IN BRACE
STUD

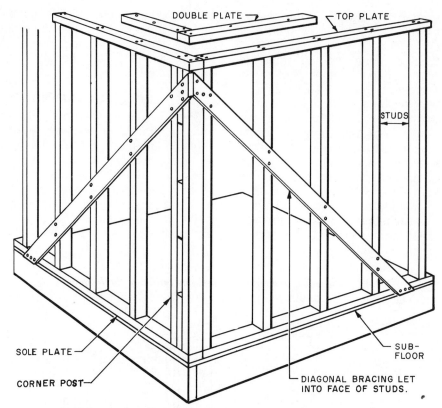

DOUBLE PLATE
TOP PLATE
STUDS
SOLE PLATE
CORNER POST
SUB-FLOOR
DIAGONAL BRACING LET INTO FACE OF STUDS.

25–54. *Diagonal bracing at the corner. Note the nailing pattern.*

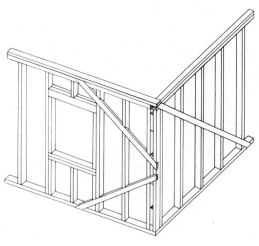

25–55. *A diagonal brace should reach from the double plate to the sole plate and cross 3 stud spaces. If this is not possible because of an opening, a K-brace should be used.*

25–56. *This carpenter has the power saw set to make the first cut into the studs. These cuts are made to a depth equal to the thickness of the brace stock. The saw blade is placed against the brace, which serves as a guide for cutting the gain.*

25–58. *The brace is nailed securely at the bottom only (near top of photo). All other nails are just started. After the wall is set up and plumbed all the nails are driven home to hold the wall plumb.*

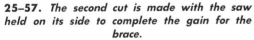

25–57. *The second cut is made with the saw held on its side to complete the gain for the brace.*

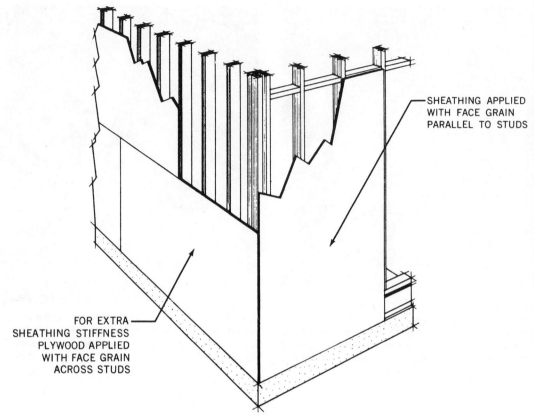

SHEATHING APPLIED
WITH FACE GRAIN
PARALLEL TO STUDS

FOR EXTRA
SHEATHING STIFFNESS
PLYWOOD APPLIED
WITH FACE GRAIN
ACROSS STUDS

25–59. *Plywood used as sheathing eliminates the need for diagonal bracing.*

Plumbing Stack Vents

Plumbing stack vents are usually installed by using a nominal 2″ × 6″ plate and placing the studs flatwise at each side. Fig. 25-62. This provides the needed wall thickness for the bell (large end) of a 4″ cast-iron soil pipe, which is larger than the thickness of a 2 × 4 stud wall. It is also possible to fur out several studs to a 6″ width at the soil stack, rather than thickening the entire wall. In areas where building regulations permit the use of 3″ vent pipe, a 2 × 4 stud wall may be used, but it requires reinforcing scabs at the top plate.

Fig. 25-63. Use 12d nails to fasten the scabs.

Bathtub Framing

The floor joists in the bathroom which support the tub or shower should be arranged so that no cutting is necessary in connecting the drainpipe. This may require only a small adjustment in spacing the joists. Fig. 25-64. When joists are parallel to the length of the tub, they are usually doubled under the outer edge. Tubs are supported at the enclosing walls by hangers or woodblocks. The wall in

25–60. *Plywood paneling will have one of the four grade stamps shown here (arrow #1). After finding the grade stamp, use the top chart for information on nailing. For example, for a 12/0 panel, ⁵/₁₆″ thick (arrow #2) the maximum stud spacing is 16″. The panel should be attached with 6d nails, with 6″ spacing around the edges of the panel and 12″ spacing on the studs in between. If the panels are to be stapled, refer to the bottom chart (arrow #3).*

Wall Sheathing[1]

Panel Identification Index	Minimum Thickness (inch)	Maximum Stud Spacing (inches) Exterior Covering Nailed to:		Nail Size [2]	Nail Spacing (inches)	
		Stud	Sheathing		Panel Edges (when over framing)	Intermediate (each stud)
¹²/₀, ¹⁶/₀, ²⁰/₀	⁵/₁₆	16	16[3]	6d	6	12
¹⁶/₀, ²⁰/₀, ²⁴/₀	³/₈	24	16 24[3]	6d	6	12
²⁴/₀, ³²/₁₆	½	24	24	6d	6	12

Notes: (1) When plywood sheathing is used, building paper and diagonal wall bracing can be omitted
      (2) Common smooth, annular, spiral thread, galvanized box or T-nails of the same diameter as common nails (0.113″ dia. for 6d) may be used. Staples also permitted at reduced spacing.
      (3) When sidings such as shingles are nailed only to the plywood sheathing, apply plywood with face grain across studs.

Recommended minimum stapling schedule for plywood/All values are for 16-gauge galvanized wire staples having a minimum crown width of ³/₈″

Plywood wall sheathing/Without diagonal bracing

Plywood Thickness	Staple Leg Length	Spacing Around Entire Perimeter of Sheet	Spacing at Intermediate Members
⁵/₁₆″	1¼″	4″	8″
³/₈″	1³/₈″	4″	8″
½″	1½″	4″	8″

Typical grade-trademarks. See Fig. 5-9, Unit 5, for complete details.

STRUCTURAL I
C-D
20/0 **APA**
INTERIOR
PS 1-74 .000
EXTERIOR GLUE

C-D
32/16 **APA**
INTERIOR
PS 1-74 .000
EXTERIOR GLUE

C-C
32/16 **APA**
EXTERIOR
PS 1-74 .000

25–61. *Using plywood for corner bracing because of a lack of room. Notice the diagonal bracing on the side wall.*

25–62. *A piece of material somewhat larger than 2″ × 4″ is used for a sole plate to provide room for a 4″ soil pipe. The 2″ × 4″ studs are then installed flatwise.*

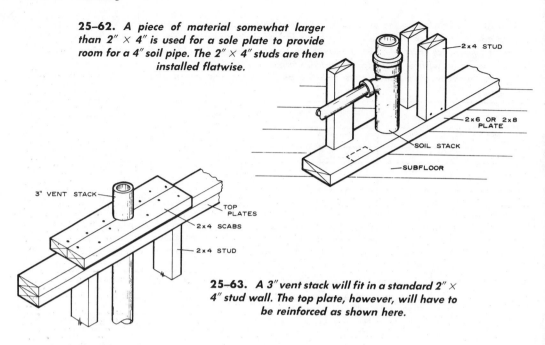

2x4 STUD

2x6 OR 2x8 PLATE

SOIL STACK

SUBFLOOR

3" VENT STACK

TOP PLATES

2x4 SCABS

2x4 STUD

25–63. *A 3″ vent stack will fit in a standard 2″ × 4″ stud wall. The top plate, however, will have to be reinforced as shown here.*

25–64. *Framing for a bathtub.*

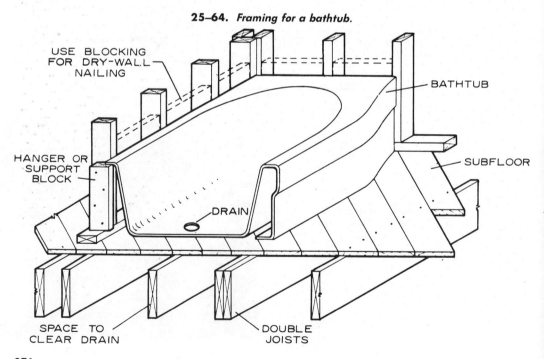

USE BLOCKING FOR DRY-WALL NAILING

BATHTUB

HANGER OR SUPPORT BLOCK

SUBFLOOR

DRAIN

SPACE TO CLEAR DRAIN

DOUBLE JOISTS

which the fixture plumbing is located should also be framed to allow for a small access panel.

CUTTING FLOOR JOISTS

Floor joists should be cut, notched, or drilled only where they will not be greatly weakened. While it is best to avoid cutting, alterations are sometimes required. Joists should then be reinforced by nailing a 2″ × 6″ scab to each side of the altered member, using 12d nails. An additional joist adjacent to the cut joist can also be used.

Notching the top or bottom of the joist should be done only in the end quarter of the span. The notch should not be more than one-sixth the depth of the joist. Thus, for a nominal 2″ × 8″ joist, 12′ long, the notch should be not more than 3′ from the end support and about 1²/₃″ deep. When a joist would require greater alteration, headers and tail beams can be used instead. Fig. 25-65. Proper planning will reduce the need for altering joists.

When necessary, holes may be bored in joists if the diameters are no greater than 2″ and the edges of the holes are not less than 2¹/₂″ from the top or bottom edges. Fig. 25-66. This usually limits a 2″-diameter hole to joists of nominal 2″ × 8″ size and larger.

Plumbing Fixtures

The weight and eventual use of plumbing fixtures requires a secure anchorage. This backing is provided at the necessary locations as shown in Fig. 25-67.

Cabinets and Utility Boxes

Special support and blocking must also be provided for most cabinets. Some cabinets are designed to fit between studs 16″ O.C. and flush with the wall covering. These are usually designed to be fastened from the inside of the cabinet, through the sides, directly into the studs. Backing must be provided at the top and bottom of the cabinet for nailing the wall covering.

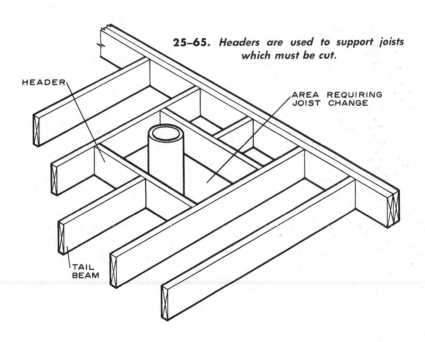

25–65. Headers are used to support joists which must be cut.

HEADER

AREA REQUIRING JOIST CHANGE

TAIL BEAM

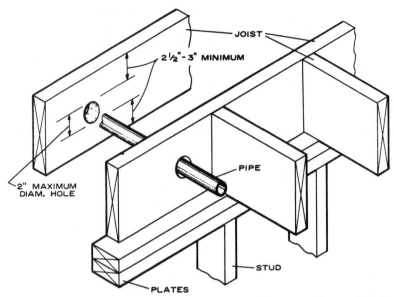

JOIST

2½"–3" MINIMUM

PIPE

2" MAXIMUM
DIAM. HOLE

STUD

PLATES

25–66. *Do not weaken the joists by cutting or boring too large a hole.*

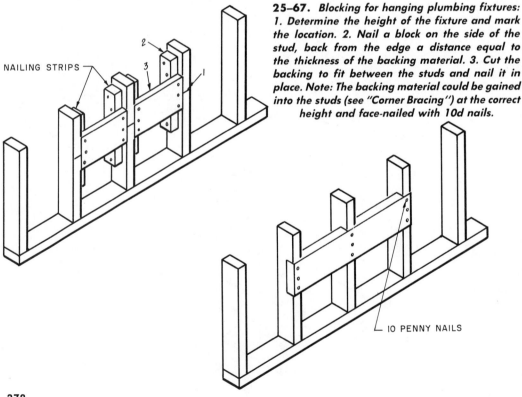

NAILING STRIPS

2

3

1

25–67. *Blocking for hanging plumbing fixtures:*
1. Determine the height of the fixture and mark
the location. 2. Nail a block on the side of the
stud, back from the edge a distance equal to
the thickness of the backing material. 3. Cut the
backing to fit between the studs and nail it in
place. Note: The backing material could be gained
into the studs (see "Corner Bracing") at the correct
height and face-nailed with 10d nails.

10 PENNY NAILS

Bathroom vanities, kitchen cabinets, and other projecting cabinets must be securely fastened. Thus they require special blocking. The location of this blocking can be determined by studying the building plans. One method of blocking for cabinets is shown in Fig. 25-68. In some cases it is advisable to install an extra full-length stud or even a 2″ × 6″ flatwise in the wall. This is especially true in kitchens in which upper and lower cabinets are to be installed.

Trim Backing

Baseboard, chair rail, ceiling, and other moldings require additional backing for nailing at the ends. Without additional backing the nails must be driven very near the ends of the molding and usually at a slight angle to anchor into the corner posts. This often results in splitting the ends of the molding. Backing such as shown in Figs. 25-25 & 25-26 will permit correct nailing procedures.

Backing for Accessories

Towel bars, shower curtain rods, soap dishes, tissue-roll holders and similar items should be supported by backing. These items do not require backing as heavy as that for cabinets or plumbing fixtures. Usually a piece of 1″ × 4″ material located and fastened as in Fig. 25-67 will provide much better support than the wall covering alone.

Heating Ducts

Heating ducts require openings in the ceiling, floor, or wall. Backing must therefore be provided for fastening the covering material. Fig. 25-69. An opening in the wall larger than the distance between the studs will require cutting off one or more studs. It also requires a header to support the shortened stud, which serves as a nailing surface for the wall covering. Fig. 25-70. An opening in the floor larger than the distance between the joists is formed as shown in Fig. 25-65.

Cabinet Soffits

The cabinet soffit is sometimes called a *bulkhead*. There are two common types. The first allows the cabinet to be constructed flush with the front of the soffit. Fig. 25-71. This type is used mainly when the cabinets are built on the job. In the second type, the cabinets are set back about 2″. This method which is more common, is used with prefabricated cabinets. Such cabinets are usually not installed until after the interior painting is completed. Details can be seen in Fig. 25-72.

CONSTRUCTING SOFFITS FOR FLUSH-MOUNTED CABINETS

Snap a chalk line along the wall studs to represent the bottom edge of the soffit. (Arrow #1, Fig. 25-71.) This line will usually be 84″ plus the thickness of the lath and plaster or dry wall above the finished floor. With 10d nails, nail a 2″ × 2″ along the top edge of this line into each wall stud.

Assume the shelves are to be 11¼″ deep. Snap a chalk line along the underside of the ceiling joists 11¼″ out from the double plate. (Arrow #2.) This allows ¾″ for the lath and plaster to be applied later. With 10d nails fasten a 2″ × 2″ along this chalk line into the bottom edge of each ceiling joist.

(If the ceiling joists run the same direction as the soffit, a ladder-type construction, similar to that used when installing nonbearing partitions, will have to be used. Refer to Fig. 25-8. The upper 2″ × 2″ for the bulkhead is nailed to the underside of the 2″ × 4″ ladder construction.)

Cut the 2″ × 4″ for the front bottom edge of the soffit to finished length. Nail 1″ × 4″ × 12″ pieces to the back (Arrow #3) and top edge (Arrow #4) of the 2″ × 4″. Temporarily tack the 2″ × 4″ in position by nailing it with 6d nails to the top 2″ × 2″. (Arrow #5.) With a level make sure that the 2″ × 4″ is level along its length and that its bottom edge is level with the bottom edge of the 2″ × 2″ nailed to the wall.

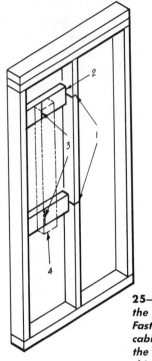

When the 2" × 4" is in position, use 6d nails to fasten the 1" × 4" pieces securely to the two 2" × 2"s. Additional 2" × 4" blocking may be required along the face or bottom edge of the soffit in order to join cabinets or install electrical outlet boxes.

CONSTRUCTING THE CONVENTIONAL CABINET SOFFIT

This type of soffit framing is used when the cabinets are prefabricated and installed later. Two 2" × 2" boards are installed, one on the wall (Fig. 25-72, arrow 1) and the second on the ceiling (arrow 3). These are positioned and nailed in place as in the previous method. The differences

25–68. *Blocking for hanging cabinets: 1. Mark the top and bottom of the cabinets on the studs. 2. Fasten blocks between the studs for attaching the cabinet backing. These blocks must be back from the edge of the studs a distance equal to the thickness of the cabinet backing. 3. Mark the position of the cabinet backing on the blocks. 4. Fasten the cabinet backing to the blocks on the location marks.*

25–69. *Recommended framing for a small heating duct opening.*

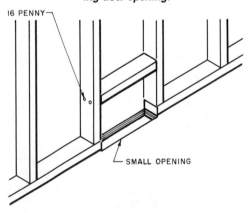

25–70. *Recommended framing for a large heating duct opening.*

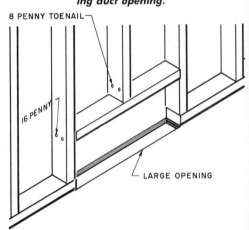

are that the 2″ × 2″ along the ceiling is positioned 14″ from the face of the wall studs to the outside edge of this 2″ × 2″, and the 2″ × 4″ at the bottom corner is turned flat rather than on edge (arrow 2). After the wall covering has been applied and painted, the cabinets are attached to the wall and to the bottom of the soffit. The top front edge of the cabinet is securely attached to the 2″ × 4″ laid flat in the soffit. Since prefabricated cabinets are usually 12″ deep, this allows a 2″ overhang of the soffit at the front edge. A piece of cove or quarter-round molding may be used to close the joint between the cabinet and the bulkhead.

Wall and Ceiling Finish Nailers

Wall and ceiling finish nailers are horizontal or vertical members to which panel-ing, dry wall, rock lath, wall board, or other covering materials are nailed. These pieces are installed at the interior corners of walls and at the junction of the wall and ceiling. The vertical nailers located at the interior corners of the walls may be made up of studs placed where they will make a good base for nailing. Fig. 25-73. This method of construction also provides a good tie between walls.

A second method of installing a vertical nailer at the interior wall corners is provided by a 2″ × 6″ nailed to the stud of the intersecting wall. In this method, a header nailed between the studs of the main wall is necessary to back up the nailer. Fig. 25-74.

At the junction of a wall and ceiling, doubling of the ceiling joists over the wall plates provides a nailing surface for the interior wall finish. Fig. 25-75. The walls are tied to the ceiling framing by toenailing through the ceiling joists into the wall plates.

25–71. Soffit construction details for flush-mounted cabinets.

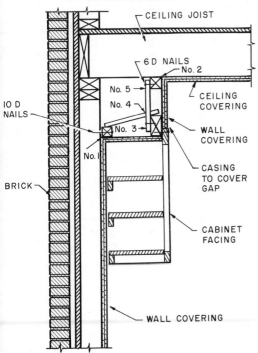

25–72. Construction details for a conventional soffit.

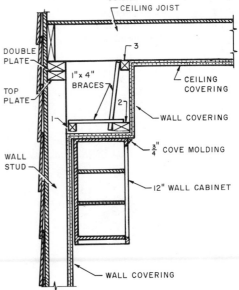

Another method of providing nailing at the ceiling line is similar to that used on the walls. A 2″ × 6″ nailer is nailed to the top of the wall plate. Fig. 25-76. A header is nailed between the ceiling joists and

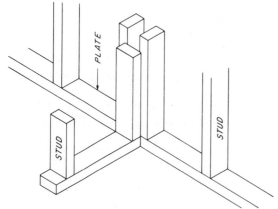

25–73. *This assembly of studs where a partition meets a wall will provide adequate nailing for the wall finish material.*

25–74. *Attach a 6″ board to the back of the last 2″ × 4″ and the wall partition to provide adequate nailing for the wall finish material. Note that a header is nailed between the wall studs so that the partition wall may be toenailed through the nailer into the header to secure the partition wall to the main wall.*

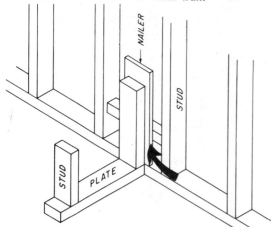

then toenailed to the wall plate through the nailer.

Nailers need not be continuous. Several short pieces of 2″ × 6″ can be used rather than new material. Nailers should be firmly secured with 16d common nails so that they will not be hammered out of position when the wall finish is nailed in place. If nailers are out of position, the wall finish material will be out of square at the corner.

Schedule for Special Framing

Usually there is a need to enclose a building quickly, to protect it from the weather. However, special framing is time consuming. Therefore during the regular framing stage a builder will take time to do only those special framing operations which must be done at that time. The rest of the special framing often is done as fill-in work during slack periods in later construction stages. The climate, other projects, and available manpower affect the scheduling of special framing.

MULTILEVEL FRAMING

The use of platform-frame construction simplifies the framing of multilevel structures. The first level of the two-story house is constructed in the same manner as for a one-story home. However, the ceiling joists

25–75a. *Nailing for the ceiling finish is provided by doubling the ceiling joists at each side of the wall plate.*

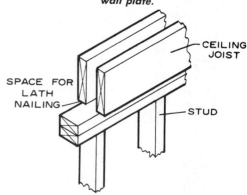

of the first story become the floor joists of the second story. Therefore these joists must be wider to support the additional load.

A subfloor is applied to the top of the joists, the second-story floor plan is laid out, and the walls are constructed and raised in the same way as for the first floor. Fig. 25-77. Some architectural designs require an overhang of the second floor joists beyond the first story walls. The framing of the floor joists can be seen in Fig. 25-78.

Split levels and other variations of architecture require a slightly different treatment of the wall framing. After consulting the sectional elevations in the building plans, make a master stud pattern for each level. This pattern will show the true lengths of the studs for the stub walls found in split-level homes. Keep in mind that each floor platform is supported by a wall section. Thus, a second-level floor platform is supported by the walls which enclose the rooms of the lower level. Fig. 25-79 (Page 387).

Bay-window framing is sometimes associated with multilevel homes. Fig. 25-80 (Page 388). A bay-window projection requires that the floor also be framed out beyond the foundation wall, as for a second-floor overhang. Fig. 25-78.

ESTIMATING WALL FRAMING
Materials and Their Costs

The number of 16" O.C. studs required for a building may be roughly estimated by figuring one for each linear foot of wall. This allows for doubling at openings and at corners. If only the material cost is to be estimated, use the charts in Figs. 25-81 and 25-82. For example, consider a building with an exterior wall 20' long and 8' high, framed with 2" × 4" studs 16" O.C. Such a wall will require 168 board feet of material.

To arrive at this answer, first determine the number of square feet in the wall (20' × 8' = 160 sq. ft.). Refer to the chart in Fig. 25-81. Find "Size of Studs" in the left column (2" × 4" for our example) and read across to the column "Spacing on Center" (16"). In the third column opposite the 16", find the factor for the board feet per square foot of area (1.05 for our example). Multiply this factor by the wall area. Thus 1.05 × 160 = 168 board feet of lumber for the wall.

25–75b. *Horizontal lath nailers at ceiling for a stud wall at right angle to joist.*

25–76. *A nailer attached to the top of a wall plate. The header nailed between the joists (arrow) can then be toenailed into the top of the nailer to secure the partition to the ceiling.*

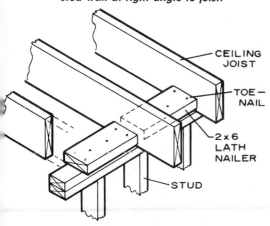

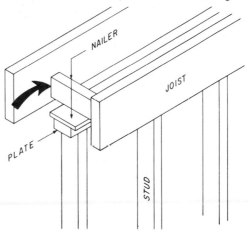

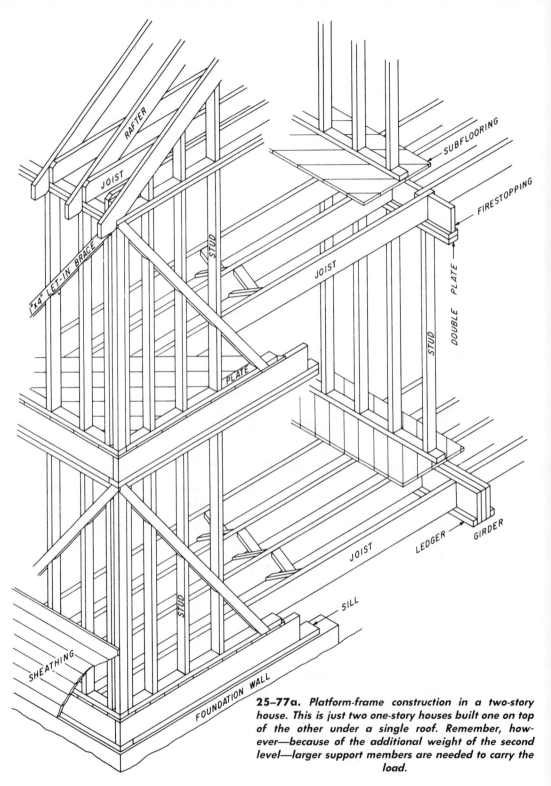

25-77a. Platform-frame construction in a two-story house. This is just two one-story houses built one on top of the other under a single roof. Remember, however—because of the additional weight of the second level—larger support members are needed to carry the load.

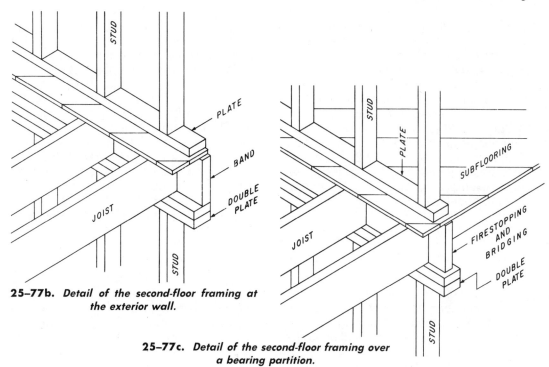

25–77b. Detail of the second-floor framing at the exterior wall.

25–77c. Detail of the second-floor framing over a bearing partition.

25–78a. Second-floor overhang on an exterior wall. Since these joists run at right angles to the wall below, they can just be extended.

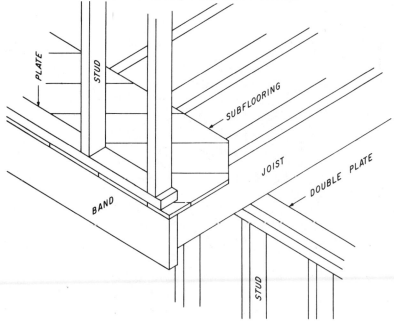

Figure the lumber costs by multiplying the cost per board foot by 168, the amount of lumber required for the wall.

To determine the amount of material for a partition wall, use Fig. 25-82 (for board feet) or Fig. 25-83 (for exact number of studs) and follow the procedure just given.

With Fig. 25-81 you can also figure the number of nails needed. For an exterior wall built with 2″ × 4″ studs, 22 lbs. of nails are needed for each 1,000 bd. ft. The wall in the example contains 168 bd. ft. and will require about 3.7 lbs. of nails:

$$\frac{168 \text{ (bd. ft.)}}{1,000} \times 22 \text{ (lbs. per} \atop 1,000 \text{ ft.)} = 3.7 \text{ pounds}$$

The size of the header for each opening will be found on the building plan. Make a list of headers and their sizes for use during construction as a cutting list. From this list determine the number of board feet of lumber, and figure the cost as described in Unit 24, "Floor Framing." Fig. 24-49 is a chart for figuring board feet content.

The amount of plate material must also be figured and added to the material list. To determine the number of lineal feet of top and bottom plates for walls having double top plates, multiply the length of the wall by three. Other materials such as gable-end studs, corner braces, fire stopping, and wall backing must also be listed. The number of board feet of lumber can be figured from this list. The total cost is determined by multiplying the total number of board feet by the cost of one board foot of lumber.

Labor Costs

The information in Fig. 25-84 can be used to determine the approximate labor cost in the same manner as described in Unit 24, page 322.

25–78b. *Second-floor overhang with wall parallel to the joists. Lookout joists are used and must be attached to a double joist. The double joist should be located a distance of twice the overhang back from the lower wall.*

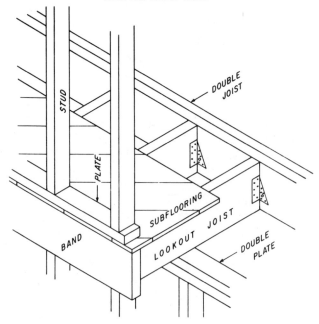

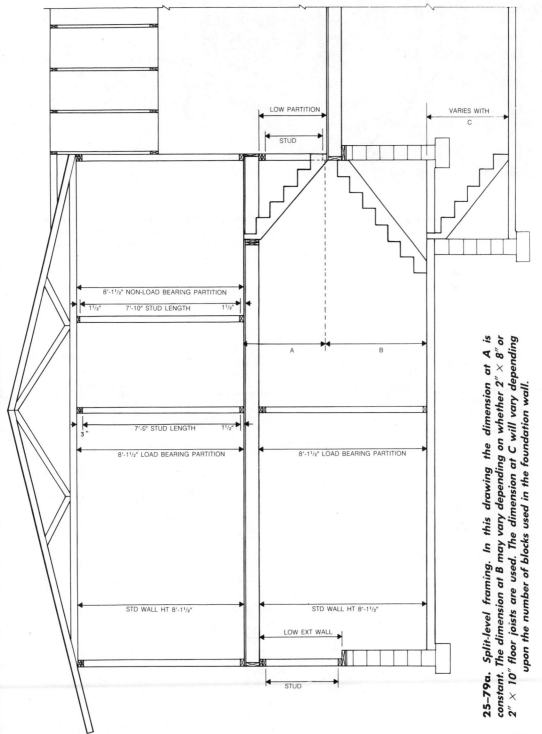

LOW PARTITION

STUD

VARIES WITH C

8'-1½" NON-LOAD BEARING PARTITION

1½"   7'-10" STUD LENGTH   1½"

A   B

3"   7'-9" STUD LENGTH   1½"

8'-1½" LOAD BEARING PARTITION   8'-1½" LOAD BEARING PARTITION

STD WALL HT 8'-1½"   STD WALL HT 8'-1½"

LOW EXT WALL

STUD

25-79a. Split-level framing. In this drawing the dimension at *A is constant.* The dimension at *B* may vary depending on whether 2″ × 8″ or 2″ × 10″ floor joists are used. The dimension at *C* will vary depending upon the number of blocks used in the foundation wall.

387

25–79b. *Floor girders are framed in for the first level. The stub wall is framed in and will help to support the floor framing for the second level of this split-level home.*

25–80a. *Bay window framing. The ceiling joists in the framing of this bay are set on top of the window headers. The top of a bay window should be kept in line with the other windows and doors in the room. Therefore the wall header will not be a standard header height. It will have to be raised so that its bottom is in line with the bottom of the bay ceiling joists.*

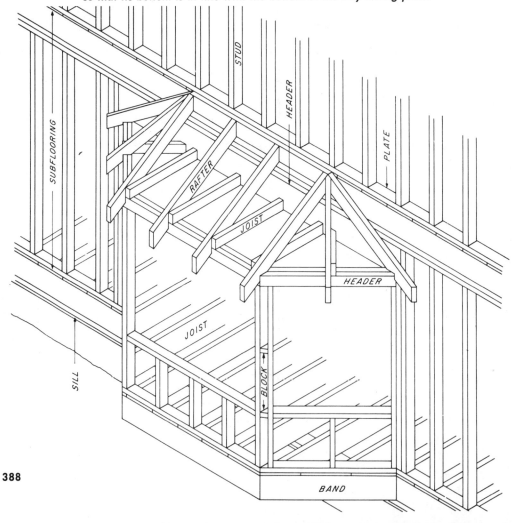

25–80b. *A bay window constructed as shown in Fig. 25–80a. Viewed from the inside of the finished room, the ceiling line is interrupted by the header in the wall. This bay window has a built-in window seat.*

25–81. *This chart tells you how you can figure the number of board feet and nails for exterior wall framing.*

Exterior Wall Studs

(Studs Including Corner Bracing.)			
Size of Studs	Spacing on Centers	Bd. Ft. per Sq. Ft. of Area	Lbs. Nails per 1000 Bd. Ft.
2″ × 3″	12″ 16″ 20″ 24″	0.83 0.78 0.74 0.71	30
2″ × 4″	12″ 16″ 20″ 24″	1.09 1.05 0.98 0.94	22
2″ × 6″	12″ 16″ 20″ 24″	1.66 1.51 1.44 1.38	15

25–82. *With this chart you can figure the board feet of lumber needed for partition studs. In addition to giving the amount of material for the partition construction, this table also gives consideration to the need for extra lumber for headers, trimmers, and other special framing.*

Partition Studs

(Studs including top and bottom plates)			
Size of Studs	Spacing on Centers	Bd. Ft. per Sq. Ft. of Area	Lbs. Nails per 1000 Bd. Ft.
2″ × 3″	12″ 16″ 24″	0.91 0.83 0.76	25
2″ × 4″	12″ 16″ 24″	1.22 1.12 1.02	19
2″ × 6″	16″ 24″	1.48 1.22	16

25–83. *With this chart you can find the number of partition studs needed for a job. The factors in the last four columns are used for figuring the number of feet of lumber. In this table they are close to the actual needs for building a partition, without consideration for special framing.*

Partition Studs

Length Partition in Feet	No. Studs Required	Number of Feet of Lumber Required Per Sq. Ft. of Wood Stud Partition Using 2″ × 4″ Studs. Studs spaced 16″ on centers, with single top and bottom plates. Ceiling Heights in Feet			
		8′ 0″	9′ 0″	10′ 0″	12′ 0″
2	3	1.25	1.167	1.13	1.13
3	3	0.833	0.812	0.80	0.80
4	4	0.833	0.812	0.80	0.80
5	5	0.833	0.812	0.80	0.80
6	6	0.833	0.812	0.80	0.80
7	6	0.833	0.75	0.75	0.80
8	7	0.75	0.75	0.75	0.70
9	8	0.75	0.75	0.75	0.70
10	9	0.75	0.75	0.75	0.70
11	9	0.75	0.70	0.70	0.67
12	10	0.75	0.70	0.70	0.67
13	11	0.75	0.70	0.70	0.67
14	12	0.75	0.70	0.70	0.67
15	12	0.70	0.70	0.70	0.67
16	13	0.70	0.70	0.70	0.67
17	14	0.70	0.70	0.70	0.67
18	15	0.70	0.70	0.67	0.67
19	15	0.70	0.70	0.67	0.67
20	16	0.70	0.70	0.67	0.67
For dbl. plate, add per sq. ft. For 2″ × 8″ studs, double above quantities. For 2″ × 6″ studs, increase above quantities 50%.		0.13	0.11	0.10	0.083

25–84.
Estimating Labor for Wall Framing

Item	Size or Kind	Est. Labor Performance	Item	Size or Kind	Est. Labor Performance
Partition plates and shoe		50 b.f. per hour	Outside wall plates and shoe	2 × 4	40 b.f. per hour
				2 × 6	50 b.f. per hour
Partition studs		50 b.f. per hour	Outside studs	2 × 4	40 b.f. per hour
Wall backing		50 b.f. per hour		2 × 6	50 b.f. per hour
Grounds		85 lin. ft. per hour	Headers for wall openings	2 × 4	40 b.f. per hour
Knee wall plates	2 × 4	40 b.f. per hour		2 × 6	50 b.f. per hour
	2 × 6	40 b.f. per hour	Gable-end studs		50 b.f. per hour
Knee wall studs	2 × 4	40 b.f. per hour	Fire-stopping		50 b.f. per hour
	2 × 6	50 b.f. per hour	Corner braces		50 b.f. per hour

QUESTIONS

1. What is the purpose of the exterior side wall in a structure?

2. What are the requirements for wall framing lumber?

3. Which of the two general types of wall framing is used most often in the United States?

4. Describe briefly the difference between balloon and platform framing.

5. List the names of several parts used in wall framing.

6. What is meant by a rough opening?

7. The width of a rough opening is measured between what two wall members?

8. What is the purpose of the top plate?

9. When laying out stud locations, why is it important for the carpenter to begin from the same side on the back and front of the house?

10. What is a story pole?

11. Describe a header and its purpose.

12. Which method of wall assembly would you prefer and why?

13. Describe how the carpenter can determine when the exterior wall is straight.

14. What is the difference between a bearing and a nonbearing wall?

15. List several examples of special framing.

16. Describe briefly the difference between the two types of cabinet soffits.

17. Why is platform framing easier than balloon framing for multilevel structures?

Wall Sheathing

Exterior coverings over wall framing commonly consist of a sheathing material to which is added finish siding, brick, stone, or other exterior wall material. The inner layer of the outside wall covering on a frame structure is called the *sheathing* (usually pronounced "sheeting"). The outer layer is called the *siding*. Siding, because it is not a structural element, is considered a part of the exterior finish. The sheathing, because it strengthens and braces the wall framing, is considered a structural element. It is therefore a part of the framing. Sheathing forms a flat base upon which the siding is applied and adds not only strength but also insulation to the house. If the type of sheathing used does not provide stiffness and rigidity, diagonal corner bracing should be used. Sheathing is sometimes eliminated from houses built in mild climates.

TYPES OF SHEATHING

The four most common types of sheathing used on modern structures are wood, plywood, fiberboard, and gypsum. Fig. 26-1.

Wood

Wood sheathing consists usually of 1" × 6" or 1" × 8" boards, but thicker and/or wider stock is sometimes used. Boards may be square-edged for ordinary edge-butt joining, or they may be shiplapped

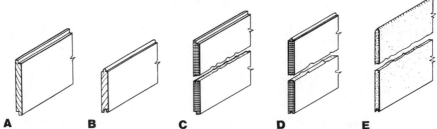

26–1. *Some common types of sheathing: A. Wood shiplap. B. Wood tongue-and-groove. C. Fiberboard shiplap. D. Fiberboard tongue-and-groove. E. Gypsum.*

or dressed-and-matched. Fig. 26-2. *Dressed-and-matched* is simply a term which is used instead of *tongue-and-groove* with reference to sheathing, siding, or flooring.

Plywood

Plywood wall sheathing covers large areas fast and adds great strength and rigidity to the structure of the house. A plywood-sheathed wall is twice as strong

26–2. *Types of edges on wood sheathing: A. Square. B. Shiplap. C. Dressed-and-matched.*

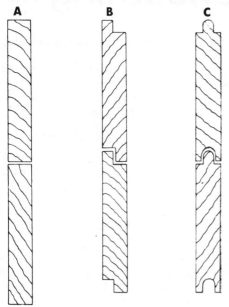

and rigid as a wall sheathed with diagonal boards. Therefore let-in corner bracing is not needed with plywood wall sheathing. Plywood also holds nails well and makes a solid nailing base for the finished siding. Plywood sheathing comes in sheets 4' wide and 8' or more long and is squared-edged. It may be applied either vertically or horizontally. Fig. 26-3. Plywood and hardboard are also used as exterior coverings without sheathing, but grades, thicknesses, and types vary from normal sheathing requirements. This phase of wall construction will be discussed in Unit 39, "Exterior Wall Coverings."

Fiberboard

Fiberboard (sometimes called insulation board) is a synthetic material usually coated or impregnated with asphalt to increase water resistance. Fiberboard sheathing is commonly used in 2' × 8' sheets, which are usually applied horizontally, and in 4' × 8' sheets, which are applied vertically. Fig. 26-4. Edges are usually shiplapped or dressed-and-matched for joining. Thickness is normally $^{25}/_{32}$".

Gypsum

Gypsum sheathing consists of a treated gypsum filler faced on both sides with lightweight paper. Gypsum sheathing comes in 2' × 8' sheets which are applied horizontally. Sheets are usually dressed-and-matched, with V-shaped grooves and tongues. This makes application easier

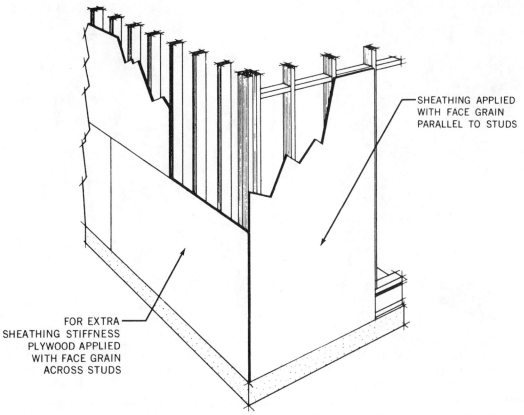

SHEATHING APPLIED
WITH FACE GRAIN
PARALLEL TO STUDS

FOR EXTRA
SHEATHING STIFFNESS
PLYWOOD APPLIED
WITH FACE GRAIN
ACROSS STUDS

26–3. *Plywood sheathing may be applied either horizontally or vertically. When applied horizontally, additional blocking should be included at the horizontal joint between the studs as a base for nailing.*

and adds a small amount of tie between sheets. E, Fig. 26-1.

APPLICATION OF SHEATHING

Sheathing may be applied at any one of several stages during construction:

• When the wall is lying on the subfloor, completely framed and squared. The advantage in applying the sheathing at this time is that it can be nailed in place while the wall sections are lying flat, thus eliminating ladders or scaffolding. The disadvantage is the added weight that must be lifted when erecting the walls.

• When the wall frames have been erected, plumbed, and braced.

• When the ceiling joists have been installed and the wall frames plumbed and braced.

• When the roof has been framed or after the roof frame has been covered.

Most carpenters apply the sheathing as soon as possible because it adds strength and rigidity to the structure. Walls that have been covered with sheathing provide a more solid structure for the ceiling and roof members. Scaffolding is usually erected as soon as the sheathing is applied. Most workers apply the sheathing before the ceiling joists or roof framing so that they will have the scaffolding to stand on when framing the roof.

26-4a. *Installing asphalt-impregnated insulating sheathing vertically.*

26-4b. *Installing 2' × 8' sheets of fiberboard (insulation board) horizontally.*

Wood

Wood sheathing may be applied either horizontally or diagonally. Fig. 26-5. Horizontal sheathing is more often used because it is easy to apply and there is less lumber waste than with diagonal sheathing. However, horizontal sheathing requires diagonal corner bracing in the wall framework. Diagonal sheathing is applied at a 45° angle. This method of sheathing adds greatly to the rigidity of the wall and eliminates the need for corner bracing. There is more lumber waste than with horizontal sheathing because of angle cuts. As stated, the application is somewhat more difficult. The building specifications will state whether horizontal or diagonal application is required.

With either method of application, 6" or 8" boards should be nailed with two 8d nails at each stud crossing. Wider boards should be nailed with three 8d nails. Unless boards are end-matched (shaped on the ends for tongue-and-groove joining), end joints must lie on the centers of studs. End joints must be broken, meaning that no two end joints may lie next to each other on the same stud. Fig. 26-6. If end-matched boards are used, end joints may lie in the spaces between studs, but end joints in adjacent courses (each strip of sheathing is called a course) must not lie in the same stud space.

Before nailing, each board should be driven tightly against the board already in place. Boards at openings should be laid or cut to bring the boards exactly flush with trimmers, headers, or subsills. Sheathing should normally be carried down over the outside floor framing members. This provides an excellent tie between wall and floor framing. Fig. 26-7.

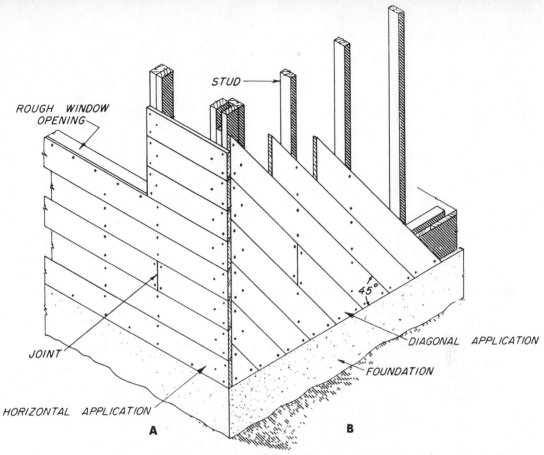

26–5. *Wood sheathing application: A. Horizontal. B. Diagonal.*

STUD

ROUGH WINDOW OPENING

JOINT

HORIZONTAL APPLICATION

A

45°

DIAGONAL APPLICATION

FOUNDATION

B

26–6. *Diagonal wood sheathing is applied at a 45° angle and is nailed with 8d nails to the framing. Note that the pieces are joined at the center of a stud.*

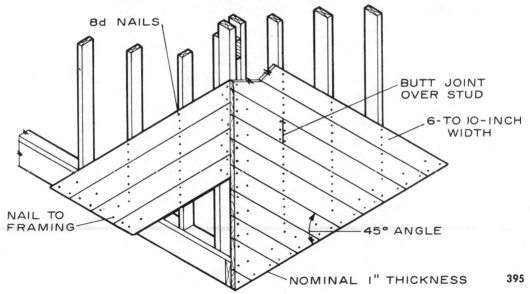

8d NAILS

BUTT JOINT OVER STUD

6- TO 10-INCH WIDTH

NAIL TO FRAMING

45° ANGLE

NOMINAL 1" THICKNESS

395

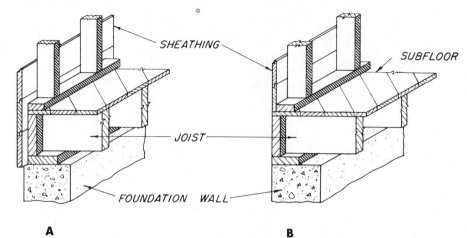

26–7. *Location of sheathing: A. Sheathing may be set on the foundation wall. B. Sheathing may be set on the subfloor.*

26–8. Horizontal application of 2′ × 8′ sheathing: A. Fiberboard. B. Gypsum.

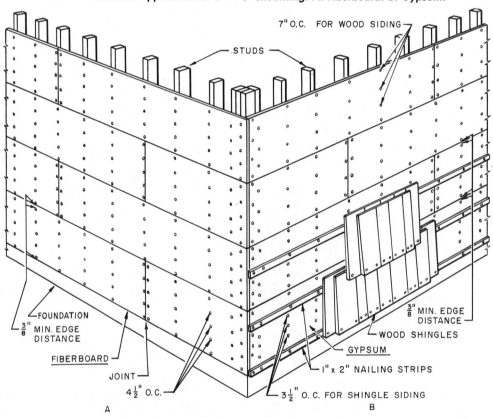

Fiberboard

The most popular size of fiberboard and gypsum sheathing is 2' × 8'. This size is applied horizontally with the vertical joints staggered. Fig. 26-8. Fiberboard sheathing should be nailed at all studs with 2" galvanized roofing nails or other types of noncorrosive nails. Space nails 4½" on center or 6 nails for every 2' of height. Nails should be kept at least 3/8" from the edge of the sheet. Fig. 26-8.

Fiberboard in 4' × 8' sheets is usually applied vertically because perimeter nailing (nailing around outside edges) is possible. When the sheathing is started at the foundation wall rather than the subfloor, use a 2" × 4" nailing strip between the studs for backing. The sheathing should be nailed with 2" galvanized roofing nails spaced 3" on center at the edges and 6" on center at intermediate framing members. The minimum edge distance is 3/8". It is also recommended that the sheets be spaced 1/8" apart to allow for expansion and avoid buckling. Joints should come on the centerline of framing members. Fig. 26-9.

Gypsum

Gypsum sheathing is nailed with 1¾" or 2" galvanized roofing nails. When the exterior finish to be used is one in which

26-9. Fiberboard in 4' × 8' sheets may be applied either vertically or horizontally. When the application is horizontal, the edge nailing is 4" on center. When the application is vertical, the edge nailing is 3" on center.

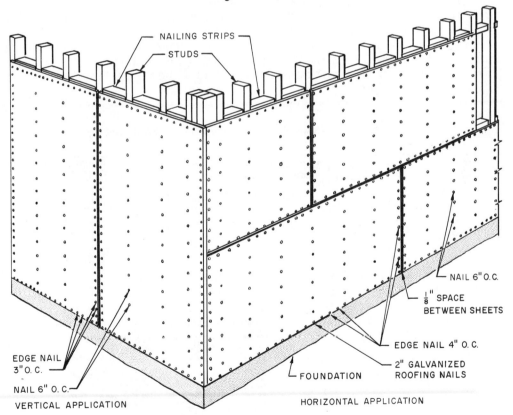

NAILING STRIPS

STUDS

NAIL 6" O.C.

1/8" SPACE BETWEEN SHEETS

EDGE NAIL 4" O.C.

2" GALVANIZED ROOFING NAILS

FOUNDATION

EDGE NAIL 3" O.C.

NAIL 6" O.C.

VERTICAL APPLICATION

HORIZONTAL APPLICATION

siding nails carry through into the studs, the roofing nails for the sheathing may be spaced either 7" on center or 4 nails for every 2' of height. Otherwise, nails should be spaced 3½" on center (7 nails in 2' height). Fig. 26-8.

Plywood

Plywood used for sheathing is usually in 4' × 8' sheets and should be a minimum of 5/16" thick for studs spaced 16". Six-penny nails should be used, spaced not more than 6" apart on edge members and 12" on intermediate members. Fig. 26-10.

Plywood is usually applied vertically, using perimeter nailing with no additional blocking. Fig. 26-11. When the sheathing is started at the foundation wall rather than the subfloor, use a 2" × 4" nailing strip between the studs for backing. Fig. 26-12.

Plywood is applied horizontally in the same way as 4' × 8' fiberboard sheathing. Fig. 26-9. However, the type of nail and nail spacing is different for plywood than for fiberboard. Fig. 26-10 a & c. Blocking is desirable at the horizontal joint between studs as a base for nailing.

When finish siding requires nailing between studs (as with wood shingles), the plywood should be 3/8" thick. If 5/16" plywood is used for sheathing, the wood shingles must either be nailed to stripping or attached with barbed nails.

BUILDING PAPER

Building paper is applied between the sheathing and the siding. It prevents the passage of air through the walls, but it is of relatively little value as a heat insulator because of its thinness.

Building paper should be used behind exterior stucco finish and also over wood sheathing. It should be provided whether or not the sheathing is tongued and grooved. For other than stucco finish it may be omitted when any of the following sheathings are used:

- Plywood.
- Fiberboard that has been treated at the factory to render it water-resistant.
- Core-treated water-repellent gypsum.
- Non-core-treated gypsum that has been treated at the factory to make it moisture-resistant.

26–10a. *Plywood wall sheathing application details.*

Wall sheathing[1]

Panel Identification Index	Minimum Thickness (inch)	Maximum Stud Spacing (inches) Exterior Covering Nailed to:		Nail Size [2]	Nail Spacing (inches)	
		Stud	Sheathing		Panel Edges (when over framing)	Intermediate (each stud)
12/0, 16/0, 20/0,	5/16	16	·16[3]	6d	6	12
16/0, 20/0, 24/0	3/8	24	16 24[3]	6d	6	12
24/0, 32/16,	½	24	24	6d	6	12

Notes: (1) When plywood sheathing is used, building paper and diagonal wall bracing can be omitted

(2) Common smooth,annular, spiral thread, galvanized box or T-nails of the same diameter as common nails (0.113" dia. for 6d) may be used. Staples also permitted at reduced spacing.

(3) When sidings such as shingles are nailed only to the plywood sheathing, apply plywood with face grain across studs.

Look for these APA grade-trademarks on wall sheathing. See Fig. 5-9, Unit 5, for complete details.

STRUCTURAL I
C-D
20/0 (APA)
INTERIOR
PS 1-74 000
EXTERIOR GLUE

C-C
32/16 (APA)
EXTERIOR
PS 1-74 000

In general, the soft, porous, relatively thick machine-finished building papers should be avoided in favor of asphalt-saturated paper or paraffined paper. These papers usually have sufficient water and air tightness and can usually stand handling in all kinds of weather. Building paper should be water-resistant but not vapor-resistant. Vapor-resistant paper might trap moisture between the walls.

The paper should be applied smoothly, and the joints should be lapped and nailed without bulges or cracks through which the air could find its way. Care should be taken around window and door openings to close all cracks.

Building paper should be applied horizontally, starting at the bottom of the wall. Succeeding layers should overlap about 4″ and should cover strips previously applied

26–10b. *Stapling plywood sheathing.*

26–11. *Installing plywood sheathing vertically.*

26–10c. *Stapling schedule.*

Recommended minimum stapling schedule for plywood / All values are for 16-gauge galvanized wire staples having a minimum crown width of ⅜″

Plywood wall sheathing / Without diagonal bracing			
Plywood Thickness	**Staple Leg Length**	**Spacing Around Entire Perimeter of Sheet**	**Spacing at Intermediate Members**
⁵/₁₆″	1¼″	4″	8″
⅜″	1⅜″	4″	8″
½″	1½″	4″	8″

around openings. Strips about 6″ wide should be installed behind all exterior trim of exterior openings.

ESTIMATING
Materials

To determine the amount of wall sheathing required for a structure, first figure the exterior wall area. On a building with walls 8′ high and front and rear walls 40′ long, the front wall area is 320 sq. ft. (8 × 40).

The front and rear walls thus have a total area of 640 sq. ft. (2 × 320). If the end walls are 8′ high and 20′ long, they contain 160 sq. ft. each, (8 × 20), for a total of 320 sq. ft. (2 × 160). The entire structure contains 960 sq. ft. of exterior wall area (320 + 640).

For a house with a hip roof this would be the exterior wall area. For a house with a gable roof, the gable ends have to be figured and added to this amount. To find the area of one gable, multiply the height of the gable by one-half the width of the

26–12. *Plywood sheathing started at the foundation wall will require a nailing strip along the top joint.*

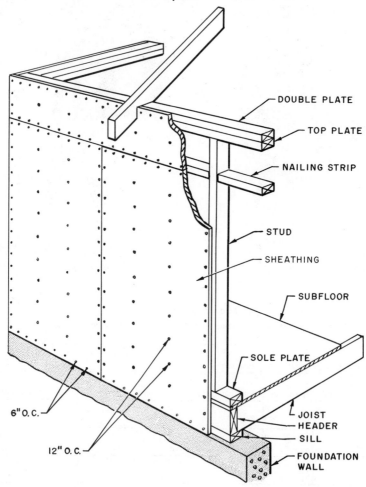

DOUBLE PLATE
TOP PLATE
NAILING STRIP
STUD
SHEATHING
SUBFLOOR
SOLE PLATE
JOIST
HEADER
SILL
FOUNDATION WALL
6″ O. C.
12″ O. C.

bottom. Multiply this area by the number of gables to determine the total gable area.

A 4' × 8' sheet of plywood contains 32 sq. ft. The number of plywood sheets required to sheathe the house is 30:

$$\frac{960 \text{ (total exterior wall area)}}{32 \text{ (sq. ft. in 1 plywood sheet)}} = 30 \text{ sheets}$$

The amount of fiberboard or gypsum sheathing is estimated in the same way.

Refer to the chart in Fig. 26-13 if lumber is used for sheathing. For example, let us say that 1 × 6, S4S material is to be used and applied horizontally. Read down the left column headed "Type" to S4S and across from size 1 × 6. The column "Board Feet per Square Foot of Area" indicates a factor of 1.15. Multiply this factor by the exterior wall area, which in this example is 960 sq. ft. A total of 1,104 board feet of S4S, 1 × 6 material is needed to sheathe the structure (960 × 1.15).

The number of nails required for applying this sheathing can also be determined from the chart in Fig. 26-13. If the wall studs are spaced on 16" centers, 31 pounds of nails are required for each thousand board feet of material. For our example, 34 pounds of nails would be required.

$$\frac{1,104 \text{ (bd. ft. of sheathing)}}{1,000} = 1.104$$

(amount of sheathing expressed in thousands of bd. ft.)

1.104 × 31 (lbs. of nails per 1,000 bd. ft.) = 34.2 lbs.

The cost of the material can be figured by multiplying the unit cost times the amount needed.

Labor

The charts in Fig. 26-14 can be used as a guide in estimating labor time for sheathing application.

26–13. *Estimating lumber sheathing and nails for horizontal application.*

Sheathing and Subflooring (Horizontal Application)

Type	Size	Bd. Ft. per Sq. Ft. of Area	12"	16"	20"	24"
			Lbs. Nails Per 1000 Bd. Ft. — Spacing of Framing Members			
T & G	1 × 4	1.32	66	52	44	36
	1 × 6	1.23	43	33	28	23
	1 × 8	1.19	32	24	21	17
	1 × 10	1.17	37	29	24	20
Shiplap	1 × 4	1.38	69	55	46	38
	1 × 6	1.26	44	34	29	24
	1 × 8	1.21	32	25	21	17
	1 × 10	1.18	37	29	25	20
S4S	1 × 4	1.19	60	47	40	33
	1 × 6	1.15	40	31	26	22
	1 × 8	1.15	30	23	20	17
	1 × 10	1.14	36	28	24	19

26–14a. *Estimating labor time for wood sheathing.*

Wood Sheathing (Diagonal Application)

Size	Estimated Labor Performance
1" × 6"	65 bd. ft. per hour
1" × 8"	70 bd. ft. per hour
1" × 10"	75 bd. ft. per hour

26–14b. *Estimating labor time for gypsum and plywood sheathing.*

Other Types of Sheathing

Type of Sheathing	Size	Estimated Labor Time per 100 sq. ft.
Gypsum board	48" × 96"	2.2 hrs.
Plywood panels	48" × 96"	1.8 hrs.

QUESTIONS

1. Why is wall sheathing considered a structural element and a part of the framing?

2. List four common types of sheathing.

3. Why isn't let-in corner bracing required when plywood wall sheathing is used?

4. Wall sheathing is usually applied at one of four stages of construction. At which stage would you apply wall sheathing? List reasons.

5. Why is building paper applied between sheathing and siding?

6. Which of the various kinds of building papers available is recommended for application between the sheathing and the siding? Why?

UNIT 27 Ceiling Framing

When the wall framing has been completed, it is ready to be tied together with the roof framing. There are two basic methods of roof framing for platform-frame construction: trussed roof construction and conventional roof construction.

Trusses are prefabricated assemblies placed on the building and attached as a unit. Fig. 27-1. The lower chords in the trusses form the ceiling of the room and support the ceiling finish. Since trusses are the completed roof frame, they will be discussed in a separate chapter on roof framing (Unit 33).

In conventional roof construction, the *ceiling joists* and *rafters* are laid out, cut, and fastened one piece at a time to the building. Fig. 27-2. *Ceiling joists* are the parallel beams which support ceiling loads. They are supported in turn by larger beams, girders, or bearing walls. *Rafters* are the inclined members of a roof framework. They support the roof loads. Conventional roof framing requires ceiling joists which serve as a tie between the exterior walls and the interior partitions. The ceiling joists also serve as floor joists for an attic or a second story.

27-1. *Trussed rafter assembly.*

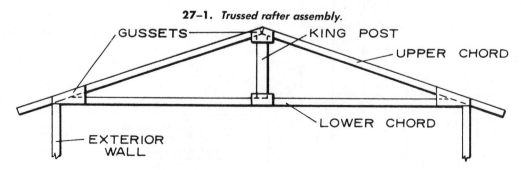

GUSSETS — KING POST
— UPPER CHORD
LOWER CHORD
EXTERIOR WALL

CEILING JOISTS

Size

The size of the ceiling joists is determined by the distance they must span and the load they must carry. The species and grade of wood are also factors to be considered. The correct size for the joists will be found on the building plans as recommended by the building code. As a general reference, the table in Fig. 27-3 shows the joist sizes and the spacing and span limitations, but be sure to confirm these with the local building code.

Layout

Ceiling joists are usually located across the width of the building and parallel to the rafters. The ends of the joists which rest on the exterior wall plates next to the rafters will usually project above the top edge of the rafter. These ends must be cut off on a slope that is equal to the roof pitch. Fig. 27-4.

The spacing of the joists is 16" or 24" on center, depending on the building specifications. Installation is begun at one end of the building and continued across the structure. Extra joists, if needed, are placed without altering the spacing of the prime joists. The first joist is located at the inside edge of the plate on an end wall. This provides edge nailing for the ceiling finish. Fig. 27-5. The second joist is usually

located over the stud in the side wall. The distance between the first two joists will thus be less than 16" or 24", depending on the center spacing used. Fig. 27-6. Each succeeding joist is spaced 16" or 24" on center.

If the layout of the ceiling joists places a joist over a stud and the studs for the two sidewalls have been laid out from the same end of the building, the ends of the joists will butt against each other over the bearing wall. Fig. 27-6. The total length

27-3. *Allowable spans for ceiling joists using nonstress-graded lumber.*

Size of Ceiling Joists (Inches)	Spacing of Ceiling Joists (Inches)	Maximum Allowable Span (Feet and Inches)			
		Group I	Group II	Group III	Group IV
2 × 4	12	11-6	11-0	9-6	5-6
	16	10-6	10-0	8-6	5-0
2 × 6	12	18-0	16-6	15-6	12-6
	16	16-0	15-0	14-6	11-0
2 × 8	12	24-0	22-6	21-0	19-0
	16	21-6	20-6	19-0	16-6

27-4. *The ends of the ceiling joists must be cut off on a slope equal to the roof pitch. It is best to cut them off about 1/8" below the top edge of the rafter.*

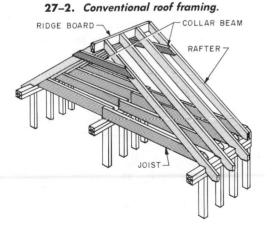

27-2. *Conventional roof framing.*

RIDGE BOARD

COLLAR BEAM

RAFTER

JOIST

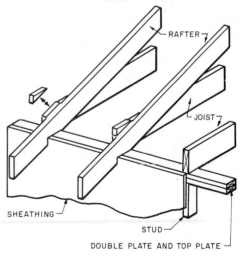

RAFTER

JOIST

SHEATHING

STUD

DOUBLE PLATE AND TOP PLATE

of the two joists will equal the width of the building, and the ends that butt will have to be squared and cut off to length. Since each of the joist ends will be resting on just half of the partition wall plate, a plywood joist splice should be nailed securely to both sides of the joists. Fig. 27-7a. Metal connectors are also available for this purpose. Fig. 27-7b.

An alternate method of joist layout is to offset the joists 1½" on the two outside walls so that they lap each other when they join over the bearing partition. Fig. 27-8. This lap is face-nailed with three 16d nails, and the joists are toenailed to the bearing partition wall plate with two 10d nails. Nonbearing partitions which run parallel to the ceiling joists are nailed to blocks installed between the joists. Fig. 27-9.

Installation

Sight down the edge of the joist to determine the crown and place the crown, or camber, up. When attaching the ceiling joist to the exterior wall plate, keep the end of the joist even with the outside edge of the plate. Fasten this end of the joist first. Toenail three 10d nails through the joist into the plate or use a metal bracket and the special nails furnished with the brack-

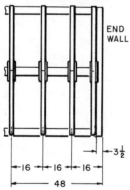

27–5. *The first ceiling joist is set on the inside edge of the end wall to permit nailing of the material for the ceiling surface. (See arrow.)*

27–6. *The distance between the first two joists is less than 16". The joists shown here are butted end to end on the bearing wall. The butt joint must be reinforced.*

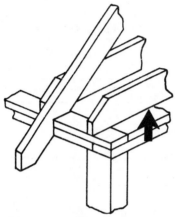

END
WALL

$3\frac{1}{2}$

16 16 16

48

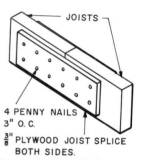

JOISTS

4 PENNY NAILS
3" O. C.

$\frac{3}{8}$" PLYWOOD JOIST SPLICE
BOTH SIDES.

27–7a. *Ceiling joists butted end to end must be spliced together for strength. When lumber is used instead of plywood for a splice, it must be 3/4" thick and at least 24" long.*

27–7b. *A metal connector may be used to reinforce the butt end joint of the ceiling joists.*

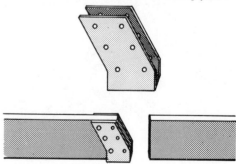

et. Fig. 27-10. Make sure the walls are straight along the top edge and plumb. Fasten the joists together where they join and then to any other double plates they may cross over. This will tie the building together at the top and make it ready for the roof framing or the sheathing.

SPECIAL FRAMING

When framing a low-pitched hip roof, the first ceiling joist will interfere with the bottom edge of the rafters. Stub joists

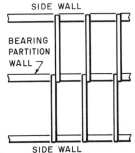

27–8. Ceiling joists lapped on the bearing wall.

27–9. A nonbearing partition wall is fastened to a block which has been nailed between the joists. Notice the backing which has been attached to the top of the partition for nailing the ceiling material.

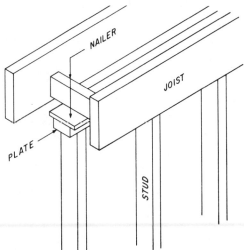

installed at right angles to the regular joists will correct this situation. Fig. 27-11. Space the stub 16″ on center for attaching the finished ceiling. Locate them so that the rafters, when installed, may be nailed directly to the side of the stubs.

Openings in the ceiling that are larger than the spacing between the joists are often necessary. They may be needed for a chimney or for access to the attic area. An enlarged opening will require the cutting of one or more joists. Such joists will need to be supported and framed as described earlier in the section "Framing Floor Openings" in Unit 24.

FRAMING FOR FLUSH CEILING AREA

A living-dining-kitchen group or kitchen-dining-family room group is often designed as one open area with a flush ceiling throughout. This makes the rooms appear much larger than they actually are. When trusses are used, there is no problem because they span from one exterior wall to the other. However, if ceiling joists and rafters are used, some type of beam is

27–10. A metal connector used to fasten the ceiling joist to the double plate.

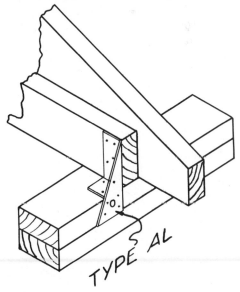

needed to support the interior ends of the ceiling joists.

This support can be provided by a flush beam, which spans from an interior cross wall to an exterior end wall. Joists are fastened to the beam with joist hangers. Fig. 27-12. These hangers are nailed to the beam with 8d nails and to the joist with nails furnished with the hangers or 1½″ roofing nails. Hangers are perhaps most easily fastened by nailing them to the end of the joist before the joist is raised into place.

An alternate method of framing utilizes a wood bracket at each pair of ceiling joists, tying them to a beam which spans the open area. Fig. 27-13. This beam is blocked up and fastened at each end to the double plate, at a height equal to the depth of the ceiling joists.

On a long span with a continuous joist, the joists sometimes need additional support. In this case, a ceiling joist strongback is constructed. Fig. 27-14. The joists should be aligned and properly spaced where they pass under the strongback before they are blocked and nailed.

ESTIMATING
Determining the Number of Joists

One way to determine the number of joists necessary for a building is to divide the length of the floor (in feet) by the joist spacing (in feet), and add one for the end joist. Then multiply by two.

27–11a. *Stub joists are securely anchored to the regular joist with metal straps.*

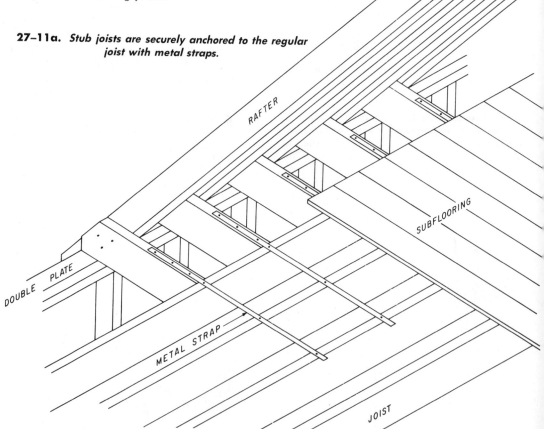

For joists on 16″ centers, take three-fourths of the length of the building, add one, and multiply the total by two. For example, if a building is 40′ long, 31 joists will be required to span from the front exterior wall to the bearing partition wall ($^3/_4 \times 40 = 30$; $30 + 1 = 31$). Another 31 joists will be needed to span from the partition to the rear exterior wall. The total number of joists will be 62 (31×2).

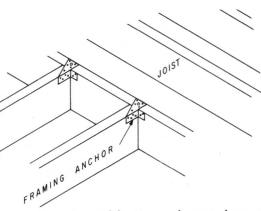

27–11b. *A metal framing anchor may be used in place of the metal straps to secure the stub joists to the regular joist.*

If the building is 20′ wide and has a bearing partition wall at the center, each joist should be 10′ long. Thus, for a building 20′ × 40′, 62 ten-foot joists will be required. Add to this amount extra joists for openings for which trimmer joists are specified.

Determining the Material Cost

An accurate estimate of the cost can be figured by multiplying the number of joists required by the cost per joist. Sometimes the builder will not make up a complete bill of materials, but will want a rough cost estimate. Without knowing the exact number of pieces needed, this estimate can be made by first finding the area of the ceiling. The area of the ceiling is the length of the building times the width of the building for each level. For example, a one-story building 20′ wide and 40′ long has a ceiling area of 20 × 40, or 800 square feet.

The number of board feet required for the joists of a building can be found in the chart in Fig. 27-15. Use the building in the example, which contains 800 sq. ft. of ceiling area. According to the chart, if the joists are 2″ × 6″ and 16″ on center, there are 88 board feet of lumber for each 100

27–12. *Ceiling joists are fastened to a flush beam with joist hangers.*

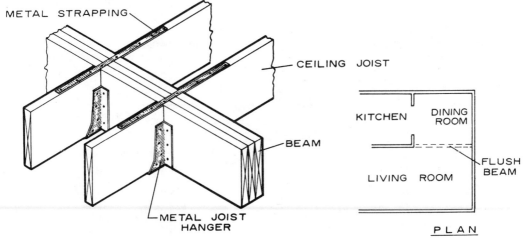

METAL STRAPPING

CEILING JOIST

BEAM

METAL JOIST HANGER

KITCHEN   DINING ROOM

LIVING ROOM

FLUSH BEAM

P L A N

sq. ft. of ceiling surface area. Divide the total ceiling area by 100 and multiply by the factor in the chart:

$$\frac{800 \text{ (ceiling area)}}{100} = 8 \text{ (ceiling area expressed in hundreds)}$$

8 × 88 (bd. ft. per 100 sq. ft. of surface) = 704 bd. ft.

The chart in Fig. 27-15 also has information for determining the number of nails necessary. For the ceiling in the example, 13 pounds of nails are needed for each 1,000 board feet. Since the ceiling in the example has 704 board feet, the ceiling will require about nine pounds of nails:

$$\frac{704 \text{ (bd. ft. of material)}}{1,000} = 0.704 \text{ (material expressed in thousands)}$$

0.704 × 13 (lbs. of nails per 1,000 bd. ft.) = 9.152, or 9 lbs. of nails

Determining the Labor Cost

The labor cost for framing a ceiling can be found by using the information in the column headed "Labor" in Fig. 27-15. To use this chart it is necessary to know how many board feet of joist material are required for a ceiling. This can be determined as described earlier in "Deter-

mining Material Cost." A more accurate method is to find the number of board feet in one joist and then multiply by the total number of joists in the building.

Use the same example described earlier: a single-story building 20' wide × 40' long. The chart in Fig. 27-16 can be used to determine the board feet content of a single joist. To use this chart, the length of the piece must be known. The length of the joists will be found on the building plan. The building in the example is 20' wide

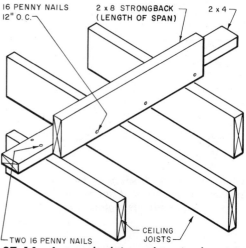

27–14. *A strongback is used to give long joists additional support.*

27–13. *Wood brackets may also be used to attach the ceiling joists to the beam.*

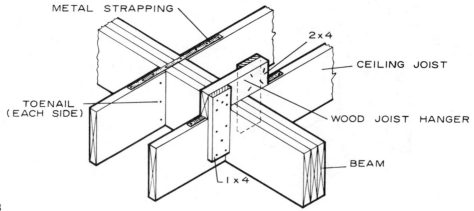

with a bearing partition centered in the building to support the ceiling joists. The ceiling joists will be 2″ × 6″ × 10′ long.

Find 2″ × 6″ in the left column of the table in Fig. 27-16. Read across to the column headed "10" for a ten-foot joist. There are 10 board feet in one 2″ × 6″ × 10′ piece of lumber. The ceiling in the example has 62 joists. If one joist contains

10 board feet, 62 joists contain a total of 620 board feet of joist material (62 × 10).

Refer to the table in Fig. 27-15. The column headed "Labor" shows that a worker can frame 65 board feet of joists per hour. The number of hours required to frame the ceiling in the example is 9.5 (620 ÷ 65). Multiply this by the hourly rate to determine the labor cost.

27-15. *Material and labor requirements for ceiling joists.*

CEILING JOIST	Material				Nails	Labor
	Board Feet Required for 100 Sq. Ft. of Surface Area				Per 1000 Bd. Ft.	Board Feet per Hr.
Size Of Joist	12″ O.C.	16″ O.C.	20″ O.C.	24″ O.C.	Pounds	Board Feet
2 x 4	78	59	48	42	19	60
2 x 6	115	88	72	63	13	65
2 x 8	153	117	96	84	9	65
2 x 10	194	147	121	104	7	70
2 x 12	230	176	144	126	6	70

27-16. *Board feet content for various sizes of timber.*

Board Feet Content

Size of Timber in Inches	Length Of Piece In Feet							
	10	12	14	16	18	20	22	24
1 × 2	1⅔	2	2⅓	2⅔	3	3⅓	3⅔	4
1 × 3	2½	3	3½	4	4½	5	5½	6
1 × 4	3⅓	4	4⅔	5⅓	6	6⅔	7⅓	8
1 × 5	4⅙	5	5⅚	6⅔	7½	8⅓	9⅙	10
1 × 6	5	6	7	8	9	10	11	12
1 × 8	6⅔	8	9⅓	10⅔	12	13⅓	14⅔	16
1 × 10	8⅓	10	11⅔	13⅓	15	16⅔	18⅓	20
1 × 12	10	12	14	16	18	20	22	24
1 × 14	11⅔	14	16⅓	18⅔	21	23⅓	25⅔	28
1 × 16	13⅓	16	18⅔	21⅓	24	26⅔	29⅓	32
1 × 20	16⅔	20	23⅓	26⅔	30	33⅓	36⅔	40
1¼ × 4	4⅙	5	5⅚	6⅔	7½	8⅓	9⅙	10
1¼ × 6	6¼	7½	8¾	10	11¼	12½	13¾	15
1¼ × 8	8⅓	10	11⅔	13⅓	15	16⅔	18⅓	20
1¼ × 10	10⅓	12½	14½	16⅔	18⅔	20⅚	22⅚	25
1¼ × 12	12½	15	17½	20	22½	25	27½	30

(Continued on page 410)

Board Feet Content (Continued from page 409)

Size of Tim- ber in Inches	Length Of Piece In Feet							
	10	12	14	16	18	20	22	24
1½ × 4	5	6	7	8	9	10	11	12
1½ × 6	7½	9	10½	12	13½	15	16½	18
1½ × 8	10	12	14	16	18	20	22	24
1½ × 10	12½	15	17½	20	22½	25	27½	30
1½ × 12	15	18	21	24	27	30	33	36
2 × 4	6⅔	8	9⅓	10⅔	12	13⅓	14⅔	16
2 × 6	10	12	14	16	18	20	22	24
2 × 8	13⅓	16	18⅔	21⅓	24	26⅔	29⅓	32
2 × 10	16⅔	20	23⅓	26⅔	30	33⅓	36⅔	40
2 × 12	20	24	28	32	36	40	44	48
2 × 14	23⅓	28	32⅔	37⅓	42	46⅔	51⅓	56
2 × 16	26⅔	32	37½	42⅔	48	53⅓	58⅔	64
2½ × 12	25	30	35	40	45	50	55	60
2½ × 14	29⅙	35	40⅚	46⅔	52½	58⅓	64⅙	70
2½ × 16	33⅓	40	46⅔	53⅓	60	66⅔	73⅓	80
3 × 6	15	18	21	24	27	30	33	36
3 × 8	20	24	28	32	36	40	44	48
3 × 10	25	30	35	40	45	50	55	60
3 × 12	30	36	42	48	54	60	66	72
3 × 14	35	42	49	56	63	70	77	84
3 × 16	40	48	56	64	72	80	88	96
4 × 4	13⅓	16	18⅔	21⅓	24	26⅔	29⅓	32
4 × 6	20	24	28	32	36	40	44	48
4 × 8	26⅔	32	17⅓	42⅔	48	53⅓	58⅔	64
4 × 10	33⅓	40	46⅔	53⅓	60	66⅔	73⅓	80
4 × 12	40	48	56	64	72	80	88	96
4 × 14	46⅓	56	65⅓	74⅔	84	93⅓	102½	112

QUESTIONS

1. What are the functions of the ceiling joists?

2. When laying out the ceiling joists, why are the first two ceiling joists less than 16″ on center?

3. What are two methods used to join the ceiling joists on the partition plate?

4. What must be done to the building before the joists are nailed together where they join to the partition plates?

5. What is a ceiling joist strongback?

6. How many joists are needed for a building 48′ long if the joists are spaced 16″ o.c.?

UNIT 28

Roof Framing

The primary function of a roof is to protect the house in all types of weather with a minimum of maintenance. Roof construction should be strong in order to withstand snow and wind loads. Roofing members should be securely fastened to each other to provide continuity across the building, and they should be anchored to exterior walls.

A second consideration is appearance. Besides being practical, a roof should add to the attractiveness of the home. Various roof styles are used to create different architectural effects. A carpenter must understand and be able to frame these various styles. Fig. 28-1.

ROOF STYLES

The basic roof styles used for homes and small buildings are as follows:
- Flat.
- Shed.
- Gable.
- Gable with dormer.
- Gable and valley.
- Hip.
- Hip and valley.

Variations of these roofs are associated with architectural styles of different countries or geographic regions. Some of these variations include the following. Fig. 28-2.
- Gambrel.
- Mansard.
- Butterfly.
- Dutch hip.

Flat Roof

Roof joists for flat roofs are laid level or at a slight slope for drainage. Sheathing and roofing are applied to the top of the joists, which in this case serve as rafters. The ceiling material is applied to the underside. Fig. 28-2.

Shed Roof

Sometimes called a lean-to, this roof is nearly flat, and its slope is in one direction only. The shed roof is used for contemporary homes and for additions to existing structures. When it is used as an addition, the roof may be attached to the side of the existing structure or to the existing roof. Figs. 28-2 and 28-3. (Page 414).

Gable Roof

The gable roof is the most common. It has two roof slopes which meet at the top, or ridge, to form a gable at each end. For variation, the gable may include dormers which add light and ventilation to second-floor rooms. Fig. 28-2.

Hip Roof

The hip roof slopes at the ends of the building as well as at the two sides. This slope to all sides makes possible an even overhang all around the building. The low appearance of this roofline and the fact that it minimizes maintenance (there is no siding above the eaves) make it a popular choice. Fig. 28-2.

Mansard Roof

The mansard is a variation of the hip roof. It has steep slopes on all four sides but these do not meet at the center as in a hip roof. Partway up on each side, a second slope is developed. The second slope

28–1. *The roof style is an important part of the architectural design of a building: A. Gable roof. B. Hip roof. C. Mansard roof.*

is almost flat and continues toward the center of the building where it meets with the slopes from the other sides. The mansard roof style was brought to this continent by the French when they settled in Quebec. Fig. 28-2.

Gambrel Roof

The gambrel roof is a variation of the gable roof. It has steep slopes on two sides. Partway up a second slope is developed which continues toward the center of the building, where it meets with the roof from the other side. This roof style was first used in the United States by German settlers in New York and Pennsylvania. Fig. 28-2.

Butterfly and Dutch Hip Roofs

A butterfly roof is an inverted gable roof. The Dutch hip roof is a hip roof with a small gable at the ridge. Construction details are provided in the architectural plans.

ROOF FRAMING TERMS

Span. The distance between the outside edge of the double plates. It is measured at right angles to the ridge board. Fig. 28-4.

Run. One-half the span distance (except when the pitch of the roof is irregular). Fig. 28-4.

Unit of Run. One unit of run is equal to 1', or 12", of run. Fig. 28-5.

Measuring line. An imaginary line running lengthwise from the outside wall to the ridge. Fig. 28-4.

Rise. The vertical distance from the top of the double plate to the upper end of the measuring line. Fig. 28-4.

Unit rise. The number of inches that a roof rises for every foot of run (unit of run). Fig. 28-5.

Pitch. The angle which the roof surface makes with a horizontal plane. It is the ratio of the rise to the span. Fig. 28-6.

For example, a roof may have a rise of 6' and a span of 24'. Fig. 28-7. Such a roof has 1/4 pitch:

$$\frac{6' \text{ (rise of rafter)}}{24' \text{ (span of building)}} = \frac{1}{4} \text{ pitch}$$

A ½ pitch roof rises one-half the distance of the span. For a 24′ span, the rise of a ½ pitch roof would be 12′. Fig. 28-8.

The common roof pitches are ¼, ½, and ⅓. Figs. 28-7, 28-8, and 28-9.

Cut of a roof. The rise in inches and the unit of run. It is used when referring to the roof pitch. Fig. 28-10. For example, the cut may be 6-12 (6″ of rise per foot of run), 8-12

28–2. *Common roof styles.*

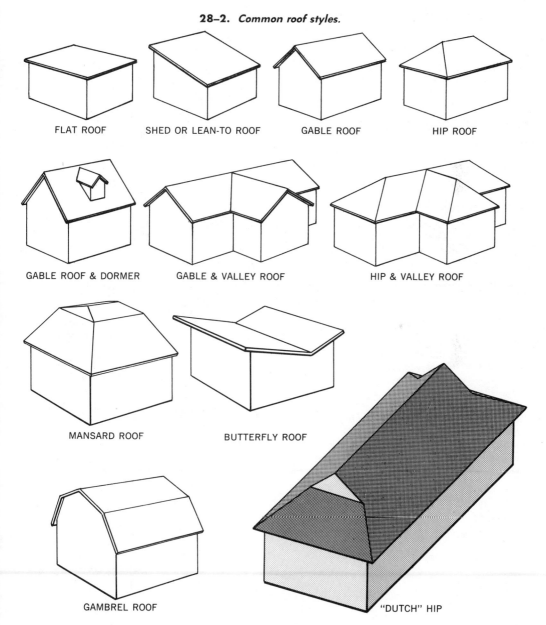

FLAT ROOF

SHED OR LEAN-TO ROOF

GABLE ROOF

HIP ROOF

GABLE ROOF & DORMER

GABLE & VALLEY ROOF

HIP & VALLEY ROOF

MANSARD ROOF

BUTTERFLY ROOF

GAMBREL ROOF

"DUTCH" HIP

413

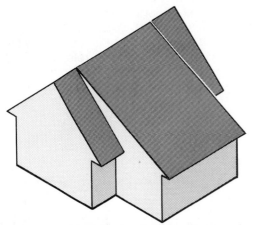

28–3. *A shed roof on an addition tied into the main roof of the existing structure.*

(8″ of rise per foot of run), or 12-12 (12″ of rise per foot of run). Figs. 28-7, 28-8, and 28-9.

Slope. The incline of a roof. Slope is the inches of vertical rise in twelve inches of horizontal run. It is expressed sometimes as a fraction (×/12), but typically as "× in 12." For example, a roof that rises at the rate of 4″ for each foot (12″) of run is designated as having a 4-in-12 slope (or 4/12). The triangular symbol above the roof in Fig. 28-7 conveys this information.

Plumb and level lines. These terms refer to the direction of a line on a rafter, not to any particular rafter cut. Any line that is vertical when the rafter is in its proper position is called a plumb line. Any line

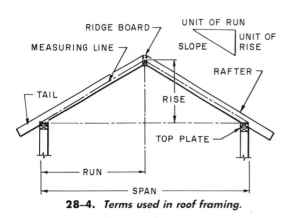

28–4. *Terms used in roof framing.*

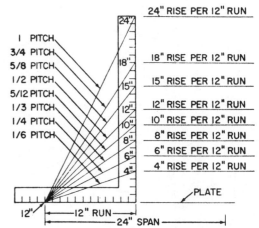

28–6. *A comparison of the two methods of expressing the pitch.*

28–5. *A comparison of total and unit terms. The unit of run is always 12″. Therefore the unit of span is always 24″. The rise in inches is variable, depending on the pitch assigned to the roof. In the example shown here, there are 8″ of rise per (foot) unit of run.*

28–7. *This roof has a 1/4 pitch, or 6″ to the foot.*

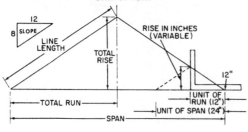

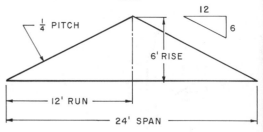

that is level when the rafter is in its proper position is called a level line. Fig. 28-11.

Ridge board. The horizontal piece that connects the upper ends of the rafters. Fig. 28-12.

Tail. The portion of the rafter which extends beyond the wall of the building to form the overhang or eave. Fig. 28-12, arrows.

Rafters. The inclined members of the roof framework. They serve the same purpose in the roof as joists in the floor or studs in the wall and are usually spaced 16" or 24" apart. They vary in width depending on their length, the distance they are spaced apart, their slope, and the kind of roof covering to be used. Rafters sometimes extend beyond the wall of the building to form eaves and protect the sides of the building.

Wood members used for roof framing should normally not exceed 19% moisture content. There are several kinds of rafters necessary for framing the many different roof styles.

Common rafters extend from the plate to the ridge board at 90° to both. Fig. 28-13.

Hip rafters extend diagonally from the corners formed by the plate to the ridge board. Fig. 28-13.

Valley rafters extend diagonally from the plates to the ridge board along the

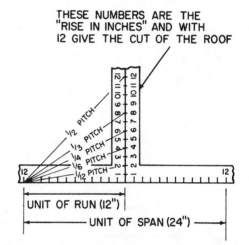

THESE NUMBERS ARE THE "RISE IN INCHES" AND WITH 12 GIVE THE CUT OF THE ROOF

28–10. *Pitch can be expressed as the rise in inches per unit of run.*

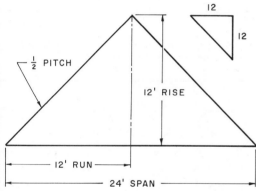

28–8. *A roof with a 1/2 pitch, or 12" to the foot.*

28–9. *A roof with a 1/3 pitch, or 8" to the foot.*

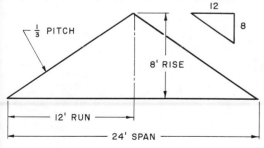

28–11. *The framing square is used to lay out the plumb and level lines on a rafter. The plumb line is drawn along the tongue of the square and the level line along the body (sometimes called the blade).*

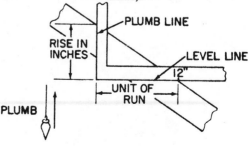

lines where two roofs intersect. Fig. 28-13.

Jack rafters never extend the full distance from the plate to the ridge board. There are three kinds of jack rafters: hip jacks extend from the plate to a hip rafter, valley jacks extend from the ridge to a valley rafter, and cripple jacks extend between a hip rafter and a valley rafter or between two valley rafters. Fig. 28-13.

LAYING OUT A ROOF FRAME PLAN

Before cutting rafters, the carpenter must determine what kinds are necessary to frame the roof. A roof framing diagram may be included among the working drawings; if not, you should lay one out for yourself.

There are four types of roofs which will usually be of concern to the carpenter:

- Gable.
- Hip.
- Gable and valley.
- Hip and valley.

The lean-to or shed roof is one-half of a gable roof. It extends from ridge board to plate on one side only. The gambrel roof plan is the same as for a gable roof. The mansard roof is a combination of a hip and a flat roof or two hip roofs. The first hip off the plate has a steep slope and the second is either flat or has a very low slope.

Always make a roof frame plan to determine the kinds of rafters that will be needed for framing. If the plan is drawn to scale, the exact number of each kind of rafter can also be determined. However, the actual length of each rafter should be figured from the dimensions taken directly off the building. The roof frame plan for each of the more common roof styles can be made as follows:

Gable Roof

1. Lay out the outline of the building. A, Fig. 28-14.

28–12. *The ridge board fastens the upper ends of the rafters together and maintains the correct spacing between them. The rafter tails are shown at the arrows.*

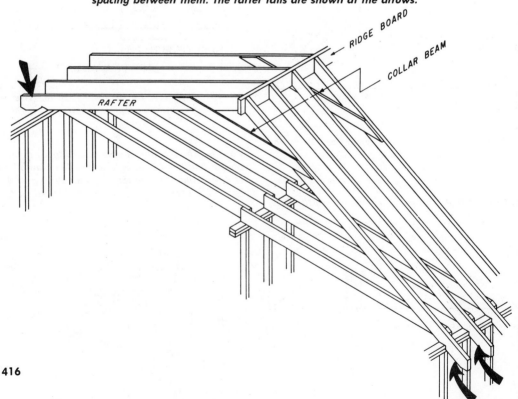

RIDGE BOARD

COLLAR BEAM

RAFTER

2. Determine the direction in which the rafters will run and draw the center line at right angles to this direction. B, Fig. 28-14.

3. The center line determines the location of the ridge line. C, Fig. 28-14.

4. Determine the distance between the rafters and lay out the roof frame plan. D, Fig. 28-14.

Hip Roof

1. Lay out the outline of the building. A, Fig. 28-15.

2. Locate and draw a center line. B, Fig. 28-15.

3. At each corner, draw a 45° line from corner to center line. This establishes location of hip rafters. C, Fig. 28-15.

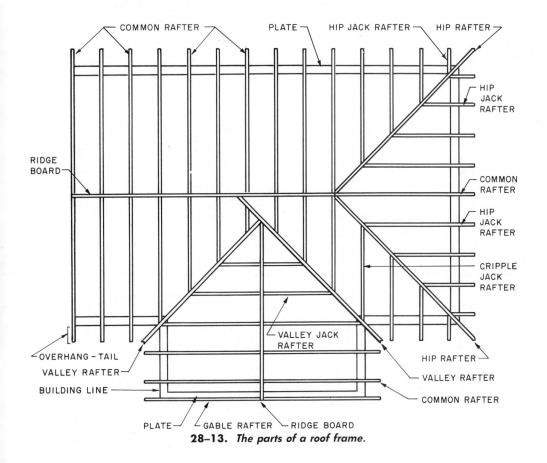

28–13. The parts of a roof frame.

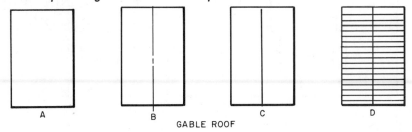

28–14. *Frame plan for gable roof. The frame plan for a shed roof would be one-half of this.*

GABLE ROOF

4. Draw the ridge line between the intersecting points of the hip rafters. D, Fig. 28-15.

5. Determine the distance between the rafters and lay out the roof frame plan. E, Fig. 28-15.

Gable and Valley Roof

1. Lay out the outline of the building. A, Fig. 28-16.

2. Draw the center line of the larger rectangle. B, Fig. 28-16 (arrow 1).

3. Draw the center line of the smaller rectangle. B, Fig. 28-16 (arrow 2).

4. Draw a line at 45° between the interior corners of the building outline and the ridge line. C, Fig. 28-16.

5. Draw in the ridge lines. D, Fig. 28-16.

6. Determine the distance between the rafters and lay out the roof frame plan. E, Fig. 28-16.

Hip and Valley Roof

1. Layout the outline of the building. A, Fig. 28-17.

2. Outline the largest possible rectangle inside the building outline. B, Fig. 28-17.

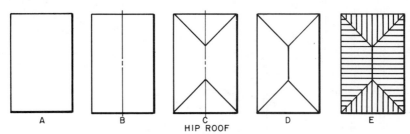

A   B   C   D   E
HIP ROOF

28–15. *Frame plan for hip roof.*

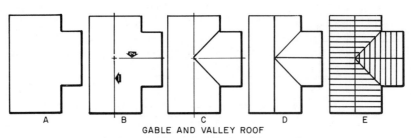

A   B   C   D   E
GABLE AND VALLEY ROOF

28–16. *Gable and valley roof frame plan.*

28–17. *Hip and valley roof frame plan.*

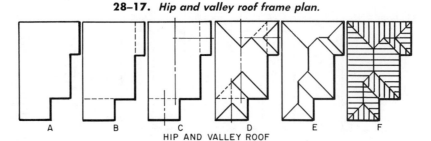

A   B   C   D   E   F
HIP AND VALLEY ROOF

3. Draw center lines for each rectangle formed inside the building outline. C, Fig. 28-17.

4. Draw a line at 45° from each corner, both inside and outside, and extend these lines to intersect with the center lines drawn in C. The solid lines indicate the location of the hip rafters on outside cor-

ners and valley rafters on inside corners. D, Fig. 28-17.

5. The center lines drawn in C connect the hip and valley rafters. Draw these in as solid lines to indicate the location of the ridges. E, Fig. 28-17.

6. Figure distance between rafters and lay out roof frame plan. F, Fig. 28-17.

QUESTIONS

1. What is the main purpose of a roof?
2. List several roof styles.
3. What is the ridgeboard's function?
4. What pieces in the roof serve the same purpose as joists in the floor?

5. What is the difference between a hip rafter and a valley rafter?
6. What is the purpose of a roof frame plan?

Conventional Roof Framing With Common Rafters

UNIT 29

There are two methods of roof framing for the pitched roof styles discussed in the previous unit: conventional and trussed roof construction. In conventional roof construction, the carpenter builds the roof with ceiling joists and rafters, a piece at a time, on the building's walls. Fig. 29-1. In trussed roof construction the trusses are usually prefabricated and are attached to the building as units. Fig. 29-2. These two framing methods are used most often for roof slopes of 4 in 12 ($\frac{1}{6}$ pitch) and greater. The framing of a low-pitched or flat roof will be discussed later.

In this unit the layout and cutting procedures for framing a conventional roof will be discussed. The procedure for constructing a pitched roof using trusses will be discussed in Unit 33.

The joist and rafter method is known by most carpenters and therefore is used fre-

quently. Common types of sheathing and finish materials are used, insulation is easily installed between the joists, and the roof load is carried on the walls without causing the ceiling to deflect.

There are some disadvantages to this type of construction. It takes longer, and

29-1. *Conventional roof framing.*

the building is therefore exposed to the weather longer. Also, the carpenters building a roof with joists and rafters must stand on scaffolding and ceiling joists.

In conventional roof construction, the rafters should not be erected until the ceiling joists have been fastened in place. The ceiling joists act as a tie and prevent the rafters from spreading and pushing out on the exterior walls.

LAYING OUT COMMON RAFTERS

The rafters are the skeleton of the roof and must be carefully made and fitted if they are to support the roof weight. The top of the rafter rests against the ridge board and is called a top or plumb cut. The bottom of the rafter rests on the plate; this is a level or seat cut. Fig. 29-3. These cuts must be made accurately if the rafter is to fit properly.

A plumb cut line is drawn with the framing square as a guide. The unit run (12″ mark) on the body of the square is aligned with the edge of the rafter. The unit rise on the tongue of the square (number used will correspond to the slope of the roof) is aligned on the same edge of the rafter. The plumb line is then drawn along the edge of

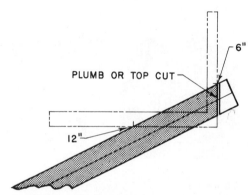

29–4. *This plumb line has been drawn for the top cut on a roof with a 6″ unit rise (¼ pitch roof).*

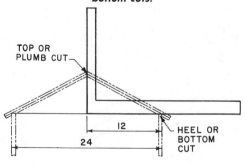

29–2. *Framing the roof with trussed rafters.*

29–5. *A level line drawn for the seat cut of a bird's-mouth. This cut is made for a roof with a 6″ unit rise.*

29–3. *This framing square was enlarged to show its relationship to the roof and to the top and bottom cuts.*

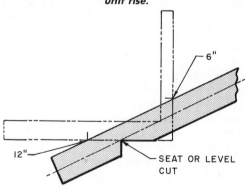

the tongue. Fig. 29-4. A level line is drawn for the same roof pitch with the square in the same position on the rafter except that the line is drawn along the body of the framing square. Fig. 29-5.

The theoretical length of a common rafter is the shortest distance between the outer edge of the plate (A) and a point where the measuring line of the rafter comes in contact with the ridge line (B). Figs. 29-6 and 29-7. This length is found

along the measuring line and may be calculated in several ways:
● By using the Pythagorean theorem.
● By applying the unit length obtained from the rafter table on the framing square.
● By stepping off the length with the framing square.

Pythagorean Theorem Method

The Pythagorean theorem states that the square of the hypotenuse of a right triangle is equal to the sum of the squares of the other two sides ($A^2 = B^2 + C^2$). The rise, the run, and the rafter of a roof form a right triangle, with the rafter as the hypotenuse. Fig. 29-8. The length of the rafter (A) can thus be calculated from the rise (B) and the run (C). Fig. 29-9.

Because of the ease and convenience of other methods, the Pythagorean theorem

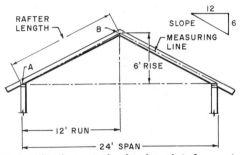

29–6. The theoretical rafter length is from point A to point B.

29–7. Actual and theoretical length of common rafter.

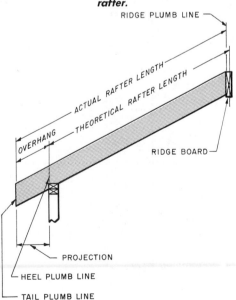

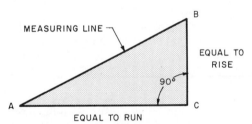

29–8. The measuring line is the hypotenuse of the right triangle and represents the length of the rafters.

29–9. The Pythagorean theorem states that the length of the hypotenuse (A) will be the square root of the sum of the squares of the other two sides ($\sqrt{B^2 + C^2}$).

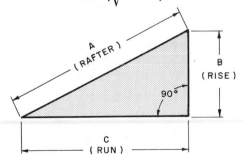

method is seldom used. However, the other methods commonly used on the job are based on this mathematical formula.

Unit Length Method

The unit length is the hypotenuse of a right triangle with the unit of run (12″) as the base and the unit rise (rise in inches per foot of run) as the altitude. Fig. 29-10a. The unit length is found on the rafter table of the framing square. Fig. 29-10b. The inch markings along the top of the table represent unit rise. The top line of the table reads: "Length Common Rafters per Foot Run." If you follow across the top line to the figure under 6 (for a unit rise of 6″), you will find the figure 13.42. This is the unit length for a roof triangle with a unit run of 12″ and a unit rise of 6″.

Let's figure the total length of a rafter for a small building with a unit rise of 5″, a span of 6′, and a run of 3′. Look at the rafter table to obtain the unit length. Fig. 29-10b.

For a unit rise of 5″, the unit length is 13″ per unit of run (one foot). The total length is the unit length times the total run. The total run of the building in this example is 3′. Therefore the total length of the common rafters is 39″. Fig. 29-11.

13 (inches per unit of run) × 3 (units of run)
= 39″

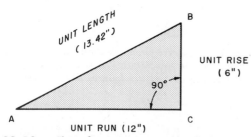

29–10a. *The rafter is represented by the line AB. The length of this line can be found in the rafter table on the framing square.*

29–10b. *To find the unit length of common rafters, check the rafter table on the face of the steel square.*

Step-Off Method

A third method for finding the theoretical rafter length is by using the framing square to "step off" the length. Fig. 29-12. Place the square on the rafter with the tongue on the plumb cut. Step off the cut of the roof (for example in Fig. 29-12, 6" on the tongue and 12" on the blade) on the rafter stock as many times as there are feet in the total run. In this case, it would be three times.

Often the run of a building will not come out in even feet. For example, the run might be 3' 4". The extra 4" is taken care of in the same manner as the full foot run. With the square at the first step position, draw a line along the edge of the tongue to represent the center line of the ridge board. At the 4" mark of the blade, make a mark on the rafter along the level line—not along the edge of the rafter. Fig. 29-13. Then, as in Fig. 29-12, step off the cut three more times, for a total run of 3' 4". This is the theoretical length of the rafter. The ridge board thickness and overhang can now be figured and laid out.

CUTTING COMMON RAFTERS
Common Rafter Ridge Allowance

The theoretical length does not take into account the thickness of the ridge board or the length of the overhang, if there is one. To cut a rafter without an overhang to its actual length, you must deduct one-half the thickness of the ridge board from the

29-11. *The total length of a rafter is the total run times the unit length. In this example the total run is 3' and the unit length is 13". Therefore the theoretical length of the rafter is 39".*

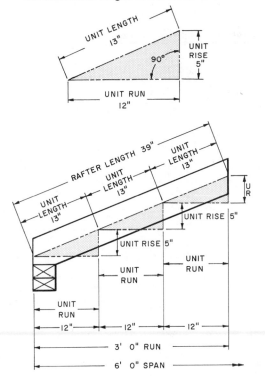

29-12. *Stepping off the length of a common rafter.*

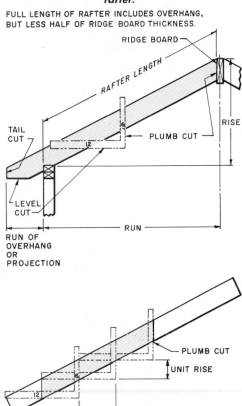

ridge end. Fig. 29-14. For example, if 2″ material is used for the ridge board, the actual thickness is 1½″. One-half of this is ³/₄″. The ³/₄″ is laid off along the level line, and the line for the actual ridge plumb cut is made. Fig. 29-15.

Common Rafter Overhang

A roof may or may not have an overhang. If not, the rafter must be cut so that its lower end is even with the outside of the exterior wall. Fig. 29-16. The portion of the rafter which rests on the plate is called the seat. To lay out the seat, place the tongue of the framing square on the heel plumb line with the rafter edge intersecting the correct seat width on the blade. Fig. 29-17. Draw a line from the heel plumb line along the blade.

A roof with a wide overhang at the cornice and the gable ends not only enhances appearance but also provides protection to side and end walls. Thus even in lower-cost houses, when style and design permit, wide overhangs are desirable. Though

it adds slightly to the initial cost, future savings on maintenance usually merit this type of roof extension.

If the roof does have an overhang, or eave, the overhanging part of the rafter is called the tail and must be added to the

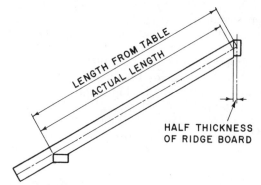

29-14. *Subtract one-half the actual thickness of the ridge board from the theoretical length of the rafter to obtain the rafter's actual length. If there is to be an overhang, this will be added later.*

29-13. *Stepping off a rafter when the run is not an even number of feet.*

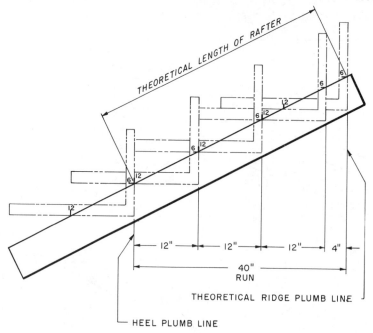

length of the rafter. The length of the tail may be calculated as if it were a separate little rafter. Any of the methods used for finding rafter length may be used to find the length of the tail. Suppose the run of the overhang (sometimes called the projection) is 2′ and the unit rise of the roof is 8″. Fig. 29-18. Look at the rafter table and find the unit length for a common rafter with a

unit rise of 8″. Fig. 29-10b. The unit length of the rafter is 14.42″. Since the total run of the overhang is 2′, the tail (length of overhang) is 28.84″, or 28 $^{27}/_{32}$″:

$$14.42 \text{ (inches per unit of run)} \times 2 \text{ (units of run)} = 28.84″$$

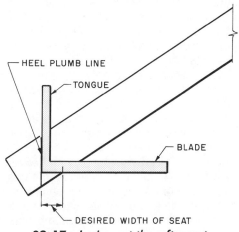

29–17. *Laying out the rafter seat.*

29–18. *The length of the rafter overhang may be found by using the rafter table on the framing square.*

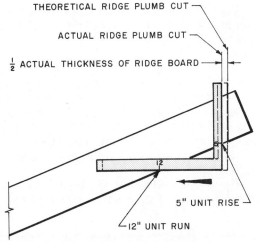

29–15. *Lay off one-half the thickness of the ridge board at right angles to the tongue of the square (along the level line). Do not lay it off along the edge of the rafter.*

29–16. *The rafter without an overhang may rest on the exterior wall plate with or without a heel. Which do you think would be stronger?*

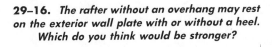

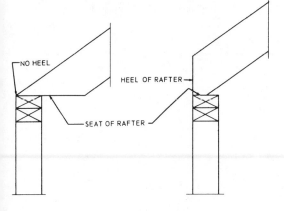

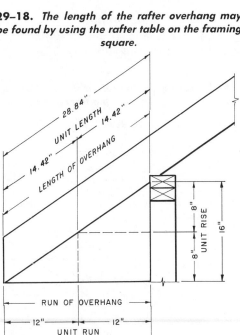

Another way to lay off the overhang is with the framing square. Suppose the run of the overhang is 10". Fig. 29-19. Start the layout by placing the tongue of the square along the heel plumb line and setting the square to the cut of the roof. In Fig. 29-19, the square is set to a unit rise of 8" and a unit run of 12". Move the square in the direction of the arrow in Fig. 29-19 until the 10" mark of the blade is on the heel plumb line. Draw a line along the tongue. This will be the tail cut.

Many carpenters do not cut the tail to the finished length until after the rafters have been fastened in place. The length of the tail is calculated, and a sufficient amount of material is left beyond the bird's-mouth for the overhang. Fig. 29-20. All other cuts except the tail plumb cut are made. After the rafters are fastened in place, the exact length of the tail is marked on the end rafters. A chalk line is snapped on the top edge of all the rafters. A tail plumb line is then drawn down from this chalk line on each rafter and the tail is cut along the line.

Bird's-mouth

A rafter with an overhang has a notch in it called a bird's-mouth. Fig. 29-20. The plumb cut of the bird's-mouth, which bears against the side of the rafter plate, is called the heel cut. The level cut, which bears on the top of the rafter plate, is called the seat cut.

The size of the bird's-mouth for a common rafter is usually stated in terms of the depth of the heel cut rather than the width of the seat cut. The bird's-mouth is laid out much the same way as the seat on a rafter without an overhang. Measure off the depth of the heel on the heel plumb line, set the square, and draw the seat line along the blade. Fig. 29-21.

Common Rafter Pattern

Calculate the actual length of a common rafter and lay one out on a piece of stock. When laying out rafters, remember to use the crown of the rafter member for the top edge. Carefully cut out the rafter. Use this rafter as a pattern for cutting a second rafter.

Try the two rafters on the building with the ridge board or a scrap piece of the

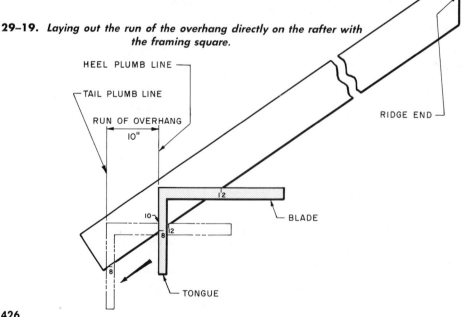

29–19. *Laying out the run of the overhang directly on the rafter with the framing square.*

HEEL PLUMB LINE

TAIL PLUMB LINE

RUN OF OVERHANG

10"

RIDGE END

BLADE

TONGUE

same size material as the ridge board between to see how the heel cut and the top cut fit. If they are all right, use one of these rafters as a pattern to cut all others needed. Distribute the rafters to their locations on the building. The rafters are usually leaned against the building with the ridge cut up. The workers on the building can then pull them up as needed and fasten them in position. Fig. 29-22.

SADDLE BRACE ROOF FRAMING

A metal bracket used for roof framing permits the use of square-end lumber for rafters, eliminating both the plumb cut at the ridge and the bird's-mouth at the plate. This bracket will adjust to any pitch. Fig. 29-23. Hip and jack rafter brackets are also available. Fig. 29-24. This "saddle brace" produces a strong roof which exceeds regulations of federal, state, and local building codes and also meets FHA and CMHC (Canada) requirements. (CMHC is the Central Mortgage and Housing Corporation. It can be compared to the FHA in the United States.) To use the metal brackets on a gable roof follow this procedure:

1. Lay out the rafter spacing on the top plates and nail the anchor brackets to the top plates with two 1½" roofing nails.

2. Install a ceiling joist alongside each anchor bracket. Fig. 29-25.

3. Set the ridge in the center of the building at the right height for the required pitch. Fig. 29-26. (Installation of the ridge board is discussed in Unit 32.)

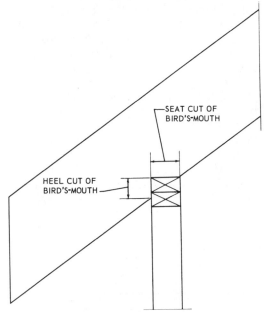

29–20. *The bird's-mouth on a rafter with an overhang.*

29–21. *Using the square to lay out a bird's-mouth. For a common rafter on a gable roof, the depth of the heel cut is laid out along the heel plumb line. This completes the layout of the bird's-mouth. The length of the line at arrow#1 will be the important dimension when laying out the bird's-mouth for a hip and valley rafter, as you will learn later.*

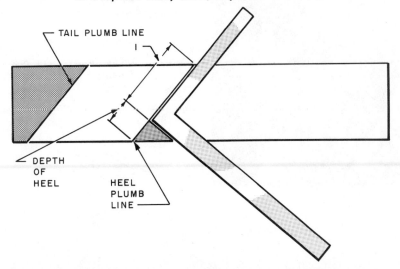

29-22. Framing a gable roof. The rafters are leaned against the building and pulled up as needed. It is best to have three workers when framing a roof: one at the ridge and one at each plate where the rafters are to be fastened. The rafters are erected alternately: one from the front, then one from the back.

29-23a. Framing a roof using saddle braces.

4. Lay out the rafter spacing on the ridge board and install the brackets over the ridge. Fig. 29-26. Nail each bracket in place with three 1½" roofing nails.

5. Insert a square-end lumber rafter into the saddle brace plate anchor and ridge bracket. Fig. 29-27. Make sure the rafter is pushed firmly against the ridge.

6. With the rafter and ceiling joist in place at the plate, drive a 16d nail from the ceiling joist side through to the rafter. Drive another 16d nail from the rafter side through to the ceiling joist. Fig. 29-28.

7. Make sure the bottom of the saddle brace at the ridge is snug against the bottom of the rafter. Nail each rafter face through the ridge bracket with two 1½" roofing nails. Fig. 29-29.

8. After installing all rafters, attach a collar tie to every fourth rafter if the spacing is 16" on center and every third rafter if

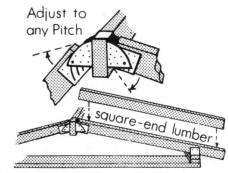

Adjust to any Pitch

square-end lumber

29-23b. Square-end lumber is used with the saddle brace, which will adjust to any pitch.

29-24. Saddle braces may also be used for framing hip and jack rafters.

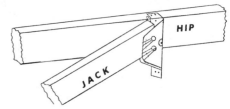

HIP

JACK

the spacing is 24" on center. This is a minimum standard. Local building codes may vary. Fig. 29-30.

Attaching a Shed Roof

A shed roof may be attached to an existing building by using these metal brackets. Cut the saddle brace in half at the ridge strap and bend the strap up. Nail it against the existing wall. Figs. 29-31.

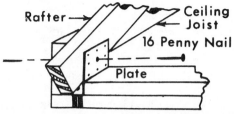

29-28. *Nail through the anchor plate into the rafter from one side. From the other side, nail through the ceiling joist into the rafter.*

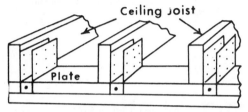

29-25. *Anchor brackets are used to attach the rafters to the ceiling joists and the double plate.*

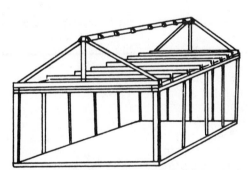

29-26. *The ridge is set in place ready to receive the rafters. Brackets are installed over the ridge.*

29-27. *Place the rafter in the saddle brace at the ridge and the anchor bracket at the plate. Push it up firmly against the ridge.*

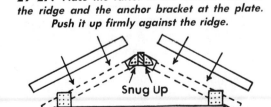

29-29. *Nail the rafter to the ridge bracket with two nails on each side.*

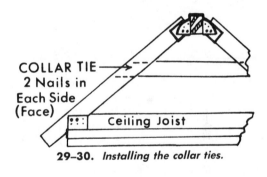

29-30. *Installing the collar ties.*

29-31. *Using the saddle brace to attach a shed roof to an existing building.*

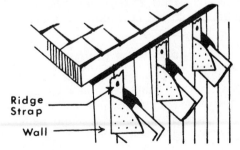

QUESTIONS

1. What are some of the disadvantages of conventional roof framing?

2. What prevents the rafters from spreading and pushing out on the exterior walls?

3. What is the cut at the top of the rafter called?

4. What is the name of the cut that rests on the plate?

5. When laying out the rafter, what line is drawn along the edge of the tongue on the framing square?

6. What are three ways that the length of a common rafter can be calculated?

7. A building is 24' wide and has a 1/3 pitch. What is the theoretical length of a common rafter?

8. Describe the step-off method for finding the length of a rafter.

9. What is meant by ridge allowance?

10. What is the bird's-mouth?

11. What is a saddle brace? If saddle braces are used, will the roof meet the FHA requirements?

Hip and Valley Rafters

The hip rafter is a roof member that forms a raised area or "hip" in the roof, usually extending from the corner of the building diagonally to the ridge. Fig. 30-1 a and b. The valley rafter is similar, but it forms a depression in the roof instead of a hip. Fig. 30-1 a and c. Like the hip rafter, it extends diagonally from plate to ridge.

The total rise of hip and valley rafters is the same as that of common rafters. Fig. 30-1a. Hip and valley rafters may be the same thickness as common rafters, but they should be 2" deeper to permit full bearing with the beveled end of the jack rafter.Fig. 30-2.

HIP RAFTER LAYOUT

The length of a hip rafter, like the length of a common rafter, is calculated on the basis of the unit run and unit rise and/or the total run and total rise. Any of the methods previously described for determining the length of a common rafter may be used.

However, some of the basic data for hip and valley rafters is different.

Fig. 30-3 shows part of a roof framing plan for a hip roof. On a hip roof framing plan, the lines which indicate the hip rafters (EC, AC, KG, and IG in Fig. 30-3) form 45° angles with the building lines. A line which indicates a rafter in the roof framing diagram corresponds to the total run (not length) of the rafter it represents. You can see from the diagram that the total run of a hip rafter is the hypotenuse of a right triangle, with the shorter sides each equal to the total run of a common rafter. Fig. 30-3. In Fig. 30-4 one corner of the roof framing plan (ABCF in Fig. 30-3) has been drawn in perspective to show the relative position of the hip rafter to the common rafter.

The unit run of a hip rafter is the hypotenuse of a right triangle with the shorter sides each equal to the unit run of a common rafter. Fig. 30-5. The unit run of a

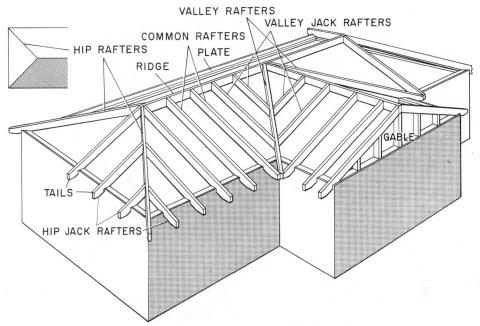

VALLEY RAFTERS
VALLEY JACK RAFTERS
COMMON RAFTERS
HIP RAFTERS
PLATE
RIDGE
GABLE
TAILS
HIP JACK RAFTERS

30–1a. *Roof frame with hip and valley rafters.*

30–1b. *Hip rafter framing at the ridge.*

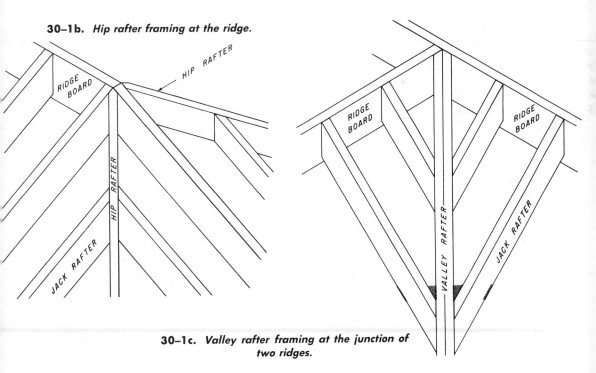

RIDGE BOARD
HIP RAFTER
HIP RAFTER
JACK RAFTER

RIDGE BOARD
RIDGE BOARD
VALLEY RAFTER
JACK RAFTER

30–1c. *Valley rafter framing at the junction of two ridges.*

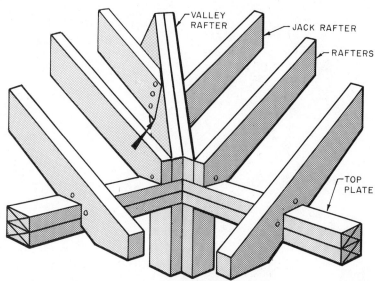

30–2. *Because the beveled cut on the ends of jack rafters creates a longer cut (see arrow), the hip and valley rafters must be 2″ deeper than common rafters to permit a full bearing surface. The valley rafter in this drawing has been cut off at the plate. Normally it is extended to become part of the overhang.*

common rafter is 12″. By the Pythagorean theorem, the unit run of a hip rafter is the square root of $12^2 + 12^2$, which is 16.97, or 17. Fig. 30-6. The unit run of a valley rafter is also 17″.

Like the unit length of a common rafter, the unit length of a hip rafter may be obtained from the rafter table on the framing square. In Fig. 29-10b, the second line

in the table is headed "Length Hip or Valley per Foot Run." This means "per foot run of a common rafter in the same roof." Another way to state this would be "per 16.97″ run of hip or valley rafter." For example, the unit length for a unit rise of 8″

30–3. *Hip roof framing diagram.*

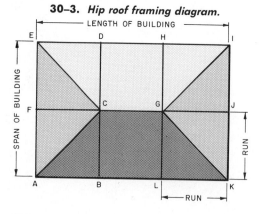

30–4. *The relative position of a hip rafter to a common rafter is shown in this perspective drawing of a corner from the roof framing diagram in Fig. 30–3.*

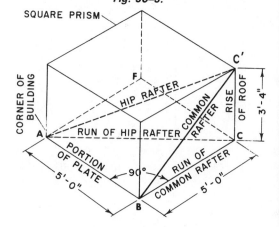

is 18.76″. To calculate the length of a hip rafter, multiply the unit length by the number of feet in the total run of a common rafter.

In Fig. 30-5, the corner of the building from Fig. 30-3 is shown. In this example the run of a common rafter is 5′. The unit rise is 8″ and the unit length of the hip rafter for this unit rise is 18.76″. The unit length multiplied by the total run in feet is the length of the hip rafter in inches (18.76″ × 5 = 93.8″, or 7′9¹³⁄₁₆″). As in the case of common rafters, this is the theoretical length. To obtain the actual length, the ridge board shortening allowance and the rafter tail will have to be calculated and laid out.

Plumb and Level Lines

The plumb and level lines on a hip or valley rafter are also referred to as the top and bottom cuts. The top cut is the plumb line and the bottom cut is the level line. To obtain the top and bottom cuts of the hip or valley rafters, set off 17″ on the body of the square. On the tongue set off the rise per foot of common rafter run. A line drawn

30–5. *The relationship between the unit run of a hip rafter and the unit run of a common rafter.*

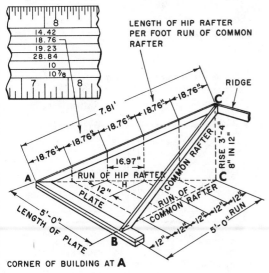

CORNER OF BUILDING AT **A**

along the body will be the level or seat cut, and a line drawn along the tongue will be the plumb or top cut. Fig. 30-7.

Ridge Allowance

As is the case with a common rafter, the theoretical length of a hip rafter does not take into account the thickness of the ridge board. The ridge-end shortening allowance for a hip rafter depends on the manner in which the ridge end of the hip rafter is joined to the other structural members. The ridge end of the hip rafter may be framed against the ridge board or against the ridge end of common rafters. Figs. 30-8 and 30-9.

If the hip rafter is framed against the ridge board, the shortening allowance is one-half the 45° thickness of the ridge piece. The 45° thickness of a piece of stock is the length of a line laid at 45° across the thickness of the stock. If the hip rafter is framed against the common rafters, the shortening allowance is one-half the 45° thickness of a common rafter.

To lay off the shortening allowance, set the tongue of the framing square to the theoretical ridge plumb cut line. Measure off the shortening allowance along the blade. Fig. 30-10a. Set the square at the mark to the cut of the rafter (unit rise and unit run) and draw the actual ridge plumb cut line. Fig. 30-10b. Remember that the

30–6. *The hypotenuse of a right triangle whose shorter sides each equal 12″ is 16.97″. This can be rounded off to 17″.*

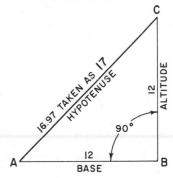

cut of the common rafter is based on a 12″ unit run whereas the unit run of the hip or valley rafter is 17″. Therefore, to set the square at the cut of the hip rafter, the tongue is set at the unit rise and the blade is set at the 17″ mark.

Side Cuts

Since a common rafter runs at 90° to the ridge board, the ridge end of a common rafter is cut square, or at 90° to the lengthwise line of the rafter. A hip rafter, however, joins the ridge piece or the ridge ends of the common rafters at an angle. The ridge end of a hip rafter must therefore be cut to a corresponding angle. This cut is called a *side cut*. Figs. 30-8 and 30-9. The side cut may be laid out in one of two ways.

One method is illustrated in Figs. 30-11 and 30-12. Place the tongue of the framing square along the actual ridge plumb cut line and measure off one-half the thickness of the hip rafter along the blade (level line). Shift the tongue to the mark, set the square to the cut of the rafter (17″ and 8″ in this example), and draw the plumb line. A, Fig. 30-11. Turn the rafter edge up, draw an edge center line, and draw in the angle of the side cut. Fig. 30-12. For a hip rafter which is to be framed against the ridge piece, there will be only a single side cut. Fig. 30-8. For a hip rafter which is to be framed against the ridge ends of the common rafters, there will be a double side cut. Fig. 30-9. In either case, the tail of the rafter must have a double side cut at the same angle, but in the reverse direction, to allow attachment of the fascia board. Fig. 30-13.

A second method of laying out the angle of the side cut on a hip rafter is by referring to the rafter table on the framing square. In Fig. 29-10b, the bottom line of the table is

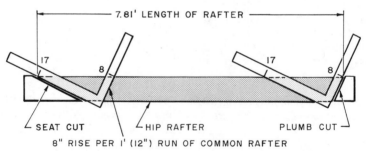

SEAT CUT   HIP RAFTER   PLUMB CUT

7.81′ LENGTH OF RAFTER

8″ RISE PER 1′ (12″) RUN OF COMMON RAFTER

30-7. *Marking the top (plumb) cut and the seat (level) cut of a hip rafter.*

30-8. *A hip rafter framed against the ridge board requires a single side cut. However, the end common rafter must have a 45° angle cut for framing against the side of the hip rafter.*

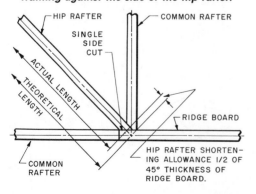

HIP RAFTER   COMMON RAFTER

SINGLE SIDE CUT

ACTUAL LENGTH

THEORETICAL LENGTH

RIDGE BOARD

COMMON RAFTER

HIP RAFTER SHORTENING ALLOWANCE 1/2 OF 45° THICKNESS OF RIDGE BOARD.

30-9. *A hip rafter framed against the ridge-end common rafters requires a double side cut.*

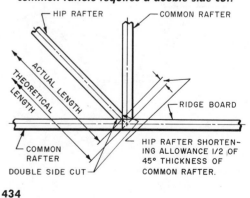

HIP RAFTER   COMMON RAFTER

ACTUAL LENGTH

THEORETICAL LENGTH

RIDGE BOARD

COMMON RAFTER

DOUBLE SIDE CUT

HIP RAFTER SHORTENING ALLOWANCE 1/2 OF 45° THICKNESS OF COMMON RAFTER.

headed "Side Cut Hip or Valley Use." Follow this line over to the column headed by the figure 8 (for a unit rise of 8"). The number shown is 10⅞. Place the framing square face up on the rafter edge, with the tongue on the ridge-end plumb cut line. (This is line A in Fig. 30-12.) Set the square to a cut of 10⅞" on the blade and 12" on the tongue. Draw the side cut angle along the tongue. Fig. 30-14.

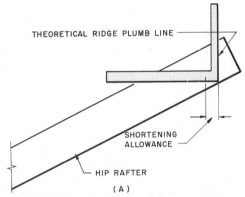

30-10a. *To lay off the shortening allowance, place the tongue of the square along the theoretical ridge plumb cut line and measure off the shortening allowance along the blade of the square (level line).*

30-10b. *Set the square to the cut of the roof (8" unit rise for this example) with the tongue on the shortening allowance mark. Draw the actual ridge plumb line along the edge of the tongue.*

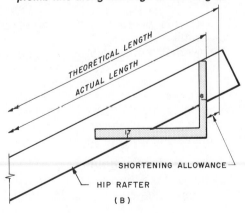

Overhang

A hip or valley rafter overhang, like a common rafter overhang, is figured as a separate rafter. The run of the overhang, however, is not the same as the run of a common rafter overhang in the same roof. The run of the hip or valley rafter overhang is the hypotenuse of a right triangle whose shorter sides are each equal to the run of a common rafter overhang. Fig. 30-15. If the run of the common rafter overhang is 2' for a roof with an 8" unit rise, the length of the hip or valley rafter tail is figured as follows:

1. Find the unit length of the hip or valley rafter on the framing square. Fig. 29-10b. For this roof it is 18.76".

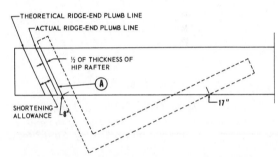

30-11. *To lay out the side cut, at a right angle to the ridge plumb cut line measure off one-half the thickness of the hip rafter from the actual ridge plumb cut line.*

30-12. *Draw a centerline on the edge of the rafter (arrow 1). Extend the plumb lines from the face of the rafter to intersect the centerline at 90°. The side cut line is drawn from line A through the intersection of the centerline and the actual ridge-end plumb line.*

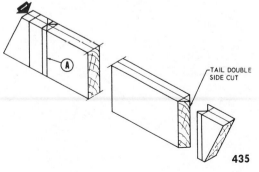

435

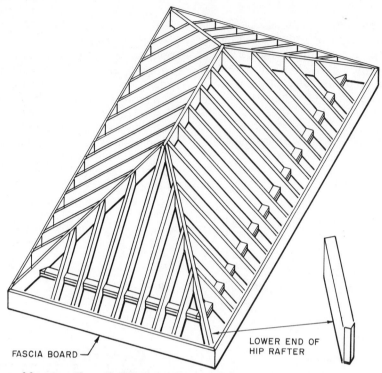

FASCIA BOARD

LOWER END OF
HIP RAFTER

30–13. *Hip roof framing. The end of the hip rafter has a double side cut. The fascia boards from the side and end will be fastened along the ends of the rafters and mitered to form an outside corner at the hip rafter. See Fig. 30–12 for detail of hip rafter tail double side cut.*

30–14. *Framing square in position on the back edge of the hip rafter for a unit rise of 8″. A single side cut will be made for framing against the ridge board.*

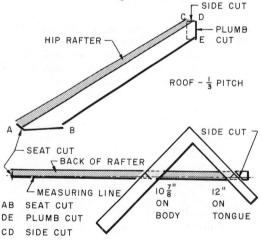

HIP RAFTER

SIDE CUT
PLUMB
CUT

ROOF – ⅓ PITCH

A   B

SEAT CUT
BACK OF RAFTER

SIDE CUT

MEASURING LINE

$10\frac{7}{8}$″
ON
BODY

12″
ON
TONGUE

AB   SEAT CUT
DE   PLUMB CUT
CD   SIDE CUT

2. Multiply the unit length of the hip or valley rafter by the run of the common rafter overhang:

18.76″ (unit length of hip or valley rafter) × 2 (feet of run in common rafter overhang) = 37.52″, or 37½″

3. Add this product to the theoretical rafter length.

The overhang may also be stepped off as described in Unit 29 for a common rafter. When stepping off the length of the overhang, set the 17″ mark on the blade of the square even with the edge of the rafter. Set the unit rise, whatever it might be, on the tongue even with the same rafter edge.

Bird's-mouth

Laying out the bird's-mouth for a hip rafter is much the same as for a common

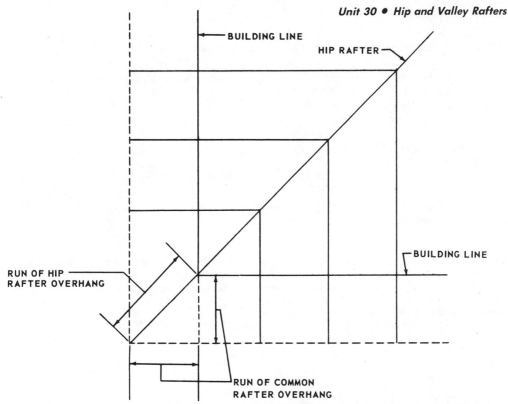

30–15 *Run of hip rafter overhang. For each unit of run (12″) of a common rafter, the unit of run for the hip rafter is 17″. Therefore if the run of the common rafter overhang in this drawing is 2′ (24″), the run of the hip rafter overhang will be 34″ (2 × 17).*

rafter. However, there are a couple of things to remember. When the plumb (heel cut) and level (seat cut) lines are laid out for a bird's-mouth on a hip rafter, set the body of the square at 17″ and the tongue to the unit rise (depending on the roof pitch). Fig. 30-7.When laying out the depth of the heel for the bird's-mouth, measure along the heel plumb line down from the top edge of the rafter a distance equal to the same dimension on the common rafter. Fig. 30-16. This must be done so that the hip rafters, which are usually wider than the common rafters, will be level with the common rafters.

Backing or Dropping a Hip Rafter

If the dimension above the bird's-mouth is exactly the same on a hip rafter as on a common rafter, the edges of the hip rafter will extend above the upper ends of the jack rafters and interfere with the application of the sheathing. A, Fig. 30-17. This can be corrected by either backing or dropping the hip rafter. *Backing* means to bevel the upper edge of the hip rafter. B, Fig. 30-17. *Dropping* means to deepen the bird's-mouth so as to bring the top edge of the hip rafter down to the upper ends of the jacks. C, Fig. 30-17.

The amount of backing or drop required is calculated as shown in A of Fig. 30-18. Set the framing square to the cut of the rafter (8″ and 17″ in this example) on the upper edge, and measure off one-half the thickness of the rafter from the edge along the blade (arrow 1). A backing line drawn through this mark, parallel to the edge, will

437

indicate the bevel angle, if the rafter is to be backed. B, Fig. 30-18. The perpendicular distance between the backing line and the edge of the rafter will be the amount of drop (arrow 2). A, Fig. 30-18. This is the amount by which the depth of the hip rafter bird's-mouth should exceed the depth of the common rafter bird's-mouth. C, Fig. 30-18.

VALLEY RAFTER LAYOUT

A valley rafter follows the line of intersection between a main roof surface and a gable roof addition or gable roof dormer surface. Most roofs which contain valley rafters are equal-pitch roofs; that is, the pitch of the addition or dormer roof is the same as the pitch of the main roof. In an equal-pitch roof the valley rafters always run at 45° to the building line and the ridge boards.

Framing an Equal-Span Roof Addition

In equal-span framing, the span of the addition is the same as the span of the main roof. Fig. 30-19. When the pitch of

30–16. *When laying out the bird's-mouth on a hip rafter, measure down from the top edge. Dimension A in the drawing must be the same for common and hip rafters so that the tops of all the rafters will be in line for the application of sheathing.*

30–17. *A. The edge of a hip rafter may extend above the upper ends of the jack rafters. B. Backing a hip rafter. C. Dropping a hip rafter.*

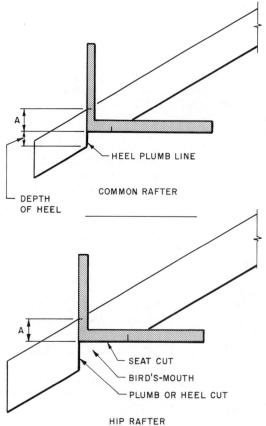

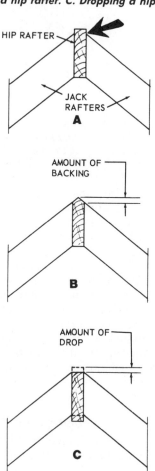

the addition roof is the same as the pitch of the main roof, equal spans bring the ridge pieces to equal heights.

Look at the roof framing diagram in Fig. 30-19. The total run of a valley rafter (indicated by AB and AD in the diagram) is the hypotenuse of a right triangle whose shorter sides are each equal to the total run of a common rafter in the main roof. The unit run of a valley rafter is therefore 16.97″, the same as the unit run for a hip rafter. Figuring the length of an equal-span addition valley rafter is thus the same as figuring the length of a hip rafter.

The ridge-end shortening allowance for an equal-span addition valley rafter is

one-half the 45° thickness of the ridge board. Fig. 30-20. Side cuts are laid out as they are for a hip rafter. The valley rafter tail has a double side cut, like the hip rafter tail, but in the reverse direction, since the tail cut on a valley rafter must form an inside rather than an outside corner. Fig. 30-21. The overhang, if any, and the bird's-mouth are figured just as they are for a hip rafter. A valley rafter, however, does not require backing or dropping.

Framing an Unequal-Span Roof Addition

Sometimes the span of the roof addition is shorter than the span of the main roof.

30–18. *Backing or dropping a hip rafter: A. Determining amount of backing or drop. B. Bevel angle for backing the rafter. C. Deepening the bird's-mouth for dropping the rafter.*

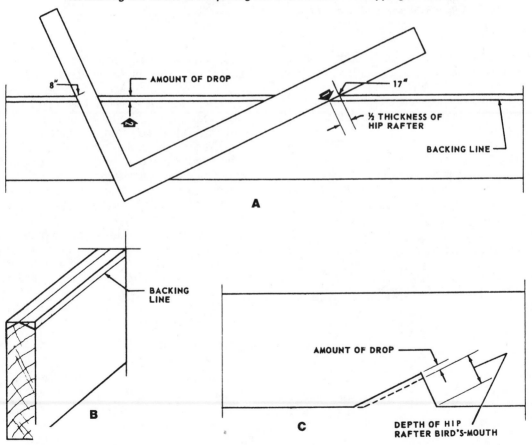

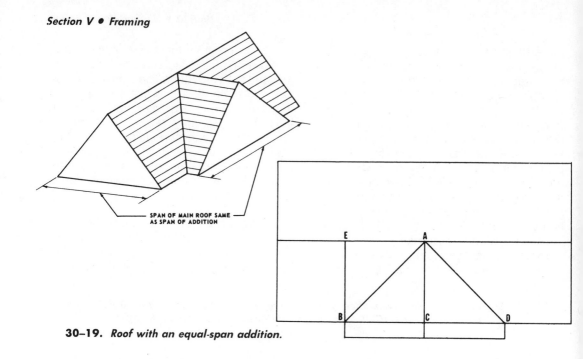

SPAN OF MAIN ROOF SAME
AS SPAN OF ADDITION

30-19. *Roof with an equal-span addition.*

Fig. 30-22. When the pitch of the addition roof is the same as the pitch of the main roof, the shorter span of the addition brings the addition ridge board down to a lower level than that of the main roof ridge board.

30-20. *Ridge-end shortening allowance for an equal-span addition valley rafter.*

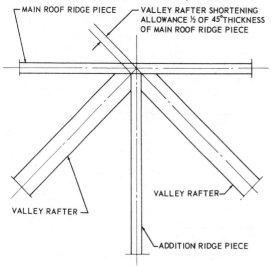

MAIN ROOF RIDGE PIECE — VALLEY RAFTER SHORTENING ALLOWANCE ½ OF 45° THICKNESS OF MAIN ROOF RIDGE PIECE

VALLEY RAFTER

VALLEY RAFTER

ADDITION RIDGE PIECE

There are two ways of framing an addition of this type. In one method, a full-length valley rafter (AD in Fig. 30-22) is framed between the rafter plate and the ridge board, and a shorter valley rafter (CB in the figure) is then framed to the longer one. The total run of the longer valley rafter is the hypotenuse of a right triangle whose shorter sides are each equal to the total run of a common rafter *in the main roof*. The total run of the shorter valley rafter, on the other hand, is the hypotenuse of a right triangle with shorter sides each equal to the total run of a common rafter *in the addition*. The total run of a common rafter in the main roof is equal to one-half the span of the main roof. The total run of a common rafter in the addition is equal to one-half the span of the addition.

DETERMINING THE LENGTH OF A VALLEY RAFTER

When the total run of any rafter is known, the theoretical length can be found by multiplying the unit length by the total run. Suppose, for example, that the addition in Fig. 30-22 has a span of 30' and that the

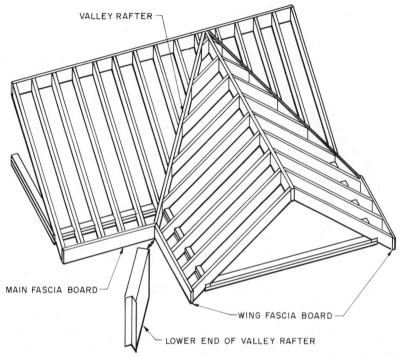

VALLEY RAFTER

MAIN FASCIA BOARD

WING FASCIA BOARD

LOWER END OF VALLEY RAFTER

30–21. *Valley rafter framing. Notice the inside corner formed by the fascia boards.*

30–22. *An addition with a span less than the main roof span. This addition is formed with a long and a short valley rafter.*

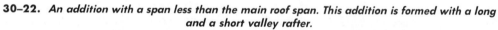

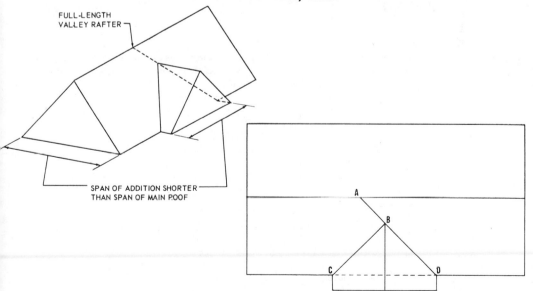

FULL-LENGTH
VALLEY RAFTER

SPAN OF ADDITION SHORTER
THAN SPAN OF MAIN ROOF

A
B
C
D

unit rise of a common rafter in the addition is 9″. The rafter table in Fig. 29-10b shows that the unit length for a valley rafter in a roof with a common rafter unit rise of 9″ is 19.21″. To find the theoretical length of the

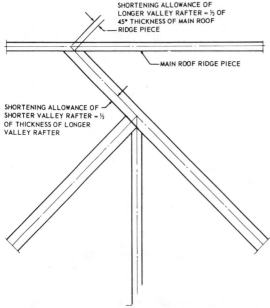

SHORTENING ALLOWANCE OF LONGER VALLEY RAFTER = ½ OF 45° THICKNESS OF MAIN ROOF RIDGE PIECE

MAIN ROOF RIDGE PIECE

SHORTENING ALLOWANCE OF SHORTER VALLEY RAFTER = ½ OF THICKNESS OF LONGER VALLEY RAFTER

30–23. *Long and short valley rafter shortening allowances.*

valley rafter, multiply its unit length by the number of feet in a common rafter of the roof to which it belongs. The total run of a common rafter is equal to one-half the span. Therefore the length of the longer valley rafter in Fig. 30-22 would be 19.21″ times one-half the span of the main roof. The length of the shorter valley rafter would be 19.21″ times one-half the span of the addition. Since one-half the span of the addition is 15′, the length of the shorter valley rafter is 19.21″ × 15, or 288.15″. Converted to feet, this is 24.01′.

The shortening allowances for the long and short valley rafters are shown in Fig. 30-23. Note that the long valley rafter has a single side cut for framing to the main roof ridge piece, while the short valley rafter is cut square for framing to the addition ridge piece.

A Second Method of Framing an Unequal-Span Addition

Another method of framing an equal-pitch, unequal-span addition is to nail the inboard end of the addition ridge piece to a piece of stock which hangs from the main roof ridge board. Fig. 30-24. This method calls for two short valley rafters, each of which extends from the rafter plate to the

30–24. *Another addition with a span less than the main roof span. This addition is framed with the addition ridge board suspended from the main roof ridge board. The two valley rafters (AB and AC) are the same length.*

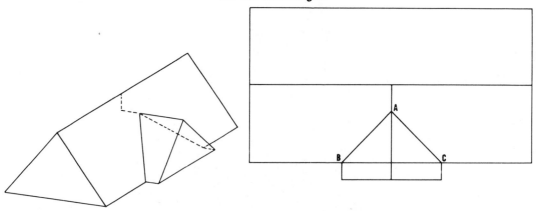

addition ridge piece. The total run of each of these valley rafters is the hypotenuse of a right triangle whose shorter sides are each equal to the total run of a common rafter in the addition.

The shortening allowance for each short valley rafter is one-half the 45° thickness of the addition ridge piece. Fig. 30-25. Each rafter is framed to the addition ridge piece with a single side cut.

Framing a Gable Dormer without Side Walls

When a gable dormer without side walls is framed, the dormer ridge piece is fastened to a header set between a couple of doubled main roof common rafters. Fig. 30-26. The valley rafters are framed between this header and a lower header. The total run of a valley rafter is the hypotenuse of a right triangle whose shorter sides are each equal to the total run of a common rafter in the dormer.

The arrangement and names of framing members in this type of dormer framing are shown in Fig. 30-27. Note that the upper edges of the headers must be beveled to the cut of the main roof.

30–25. *Shortening allowance of valley rafters in suspended-ridge method of addition roof framing.*

In this framing method, the shortening allowance for the upper end of a valley rafter is one-half the 45° thickness of the inside member in the upper doubled header. Fig. 30-28. The shortening allowance for the lower end is one-half the 45° thickness of the inside member in the doubled

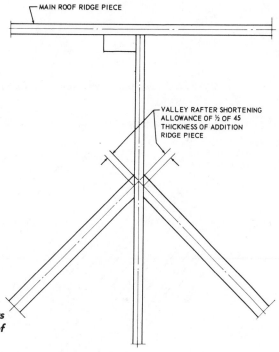

MAIN ROOF RIDGE PIECE

VALLEY RAFTER SHORTENING ALLOWANCE OF ½ OF 45 THICKNESS OF ADDITION RIDGE PIECE

30–26. *Framing a dormer without side walls.*

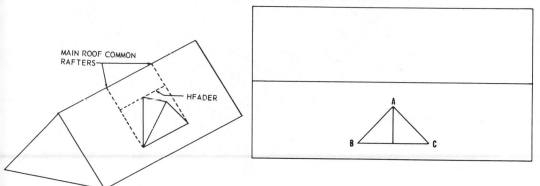

MAIN ROOF COMMON RAFTERS

HEADER

A
B   C

common rafter. Each valley rafter has a double side cut at the upper and the lower end.

Framing a Gable Dormer with Side Walls

A method of framing a gable dormer with side walls is illustrated in Fig. 30-29. As indicated in the framing diagram, the total run of a valley rafter is again the hypote-nuse of a right triangle whose shorter sides are each equal to the run of a common rafter in the dormer. Figure the lengths of the dormer corner posts and side studs just as you do the lengths of gable-end studs (see p. 462). Lay off the lower-end cutoff angle by setting the square to the cut of the main roof. The valley rafter shortening allowances for this method of framing are shown in Fig. 30-30.

30–27. *Arrangement and names of framing members for a dormer without side walls.*

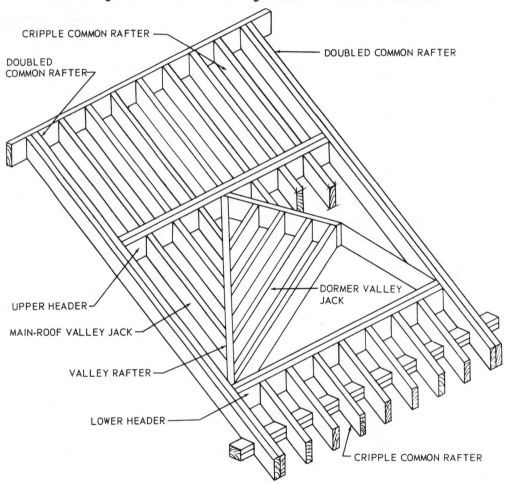

CRIPPLE COMMON RAFTER

DOUBLED COMMON RAFTER

DOUBLED COMMON RAFTER

DORMER VALLEY JACK

UPPER HEADER

MAIN-ROOF VALLEY JACK

VALLEY RAFTER

LOWER HEADER

CRIPPLE COMMON RAFTER

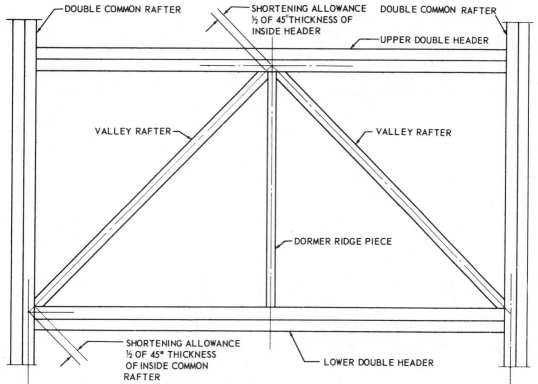

DOUBLE COMMON RAFTER

SHORTENING ALLOWANCE
½ OF 45°THICKNESS OF
INSIDE HEADER

DOUBLE COMMON RAFTER

UPPER DOUBLE HEADER

VALLEY RAFTER

VALLEY RAFTER

DORMER RIDGE PIECE

SHORTENING ALLOWANCE
½ OF 45° THICKNESS
OF INSIDE COMMON
RAFTER

LOWER DOUBLE HEADER

30–28. *Valley rafter shortening allowances for a dormer without side walls.*

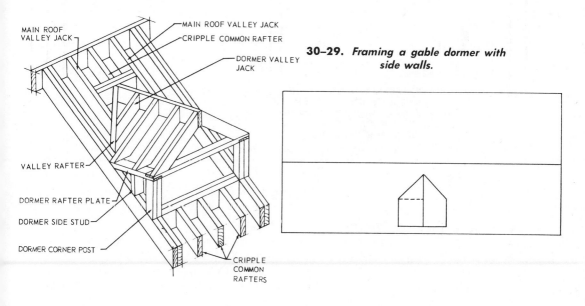

MAIN ROOF
VALLEY JACK

MAIN ROOF VALLEY JACK

CRIPPLE COMMON RAFTER

DORMER VALLEY
JACK

VALLEY RAFTER

DORMER RAFTER PLATE

DORMER SIDE STUD

DORMER CORNER POST

CRIPPLE
COMMON
RAFTERS

30–29. *Framing a gable dormer with side walls.*

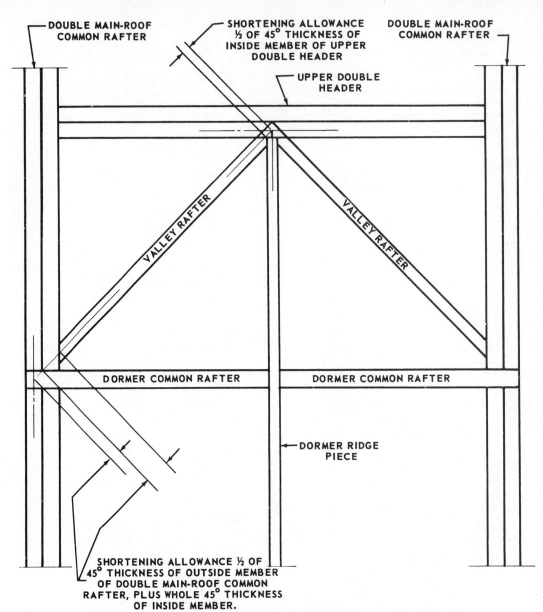

DOUBLE MAIN-ROOF
COMMON RAFTER

SHORTENING ALLOWANCE
½ OF 45° THICKNESS OF
INSIDE MEMBER OF UPPER
DOUBLE HEADER

DOUBLE MAIN-ROOF
COMMON RAFTER

UPPER DOUBLE
HEADER

VALLEY RAFTER

VALLEY RAFTER

DORMER COMMON RAFTER

DORMER COMMON RAFTER

DORMER RIDGE
PIECE

SHORTENING ALLOWANCE ½ OF
45° THICKNESS OF OUTSIDE MEMBER
OF DOUBLE MAIN-ROOF COMMON
RAFTER, PLUS WHOLE 45° THICKNESS
OF INSIDE MEMBER.

30–30. *Valley rafter shortening allowances for a dormer with side walls.*

QUESTIONS

1. What is a hip rafter?
2. What is a valley rafter?
3. Explain why the unit run of a hip rafter is 16.97 when the unit run of a common rafter is 12.
4. What number is used on the body of the square when making a cut for a hip or valley rafter?
5. What is the shortening allowance at the ridge for a hip rafter when the ridge end is framed against the ridge board?
6. Why is the run of a hip rafter over-

hang greater than the run of a common rafter overhang in the same roof?

7. When laying out the depth of the heel for the bird's-mouth on a hip rafter, why must you measure down from the top edge of the rafter a distance equal to the same dimension used on the common rafter?

8. Why must the upper edge of the hip rafter be beveled or the bird's-mouth cut deeper?

9. Describe two methods of framing an unequal-span roof addition.

10. In a right triangle, what is the side opposite the right angle called?

UNIT
31

Jack Rafters

A jack rafter is a shortened common rafter that may be framed to a hip rafter, a valley rafter, or both. This means that in an equal-pitch framing situation, the unit rise of a jack rafter is always the same as the unit rise of a common rafter.

TYPES OF JACK RAFTERS

A *hip jack* rafter extends from a hip rafter to a rafter plate. Fig. 31-1.

A *valley jack* rafter extends from a valley rafter to a ridge board. Fig. 31-1.

A *cripple jack* rafter does not contact either a plate or a ridge piece. There are two kinds of cripple jack rafters: (1) the *valley cripple jack* extends between two valley rafters in the long-and-short-valley-rafter method of addition framing and (2) the *hip-valley cripple jack* extends from a hip rafter to a valley rafter. Fig. 31-2.

LENGTHS OF HIP JACK RAFTERS

A roof framing diagram for a series of hip jack rafters is shown in Fig. 31-3. The jacks are always on the same spacing as the common rafters. The spacing in this instance is 16" on center. You can see from the arrow in the diagram that the total run of the shortest jack is also 16".

Suppose the unit rise of a common rafter in this roof is 8" per 12" or run. The jacks have the same unit rise as a common rafter. The unit length of a rafter is the hypotenuse of a right triangle with the unit run as base and the unit rise as height. The unit length of a jack rafter in the example is therefore the square root of ($12^2 + 8^2$), or 14.42. This means that a jack is 14.42" long for every 12" of run.

The theoretical total length of the shortest jack rafter can now be calculated:

$$\frac{12\text{" (unit run)}}{14.42\text{" (unit length)}} = \frac{16\text{" (total run)}}{\times \quad \text{(total length)}}$$

$$x = 19.23\text{"}$$

This is the length of the shortest hip jack when the jacks are spaced 16" on center and the unit rise is 8". It is also the *common difference* of these jacks. This means that the next hip jack will be 2 × 19.23" long, the one after that 3 × 19.23" long, and so on.

The common difference for hip jacks spaced 16" on center and for hip jacks spaced 24" on center can be found in the rafter table on the framing square. Fig. 29-10b. The third line of the table reads

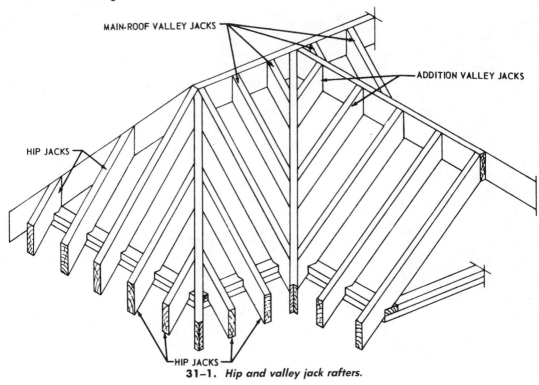

MAIN-ROOF VALLEY JACKS

ADDITION VALLEY JACKS

HIP JACKS

HIP JACKS

31-1. *Hip and valley jack rafters.*

31-2. *Valley cripple jack and hip-valley cripple jacks.*

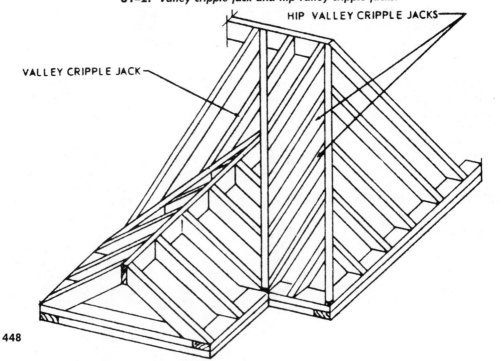

HIP VALLEY CRIPPLE JACKS

VALLEY CRIPPLE JACK

"Difference in Length of Jacks 16 Inches Centers." Follow this line to the column headed 8 (for a unit rise of 8") to find the length of the first jack rafter and the common difference, 19¼".

LENGTHS OF VALLEY AND CRIPPLE JACKS

The best way to figure the total lengths of valley jacks and cripple jacks is to lay out a framing diagram. Fig. 31-4 shows part of a framing diagram for a main hip roof with a long-and-short-valley-rafter gable addition. By studying the diagram you can figure the total lengths of the valley jacks and cripple jacks as follows:

The run of valley jack No. 1 is the same as the run of hip jack No. 8, which is the shortest hip jack. The length of valley jack No. 1 is therefore equal to the common difference of jacks.

The run of valley jack No. 2 is the same as the run of hip jack No. 7, and the length is therefore twice the common difference of jacks.

The run of valley jack No. 3 is the same as the run of hip jack No. 6. The length is therefore three times common difference of jacks. The run of hip-valley cripple No. 4, and also of hip-valley cripple No. 5, is the same as the run of valley jack No. 3. The length of these rafters is thus the same as the length of No. 3.

The run of valley jack No. 9, and also of valley jack No. 10, is equal to the spacing of jacks on center. Therefore the length of each of these jacks is equal to the common difference of jacks. The run of valley jacks Nos. 11 and 12 is twice the run of valley jacks Nos. 9 and 10. The length of each of these jacks is therefore twice the common difference of jacks.

The run of valley cripple No. 13 is twice the spacing of jacks on center, and the length is therefore twice the common difference of jacks. The run of valley cripple

31–3. *Hip jack framing diagram.*

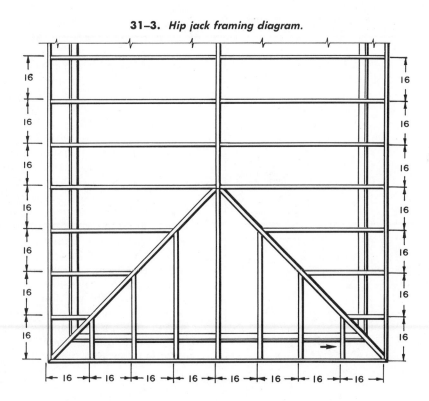

No. 14 is twice the run of valley cripple No. 13, and the length is therefore twice the common difference of jacks.

JACK RAFTER SHORTENING ALLOWANCES

A hip jack rafter has a shortening allowance at the upper end consisting of one-half the 45° thickness of the hip rafter. Fig. 31-5. A valley jack rafter has a shortening allowance at the upper end, consisting of one-half the thickness of the ridge board (Figs. 30-11 and 30-12) and another at the lower end, consisting of one-half the 45° thickness of the valley rafter. Fig. 31-5. A hip-valley cripple has a shortening allowance at the upper end, consisting of one-half the 45° thickness of the hip rafter, and another at the lower end, consisting of one-half the 45° thickness of the valley rafter. A valley cripple has a shortening allowance at the upper end, consisting of one-half the 45° thickness of the long val-

ley rafter, and another at the lower end, consisting of one-half the 45° thickness of the short valley rafter.

JACK RAFTER SIDE CUTS

The side cut on a jack rafter can be laid out by the method illustrated in Fig. 30-11 and 30-12 for laying out the side cut on a hip rafter. Another method is to use the rafter table on the framing square. Fig. 29-10b. Find the line headed "Side Cut of Jacks Use" and read across to the figure under the unit rise. For a unit rise of 8, the figure given is 10. To lay out the side cut on a jack with this unit rise, set the square face-up on the edge of the rafter to 12" on the tongue and 10" on the blade. Draw the side cut line along the tongue, as was described earlier for side cuts on hip rafters. Fig. 30-14.

JACK RAFTER BIRD'S-MOUTH AND OVERHANG

A jack rafter is a shortened common rafter. Consequently the bird's-mouth and overhang on a jack rafter are laid out just as they are on a common rafter.

JACK RAFTER PATTERN

Lay out and cut the longest jack rafter first. Be careful to calculate and make all necessary allowances to determine the actual length. Set the rafter in place on the building and check the fit of all the cuts. See that the spacing between the centers of the rafters is correct. When everything is correct, use this rafter as a pattern. On the top edge of the rafter, measure down the center line from the ridge end a distance equal to the common difference measurement (found on the framing square rafter table). This is the length of the second longest jack rafter. Continue to mark the common difference measurements along the top edge until the lengths of all the jacks have been laid out. Fig. 31-6. Using this pattern, mark off all the jack rafters. When all the rafters have been cut, the pattern is used as a part of the roof frame.

31–4. *Jack rafter framing diagram for a hip roof with a gable addition.*

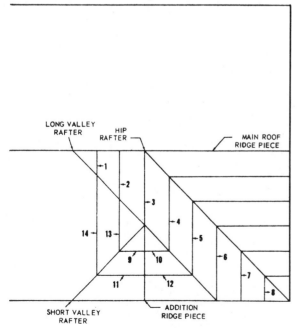

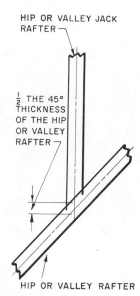

HIP OR VALLEY JACK RAFTER

½ THE 45° THICKNESS OF THE HIP OR VALLEY RAFTER

HIP OR VALLEY RAFTER

31–5. *The shortening allowance for the upper end of a hip jack or the lower end of a valley jack rafter is one-half the 45° thickness of the hip or valley rafter, whichever the jack rafter intersects.*

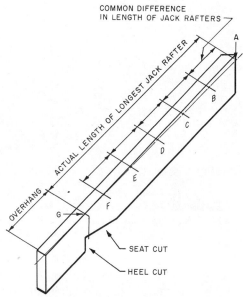

COMMON DIFFERENCE IN LENGTH OF JACK RAFTERS

A
B
C
D
E
F
G

ACTUAL LENGTH OF LONGEST JACK RAFTER

OVERHANG

SEAT CUT

HEEL CUT

31–6. *Use the longest jack rafter as a pattern. The second jack rafter is BG, the third jack rafter is CG, and so on.*

QUESTIONS

1. What is a jack rafter?
2. What is a hip jack rafter?
3. What is a valley jack rafter?
4. What is a cripple jack rafter?

5. If the shortest hip jack rafter is 19¼″ long, how long will the third hip jack rafter be?

UNIT 32 Layout and Erection of the Roof Frame

When the building has been framed, plumbed, and squared and the ceiling joists are in place, the structure is ready for roof framing. Lay out and cut two common rafters for trial purposes as discussed on pages 420-427. With a scrap piece of material the same thickness as the ridge board, set the two common rafters on the building and check all the cuts to make certain the rafters fit properly. If necessary, make corrections on the trial rafters. When they fit properly, use one as a pattern and

32–1. *Framing a gable roof with an unequal-span addition and a Dutch hip on the front of the addition. The workers are framing the overhang on the gable end.*

cut the required number of common rafters for the roof frame. Lean the rafters in position against the building with the ridge end up. Calculate and lay out the actual length of the ridge board. Then lay out the rafter locations on the double plates and ridge board in preparation for the roof frame erection.

LAYING OUT THE RIDGE BOARD
Gable Roof

Laying out the ridge piece for a gable main roof presents no particular problem, since the theoretical length of the ridge piece is equal to the length of the building. The actual length would include any overhang . Fig. 32-1.

Hip Roof

For a hip main roof, the ridge piece layout requires a certain amount of calculation. In an equal-pitch hip roof the theoretical length of the ridge piece amounts to the length of the building minus twice the total run of a main roof common rafter. The actual length, however, depends upon the way in which the hip rafters are framed to the ridge.

The theoretical ends of the ridge board are at the points where the ridge center

line and the hip rafter center lines cross. Fig. 32-2. If the hip rafter is framed against the ridge board, the actual length of the ridge board exceeds the theoretical length, at each end, by one-half the thickness of the ridge board plus one-half the 45° thickness of the hip rafter. A, Fig. 32-2. If the hip rafter is framed between the common rafters, the actual length of the ridge board exceeds the theoretical length, at each end, by one-half the thickness of a common rafter. B, Fig. 32-2.

Equal-Span Addition

For an equal-span addition, the length of the ridge board is equal to the length the addition projects beyond the building, plus one-half the span of the building, minus the shortening allowance at the main roof ridge. The shortening allowance amounts to one-half the thickness of the main roof ridge piece. Fig. 32-3.

Unequal-Span Addition

The length of the ridge board for an unequal-span addition varies with the method of framing the ridge piece. Fig. 32-4. If the addition ridge board is suspended from the main roof ridge board, the length is equal to the length the addition

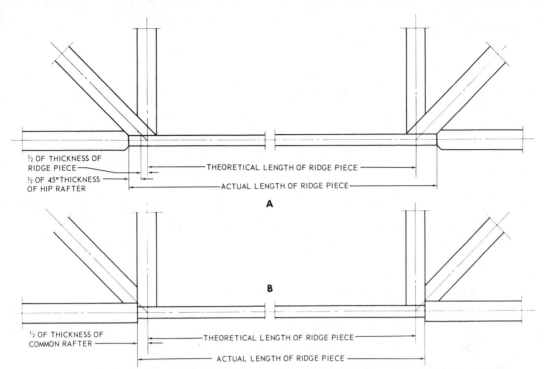

½ OF THICKNESS OF
RIDGE PIECE

½ OF 45° THICKNESS
OF HIP RAFTER

THEORETICAL LENGTH OF RIDGE PIECE

ACTUAL LENGTH OF RIDGE PIECE

A

B

½ OF THICKNESS OF
COMMON RAFTER

THEORETICAL LENGTH OF RIDGE PIECE

ACTUAL LENGTH OF RIDGE PIECE

32–2. *Theoretical and actual lengths of hip roof ridge boards. A. Hip rafter framed against ridge board. B. Hip rafter framed between common rafters. In this drawing the ridge board is 1″ material. Usually it is 2″, the same as the ridge-end common rafters, and the side cuts on these rafters (A) are not needed.*

32–3. *Determining the length of the ridge board for an equal-span addition.*

32–4. *Determining the length of the ridge board for an unequal-span addition.*

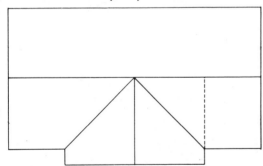

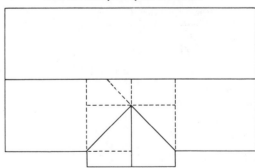

projects beyond the building, plus one-half the span of the main roof.

If the addition ridge board is framed by the long-and-short-valley-rafter method, the length is equal to the length the addition projects beyond the building, plus

one-half the span of the addition, minus a shortening allowance consisting of one-half the 45° thickness of the long valley rafter.

If the addition ridge piece is framed to a double header set between a couple of double main roof common rafters, the length of the ridge piece is equal to the length the addition projects beyond the building, plus one-half the span of the addition, minus a shortening allowance consisting of one-half the thickness of the inside member of the double header.

Dormer without Side Walls

The length of the ridge piece on a dormer without side walls is equal to one-half the span of the dormer, less a shortening allowance consisting of one-half the thickness of the inside member of the upper double header. Fig. 32-5.

Dormer with Side Walls

The length of the ridge board on a dormer with side walls amounts to the length of the dormer side-wall top plate, plus one-half the span of the dormer, minus a shortening allowance consisting of one-half the thickness of the inside member of the upper double header. Fig. 32-6.

LAYING OUT THE RAFTER LOCATIONS

The layout of the rafter spacing on the wall plates and ridge board is determined by checking either the building plans or the roof frame plan. Rafter locations are laid out on plates, ridge board, and other rafters with the same lines and X's used to lay out stud and joist locations. (See Units 25 and 27). In most cases the rafters are

located next to the ceiling joists. The rafters can then be fastened to the side of the joists to tie the building together.

Gable Roof

For a gable roof the rafter locations are laid out on the top plates first. The locations are then transferred to the ridge piece by matching the ridge board against a top plate. Fig. 32-7.

On a gable roof the first rafters on each end are usually set even with the outside wall to permit a smooth unbroken surface for the application of the sheathing. Since the first ceiling joist was set on the inside edge of the wall, it will be necessary to place a spacer block between the first rafter and the first ceiling joist. Fig. 32-8.

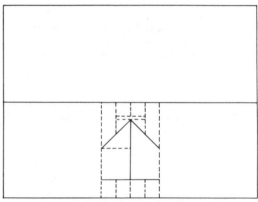

32–6. *Determining length of ridge board on dormer with side walls.*

32–5. *Determining length of ridge board on dormer without side walls.*

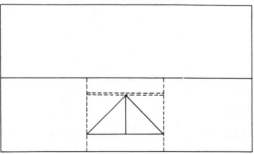

32–7. *Lay the ridge board on edge on the top plate and extend the layout lines from the plate onto the ridge board.*

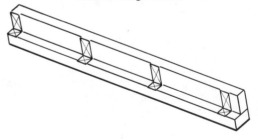

The other rafters are fastened to the side of the joists along the length of the building.

If the rafters are on 24″ centers and the ceiling joists are on 16″ centers, the first rafter will be placed as shown in Fig. 32-8. The second rafter will rest on the plate between the second and third joists. The third rafter will fasten to the side of the fourth joist. The rafters will continue to alternately rest on the plate between two joists and fasten to the side of a joist along the remaining length of the building. Fig. 32-9.

Always begin the rafter layout on the plates from the same end of the building for the two opposite walls, and continue along the length of the building. Fig. 32-10. This will insure direct bearing down through the walls to the foundation wall. It will also make the rafters butt directly opposite each other on the ridge board. Fig. 32-11.

Hip Roof

The top plate locations of the ridge-end common rafters in an equal-pitch hip roof

32–8. *Gable roof rafter locations.*

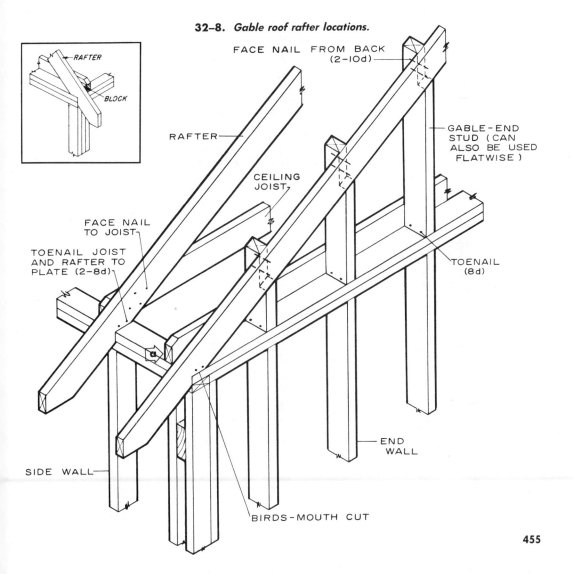

measure one-half of the span (or the run of a main roof common rafter) away from the building corners. These locations, plus the top plate locations of the rafters lying between the ridge-end common rafters, can be transferred to the ridge board by matching the ridge board against the top plates. Fig. 32-12.

Addition Roofs

In an equal-span addition the valley rafter locations on the main roof ridge

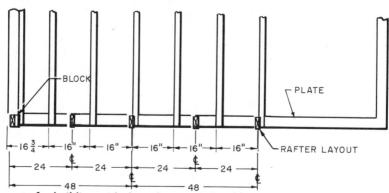

32–9. *Layout of a building with the rafters on 24" centers and the ceiling joists on 16" centers.*

32–10. *Begin the layout of the rafters from the same end of the building as the layout of the floor joists, wall studs, and ceiling joists. In this drawing the layout for each phase began at arrow A on the two side walls.*

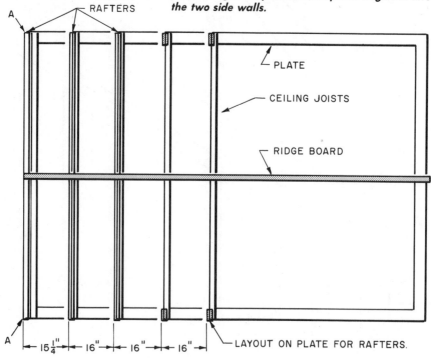

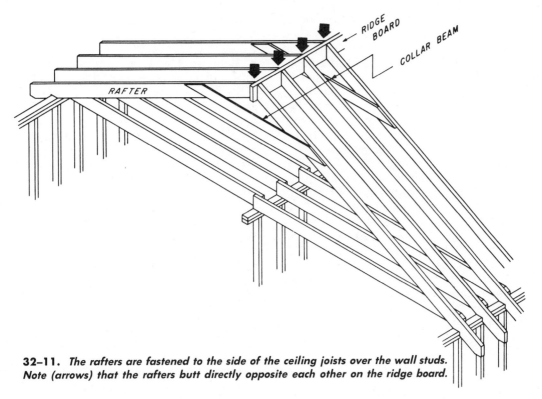

32–11. *The rafters are fastened to the side of the ceiling joists over the wall studs. Note (arrows) that the rafters butt directly opposite each other on the ridge board.*

32–12. *The locations of the rafters in the area "A" are transferred to the ridge board from the top plate.*

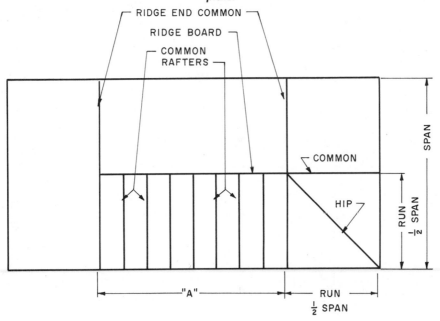

457

board lie alongside the addition ridge board location. In Fig. 32-13 the distance between the end of the main roof ridge board and the addition ridge piece location is equal to distance A plus distance B, distance B being one-half the span of the addition. In Fig. 32-14 the distance between the *theoretical* end of the main roof ridge board and the addition ridge board location is the same as distance A.

In an unequal-span addition, if framing is by the long-and-short-valley-rafter method, the distance from the end of the main roof ridge board to the upper end of the longer valley rafter is equal to distance A plus distance B, distance B being one-half the span of the main roof. Fig. 32-15. The location of the inboard end of the shorter valley rafter on the longer valley rafter can be determined as follows:

• Obtain the unit length of the longer valley rafter from the rafter table. Fig. 29-

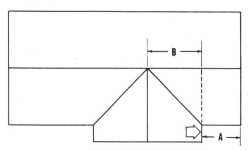

32–13. *Ridge board location for equal-span addition on a gable roof.*

32–14. *Ridge board location for equal-span addition on a hip roof.*

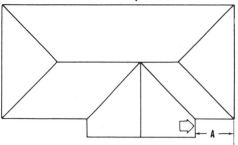

10b. Suppose that the common rafter unit rise is 8″. In that case the unit length of a valley rafter is 18.76″.

• Between the point where the shorter rafter ties in and the top plate, the total run of the longer valley rafter is the hypotenuse of a right triangle whose shorter sides are each equal to the total run of a common rafter in the addition. The total run of a common rafter in the addition is one-half the span. Suppose the addition is 20′ wide; the run of a common rafter would be 10′. C, Fig. 32-15.

• You know that the valley rafter is 18.76″ long for every foot of common rafter run. The location mark for the inboard end of the shorter valley rafter on the longer valley rafter can thus be calculated:

18.76 (in. per ft. of run) × 10 (ft. of run) = 187.6″; 187.6″ = 15.63′, or 15′ 7⁹/₁₆″

This is the distance from the heel plumb cut line of the longer valley rafter to the location mark.

If framing is by the suspended-ridge method, the distance between the suspension point on the main roof ridge board and the end of the main roof ridge piece is equal to distaince A plus distance C. Fig. 32-15. Distance C is one-half the span of the addition. The distance between the outboard end of the addition ridge board and the valley rafters (both short in this method of framing) tie into the addition ridge board is equal to one-half the span of

32–15. *Ridge board and valley rafter locations for unequal-span addition.*

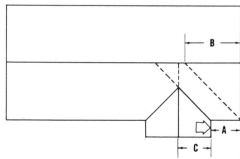

the addition plus the length of the addition side-wall top plate.

ERECTING THE RIDGE BOARD

Many carpenters raise the ridge board and the gable-end rafters all at one time. Each member supports the other. However, with the gable-end rafters nailed in place, it is difficult to make adjustments.

It is possible to put the ridge board in place before raising any rafters. Nail uprights on the walls and cross partitions below the center line of the ridge board for support. Erect the ridge board, level and align it, and nail it in place. Fig. 32-16a. The ridge board should also be braced longitudinally to prevent the roof from swaying. This is particularly important on a gable roof. Fig. 32-16b. After the ridge board is nailed in position, begin the erection of the rafters.

ERECTING THE RAFTERS

Roof framing should be done from a scaffold with planking not less than 4' below the level of the main roof ridge board. The usual type of roof scaffold consists of diagonally braced, two-legged horses, spaced about 10' apart and extending the full length of the ridge piece.

If the building has an addition, as much as possible of the main roof is framed before the addition framing is started. All types of jack rafters are usually left out until after the headers, hip rafters, valley rafters, and ridges to which they will be framed have been installed.

Gable Roof

For a gable roof the two pairs of gable-end rafters and the ridge piece are usually erected first. Two people, one at either end of the scaffold, hold the ridge board in position, while a third person sets the gable-end rafters in place and toenails them at the top plate with 8d nails, two on one side and one on the other side. Make certain the person standing at the ridge pulls the rafter up so that the (plumb) heel cut of the bird's-mouth is tight against the side of the building when the rafter is nailed at the plate. Each worker on the scaffold then end-nails the ridge piece to one of the rafters with three 10d nails driven through the ridge piece into the end of the rafter. The other rafter is toenailed to the ridge piece and to the first rafter with four 8d nails, two on each side of the rafter.

32–16b. *Longitudinal bracing for the ridge board.*

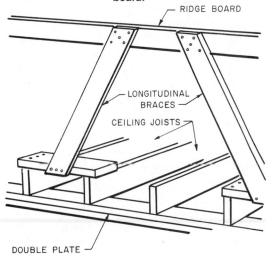

32–16a. *An upright (leg) supports the ridge board in position for the rafter erection.*

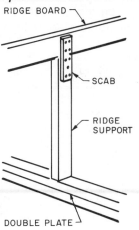

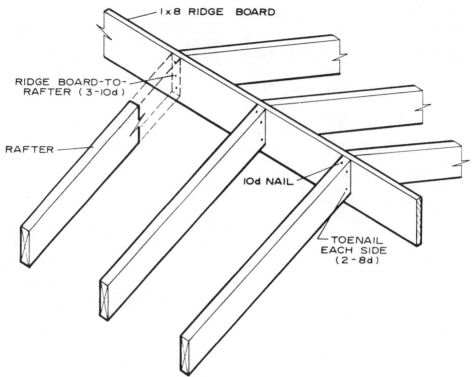

32–17. *Rafter nailing procedure at the ridge board.*

If the ridge board has not been previously erected and braced, temporary braces like those for a wall should be installed at the ridge ends to hold the rafters approximately plumb, after which the rafters between the end rafters should be erected. Figs. 32-17, 18, and 19. The braces should then be released, and the pair of rafters at one end should be plumbed. The braces are then reset and left in place until enough sheathing has been installed to hold the rafters plumb.

Ceiling-joist ends are nailed to adjacent rafters with four 10d nails, two to each side. Metal brackets may also be used to attach the rafters to the plate. Fig. 32-20.

Hip Roof

On a hip roof the ridge board and the common rafters extending from the ridge ends to the side walls are erected first, in about the same manner as for a gable roof. The intermediate common rafters are then filled in. After that, the ridge-end common rafters extending from the ridge ends to the mid-points on the end walls are erected. The hip rafters and hip jacks are installed next.

The common rafters in a hip roof do not require plumbing. If the hip rafters are correctly cut, installing the hip rafters and the common rafter which projects from the end of the ridge board to the end wall will bring the common rafters plumb.

Hip rafters are toenailed to plate corners with 10d nails, two to each side. At the ridge board, they are toenailed with four 8d nails. After the hip rafters are fastened in place, partially drive a nail in the center of the top edge of the hip rafter at the ridge

32–18. *Toenailing a rafter to the ridge board and to the rafter opposite it.*

32–19. *Face-nailing a jack rafter through the ridge board.*

end and at the plate end. Pull a line taut between these nails and as the hip jacks are nailed to the hip rafter, keep the string centered on the top edge of the hip rafter to insure a straight hip line.

The hip jacks should be nailed in pairs, one opposite the other. *Do not nail* all the jacks on one side of the hip and then all the jacks on the opposite side as this would push the hip out of alignment and cause a bow. Hip jacks are toenailed to hip rafters with 10d nails, three to each jack, and to the plate with 10d nails, two to each side.

Additions and Dormers

For an addition or dormer the valley rafters are usually erected first. Valley rafters are toenailed to plates with 10d nails, two to each side, and to ridge pieces and headers with three 10d nails. Ridge pieces and ridge-end common rafters are erected

next, then other addition common rafters, and last, valley and cripple jacks. As with hip rafters, pull a line along the top edge of the valley rafter and nail the jacks in pairs. A valley jack should be held in position for nailing as shown in Fig. 32-21. When properly nailed, the end of a straightedge laid along the top edge of the jack should contact the center line of the valley rafter as shown.

COLLAR BEAM FRAMING

Gable or double-pitch roof rafters are often reinforced by horizontal members called *collar beams*. In a finished attic the collar beams also function as ceiling joists.

The length of a collar beam is calculated on the basis of the height of the beam above the level of the side-wall top plates. The theoretical length of a beam in feet is found by dividing this height in inches by the unit rise of a common rafter in the roof, and subtracting twice the result from the span of the building. For example, in the roof shown in Fig. 32-22, the collar beam is 3' 6", or 42", above the rafter plate. The unit rise of a common rafter in the roof is 10. Forty-two divided by 10 is 4.2, and twice 4.2 is 8.4. This is subtracted from the span of the building: $16-8.4 = 7.6'$, or about 7' 7³/₁₆", which is the theoretical length of the beam.

To bring the ends of the collar beam flush with the upper edges of the common rafters, you must add to the theoretical length of the beam, at each end, an amount equal to the *level width* of a rafter minus the width of the rafter seat cut. The level width is obtained by setting the square on the rafter to the cut of the roof, drawing a level line from edge to edge, and measuring the length of this line.

Lay out the end cuts on a collar beam by setting the framing square on the beam to the cut of the roof. Fig. 32-23.

32-20. *Metal brackets are sometimes used to fasten the rafter to the plate. These brackets are fastened with special nails (11 gauge-1 1/4" long) which are furnished with the bracket.*

Collar beams are nailed to common rafters with four 8d nails to each end of a one-inch beam. If two-inch material is used for the beams, they are nailed with three 16d nails at each end.

GABLE-END FRAMING

Gable-end studs are members which rest on the top plate and extend to the rafter line in the ends of a gable roof. Fig. 32-24a & b. They may be placed with the edge of the stud even with the outside wall and the top notched to fit the rafter (Fig. 32-24c), or they may be installed flatwise with a cut on the top of the stud to fit the slope of the rafter.

The position of the first gable-end stud is located by making a mark on the double plate directly above the wall stud nearest the ridge line. A, Fig. 32-25. Plumb the gable-end stud on this mark and mark the stud where it hits the bottom of the rafter. B, Fig. 32-25. Mark the cut of the roof across the edge of the gable stud and notch the stud to a depth equal to the thickness of the rafter. C, Fig. 32-25.

The lengths of the other gable studs will depend on the spacing. For studs 24" on

32-21. *Correct position for nailing a valley jack rafter.*

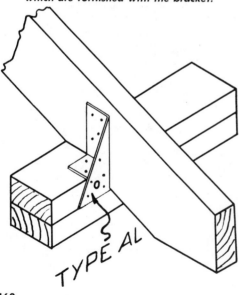

STRAIGHTEDGE

VALLEY JACK

VALLEY JACK

VALLEY RAFTER

TYPE AL

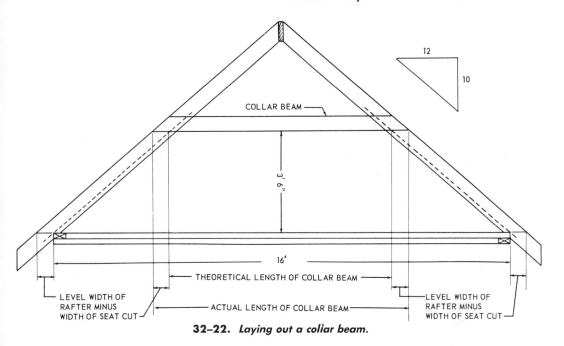

32-22. *Laying out a collar beam.*

32-23. *Laying out the end cut on a collar beam for a roof with a unit rise of 10".*

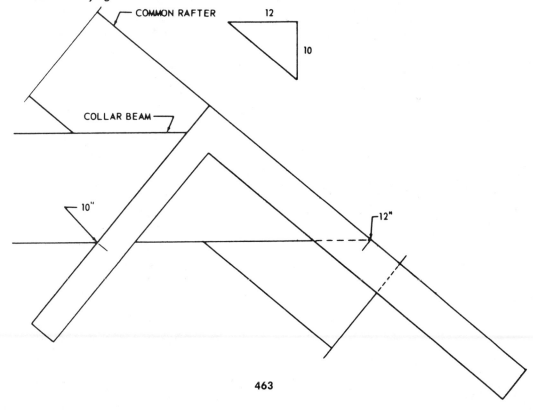

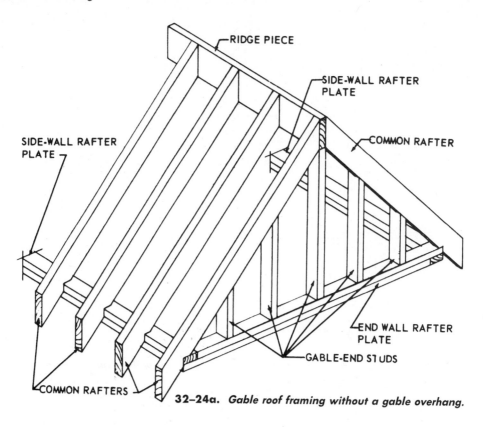

RIDGE PIECE

SIDE-WALL RAFTER PLATE

COMMON RAFTER

SIDE-WALL RAFTER PLATE

END WALL RAFTER PLATE

GABLE-END STUDS

COMMON RAFTERS

32–24a. *Gable roof framing without a gable overhang.*

32–24b. *Roof framing for overhang at gable end.*

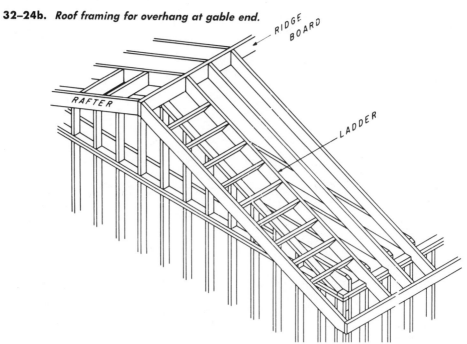

RIDGE BOARD

RAFTER

LADDER

center, the line DE in Fig. 32-25 represents 2 units of run (one unit is 12″). For a roof with a unit rise of 6″ and studs 24″ O.C., the second gable stud will be 12″ shorter.

The common difference in the length of the gable studs may be figured by the following method:

$$\frac{24'' \ (\text{O.C. spacing})}{12'' \ (\text{unit run})} = 2$$

2 × 6″ (unit rise) = 12″ (common difference)

A common difference of 12″ means that each stud will be 12″ shorter than the first, the third stud 24″ shorter than the first, the fourth stud 36″ shorter, and so on. If the studs are spaced 16″ O.C. for the same roof, the common difference in length is 8″:

$$\frac{16'' \ (\text{O. C. spacing})}{12'' \ (\text{unit run})} = 1\frac{1}{3}$$

$1\frac{1}{3}$ × 6″ (unit rise) = 8″ (common difference)

The common difference in the length of the gable studs may also be laid out directly with the framing square. Fig. 32-26. Place the framing square on the stud to the cut of the roof (6 and 12 for this example). Draw a line along the blade at A. Slide the square along this line in the direction of the arrow at B until the spacing desired between the studs (16 for this example) is at the intersection of the line drawn at A and the edge of the stud. C, Fig. 32-26. Read the dimension on the tongue which is aligned with the same edge of the stud. This is the common difference (8″ for this example) between the gable studs.

Toenail the studs to the plate with two 8d nails from each side. Fig. 32-8. As the studs are nailed in place, care must be taken not to force a crown into the top of the rafter.

FRAMING A GAMBREL ROOF

The gambrel roof is a gable roof with two slopes. It has the advantage of providing additional space for rooms in the attic

32–24c. Gable-end studs notched to fit the rafter.

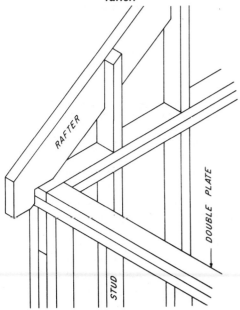

32–25. Locating the position of the gable-end studs and determining the common difference in length.

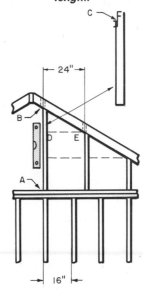

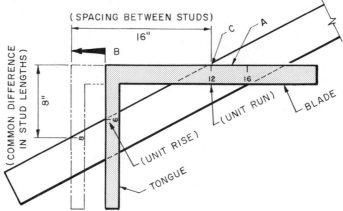

32-26. *Using the framing square to determine the common difference in the length of gable-end studs.*

area. It also minimizes the roof area exposed to snow loads. The framing of this roof style is simply a combination of two common rafters, the lower one having a steep pitch and the upper one a low pitch. If the pitches are known, the rafters may be laid out in the same manner as any common rafter.

The roof may also be laid out full size on the subfloor. Use the run of the building (AB) as a radius and draw a semicircle. Fig. 32-27. Draw a perpendicular line from point A to intersect the semicircle at E. This

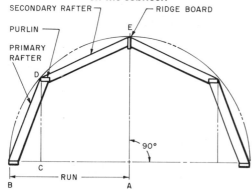

32-27. *The patterns for the rafters in a gambrel roof may be made by laying the roof out full size on the subfloor.*

locates the ridge line. Find the height of the partition walls from the plans. Draw a perpendicular line this length, between the plate and the semicircle. Line CD, Fig. 32-27. Connect the points B and D and the points D and E. This gives the location and pitch of the primary rafter BD and the secondary rafter DE. From this layout the rafter patterns can be made and cut for trial on the building.

FRAMING A SHED ROOF

A shed roof is essentially one-half of a gable roof. Like the full-length rafters in a gable roof, the full-length rafters in a shed roof are common rafters. However, the total run of a shed roof common rafter is equal to the span of the building *minus the width of the top plate on the higher rafter-end wall.* Fig. 32-28. Also, the run of the overhang on the higher wall is measured from the *inner edge* of the top plate. With these exceptions, shed roof common rafters are laid out like gable roof common rafters. A shed roof common rafter has two bird's-mouths, but they are laid out just like the bird's-mouth on a gable roof common rafter.

For a shed roof, the height of the higher rafter-end wall must exceed the height of the lower by an amount equal to the total rise of a common rafter.

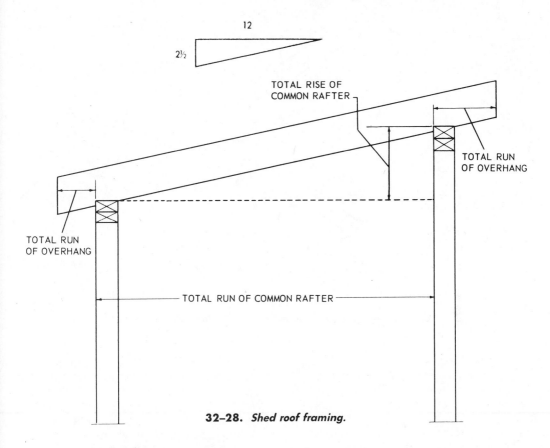

32–28. *Shed roof framing.*

FRAMING A SHED DORMER

When framing a shed dormer (Fig. 32-29), there are three layout problems to be solved:

• Determining the total run of a dormer rafter.

• Determining the angle of cut on the inboard ends of the dormer rafters.

• Determining the lengths of the dormer side-wall studs.

To determine the total run of a dormer rafter, divide the height of the dormer end wall, in inches, by the difference between the unit rise of the dormer roof and the unit rise of the main roof. For example, suppose the height of the dormer end wall is 9′, or 108″. A, Fig. 32-30. The unit rise of the main roof is 8; the unit rise of the dormer

32–29. *Framing a shed dormer.*

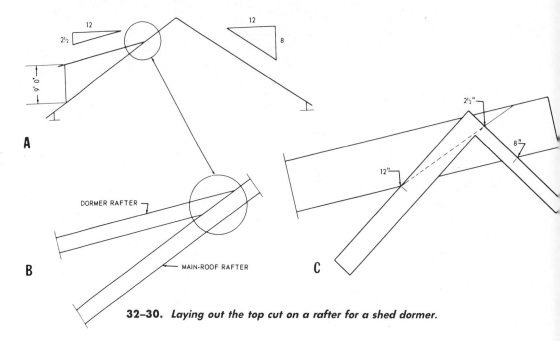

A

B

C

DORMER RAFTER

MAIN-ROOF RAFTER

32–30. Laying out the top cut on a rafter for a shed dormer.

roof is 2½. The difference between them is 5½. The total run of a dormer rafter is therefore 108 divided by 5½, or 19.63 feet. Knowing the total run and the unit rise, you can figure the length of a dormer rafter by any of the methods already described.

The inboard ends of the dormer rafters must be cut to fit the slope of the main roof.

32–31. Determining the common difference in the length of dormer side-wall studs by direct layout.

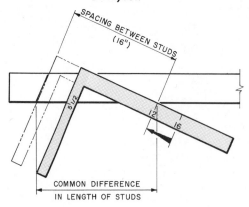

SPACING BETWEEN STUDS
(16")

COMMON DIFFERENCE
IN LENGTH OF STUDS

B, Fig. 32-30. To get the angle of this cut, set the square on the rafter to the cut of the main roof. C, Fig. 32-30. Measure off the unit rise of the dormer roof from the heel of the square along the tongue. Make a mark at this point and draw the cut off line through this mark from the 12″ mark.

The lengths of the side-wall studs on a shed dormer are determined as follows: Suppose a dormer rafter rises 2½″ for every 12″ of run and a main roof common rafter rises 8″ for every 12″ of run. A, Fig. 32-30. If the studs were spaced 12″ O.C., the length of the shortest stud (which is also the common difference of studs) would be the difference between 8″ and 2½″, or 5½″. This being the case, if the stud spacing is 16″, the length of the shortest stud is the value of x in the proportional equation $12{:}5\frac{1}{2} :: 16{:}x$. Thus $x = 7\frac{5}{16}$. The shortest stud will be $7\frac{5}{16}$″ long. The next stud will be $2 \times 7\frac{5}{16}$″ long, or $14\frac{5}{8}$″, and so on.

A second method of determining the length of the shortest stud (the common difference of the studs) is to make the layout directly on a stud with the framing square. Fig. 32-31. The difference in the rise of the two roofs is 5¹/₂″. Find the 5¹/₂″ mark on the tongue of the square and place it on the edge of a stud. Place the 12″ mark of the body of the square on the same edge of the stud. Draw a line along the body of the square onto the stud. Slide the square along this line until the 16″ mark (the on-center spacing between the studs) is over the point of the 12″ mark. Draw a line along the tongue of the square. This completes the layout for the shortest stud; the second stud will be twice as long, and so on.

To get the lower-end cut off angle for studs, set the square on the stud to the cut of the main roof. To get the upper-end cut off angle, set the square to the cut of the dormer roof.

FRAMING A FLAT OR LOW-PITCH ROOF

The two basic types of roofs--flat and pitched--have numerous variations. The so-called flat roof may actually have some slope for drainage. As discussed earlier, the slope is generally expressed as the inches of vertical rise in 12 inches of horizontal run. For purposes of definition, flat roofs might be classed as those having less than a 3-in12-slope. Fig. 32-32.

Post-and-beam construction is frequently used with flat or low-slope roofs. Fig. 32-33. In conventional stud wall framing for buildings with flat or low-slope roofs, the rafters or roof joists usually serve as ceiling joists for the space below.

The flat or low-slope roof sometimes combines ceiling and roof elements in one system. This system serves as an interior finish, or as a fastening surface for the finish, and as an outer surface for application of the roofing. Fig. 32-34. In mild climates flat or low-pitch roofs may be built with 2″ matched planks for roof sheathing supported on large beams spaced about 6′ apart. The planking and beams are exposed on the underside. Fig. 32-35. The

exposed material may be dressed smooth and finished with varnish or otherwise decorated.

The structural elements can be arranged in several ways by the use of ceiling

32–32. *Variations of flat roof styles.*

Flat Single Span

Flat Double Span

Shed

Combination

Ridge & Plate Beams

Rafter Beams

Cantilever Beam

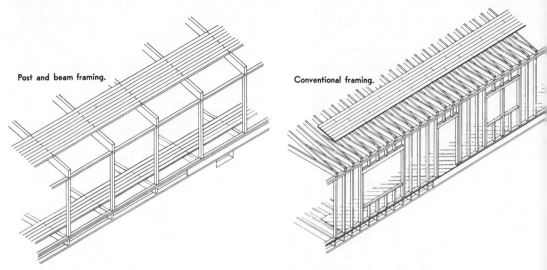

32–33. *Comparison of post-and-beam system with conventional framing.*

32–34. *Rafter-joist construction for a flat roof: A. Detail at exterior wall. B. Detail at interior wall.*

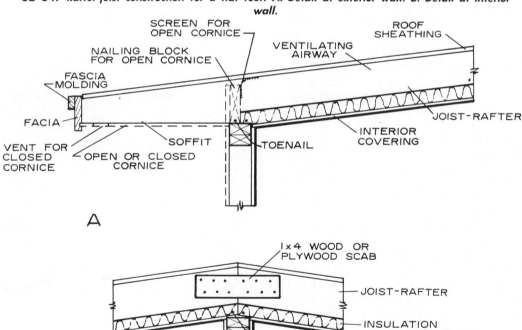

32–35. *In this home the ceiling is extended beyond the exterior wall to become the overhang. The roofing is applied on top and a finish is applied to the bottom. The single system serves as a ceiling and a roof.*

beams or thick roof decking which spans from the exterior walls to the ridge beam or center bearing partition. Fig. 32-36. The roof is generally covered with a fiberboard insulation, and this in turn with a composition roof.

Roof joists for flat roofs are commonly laid level, with roof sheathing and roofing on top and with the underside utilized to support the ceiling. Sometimes a slight roof slope may be provided for roof drainage by tapering the joist or adding a cant strip (a triangular piece of lumber) to the top. Insulation may be added just above the ceiling, and the space above the insulation should be ventilated to remove hot air in the summer and to provide protection against condensation in the winter.

Flat and low-pitch roofs generally require larger-sized rafters than pitched roofs, but the total amount of framing lumber required is usually less. In flat roof construction where rafters also serve as ceiling joists, the size of the rafters is based on both roof and ceiling loads. The size is given on the plans or determined from rafter span tables.

When there is an overhang on all sides of the house, lookout rafters are ordinarily used. Fig. 32-37. The lookout rafters are nailed to a double header and toenailed to the wall plate. The distance from the double header to the wall line is usually twice the overhang. Rafter ends may be finished with an outside header, which will serve as a nailing surface for trim.

32–36. *Wood-deck construction: A. Installing wood decking. B. Toenailing horizontal joint. C. Edge-nailing 3″ × 6″ solid decking.*

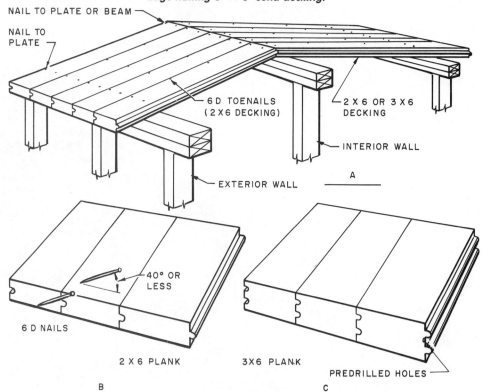

ROOF OPENINGS

Roof openings are those which require interruption of the normal run of rafters or other roof framing. Such openings may be required for a ventilator, chimney, skylight, or for dormer windows. Fig. 32-38.

Roof openings, like floor openings, are framed by headers and trimmers. Double headers are used at right angles to the rafters, which are set into the headers in the same manner as joists in floor opening construction. Just as trimmers are double

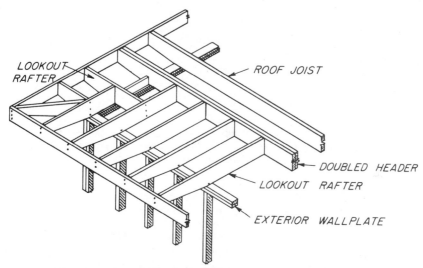

LOOKOUT RAFTER

ROOF JOIST

DOUBLED HEADER

LOOKOUT RAFTER

EXTERIOR WALLPLATE

32-37a. *Typical construction of flat roof with overhang.*

32-37b. *Corner framing for flat roofs. Note the use of metal brackets to fasten the lookout rafters to the main roof rafter joist.*

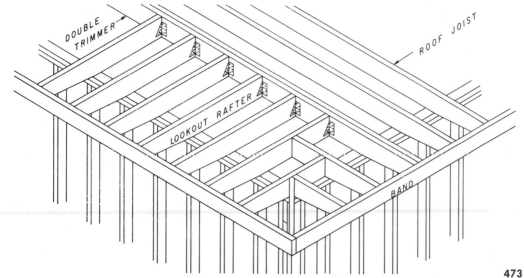

DOUBLE TRIMMER

ROOF JOIST

LOOKOUT RAFTER

BAND

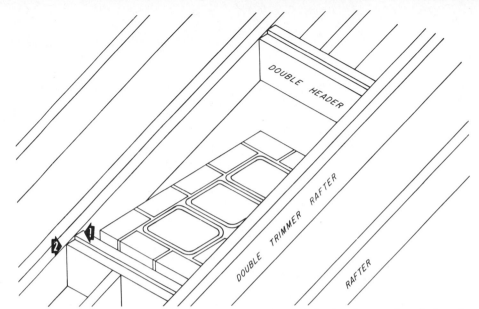

32–38. *Roof framing around the chimney. The top edges of the headers are kept below the top edge of the rafter (arrow 1). The lower edges of the headers are kept even with the top edge of the rafter (arrow 2).*

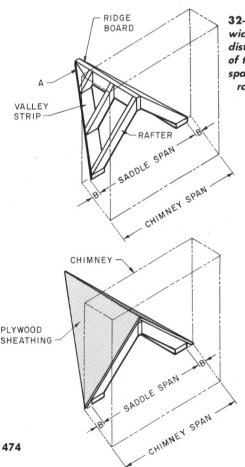

32–39. *The saddle span is less than the chimney width, as shown at "B" in the drawing. This distance ("B") must be subtracted from each side of the chimney width to obtain the actual saddle span. When the sheathing is applied to the saddle rafters, it will project beyond the valley strip.*

joists in floor construction, they are double rafters in roof openings.

CHIMNEY SADDLE

The chimney saddle sheds water and snow and prevents ice from building up behind the chimney on the roof. The saddle may be constructed on the roof. If the chimney span and roof pitch are known, it can also be fabricated on the ground and the completed assembly nailed to the roof framing. Fig. 32-39.

The valley strips are l″ × 4″ or l″ × 6″. The length is determined in the same way as for a valley rafter. Use the framing square to lay out the valley strips. Lay out the top and bottom cuts along the tongue of the square. When measuring off the length of the strip, use the unit length of a common rafter from the roof on which the saddle is to be framed.

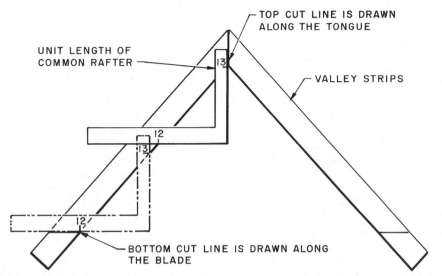

UNIT LENGTH OF COMMON RAFTER

TOP CUT LINE IS DRAWN ALONG THE TONGUE

VALLEY STRIPS

BOTTOM CUT LINE IS DRAWN ALONG THE BLADE

32–40. *Using the framing square to lay out the top and bottom cuts on the valley strips.*

For example, a roof with a unit rise of 5" has a unit length of 13". Fig. 29-10b. To lay out the valley strip, position the square with the 13" mark of the tongue and the 12" mark of the blade on the edge of the strip. Draw a line along the tongue for the top cut. Fig. 32-40. Measure and lay off the length of the valley strip. With the square set the same as for the top cut, place the edge of the blade on the length mark and draw a line along the blade for the bottom cut.

The end of the ridge rests on the valley strips. A, Fig. 32-39. This cut is the same as the seat cut for a common rafter in the main roof. Place the square on the ridge board for the cut of the roof (5" on the tongue and 12" on the blade for the example), and draw a line along the blade. The length of the ridge is equal to the run of the common rafter in the saddle span minus the allowance for the drop of the ridge, which is approximately 3/4".

The length of the longest rafter is determined by multiplying the saddle run (half the saddle span) by the unit length of the common rafter. Fig. 32-39. Deduct the ridge shortening allowance to obtain the actual length. The top and bottom cuts

are the same as for a common rafter in the main roof. However, there is a side cut on the bottom where the rafter rests on the valley strip. This cut is the same as for regular valley jacks. On the rafter table of the framing square, the side cut figure for a valley jack in a roof with a 5" unit rise is 11 1/2". Fig. 29-10b. Lay out and make the cut as described on p. 450 for jack rafters.

The cuts are the same for all the rafters in the chimney saddle. However, the rafter lengths differ. The difference in the length of the rafters can be found on the rafter table of the framing square under "Difference in Length of Jacks." For rafters 16" on center in a roof with a unit rise of 5", the second rafter will be 17 5/16" shorter than the first rafter. Fig. 29-10b. The third rafter will be 34 5/8" (2 × 17 5/16) shorter than the first rafter, and so on. When the saddle framing is complete, the sheathing, flashing (to prevent water seepage), and roofing are applied.

ESTIMATING
Number of Rafters

The number of rafters necessary for a building may be counted directly from the roof framing plan. The number of rafters may also be obtained in the same way the number of floor joists is estimated. For rafters on 16″ centers, take three-fourths of the building's length and add one. For example, if a rectangular building is 40′ long, 31 rafters will be required for each of the longer sides ($\frac{3}{4} \times 40 = 30$; $30 + 1 = 31$). A total of 62 rafters would thus be needed. Add to this amount extra rafters for the required trimmers and any other special framing.

Material Cost

An accurate estimate of the cost can be figured by multiplying the number of rafters required by the cost per rafter. Sometimes the builder will not make up a complete bill of materials, but wants a rough cost estimate. This can be figured without knowing the exact number of pieces needed by first finding the area of the roof.

The area of the roof is the length of the building times the width of the building

32–41.
Obtaining Roof Area From Plan Area

When a roof has to be figured from a plan only, and the roof pitch is known, the roof area may be fairly accurately computed from the table. The horizontal or plan area (including overhangs) should be multiplied by the factor shown in the table opposite the rise, which is given in inches per horizontal foot. The result will be the roof area.

Rise	Factor	Rise	Factor
3″	1.031	8″	1.202
3½″	1.042	8½″	1.225
4″	1.054	9″	1.250
4½″	1.068	9½″	1.275
5″	1.083	10″	1.302
5½″	1.100	10½″	1.329
6″	1.118	11″	1.357
6½″	1.137	11½″	1.385
7″	1.158	12″	1.414
7½″	1.179		

times the factor from the table in Fig. 32-41. For example, a building 20′ wide and 40′ long with no overhang, will have a roof area of 20 times 40, or 800 (square feet), times the factor from the table. For a roof with a unit rise of 5″, the factor is 1.083. The area of the roof, then, is 800 × 1.083, which equals 866.4, or 867 square feet of roof area.

The number of board feet of material for rafters, ridge board, and collar beams required for a building is next determined. Use the same dimensions of the building from the previous example, which contained 867 square feet of roof area, and refer to the chart in Fig. 32-42. If the rafters are 2″ × 6″ and 16″ on center, the chart indicates 102 board feet of lumber for each 100 square feet of roof surface area. Divide the total roof area by 100 and multiply by the factor in the chart:

$$\frac{867 \text{ (total sq. ft. of roof area)}}{100} = 8.67 \text{ (roof area expressed in hundreds of sq. ft.)}$$

$$8.67 \times 102 \text{ (bd. ft. per sq. ft.)} = 884.3 \text{ bd. ft.}$$

Multiply this figure by the cost per board foot to find the total cost of lumber for the roof.

The chart in Fig. 32-42 also has information for determining the number of nails necessary. For the roof in the example, 12 pounds of nails are needed for each 1,000 board feet. The roof in the example has about 884 board feet, requiring about 10½ pounds of nails:

$$\frac{884 \text{ (bd. ft. of material)}}{1000} = 0.884$$

(material expressed in thousands of bd. ft.)

$$0.884 \times 12 \text{ (lbs. of nails per 1000 bd. ft.)} = 10.6, \text{ or } 10\frac{1}{2} \text{ lbs. of nails}$$

The cost of the nails for the roof framing is determined by multiplying the cost for one pound by the total number needed.

Labor Cost

The cost of labor for framing a roof can also be found by using the information in Fig. 32-42. To use this chart it is necessary to know how many board feet of rafter material are required for a roof. Use the same examples as previously described: a building 20′ wide and 40′ long with a gable roof having a unit rise of 5″. Find the number of board feet in one rafter and then multiply this by the total number of rafters in the building.

The chart in Fig. 32-43 can be used to determine the board feet content of a single piece. To use this chart the length of a piece must be known. To find the length of the rafter for figuring board feet, refer to the chart in Fig. 32-44. The roof in the example has a unit rise of 5″ and the building width

is 20′. The chart in Fig. 32-44 indicates the rafter will be approximately 10′ 10″ long.

A 2″ × 6″ rafter, 10′ 10″ long, would be cut from a 2″ × 6″ × 12′ member. Find 2 × 6 in the left column of the table in Fig. 32-43 and read across to the column headed 12, for a 12′ piece. There are 12 board feet in one 2″ × 6″ × 12′ piece of lumber. The roof in the example has 62 rafters (see "Number of Rafters"). Therefore, if one rafter contains 12 board feet of material, 62 rafters contain 62 × 12, or 744 board feet of rafter material.

The cost of framing a roof can be found by using the factor from the table in Fig. 32-42. There are 744 board feet of common rafter material in the building used for the example. The table shows a worker can frame 35 board feet of common rafters per hour. The number of hours required to frame the roof is 744 divided by 35, or 21.2 hours. To find the cost of labor, multiply the number of hours by the hourly labor rate.

32–42. *Estimating materials and labor for roof framing.*

RAFTERS	Board Feet Required 100 Square Feet Surface Area			Nails	Labor
	12″ O.C.	16″ O.C.	24″ O.C.	Per 1000 Board Feet	Board Feet Per Hour
2 x 4	89	71	53	17	
2 x 6	129	102	75	12	See
2 x 8	171	134	112	9	Table
2 x 10	212	197	121	7	Below
2 x 12	252	197	143	6	

Note: Includes common, hip and valley rafters, ridge boards and collar beams.

RAFTERS			Labor			
	Common	Hip	Jack	Valley	Ridge	Collars
Board Feet Per Hour	35	35	25	35	35	65

32–43. *Board feet content of various timber sizes.*
Board Feet Content

Size of Timber in Inches	Length Of Piece In Feet							
	10	12	14	16	18	20	22	24
1 × 2	1⅔	2	2⅓	2⅔	3	3⅓	3⅔	4
1 × 3	2½	3	3½	4	4½	5	5½	6
1 × 4	3⅓	4	4⅔	5⅓	6	6⅔	7⅓	8
1 × 5	4⅙	5	5⅚	6⅔	7½	8⅓	9⅙	10
1 × 6	5	6	7	8	9	10	11	12
1 × 8	6⅔	8	9⅓	10⅔	12	13⅓	14⅔	16
1 × 10	8⅓	10	11⅔	13⅓	15	16⅔	18⅓	20
1 × 12	10	12	14	16	18	20	22	24
1 × 14	11⅔	14	16⅓	18⅔	21	23⅓	25⅔	28
1 × 16	13⅓	16	18⅔	21⅓	24	26⅔	29⅓	32
1 × 20	16⅔	20	23⅓	26⅔	30	33⅓	36⅔	40
1¼ × 4	4⅙	5	5⅚	6⅔	7½	8⅓	9⅙	10
1¼ × 6	6¼	7½	8¾	10	11¼	12½	13¾	15
1¼ × 8	8⅓	10	11⅔	13⅓	15	16⅔	18⅓	20
1¼ × 10	10⅓	12½	14½	16⅔	18⅔	20⅚	22⅚	25
1¼ × 12	12½	15	17½	20	22½	25	27½	30
1½ × 4	5	6	7	8	9	10	11	12
1½ × 6	7½	9	10½	12	13½	15	16½	18
1½ × 8	10	12	14	16	18	20	22	24
1½ × 10	12½	15	17½	20	22½	25	27½	30
1½ × 12	15	18	21	24	27	30	33	36
2 × 4	6⅔	8	9⅓	10⅔	12	13⅓	14⅔	16
2 × 6	10	12	14	16	18	20	22	24
2 × 8	13⅓	16	18⅔	21⅓	24	26⅔	29⅓	32
2 × 10	16⅔	20	23⅓	26⅔	30	33⅓	36⅔	40
2 × 12	20	24	28	32	36	40	44	48
2 × 14	23⅓	28	32⅔	37⅓	42	46⅔	51⅓	56
2 × 16	26⅔	32	37½	42⅔	48	53⅓	58⅔	64
2½ × 12	25	30	35	40	45	50	55	60
2½ × 14	29⅙	35	40⅚	46⅔	52½	58⅓	64⅙	70
2½ × 16	33⅓	40	46⅔	53⅓	60	66⅔	73⅓	80
3 × 6	15	18	21	24	27	30	33	36
3 × 8	20	24	28	32	36	40	44	48
3 × 10	25	30	35	40	45	50	55	60
3 × 12	30	36	42	48	54	60	66	72
3 × 14	35	42	49	56	63	70	77	84
3 × 16	40	48	56	64	72	80	88	96
4 × 4	13⅓	16	18⅔	21⅓	24	26⅔	29⅓	32
4 × 6	20	24	28	32	36	40	44	48
4 × 8	26⅔	32	17⅓	42⅔	48	53⅓	58⅔	64
4 × 10	33⅓	40	46⅔	53⅓	60	66⅔	73⅓	80
4 × 12	40	48	56	64	72	80	88	96
4 × 14	46⅓	56	65⅓	74⅔	84	93⅓	102½	112

32–44. *The length of the rafters for some of the more common roofs can be found in this chart.*

Rise Of Rafter	Building Width (In Feet)											
	10	12	14	16	18	20	22	24	26	28	30	32
3″	5′2″	6′2″	7′3″	8′3″	9′3″	10′4″	11′4″	12′4″	13′5″	14′5″	15′6″	16′6″
4″	5′3″	6′4″	7′5″	8′5″	9′6″	10′7″	11′7″	12′8″	13′8″	14′9″	15′10″	16′10″
5″	5′5″	6′6″	7′7″	8′8″	9′9″	10′10″	11′11″	13′0″	14′1″	15′2″	16′3″	17′4″
6″	5′7″	6′8″	7′10″	9′0″	10′1″	11′2″	12′4″	13′5″	14′6″	15′8″	16′9″	17′11″

Note: Tables accurate only to nearest inch.

QUESTIONS

1. Why is the ridge board for a hip roof shorter in length than for a gable roof?

2. Why is it necessary that the rafter locations on the ridge board be laid out exactly from the top plate?

3. Why is it best to erect the ridge board in its proper position before beginning the installation of the rafters?

4. When nailing the common rafters in place, why must the rafter be nailed at the bird's-mouth first?

5. When erecting the rafters for a hip roof, which rafters are erected first?

6. Why must hip jack rafters be installed in pairs?

7. What special treatment is required around roof openings?

8. What is a collar beam?

9. Describe the two methods of installing gable end studs.

10. How does a gambrel roof differ from a gable roof?

11. When figuring the length of the rafters for a shed roof, how do they differ from the rafters for a gable roof?

12. Why is a flat or a low-slope roof usually less expensive to construct than a pitched roof?

13. What is the purpose of a chimney saddle?

Roof Trusses

Much modern roof framing of residential and commercial buildings is done with roof trusses. The *simple truss*, or *trussed rafter*, is an assembly of members forming a rigid framework of triangular shapes. These members are usually connected at the joints by gussets. *Gussets* are flat wood, plywood, or similar type members. They are fastened to the truss by nails, screws, bolts, or adhesives. The roof truss is capable of supporting loads over long spans without intermediate support. Fig. 33-1. It has been greatly refined during its development over the years, and the gus-

33-1a. *Through the use of trusses, this barn has a clear span of 80'.*

33-2. *Testing a nail-glued king-post truss.*

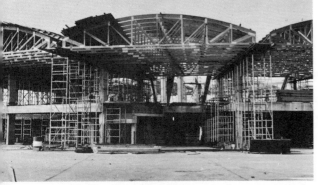

33-1b. *Bowstring trusses used as supports for concrete forms.*

set and other preassembled types of roof trusses are being used extensively in the housing field. Figs. 33-2 and 33-3.

Roof trusses save material and on-site labor costs. It is estimated that a material savings of about 30% is made on roof members and ceiling joists. The double top plate on interior partition walls and the double floor joists under interior bearing partitions are not necessary. Roof trusses also eliminate interior bearing partitions because trusses are self-supporting.

Trusses can be erected quickly, and therefore the house can be enclosed in a short time. The roof frame can be ready for sheathing in less than an hour. A long boom mobile crane with a spreader bar can be used to lift the trusses up to the top plates. The trusses are hung on the bar at 24" centers while on the ground. Six to ten trusses are held in position on the bar by special blocking or by 1" × 4" ribbons nailed to the top chord. The assembly is swung to the top wall plates, and a worker on each wall plate positions the trusses and nails them. The bar is then removed and returned for another load. Fig. 33-4.

Trusses are usually designed to span from one exterior wall to the other with lengths of 20' to 32' or more. Because no interior bearing walls are required, the interior of the building becomes one large workroom. This allows increased flexibility for interior planning, since partitions can be placed without regard to structural requirements. Fig. 33-5.

Most trusses are fabricated in a shop and then delivered to the job site. Fig.

33-3. *Both roof and floor trusses are used in this southern California home.*

33-6. Some, however, are constructed at the job site.

The following wood trusses are most commonly used for houses. Fig. 33-7.

- King-post.
- W-type.
- Scissors.

These and similar trusses are most adaptable to rectangular houses because the constant width requires only one type of truss. However, trusses can also be used for L-shaped houses. For hip roofs, hip trusses can be provided for each end and valley area. Fig. 33-8 (Page 484).

Trusses are commonly designed for 24" spacing. This spacing requires somewhat thicker interior and exterior sheathing or finish material than is needed for conventional joist and rafter construction using 16" spacing. Truss designs, lumber grades, and construction details are available from several sources, including the American Plywood Association.

KING-POST TRUSS

The king-post is the simplest form of truss used for houses. It consists of upper and lower chords and a center vertical post. B, Fig. 33-7. Allowable spans are somewhat less than for the W-truss when the same size members are used because of the unsupported length of the upper chord. For example, under the same conditions, a plywood gusset king-post truss with 4-in-12 slope and 2' spacing is limit-

33-4. *Setting a series of eight prespaced trusses on the exterior walls with a long boom mobile crane.*

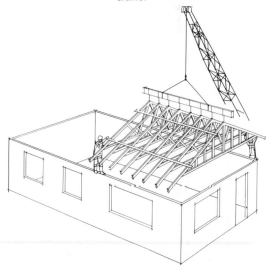

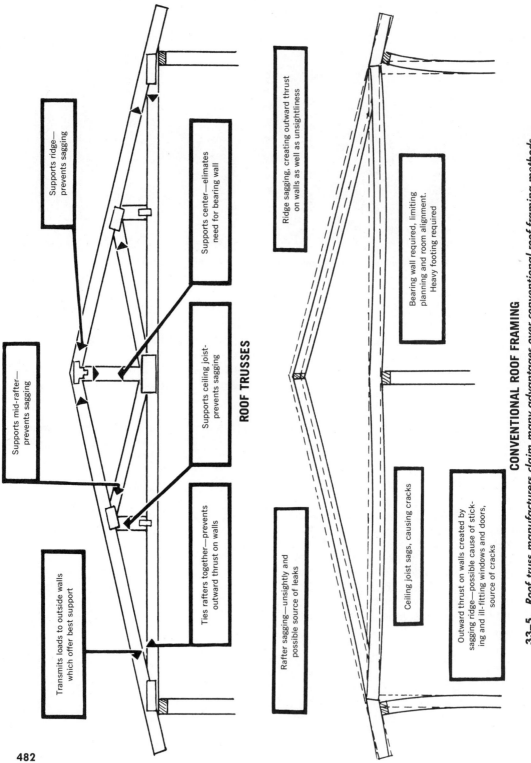

ROOF TRUSSES

Supports ridge—prevents sagging

Supports center—eliminates need for bearing wall

Supports mid-rafter—prevents sagging

Supports ceiling joist—prevents sagging

Ties rafters together—prevents outward thrust on walls

Transmits loads to outside walls which offer best support

CONVENTIONAL ROOF FRAMING

Ridge sagging, creating outward thrust on walls as well as unsightliness

Bearing wall required, limiting planning and room alignment. Heavy footing required

Ceiling joist sags, causing cracks

Rafter sagging—unsightly and possible source of leaks

Outward thrust on walls created by sagging ridge—possible cause of sticking and ill-fitting windows and doors, source of cracks

33–5. Roof truss manufacturers claim many advantages over conventional roof framing methods.

ed to about a 26′ span for 2″ × 4″ members. A W-truss with the same size members and spacing could be used for a 32′ span. Furthermore, the grades of lumber used for the two types might also vary.

For short and medium spans, the king-post truss is probably more economical than other types because it has fewer pieces and can be fabricated faster. However, local prices and design load require-

33-6. *These trusses are bundled in sets with 3/4″ steel tape for transporting to the building site.*

33-7. *Light wood trusses: A. W-type. B. King-post. C. Scissors.*

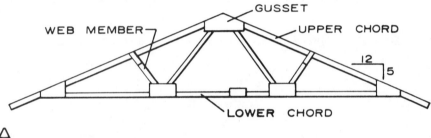

GUSSET

WEB MEMBER

UPPER CHORD

12
5

LOWER CHORD

A

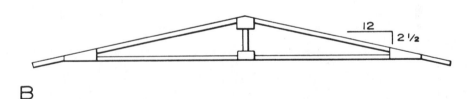

12
2 ½

B

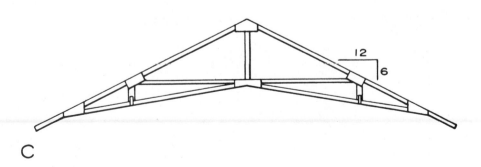

12
6

C

ments (for snow, wind, etc.) as well as the span should govern the type of truss to be used.

W-TYPE TRUSS

The W-Type truss is perhaps the most popular and most widely used of the light wood trusses. A, Fig. 33-7. Its design includes the use of three more members than the king-post truss, but distances between connections are less. This usually allows the use of lower grade lumber and somewhat greater spans for the same member size.

SCISSORS TRUSS

The scissors truss is used for houses with a sloping living room ceiling. C, Fig.

33-7. Somewhat more complicated than the W-type truss, it provides good roof construction for a "cathedral" ceiling with a saving in materials over conventional framing methods.

DESIGN

Plans for the fabrication of trussed rafters must include the preparation of an engineered design. For use on FHA-insured projects, the design must be approved by the FHA. Some building codes require approval and a certificate of inspection on trussed rafters.

The design of a truss includes consideration of not only snow and wind loads but the weight of the roof itself. Design also takes into account the slope of the roof.

33-8. *Special trusses are available for hip and valley areas.*

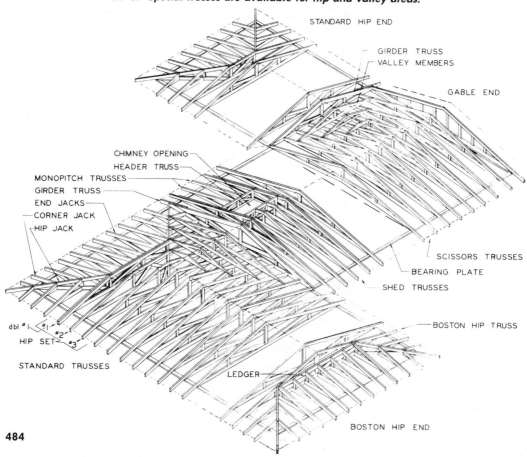

STANDARD HIP END

GIRDER TRUSS
VALLEY MEMBERS

GABLE END

CHIMNEY OPENING
HEADER TRUSS
MONOPITCH TRUSSES
GIRDER TRUSS
END JACKS
CORNER JACK
HIP JACK

SCISSORS TRUSSES
BEARING PLATE
SHED TRUSSES

dbl. #1 #1
HIP SET #2
         #3

BOSTON HIP TRUSS

STANDARD TRUSSES

LEDGER

BOSTON HIP END

Generally, the flatter the slope, the greater the stresses. Flatter slopes therefore require larger members and stronger connections in roof trusses.

A great majority of the trusses used are fabricated with gussets of plywood (nailed, glued, or bolted in place) or with metal gusset plates. Fig. 33-9. Others are assembled with split-ring connectors. Fig. 33-10. Some trusses are designed with a 2" × 4" soffit return at the end of each upper chord to provide nailing for the soffit of a wide box cornice.

Designs for standard W-type and king-post trusses with plywood gussets are usually available through the American Plywood Association or a local lumber dealer. Fig. 33-11. Many lumber dealers are able to provide the builder with completed trusses ready for erection.

To illustrate the design and construction of a typical wood W-truss more clearly, the following example is given:

The span for the gusset truss is 26', the slope 4 in 12, and the spacing 24". Fig. 33-12 (Page 491). The gussets are nail-glued; that is, nails or staples are used to

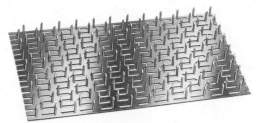

33-9b. *Metal gusset plate.*

33-9c. *This jig holds the truss firmly in place during application of the metal gusset plate.*

33-9a. *Trussed rafter assembled with gusset plates, glue, and nails.*

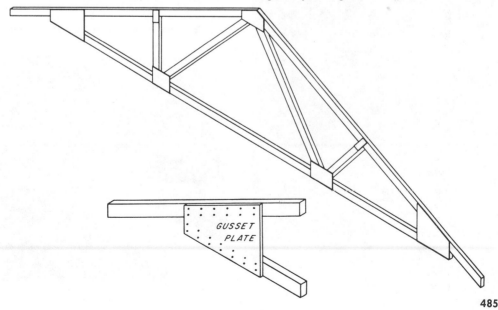

GUSSET
PLATE

485

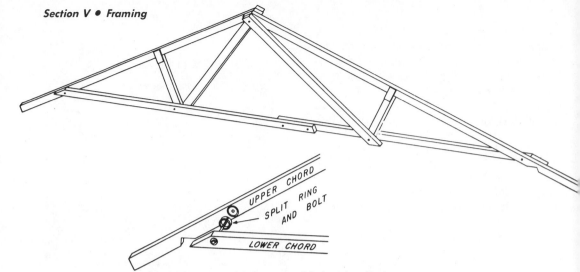

33-10a. *Trussed rafter assembled with split-ring connectors.*

supply pressure while the glue sets. Total roof load is 40 pounds per square foot, which is usually sufficient for areas with moderate to heavy snow. The upper and lower chords can be 2″ × 4″ in size, but the upper chord requires a slightly higher

grade of material. It is often deirable to use dimension material with a moisture content of about 15%. The moisture content should not exceed 19%.

FABRICATION
Applying Gussets

Plywood gussets can be made from ³/₈″ or ¹/₂″ standard plywood with exterior glueline or exterior sheathing grade plywood. The cutout size of the gussets and

33-10b. *Cutting a hole for a split metal ring and bolt.*

33-10c. *Assembling the trussed rafter on the job using the split-ring connector.*

33–11. *An example of a nail-glued truss plan for a king-post truss with a 4-in-12 slope. This is one of the designs available from the American Plywood Association.*

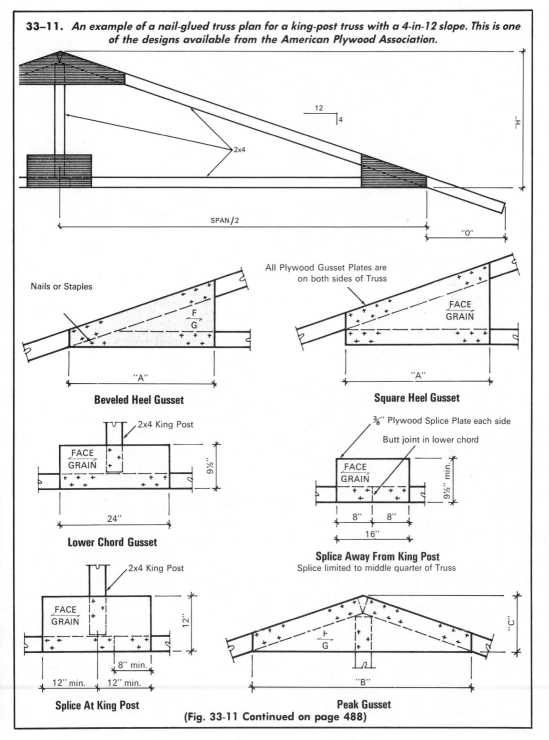

Beveled Heel Gusset

Square Heel Gusset

Lower Chord Gusset

Splice Away From King Post
Splice limited to middle quarter of Truss

Splice At King Post

Peak Gusset

(Fig. 33-11 Continued on page 488)

(Fig. 33-11 Continued from page 487)

SELECTING THE TRUSS DESIGN

☐ The first step in selecting a design from this plan is to determine the design load used in your area. Consult your local building official for this information; or the nearest FHA office, for FHA-insured construction.

☐ Two loading conditions are given—30 and 40 lbs. per sq. ft. total roof load. Allowance has been made for ceiling and attic loads in addition to these roof loads. Enter the TRUSS DIMENSION TABLE with the design load for your area.

☐ These designs are based on a 24″ O.C. spacing. Where trusses are spaced 16″ O.C. they can carry greater loads. For instance, on the 16″ spacing, total allowable roof loads are 45 lbs. per sq. ft. for the tabulated 30 lbs. per sq. ft. loading condition, and 60 lbs. per sq. ft. for the tabulated 40 lbs. per sq. ft. loading condition.

☐ A choice of two heel gusset arrangements is offered. The beveled heel gusset provides the lowest roof line. The square heel joint offers the most economical fabrication.

☐ With the loading condition and heel gusset arrangement determined, gusset plate sizes and truss dimensions may be selected for the applicable span.

☐ Lumber grades may be chosen from the CHORD CODE TABLE. For each span in the TRUSS DIMENSION TABLE the lumber species and grades for the upper and lower chord members may be picked from those listed under the corresponding code in the CHORD CODE TABLE.

GENERAL NOTES

1. See page 487 for detailed material and fabrication recommendations.

2. Use ⅜″ minimum thickness plywood gusset plates nail-glued in accordance with fabrication recommendations. All plywood shall bear the APA grade-trademark of the American Plywood Association.

3. Use stress grade lumber for chord members from the CHORD CODE TABLE. The king-post may be cut from Construction grade Douglas fir, larch, hemlock, or No. 2 Southern pine.

4. Trusses with pitches over 4 in 12 and not more than 5 in 12, require the same gusset plate sizes and chord lumber grades as specified for the trusses on this sheet.

5. Trusses for spans intermediate between those listed in the TRUSS DIMENSION TABLE, require gusset plate sizes and chord codes for the next longer span. For example, a 27′ span truss would require the gusset plate sizes and chord codes listed for the 28′8″ design.

Nail-Glued Truss Plans

	Chords			King Post KP-2
Upper	Lower	Pitch	
2 x 4	2 x 4	4 in 12	

MATERIALS

Plywood

Use only plywood bearing the APA grade-trademark of the American Plywood Association, in the thicknesses specified on the drawings. In normal situations where the moisture content of the trusses in service will not exceed 18%, use the regular Interior-APA-PlyScord or WSP-1 or WSP-2 CD sheathing grades.

The above grades manufactured with Exterior glue or Exterior-APA type plywood may be used for added assurance of durability. These premium grades manufactured with Exterior glue are required in FHA insured conetruction.

Where the moisture content of the wood is likely to exceed 18% in service, use only Exterior-APA grades.

Plywood for gusset plates shall have a moisture content of 16% or less. Normally, plywood may be used as received unless it has been stored out of doors. Surfaces to be

(Fig. 33-11 Continued)

glued must be clean and free from oil, dust, and paper tape.

Lumber

Lumber must be of the stress grade called for in the CHORD CODE TABLE, as indicated by an approved grading agency. At time of gluing it should be conditioned to a moisture content approximately that which it will attain in service, but in any case between 7% and 16%. Surfaces to be glued should be clean and free from oil, dust, and other foreign matter. Each piece should be machine finished, but not sanded.

Use no lumber which has in the area of the gussets any roughness, cup, or twist which might prevent good contact between gusset and lumber.

Keep surfaces of intersecting lumber members flush within 1/32".

Glue

Use casein type, conforming with Federal Specification MMM-A-125, Type II for dry, indoor exposures. For wet conditions, or if any glue joint is exposed, even at a soffit, use resorcinol-type glue, conforming with Military Specification MIL-A-46051.

KP-2 Designs (when using STANDARD sheathing or C-C EXT-APA grade plywood

Loading Condition Total Roof Load (lbs. per sq. ft.)	Span	Beveled Heel Gusset							Square Heel Gusset						
		Dimensions in Inches					Chord Code		Dimensions in Inches					Chord Code	
		A	B	C	H	O	Upper	Lower	A	B	C	H	O	Upper	Lower
30 (a) Meets FHA requirements	20'8"	32	48	12	45⅛	44	2	3	19	32	12	48¾	44	2	3
	22'8"	32	48	12	49⅛	48	1	2	19	32	12	52¾	48	1	3
	24'8"	48	60	16	53⅛	48	1	2	24	48	12	56¾	48	1	3
	26'8"	48	72	16	57⅛	48	1	2	32	60	16	60¾	48	1	2
40 (b)	20'8"	32	48	12	45⅛	43	8	9	19	32	12	48¾	48	7	9
	22'8"	32	60	16	49⅛	48	7	8	19	48	12	52¾	48	7	9
	24'8"								32	60	16	56¾	48	7	9

Truss Dimension Table KP-2

(a) 30 psf (20 psf live load, 10 psf dead load; on upper chord and 10 psf dead load on lower chord
(b) 40 psf (30 psf live load, 10 psf dead load) on upper chord and 10 psf dead load on lower chord

Chord Code Table

Chord Code	Size	Grade and Species meeting Stress requirements	Grading Rules	f	t//	c//
1	2 x 4	Select Structural Light Framing WCDF	WCLIB	1950	1700	1400
		No. 1 Dense Kiln Dried Southern Pine	SPIB	2000	2000	1700
		1.8E	WWPA	2100	1700	1700
2	2 x 4	1500f Industrial Light Framing WCDF	WCLIB	1500	1300	1200
		1500f Industrial Light Framing WCH	WCLIB	1450	1250	1100
		No. 1 2" Dimension Southern Pine	SPIB	1450	1450	1350
		1.4E	WWPA	1500	1200	1200
3	2 x 4	1200f Industrial Light Framing WCDF	WCLIB	1200	1100	1000
		1200f Industrial Light Framing WCH	WCLIB	1150	1000	900
		No. 2 2" Dimension Southern Pine	SPIB	1200	1200	900

(Continued on page 490)

Chord Code	Size	Grade and Species meeting Stress requirements	Grading Rules	f	t//	c//
		(Fig. 33-11 Continued from page 489)				
4	2 x 6	Select Structural J & P WCDF	WCLIB	1950	1700	1600
		Select Structural J & P Western Larch	WWPA	1900	1600	1500
		No. 1 Dense Kiln Dried Southern Pine	SPIB	2000	2000	1700
		1.8E	WWPA	2100	1700	1700
5	2 x 6	Construction Grade J & P WCDF	WCLIB	1450	1300	1200
		Construction Grade J & P WCH	WCLIB	1450	1250	1150
		Structural J & P Western Larch	WWPA	1450	1300	1200
		No. 1 2" Dimension Southern Pine	SPIB	1450	1450	1350
		1.4E	WWPA	1500	1200	1200
6	2 x 6	Standard Grade J & P WCDF	WCLIB	1200	1100	1050
		Standard Grade J & P WCH	WCLIB	1150	1000	950
		Standard Structural Western Larch	WWPA	1200	1100	1050
		No. 2 2" Dimension Southern Pine	SPIB	1200	1200	900
7	2 x 4	Select Structural Light Framing WCDF	WCLIB	1900	1900	1400
		Select Structural Light Framing Western Larch	WWPA	1900	1900	1400
		No. 1 Dense Kiln Dried Southern Pine	SPIB	2050	2050	1750
		1.8E	WWPA	2100	1700	1700
8	2 x 4	1500f Industrial Light Framing WCDF	WCLIB	1500	1500	1200
		Select Structural Light Framing WCH	WCLIB	1600	1600	1100
		Select Structural Light Framing WH	WWPA	1600	1600	1100
		1500f Industrial Light Framing Western Larch	WWPA	1500	1500	1200
		No. 1 2" Dimension Southern Pine	SPIB	1500	1500	1350
		1.4E	WWPA	1500	1200	1200
9	2 x 4	1200f Industrial Light Framing WCDF	WCLIB	1200	1200	1000
		1200f Industrial Light Framing Western Larch	WWPA	1200	1200	1000
		1500f Industrial Light Framing WCH	WCLIB	1500	1500	1000
		1500f Industrial Light Framing WH	WWPA	1500	1500	1000
		No. 2 2" Dimension Southern Pine	SPIB	1200	1200	900
10	2 x 6	Select Structural J & P WCDF	WCLIB	1900	1900	1500
		Select Structural J & P Western Larch	WWPA	1900	1900	1500
		No. 1 Dense Kiln Dried Southern Pine	SPIB	2050	2050	1750
		1.8E	WWPA	2100	1700	1700
11	2 x 6	Construction Grade J & P WCDF	WCLIB	1500	1500	1200
		Construction Grade J & P Western Larch	WWPA	1500	1500	1200
		Select Structural J & P WCH	WCLIB	1600	1600	1200
		Select Structural J & P WH	WWPA	1600	1600	1200
		No. 1 2" Dimension Southern Pine	SPIB	1500	1500	1350
		1.4E	WWPA	1500	1200	1200
12	2 x 6	Standard Grade J & P WCDF	WCLIB	1200	1200	1000
		Standard Grade J & P Western Larch	WWPA	1200	1200	1000
		Standard Grade J & P WCH	WCLIB	1200	1200	1000
		Standard Grade J & P WH	WWPA	1200	1200	1000
		No. 2 2" Dimension Dense Southern Pine	SPIB	1400	1400	1050

the general nailing pattern for nail-gluing are shown in Figs. 33-12 and 33-13. More specifically, 4d nails should be used for plywood gussets up to $3/8''$ thick and 6d for plywood $1/2''$ to $7/8''$ thick. Three-inch spacing should be used when plywood is no more than $3/8''$ thick and $4''$ spacing should be used for thicker plywood. When wood truss members are a nominal $4''$ wide, use two rows of nails with a $3/4''$ edge distance. Use three rows of nails when truss members are $6''$ wide. Gussets are used on both sides of the truss. Fig. 33-14.

For normal conditions and where relative humidities in the attic area tend to be high, such as might occur in the southern and southeastern United States, resorcinol glue should be used for the gussets. In dry and arid areas where conditions are more favorable, a casein or similar glue might

be considered. For estimating purposes, approximately $1/10$ pound of glue is required per square foot of gusset.

Glue should be spread on the clean surfaces of the gusset, truss member, or both. When mixing and using the glue, be sure to follow the glue manufacturer's recommendations. If the glue is to be spread on the lumber, it is a good idea to mark the ends of the gussets on the jig table so that the glue will be spread only over the area to be covered by the gusset. To prevent interference with the sheathing or ceiling, the edge of the gussets may be held back $1/8''$ from the edge of the lumber.

Either nails or staples might be used to supply pressure until the glue has set, although only nails are recommended for plywood $1/2''$ and thicker. Use the nail spacing previously outlined. Closer or in-

33–12. Construction details of a 26′ W-truss: A. Beveled heel gusset. B. Peak gusset. C. Upper chord intermediate gusset. D. Splice of lower chord. E. Lower chord intermediate gusset.

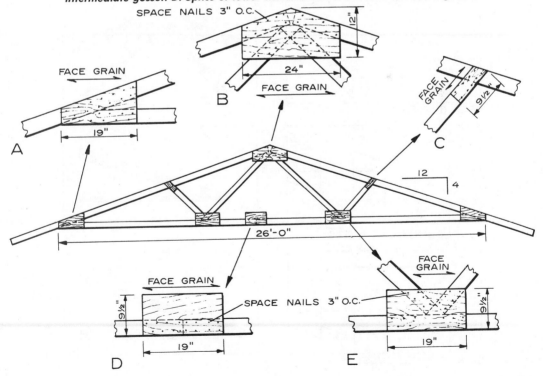

termediate spacing may be used to insure "squeezeout" at all visible edges.

Gluing should be done under closely controlled temperature conditions. This is especially true if using the resorcinol adhesives. Follow the assembly temperatures recommended by the manufacturer.

The complete truss should be set aside immediately after assembly. It should not be disturbed until the glue has set. A holding frame which stores the trusses in an upright, inverted position is a convenient way to store the trusses during curing.

Assembly

Plywood trusses can be built without expensive equipment or facilities. The use of trusses is therefore possible for small and medium operations or where a special truss design is needed. A glue spreader or

33–13. *Plywood gusset cutting layouts. Make sure the grain of the plywood runs parallel with the lower chord, except for the upper chord intermediate gusset. For this gusset the grain should run parallel to the compression web.*

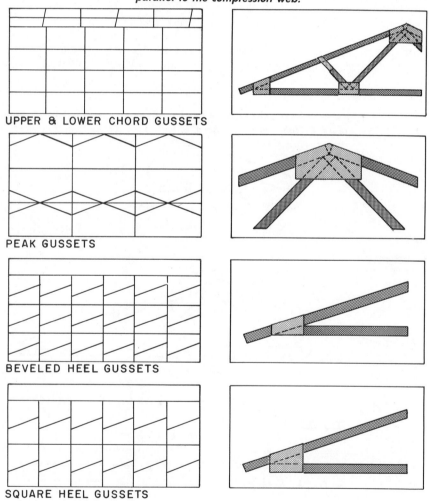

UPPER & LOWER CHORD GUSSETS

PEAK GUSSETS

BEVELED HEEL GUSSETS

SQUARE HEEL GUSSETS

roller and hammers are all the equipment required. Power nailers or staplers are often used. Jig tables are more convenient for hand or power nailing. Fig. 33-15. It is possible, however, to lay out the truss on the subfloor and then cut and assemble the trusses before the wall is erected.

To insure proper temperature control, nail-glued trusses should be fabricated in an enclosed building. For normal truss assembly with moisture-resistant glues such as casein, a minimum room temperature of 50° F is needed. Where exterior glues such as resorcinol-resin glue are used, a 70° F temperature should be maintained during the assembly and cure.

Therefore an area at least big enough to construct the largest truss, with sufficient room for turning and storage after assembly, is recommended.

A simple jig can be made by nailing blocks to a table. Lay out the overall truss dimensions on the table with a chalk line. Position the upper and lower chords. Fig. 33-16. For best appearance don't forget to provide for camber (a slight curvature). Camber of $1/4''$ for a 30' truss with proportional amounts for other lengths is recommended.

For the W-type trusses lay out the long diagonals with the diagonal centerline intersecting the lower chord centerline at $1/3$

33–14a. *Nailing plywood gussets in assembling trussed rafters. The operator is using a portable pneumatic (air operated) nailing machine.*

33–14b. *Nail and staple schedule.*

Plywood Thickness	Nails (1)		Staples (2)	
	Size	Spacing (3)	Size	Spacing (3)
$3/8''$	4d	3″ o.c.	$1 1/8''$	3″ o.c.
$1/2''$-$3/4''$	6d	4″ o.c.	Not recommended	

(1) Nails—Box, common, cement coated or T-nails.
(2) Staples—16 gauge with $7/16''$ crown width.
(3) Use two rows nails or staples for 4″ wide lumber, three rows for 6″ wide lumber and set in $3/4''$ from lumber edge. Stagger nails from opposite sides. Nails or staples provide gluing pressure.

33–15a. *A truss jig table designed to allow the worker to walk through while setting the pieces in place for assembly.*

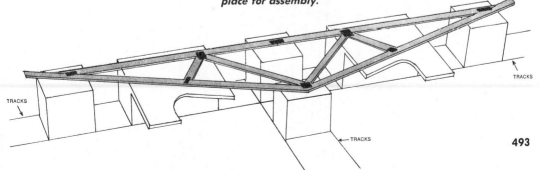

TRACKS

TRACKS

TRACKS

SUGGESTED ASSEMBLY TABLE

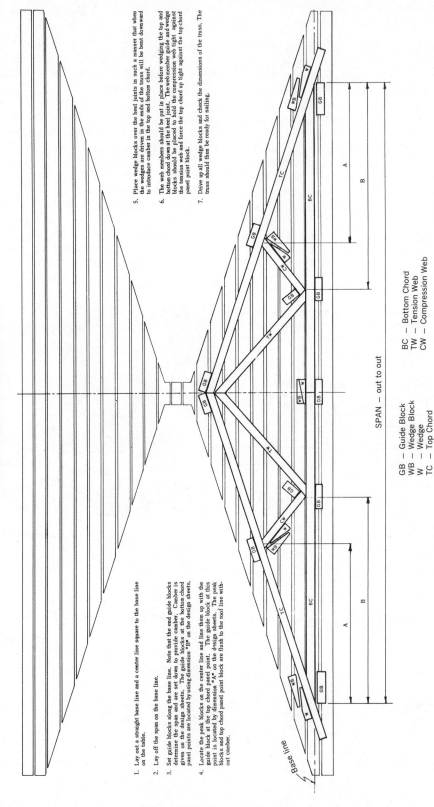

1. Lay out a straight base line and a center line square to the base line on the table.

2. Lay off the span on the base line.

3. Set guide blocks along the base line. Note that the end guide blocks determine the span and are set down to provide camber. Camber is given on the design sheets. The guide blocks at the bottom chord panel points are located by using dimension "B" on the design sheets.

4. Locate the peak blocks on the center line and line them up with the guide block at the top chord panel point. The guide block at this point is located by dimension "A" on the design sheets. The peak blocks and top chord panel point block are flush to the roof line without camber.

5. Place wedge blocks over the heel joints in such a manner that when the wedges are driven in the ends of the truss will be bent downward to introduce camber in the top and bottom chord.

6. The web members should be put in place before wedging the top and bottom chord down at the heel joint. The web-member guide and wedge blocks should be placed to hold the compression web tight against the tension web and force the top chord up tight against the top chord panel point block.

7. Drive up all wedge blocks and check the dimensions of the truss. The truss should then be ready for nailing.

SPAN — out to out

GB – Guide Block
WB – Wedge Block
W – Wedge
TC – Top Chord

BC – Bottom Chord
TW – Tension Web
CW – Compression Web

Base line

33–15b. *Building a truss assembly table.*

494

the span of the lower chord. Fig. 33-17. The short diagonals may then be positioned with the upper end centered midway between the clear distance from the peak to the heel gusset. Another way to determine the points at which the diagonals intersect the upper chord is to divide the bottom chord into four equal parts. Perpendiculars extended from these divisions will intersect the upper chord at the locations for the diagonals (the "1/4 point" in Fig. 33-17).

For the king-post truss, simply place the king post in position, and recheck alignment of all members to insure accurate dimensions. When all members of a truss are in position, nail the holding blocks into place. Fig. 33-18.

Commercially, trusses are made on specially designed jig tables that are adjustable to various sizes and styles of trusses. Fig. 33-19. Truss plates similar to those shown in Fig. 33-9b & c are used in place of plywood gussets on most commercially made trusses. The plates are pressed into place with a 40-ton moving press. This press runs on a track and moves over a fixed table which may be of any length.

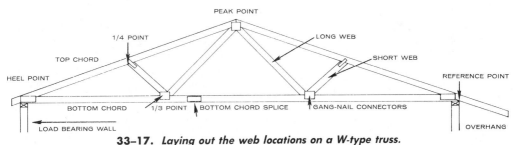

33–16. *A king-post truss laid out and blocked on an assembly table.*

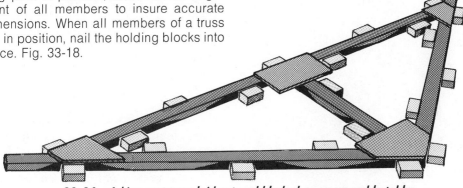

33–17. *Laying out the web locations on a W-type truss.*

33–18. *Nail the holding blocks into position on the jig table. This is a W-type truss.*

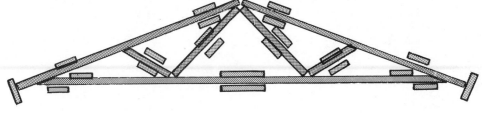

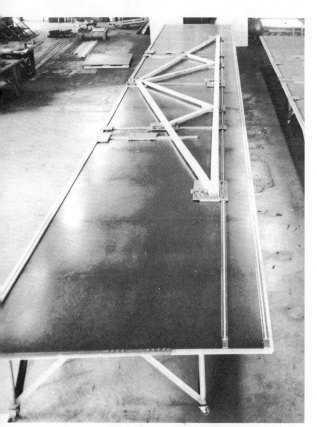

33–19. *A commercial jig table adjustable for various sizes and styles of trusses.*

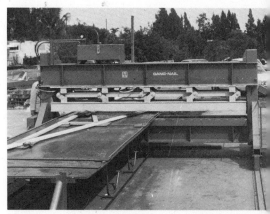

33–20. *A 40-ton moving press. The press moves on a track over the jig table. It is stopped as it passes over the metal truss plate, and the truss plate is pressed into the truss members.*

Fig. 33-20. It can be used for the production of trussed joists as well as trussed rafters. Fig. 33-21.

Handling

In handling and storage of completed trusses, avoid placing unusual stresses on them. They were designed to carry roof loads in a vertical position, and it is important that they be lifted and stored upright. If they must be handled in a flat position, enough support should be used along their length to minimize bending deflections. Never support the trusses only at the center or only at each end when they are in a flat position.

TRUSS ERECTION

Completed trusses can be raised in place with a mobile crane after delivery to the building site. They can also be placed by hand over the exterior walls in an inverted position and then rotated into an upright position. Fig. 33-22.

The top plates of the two side walls should be marked for the location of each set of trusses. The trusses are fastened to the outside walls and two 1″ × 4″ or 1″ × 6″ temporary horizontal braces (ribbons) are located near the ridge line to space and align them until the roof sheathing has been applied.

When fastening trusses, resistance to uplift stresses as well as thrust must be considered. Trusses are fastened to the outside walls with nails or framing anchors. The ring-shank nail provides a simple connection which will resist wind uplift forces. Toenailing is sometimes done, but this is not always the most satisfactory method. The heel gusset and a plywood gusset or metal gusset plate are located at the wall plate and make toenailing difficult. However, two 10d nails on each side of the truss can be used in nailing the lower

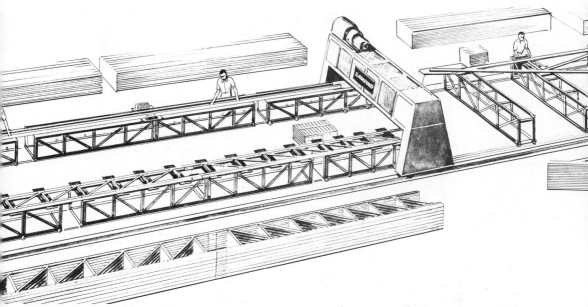

33-21. *On this assembly line a single press is used to assemble both roof trusses and trussed joists. With this system a three-member crew can produce up to 300 trusses daily.*

chord to the plate. Fig. 33-23. Predrilling may be necessary to prevent splitting. Because of the single-member thickness of the truss and the presence of gussets at the wall plates, it is usually a good idea to use some type of metal connector to supplement the toenailings.

A better system of fastening trusses involves the use of a metal connector or bracket. These brackets are available commercially or can be formed from sheet metal. Fig. 33-24. The brackets are nailed to the wall plates at side and top with 8d nails and to the lower chords of the truss with 6d or 1½" roofing nails. Fig. 33-23.

INTERIOR PARTITION INSTALLATION

Where partitions run parallel to but between the bottom truss chords and the partitions are erected *before* the ceiling finish is applied, install 2" × 4" blocking between the lower chords. Fig.33-25. This blocking should be spaced not over 4' on center. Nail the blocking to the chords with two 16d nails in each end. To provide nailing for lath or wallboard, nail a 1" × 6" or 2" × 6" continuous backer to the blocking. Set the bottom face level with the bottom of the lower truss chords.

When partitions are erected *after* the ceiling finish is applied, 2" × 4" blocking is set with the bottom edge level with the bottom of the truss chords. Nail the blocking with two 16d nails in each end.

If the partitions run at right angles to the bottom of the truss chords, the partitions are nailed directly to lower chord members. A 2" × 6" backer for the application of the ceiling finish is nailed on top of the partition plates between the trusses. Fig. 33-26.

ESTIMATING
Materials

To determine the amount of material necessary for building a roof truss on the

33–22a. *Fabricated roof trusses are lightweight and may be placed by hand on the exterior walls.*

33–22b. *Using special poles to set up roof trusses while standing on the ground.*

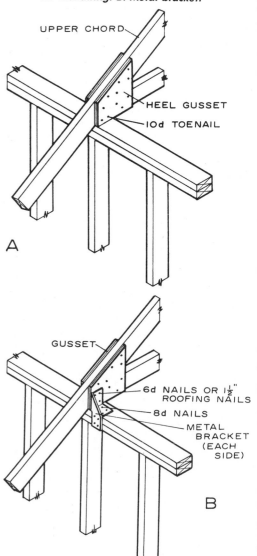

33–22c. *Roof trusses may be tipped up as they are nailed in place.*

33–23. *Fastening trusses to the wall plate: A. Toenailing. B. Metal bracket.*

UPPER CHORD

HEEL GUSSET

10d TOENAIL

A

GUSSET

6d NAILS OR 1½" ROOFING NAILS

8d NAILS

METAL BRACKET (EACH SIDE)

B

job site, figure the material as described for estimating conventional roof framing, page 475. To this add the cost of any special connectors, plywood gussets, or metal gusset plates.

Most builders, however, buy trusses ready-made. A manufacturer may be able to fabricate a roof truss, deliver it to the job site, and set it on the rafter plate with a crane, ready for the worker to tip it into place, for less than the cost of the material alone in conventional framing.

Labor

Installing the roof trusses on an average size home, with attached garage, containing about 2,000 square feet of ceiling area will take 3 workers about 2 hours. This includes laying out the locations, setting up, and attaching the roof trusses to the rafter plates.

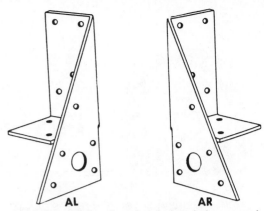

33-24b. *Left-hand and right-hand sheet-metal brackets are available for installation on both sides of the rafter truss.*

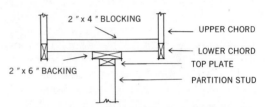

2 " x 4 " BLOCKING
UPPER CHORD
LOWER CHORD
TOP PLATE
2 " x 6 " BACKING
PARTITION STUD

33-25. *Construction details for partitions that run parallel to the roof truss.*

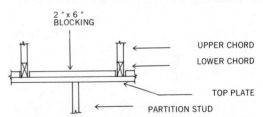

2 " x 6 "
BLOCKING
UPPER CHORD
LOWER CHORD
TOP PLATE
PARTITION STUD

33-26. *Construction details for partitions that run at right angles to the roof truss.*

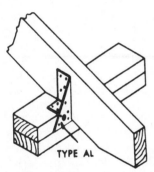

TYPE AL

33-24a. *Sheet-metal brackets are ideal for attaching the rafter truss to the wall plate.*

QUESTIONS

1. What are some of the advantages of using roof trusses in residential construction?

2. What three types of trusses are most commonly used for residential building?

3. Which of the three types of trusses used in building construction is the most popular and extensively used?

4. What are gussets usually made from?

5. What is a truss plate?

6. What is the best method of attaching the truss to the wall plates?

Roof Sheathing

Roof sheathing covers the rafters or roof joists. The roof sheathing, like the wall sheathing and the subflooring, is a structural element. Therefore it is a part of the framing. Sheathing provides a nailing base for the finish roof covering and gives rigidity and strength to the roof framing. Plywood or lumber roof sheathing is most commonly used for pitched roofs. Lumber or laminated roof decking is sometimes used in homes with exposed ceilings. Fig. 34-1. A manufactured wood fiber roof decking is also adaptable to exposed ceiling applications.

LUMBER ROOF SHEATHING

Roof sheathing boards are generally No. 3 common or better. The species used are the pines—Douglas fir, redwood, hemlocks, western larch, the firs, and the spruces. If the roof is to be covered with asphalt shingles, it is important that thoroughly seasoned material be used for the sheathing. Unseasoned wood will dry out and shrink in width. This shrinkage will cause buckling or lifts of the shingles which may extend along the full length of the board.

Nominal 1" boards are used for both flat and pitched roofs. Where flat roofs are to be used for a deck or a balcony, thicker sheathing boards will be required. Board roof sheathing, like board wall sheathing and subflooring, may be laid either horizontally or diagonally. Fig. 34-2. Horizontal board sheathing may be either closed (laid with no spaces between the courses) or open (laid with spaces between the cours-

es). In areas where wind-driven snow conditions prevail, a solid roof deck is recommended.

Installation
CLOSED SHEATHING

Roof boards used for sheathing under materials that require solid and continuous support—such as asphalt shingles, composition roofing, and sheet-metal roofing—must be laid closed. Fig. 34-3. Closed roof sheathing may also be used for wood shingles. The boards are a nominal 1" × 8" and may be square-edged, dressed and matched, or shiplapped.

OPEN SHEATHING

Open sheathing is used under wood shingles or shakes as a roof covering in blizzard-free areas or damp climates. Open sheathing usually consists of 1" × 4" strips with the on-center spacing equal to the shingle weather exposure but not over 10". (A 10" shingle which is lapped 4" by the shingle above it is said to be laid 6" to the weather.) When applying open sheathing, the boards should be laid up without spacing to a point on the roof above the overhang. Fig. 34-3.

Nailing

Lumber roof sheathing is nailed to each rafter with two 8d nails. Joints must be made on the rafters, just as wall sheathing joints must be made over studs. When end-matched boards are used, joints may be made between rafters, but in no case should the joints of adjoining boards be

34–1. *Lumber roof decking is available in a variety of patterns and is used in homes with exposed ceilings.*

34–2. *Roof board sheathing laid horizontally.*

34–3. *Installation of board roof sheathing, showing both closed and spaced types.*

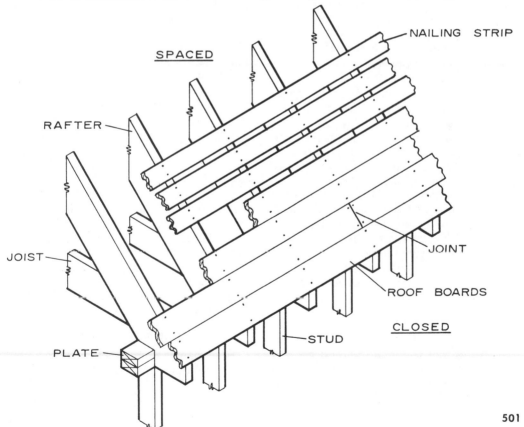

SPACED

NAILING STRIP

RAFTER

JOINT

JOIST

ROOF BOARDS

CLOSED

STUD

PLATE

34-4. *Both the roof and sidewalls of this large home are sheathed with plywood.*

made over the same rafter space. Each board should bear on at least two rafters.

PLYWOOD ROOF SHEATHING

Many different roof forms are possible with plywood construction. Plywood offers flexibility in design, ease of construction, economy, and durability. Fig. 34-4.

It can be installed quickly over large areas and provides a smooth, solid base with a minimum of joints. A plywood deck is also equally effective under any type of shingles or built-up roofing. Waste is minimal, contributing to the low in-place cost. It is frequently possible to cut costs still further by using fewer rafters with a somewhat thicker panel for the decking, for example, $3/4''$ plywood over framing 4' on center. Plywood and trusses are often combined in this manner. For recommended spans and plywood grades, see Fig. 34-5.

Installation

Plywood roof sheathing should be laid with the face grain perpendicular to the rafters. Fig. 34-6. Sheathing grade (unsanded) plywood is ordinarily used. Joints should be made over the centers of the rafters.

For wood or asphalt shingles with a rafter spacing of 16", $5/16''$ plywood may be used. For a 24" span $3/8''$ plywood should be used. For slate, tile, and asbestos cement shingles $1/2''$ plywood is recommended for 16" rafter spacing, and $5/8''$ plywood for 24" spacing. Fig. 34-5. If wood shingles are used and the plywood sheathing is less than $1/2''$ thick, 1" × 2" nailing strips spaced according to the shingle exposure should be nailed to the plywood.

Plywood roof sheathing, unless it is of the exterior type, should have no surface or edge exposed to the weather. To reduce handling costs and help the worker apply plywood roof sheathing, a roof platform may be constructed. Fig. 34-7 (Page 506). This platform supports a supply of plywood sheathing on the roof that is readily accessible to the worker. Another aid which may be constructed is a plywood ladder. It will eliminate the need for a second person to hand the plywood up to the one on the roof. Fig. 34-8. This ladder is leaned against the building in the normal position of a ladder. The plywood sheets are then set on end on the $3/8''$ plywood gusset. From there they can be pulled onto the roof as needed for application.

NAILING

Nails should be spaced 6" at the panel edges and 12" at intermediate supports except where the spans are 48" or more. Then the nails should be spaced 6" on all supports. Use 6d common smooth, ringshank, or spiral thread nails for plywood $1/2''$ thick or less. For plywood 1" thick or less use 8d common smooth, rink-shank, or spiral thread nails. Fig. 34-5.

DECKING OR PLANKING

Roof decking provides a solid permanent roof deck and an attractive ready-to-finish interior ceiling. It serves as an excellent base for any roofing material. Decking, with tongue-and-groove edges and decorative face patterns, is a standard building product for residential, commercial, and institutional construction. Fig. 34-9 (Page 508). Although known and used as roof decking, its load bearing capacities also make it useful as floor decking and for solid sidewall construction. This material is available in grades, patterns, and sizes suitable for both residential and commercial construction.

Grades. The *commercial* grade is designed for use in buildings where appearance and strength requirements are not a prime factor. The *select* grade decking is ideally suited for homes, schools, churches, motels, and restaurants, or wherever an attractive surface appearance is important.

34–5. *Plywood roof sheathing application specifications.*

Plywood roof sheathing[1][2][3] / (Plywood continuous over two or more spans; grain of face plies across supports)

Panel Identification Index	Plywood Thickness (inch)	Max. Span (inches)[4]	Unsupported Edge-Max. Length (inches)[5]	Allowable Roof Loads (psf)[6][7]										
				Spacing of Supports (inches center to center)										
				12	16	20	24	30	32	36	42	48	60	72
12/0	5/16	12	12	100 (130)										
16/0	5/16, 3/8	16	16	130 (170)	55 (75)									
20/0	5/16, 3/8	20	20		85 (110)	45 (55)								
24/0	3/8, 1/2	24	24		150 (160)	75 (100)	45 (60)							
30/12	5/8	30	26			145 (165)	85 (110)	40 (55)						
32/16	1/2, 5/8	32	28				90 (105)	45 (60)	40 (50)					
36/16	3/4	36	30				125 (145)	65 (85)	55 (70)	35 (50)				
42/20	5/8, 3/4, 7/8	42	32					80 (105)	65 (90)	45 (60)	35 (40)			
48/24	3/4, 7/8	48	36						105 (115)	75 (90)	55 (55)	40 (40)		
2-4-1	1 1/8	72	48							175 (175)	105 (105)	80 (80)	50 (50)	30 (35)
1 1/8 G1&2	1 1/8	72	48							145 (145)	85 (85)	65 (65)	40 (40)	30 (30)
1 1/4 G3&4	1 1/4	72	48							160 (165)	95 (95)	75 (75)	45 (45)	25 (35)

Notes: (1) Applies to Standard, Structural I and II and C-C grades only.
(2) For applications where the roofing is to be guaranteed by a performance bond, recommendations may differ somewhat from these values. Contact American Plywood Association for bonded roof recommendations.
(3) Use 6d common smooth, ring-shank or spiral thread nails for 1/2″ thick or less, and 8d common smooth, ring-shank or spiral thread for plywood 1″ thick or less (if ring-shank or spiral thread nails same diameter as common). Use 8d ring-shank or spiral thread or 10d common smooth shank nails for 2.4.1, 1 1/8″ and 1 1/4″ panels. Space nails 6″ at panel edges and 12″ at intermediate supports except that where spans are 48″ or more, nails shall be 6″ at all supports.
(4) These spans shall not be exceeded for any load conditions.
(5) Provide adequate blocking, tongue and grooved edges or other suitable edge support such as PlyClips when spans exceed indicated value. Use two PlyClips for 48″ or greater spans and one for lesser spans.
(6) Uniform load deflection limitation: 1/180th of the span under live load plus dead load, 1/240th under live load only. Allowable live load shown in boldface type and allowable total load shown within parenthesis.
(7) Allowable roof loads were established by laboratory test and calculations assuming evenly distributed loads.

Patterns. Lumber roof decking with a double tongue and groove is available in several patterns. Some of the more common are the regular V-joint, grooved, striated, and eased joint (bullnosed) patterns. Single tongue-and-groove decking in nominal 2″ × 6″ and 2″ × 8″ sizes is available with the V-joint pattern only. Fig. 34-10 (Page 509).

Sizes. Decking comes in nominal widths of 4″ to 12″ and in nominal thicknesses of 2″ to 4″. The 3″ and 4″ roof decking is

34–6a. *The grain of plywood sheathing should be at right angles to the supporting members.*

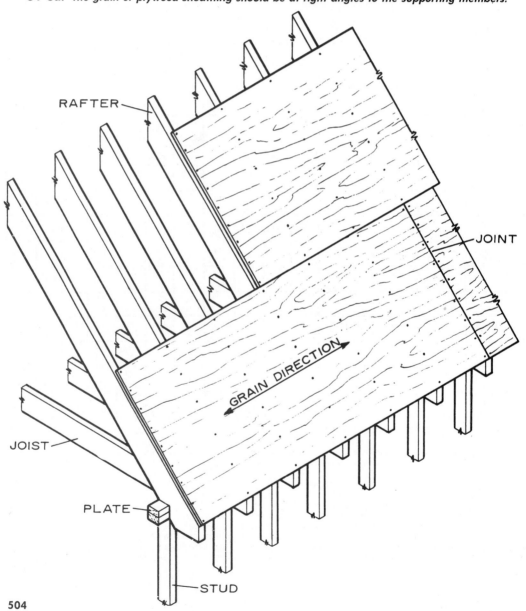

available in random lengths of 6' to 20' or longer (odd and even).

Decking is also available laminated. It comes in six different species of softwood lumber: Idaho white pine, inland red cedar, Idaho white fir, ponderosa pine, Douglas fir, larch, and southern pine. Because of the laminating feature, this material may have a facing of one wood species and back and interior laminations of different woods. It is also available with all laminations of the same species. For all types of decking make sure the material is the correct thickness for the span by checking the manufacturer's recommendations.

Installation

Roof decking that is to be applied to a flat roof should be installed with the tongue away from the worker. Roof decking that is being applied to a sloping roof should be installed with the tongues up. The butt ends of the pieces are cut at approximately a 2° angle. This provides a bevel cut from the face to the back to insure a tight face butt joint when the decking is laid in a random length pattern. Fig. 34-11. If there are three or more supports for the decking, a controlled random laying pattern may be used. Fig. 34-12 (Page 510). This is an economical pattern because it makes use of random plank lengths, but the following rules must be observed:
• Stagger the end joints in adjacent planks as widely as possible and not less than 2'.
• Separate the joints in the same general line by at least two courses.

34–6b. *Plywood roof sheathing details.*

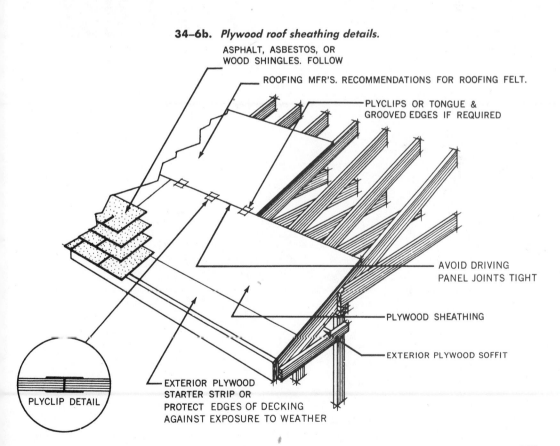

ASPHALT, ASBESTOS, OR WOOD SHINGLES. FOLLOW

ROOFING MFR'S. RECOMMENDATIONS FOR ROOFING FELT.

PLYCLIPS OR TONGUE & GROOVED EDGES IF REQUIRED

AVOID DRIVING PANEL JOINTS TIGHT

PLYWOOD SHEATHING

EXTERIOR PLYWOOD SOFFIT

PLYCLIP DETAIL

EXTERIOR PLYWOOD STARTER STRIP OR PROTECT EDGES OF DECKING AGAINST EXPOSURE TO WEATHER

34–6c. *Power-driven fasteners speed plywood installation. Be sure to use a fastener which will provide the same holding power as the nail size recommended.*

34–7b. *Having the plywood readily available on the roof increases efficiency.*

34–7a. *Details for adjustable roof platform.*

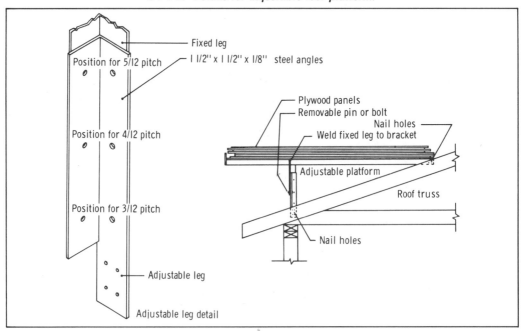

Fixed leg

1 1/2" x 1 1/2" x 1/8" steel angles

Position for 5/12 pitch

Position for 4/12 pitch

Position for 3/12 pitch

Adjustable leg

Adjustable leg detail

Plywood panels

Removable pin or bolt

Nail holes

Weld fixed leg to bracket

Adjustable platform

Roof truss

Nail holes

• Minimize joints in the middle one-third of all spans, make each plank bear on at least one support, and minimize the joints in the end span.

The ability of the decking to support specific loads depends on the support spacing, plank thickness, and span arrangement. Although two-span continuous layout offers structural efficiency, use of random-length planks is the most economical. Random-length double tongue-and-groove decking is used when there are three or more spans. It is not intended for use over single spans and it is not recommended for use over double spans. Fig. 34-12. Each piece should bear on at least one support.

NAILING

Fasten the decking with common nails twice as long as the nominal plank thickness. For widths 6″ or less, toenail once and face-nail once at each support. For widths over 6″, toenail once and face-nail

34-8b. *This ladder is made from scrap plywood nailed to a pair of 2 × 4s. A worker can load this ladder before getting on the roof.*

34-8a. *Plywood ladder detail.*

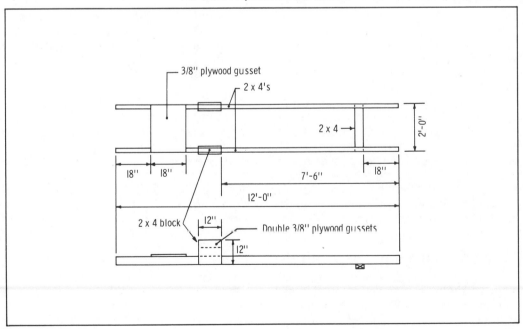

34-8c. *When applying the sheathing, the worker can pull it from the ladder onto the roof.*

34-9. *Decking with tongue-and-groove edges and decorative face patterns provides a durable roof and an attractive ceiling for residential and commercial buildings.*

twice. Decking 3″ and 4″ thick must be predrilled and toenailed with 8″ spikes. Fig. 34-13. Some manufacturers provide the 3″ and 4″ thick roof decking with predrilled nail holes on 30″ centers. Bright common nails may be used but dipped galvanized common nails have better holding power and reduce the possibility of rust streaks. End joints not over a support should be side-nailed within 10″ of each plank end. Metal splines are recommended on end joints of 3″ and 4″ material for better alignment, appearance, and strength.

WOOD FIBER ROOF DECKING

The all-wood fiber roof decking combines strength and insulation advantages that make possible quality construction with economy. This type of decking is weatherproof and protected against termites and rot. It is ideally suited for built-up roofing, as well as for asphalt and wood shingles on all types of buildings. It is available in four thicknesses: 2³/₈″, 1⁷/₈″, 1³/₈″, and ¹⁵/₁₆″. The standard panels are 2′ × 8′ with tongue-and-groove edges and square ends. Fig. 34-14. The surfaces are coated on one or both sides at the factory in a variety of colors.

Installation

Wood fiber roof decking is laid with the tongue-and-groove joint at right angles to the support members. Begin laying at the eave line with the groove edge away from the applicator. Staple wax paper in position over the rafter before installing the roof deck. Fig. 34-15. The wax paper protects the exposed interior finish of the decking when the beams are stained later. Calk the end joints with a nonstaining calking compound. Fig. 34-16. Butt the adjacent piece up against the calked joint. Fig. 34-17 (Page 512). Drive the tongue-and-groove edges of each of the units firmly together with a wood block cut to fit the grooved

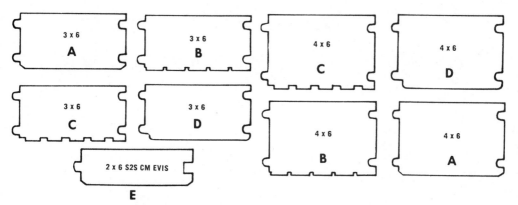

34-10. *Some lumber roof decking patterns and sizes: A. Regular V-jointed. B. Striated. C. Grooved. D. Eased joint (bullnosed). E. Single tongue-and-groove V-joint.*

edge of the decking. Fig. 34-18. End joints must be made over a support member.

NAILING

These panels are tongued and grooved but are nailed through the face into the wood, rafters, or trusses. Face-nail 6" on center with 6d nails for $^{15}/_{16}$", 8d for $1^3/_8$", 10d for $1^7/_8$", and 16d for $2^3/_8$" thickness.

SHEATHING AT THE ENDS OF THE ROOF

Where the gable ends of the roof have little or no extension other than the molding and trim, the roof sheathing is usually sawed flush with the outer face of the side wall sheathing. Cuts should be made even so that the trim and molding can be properly installed. See Unit 36, "Roof Trim." Roof sheathing that projects beyond the end walls should span not less than three rafter spaces to insure proper anchorage to the rafters and to prevent sagging. Fig. 34-19. In general, it is desirable to use the longest boards at overhangs to secure good anchorage.

SHEATHING DETAILS AT CHIMNEY OPENINGS

Where chimney openings occur in the roof structure, the roof sheathing should have a clearance of $^3/_4$" from the finished

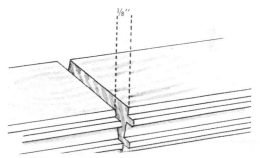

34-11. *The ends of lumber decking are cut at a 2° angle to insure a tight face joint on the exposed ceiling.*

masonry on all sides. Fig. 34-20. Framing members should have a 2" clearance for fire protection. The sheathing should be securely nailed to the rafters and to the headers around the opening.

SHEATHING AT VALLEYS AND HIPS

The sheathing at the valleys and hips should be fitted to give a tight joint. It should be securely nailed to the valley or the hip rafter. Fig. 34-20. This will give a solid and smooth base for the flashing.

ESTIMATING
Material

Determine the total area to be covered. To figure the roof area without actually

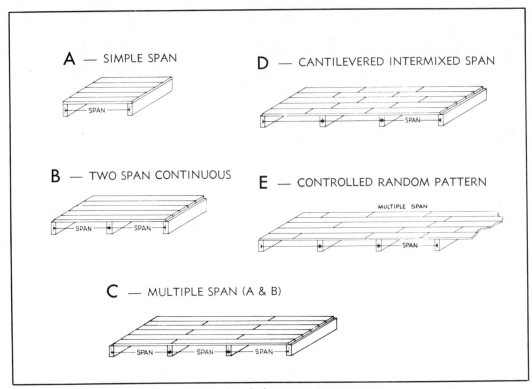

34–12. *Lumber decking span arrangments.*

34–13. *Nailing details for lumber decking.*

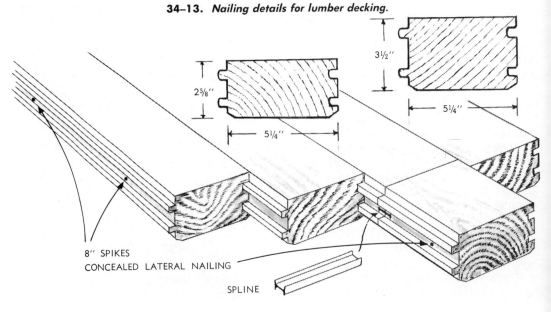

getting on the roof and measuring, find the dimensions of the roof on the plans. Multiply the length times the width of the roof including the overhang. Then multiply by the factor shown opposite the rise of the roof in Fig. 34-21 (Page 514). The result will be the roof area.

For example, assume that a home is 70' long and 30' wide including the overhang and that the roof has a rise of 5½". 70' ×

34–14.

Wood Fiber Roof Decking Specifications

Thickness	Rafter spacing, with allowable live load* of		Size	Finish	Attachment
	40 psf.	50 psf.			To wood rafters with galvanized common nail
15/16"	24" o.c.	16" o.c.	Standard panel 2' x 8' (nominal). Actual is 23⅛" x 95⅞"	Ends: square Long edges; T&G and V-jointed on ceiling side; Special order: All 4 edges T&G up to 4' x 8'	6d.
1⅜"	32" o.c.	24" o.c.			8d.
1⅞"	48" o.c.	32" o.c.	Special order; Widths to 4' Lengths to 12'		10d.
2⅜"	60" o.c. +	48" o.c.			16d.

*Based on spacing and thickness indicated.

34–15. *Stapling wax paper to the top of the beams. The paper will protect the exposed ceiling later when the beams are being finished.*

34–16. *Calking the end joints.*

34–17. *Butt the adjacent pieces together at the calking.*

34–18. *Protect the edges of the roof decking with a wood block when driving the joint up tightly.*

34–19. *Sheathing at the ends of a gable roof.*

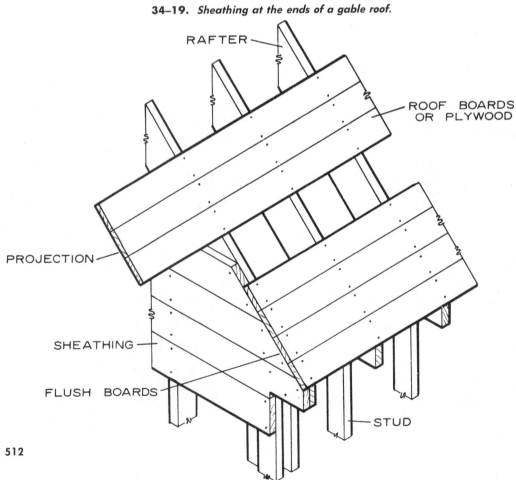

RAFTER

ROOF BOARDS OR PLYWOOD

PROJECTION

SHEATHING

FLUSH BOARDS

STUD

34-20. *Sheathing details at the valley and at the chimney opening. Section A-A shows the clearance between the masonry chimney and the wood structure.*

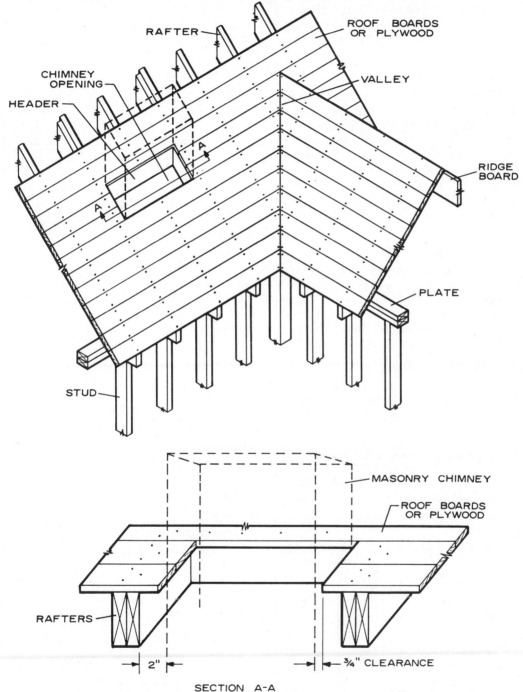

RAFTER

CHIMNEY OPENING

HEADER

ROOF BOARDS OR PLYWOOD

VALLEY

RIDGE BOARD

PLATE

STUD

MASONRY CHIMNEY

ROOF BOARDS OR PLYWOOD

RAFTERS

2"

¾" CLEARANCE

SECTION A-A

30′ = 2100 sq. ft. For a rise of 5½″, the factor on the chart is 1.100. Fig. 34-21. 2100 sq. ft. × 1.100 = 2310 sq. ft. Use this total area for figuring roofing needs, such as sheathing, felt underlayment, or shingles.

LUMBER SHEATHING

First figure the total area to be covered. Determine the size boards to be used and refer to the chart in Fig. 34-22. Multiply the total area to be covered by the factor from the chart. For example, if 1″ × 8″ tongue-and-groove sheathing boards are used, the total roof area is multiplied by 1.16. To determine the total number of board feet needed, add 5% for a trim and waste factor.

PLYWOOD SHEATHING

Determine the total roof area to be covered and divide by 32 (the number of square feet in one sheet of 4′ × 8′ plywood). This will give the number of sheets required for covering the area. Be sure to add 5% for a trim and waste allowance.

DECKING OR PLANKING

Determine the area to be covered. Then refer to the chart in Fig. 34-23, and read in the left column the size planking to be

34-21. *When a roof has to be figured from a plan only and the roof pitch is known, the roof area may be computed from the table below.*

Determining Roof Area from a Plan

Rise	Factor	Rise	Factor
3″	1.031	8″	1.202
3½″	1.042	8½″	1.225
4″	1.054	9″	1.250
4½″	1.068	9½″	1.275
5″	1.083	10″	1.302
5½″	1.100	10½″	1.329
6″	1.118	11″	1.357
6½″	1.137	11½″	1.385
7″	1.158	12″	1.414
7½″	1.179		

applied. For example, if 2″ × 6″ material is selected, the factor given is 2.40. Multiply the area to be covered by this factor and add a 5% trim and waste allowance to arrive at the amount of material required.

34-22.
Roof Board Sheathing Specifications

	Nominal Size	Width Dress	Width Face	Area Factor
Shiplap	1 x 6	5⁷/₁₆	4¹⁵/₁₆	1.22
	1 x 8	7⅛	6⅝	1.21
	1 x 10	9⅛	8⅝	1.16
	1 x 12	11⅛	10⅝	1.13
Tongue And Groove	1 x 4	3⁷/₁₆	3³/₁₆	1.26
	1 x 6	5⁷/₁₆	5³/₁₆	1.16
	1 x 8	7⅛	6⅞	1.16
	1 x 10	9⅛	8⅞	1.13
	1 x 12	11⅛	10⅞	1.10
S4S	1 x 4	3½	3½	1.14
	1 x 6	5½	5½	1.09
	1 x 8	7¼	7¼	1.10
	1 x 10	9¼	9¼	1.08
	1 x 12	11¼	11¼	1.07

34-23.
Area Factors for Estimating Decking

Size	Area Factor
2″ x 6″	2.40
2″ x 8″	2.29
3″ x 6″	3.43
4″ x 6″	4.57

Waste allowance not included in above factors.

34-24.
Labor Time for Roof Sheathing

1 x 6 boards S4S	65 b.f. per hour
1 x 6 center match	55 b.f. per hour
1 x 8 shiplap	60 b.f. per hour
1 x 10 shiplap	75 b.f. per hour

WOOD FIBER ROOF DECKING

Determine the total roof area to be covered. For every 100 square feet of area 6.25 panels, 2′ × 8′ in size, will be required. Therefore divide the roof area by 100 and multiply by 6.25. Using our previous example with a roof area of 2310 square feet, 145 panels will be needed, as shown by the following:

$$\frac{2310\ (\text{roof area})}{100} = 23.1 \text{ (roof area expressed in hundreds of sq. ft.)}$$

$$23.1 \times 6.25 = 144.37, \text{ or } 145$$

Labor

Use the chart in Fig. 34-24 when estimating labor for roof sheathing application.

QUESTIONS

1. What are the most commonly used sheathing materials for pitched roofs?

2. List several materials that are commonly used for roof sheathing.

3. Why is it important that lumber roof sheathing be thoroughly seasoned when used with asphalt shingles?

4. When installing plywood roof sheathing, in which direction should the grain run in relation to the rafters?

5. What is the minimum thickness of plywood that may be used for roof sheathing under asphalt shingles with a rafter spacing of 16″?

6. What are some of the advantages of decking or planking in contemporary architecture?

7. What are some of the advantages of wood fiber roof decking?

8. What clearance is recommended between roof sheathing and finished masonry, such as a chimney?

UNIT 35

Roof Coverings*

The roof covering, or roofing, is a part of the exterior finish. It should provide long-lasting waterproof protection for the building and its contents from rain, snow, wind and, to some extent, heat and cold. Fig. 35-1. Materials used for pitched roofs include shingles of wood, asphalt, and asbestos. Tile and slate are also popular. Sheet materials such as roll roofing, galvanized iron, aluminum, copper, and tin are sometimes used. For flat or low-pitched roofs, composition or built-up roofing with a gravel topping or cap sheet are frequent combinations. Built-up roofing consists of a number of layers of asphalt-saturated felt mopped down with hot asphalt or tar. Metal roofs are sometimes used on flat decks of dormers, porches, or entryways.

The choice of materials and method of application is influenced by cost, roof slope, expected service life of the roofing, wind resistance, fire resistance, and local

*Some of the material from this unit was adapted from *Construction: Principles, Materials & Methods* by courtesy of the American Savings & Loan Institute Press.

35–1. *Applying wood shake roof covering.*

climate. Due to the large amount of exposed surface, appearance is also important. Shingles, for example, add color, texture, and pattern to the roof surface. All shingles are applied to roof surfaces in some overlapping fashion to shed water.

35–2. *Terminology used in roofing: E=exposure, TL=toplap, HL=headlap, W=width of strip shingles or length of individual shingles.*

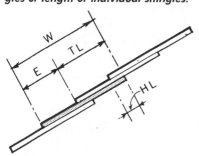

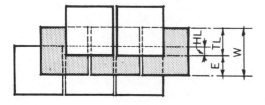

Therefore they are suitable for any roof with enough slope to insure good drainage.

ROOFING TERMINOLOGY

Square. Roofing is estimated and sold by the square. A square of roofing is the amount required to cover 100 sq. ft. of roof surface.

Coverage. This term indicates the amount of weather protection provided by the overlapping of shingles. Depending on the kind of shingle and method of application, shingles may furnish one (single coverage), two (double coverage), or three (triple coverage) thicknesses of material over the surface of the roof.

Shingles providing single coverage are suitable for reroofing over existing roofs. Shingles providing double and triple coverage are used for new construction, both having increased weather resistance and a longer service life.

Exposure. The shortest distance in inches between exposed edges of overlapping shingles. Fig. 35-2.

Toplap. The shortest distance in inches from the lower edge of an overlapping shingle or sheet to the upper edge of the lapped unit in the first course below (that is, the width of the shingle minus the exposure). Fig. 35-2.

Headlap. The shortest distance in inches from the lower edges of an overlapping shingle or sheet to the upper edge of the unit in the second course below. Fig. 35-2.

Side- or Endlap. The shortest distance in inches by which adjacent shingles or sheets horizontally overlap each other. Fig. 35-2.

Shingle Butt. The lower exposed edge of the shingle.

Slope and Pitch

These terms are often incorrectly used synonymously when referring to the incline of a sloped roof. Both are defined on the next page. Fig. 35-3 also compares some common roof slopes to corresponding roof pitches.

Slope. Slope indicates the incline of a roof as a ratio of vertical rise to horizontal run. It is expressed sometimes as a fraction but typically as *X* in 12. For example, a roof that rises at the rate of 4" for each foot (12") of run, is designated as having a 4-in-12 slope. The triangular symbol above the roof in Fig. 35-3 conveys this information.

Pitch. Pitch indicates the incline of a roof as a ratio of the vertical rise to *twice* the horizontal run. It is expressed as a fraction. For example, if the rise of a roof is 4' and the run 12', the roof is designated as having a pitch of $^1/_6$ ($^4/_{24}$ = $^1/_6$).

ROOFING ACCESSORIES

In addition to the shingles, many accessory materials are required to prepare the roof deck and to apply the shingles. These accessories include: underlayment, flashing, roofing cements, eaves flashing, drip edge, and roofing nails or fasteners. With some kinds of shingles, other accessories may be required, such as starter shingles and hip and ridge units. Regardless of the type of shingle to be installed, always check the instructions and recommendations of the shingle manufacturer to insure proper performance.

We will assume in this unit that the roof is correctly and adequately ventilated. See Unit 53 for prevention of water vapor condensation. When applying shingles, the exposure distance is important. This distance depends mostly on roof slope and shingle type. Fig. 35-4. The minimum slope on main roofs is 4 in 12 for wood, asphalt, asbestos, and slate shingles. For built-up roofs the maximum slope is 3 in 12.

Underlayment

Underlayment is normally required for asphalt, asbestos, and slate shingles and for tile roofing, but it may be omitted for wood shingles. In areas where snow is common and ice dams occur (melting snow freezes at the eave line), it is a good practice to apply one course of 55-pound smooth-surfaced roll roofing at the eaves.

35–4. *Minimum pitch requirements for asphalt roofing products.*

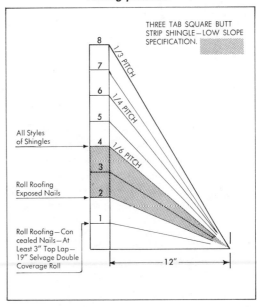

35–3. *Slope and pitch.*

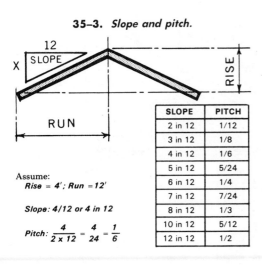

SLOPE	PITCH
2 in 12	1/12
3 in 12	1/8
4 in 12	1/6
5 in 12	5/24
6 in 12	1/4
7 in 12	7/24
8 in 12	1/3
10 in 12	5/12
12 in 12	1/2

Assume:
Rise = 4'; Run = 12'

Slope: 4/12 or 4 in 12

Pitch: $\dfrac{4}{2 \times 12} = \dfrac{4}{24} = \dfrac{1}{6}$

$$Slope = \frac{Rise}{Run;} \qquad Pitch = \frac{Rise}{2 \times Run}$$

35–5. *A. Snow and ice dams can build up on the overhang of roofs and gutters, causing melting snow to back up under shingles and under the fascia board of closed cornices. Damage to interior ceilings and walls and to exterior paint results from this water seepage. B. Protection from snow and ice dams is provided by eave flashing. Ventilation of the cornice by means of vents in the soffit and sufficient insulation will minimize the melting.*

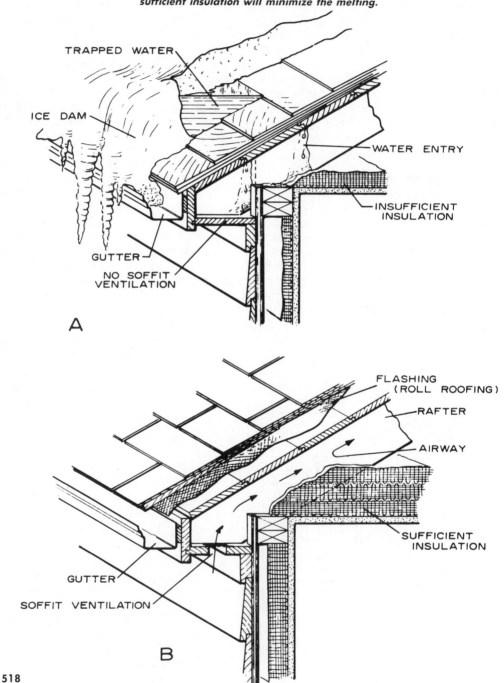

Fig. 35-5. Roof underlayment generally has three purposes:

- It protects the sheathing from moisture absorption until the shingles can be applied.
- It provides important additional weather protection by preventing the entrance of wind-driven rain below the shingles onto the sheathing or into the structure.
- In the case of asphalt shingles, it prevents direct contact between the shingles and resinous areas in wood sheathing which may be damaging to the shingles because of chemical incompatibility.

Underlayment should be a material with low vapor resistance, such as asphalt-saturated felt. Do not use materials such as coated felts or laminated waterproof papers which act as a vapor barrier. These allow moisture or frost to accumulate between the underlayment and the roof sheathing. Underlayment requirements for different kinds of shingles and various roof slopes are shown in Fig. 35-6.

35–6.

Summary Of Underlayment Recommendations For Shingle Roofs

Type of Roofing	Sheath-ing	Type of Underlayment	Normal Slope		Low Slope	
Asbestos-Cement Shingles	Solid	No. 15 asphalt saturated asbestos (inorganic) felt, OR No. 30 asphalt saturated felt	$5/12$ and up	Single layer over entire roof	$3/12$ to $5/12$	Double layer over entire roof[1]
Asphalt Shingles	Solid	No. 15 asphalt saturated felt	$4/12$ and up	Single layer over entire roof	$2/12$ to $4/12$	Double layer over entire roof[2]
Wood Shakes	Spaced	No. 30 asphalt saturated felt (interlayment)	$4/12$ and up	Underlayment starter course; interlayment over entire roof	Shakes not recommended on slopes less than $4/12$ with spaced sheathing	
	Solid[3,5]	No. 30 asphalt saturated felt (interlayment)	$4/12$ and up	Underlayment starter course; interlayment over entire roof	$3/12$ to $4/12$[4]	Single layer underlayment over entire roof; interlayment over entire roof
Wood Shingles	Spaced	None required.	$5/12$ and up	None required	$3/12$ to $5/12$[4]	None required
	Solid[5]	No. 15 asphalt saturated felt	$5/12$ and up	None required[6]	$3/12$ to $5/12$[4]	None required[6]

1. May be single layer on 4 in 12 slope in areas where outside design temperature is warmer than 0° F.
2. Square-Butt Strip shingles only; requires Wind Resistant shingles or cemented tabs.
3. Recommended in areas subject to wind driven snow.
4. Requires reduced weather exposure.
5. May be desirable for added insulation and to minimize air infiltration.
6. May be desirable for protection of sheathing.

INSTALLING UNDERLAYMENT

Apply the underlayment as soon as the roof sheathing has been completed. For single underlay, start at the eave line with the 15-pound felt. Roll across the roof with a toplap of at least 2″ at all horizontal joints and a 4″ sidelap at all end joints. Fig. 35-7. Lap the underlayment over all hips and ridges 6″ on each side. A double underlay can be started with two layers at the eave line, flush with the fascia board or molding. The second and remaining strips have 19″ headlaps with 17″ exposures. Fig. 35-8. Cover the entire roof in this manner, making sure that all surfaces have double coverage. Use only enough fasterners to hold the underlayment in place until the shingles are applied. Do not apply shingles over wet underlayment.

Flashings

Flashing is a special construction of sheet metal or other material used to protect the building from water seepage. Flashing must be made watertight and water-shedding. Metal used for flashing must be corrosion-resistant. It should be galvanized steel (at least 26-gauge), 0.019″ thick aluminum, or 16 oz. copper.

Flashing is required at the point of intersection between roof and soil stack or ventilator, in the valley of a roof, around

35–7b. *Stapling the underlayment into place.*

chimneys, and at the point where a wall intersects a roof.

SOIL STACKS

Apply the roofing up to the stack, cutting it to fit. Fig. 35-9. Then install a corrosion-resistant metal sleeve which slips over the stack and has an adjustable flange to fit any roof slope. Figs. 35-10 and 35-11. A piece of 55-pound roll roofing can also be used. For roll roofing, lay out and cut an opening for the stack as shown in Fig. 35-12. Slip this flange in place over the stack and apply a roof cement 2″ up on the stack and 2″ out on the flange. Continue shingling over the flange. Cut the shingles to fit around the stack, pressing them firmly into the cement. Fig. 35-13.

VALLEYS

The open or closed method may be used to construct valley flashing. A valley underlayment strip of No. 15 asphalt-saturated felt, 36″ wide, is applied first.

35–7a. *Details for applying the underlayment for single coverage.*

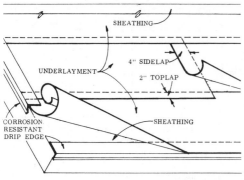

SHEATHING

4″ SIDELAP

UNDERLAYMENT

2″ TOPLAP

CORROSION RESISTANT DRIP EDGE

SHEATHING

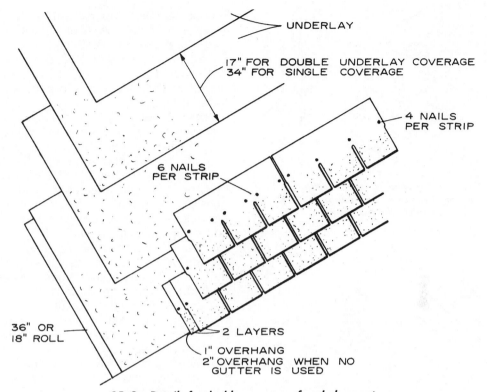

UNDERLAY

17" FOR DOUBLE UNDERLAY COVERAGE
34" FOR SINGLE COVERAGE

4 NAILS
PER STRIP

6 NAILS
PER STRIP

36" OR
18" ROLL

2 LAYERS

1" OVERHANG
2" OVERHANG WHEN NO
GUTTER IS USED

35–8. *Details for double coverage of underlayment.*

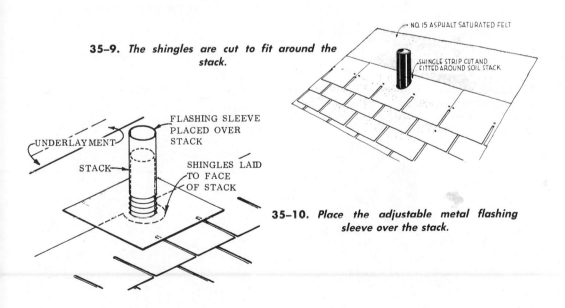

35–9. *The shingles are cut to fit around the stack.*

NO. 15 ASPHALT SATURATED FELT

SHINGLE STRIP CUT AND
FITTED AROUND SOIL STACK

UNDERLAYMENT

FLASHING SLEEVE
PLACED OVER
STACK

STACK

SHINGLES LAID
TO FACE
OF STACK

35–10. *Place the adjustable metal flashing sleeve over the stack.*

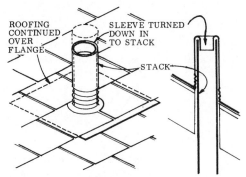

35–11. Lay the shingles over the flange. Turn the top of the sleeve down into the stack to complete the installation.

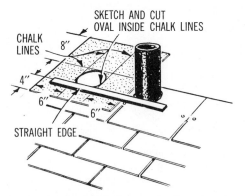

35–12. Laying out a flange of roll roofing for a soil stack.

35–13. Completing the installation of the flange. Lay shingles over the flange, fit them around the stack, and press the shingles firmly into the cement.

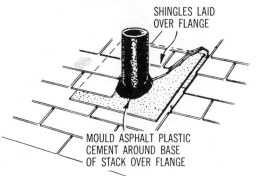

Fig. 35-14. The strip is centered in the valley and secured with enough nails to hold it in place. The horizontal courses of underlayment are cut to overlap this valley strip a minimum of 6". Where eaves flashing is required, it is applied over the valley underlayment strip.

Open Valleys. Open valleys may be flashed with metal or with 90 lb. mineral-surfaced asphalt roll roofing in a color to match or to contrast with the roof shingles. The method is illustrated in Fig. 35-15. An 18" wide strip of mineral-surfaced roll roofing is placed over the valley underlayment. It is centered in the valley with the surfaced side down and the lower edge cut to conform to and be flush with the eaves flashing. When it is necessary to splice the material, the ends of the upper segments are laid to overlap the lower segments 12" and are secured with asphalt plastic cement. Only enough nails are used in rows 1" in from each edge to hold the strip smoothly in place.

Another strip, 36" wide, is placed over the first strip. It is centered in the valley with the surfaced side up and secured with nails. It is lapped if necessary the same way as the underlying 18" strip.

35–14. Applying the underlayment in the valley.

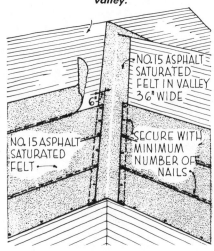

Before shingles are applied, a chalk line is snapped on each side of the valley for its full length. The lines should start 6" apart at the ridge and spread wider apart (at the rate of $1/8$" per foot) to the eave. The chalk lines serve as a guide in trimming the shingle units to fit the valley, insuring a clean, sharp edge. The upper corner of each end shingle is clipped to direct water

into the valley and prevent water penetration between courses. Fig. 35-16. Each shingle is cemented to the valley lining with asphalt cement to insure a tight seal. No exposed nails should appear along the valley flashing.

Closed Valleys. Closed (woven) valleys can be used only with strip shingles. This method has the advantage of doubling the coverage of the shingles throughout the length of the valley, increasing weather resistance at this vulnerable point. A valley lining made from a 36" wide strip of 55-pound (or heavier) roll roofing is placed over the valley underlayment and centered in the valley. Fig. 35-17.

Valley shingles are laid over the lining by either of two methods:

• They may be applied on both roof surfaces at the same time, with each course in turn woven over the valley.

• Each surface may be covered to the point approximately 36" from the center of the valley and the valley shingles woven in place later.

In either case, the first course at the valley is laid along the eaves of one surface over the valley lining and extended

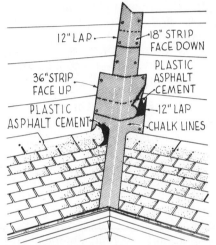

35–15. *Open valley flashing details using roll roofing.*

35–16. *Begin laying each course at the chalk line in the valley. Clip the top corners of the shingles as shown at the arrow to prevent water penetration.*

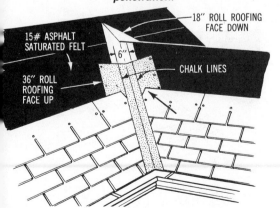

35–17. *Closed valley flashing using woven strip shingles.*

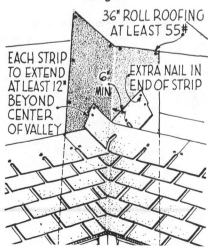

along the adjoining roof surface for a distance of at least 12″. The first course of the adjoining roof surface is then carried over the valley on top of the previously applied shingle. Succeeding courses are then laid alternately, weaving the valley shingles over each other.

The shingles are pressed tightly into the valley and nailed in the usual manner (see "Fastening Strip Shingles," page 530) except that no nail should be located closer than 6″ to the valley center line and two nails are used at the end of each terminal strip. Fig. 35-17.

CHIMNEYS

Apply the shingles over the felt up to the chimney face. If 90-pound roll roofing is to be used for flashing, cut wood cant strips and install them above and at the sides of the chimney. Fig. 35-18. The roll roofing flashing should be cut to run 10″ up the chimney. Working from the bottom up, fit metal flashing over the base flashing and insert it 1½″ into the mortar joints. Refill the joints with mortar or roofing cement.

WALL INTERSECTIONS

Start at the eave line and work upward. Apply the metal flashing over the felt and up onto the wall sheathing but under the roofing and the siding. The siding should be cut so that it clears the roof by at least ³/₄″. Fig. 35-19.

Roofing Cements

Roofing cements are used for installing eaves flashing, for flashing assemblies, for cementing tabs of asphalt shingles and laps in sheet material, and for roof repairing. There are several types of cement, including plastic asphalt cements, lap cements, quick-setting asphalt adhesives, roof coatings, and primers. The type and quality of materials and methods of application on a shingle roof should be those recommended by the manufacturer of the shingle roofing.

Eaves Flashing

Eaves flashing is recommended in areas where the temperature goes below 0° F, or wherever there is a possibility of ice forming along the eaves. This ice forms a dam which allows water to back up under the shingles.

Eaves flashing is usually formed by an additional course of underlayment or roll roofing applied over the underlayment. For normal slopes this extends up the roof to cover a point at least 12″ inside the interior wall line of the building. A, Fig. 35-20.

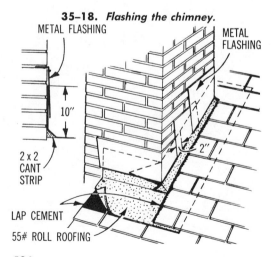

35–18. *Flashing the chimney.*

METAL FLASHING

METAL FLASHING

10″

2 x 2 CANT STRIP

2″

LAP CEMENT

55# ROLL ROOFING

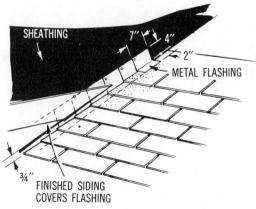

35–19. *Flashing the intersection of a roof and wall.*

SHEATHING

7″   4″

2″

METAL FLASHING

³/₄″   FINISHED SIDING COVERS FLASHING

For low slopes or in areas subject to severe icing, eaves flashing formed by cementing an additional course of underlayment over the first underlayment as for normal conditions. However, it extends up the roof to cover a point at least 24″ inside the interior wall line of the building. B, Fig. 35-20 and Fig. 35-21.

Drip Edge

Drip edges are designed and installed to protect the edges of the roof. They prevent leaks at this point by causing water to drip free of underlying eave and cornice construction. Some shapes of pre-formed drip edges are shown in Fig. 35-22. A drip edge is recommended for most

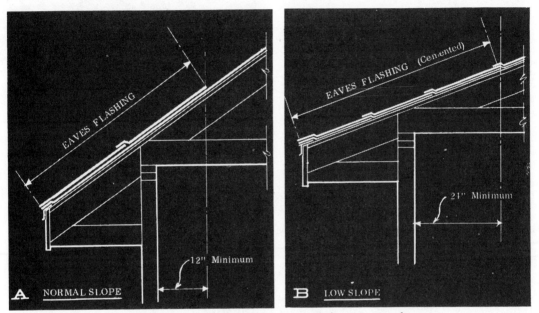

35–20. *Eaves flashing: A. Normal slope. B. Low slope.*

35–21. *Details for applying the underlayment for a low slope.*

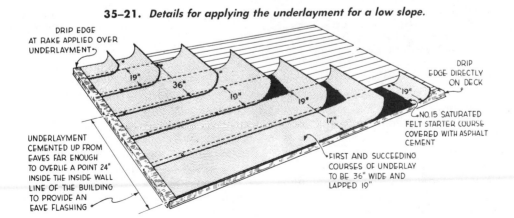

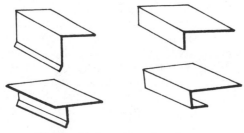

35-22. *Various drip edge shapes.*

35-23b. *Using a pneumatic stapler to install the metal drip edge at the rake.*

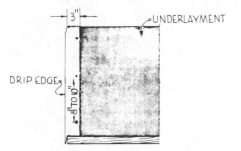

A RAKE

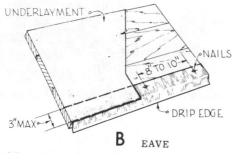

B EAVE

35-23a. *Drip edge application: A. Along the rake. B. At the eave. Note that the underlayment goes under the drip edge at the rake and over the drip edge at the eave.*

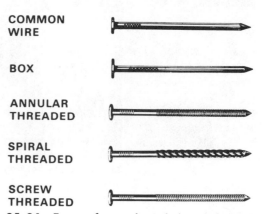

COMMON WIRE

BOX

ANNULAR THREADED

SPIRAL THREADED

SCREW THREADED

35-24. *Types of smooth and threaded shank nails recommended for the application of shingle roofing.*

shingle roofs. It is applied to the sheathing and under the underlayment at the eaves, but over the underlayment up the rake. Fig. 35-23.

Roofing Nails

No single step in applying roof shingles is more important than proper nailing. Suitability is dependent on several factors:

● Selecting the correct nail for the kind of shingle and type of roof sheathing. Fig. 35-24.
● Using the correct number of nails.
● Locating them in the shingle correctly.
● Choosing nail metal compatible with metal used for flashing.

Roofing nails should be long enough to penetrate through the shingle and through the roof sheathing. They should penetrate

35-25. *Asphalt roof shingles.*

SHINGLE TYPE*	SHIPPING WEIGHT PER SQUARE	PACKAGES PER SQUARE	LENGTH	WIDTH	UNITS PER SQUARE	SIDE-LAP	TOP-LAP	HEAD-LAP	EXPOSURE
STRIP SHINGLES — 2 & 3 TAB SQUARE BUTT	235 Lb	3	36"	12"	80		7"	2"	5"
STRIP SHINGLES — 2 & 3 TAB HEXAGONAL	195 Lb	3	36"	11⅓"	86		2"	2"	5"
INDIVIDUAL — STAPLE / LOCK	145 Lb	2	16"	16"	80	2½"			
GIANT INDIVIDUAL — AMERICAN	330 Lb	4	16"	12"	226		11"	6"	5"
GIANT INDIVIDUAL — DUTCH LAP	165 Lb	2	16"	12"	113	3"	2"		10"

at least 1" into plank decking. Nails for applying shingles over plywood sheathing should have threaded shanks.

Specific recommendations for the type, size, number, and spacing of roofing nails are given later in conjunction with the information on asphalt and wood shingles and wood shakes.

ASPHALT SHINGLES

Asphalt roof shingles are manufactured in three basic kinds of units. Fig. 35-25.

• Strip shingles of the square-butt or hexagonal type.

• Individual shingles of the interlocking or staple-down type.

• Giant individual shingles for application by either the American or Dutch lap methods.

In areas where high winds prevail, wind resistant strip shingles with factory-applied adhesive or integral locking tabs are recommended.

Shingles are laid so that they overlap and cover each other to shed water. Before applying shingles make sure that:

• The underlayment, drip edge, and flashings are in place.

• The roof deck is tight and provides a suitable nailing base.

• The chimney is completed and the counter flashing installed.

• Stacks and other equipment requiring openings in the roof are in place with counter flashing where necessary.

Strip Shingles

On small roofs strip shingles may be laid from either rake. Fig. 35-26. On roofs 30' and longer, shingles should be started at the center and applied both ways from a vertical line. This will assure more accurate vertical alignment and will provide for meeting and matching above a projection such as a dormer or chimney. To assure accurate alignment of shingles, use horizontal and vertical chalk lines.

The first course of shingles, called the *starter* course, is applied over the eaves flashing strip and even with its lower edge along the eave. The starter course may be a 9" wide (or wider) starter strip of

35-26a. *Shingles laid out on a roof. Some of them have been opened up in preparation for the roofer, who is stapling the shingles in place. The helper is cutting the starter shingles on the edge of the roof.*

35-26b. *This ladder is equipped with an electric hoist used to raise bundles of shingles and other materials up to the roof.*

mineral-surfaced roll roofing of a color to match the shingles. A row of inverted shingles may also be used for the starter course. Fig. 35-27. Fasten the starter strip with roofing nails placed about 3" or 4" above the eave edge and spaced so that the nailheads will not be exposed at the cutouts between the tabs on the first course. If square-butt strip shingles are used as a starter strip, cut off 3" of the first inverted starter course shingle to be laid at the rake. Then the first course laid right side up is started with a full shingle. Succeeding courses are started with full or cut strips depending upon the pattern desired. Three variations for laying square-butt strip shingles are as follows:

• *Cutouts breaking joints on halves.* When the cutouts break the joints on halves, the second course is started with a full shingle cut 6" short (half a tab). The third is started with a full shingle minus the entire first tab. The fourth is started with $1\frac{1}{2}$ tabs cut from the shingle, and so on, causing the cutouts to be centered on the tabs of the course below. Fig. 35-28. This pattern can also be made with less cutting by starting a third course with a full shingle.

• *Cutouts breaking joints on thirds.* When the cutouts break the joints on thirds, the second course is started with a full shingle cut 4" short ($\frac{1}{3}$ of a tab). The third course is started with a full shingle cut 8" short ($\frac{2}{3}$ of a tab), and the fourth with a full shingle. Fig. 35-29.

528

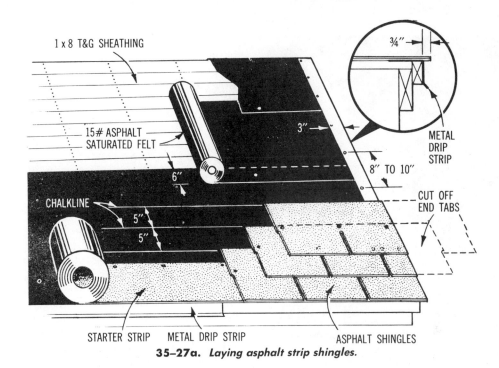

1 x 8 T&G SHEATHING

15# ASPHALT
SATURATED FELT

6"

CHALKLINE

5"

5"

3"

8" TO 10"

¾"

METAL
DRIP
STRIP

CUT OFF
END TABS

STARTER STRIP   METAL DRIP STRIP   ASPHALT SHINGLES

35–27a. *Laying asphalt strip shingles.*

35–27b. *This staple gun is equipped with a guide which is placed at the lower edge of the row of shingles previously laid. This positions the shingle being installed for the correct exposure while the shingle is held in position with the left hand.*

35–27c. *After the shingle is located in position, it is stapled from left to right, with four staples.* **529**

35–27d. *Nailing the courses of shingles.*

● *Random spacing* is achieved by removing different amounts from the starting tab of succeeding courses in accordance with the following general principles:

1. The width of any starting tab should be at least 3″.

2. Cutout centerlines of any course should be located at least 3″ laterally from the cutout centerlines in both the course above and the course below.

3. Starting tab widths should be varied sufficiently so that the eye will not follow a cutout alignment. Fig. 35-30.

Regardless of the laying pattern, each succeeding course of shingles is placed so that the lower edges of the butts are aligned with the top of the cutouts on the underlying course.

FASTENING STRIP SHINGLES

Nails for applying asphalt roofing should be corrosion-resistant. Hot-dipped galvanized steel or aluminum nails with sharp points and flat heads that are $3/8″$ to $7/16″$ in diameter are recommended. Shanks should be 10- to 12-gauge wire and may be smooth or threaded. Threaded nails are preferred because of their increased holding power. Aluminum nails should have screw threads with approximately a $12^1/2°$ thread angle. Fig. 35-31. Galvanized steel nails, if threaded, should have annular threads. Nail lengths typically required are given in Fig. 35-32.

The number and the placement of nails are important for good roof application.

35–28. *Laying asphalt square-butt strip shingles with the cutouts breaking joints on halves.*

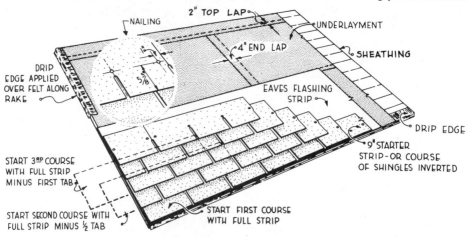

Nailing should start at the end nearest the shingle last applied and proceed to the opposite end. To prevent buckling be sure each shingle is in perfect alignment before driving any nail. Drive the nail straight to avoid cutting the shingle with the edge of the nailhead. The nailhead should be driv-en flush, not sunk below the surface of the shingle.

If the shingles are laid in windy areas, they will require additional protection. A spot of quick-setting cement about 1" square for each tab is applied on the underlying shingle with a putty knife or

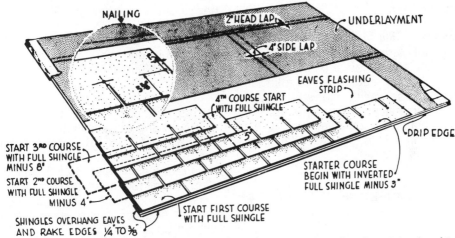

35-29. *Laying asphalt square-butt strip shingles with the cutouts breaking joints on thirds.*

35-30. *Random spacing of asphalt square-butt strip shingles. The first course was started with a full-length strip.*

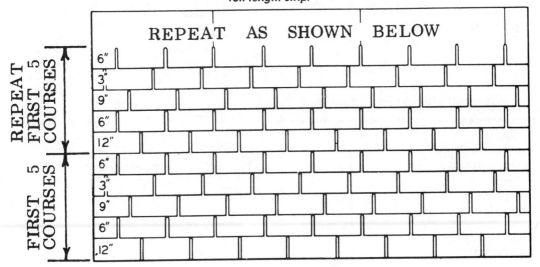

calking gun. The free tab is then pressed against the cement. Fig. 35-33.

Three-tab square-butt shingles require four nails for each strip. When the shingles are applied with a 5″ exposure, the four nails are placed ⅝″ above the top of the cutouts and located horizontally with one nail 1″ back from each end, and one nail on the center line of each cutout. Fig. 35-28. Two-tab square-butt shingles are nailed in a similar manner.

HEXAGONAL STRIP SHINGLES

Hexagonal strip shingles permit no spacing variations as do square-butt strips. Application begins with a roll roofing starter course or with inverted shingles. The first course starts with a full strip. The remaining courses begin alternately with a full strip minus ½ tab and then a full strip.

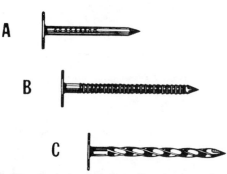

35–31. *Asphalt shingle nails: A. Smooth. B. Annular threaded. C. Screw threaded.*

35–32.
Asphalt Shingle Nail Lengths

Application	1″ Sheathing	⅜″ Plywood
Strip or Individual Shingle (new construction)	1¼″	⅞″
Over Asphalt Roofing (reroofing)	1½″	1″
Over Wood Shingles (reroofing)	1¾″	—

Fig. 35-34. Each course is applied so that the lower edge of the tabs is aligned with the top of the cutouts on the preceding course.

Two- and three-tab hexagonal shingles require four nails per strip located in a line 5¼″ above the exposed butt-edge and horizontally as follows:
● For two-tab shingles, one nail 1″ back from each end of the strip and one nail ¾″ back from each angle of the cutouts. Fig. 35-34a.
● For three-tab shingles, one nail 1″ back from each end and one nail centered above each cutout. Fig. 35-34b.

HIPS AND RIDGES

Hips and ridges may be finished by using hip and ridge shingles furnished by the manufacturer or by cutting pieces at least 9″ × 12″ either from 12″ × 36″ square-butt shingle strips or from mineral-surfaced roll roofing of a color to match the shingles. They are applied by bending each shingle lengthwise down the center with an equal amount on each side of the hip or ridge. Proper alignment can best be

35–33. *Shingle tabs are cemented down for wind protection.*

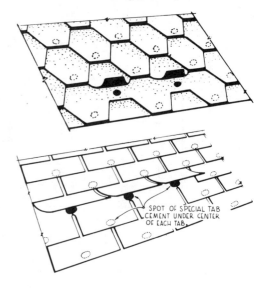

SPOT OF SPECIAL TAB CEMENT UNDER CENTER OF EACH TAB.

maintained by snapping a chalk line down one side of the ridge on which the edge of the shingle is aligned as it is nailed in place.

Apply the hip and ridge shingles by beginning at the bottom of a hip or one end of the ridge. Use a 5″ exposure. Each shingle is secured with one nail at each side 5½″ back from the exposed end and 1″ up from the edge. Fig. 35-35. When laying the shingles on the ridge, always lay the exposed edge away from the prevailing winds.

VALLEYS

Valley treatment may be open or closed. See pages 522-524 for details about shingle application at the valleys.

Strip Shingles on a Low-Pitch Roof

Square-tab strip shingles may be used on roof slopes less than 4 in 12 but not less than 2 in 12. Low-slope application methods require:

- Double underlayment.
- Cemented eaves flashing strip.

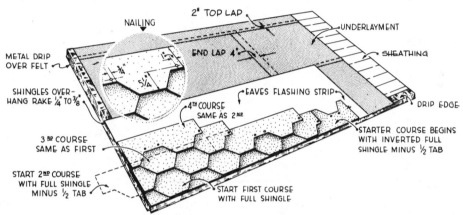

35–34a. *Laying two-tab hexagonal strip shingles.*

35–34b. *Laying three-tab hexagonal strip shingles.*

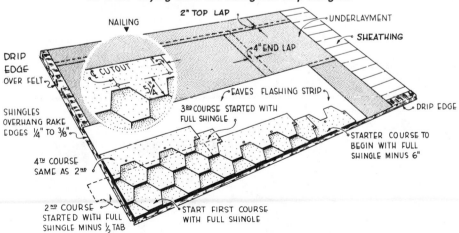

● Shingles provided with factory-applied adhesives. Or, each free tab of square-butt shingles should be cemented. Fig. 35-33. Application of strip shingles over double underlayment and cemented eaves flashing is shown in Fig. 35-36. Any shingle laying pattern described under normal slope application may be used.

Interlocking Shingles

Interlocking (lock-down) shingles are designed to provide resistance to strong winds. They have integral locking devices that vary in detail but which can be classified into five general groups. A, Fig. 35-37. Types 1, 2, 3, and 4 are individual shingles while Type 5 is a strip shingle usually having two tabs per strip. Interlocking shingles do not require use of adhesives although cement may occasionally be needed along rakes and eaves where the locking devices may have to be removed. The roof slope should not be less than the minimum specified by the shingle manufacturer. The individual shingles (Types 1 through 4) are intended for roof slopes of 4 in 12 and greater.

INSTALLING INTERLOCKING SHINGLES

Due to the number of designs available, the manufacturer's instructions should be studied carefully. Interlocking shingles are self-aligning but are sufficiently flexible to allow for a limited amount of adjustment to save time, especially on long roofs. It is recommended that the roofs be laid out with horizontal chalk lines to provide guides for the meeting, matching, and locking of courses above dormers and other projections through the roof. The locking devices should be engaged correctly. B, Fig. 35-37.

The proper location of nails is essential to the performance of the locking device. The shingle manufacturer's instructions will specify where the nail should be located to insure the best results.

The procedure for finishing hips and ridges is the same as for strip shingles. Fig. 35-35.

WOOD SHAKES

There are three types of wood shakes: handsplit-and-resawn, tapersplit, and straightsplit. Shakes are produced in three

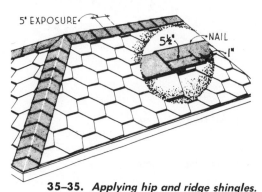

35–35. *Applying hip and ridge shingles.*

35–36. *Applying strip shingles on a low slope over a double underlayment.*

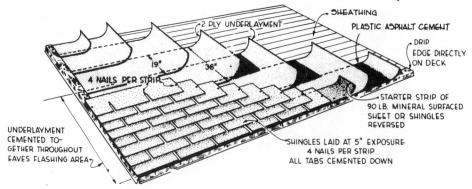

lengths: 18″, 24″, and 32″. Fig. 35-38a. The maximum exposure recommended for double coverage on a roof is 13″ for 32″ shakes, 10″ for 24″ shakes, and 7½″ for 18″ shakes. A triple coverage roof can be achieved by reducing these exposures to 10″ for 32″ shakes, 7½″ for 24″ shakes, and 5½″ for 18″ shakes. Fig. 35-38b.

Shakes are recommended on slopes of 4 in 12 or steeper. By taking special precautions, installations may be made on slopes as low as 3 in 12. These precautions are:

• Reduce the exposure to provide triple coverage.
• Use solid sheathing with an underlayment of No. 30 asphalt-saturated felt applied over the entire roof with a No. 30 asphalt-saturated felt interlayment between each course.

Shakes may be applied over either spaced or solid sheathing depending on climate conditions. See "Roof Sheathing," Unit 34.

Eaves Flashing

In areas where the outside design temperature is 0° F or colder, or where there is a possibility of ice forming along the eaves and causing a backup of water, eaves flashing is recommended. Fig. 35-39. In these areas shakes should be applied over solid sheathing.

On slopes 4 in 12 or steeper, eaves flashing is formed by applying an additional course of No. 30 asphalt-saturated felt over the underlayment starting course at the eaves. The eaves flashing should extend up the roof to cover a point at least 24″ inside the exterior wall line of the

35–37. *Interlocking shingles: A. Various styles of interlocking shingle tabs. B. Methods of locking shingles.*

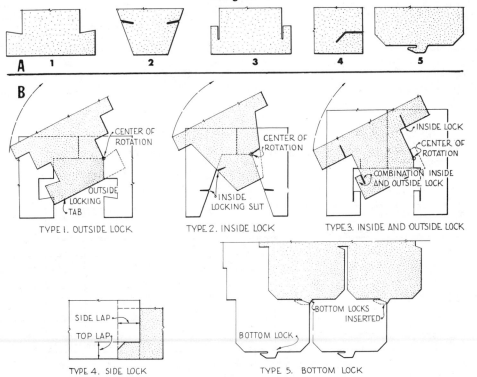

building. When the eave overhang requires flashing to be wider than 36″, the
necessary horizontal joint is cemented and
located outside the exterior wall line of the
building. A, Fig. 35-20.

For slopes 3 in 12 to 4 in 12 or in areas
subject to severe icing, eaves flashing
may be formed as described for 4 in 12
slopes except that a double layer of No. 30
asphalt-saturated felt underlayment is cemented together with a continuous layer of
plastic asphalt cement. Cement is also

applied to the 19″ underlying portion of
each succeeding course which lies within
the eaves flashing area, before applying
the next course of asphalt felt.

Drip Edge

Wood shakes should extend out over the
eave and rake a distance of 1″ to 1¹/₂″ to
form a drip. To align this drip edge, nail a
shingle at each end of the roof line and
allow it to project. Attach a chalk line to the
butt edge and pull it taut between the two

35–38a. *Types of red cedar shakes.*

Grade	Length and Thickness	Bundles Per Square	Weight (lbs. per square)		Description
No. 1 Handsplit & Resawn	18″ x ½″ to ¾″ 18″ x ¾″ to 1¼″ 24″ x ½″ to ¾″ 24″ x ¾″ to 1¼″ 32″ x ¾″ to 1¼″	4 5 4 5 6	220 250 280 350 450		 These shakes have split faces and sawn backs. Cedar blanks or boards are split from logs and then run diagonally through a bandsaw to produce two tapered shakes from each.
No. 1 Tapersplit	24″ x ½″ to ⅝″	4	260		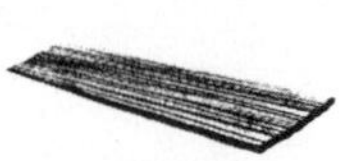 Produced largely by hand, using a sharp-bladed steel froe and a wooden mallet. The natural shingle-like taper is achieved by reversing the block, end-for-end, with each split.
No. 1 Straight-Split (Barn)	18″ x ⅜″ 24″ x ⅜″	5 5	200 260		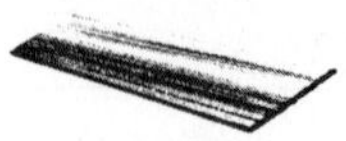 Produced in the same manner as taper-split shakes except that by splitting from the same end of the block, the shapes acquire the same thickness throughout.

35–38b.
Roof Coverage Of Shakes At Varying Weather Exposures

Length and Thickness	Type of Shake	No. of Bundles	Approximate Coverage (sq. ft.)								
			Weather Exposures								
			5½″	6½″	7″	7½″	8″	8½″	10″	11½″	13″
18″ x ½″ to ¾″	Handsplit-and-Resawn	4	55*	65	70	75**	—	—	—	—	—
18″ x ¾″ to 1¼″	Handsplit-and-Resawn	5	55*	65	70	75**	—	—	—	—	—
24″ x ½″ to ¾″	Handsplit-and-Resawn	4	—	65	70	75*	80	85	100**	—	—
24″ x ¾″ to 1¼″	Handsplit-and-Resawn	5	—	65	70	75*	80	85	100**	—	—
32″ x ¾″ to 1¼″	Handsplit-and-Resawn	6	—	—	—	—	—	—	100**	115	130**
24″ x ½″ to ⅝″	Tapersplit	4	—	65	70	75*	80	85	100**	—	—
18″ x ⅜″	Straight-Split	5	65*	—	—	—	—	—	—	—	—
24″ x ⅜″	Straight-Split	5	—	65	70	75*	—	—	—	—	—
15″ Starter-Finish Course			Use supplementary with shakes applied not over 10″ exposure								

*Recommended maximum weather exposure for 3-ply roof construction.
**Recommended maximum weather exposure for 2-ply roof construction.

shingles. The butt edge of each wood shingle applied as a drip edge can then be aligned to the chalk line.

Flashing

Unless special precautions are taken, copper flashing materials are not recommended for use with red cedar shakes. Fig. 35-40.

For valley flashing the *open method* (roofing, felt, and sheet metal) or the *closed method* (hand-fitted shakes) may be used. The open method is highly recommended for longer service life. Open valleys are first covered with a valley underlayment strip of No. 30 asphalt-saturated felt at least 20″ wide. The strip is centered in the valley and secured with enough nails to hold it in place. Metal flashing strips 20″ wide are then nailed over the underlayment. If the flashing is galvanized steel, it should be preferably 18-gauge, but not less than 26-gauge. Flashing that is center crimped and painted on both surfaces is preferred. The valley edges should be edge crimped to

provide an additional water stop. This is done by turning the edges up and back approximately 1/2″ toward the valley center line. The shakes laid to finish at the valley are trimmed parallel with the valley to form a 6″ wide gutter.

Closed valleys are first covered with a 1″ × 6″ wood strip. This strip is nailed flat into the saddle and covered with roofing felt as specified previously for open valleys. Shakes in each course are edge trimmed

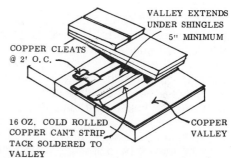

35–40. A special cant strip is available to minimize the surface contact of copper with red cedar.

35–39. Outside design temperatures.

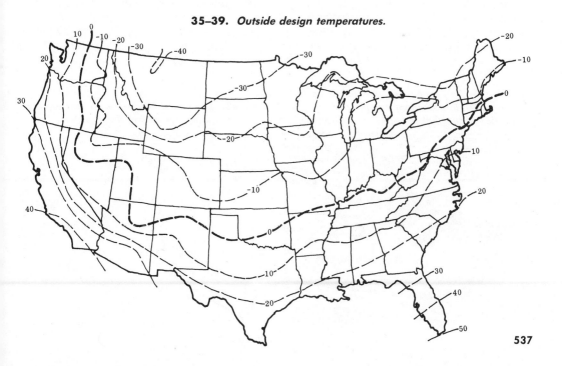

to fit into the valley, then laid across the valley with an undercourse of metal flashing having a 2″ headlap and extending 10″ under the shakes on each side of the saddle.

Application of Shakes

Apply a starter strip of No. 30 asphalt-saturated felt underlayment 36″ wide over the sheathing. The starter course of shakes at the eave line should be doubled, using an undercourse of 24″, 18″, or 15″ shakes. The latter is made expressly for this purpose.

After each course of shakes is applied, an 18″ wide strip of No. 30 asphalt-saturated felt interlayment should be applied over the top portion of the shakes extending onto the sheathing. Fig. 35-41. The bottom edge of the interlayment should be positioned at a distance above the butt edge of the shake equal to twice the exposure. For example, if 24″ shakes are being laid at 10″ exposure, the bottom edge of the felt should be positioned at a distance 20″ above the shake butts. The 18″ felt strip will then cover the top 4″ of the shakes and extend 14″ onto the sheathing.

Individual shakes should be spaced approximately ¼″ to ⅜″ apart to allow for possible expansion due to moisture ab-sorption. The joints between shakes should be offset at least 1½″ in adjacent courses. The joints in alternate courses also should be kept out of direct alignment when a three-ply roof is being built.

When straightsplit shakes, which are of equal thickness throughout, are applied, the froe-end of the shake (the smoother end from which it has been split) should be laid undermost. The application of shakes can be speeded by use of a shingler's or lather's hatchet. Fig. 35-42. The hatchet is used for nailing, carries an exposure gauge, and has a sharpened heel which can be used for trimming.

NAILING

Only two nails should be used to apply each shake, regardless of its width. The nails should be placed approximately 1″ in from each edge and from 1″ to 2″ above the butt line of the succeeding course. Nails should be driven until their heads meet the shake surface but no farther. Fig. 35-43.

Nail lengths typically required are 6d (2″) for shakes with ¾″ to 1¼″ butt thickness and 5d (1¾″) for shakes with ½″ to ¾″ butt thickness. Nails two-penny (2d) sizes larger should be used to apply hip and ridge units. The nail should always be long enough to penetrate through the sheathing.

35–41. *An 18″ wide strip of No. 30 asphalt-saturated felt interlayment is applied over the top portion of the shakes between each course.*

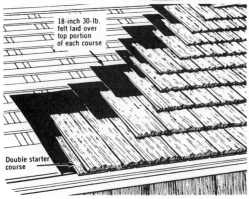

35–42. *Shingler's hatchet with an adjustable gauge for setting the exposure.*

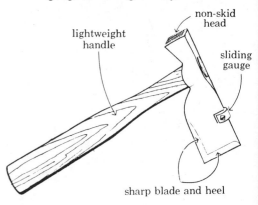

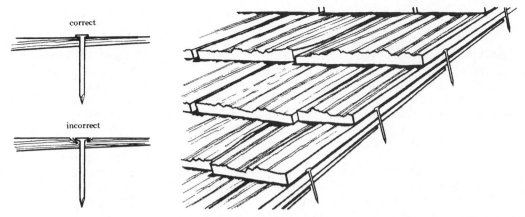

35–43. Nails should be driven flush with the surface of the wood shake but no farther. Place the nails from 1″ to 2″ above the butt line of the succeeding course.

HIPS AND RIDGES

The final shake course at the ridge line, as well as shakes that terminate at hips, should be secured with additional nails. This final shake course should also be composed of smoother textured shakes. A strip of No. 30 asphalt-saturated felt, at least 12″ wide, should be applied over the crown of all hips and ridges, with an equal exposure of 6″ on each side.

Prefabricated hip and ridge units can be used, or the hips and ridges can be cut and applied on the site. In site-construction of hips, shakes approximately 6″ wide are sorted out. Two wooden straight-edges are tacked on the roof 6″ from the center line of the hip, one on each side. The starting course of shakes should be doubled. The first shake on the hip is nailed in place with one edge resting against the guide strip. The edge of the shake projecting over the center of the hip is cut back on a bevel. The shake on the opposite side is then applied and the projecting edge cut back to fit. Shakes in the following courses are applied alternately in reverse order. Fig. 35-44.

Ridges are constructed in a similar manner. Exposure of the hip and ridge shakes normally is the same as the shakes on the roof. Ridge shakes are laid along an unbroken ridge that terminates in a gable at each end. They should be started at each gable end and terminate in the middle of the ridge. At that point, a small saddle is face-nailed to splice the two lines. The first

35–44. Hip and ridge construction using wood shakes.

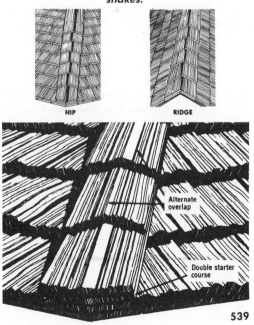

HIP   RIDGE

Alternate overlap

Double starter course

course of shakes should always be doubled at each end of the ridge.

GABLE RAKES

Dripping water may be eliminated by inserting a single strip of bevel siding the full length of each gable rake with the thick edge flush with the sheathing edge. The inward pitch of the roof surface will then divert the water away from the gable edge.

ALTERNATE APPLICATION METHODS

• **Graduated exposures.** By reducing the exposure of each course from eaves to ridge, a variation in roof appearance may be achieved with handsplit shakes. This requires shakes of several lengths. Fig. 35-45.

• **Staggered lines.** Irregular and random roof patterns can be achieved by laying shakes with butts placed slightly above or below the horizontal lines governing each course. Fig. 35-46. For an extremely irregular pattern, longer shakes may be interspersed over the roof with their butts several inches lower than the course lines.

WOOD SHINGLES

Wood shingles are manufactured in 24", 18", and 16" lengths conforming to three grades: #1, #2, and #3. Fig. 35-47. Preformed, factory-built hip and ridge units are available.

The exposure of wood shingles is dependent on the slope of the roof. Standard exposures of 5", 5$\frac{1}{2}$", and 7$\frac{1}{2}$" for shingle lengths of 16", 18", and 24" respectively are used on slopes 5 in 12 or greater. On 4-in-12 slopes, exposures should be reduced to 4$\frac{1}{2}$", 5", and 6$\frac{3}{4}$" for 16", 18", and 24" shingles respectively. On 3-in-12 slopes they should be reduced further to 3$\frac{3}{4}$", 4$\frac{1}{4}$", 5$\frac{3}{4}$" for 16", 18", and 24" singles respectively. This will assure four layers of shingles over the roof area. Fig. 35-48. Wood shingles are not recommended on slopes less than 3 in 12.

Underlayment

Underlayment is not usually used between shingles on spaced or solid sheathing. However, it may be desirable for the protection of sheathing and to insure against air infiltration. For underlayment, No. 15 asphalt-saturated felt may be used.

Drip Edge

Wood shingles should extend out over the eave and rake a distance of 1" to 1$\frac{1}{2}$" to form a drip. This edge is aligned in the same manner as described for wood shakes.

Eaves Flashing

In areas where the outside design temperature is 0° F or colder, or where there is possibility of ice forming along the eaves and causing a backup of water, eaves flashing is recommended. Fig. 35-39. Sheathing should be applied solidly above the eave line to cover a point at least 24" inside the interior wall line of the building. Fig. 35-20.

35–45. *A roof with graduated exposure is created by using all three shake sizes.*

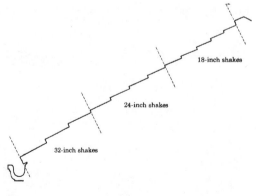

18-inch shakes

24-inch shakes

32-inch shakes

35–46. *Irregular roof patterns can be made by laying the shakes with butts slightly above or below the horizontal lines.*

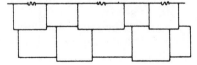

For 4-in-12 slopes, the eaves flashing is formed by applying a double layer of No. 15 asphalt-saturated felt to cover this section of solid sheathing. When the eave overhang requires the flashing to be wider than 36", the necessary horizontal joint between the felt strips is cemented and located outside the exterior wall line.

For slopes from 3 in 12 up to 4 in 12, or in areas subject to severe icing, eaves flashing may be formed as described 4-in-12 slopes, except that the double layer of No. 15 asphalt-saturated underlayment is cemented. The eaves flashing is formed by applying a continuous layer of plastic asphalt cement, at the rate of 2 gals. per 100 sq. ft., to the surface of the underlayment starter course before the second layer of underlayment is applied.

Cement is also applied to the 19" underlying portion of each succeeding course which lies within the eaves flashing area, before placing the next course. It is important to apply the cement uniformly with a comb trowel, so that at no point does underlayment touch underlayment when the application is completed. The overlying sheet is pressed firmly into the entire cemented area. Fig. 35-21.

Flashing

If copper flashing is used with wood shingles, take special precautions. Fig. 35-40. Premature deterioration of the copper may occur when the metal and wood are in intimate contact in the presence of moisture.

Only the open method should be used to construct valley flashing. A closed valley is not recommended.

On slopes up to 12 in 12, metal valley sheets should be wide enough to extend at

35–47. Types of red cedar shingles.

Grade	Size	Bundles Per Square	Weight Per Square		Description
No. 1	24″	4 bdls.	192 lbs.		The premium grade of shingles for roofs and side-walls. These shingles are 100% heartwood, 100% clear and 100% edge-grain.
	18″	4 bdls.	158 lbs.		
	16″	4 bdls.	144 lbs.		
No. 2	24″	4 bdls.	192 lbs.		A good grade for all applications. Not less than 10″ clear on 16″ shingles, 11″ clear on 18″ shingles and 16″ clear on 24″ shingles. Flat grain and limited sapwood are permitted.
	18″	4 bdls.	158 lbs.		
	16″	4 bdls.	144 lbs.		
No. 3	24″	4 bdls.	192 lbs.		A utility grade for economy applications and secondary buildings. Guaranteed 6″ clear on 16″ and 18″ shingles, 10″ clear on 24″ shingles.
	18″	4 bdls.	158 lbs.		
	16″	4 bdls.	144 lbs.		

35–48. Roof Coverage Of Wood Shingles At Varying Exposures

Length and Thickness*	Approximate Coverage (sq. ft.) of Four Bundles									
	Weather Exposures									
	3½″	4″	4½″	5″	5½″	6″	6½″	7″	7½″
16′ x 5/2″	70	80	90	100**	——	——	——	——	——
18″ x 5/2¼″	——	72½	81½	90½	100**	——	——	——	——
24″ x 4/2″	——	——	——	——	——	80	86½	93	100**

*Sum of the thickness e.g. 5/2″ means 5 butts = 2″
**Maximum exposure recommended for roofing

least 10″ on each side of the valley center line. Fig. 35-49. On roofs of steeper slope, narrower sheets may be used extending on each side of the valley center line for a distance of at least 7″. The open portion of the valley should be at least 4″ wide, but valleys may begin 2″ wide and increase at the rate of $1/2$″ per 8′ of length as they descend.

In areas where the outside design temperatures is 0° F or colder, underlayment should be installed under metal valley sheets.

Application of Wood Shingles

The first course of shingles at the eaves should be doubled or tripled. It should project 1″ to $1^1/2$″ beyond the eaves to provide a drip.

The second layer of shingles in the first course should be nailed over the first layer to provide a minimum sidelap of at least $1^1/2$″ between joints. Fig. 35-50. If possible, joints should be "broken" by a greater margin. A triple layer of shingles in the first course provides additional insurance against leaks at the cornice. No. 3 grade shingles frequently are used for the starter course.

Shingles should be spaced at least $1/4$″ apart to provide for expansion. Joints between shingles in any course should be separated not less than $1^1/2$″ from joints in the adjacent course above or below. Joints in alternate courses should not be in direct alignment. When shingles are laid with the recommended exposure, triple coverage of the roof results. Fig. 35-50.

35–49. *Open valley flashing construction with wood shingles.*

On Roofs Flatter than Half Pitch, Valley Sheets should extend at least 10″ from Valley Center

On Half Pitch and steeper, Valley Sheets should extend at least 7″ from Valley Center

When the roof terminates in a valley, the shingles for the valley should be carefully cut to the proper miter at the exposed butts. These shingles should be nailed in place first so that the direction of shingle application is away from the valley. This permits valley shingles to be carefully selected and insures shingle joints will not break over the valley flashing.

NAILING

To insure that shingles will lie flat and give maximum service, only two nails should be used to secure each shingle. Nails should be placed not more than 3/4″ from the side edge of shingles, at a distance of not more than 1″ above the exposure line. Fig. 35-50. Nails should be driven flush but not so that the nailhead crushes the wood. Fig. 35-43. The recom- mended nail sizes for the application of wood shingles are shown in Fig. 35-51. As with shakes, the application of wood shingles can be speeded by the use of a shingler's or lather's hatchet. Fig. 35-42.

35–51.
Nail Sizes Recommended
For Application Of
Wood Shingles

Size	Length	Gauge	Head	Shingles
3d*	1¼″	14½	7/32″	16″ & 18″
4d*	1½″	14	7/32″	24″
5d**	1¾″	14	7/32″	16″ & 18″
6d**	2″	13	7/32″	24″

*3d and 4d nails are used for new construction.
**5d and 6d nails are used for reroofing.

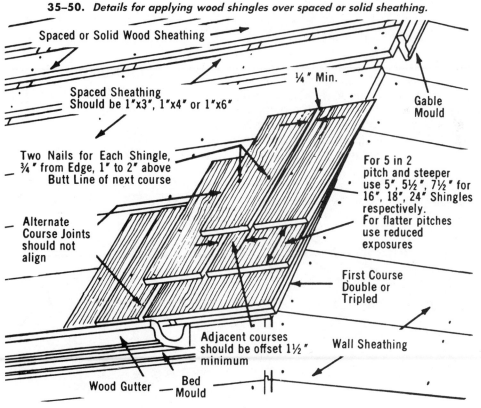

35–50. *Details for applying wood shingles over spaced or solid sheathing.*

Spaced or Solid Wood Sheathing

Spaced Sheathing
Should be 1″x3″, 1″x4″ or 1″x6″

¼″ Min.

Gable Mould

Two Nails for Each Shingle, ¾″ from Edge, 1″ to 2″ above Butt Line of next course

For 5 in 2 pitch and steeper use 5″, 5½″, 7½″ for 16″, 18″, 24″ Shingles respectively. For flatter pitches use reduced exposures

Alternate Course Joints should not align

First Course Double or Tripled

Adjacent courses should be offset 1½″ minimum

Wall Sheathing

Wood Gutter — Bed Mould

HIPS AND RIDGES

Hips and ridges should be of the modified "Boston" type with protected nailing. Fig. 35-52a. Nails at least two sizes larger than the nails used to apply the shingles are required.

Hips and ridges should begin with a double starter course. Either site-applied or preformed factory-constructed hip and ridge units may be used. Fig. 35-52b.

GABLE RAKES

Shingles should project 1" to 1½" over the rake. End shingles may be canted to eliminate drips as discussed in the application of wood shakes.

ALTERNATE APPLICATION METHODS

Several alternate methods of applying shingles, giving a different appearance to the roof, are illustrated in Fig. 35-53.

ROLL ROOFING

When economy is a factor in construction, the use of mineral-surfaced roll roofing might be considered. While this type of roofing will not be as attractive as an asphalt shingle roof and perhaps not as durable, it may cost up to 15% less than standard asphalt shingles. Roll roofing is excellent over old roofs as well as for new decks. The 19" selvage, double coverage rolls (65 pounds minimum weight with a

mineral surface) are designed for flat decks with a slope of 1" or more per foot.

Roll roofing should be installed over a double coverage underlay. First nail the metal drip edges at eaves and rakes. Use the 19" selvage cut, trimmed out of a full sheet, as a starter strip. Save the remaining 17" strip for the last course. Nail the strip so that it overhangs eaves and rakes ¼" to ⅜". With galvanized roofing nails, apply the first course. Fig. 35-54. Apply the second and following courses with a full 19" overlap, leaving just the mineral surface exposed.

Next, lift the mineral surface of each course and apply a quick-setting lap cement to the underlying sheet to within ¼" of the exposed edge. Apply firm pressure over the entire cemented area, using a light roller or broom.

After you have finished the whole roof, check for any loose laps, and re-cement to insure complete bond. The ridge can be finished with a Boston-type covering or by 12" wide strips of the roll roofing, using at least 6" on each side.

BUILT-UP ROOFS

Built-up roof coverings are installed by roofing companies that specialize in this

35–52b. *Wood shingle hip and ridge construction.*

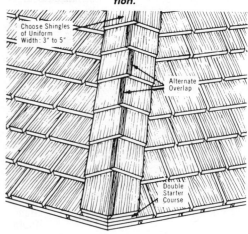

35–52a. *Boston ridge with wood shingles.*

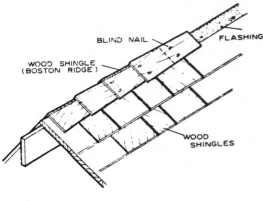

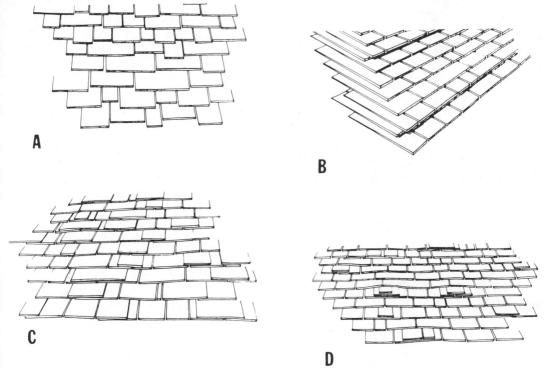

A

B

C

D

35–53. *Alternate application methods for wood shingles: A. Thatch. Shingles are positioned above and below a hypothetical course line, with deviation from the line not to exceed 1". B. Serrated. Courses are doubled every 3rd, 4th, 5th, or 6th course. Doubled courses can be laid butt-edge flush or with a slight overhang. C. Dutch weave. Shingles are doubled or superimposed at random throughout the roof area. D. Pyramid. Two extra shingles, narrow shingle over a wide one, are superimposed at random.*

35–54. *Application details for roll roofing.*

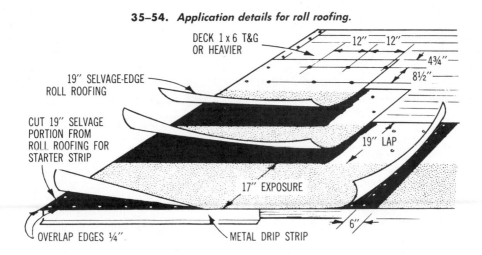

DECK 1 x 6 T&G
OR HEAVIER

19" SELVAGE-EDGE
ROLL ROOFING

CUT 19" SELVAGE
PORTION FROM
ROLL ROOFING FOR
STARTER STRIP

12" 12"

4¾"

8½"

19" LAP

17" EXPOSURE

OVERLAP EDGES ¼".

METAL DRIP STRIP

6"

35-55. *Preparing a roof deck for a built-up roof.*

field. Fig. 35-55. Roofs of this type may have three, four, or five layers of roofer's felt, each mopped down with tar or asphalt. The final surface may be coated with asphalt and covered with gravel embedded in asphalt or tar, or covered with a cap sheet. For convenience, it is customary to refer to built-up roofs as 10-year, 15-year, or 20-year roofs, depending upon the method of application. Fig. 35-56.

For example, a 15-year roof over a wood deck may have a base layer of 30-pound saturated roofer's felt laid dry, with edges lapped and held down with roofing nails. All nailing is done with either roofing nails driven through flat tin caps or with 10-gauge roofing nails having heads of not less than 5/8″ diameter. The dry sheet is intended to prevent tar or asphalt from entering the rafter spaces. Three layers of 15-pound saturated felt follow, each of which is mopped on with hot tar rather than being nailed. The final coat of tar or as-

phalt may be covered with roofing gravel or a cap sheet of roll roofing. Fig. 35-57.

The cornice or eave line of projecting roofs is usually finished with metal edging or flashing, which acts as a drip edge. A metal gravel strip is used in conjunction with the flashing at the eaves when the roof is covered with gravel. Fig. 35-58. Where built-up roofing is finished against another wall, the roofing is turned up on the wall sheathing over a cant strip and is flashed with metal. Fig. 35-59. This flashing is generally extended up about 4″ above the bottom of the siding.

METAL ROOFING

Corrugated metal roofing comes in widths up to 4′ and lengths up to 24′ and therefore covers large areas quickly. It is ideal for utility buildings such as garages and sheds. It can be used on slopes as low as 4 in 12, or as low as 3 in 12 if a single panel will cover from eave to ridge.

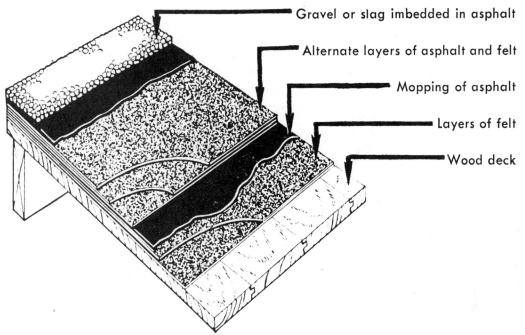

35–56. *Cross section of a 20-year, 5-ply built-up roof.*

35–57. *Application detail for a built-up roof.*

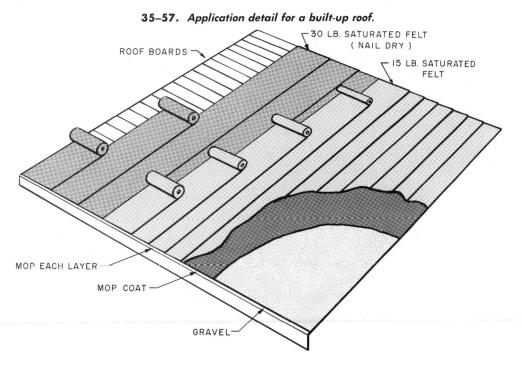

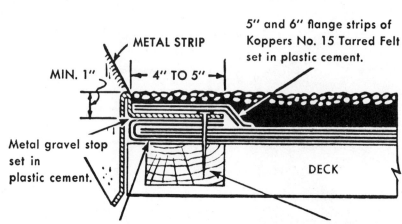

METAL STRIP

5" and 6" flange strips of Koppers No. 15 Tarred Felt set in plastic cement.

MIN. 1"

4" TO 5"

Metal gravel stop set in plastic cement.

DECK

Two lower plies of roofing felt are turned up over top plies of felt and run back on deck at least five inches.

Nails spaced 1" from edge of metal strip and 3" apart. Creosoted wood nailing strip furnished when deck does not permit nailing.

35–58a. *A cross section of a built-up roof without insulation showing the metal gravel strip at the roof edge.*

35–58b. *A cross section of a built-up roof with insulation showing the metal gravel strip at the roof edge.*

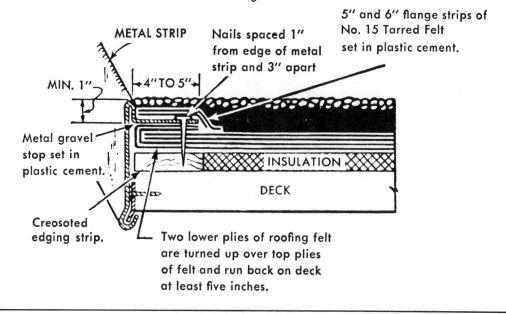

METAL STRIP

Nails spaced 1" from edge of metal strip and 3" apart

5" and 6" flange strips of No. 15 Tarred Felt set in plastic cement.

MIN. 1"

4" TO 5"

Metal gravel stop set in plastic cement.

INSULATION

DECK

Creosoted edging strip.

Two lower plies of roofing felt are turned up over top plies of felt and run back on deck at least five inches.

Apply purlins across the roof rafters to support the roofing, spacing according to roofing manufacturer's instructions. No sheathing is required. Set closure strips (they come with the roofing) at eave. Cut panels to length with tin snips, the length being the dimension from ridge to eave plus 2½″. Nail through tops of ribs, using 1¾″ screwshank nails with neoprene washers to prevent leaks. Next, set closure strips (inverted) at ridge and apply metal cap. Finally, flash at eaves as shown in Fig. 35-60.

Sheet-metal roofing must be laid over sheathing. The joints should be watertight, and the deck should be properly flashed where it joins with a wall. Nails should be of the same metal as that used on the roof,

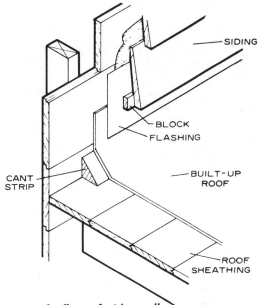

35–59. *The intersection of a flat roof with a wall showing the cant strip and flashing.*

35–60a. *Installation details for corrugated roofing.*

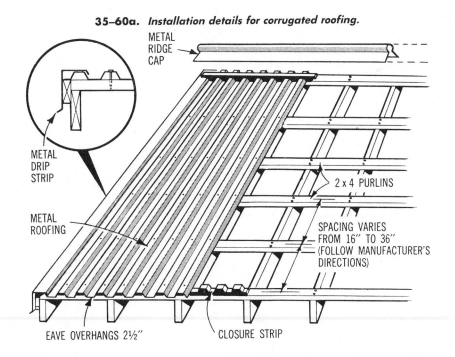

35–60b. *Installing corrugated aluminum roofing.*

except with tin roofs, where steel nails may be used. All exposed nailheads on tin roofs should be soldered.

GUTTERS AND DOWNSPOUTS

Various types of gutters are on the market. Fig. 35-61. The two general types are the formed metal and the half-round. Fig. 35-62a. Downspouts or leaders are rectangular or round, with the round leader being ordinarily used with the half-round gutter. Both the round and the rectangular leaders are usually corrugated for added strength. Fig. 35-62b. The corrugated patterns are less likely to burst when plugged with ice.

35–61. *Parts of a formed metal gutter system.*

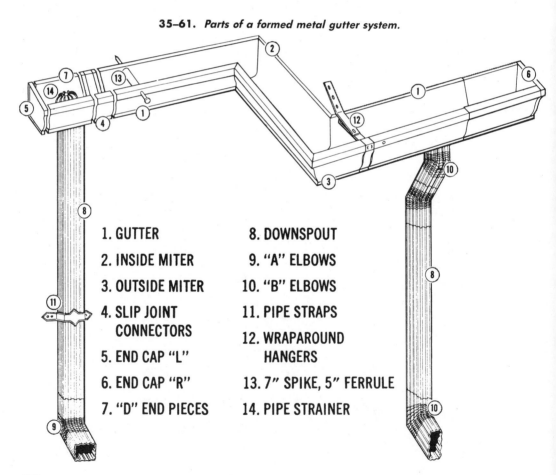

1. GUTTER
2. INSIDE MITER
3. OUTSIDE MITER
4. SLIP JOINT CONNECTORS
5. END CAP "L"
6. END CAP "R"
7. "D" END PIECES
8. DOWNSPOUT
9. "A" ELBOWS
10. "B" ELBOWS
11. PIPE STRAPS
12. WRAPAROUND HANGERS
13. 7" SPIKE, 5" FERRULE
14. PIPE STRAINER

Wood gutters may be used in place of metal gutters and are usually fastened by means of rust-resistant screws or nails. Fig. 35-63. Nailing or spacing blocks are placed between the gutter and the fascia or frieze board about 16" on center. Wood gutters are given very little pitch because they usually are part of the architectural treatment. Joints in wood gutters are best made by dowels or splines. The joints are covered with heavy fabric tacked in place and covered with a mastic. It is recommended that 1 or 2 coats of water repellent containing a preservative be applied to the bare wood.

Hanging metal gutters are held with flat or wire metal hangers that are so installed that a pitch is formed for drainage. Fig. 35-64. Joints in metal gutters and downspouts should be soldered. Gutters should be mounted so that the shingle extension is over the center of the gutter. Hangers should be spaced 3' to 4' on center.

Another type of formed metal gutter has an extension or flashing strip that is fastened to the roof boards. Fig. 35-65. These gutters are usually made so that the back varies in height to allow a pitch for drainage. Hangers are used at the outer rim to add stiffness.

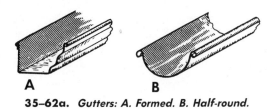

35-62a. *Gutters: A. Formed. B. Half-round.*

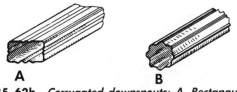

35-62b. *Corrugated downspouts: A. Rectangular. B. Round.*

35-63. *A wood gutter.*

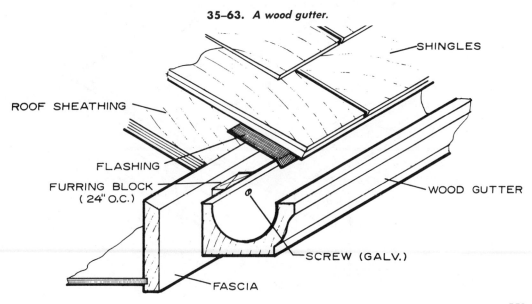

SHINGLES

ROOF SHEATHING

FLASHING

FURRING BLOCK (24" O.C.)

WOOD GUTTER

SCREW (GALV.)

FASCIA

Metal gutters are placed to drain toward the downspouts. With a slope of approximately 1" in 10', the maximum run between the high point and the downspout should not ordinarily exceed 25'. A gooseneck is used to bring the downspout in line with the wall. The form of this gooseneck will vary according to the extent of the cornice overhang.

Downspouts are fastened to the wall by means of leader straps or hooks. Fig. 35-66. Many patterns of straps are made to allow a space between the wall and the downspout. A minimum of two straps should be used in an 8' length of leader: one at the gooseneck and one at the bottom elbow. The elbow is used to lead the water to a splash block that carries the water away from the foundation. The minimum length of the splash block should be 3'. It is the practice, in some areas, to carry the water to a storm sewer by means of tile lines. Fig. 35-67. In final grading the slope should be such as to insure positive drainage of water away from the foundation walls.

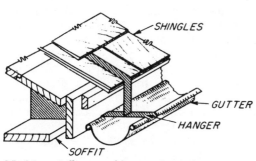

35–64a. *A flat metal hanger is used to support this half-round gutter.*

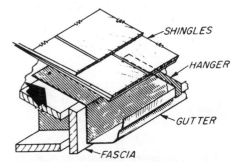

35–65. *This formed metal gutter has an extension which also serves as a flashing strip. It is used to fasten the gutter to the roof boards.*

35–64b. *A flat metal hanger used to support a formed gutter.*

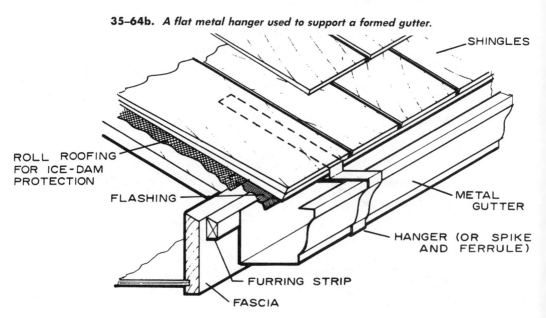

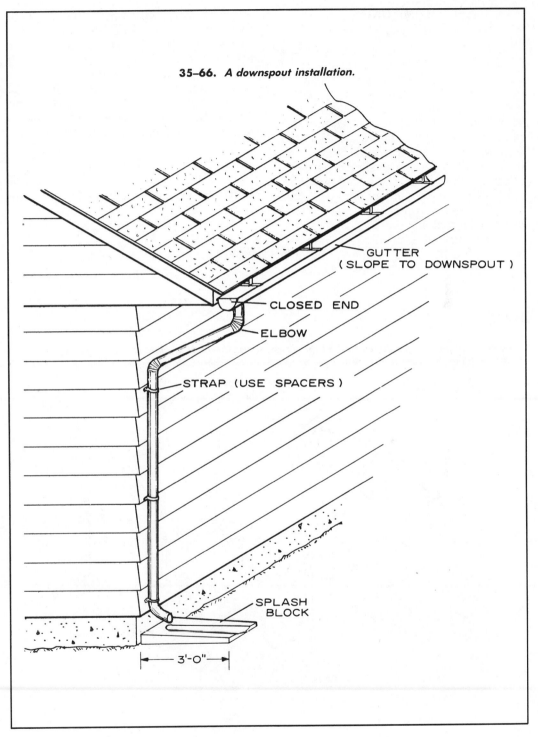

35–66. *A downspout installation.*

GUTTER
(SLOPE TO DOWNSPOUT)

CLOSED END

ELBOW

STRAP (USE SPACERS)

SPLASH
BLOCK

3'-0"

ESTIMATING

Materials

To determine the number of shingles needed, the total area to be covered must be figured first. To figure the roof area without actually getting on the roof to measure the roof line, use the chart in Fig. 35-68. When the roof slope is known, this chart may be used to figure the roof area from the plan. The plan area of the roofline (including the overhang) should be multiplied by the factor shown in the table opposite the rise of the roof, which is given in inches per horizontal foot. The result will be the total roof area.

For example, if a home is 70′ long and 30′ wide including the overhang, the area is 2100 square feet. If the rise of the roof is 5½″, multiply the area by 1.100 for a total roof area of 2310 square feet. Use this total roof area for figuring the amounts of roofing materials, such as the felt underlayment or shingles, that will be needed.

One square of shingles covers 100 square feet of roof surface. To determine the number of squares needed to cover the roof, divide the total area by 100.

$$\frac{2310}{100} \times 23.1 \text{ squares of shingles}$$

A 10% waste and cutting factor must be added to this amount.

$$23.1 \times 0.10 = 2.31$$

$$23.1 + 2.31 = 25.41, \text{ or 25 squares of shingles}$$

35-67. *All gutters are sloped toward the downspouts. The final grading should slope away from the building to insure proper drainage. Sometimes downspouts are connected to a storm sewer.*

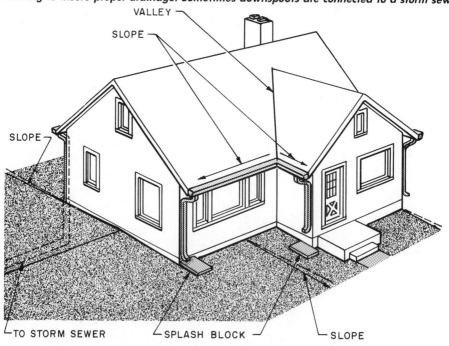

VALLEY

SLOPE

SLOPE

SLOPE

TO STORM SEWER   SPLASH BLOCK   SLOPE

Another method of figuring the area of a plain gable roof is to multiply the length of the ridge by the length of a rafter. This will give you one-half the roof area. Multiply by 2 to obtain the total square feet of roof surface.

A similar method may be used to find the area of a hip roof. Multiply the length of the eaves by ½ the length of the common rafter at the end. Multiply this by 2 to obtain the area of both ends. To find the area of the sides, add the length of the eave to the length of the ridge and divide by 2. Multiply this by the length of the common rafter to obtain the area of one side of the roof. Multiply by 2 to find the number of square feet on both sides of the roof. Add this to the area of the two ends and divide the total area by 100 to get the number of squares.

The area of a plain hip roof running to a point at the top is obtained by multiplying the length of the eaves at one end by one-half the length of the rafter. This gives the area of one end of the roof. To obtain the total area, multiply by 4.

35–68. *When a roof area has to be figured from a plan only and the roof slope is known, the roof area may be computed from this table.*

Determining Roof Area from a Plan

Rise	Factor	Rise	Factor
3″	1.031	8″	1.202
3½″	1.042	8½″	1.225
4″	1.054	9″	1.250
4½″	1.068	9½″	1.275
5″	1.083	10″	1.302
5½″	1.100	10½″	1.329
6″	1.118	11″	1.357
6½″	1.137	11½″	1.385
7″	1.158	12″	1.414
7½″	1.179		

Be sure to subtract for openings in the roof such as the chimney, and where the dormer intersects the roof line so that the dormer area is not figured twice. Quantities of starter strips, eaves flashings, valley flashings, and ridge shingles all depend upon linear measurements along the hips, rakes, valleys, eaves, and ridge. Measurements for horizontal elements can be taken directly off the roof plan. The rakes, hips, and valleys run on a slope. The actual length of rakes, hips, and valleys must therefore be measured on the roof. The length of the rafters can also be figured as described in Unit 28.

The number of nails needed for asphalt roofing can be determined from the chart in Fig. 35-69. For the example, 25 squares of three-tab square-butt shingles are required to cover the roof area. Read down the chart from the heading "Pounds per Square" to the line "3 Tab Square-Butt on New Deck." Using 11-gauge nails, 1.44 pounds are required for each square. The total number of pounds needed then would be 36.

25 (squares of shingles) × 1.44 (lbs. per square) = 36 lbs. of nails.

The number of nails required for wood shingles can be determined from the information on the chart in Fig. 35-70.

Labor

To figure the labor for installing the roof in the example, refer to Fig. 35-69. Read down the column headed "Labor Hours per Square" to the line "3 Tab Square-Butt on New Deck." Each square requires 1½ hours. Twenty-five squares would require 37½ hours.

25 (squares to be laid) × 1.5 (hours per square) = 37.5 hours

To figure the labor for laying wood shingles, use the information from the chart in Fig. 35-70.

35–69.
Nail Requirements For Asphalt Roofing Products

Type of Roofing	Shingles per Sq.	Nails per Shingle	Length of Nail*	Nails per Square	Pounds per Square (approximate)		Labor Hours per Square
					12 ga. by 7/16" head	11 ga. by 7/16" head	
Roll Roofing on new deck			1"	252**	0.73	1.12	1
Roll Roofing over old roof'g			1¾"	252**	1.13	1.78	1¼
19" Selvage over old shing.			1¾"	181	0.83	1.07	1
3 Tab Sq. Butt on new deck	80	4	1¼"	336	1.22	1.44	1½
3 Tab Sq. Butt reroofing	80	4	1¾"	504	2.38	3.01	1⁵/₆
Hex Strip on new deck	86	4	1¼"	361	1.28	1.68	1½
Hex Strip reroofing	86	4	1¾"	361	1.65	2.03	2
Giant Amer.	226	2	1¼"	479	1.79	2.27	2½
Giant Dutch Lap	113	2	1¼"	236	1.07	1.39	1½
Individ. Hex	82	2	1¾"	172	0.79	1.03	1½

(*)   Length of nail should always be sufficient to penetrate at least ¾" into sound wood. Nails should show little, if any, below underside of deck.
(**) This is the number of nails required when spaced 2" apart.

35–70. *Estimating wood shingles.*

WOOD SHINGLES	Material			Nails		Labor
	Per 100 Square Feet of Surface			per 100 Square Feet		Hours
Laid To Weather	Shingles per 100 Sq. Feet	Waste	Shingles per 100 Sq. Ft. with Waste	3d Nails	4d Nails	per 100 Square Feet
4"	900	10%	990	3¾ Pounds	6½ Pounds	3¾
5"	720	10%	792	3 Pounds	5¼ Pounds	3
6"	600	10%	660	2½ Pounds	4¼ Pounds	2½

Note: Nails based on using 2 nails per shingle. Increase time factor 25% for hip roofs.

QUESTIONS

1. List several materials used for covering pitched roofs.

2. List several materials which are considered roofing accessories.

3. What is the minimum slope on main roofs for the application of wood, asphalt, asbestos, and slate shingles?

4. What is the maximum slope recommended for built-up roofs?

5. Why is roof underlayment used under shingles?

6. What is flashing?

7. On roofs longer than 30', why are strip shingles started at the center and applied toward the ends?

8. List 3 variations for laying square-butt strip shingles.

9. When laying shingles on the ridge, in what direction should the exposed edge by laid?

10. What are the two methods used to construct valley flashing? Describe the difference.

11. What special precautions should be taken when laying strip shingles on a low-pitch roof?

12. What is the main advantage of interlocking shingles?

13. What is the difference between wood shakes and wood shingles?

14. What is considered the least expensive type of roofing?

15. How is the quality of a built-up roof designated?

16. What is the recommended slope for metal gutters?

17. Describe one method for determining the area of a plain gable roof. For determining the area of a plain hip roof.

Section VI

COMPLETING THE EXTERIOR

Roof Trim

As explained in Unit 28, the rafter-end overhangs of a roof are called the *eaves.* For example, on a hip roof, all four edges—the sides and ends—of the roof have eaves. A gable roof, however, has eaves only on the side-wall edges. The gable-end (end-wall) edges are called *rakes.* The exterior finish at and just below the eaves is called the *cornice.*

The cornice work may be done as soon as the roof has been framed. It may also, with the exception of the fascia, be done after the roofing has been applied. In most geographical areas, workers will put on the roof covering first to protect the structure from the weather. The rake molding on a gable roof, however, must be installed before the roofing.

TYPES OF CORNICES

The type of cornice required for a particular structure is shown on the wall sections of the house plans, and there are usually cornice detail drawings as well. Basically, there are three types of cornices:

- Close.
- Open.
- Box.

A roof with no rafter overhang normally has a close cornice. Fig. 36-1. This cornice consists of a single strip called a *frieze.* The frieze is beveled on its upper edge to fit close under the overhang of the eaves and rabbeted on its lower edge to overlap the upper edge of the top siding course. If trim is used, it usually consists of molding installed as shown in Fig. 36-1. Molding trim in this position is called *crown* or *shingle molding.*

A roof with a rafter overhang may have either an open cornice or a box cornice.

The simplest type of open cornice consists of only a frieze, which must be notched to fit around the rafters. Fig. 36-2. If trim is used, it usually consists of molding cut to fit between the rafters. Molding in this position is called *bed molding.*

Another type of open cornice consists of a frieze and a fascia. Fig. 36-3. A *fascia* is a strip nailed to the tail plumb cuts of the rafters. Shingle molding can be attached to the top of the fascia, but it is seldom used.

With a box cornice, the rafter overhang is entirely boxed in by the roof covering, the fascia, and a bottom strip called a *plancier,* or *soffit.* Figs. 36-4 (Page 562) and 36-5. The soffit can be nailed to the rafters. It can also be nailed to lookouts. The *lookouts* are a series of horizontal members which are nailed to the rafters and extend from the rafter ends to the face of the sheathing. Fig. 36-5b. The frieze, if any, is set just below the lookouts. If trim is used, it is placed at the intersection of the frieze and the soffit.

WOOD CORNICE CONSTRUCTION

If building paper is used on the sidewalls, the top course of paper must be applied before beginning work on the cornice. For an open or a box cornice the paper must be slit to fit around the rafters.

Open Cornice

One method of constructing an open cornice is to measure the distance between the rafter at either the top or bottom edge. Fig. 36-6 (Page 564). Cut material to this length, making sure that both ends are cut square. Do this for each rafter spacing. Nail the material in position, as shown at A in Fig. 36-7 (Page 565). The nails can be driven through the side of the rafter into the end of the block on one side. Nails will have to be toenailed on the other side, as the block installed previously gets in the way for nailing. Fig. 36-8.

If vents are needed, determine their location and bore the necessary holes. Sta-

ple or tack a piece of window screen on the back of the vent openings. Fig. 36-6.

If tongued and grooved material is to be used on the roof, nail the material on the rafters with the good surface down since it will be visible from below. Fig. 36-9. Remove the groove from the starter board. Bevel the edge if desired. Place the starter

36–1. *A close cornice.*

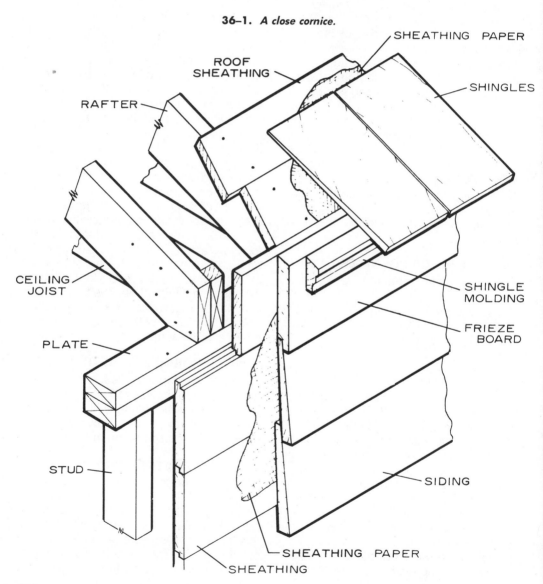

SHEATHING PAPER

ROOF SHEATHING

SHINGLES

RAFTER

CEILING JOIST

SHINGLE MOLDING

FRIEZE BOARD

PLATE

STUD

SIDING

SHEATHING PAPER

SHEATHING

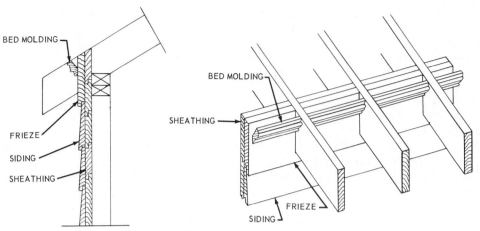

BED MOLDING

FRIEZE

SIDING

SHEATHING

BED MOLDING

SHEATHING

FRIEZE

SIDING

36–2. *Simplest type of open cornice.*

36–3a. *An open cornice with a fascia board.*

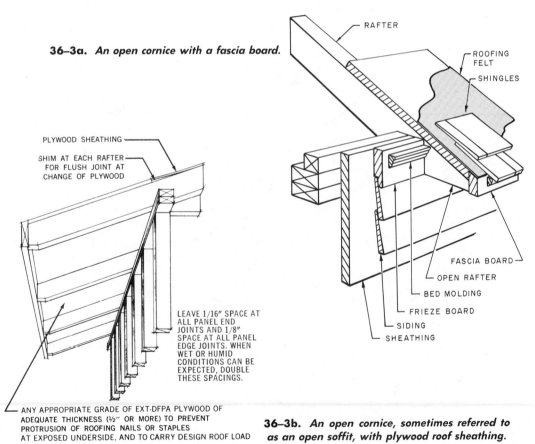

RAFTER

ROOFING FELT

SHINGLES

PLYWOOD SHEATHING

SHIM AT EACH RAFTER FOR FLUSH JOINT AT CHANGE OF PLYWOOD

LEAVE 1/16″ SPACE AT ALL PANEL END JOINTS AND 1/8″ SPACE AT ALL PANEL EDGE JOINTS. WHEN WET OR HUMID CONDITIONS CAN BE EXPECTED, DOUBLE THESE SPACINGS.

ANY APPROPRIATE GRADE OF EXT-DFPA PLYWOOD OF ADEQUATE THICKNESS (½″ OR MORE) TO PREVENT PROTRUSION OF ROOFING NAILS OR STAPLES AT EXPOSED UNDERSIDE, AND TO CARRY DESIGN ROOF LOAD

FASCIA BOARD

OPEN RAFTER

BED MOLDING

FRIEZE BOARD

SIDING

SHEATHING

36–3b. *An open cornice, sometimes referred to as an open soffit, with plywood roof sheathing.*

board in position along the top of the rafter tail and align it before nailing it in place. B, Fig. 36-7. The type of sheathing used above the blocking will depend on the kind of roof covering. Plywood or lumber sheathing may be chosen.

All joints in the construction of an open cornice should be planed smooth and fitted together tightly. All moldings must be mitered for joining on outside corners and mitered or coped (see p. 941) for joining on inside corners. Care should be taken in this type of cornice construction because the workmanship is readily visible from the ground.

Box Cornice

Before adding a box cornice, check the plumb cuts on the rafter tails to make certain they are all in line. This can be done by stretching a line along the top ends of the rafters from one corner of the building to the other. Many carpenters do not make the plumb cut on the rafter tails when the rafter is cut. Instead, the rafters are nailed in place with the tails projecting beyond the exterior wall at various lengths. Then the point at which the tails are to be

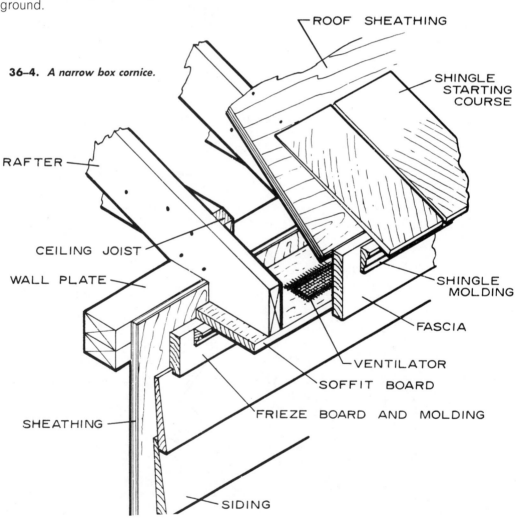

36-4. A narrow box cornice.

cut off is determined, and a chalk line is snapped along the top edge of the rafter tails. The plumb line is drawn down the side of each rafter from this line. Each rafter tail is then cut off along this plumb line. Fig. 36-10 (Page 566).

LOOKOUTS

1. Use a piece of 1″ × 4″ material to serve as a ledger and nail it temporarily against the exterior wall tight up under the rafters and aligned with the inside edge of the first rafter. Fig. 36-11. With a straight-edge against the side of the rafter, make a line on the surface of the ledger. Place an X on the side of the line away from the underside of the rafter to indicate the location of the lookout. Do this along the entire length of the building.

2. Determine the length of the lookouts. Measure on a level line from the plumb cut on the rafter tail to the wall of the building. Subtract ³/₄″ from this measurement to allow for the thickness of the nominal 1″ × 4″ ledger to which the lookouts will be nailed. Subtract another ³/₄″ to make sure that the lookouts do not project beyond the end of the rafters. Otherwise, if there is any deviation in the alignment of the exterior wall, such as a slight bow or crooked stud, the lookout may extend beyond the end of the rafter tail. This will interfere later with the proper installation and alignment of the fascia board.

3. After the lookouts have been cut to length, remove the ledger from its tempo-rary nailing and nail the lookouts to the ledger over the Xs. Nail through the back of

36–5a. *A box cornice with a sloping soffit.*

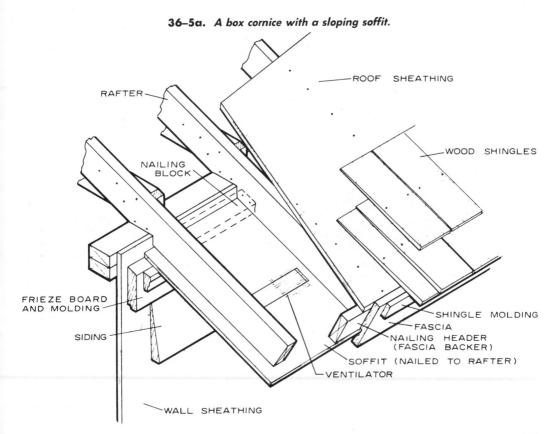

RAFTER

NAILING BLOCK

ROOF SHEATHING

WOOD SHINGLES

FRIEZE BOARD AND MOLDING

SIDING

SHINGLE MOLDING

FASCIA

NAILING HEADER (FASCIA BACKER)

SOFFIT (NAILED TO RAFTER)

VENTILATOR

WALL SHEATHING

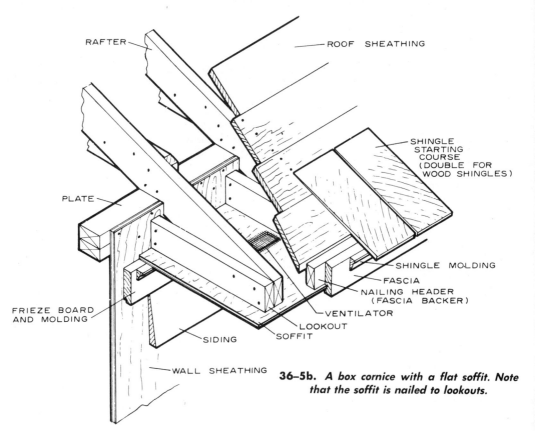

RAFTER

ROOF SHEATHING

SHINGLE STARTING COURSE (DOUBLE FOR WOOD SHINGLES)

PLATE

SHINGLE MOLDING

FASCIA

NAILING HEADER (FASCIA BACKER)

FRIEZE BOARD AND MOLDING

VENTILATOR

SIDING

LOOKOUT

SOFFIT

WALL SHEATHING

36–5b. *A box cornice with a flat soffit. Note that the soffit is nailed to lookouts.*

36–5c. *The box cornice on this flat roof overhang has a tapered soffit.*

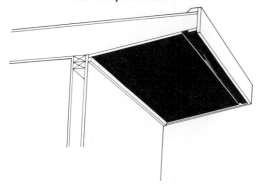

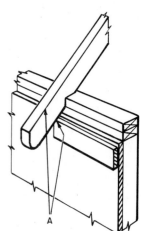

A

36–6. *Determining the length of blocking needed for an open cornice. Measuring may be done at either the top or bottom edge of the rafter, as shown by the arrows at A. For ventilation, the blocking should be drilled as shown at B and then screening stapled behind the holes as shown at C.*

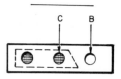

C B

564

the ledger into the end of the lookout with two 8d coated nails. Nail the last lookout into the end of the strip. Fig. 36-12. The lookouts may be made from either 2 × 4s or 1 × 4s. If 1 × 4s are used, place a 2 × 4 for additional nailing surface wherever the soffit pieces must be joined. Fig. 36-13.

4. Locate the position of the ledger on the exterior wall by leveling from the rafter tail in toward the wall and placing a mark on the sheathing. Point B, Fig. 36-14 (Page 568). Do this at each end of the building. Then snap a chalk line along the full length of the building on the sheathing.

36–8. *Nailing the blocking in place on an open cornice. Notice the vent holes which have been drilled in some of the blocks.*

36–7. *The location of the blocking used in the open cornice is determined by the bird's-mouth in the rafter as shown at A in both drawings. Notice that the roof board at B in the first drawing has been beveled on the edge to conform to the roof pitch.*

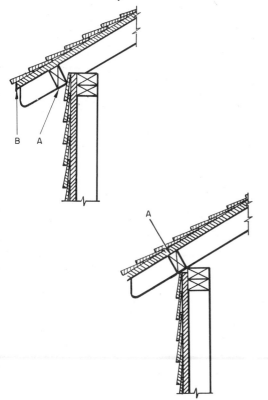

36–9. *This open cornice was extended to serve as a roof for the porch area. Notice that the roof board material needs to be carefully selected and applied because it is clearly visible from below.*

36-10. *Cutting off the rafter tails after the rafters have been nailed in place.*

5. Place bottom edge of ledger on this line. Nail it to studs through sheathing. Nail each lookout to side of rafter tail, except end lookout, which is nailed under rafter. Level each lookout as it is nailed.

FASCIA BOARD

Lay out, rip (if necessary), and groove the fascia board. The groove is made in the fascia board to receive the soffit. It should be cut about $3/8''$ up from the bottom edge of the fascia. This is done to provide a drip edge which prevents water from backing up into the groove.

Nail the fascia board in position along the ends of the rafter tails with the top of the groove even with the bottom edge of the lookouts. Fig. 36-15. If the fascia board must be spliced, it should be done with the joint on the end of a rafter tail, and the joint should be mitered. Fig. 36-16 (Page 570). The top edge of the fascia board may be beveled to the same angle as the pitch of the roof. If it is not, make certain that the fascia board is installed with its top outer edge in line with the top surface of the roof sheathing. Fig. 36-15. Make certain that the fascia is straight along its length. If necessary, straighten the fascia by driving

36-11. *Temporarily nail the ledger strip up under the rafters. Then mark the location of the lookouts.*

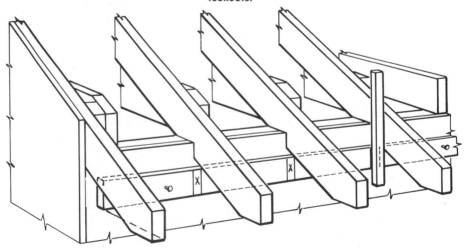

shims between the rafter tail ends and the inside of the fascia board.

SOFFIT

Several materials may be used for the soffit on a box cornice. Because of the popularity of wide overhangs, materials which are available in large sheets are frequently used. These include plywood, gypsum board, and hardboard.

Plywood. This is one of the most popular materials for a box cornice. It simplifies

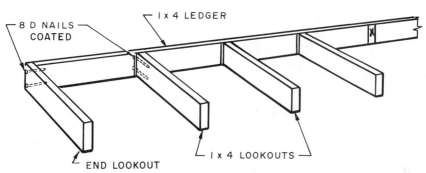

36–12. *The lookouts are nailed to the ledger strip next to the line and over the X made earlier. Note that the end lookout is nailed into the end of the ledger strip. This means that the end lookout has to be of the same thickness as the rafter and longer than the rest of the lookouts. It will have to be cut to fit under the rafter tail.*

36–13. *The soffit material must be supported wherever it is joined together. If possible, the joints are usually located under a lookout. However, rather than cut the soffit off, nail a 2 × 4 laid flat from the ledger strip to the fascia over the soffit joint.*

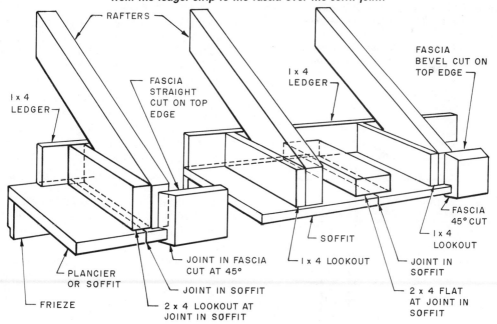

construction and presents a smooth, attractive surface. Plywood also has the advantage of matching other wood surfaces when a stained wood grain finish is desirable.

The recommended spans for box soffits are shown in Fig. 36-17. Exterior plywood should be used wherever the underside of the roof deck is exposed to the weather. Fig. 36-18.

To install plywood soffit, rip the soffit to width and slip the outer edge into the fascia groove. Then push the inside edge up against the lookouts and ledger strip. Nail the soffit securely to the ledger and to each lookout with 4d nails. The nails should be spaced about 6″ apart. If the soffit has to be made up of several pieces, join it under a lookout. If this is not possible, a 2 × 4 can be laid flat, toenailed into the ledger, and face-nailed through the fascia. The two pieces of soffit can then be joined under the center of the flat side of

36–14. *To locate the position of the ledger strip on the side wall of the building (point B), level a line in from the rafter tail as shown at A. The rafters are usually cut off before the ledger strip is located on the building. However, in some cases, point B is located on the building first. The cutoff line on the bottom of the rafter tail is then marked by leveling out from point B.*

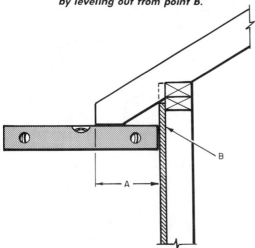

the 2 × 4. This will give adequate nail backing for the joint. Fig. 36-13.

Gypsum Board. Gypsum soffit board is a noncombustible product developed for use in soffits, carports, and similar installations where there is no direct exposure to the weather.

This soffit board has a water-resistant core and blue face paper. It is cut and scored instead of sawed, and the joints are taped and finished to provide a smooth surface. Gypsum soffit board is available in 8′ and 12′ lengths, ½″ thick and 4′ wide.

Gypsum soffit boards are installed in the same manner as regular wallboard over framing members spaced a maximum of 24″ on center. Space nails 7″ on center. Use trim around the edges that abut the building and fascia. The fascia must come at least ³/₈″ below the level of the soffit board to provide a drip edge. The round-edge joints should be prefilled with a joint compound. The application should then be taped and finished in the conventional manner. (See Unit 42 for detailed information about taping and finishing gypsum board.) Apply two coats of compound over the tape and three coats over the nail-heads. Provide adequate ventilation for the space above the soffit (see *FHA Minimum Property Standards*). Install control joints (to allow for expansion and contraction) at a maximum spacing of 40′. Fig. 36-19.

Hardboard. Hardboard panels are frequently used for soffits on the undersides of eaves and the ceilings of porches, breezeways, and carports. If the hardboard does not have a factory-applied primer, the panel should be conditioned prior to use. Follow the manufacturer's recommendations included in each product bundle. The soffit framing must provide continuous support at the edges, ends of panels, and joints.

The installation of hardboard soffits is similar to plywood soffit installation. Fasten the panels with 5d galvanized box, siding, or sinker nails. Space the nails 4″ on center around edges and approximate-

36–15. *The fascia board may be nailed to the ends of the rafters using either of the methods shown here. Bevel the top edge of the fascia board to conform to the roof pitch, as at A. Or rip the fascia board to width so that the outside top corner is in line with the top edge of the rafter, as at B. In either case, when the roof sheathing is applied it must lie flat. Notice also that the top edge of the groove in the fascia must be in line with the bottom edge of the lookout for proper installation of the soffit material.*

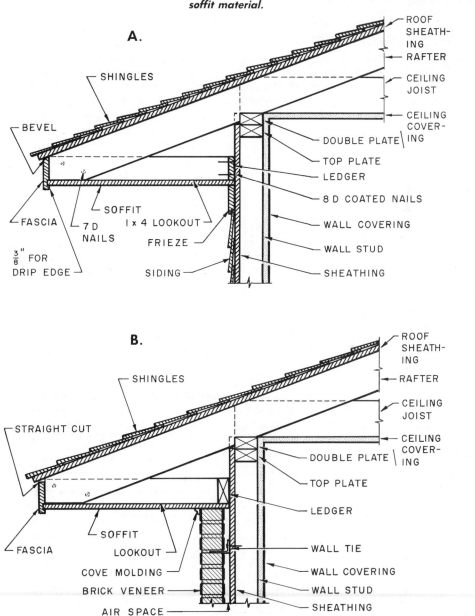

A.

ROOF SHEATHING
RAFTER
SHINGLES
CEILING JOIST
BEVEL
CEILING COVERING
DOUBLE PLATE
TOP PLATE
LEDGER
8 D COATED NAILS
SOFFIT
1 x 4 LOOKOUT
FASCIA
7 D NAILS
FRIEZE
WALL COVERING
WALL STUD
$\frac{3}{8}$" FOR DRIP EDGE
SIDING
SHEATHING

B.

ROOF SHEATHING
SHINGLES
RAFTER
CEILING JOIST
STRAIGHT CUT
CEILING COVERING
DOUBLE PLATE
TOP PLATE
LEDGER
SOFFIT
LOOKOUT
WALL TIE
FASCIA
COVE MOLDING
WALL COVERING
BRICK VENEER
WALL STUD
AIR SPACE
SHEATHING

ly 6" on center at intermediate supports. Never nail closer than ³/₈" to the edge. Fig. 36-20a. Metal moldings are available for use in the installation of hardboard soffit material. With these moldings, there are no exposed nailheads on soffits 2' or less in width. Fig. 36-20b.

METAL CORNICE CONSTRUCTION

Metal cornice material may be used for box cornices on most roofs, entryways, porches, and carport ceilings. This aluminum system requires little maintenance, is self-ventilating and self-supporting, and

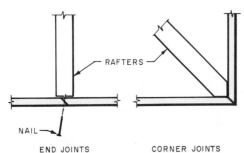

END JOINTS   CORNER JOINTS

36–16. *On a hip roof the fascia is mitered at the corner on the end of the hip rafter. On any roof where the fascia must be joined, join it on a rafter end. Miter the joint as shown.*

will not rust. It is entirely prefinished. The ribbed soffit material may be nonperforated or it may be perforated to give approximately 8% open area. It comes in coils usually 50' long and of various widths. Fig. 36-21a.

Installation

It is important to plan the sequence of operation. The soffit material is pulled from coils and fed between the fascia and frieze runner guides into its proper position. Therefore one end of a soffit run cannot be closed off with enclosures, fascia runners, or extensions of frieze runners until the respective runs of soffit material are in place. Fig. 36-21b. To illustrate metal cornice construction, this section describes the installation procedure for a box cornice on a hip roof.

FASCIA RUNNERS

Hang the fascia level, using a chalk line for alignment. Do not force the aluminum or metal fascia to conform to the wood to which it is being attached.

Secure the top edge of the fascia runner to the fascia with 1¹/₂" spiral-shank aluminum nails. Place the nails no more than 2' apart along the length of the fascia.

36–17. *This chart indicates the maximum spacing of supports for various thicknesses of box (closed) plywood soffits. It also gives the correct nail sizes and spacing to be used.*

Exterior plywood soffits (closed) / (Plywood continuous over two or more spans; grain of face plys across supports)

	Closed Soffits		Nail		Nail Spacing (inches)	
Plywood Thickness (inch)	Group	Max. Spacing of Supports c. to c. (inches)	Size	Type	Panel Edges	Intermediate (each support)
³/₈	1, 2, 3, or 4	24	6d	Non-corrosive type (galv. or alum.) box or casing.	6 (or one nail each support)	12
⁵/₈		48	8d			12

Use plywood with these typical APA grade-trademarks. See Fig. 5-9, Unit 5, for complete details.

Sanded Grades

A-C (APA) GROUP 2 EXTERIOR PS 1-74 000

Specialty Panels

303 SIDING 16 oc GROUP 3 (APA) EXTERIOR PS 1-74 000

M.D. OVERLAY GROUP 1 (APA) EXTERIOR PS 1-74 000

Make two cuts 1″ to 1¹/₂″ apart and about ⁵/₈″ deep into the top flange of the runner's guide. Fig. 36-22 (Page 574) at b. While holding the fascia runner flush against the outside face of the fascia board, bend the 1″ to 1¹/₂″ tab up against the inside face of the fascia board. Fig. 36-22 at c. When bending the tab, make sure that the top flange of the runner's guide is still straight. If not, straighten as necessary. Cut and bend as many of these tabs as necessary to hold the fascia runner in place. Space the tabs not more than 2′ apart.

If the fascia board is thicker than 1″, the runner will not be wide enough for a tab. In that case, cut some nailing tabs from scrap material about 1″ wide. Insert the nailing tabs into the slot between the fascia and channel runner. Fig. 36-23 at e. Secure the fascia runner as described previously. Bend the tabs up against the inside of the

fascia board. Secure with a 1¹/₂″ spiral-shank aluminum nail. Refer to Fig. 36-23 at g.

Be sure to allow for expansion and contraction between the fascia channels at the lap joints. Do not secure the channels to each other by nailing or any other fastening device and do not nail along the bottom edge of fascia runners. If this is done it

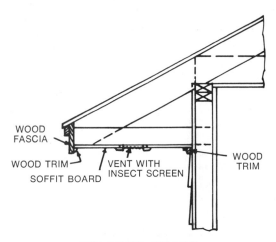

FRAME WALL

36–19a. *Gypsum board provides a flat, smooth surface for the soffit of a box cornice.*

36–19b. *A gypsum-board soffit on a masonry exterior wall should be supported along its entire length. This support is provided by a piece of wood trim attached to the exterior wall. Also be sure to provide a minimum of ¹/₄″ of space behind the soffit for expansion.*

36–18. *Plywood is frequently used for soffit material on a box soffit, but certain precautions should be taken, as shown here.*

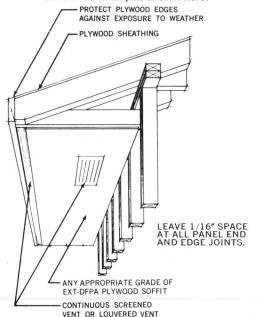

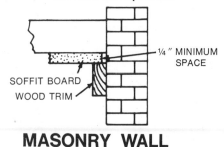

MASONRY WALL

will cause buckling on the face when the aluminum expands.

CUTTING AND FITTING FASCIA AND FRIEZE RUNNERS

A 1¼" notch is made in the fascia and frieze runners for end lap joints. To make lap cuts, use a pair of snips to cut the bottom flange of the runner through to the back of the runner. Fig. 36-24a. Then cut through both thicknesses of the runner groove back, removing piece A. On the top flange of the runner, cut to the inside of the runner groove back. Now bend the piece

36–20a. *Hardboard soffits should be supported as shown. Nail them 4" on center around the edges and 6" on center at the intermediate supports.*

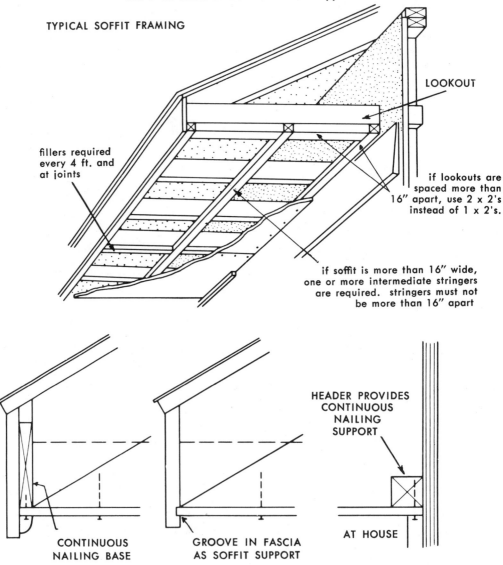

TYPICAL SOFFIT FRAMING

LOOKOUT

fillers required
every 4 ft. and
at joints

if lookouts are
spaced more than
16" apart, use 2 x 2's
instead of 1 x 2's.

if soffit is more than 16" wide,
one or more intermediate stringers
are required. stringers must not
be more than 16" apart

CONTINUOUS
NAILING BASE

GROOVE IN FASCIA
AS SOFFIT SUPPORT

HEADER PROVIDES
CONTINUOUS
NAILING
SUPPORT

AT HOUSE

back and forth until it breaks off as shown at B in Fig. 36-24b. Trim the rough edges and reshape the flanges as necessary. To make end cuts, cut through flanges a, b, and c to the runner groove back as shown in Fig. 36-25. Bend the piece back and forth until the groove back breaks. Trim off rough edges as necessary.

If the metal soffit system is to be used on a home which is of brick veneer, block out at the exterior wall above the brick line to permit the frieze runner to be installed flush with the top edge of the brick. Fig. 36-26. (Note in the detail B that a quarter-

round frieze runner may also be used for this installation with a slightly different method of blocking.)

Where overhangs exceed the width of the soffit material available, a double channel runner can be used so that two pieces of soffit can be installed. Fig. 36-27.

Trimming the corners of a horizontal soffit is slightly different from a sloping soffit. In the case of a horizontal soffit such

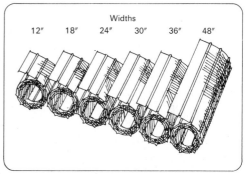

36–21a. *Aluminum soffit material is available in 50' coils, perforated or nonperforated, in various widths.*

36–20b. *The channel molding (A) is attached to the exterior sidewall and to the inside of the fascia board to support the hardboard soffit. The H molding (B) is a divider strip. If desired, a vent strip (C) may be used in place of or in addition to the H molding.*

36–21b. *Aluminum soffit material is pulled from coils and fed into the fascia and frieze runner guides.*

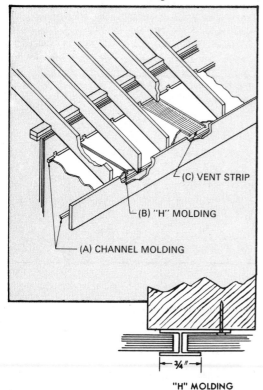

(C) VENT STRIP
(B) "H" MOLDING
(A) CHANNEL MOLDING

"H" MOLDING JOINT

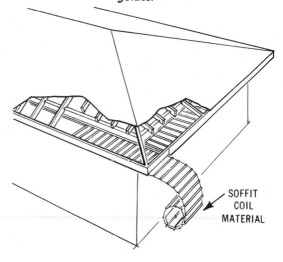

SOFFIT COIL MATERIAL

as would be used on a hip roof, miter corner trim is available and is installed as shown in Fig. 36-28a. For a sloping soffit such as the end of a gable, another type of corner trim is available and can be installed as shown in Fig. 36-28b.

CUTTING AND FITTING QUARTER–ROUND FRIEZE MOLDING

End closures are made by cutting away the area shown by the dotted line on the flanges in Fig. 36-29. Bend the flange down over the end and cut the rounded contour with double action snips to complete the closure. Corners are fitted by mitering in the usual manner. End-laps are made by cutting away the area shown by the broken lines in Fig. 36-30 and shoving the ends together. Fig. 36-31. On the gable fascia trim, the end laps are made by notching. Fig. 36-32.

INSTALLATION SEQUENCE

1. Fascia and frieze runners are available in a variety of sizes. Fig. 36-33. Select

36–22. *The fascia runner is attached to the fascia board by notching (b) and then bending the tab up (c). This will hold the fascia runner in tightly against the fascia board at the bottom.*

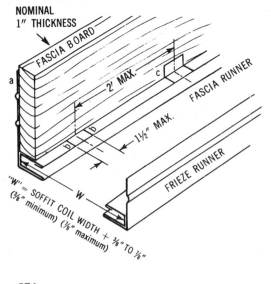

the correct size for the job and apply the fascia and frieze runners for Unit 1 as shown in Fig. 36-34 (Page 578).

2. Pull the soffit coil into place in the direction of the arrow as shown in Fig. 36-34 for Unit 1.

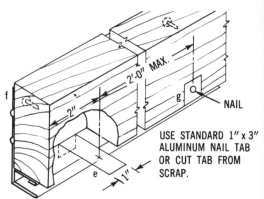

USE STANDARD 1″ x 3″ ALUMINUM NAIL TAB OR CUT TAB FROM SCRAP.

36–23. *When the fascia board is thicker than 1″, special tabs about 1″ wide are inserted (e), bent, and nailed (g). To provide for expansion and contraction, allow $1/16″$ minimum spacing between the fascia channels at lap joints. Do not secure the fascias to each other by nails, pop rivets, or screws. The tabs will hold the fascia runner in tightly against the bottom of the fascia board.*

36–24a. *Cutting a notch for end lap joints. On the bottom flange of the runner make a cut through to the back of the runner. Then cut through both thicknesses along the back edge of the runner groove back and remove piece A.*

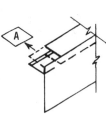

36–24b. *On the top flange of the runner, cut to the inside of the runner groove back and bend piece B back and forth until it breaks.*

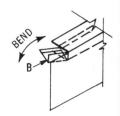

3. Install end closure runners to Units 2 and 3 where they intersect with Unit 1. Fig. 36-35a.

4. Install the fascia and frieze runners of Units 2 and 3. Fig. 36-34. Note that you must leave out one section of fascia runner equal to the soffit width to allow Unit 4 to be installed later.

5. Pull in the soffit coil for Units 2 and 3. Fig. 36-34.

6. Apply end closure runners to Units 2 and 3 where they intersect with Unit 4. Fig. 36-34.

7. Apply the fascia and frieze runners of Unit 4 and pull in the soffit coil.

8. Install the fascia closure runner referred to in Step 4. Figs. 36-34 and 36-35a.

9. Apply the corner trim angles or mitered corner trim. Fig. 36-35b.

10. Use a spline tool to apply a polyethylene spline along all sides and ends of the soffit sheet. Fig. 36-36.

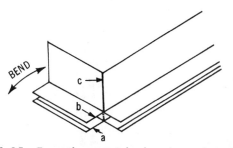

36–25. *To cut the material to length, cut through the flanges at a and b and then bend the piece back and forth at c until the back breaks.*

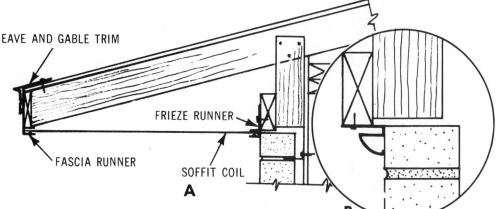

A

36–26. *On a brick veneer exterior wall, a frieze runner may be nailed to blocking material to bring it out flush with the wall (A). Or a quarter-round frieze runner can be used and blocked out (B).*

B

ALTERNATE USE OF
¼ ROUND FRIEZE RUNNER

36–27. *On a flat roof installation where the overhang is wider than 48", a double-channel runner (H molding) may be used to support two widths of the soffit coil material.*

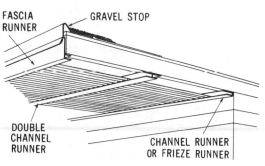

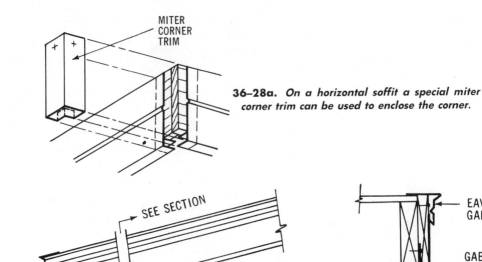

MITER
CORNER
TRIM

36–28a. *On a horizontal soffit a special miter corner trim can be used to enclose the corner.*

SEE SECTION

EAVE AND
GABLE TRIM

GABLE
FASCIA
TRIM

SOFFIT
COIL

END CLOSURE
RUNNER

CORNER
TRIM

GABLE FASCIA TRIM

SECTION

36–28b. *On a sloping soffit for a gable, special corner trim can be applied as shown here.*

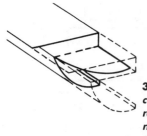

36–29. To make an end closure on a quarter-round frieze molding, notch as shown. Bend the tab down, then cut the tab to conform to the profile of the quarter-round.

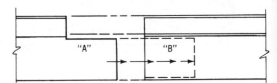

"A"   "B"

36–31. When joining two pieces of quarter-round end to end, the pieces can be cut and telescoped as shown. Part A will fit over B, and the frieze runner will butt together with the cove overlapping.

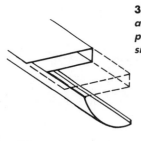

36–30. When making an end lap, notch the piece as shown. With tin snips, cut away the dotted line area.

36–32. Notch the gable fascia trim for end laps as shown by the broken lines.

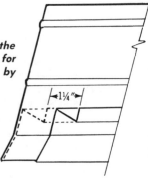

1¼"

36–33. *Fascia and frieze runners are available either straight or sloping, in a variety of sizes.*

Part Name		Part Description
Straight Fascia and Frieze Runners (Prenotched)		
1″ Leg	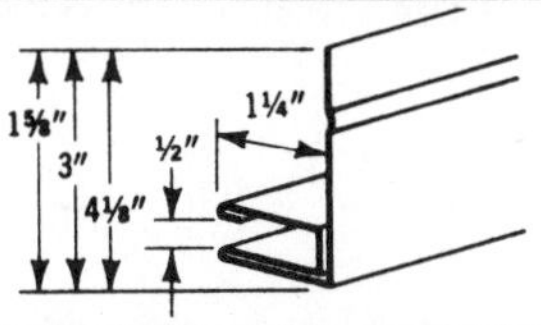	1⅝″ wide x 121¼″ long
3″ Leg		3″ wide x 121¼″ long
4″ Leg		4⅛″ wide x 121¼″ long
6″ Leg	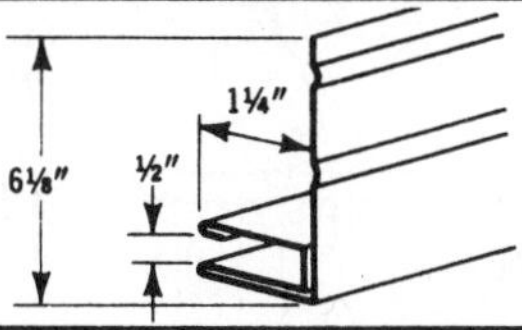	6⅛″ wide x 121¼″ long
8″ Leg	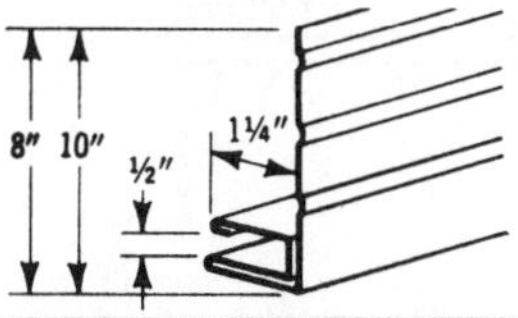	8″ wide x 121¼″ long
10″ Leg		10″ wide x 121¼″ long
Sloping Fascia and Frieze Runners, Type #1 (Prenotched)		
3″ Leg	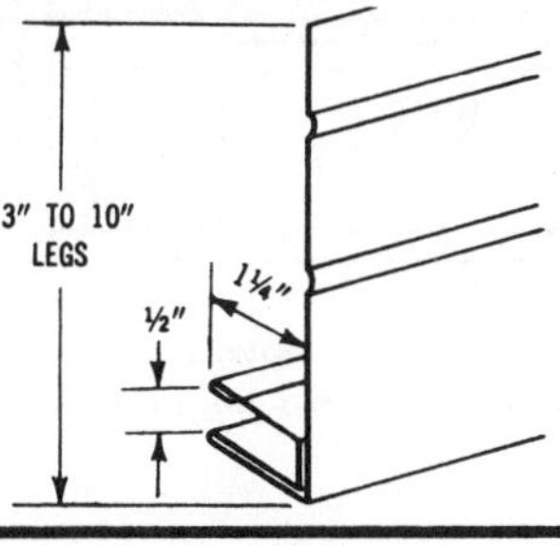	3″ wide x 121¼″ long
4″ Leg		4⅛″ wide x 121¼″ long
6″ Leg		6⅛″ wide x 121¼″ long
8″ Leg		8″ wide x 121¼″ long
10″ Leg		10″ wide x 121¼″ long
Sloping Fascia and Frieze Runners, Type #2 (Prenotched)		
3″ Leg	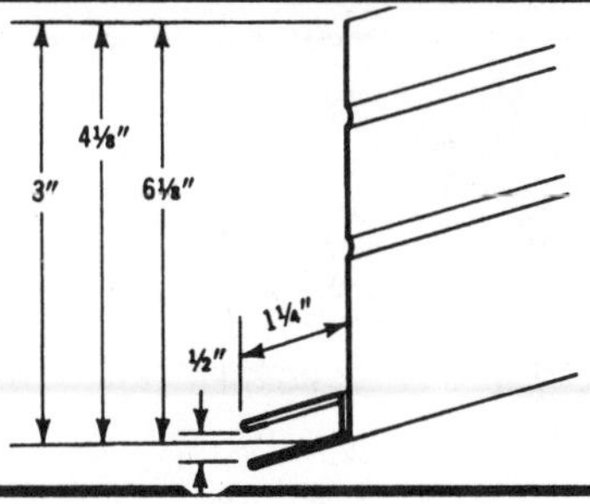	3″ wide x 121¼″ long
4″ Leg		4⅛″ wide x 121¼″ long
6″ Leg		6⅛″ wide x 121¼″ long

RAKE OR GABLE-END FINISH

The extension of a gable roof beyond the end wall is called the *rake section.* This detail may be classed as being:
- A close rake with little projection.
- A boxed or open extension varying from 6″ to 2′ or more.

When the rake extension is only 6″ to 8″, the fascia and soffit can be nailed to a series of short lookout blocks. A, Fig. 36-37. In addition, the fascia is further secured by nailing through the projecting roof sheathing. A frieze board and appropriate moldings complete the construction.

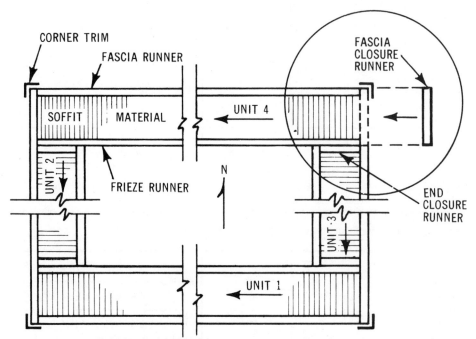

CORNER TRIM

FASCIA RUNNER

FASCIA CLOSURE RUNNER

SOFFIT | MATERIAL

UNIT 4

UNIT 2

FRIEZE RUNNER

N

UNIT 3

END CLOSURE RUNNER

UNIT 1

36–34. *This diagram shows the soffit of a hip roof from below. The arrows indicate the direction and sequence in which to pull the soffit coil into place.*

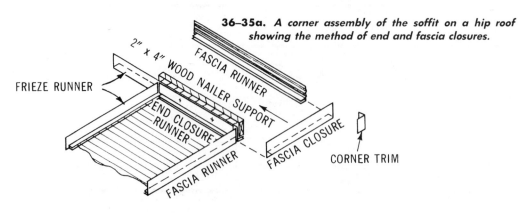

36–35a. *A corner assembly of the soffit on a hip roof showing the method of end and fascia closures.*

FRIEZE RUNNER

2″ x 4″ WOOD NAILER SUPPORT

FASCIA RUNNER

END CLOSURE RUNNER

FASCIA RUNNER

FASCIA CLOSURE

CORNER TRIM

In a moderate overhang of up to 20", both the extending sheathing and a *fly rafter* aid in supporting the rake section. B, Fig. 36-37. The fly rafter extends from the ridge board to the nailing header which connects the ends of the rafters. The roof sheathing boards or the plywood should extend from inner rafters to the end of the gable projection to provide rigidity and strength.

The roof sheathing is nailed to the fly rafter and to the lookout blocks which aid in supporting the rake section and also serve as a nailing area for the soffit. Additional nailing blocks against the sheathing are sometimes required for thinner soffit materials.

Wide gable extensions (2' or more) require rigid framing to resist roof loads and prevent deflection of the rake section. This is usually accomplished by a series of *purlins* or lookout members nailed to a fly rafter at the outside edge and supported by the end wall and a doubled interior rafter. Fig. 36-38. This framing is often called a "ladder" and may be constructed in place or on the ground or other convenient area and hoisted in place.

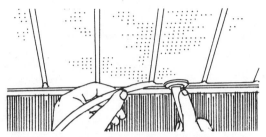

36–36. *Use the spline tool to insert the polyethylene spline along all sides and ends of the soffit sheet.*

36–35b. *After all of the soffit coils have been pulled into place, the last soffit to be pulled in must be closed off by installing the fascia runner and the corner trim.*

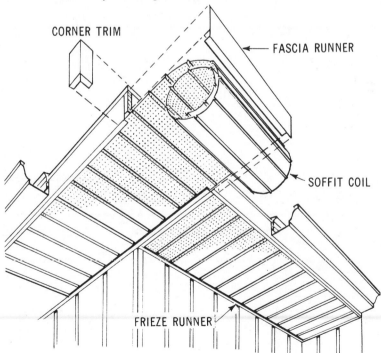

CORNER TRIM

FASCIA RUNNER

SOFFIT COIL

FRIEZE RUNNER

When ladder framing is preassembled, it is usually made up with a header rafter on the inside and a fly rafter on the outside. Each is nailed to the ends of the lookouts which bear on the gable-end wall. When the header is the same size as the rafter, be sure to provide a notch for the wall plates the same as for the regular rafters. In moderate width overhangs, nailing the header and fly rafter to the lookouts with supplemental toenailing is usually sufficiently strong to eliminate the need for the metal hangers shown in B, Fig. 36-38. The header rafters can be face-nailed directly to the end rafters with 12d nails spaced 16″ to 20″ apart.

36–37. *Normal gable-end extensions: A. Narrow overhang. B. Moderate overhang.*

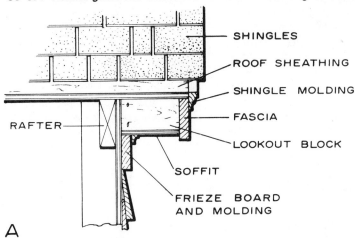

SHINGLES
ROOF SHEATHING
SHINGLE MOLDING
FASCIA
RAFTER
LOOKOUT BLOCK
SOFFIT
FRIEZE BOARD AND MOLDING

A

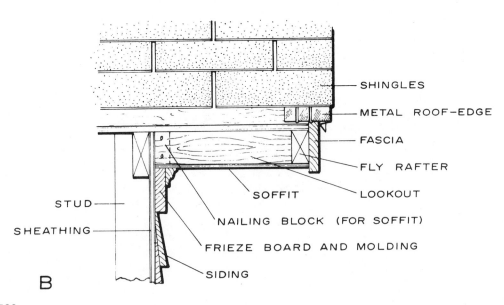

SHINGLES
METAL ROOF-EDGE
FASCIA
FLY RAFTER
LOOKOUT
STUD
SOFFIT
SHEATHING
NAILING BLOCK (FOR SOFFIT)
FRIEZE BOARD AND MOLDING
SIDING

B

Other details of soffits, fascia, frieze board, and moldings can be similar to those used for a wide gable overhang. Lookouts should be spaced 16″ to 24″ apart, depending on the thickness of the soffit material.

A close rake has no extension beyond the end wall other than the frieze board and moldings. Some additional protection and overhang can be provided by using a 2″ × 3″ or 2″ × 4″ fascia block over the sheathing. Fig. 36-39. This member acts as

36–38. *Wide gable-end extension: A. Wide overhang. B. Ladder framing for a wide overhang.*

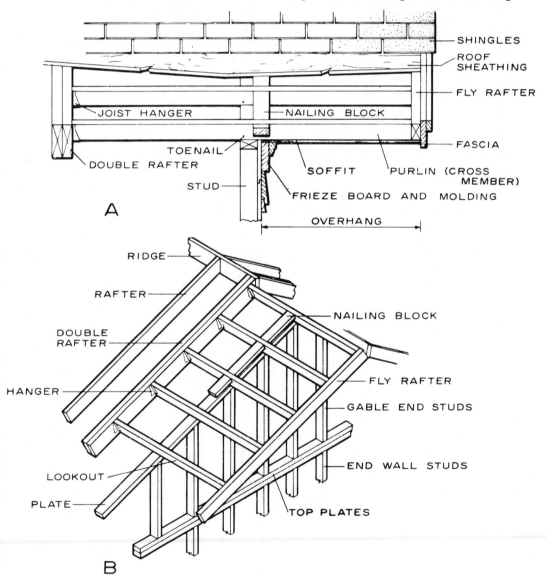

A

SHINGLES
ROOF SHEATHING
FLY RAFTER
JOIST HANGER
NAILING BLOCK
FASCIA
TOENAIL
DOUBLE RAFTER
SOFFIT
PURLIN (CROSS MEMBER)
STUD
FRIEZE BOARD AND MOLDING
OVERHANG

B

RIDGE
RAFTER
DOUBLE RAFTER
NAILING BLOCK
HANGER
FLY RAFTER
GABLE END STUDS
END WALL STUDS
LOOKOUT
PLATE
TOP PLATES

a frieze board, as the siding can be butted against it. The fascia, often 1" × 6", serves as a trim member. Metal roof edging is often used along the rake section as flashing.

CORNICE RETURN

The cornice return is the end finish of the cornice on a gable roof. On hip roofs and flat roofs, the cornice is usually continuous around the entire house. On a gable roof, however, it must be terminated or joined with the gable ends. The method selected depends to a great extent on the type of cornice and the projection of the gable roof beyond the end wall.

A narrow box cornice, often used in houses with Cape Cod or Colonial details, has a boxed return when the rake section has some projection. A, Fig. 36-40. The fascia board and shingle molding of the cornice are carried around the corner of the rake projection.

When a wide box cornice has no horizontal lookout members, the soffit of the gable-end overhang is at the same slope and coincides with the cornice soffit. B, Fig. 36-40. This is a simple system and is often used when there are wide overhangs at both sides and ends of the house.

A close rake (a gable end with little projection) may be used with a narrow box cornice or a close cornice. In this type, the frieze board of the gable end, into which the siding butts, joins the frieze board or fascia of the cornice. C. Fig. 36-40.

While close rakes and cornices with little overhang are lower in cost, the extra material and labor required for good gable and cornice overhangs are usually justified. Better sidewall protection and lower paint maintenance costs are only two of the benefits derived from good roof extensions.

GUTTERS

Wooden gutters are either built into the cornice or prefabricated and attached on the job site. They were once used extensively but are now almost obsolete. Most modern gutters are of prefabricated metal, equipped with metal straps for attaching to the roof boards. See Unit 35.

ESTIMATING
Material

The materials for the cornice are strictly linear measure. All moldings and lumber or other materials which are attached to the house can be figured by determining the

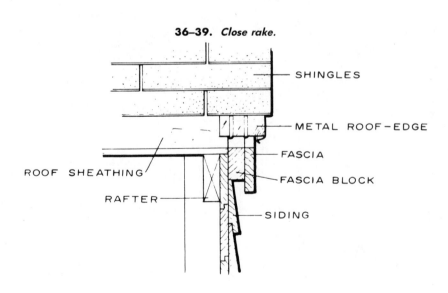

36–39. *Close rake.*

SHINGLES

METAL ROOF–EDGE

FASCIA

ROOF SHEATHING

FASCIA BLOCK

RAFTER

SIDING

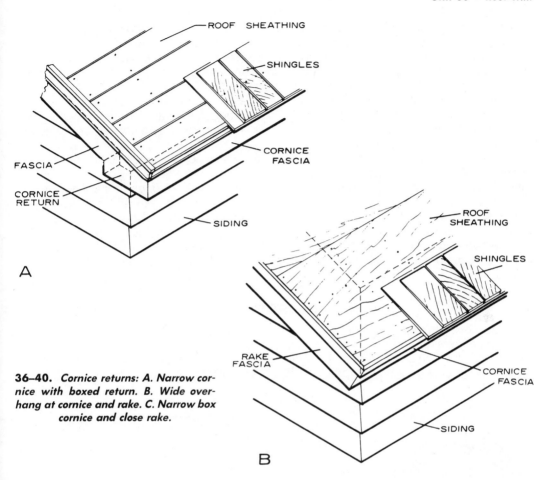

ROOF SHEATHING

SHINGLES

CORNICE
FASCIA

FASCIA

CORNICE
RETURN

SIDING

A

ROOF
SHEATHING

SHINGLES

RAKE
FASCIA

CORNICE
FASCIA

SIDING

B

36–40. *Cornice returns: A. Narrow cornice with boxed return. B. Wide overhang at cornice and rake. C. Narrow box cornice and close rake.*

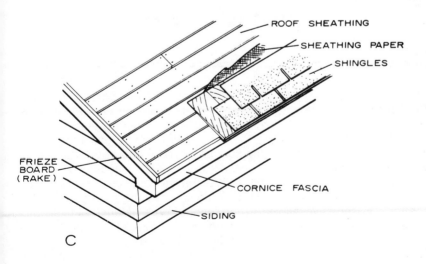

ROOF SHEATHING

SHEATHING PAPER

SHINGLES

FRIEZE
BOARD
(RAKE)

CORNICE FASCIA

SIDING

C

perimeter of the house. This includes material for ledger strips, frieze boards, and bed moldings. The amount of soffit material required is also determined by figuring the perimeter of the house and referring to the specifications of house plans for the width and thickness of the material required. The amount of material required for the fascia board and any crown molding attached to the fascia is figured by determining the perimeter of the house at the rafter ends. The linear footage for rake moldings is figured in the same way as the length of the gable-end rafter. (See Unit 28, p. 411). The amount of material necessary for the lookouts may be determined by multiplying the projection by the number of rafters.

Labor

It is possible to give a general estimate of the cost for installing a cornice. The rate of installation varies with the type of cornice.

• For a close cornice, it is estimated that workers can install 20 linear feet per hour.
• For an open cornice, 12 linear feet per hour.
• For a box cornice, 10 linear feet per hour.

A house with an open cornice and a 148′ perimeter would take about 12 hours and 20 minutes: 148′ ÷ 12 = 12$\frac{1}{3}$ hours, or 12 hours and 20 minutes. In order to find the approximate cost of installing a cornice, multiply the estimated hours of labor by the cost per hour.

QUESTIONS

1. List several types of cornices.
2. When constructing a closed cornice, what procedure is used to insure alignment of the plumb cuts on the rafter tails?

3. What is a fascia?
4. List three materials that are commonly used for the soffit on a closed cornice.
5. What is a cornice return?

UNIT 37

Windows

Windows are millwork items and are usually fully assembled at the factory. Window units often come with the sash fitted and weather-stripped, the frame assembled, and the exterior casing in place. Fig. 37-1. Standard combination storms and screens or separate units can also be included. All wood components are treated with a water-repellent preservative at the factory to provide protection before and after they are placed in the walls.

Besides letting in light and air, windows are an important part of the architectural design. Fig. 37-2. Generally the glass area of a room should be not less than 10% of the floor area. The window area that can be opened for ventilation should be not less than 4% of the floor area unless a complete air conditioning system is used.

There should be a balance of fixed picture windows and operating windows. An operating window can always be closed to

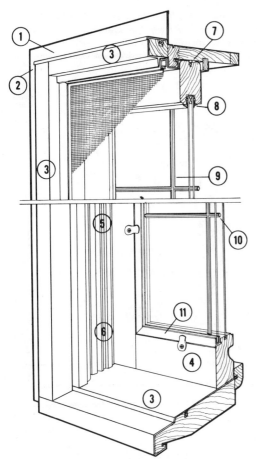

37-1. *Parts of an assembled double-hung window: 1. Head flashing. 2. Blind stops. 3. Casing. 4. Sash. 5. Counterbalancing unit. 6. Tracks. 7. Weather stripping. 8. Glazing. 9. Grill (installed on the inside when insulating glass is used). 10. Grill (installed between the glass when storm panels are used). 11. Storm panel.*

37-2. *Windows are an important part of the design in this home.*

seal out unpleasant weather or opened to a cooling breeze, but a fixed window cannot be opened. Local climate and prevailing winds determine the best window placement and the degree of ventilation required.

The type of window specified in the plans will vary with the room requirements. Not every room will need the same size and type of window. In bedrooms, light and

ventilation are a necessity, but privacy and wall space for furniture are also important factors. A row of narrow operating windows placed high on two walls of the room will provide light and ventilation as well as privacy and wall space. In the kitchen, windows should provide good ventilation of cooking odors. For the area over the sink and other hard-to-reach spots, a casement window or awning type that opens with a crank or lever action would be a good choice. The living areas are an ideal location for large picture windows which bring in scenic views.

The window style and size should be such that it is convenient to look through the window whether a person is seated or standing. Eye-level (seated and standing) height charts for various window types are shown in Fig. 37-3.

TYPES OF WINDOWS

These are the principal types of windows. Fig. 37-4.
- Double-hung.
- Casement.
- Stationary (fixed).
- Awning.
- Hopper.
- Horizontal-sliding.
- Jalousie.

37–3. *Recommended window heights for various rooms. Note that the header height of the window is standard and corresponds to the 6'8" door height. The sill height, however, will vary with the style and size window to be installed.*

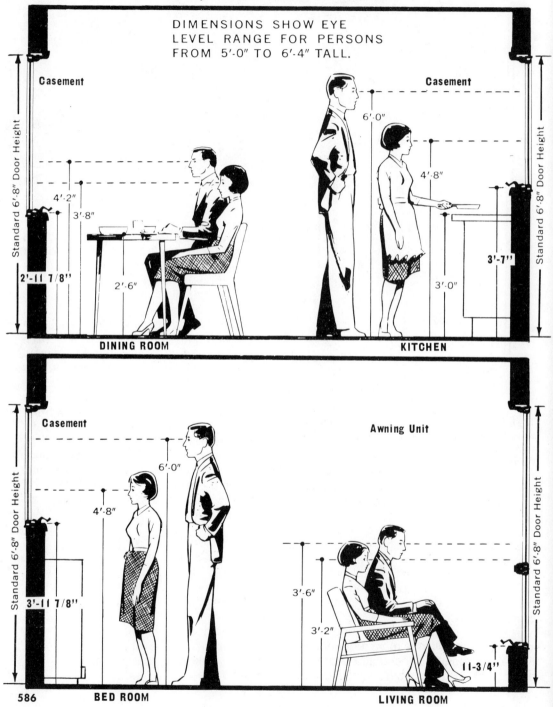

DIMENSIONS SHOW EYE LEVEL RANGE FOR PERSONS FROM 5'-0" TO 6'-4" TALL.

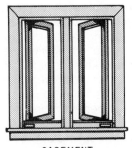

CASEMENT

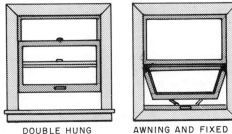

DOUBLE HUNG   AWNING AND FIXED

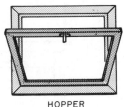

HOPPER

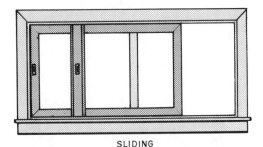

SLIDING

37-4. Six of the basic window styles.

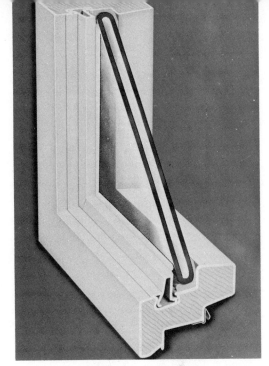

37-5. Insulated glass.

Glass blocks are sometimes used for admitting light in places where transparency or ventilation is not required.

Windows may be of wood or metal. Heat loss through metal frames and sash is much greater than through similar wood units. On the other hand, metal frames and sash require less maintenance. They also can be made narrower and thus allow larger glass areas.

Wooden window frames and sash should be made from a clear grade of all-heartwood stock of a decay-resistant wood species or from wood which has been given a preservative treatment. Species commonly used include ponderosa and other pines, cedar, cypress, redwood, and spruce. Metal window frames and sash are made of aluminum or steel.

Insulated glass, used for both stationary and movable sash, consists of two or more sheets of glass with air space between the sheets. The edges are hermetically sealed to keep the air in. This type of glass has more resistance to heat loss than a single thickness and is often used without a storm sash. Fig. 37-5.

Window jambs (sides and tops of the frames) must be the same width as the wall section, including the exterior sheathing and the interior finished wall covering. Jambs are made of nominal 1" lumber;

jamb liners are used to adapt the window unit to various wall thicknesses. Fig. 37-6. Sills (bottoms of frame) are made from nominal 2″ lumber and are sloped at about 3 in 12 for good drainage. D, Fig. 37-7. Sash are normally 1³/₈″ thick, and wood combination storm and screen windows are usually 1¹/₈″ thick.

Double-hung Windows

The double-hung window is perhaps the most familiar window type. It consists of an upper and a lower sash that slide vertically in separate grooves in the side jambs or in full-width metal weather stripping. Fig. 37-7. This type of window provides a maximum face opening for ventilation of one-half the total window area. Each sash is provided with springs, balances, or compression weather stripping to hold it in place in any location. Compression weather stripping, for example, prevents air infiltration, provides tension, and acts as a counterbalance. Several types allow the

37–6. *Some manufacturers include a jamb liner that can be repositioned and adapted to various wall thicknesses as shown at arrow No. 1 in the jamb section. The jamb liner is shown at the No. 2 arrows with the window installed in a framed wall with lath and plaster.*

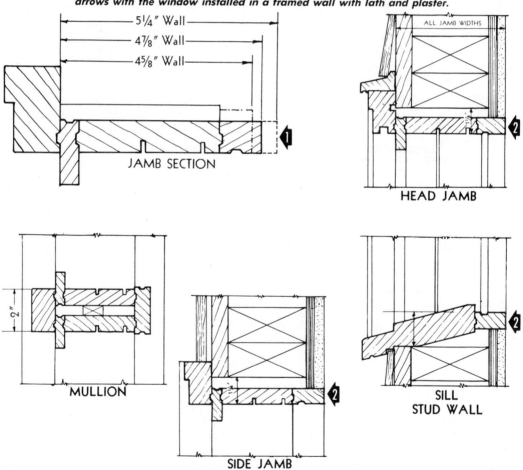

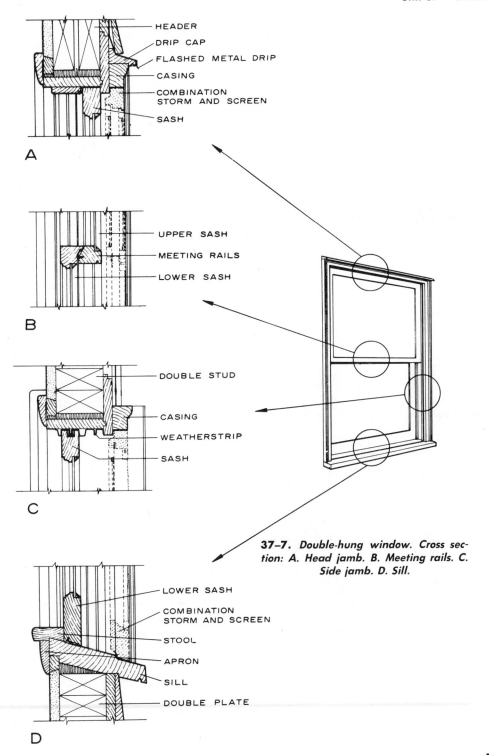

HEADER
DRIP CAP
FLASHED METAL DRIP
CASING
COMBINATION
STORM AND SCREEN
SASH

A

UPPER SASH
MEETING RAILS
LOWER SASH

B

DOUBLE STUD
CASING
WEATHERSTRIP
SASH

C

LOWER SASH
COMBINATION
STORM AND SCREEN
STOOL
APRON
SILL
DOUBLE PLATE

D

37-7. *Double-hung window. Cross section: A. Head jamb. B. Meeting rails. C. Side jamb. D. Sill.*

37-8. *Most double-hung windows are designed so that the sash can be easily removed.*

37-9a. *Preassembled dividers of various styles for subdividing window lights are easily snapped into place.*

Horizontal Bar

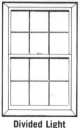
Divided Light

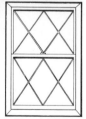
Diamond Light

sash to be removed for easy cleaning, painting, or repair. Fig. 37-8.

Sash may be divided into a number of compartments, or lights, by small wood members called muntins. A ranch-type house may provide the best appearance with top and bottom sash divided into two horizontal lights. A Colonial or Cape Cod house usually has each sash divided into six or eight lights. Some manufacturers provide preassembled dividers which snap in place over a single light, dividing it into six or eight lights. This simplifies painting and other maintenance. Fig. 37-9.

Hardware for double-hung windows includes the sash lifts (for opening the window) that are fastened to the bottom rail, although they are sometimes eliminated by providing a finger groove in the rail. Other hardware consists of sash locks or fasteners located at the meeting rails. They not only lock the window, but also draw the sash together to provide a "windtight" fit.

Double-hung windows can be arranged in a number of ways—as a single unit, doubled (or mullion) type, or in groups of three or more. One or two double-hung windows on each side of a large stationary insulated window are often used to effect a

37-10. *Double-hung windows are frequently used in combination with a large stationary window.*

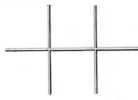

37-9b. *The use of various grill patterns can change the architectural style of a window. A divided-light grill pattern installed in conventional two-light double-hung windows enhances the traditional styling of this house.*

window wall. Fig. 37-10. Such large openings must be framed with headers large enough to carry roof loads.

Casement Windows

Casement windows have side-hinged sash, usually designed to swing outward because this type can be made more weathertight than the type that swings inward. Fig. 37-11. Screens are located inside these outward-swinging windows, and winter protection is obtained with a storm sash or by using insulated glass in the sash. One advantage of the casement window over the double-hung type is that the entire window area can be opened for ventilation.

Weather stripping is also provided for this type of window, and units are usually received from the factory entirely assembled with hardware in place. Closing hard-

ware consists of a rotary operator and sash lock. Fig. 37-12.

As in the double-hung units, casement sash can be used in a number of ways—as a pair or in combinations of two or more pairs. Style variations are achieved by divided lights. For example, snap-in muntins provide a small multiple-pane appearance for traditional styling.

Stationary Windows

Stationary windows, used alone or in combination with other types of windows, usually consist of a wood sash with a large single light of insulated glass. They are designed for providing light, as well as for attractive appearance, and are fastened permanently into the frame. Fig. 37-13 (Page 594). Because of their size (sometimes 6 to 8 feet in width) and because of the thickness of the insulating glass, a $1^{3}/_{4}"$ thick sash is usually used to provide strength.

Stationary windows may also be installed without a sash. The glass is set directly into rabbeted frame members and held in place with stops. As with the window-sash units, back puttying and face

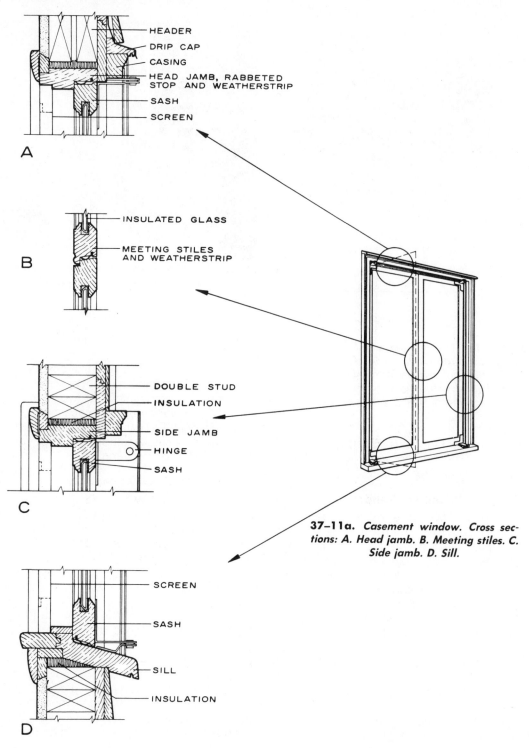

A

- HEADER
- DRIP CAP
- CASING
- HEAD JAMB, RABBETED STOP AND WEATHERSTRIP
- SASH
- SCREEN

B

- INSULATED GLASS
- MEETING STILES AND WEATHERSTRIP

C

- DOUBLE STUD
- INSULATION
- SIDE JAMB
- HINGE
- SASH

D

- SCREEN
- SASH
- SILL
- INSULATION

37–11a. *Casement window. Cross sections: A. Head jamb. B. Meeting stiles. C. Side jamb. D. Sill.*

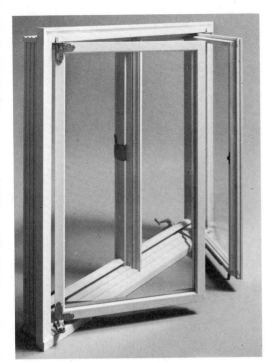

37–11b. *Wood casement window.*

37–11c. *An attractive window wall made up of casement windows with fixed units above.*

37–12a. *Casement window roto-gear operator.*

37–12b. *Casement window sash lock.*

puttying of the glass (with or without a stop) will assure moisture-resistance. Fig. 37-14 (Page 596).

Awning and Hopper Windows

Awning window units have a frame in which one or more operative sash are installed. Fig. 37-15. They often are made up for a large window wall and consist of three or more units in width and height.

Sash of the awning type are made to swing outward at the bottom. A similar unit, the hopper type, is one in which the top of the sash swings inward. Both types provide protection from rain when open.

Jambs are usually $1^1/_{16}$" or more thick because they are rabbeted, while the sill is at least $1^5/_{16}$" thick when two or more sash are used in a complete frame. Each sash may also be provided with an individual frame so that any combination in width and height can be used. Awning or hopper window units may consist of a combination of one or more fixed sash with the remainder being the operable type. Fig. 37-16 (Page 598). Operable sash are provided with hinges, pivots, and sash supporting arms. Fig. 37-17. There are three types of operating hardware available for awning windows: the standard push bar, the lever lock, and the rotary gear. Fig. 37-18. Weather stripping and storm sash and screens are usually provided. The storm sash is eliminated when the windows are glazed with insulated glass.

Horizontal-Sliding Window Units

Horizontal-sliding windows appear similar to casement sash. However, the sash (in pairs) slide horizontally in separate tracks or guides located on the sill and head jamb. Fig. 37-19. Multiple window openings consist of two or more single units and may be used when a window-wall effect is desired. Fig. 37-20. As in

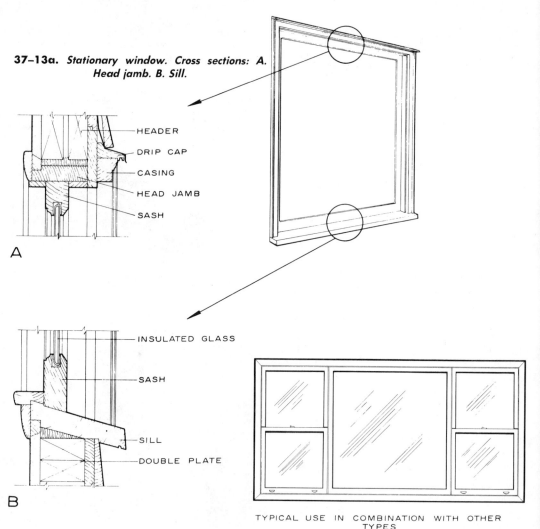

37–13a. Stationary window. Cross sections: A. Head jamb. B. Sill.

HEADER
DRIP CAP
CASING
HEAD JAMB
SASH

A

INSULATED GLASS
SASH
SILL
DOUBLE PLATE

B

TYPICAL USE IN COMBINATION WITH OTHER TYPES

37–13b. *A window wall made up of a station-ary unit and two double-hung units with a third double-hung unit on the adjacent wall. Notice when a window is positioned too close to a corner, there is no room for the drapes to hang clear. The drapes then restrict the available light and venti-lation.*

37–13c. *A ranch-type home with a stationary window used in conjunction with two double-hung window units. Note the colonial entrance and window grills.*

37–13d. *In this window, the upper sash is sta-tionary and the lower sash is an awning vent.*

595

37–14a. *Puttying (glazing) of a metal edge insulated glass unit: 1. Glazing compound. 2. Setting blocks (two, treated wood, 4" long, located ¼ of the width in from each end of the unit). 3. Edge clearance (⅛", ³⁄₁₆", or ¼" at all edges, depending on the size; distribute proportionately). 4. Lateral clearance (a minimum of ¹⁄₁₆" for glazing compound). 5. Edge coverage (cover the channel with glazing compound for uniform appearance and maximum edge insulation). 6. Glazing stop (must not bear on the unit).*

37–14b. *Glazing of insulated glass with a glass edge: 1. Glazing compound. 2. Setting blocks (two, treated wood, 2" to 3" long, located ¼ of the width in from each end of the unit). 3. Edge clearance (avoid glass to metal sash contact, allow ⅛" at all edges, and distribute proportionately). 4. Lateral clearance (a minimum of ¹⁄₁₆" for glazing compound). 5. Clips (use special clips in metal sash). 6. Glazing stop (do not allow to bear on the unit).*

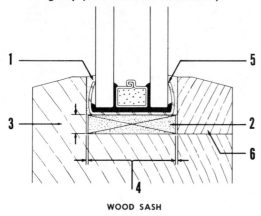

WOOD SASH

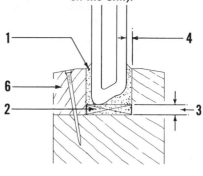

WOOD SASH

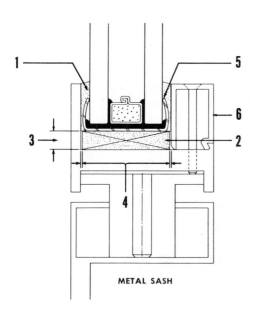

METAL SASH

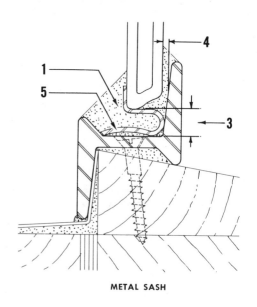

METAL SASH

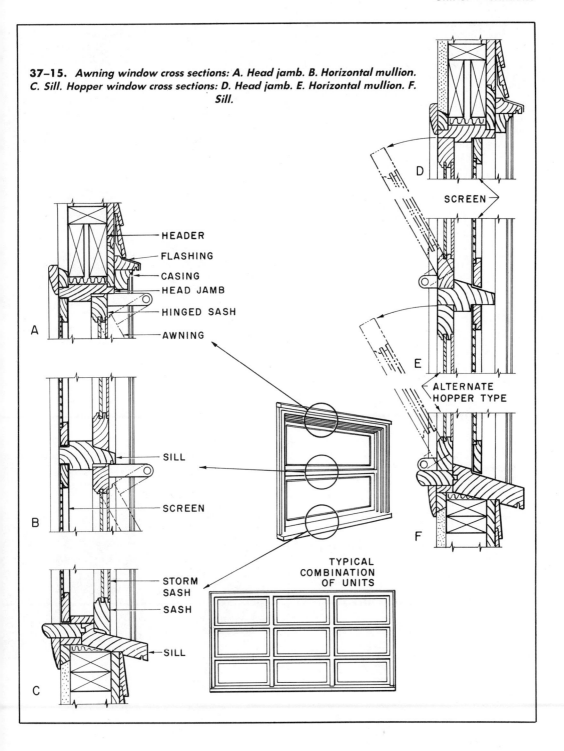

37–15. *Awning window cross sections: A. Head jamb. B. Horizontal mullion. C. Sill. Hopper window cross sections: D. Head jamb. E. Horizontal mullion. F. Sill.*

HEADER

FLASHING

CASING

HEAD JAMB

HINGED SASH

AWNING

A

SILL

SCREEN

B

STORM SASH

SASH

SILL

C

SCREEN

D

ALTERNATE HOPPER TYPE

E

F

TYPICAL COMBINATION OF UNITS

37–16a. In the cross section of the mullion, notice the spline, A, inserted in the groove provided in the side jambs. This aligns the window units when they are used in combination. A mullion strip, B, is then installed on the front edge for trim. In the vertical section, notice that the head and sill jambs also contain grooves in which a spline may be inserted to enable stacking these window units for other possible combinations.

VERTICAL SECTION

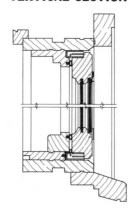

MULLION SECTION

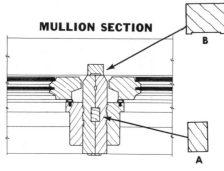

HORIZONTAL SECTION

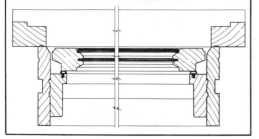

37–16b. A window wall of hopper windows with stationary windows above.

37–17. Roto-lock hardware for an awning window. As the window is closed, the bars slide toward the corners of the sash on a track. This provides a tight seal when the sash is closed.

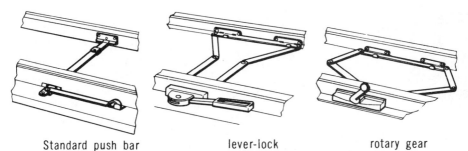

Standard push bar   lever-lock   rotary gear

37–18. *The three types of operating hardware for awning windows.*

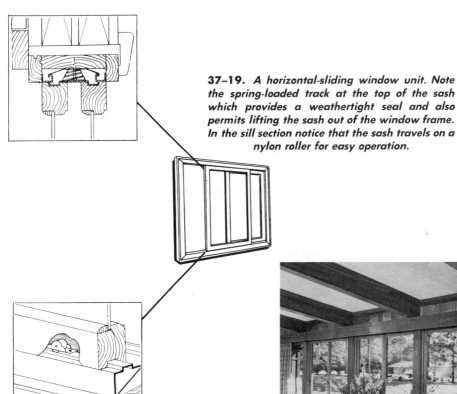

37–19. *A horizontal-sliding window unit. Note the spring-loaded track at the top of the sash which provides a weathertight seal and also permits lifting the sash out of the window frame. In the sill section notice that the sash travels on a nylon roller for easy operation.*

37–20. *A window wall of horizontal-sliding window units.*

37-21. A jalousie window.

most modern window units of all types, weather stripping, water-repellent preservative treatments, and sometimes hardware are included in these fully factory-assembled units.

Jalousie Windows

Jalousie windows consist of a series of small, horizontal glass pieces that are held by an end frame of metal. They usually open outward. Fig. 37-21. Metal sash are sometimes used but, because of low insulating value, should be installed carefully to prevent condensation and frosting on the interior surfaces during cold weather. A full storm window unit is sometimes necessary to eliminate this problem in cold climates.

Metal Sash

Metal sash are available in units made of aluminum or steel, and the principal types are casement (Fig. 37-22), double-hung (Fig. 37-23), sliding, and stationary. Lights in the sash are divided in various patterns. The aluminum sash and frames are generally made of solid extruded aluminum alloy, welded at the joints. Steel sash are made of rolled shapes about ⅛" thick, most parts being Z-shaped or T-shaped, with welded butt joints. Steel sash should be treated to make them rust-resistant. Hardware, such as hinges, latches, and operators, is special and provided with the window. Screens should be ordered with the windows.

Actual details for installation vary according to the method of manufacture. It is common practice in frame construction to use a wood buck in the window opening to hold the metal frame. Fig. 37-22. The space between the metal and the wood is filled with calking compound.

Skylights

Skylights are installed on either pitched or flat roofs for ventilation and light. The dome is double plastic for insulation. Rather than building a dormer on a room directly under a roof and enclosed with interior walls, a skylight can be installed. This is a good solution to the problem of light in a remodeled attic also. Fig. 37-24.

WINDOW SCHEDULES

A window schedule will usually contain descriptions of the various windows, plus sash openings, glass sizes, and sometimes the rough opening sizes. The location of each window in a house is found by matching the number of the window in the window schedule with the corresponding number on the house plan. Fig. 37-25 (Page 604).

Figuring Rough Opening Sizes

When the rough opening size is not provided, it will have to be figured by the builder or obtained from the window manufacturer's catalog. Tables showing glass size, sash size, and rough opening size for

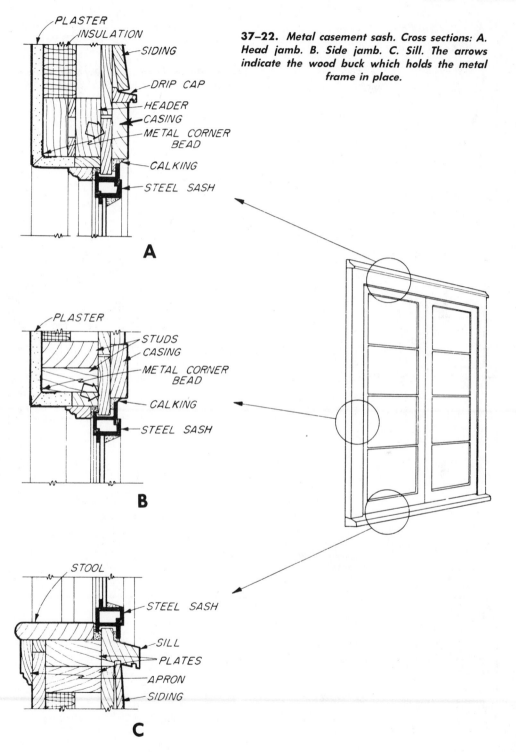

PLASTER
INSULATION
SIDING
DRIP CAP
HEADER
CASING
METAL CORNER BEAD
CALKING
STEEL SASH

A

37-22. *Metal casement sash. Cross sections: A. Head jamb. B. Side jamb. C. Sill. The arrows indicate the wood buck which holds the metal frame in place.*

PLASTER
STUDS
CASING
METAL CORNER BEAD
CALKING
STEEL SASH

B

STOOL
STEEL SASH
SILL
PLATES
APRON
SIDING

C

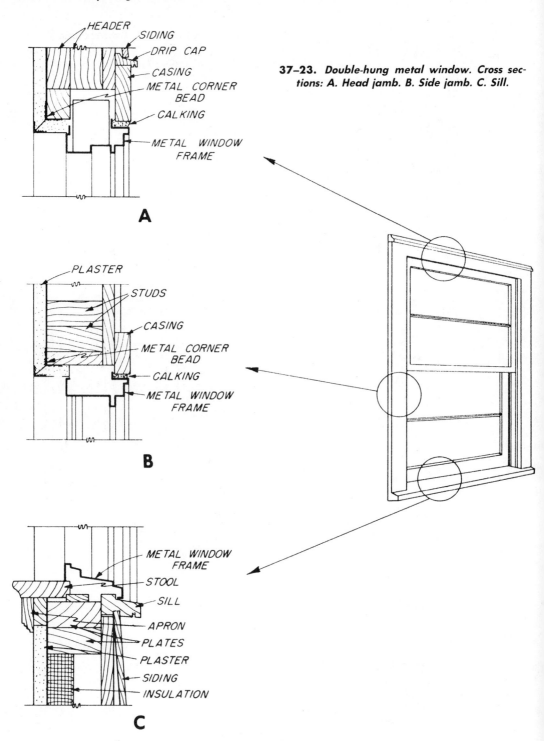

HEADER
SIDING
DRIP CAP
CASING
METAL CORNER BEAD
CALKING
METAL WINDOW FRAME

A

37–23. *Double-hung metal window. Cross sections: A. Head jamb. B. Side jamb. C. Sill.*

PLASTER
STUDS
CASING
METAL CORNER BEAD
CALKING
METAL WINDOW FRAME

B

METAL WINDOW FRAME
STOOL
SILL
APRON
PLATES
PLASTER
SIDING
INSULATION

C

windows from various manufacturers are available from suppliers. In this book, typical openings for double-hung windows are shown in Unit 25, "Wall Framing."

The rough opening size can also be figured if the glass size is known. Make the rough opening at least 6″ wider and 10″ higher than the window glass size. In specifying a window, the width of the glass is always given first, then the height, then the number of pieces of glass (or lights) and the window style. For example, 28$\frac{1}{2}$″ × 24″, 2 lights D. H. means that the glass itself is 28$\frac{1}{2}$″ wide and 24″ high and that there are two pieces of glass in a double-hung unit.

To figure the rough opening width, add 6″ to the given width: 28$\frac{1}{2}$″ + 6″ = 34$\frac{1}{2}$″, or 2′ 10$\frac{1}{2}$″. To obtain the rough opening height, add the upper and lower glass height together, and then add another 10″: 24″ + 24″ + 10″ = 58″, or 4′10″. These allowances are fairly standard and provide for the weights, springs, balances, room for plumbing and squaring, and for the normal adjustments. However, when the window manufacturer is known, use his recommended rough opening sizes.

The rough opening sizes vary slightly among manufacturers, as can be seen by comparing sample tables from two typical manufacturers' catalogs designating the sizes of standard units. Fig. 37-26. The window schedule in Fig. 37-25 indicates a glass size of 28$\frac{1}{2}$″ × 24″ for window No. 4. In Fig. 37-26a, read the width of the glass size across the top (28$\frac{1}{2}$″) and then follow down that column to the glass length (24″). The window is designated 2846, which is the manufacturer's catalog number. Note, in the case of this particular manufacturer, the number is also the sash opening. Window 2846 has a sash opening width of 2′8″ and a length of 4′6″. The rough opening given on this chart is 2′10″ × 4′9″ for a framed wall. (The unit size, which is the masonry opening for a brick veneer wall, is also given on these tables.) The rough opening figured earlier (2′10$\frac{1}{2}$″ × 4′10″) is

37–24. *A skylight installed on a pitched roof.*

larger than the manufacturer's recommended opening. Using the method for figuring a rough opening will insure getting the window unit into the opening, but it may require additional blocking or shimming. Therefore, it is always best to use the specific manufacturer's recommendations for a rough opening size.

FIGURING ROUGH OPENINGS FOR COMBINATION UNITS

Many times, window units of various styles and sizes are combined to make up larger units for a particular room and use. These combined units are separated only

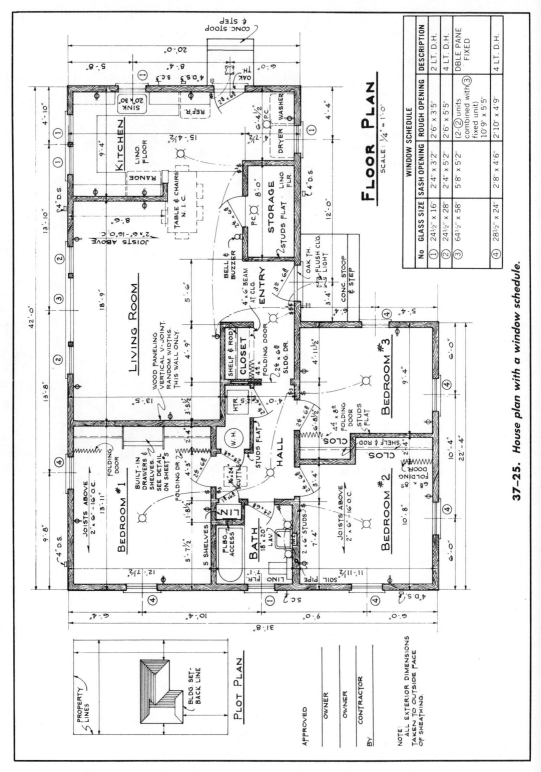

37-25. House plan with a window schedule.

604

37–26a. *A manufacturer's table for double-hung windows.*

Window Sizes

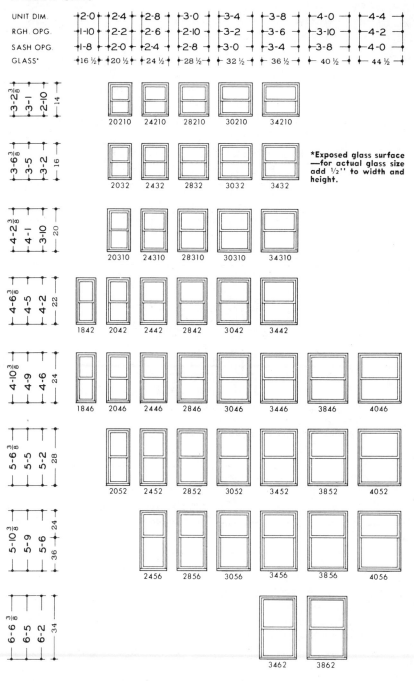

UNIT DIM.	2-0 2-4 2-8 3-0 3-4 3-8 4-0 4-4	
RGH. OPG.	1-10 2-2 2-6 2-10 3-2 3-6 3-10 4-2	
SASH OPG.	1-8 2-0 2-4 2-8 3-0 3-4 3-8 4-0	
GLASS*	16½ 20½ 24½ 28½ 32½ 36½ 40½ 44½	

*Exposed glass surface —for actual glass size add ½'' to width and height.

20210 24210 28210 30210 34210

2032 2432 2832 3032 3432

20310 24310 28310 30310 34310

1842 2042 2442 2842 3042 3442

1846 2046 2446 2846 3046 3446 3846 4046

2052 2452 2852 3052 3452 3852 4052

2456 2856 3056 3456 3856 4056

3462 3862

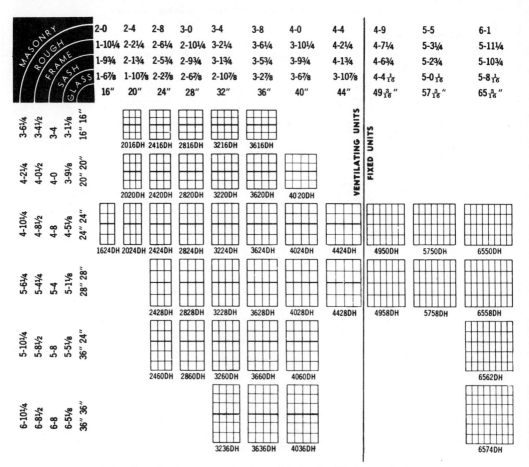

37–26b. *Another example of a table for double-hung windows.*

37–27. *Typical combinations for contemporary or traditional treatment with grills and mullions.*

A   **B**   **C**

by vertical piers called mullion strips. Fig. 37-27. Therefore, the rough opening for the combined unit will be less than the total of the rough openings for the units if they were used individually.

In the house plan in Fig. 37-25, note the window schedule on the plan calls for a combination unit in the living room consisting of two window units No. 2 and one window unit No. 3. If this unit is to be made up to look like C in Fig. 37-27, figure the width of the multiple opening by adding the individual sash openings to the width of the mullions, plus 2" for the overall rough opening. In this example the sash opening for the No. 2 unit in the window schedule is 2'4". The sash opening for the No. 3 unit is 5'8". Figs. 37-25 and 37-28. Since there are two No. 2 units, add 2'4" twice to the 5'8": 2'4" + 2'4" + 5'8" = 9'16", or 10'4". There are two mullions in our example. If each mullion is 1½" wide, the total mullion width would be 3". Add this to the sash width and then add 2" for the rough opening: 10'4" + 3" + 2" = 10'9". Note that this is the rough opening width listed for the combination window unit (No. 3) in the window schedule. Fig. 37-25. The rough opening height is figured the same way for combination units as for individual units.

INSTALLATION

Window frames are generally assembled in a mill. The preassembled window frame is easily installed, but care should be taken. Regardless of the quality of the window purchased, it is only as good as its installation. Before actually installing a window, apply a primer coat of paint to all wood members to prevent undue warpage.

The general procedure for the installation of a window frame is very similar regardless of style or manufacturer. Fig. 37-29. However, always refer to the manufacturer's instructions for any specific recommendations. For example, some manufacturers recommend that the sash be removed from the frame to prevent breakage and provide easier handling of the unit. Fig. 37-30 (Page 612). Others specify not only that the sash be left in the frame but also that diagonal braces and, in some cases, reinforcing blocks be left in place to insure that the frame remain square and in proper alignment.

When a siding material is applied over sheathing, the windows are installed first and the siding applied later. Strips of 15-pound asphalt felt should be put over the sheathing around the openings. Fig. 37-31.

Place the frame in the opening from the outside, allowing the subsill to rest on

PICTURE WINDOWS

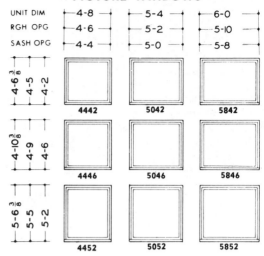

UNIT NO.	1" GLASS RABBET	1" GLASS SIZE
4442	49 x 46½	48½ x 46
5042	57 x 46½	56½ x 46⅛
5842	65 x 46½	64½ x 46
4446	49 x 50½	48½ x 50
5046	57 x 50½	56½ x 50
5846	65 x 50½	64½ x 50
4452	49 x 58½	48½ x 58
5052	57 x 58½	56½ x 58⅛
5852	65 x 58½	64½ x 58

37–28. A manufacturer's table for stationary windows.

the rough frame at the bottom, and hold the unit up tightly against the building. Fig. 37-32a (Page 613). Level and plumb the window frame, then wedge with shingles and tack in place. Fig. 37-32b. Check the sill and jamb with the level and square. Fig. 37-32c-e. When everything is in order, use 16d galvanized nails through the outside casings; or, if blind stops are provided, use 8d common nails. Fig. 37-32f-i. (When window frames are furnished with blind stops for installation, the sheathing should be installed 1½" back from the window rough openings. Fig. 37-33 and 37-34 page 616.) The nails should be spaced about 12" apart. They should penetrate the sheathing and/or the trimmer

37–29a. *Installation details for double-hung windows.*

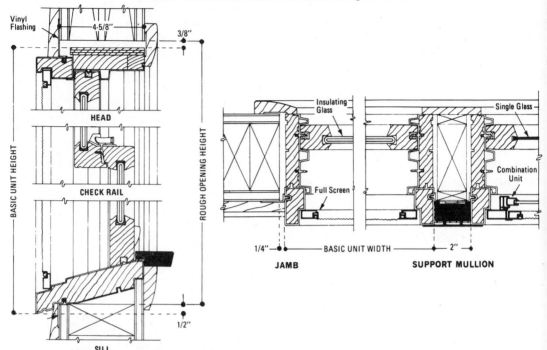

BRICK VENEER AND SOLID MASONRY INSTALLATION DETAILS

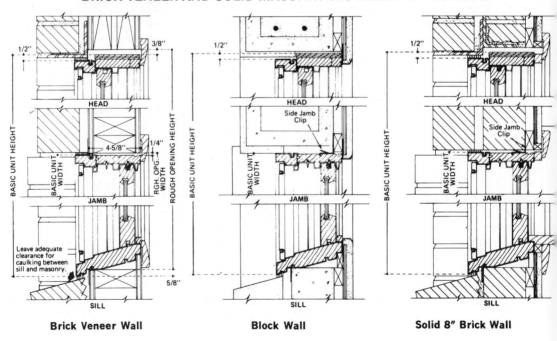

Brick Veneer Wall   **Block Wall**   **Solid 8" Brick Wall**

37-29b. *Installation details for casement windows.*

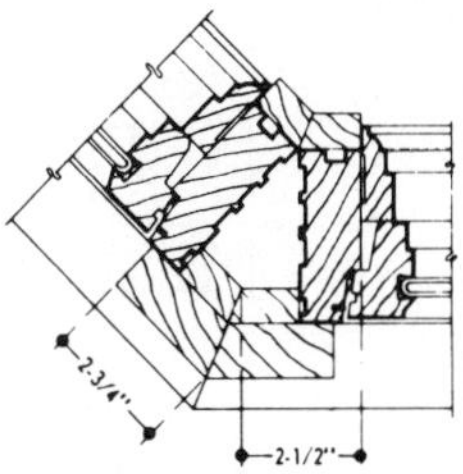

2-3/4"

2-1/2"

NON-SUPPORT MULLION

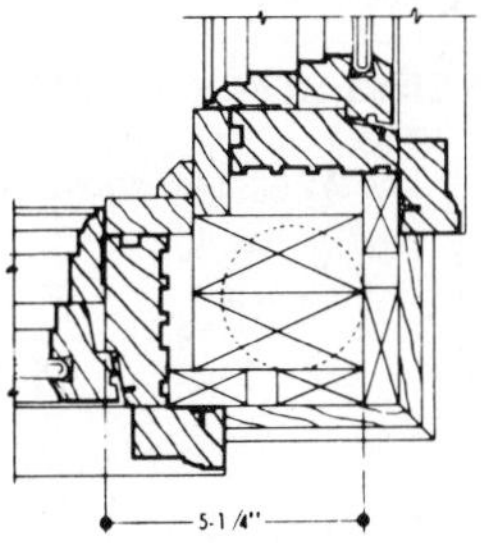

2"
S.O. WIDTH

2-5/8"

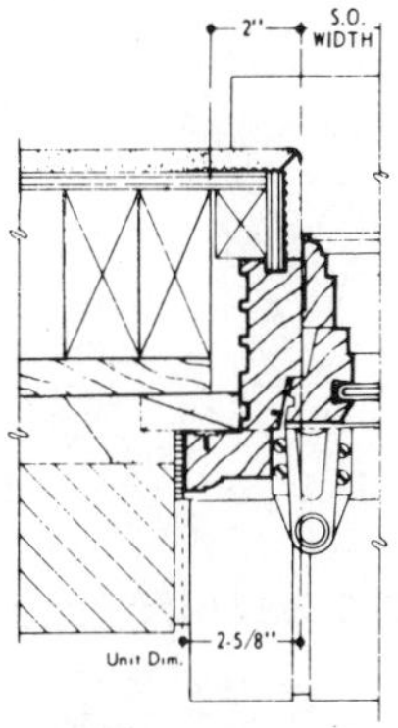

BRICK VENEER

Unit Dim.

5-1/4"

CORNER WINDOW

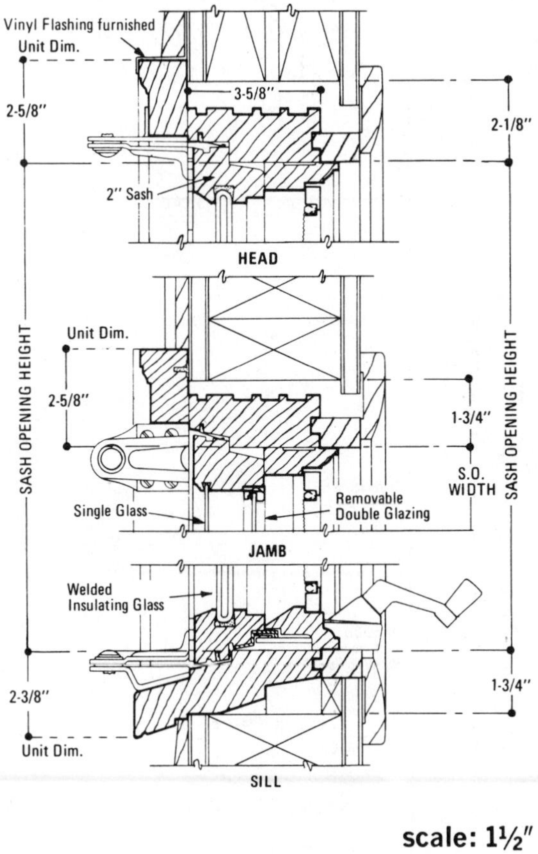

Vinyl Flashing furnished
Unit Dim.

3-5/8"

2-5/8"

2-1/8"

2" Sash

HEAD

SASH OPENING HEIGHT

Unit Dim.

2-5/8"

1-3/4"

S.O. WIDTH

SASH OPENING HEIGHT

Single Glass

Removable
Double Glazing

JAMB

Welded
Insulating Glass

2-3/8"

1-3/4"

Unit Dim.

SILL

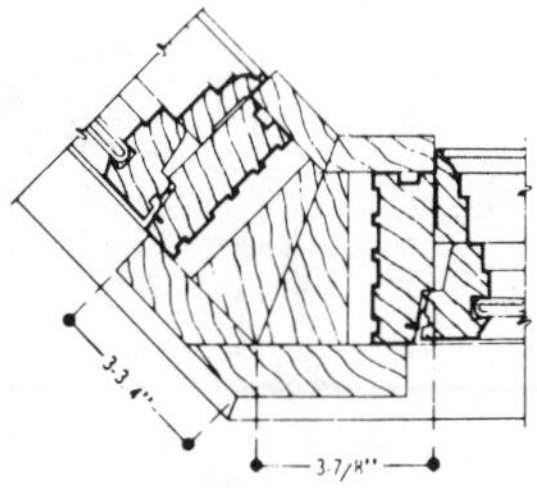

3-3/4"

3-7/8"

SUPPORT MULLION

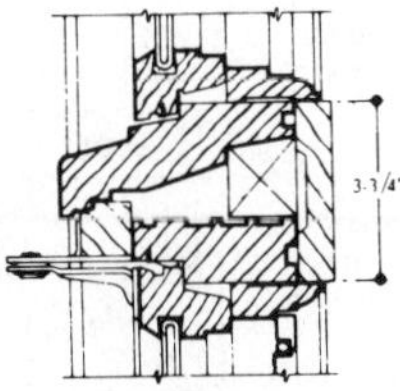

3-1/4"

TRANSOM

scale: 1½" = 1'0"

Details: typical combinations

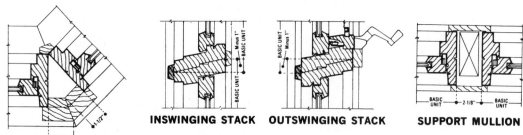

INSWINGING STACK   **OUTSWINGING STACK**   **SUPPORT MULLION**

45° ANGLE BAY

37–29c. *Installation details for awning windows.*

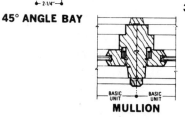

MULLION

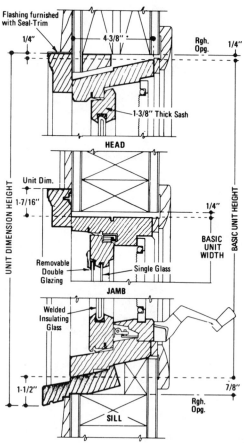

Flashing furnished with Seal-Trim

1/4"   4-3/8"   Rgh. Opg.   1/4"

1-3/8" Thick Sash

HEAD

Unit Dim.

1-7/16"   1/4"

UNIT DIMENSION HEIGHT

BASIC UNIT WIDTH

BASIC UNIT HEIGHT

Removable Double Glazing   Single Glass

JAMB

Welded Insulating Glass

1-1/2"   7/8"

Rgh. Opg.

SILL

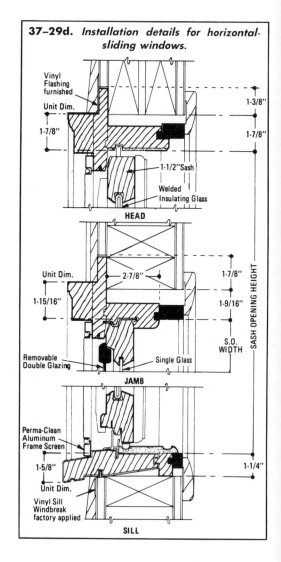

37–29d. *Installation details for horizontal-sliding windows.*

Vinyl Flashing furnished

Unit Dim.

1-7/8"   1-3/8"

1-7/8"

1-1/2"Sash

Welded Insulating Glass

HEAD

Unit Dim.

1-15/16"   2-7/8"   1-7/8"

1-9/16"

S.O. WIDTH

SASH OPENING HEIGHT

Removable Double Glazing   Single Glass

JAMB

Perma-Clean Aluminum Frame Screen

1-5/8"   1-1/4"

Unit Dim.

Vinyl Sill Windbreak factory applied

SILL

37–29e. *Installation details for a casement bow window.*

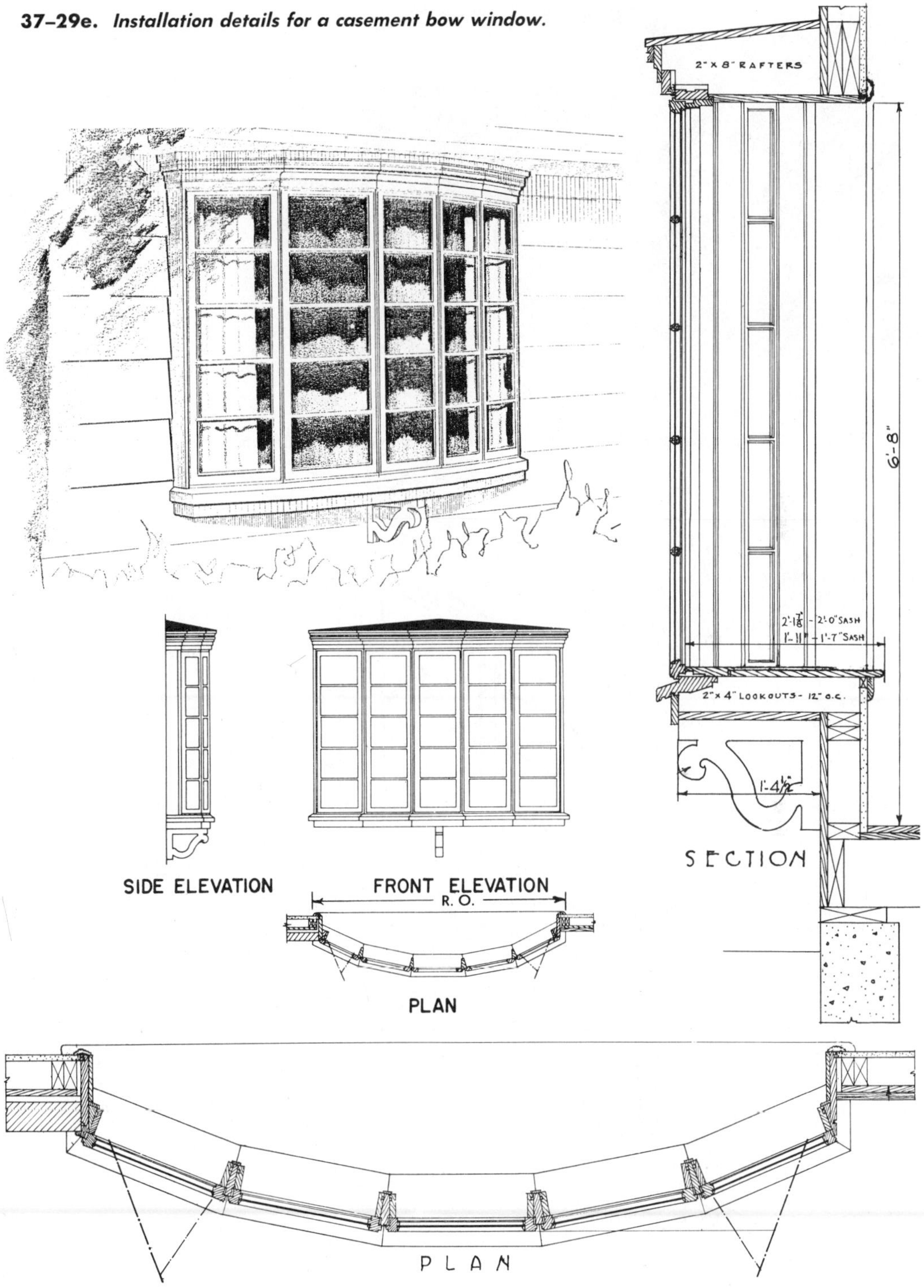

2" X 8" RAFTERS

6'-8"

2'-1⅛" – 2'-0" SASH
1'-11" – 1'-7" SASH

2" X 4" LOOKOUTS – 12" o.c.

1'-4½"

SECTION

SIDE ELEVATION

FRONT ELEVATION
R.O.

PLAN

PLAN

37-30. *These windows have been installed with the sash removed for easier handling.*

37-31. *Window frame installation details for a framed wall with horizontal lap siding.*

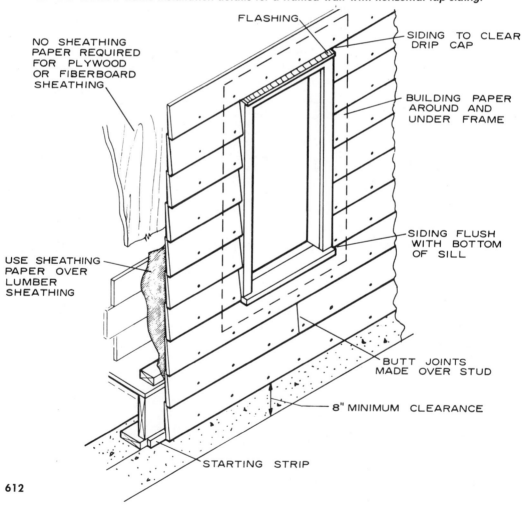

FLASHING

SIDING TO CLEAR DRIP CAP

NO SHEATHING PAPER REQUIRED FOR PLYWOOD OR FIBERBOARD SHEATHING

BUILDING PAPER AROUND AND UNDER FRAME

SIDING FLUSH WITH BOTTOM OF SILL

USE SHEATHING PAPER OVER LUMBER SHEATHING

BUTT JOINTS MADE OVER STUD

8" MINIMUM CLEARANCE

STARTING STRIP

612

37-32a. *Install the window frame in the rough opening.*

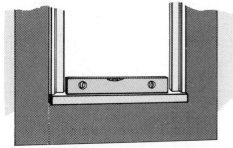

37-32b. *Level the sill, wedge the frame with shingles, and tack it in place.*

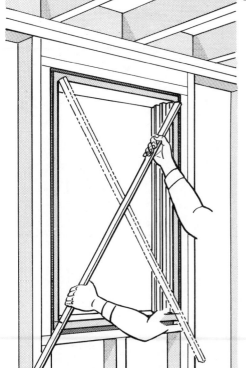

37-32c. *When the diagonal measurements are equal, the unit is square.*

37-32d. *Shim the side jambs and recheck the diagonals.*

37–32e. *Measure the distance between the side jambs to be sure they are equidistant at all points. Shim as necessary to correct.*

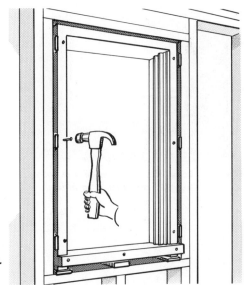

37–32f. *Nail the side jambs through the shims.*

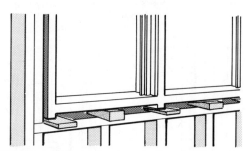

37–32g. *Shim under the raised jamb legs and at the center of long sills and mullions.*

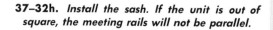

37–32h. *Install the sash. If the unit is out of square, the meeting rails will not be parallel.*

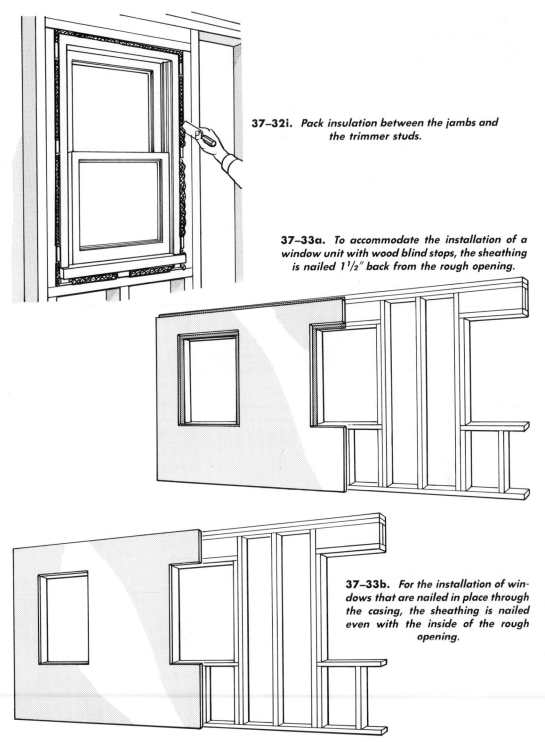

37–32i. *Pack insulation between the jambs and the trimmer studs.*

37–33a. *To accommodate the installation of a window unit with wood blind stops, the sheathing is nailed 1 1/2" back from the rough opening.*

37–33b. *For the installation of windows that are nailed in place through the casing, the sheathing is nailed even with the inside of the rough opening.*

studs and the header over the window. While nailing, open and close the sash to see that it works freely. The side and head casings are fastened in the same manner.

When a panel siding is used without sheathing, the windows are usually installed after the siding is in place. Before installing the window, place a ribbon of calking sealant over the siding at the loca-

tion of side and head casings. Fig. 37-35. Install the window as described earlier. Then place some calking at the junction of the siding and the sill and install a small molding such as a quarter-round. Fig. 37-36. If required, place metal flashing over each window.

On the back of the casing of some windows the manufacturer provides a

37–34a. *Installation details for a double-hung window showing nailing through the blind stops.*

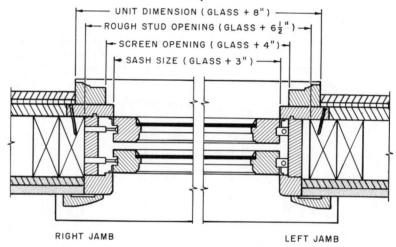

UNIT DIMENSION (GLASS + 8")

ROUGH STUD OPENING (GLASS + $6\frac{1}{2}$")

SCREEN OPENING (GLASS + 4")

SASH SIZE (GLASS + 3")

RIGHT JAMB   LEFT JAMB

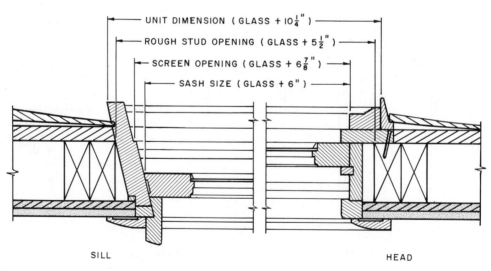

UNIT DIMENSION (GLASS + $10\frac{1}{4}$")

ROUGH STUD OPENING (GLASS + $5\frac{1}{2}$")

SCREEN OPENING (GLASS + $6\frac{7}{8}$")

SASH SIZE (GLASS + 6")

SILL   HEAD

37–34b. *Some window units are available with a vinyl covering to eliminate painting. A vinyl anchorage flange and windbreak then serves as a blind stop (see arrows).*

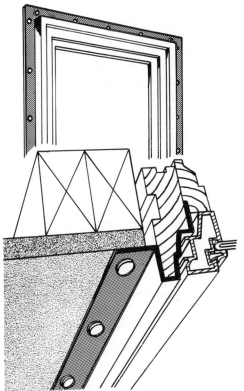

37–34c. *The window is installed with 1³/₄″ galvanized nails through the vinyl anchoring flange. The outside wall covering is applied over this flange. With this method, there are no exposed nails on the exterior of the window.*

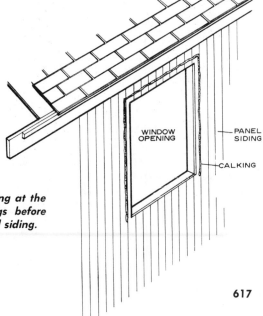

WINDOW OPENING

PANEL SIDING

CALKING

37–35. *Calk around the window opening at the location of the head and side casings before installing the window frame over panel siding.*

built-in calking. This is a bulb-type vinyl that remains pliable and provides a weathertight fit. Fig. 37-37.

Basement Windows

Basement window units are made of wood, plastic, or metal. Figs. 37-38 and 37-39. In most cases the sash is removed from the frame. The frame is set into the concrete forms for a poured wall, and the wall is poured with the window frame in place. If the windows are to be set into a concrete block wall, special blocks are available to accommodate the various types of frames. The floor framing is then constructed on the foundation wall with the window frames already set in place. The sills are usually installed later.

WINDOW SCREENS AND STORM PANELS

Window screens and storms may be an integral part of the window frame, or they may be separate units of wood, metal, or plastic. Fig. 37-40. For double-hung windows, separate units are designed so that the screens and storm panels may be stored within the unit. In this type the lower sash is usually the only one that is screened. The unit is provided with three tracks; the upper window remains in the upper position, the lower window may be slid up out of the way and the screen brought down for warm weather. For cold weather the screen is stored in the upper position, and the storm panel is lowered. Fig. 37-41.

37–36. *Installing a double-hung window frame over panel siding. This frame is installed by nailing through the casing into the studs. Note the molding under the sill.*

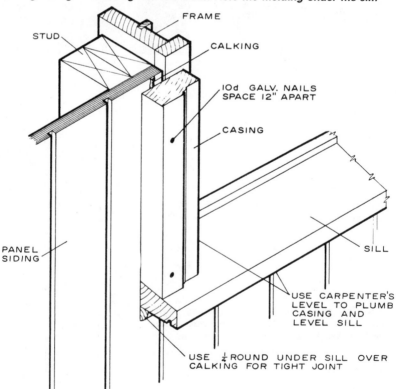

FRAME

STUD

CALKING

IOd GALV. NAILS SPACE 12" APART

CASING

PANEL SIDING

SILL

USE CARPENTER'S LEVEL TO PLUMB CASING AND LEVEL SILL

USE ¼ ROUND UNDER SILL OVER CALKING FOR TIGHT JOINT

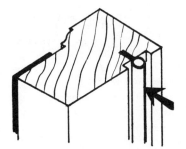

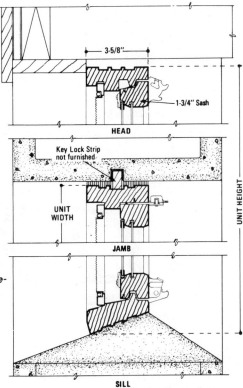

37–37. *A built-in calking on the back of window casings is sometimes provided by the manufacturer.*

HEAD

Key Lock Strip not furnished

UNIT WIDTH

JAMB

UNIT HEIGHT

3-5/8"

1-3/4" Sash

37–38a. *Installation details for a wood basement window unit.*

SILL

Typical basement installation in concrete block wall.

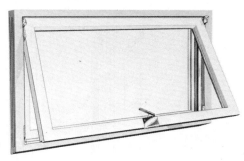

37–38b. *Some wood basement window units are dual hinged so that they may be opened from either the top or the bottom. The insert shows typical sizes of wood basement window units.*

SIZES

| UNIT | 2-8 1/8 |
| GLASS | 28" |

1-3 3/8 / 10"	2813
1-7 3/8 / 14"	2817
1-11 3/8 / 18"	2820

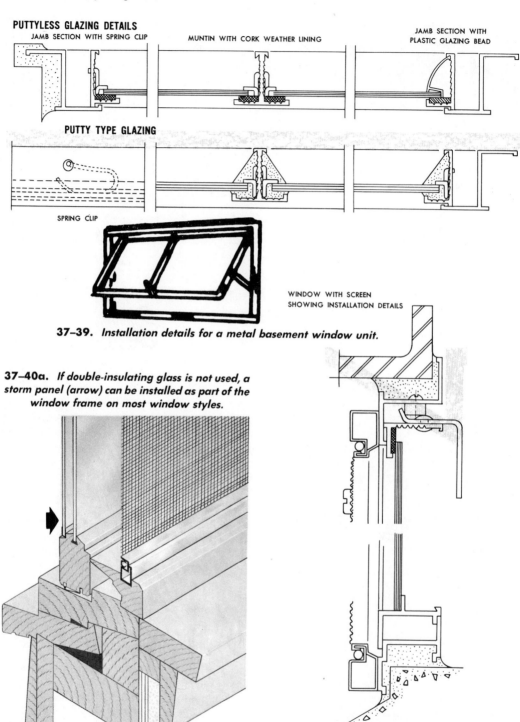

PUTTYLESS GLAZING DETAILS

JAMB SECTION WITH SPRING CLIP   MUNTIN WITH CORK WEATHER LINING   JAMB SECTION WITH PLASTIC GLAZING BEAD

PUTTY TYPE GLAZING

SPRING CLIP

WINDOW WITH SCREEN SHOWING INSTALLATION DETAILS

37-39. Installation details for a metal basement window unit.

37-40a. If double-insulating glass is not used, a storm panel (arrow) can be installed as part of the window frame on most window styles.

37-40b. *On most window styles the storm panel can be easily removed for cleaning.*

37-41a. *Operation of combination storm and screen panels.*

37-40c. *A combination unit showing the frame and the screen and storm panels.*

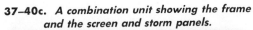

37-41b. *Some combinations of screen and storm sash can be released to allow cleaning of the sill.*

37–42. *Screens for casement windows that open out are installed on the inside.*

With Relative Humidity Conditions as shown in this Table there will be no Condensation

Outside temperature as shown		Inside temperature 70°F	
		Relative Humidity	
Below	− 20°F	Not over	15%
− 20 to	− 10°F	" "	20%
− 10 "	− 0°F	" "	25%
− 0 "	10°F	" "	30%
10 "	20°F	" "	35%
Above	20°F	" "	40%
It is important to prevent excess humidity			

Casement, awning, hopper, and sliding window units have the storm panel set into a rabbet in the outside of the window sash. The screen is a separate insert which is installed for the summer months. Screens used with double-hung, sliding, and hopper windows are installed on the outside. For casement and awning type windows that open out, the screen is installed on the inside with a projecting handle for opening and closing the window sash. Fig. 37-42.

WINDOW CONDENSATION

Condensation (formation of moisture) on windows is the result of improved heating systems, better insulation, and houses that are tightly built. Homes today do not breathe through the many small openings that existed before the extensive use of storm sash, weather stripping, insulation, and automatic heating systems. All of these improvements are fuel savers and add to people's comfort, but they do promote annoying and damaging condensation and steps should be taken to prevent it.

Condensation on the inside window surfaces results from differences in outside and inside temperatures and from the humidity conditions inside the home. Warm, humid air in the interior of the home, when temperatures are cold outside, reaches the dew-point necessary to condensation. Keeping the relative humidity within the home at a point lower than necessary for condensation to occur is the most effective way of preventing condensation on windows. The recommended indoor relative humidity for various outside temperatures is shown in Fig. 37-43. These maximum safe humidities for the home are not only better for the windows, but they will also improve paint performance and insulation and will eliminate problems with structural members.

There are three ways to reduce humidity:
• Controlling sources of humidity. For instance, venting all gas burners and clothes dryers to the outdoors and using kitchen and bathroom exhaust fans helps remove excess moisture from the air.
• Winter ventilation of homes. Because outside air usually contains less water

vapor, it will "dilute" the humidity of inside air. This takes place automatically in older houses through constant infiltration of outside air.

• Proper heating. Dry heat will reduce the relative humidity. It will counterbalance most or all of the moisture produced by modern living.

Fog on the lower corners of windows now and then is not serious. However, *excessive* condensation, condensation that blocks entire windows with fog or frost and produces water droplets, can stain woodwork and in some cases, even damage the wallpaper or plaster. Condensation on windows is easily seen and can be removed. More serious is excessive moisture in the walls and insulation, where it cannot be seen. High humidity resulting in condensation can contribute greatly to the deterioration of a house and to the discomfort of its occupants.

ESTIMATING

The cost of the individual window unit will depend on the quality, the style of the window, the glass to be installed in the unit, the material from which the unit is made, and whether or not it has a factory applied finish. To determine an accurate cost, submit a complete list of the windows that are to be installed to the supplier for pricing.

Labor

The labor required for installing and setting windows will vary considerably, depending on the size and style of the unit. The approximate time can be estimated as follows:

• For a window that contains 10 square feet or less of glass area, figure 1 hour of labor.

• For windows containing up to 20 square feet of glass area, allow an additional half hour.

• For anything over 20 square feet, figure about 2 hours.

These estimates do not include the interior trim, just the preparation of the opening and the actual installation of the window unit.

QUESTIONS

1. What are the principal types of windows?

2. How does a casement window differ from an awning window?

3. What is the main purpose of a stationary window?

4. What is a jalousie window?

5. Where is a window schedule usually found?

6. What information is contained in a window schedule?

7. What type of window installation requires that the sheathing be installed 1½" back from the window rough opening?

8. Why are basement windows the first windows to be installed in a house?

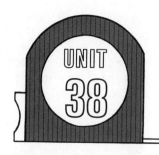
Exterior Doors and Frames

Exterior doors are made from wood or metal and are available in several styles. Figs. 38-1 and 38-2. Care should be taken to select a door that is correct for the architectural style of the house. The exterior trim around the main entrance door can vary in architectural design from a simple casing to a molded or plain pilaster with a decorative head casing. Decorative designs should always be in keeping with the architecture of the house. Figs. 38-3. Many combinations of door and entry designs for every kind of house are available along with millwork items which are adaptable to many styles. Figs. 38-4 and 38-5 (Page 628). For a house with an entry hall having no windows, it is usually desirable to have glass in the main door.

38–1. *Wood doors are available in a large number of styles: A. Three types of panel doors. B. Two types of flush doors.*

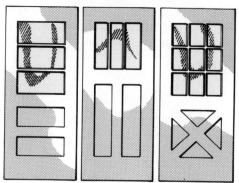

A

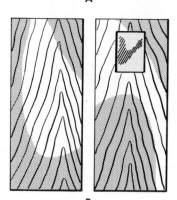

B

TYPES OF EXTERIOR DOORS
Flush Doors

Flush doors are made with plywood or other suitable facing applied over light framework onto a core of suitable thickness Figs. 38-6 and 38-7. There are two types of cores: hollow and solid. Fig. 38-8 (Page 630). Solid core construction is woodblock or particle board and is generally preferred for exterior doors. Solid core construction minimizes warping, particularly in cold climates where differences in humidity occur on opposite sides of a door.

Panel Doors

Panel doors consist of stiles (solid vertical members), rails (solid cross-members), and panels (thinner parts filling spaces between the stiles and the rails). Fig. 38-9. Many types with various wood or glass panels are available. Fig. 38-10.

Glazed Doors

French, or glazed, doors consist of stiles and rails with a space divided into lights by bars called muntins. Such doors are

38-2. *One manufacturer of prehung steel doors offers a flush and three panel door styles which can be adapted to various architectural styles.*

38-3. Many styles of wood doors can be set into a variety of decorative frames to match the architecture of the home.

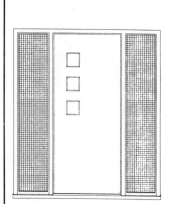

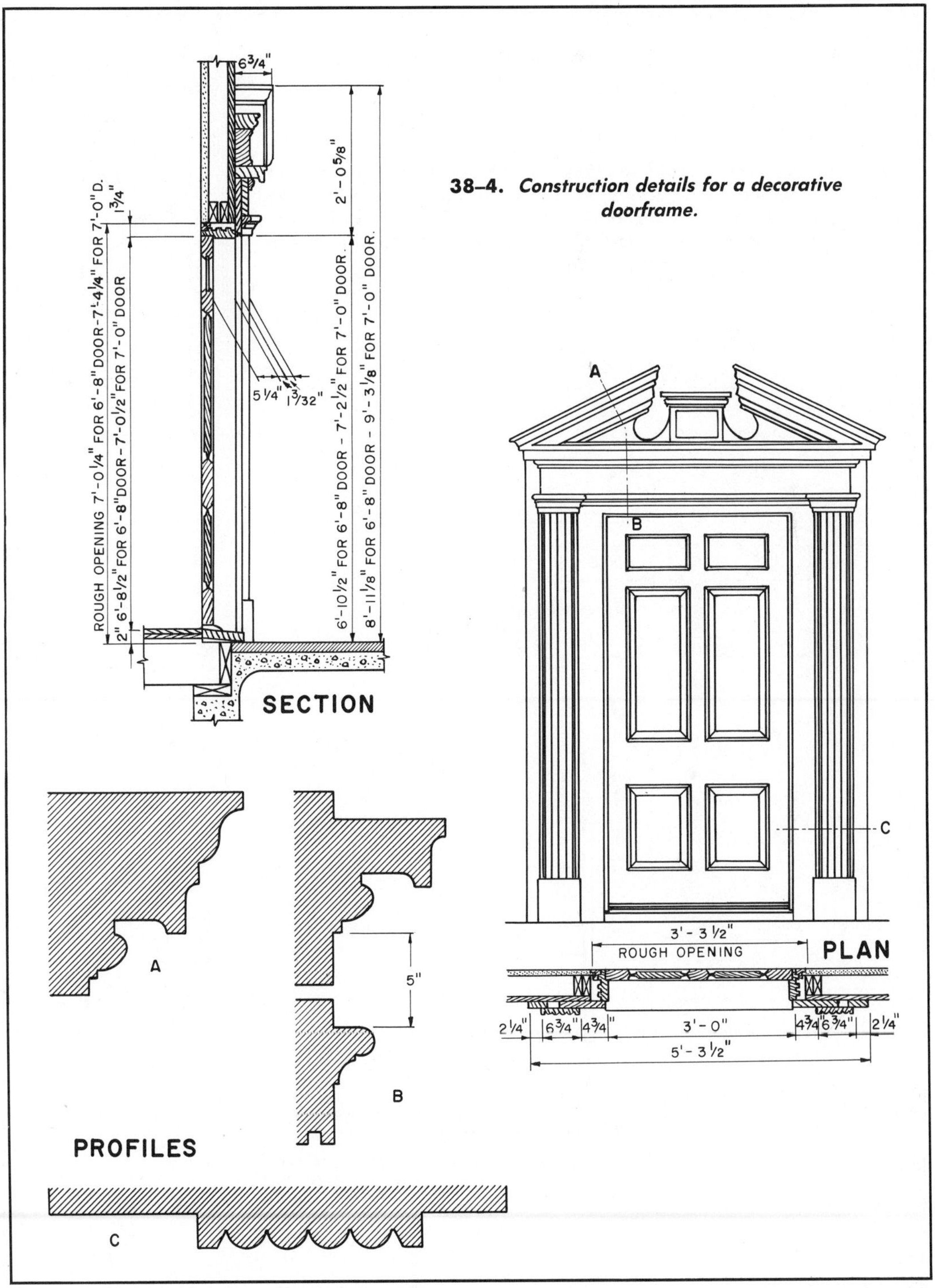
6 3/4"
2' - 0 5/8"
ROUGH OPENING 7'-0 1/4" FOR 6'-8" DOOR - 7'-4 1/4" FOR 7'-0" D.
1 3/4"
2" 6'-8 1/2" FOR 6'-8" DOOR - 7'-0 1/2" FOR 7'-0" DOOR
5 1/4"
1 3/32"
6'-10 1/2" FOR 6'-8" DOOR - 7'-2 1/2" FOR 7'-0" DOOR.
8'-11 1/8" FOR 6'-8" DOOR - 9'-3 1/8" FOR 7'-0" DOOR.
SECTION
A
B
C
3' - 3 1/2"
ROUGH OPENING
PLAN
2 1/4"
6 3/4"
4 3/4"
3' - 0"
4 3/4
6 3/4"
2 1/4"
5' - 3 1/2"
A
5"
B
PROFILES
C

38–4. *Construction details for a decorative doorframe.*

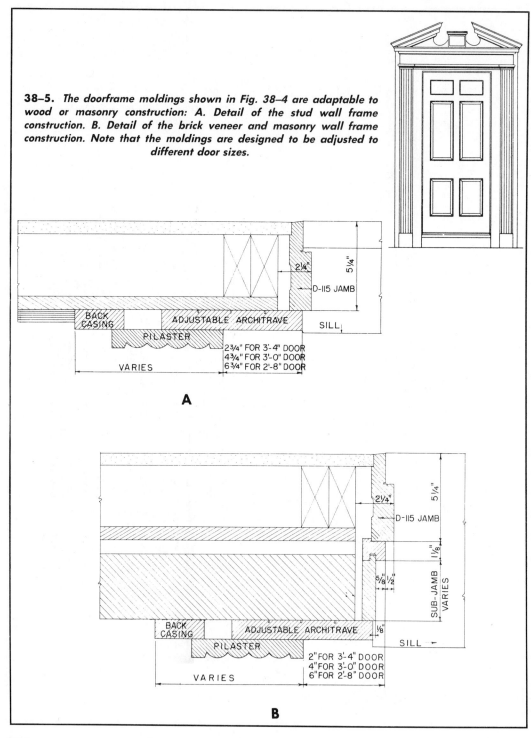

38–5. *The doorframe moldings shown in Fig. 38–4 are adaptable to wood or masonry construction: A. Detail of the stud wall frame construction. B. Detail of the brick veneer and masonry wall frame construction. Note that the moldings are designed to be adjusted to different door sizes.*

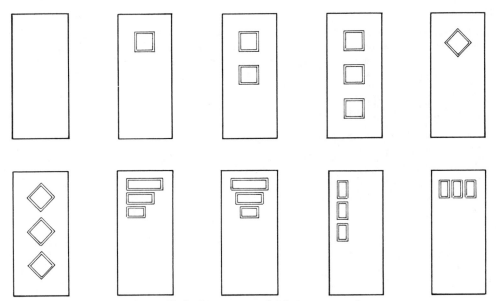

38–6. *Flush doors are available in many styles.*

38–7. *Flush doors can be individually styled by selecting from a variety of insert panels. The panels in B can be mounted on the shaded areas of the door at A. C shows one possibility.*

A

B

C

38–8. *The core construction of a flush door will vary considerably with the manufacturer. The construction details shown here are an example of the techniques used by one manufacturer. Note the built-up areas at the edges near the center for installation of the lock set.*

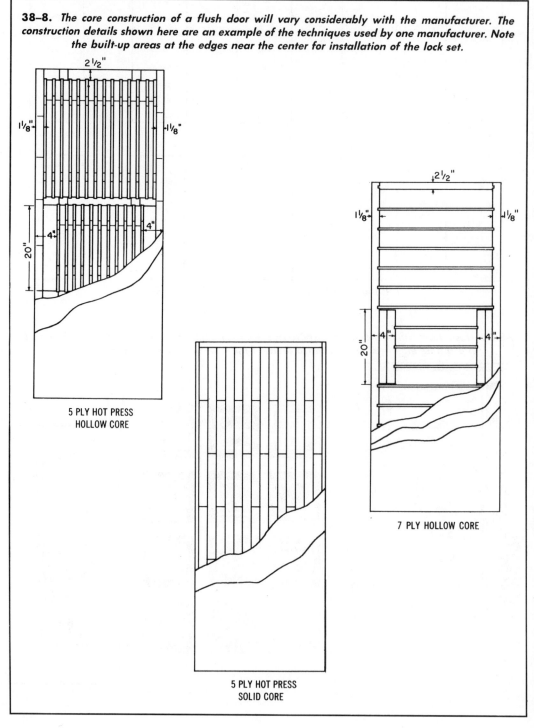

5 PLY HOT PRESS
HOLLOW CORE

7 PLY HOLLOW CORE

5 PLY HOT PRESS
SOLID CORE

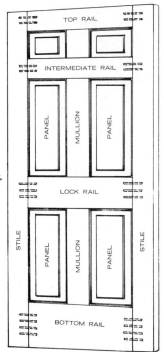

38-9. *Parts of a six-panel door.*

38-10. *Panel doors are available in many styles.*

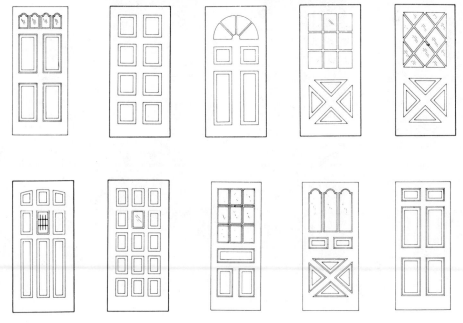

often hung in openings leading to porches or terraces. French doors may be hung singly or in pairs with a half-round molding stop between them. Fig. 38-11.

Sliding Glass Doors

Sliding glass doors are available with either wood or metal frames. Fig. 38-12. The glass may be 1″, ⅝″, or ¼″ thick insulating plate glass, depending on the local climate. These units are available in various combinations of stationary or operating doors in widths from 30″ to 120″. The door operation may be specified as right- or left-hand sliding (as viewed from the outside). Fig. 38-13. Snap-in muntins can be added to create a traditional appearance. Fig. 38-14 (Page 636).

Combination Doors

Combination storm and screen doors of wood or metal are available in several styles. Figs. 38-15 and 38-16. Panels which include screen and storm inserts are normally located in the upper portion of the door. Some types have self-storing features similar to window combination units. Heat loss through metal combination doors is greater than through similar wood doors. Weather-stripping an exterior door will reduce both air infiltration and frosting of the glass on the storm door during cold weather.

DOORFRAMES

A doorframe surrounds a door to conceal or beautify structural building parts. The doorframe consists of the doorjamb, the sill, interior trim, exterior trim, and other molding, depending on the architectural

38-12. *A wood sliding glass door with regular muntins.*

38-11. *French doors.*

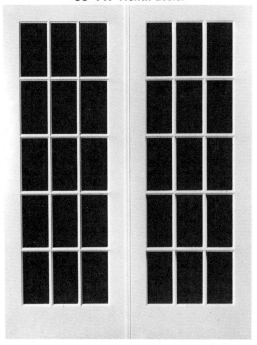

38–13a. *A table from a manufacturer's catalog illustrating wood sliding glass doors.*

WOOD DOOR SIZES

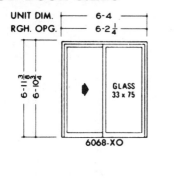

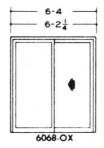

UNIT DIM. 6-4   6-4   3-5½
RGH. OPG. 6-2¼   6-2¼   3-3⅞

GLASS 33 x 75

SIDE LIGHT

6068-XO   6068-OX   3068

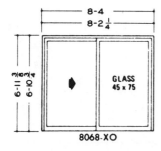

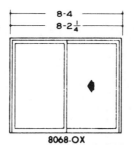

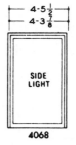

8-4   8-4   4-5½
8-2¼   8-2¼   4-3⅞

GLASS 45 x 75

SIDE LIGHT

8068-XO   8068-OX   4068

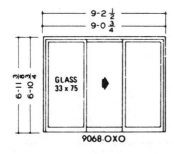

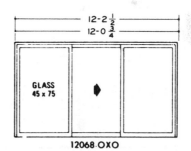

9-2½   12-2½
9-0¾   12-0¾

GLASS 33 x 75

GLASS 45 x 75

9068-OXO   12068-OXO

NUMBERING SYSTEM

No. 30, 40, 60, 80, 90 & 120—Unit Width
No. 68—Unit Height
X—Operating Door

O—Stationary Door
(Numbering figured as viewed from outside)

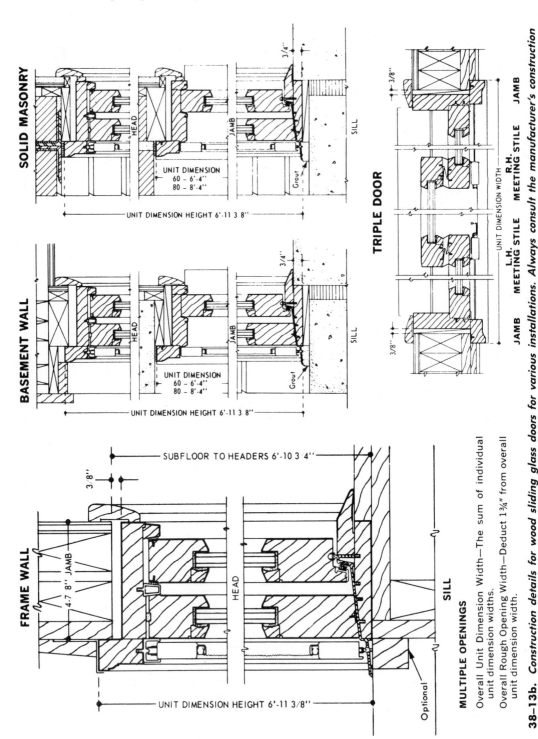

SOLID MASONRY

HEAD

JAMB

SILL

UNIT DIMENSION
60 - 6'-4''
80 - 8'-4''

Grout

3/4''

UNIT DIMENSION HEIGHT 6'-11 3 8''

BASEMENT WALL

HEAD

JAMB

SILL

UNIT DIMENSION
60 - 6'-4''
80 - 8'-4''

Grout

3/4''

UNIT DIMENSION HEIGHT 6'-11 3 8''

TRIPLE DOOR

3/8''

UNIT DIMENSION WIDTH

L.H.   R.H.

JAMB   MEETING STILE   MEETING STILE   JAMB

3/8''

3/8''

FRAME WALL

4-7 8'' JAMB

HEAD

SILL

3/8''

SUBFLOOR TO HEADERS 6'-10 3 4''

UNIT DIMENSION HEIGHT 6'-11 3/8''

Optional

MULTIPLE OPENINGS

Overall Unit Dimension Width—The sum of individual unit dimension widths.

Overall Rough Opening Width—Deduct 1¾" from overall unit dimension width.

38–13b. Construction details for wood sliding glass doors for various installations. Always consult the manufacturer's construction details for the specific door to be installed.

38-13c. *Construction details for aluminum sliding glass doors.*

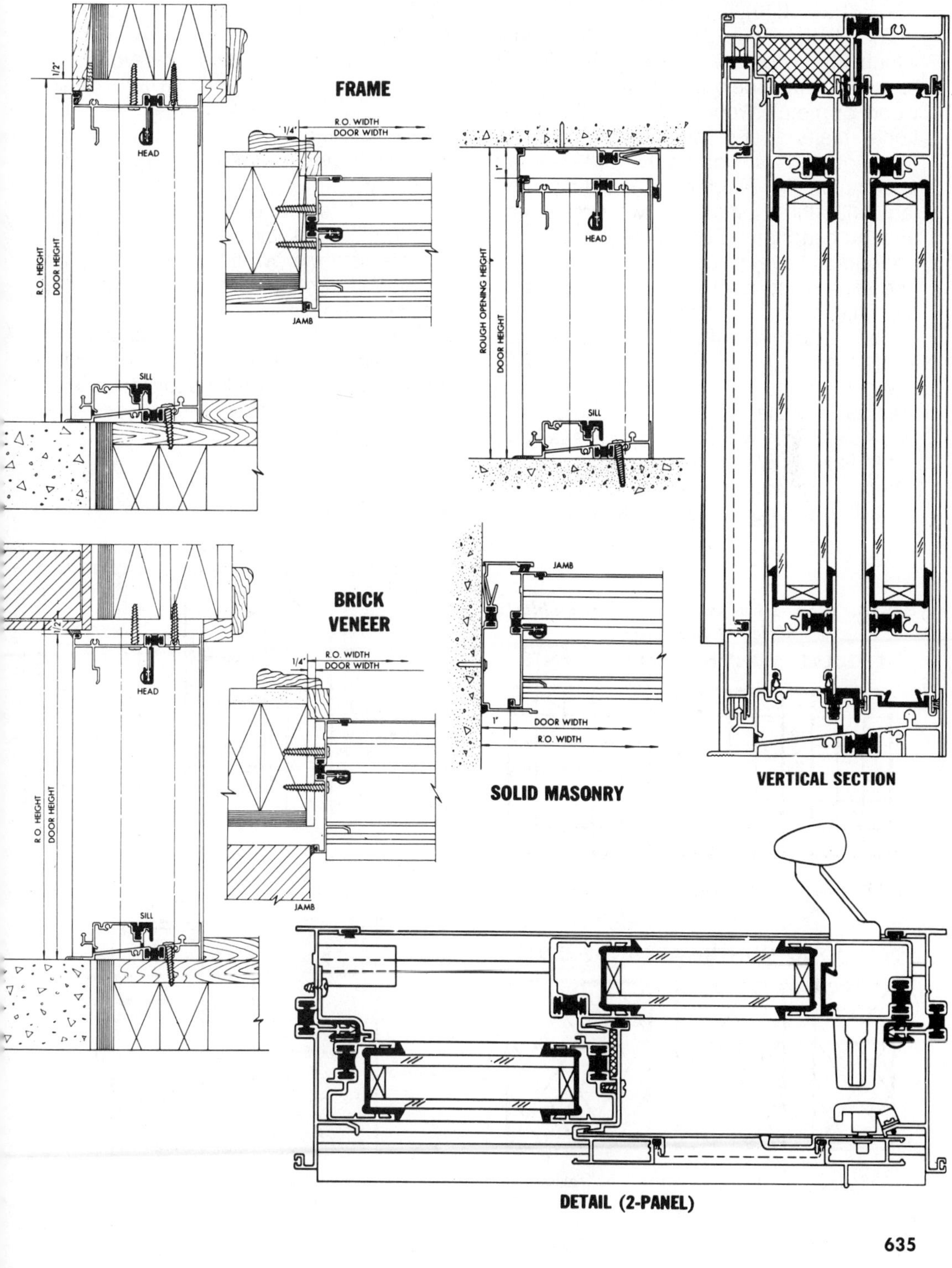

FRAME

HEAD

JAMB

SILL

R.O. WIDTH
DOOR WIDTH

1/4"

ROUGH OPENING HEIGHT
DOOR HEIGHT

HEAD

SILL

R.O. HEIGHT
DOOR HEIGHT

1/2"

**BRICK
VENEER**

HEAD

JAMB

SILL

R.O. WIDTH
DOOR WIDTH

1/4"

1/2"

R.O. HEIGHT
DOOR HEIGHT

SOLID MASONRY

JAMB

DOOR WIDTH
R.O. WIDTH

1"

VERTICAL SECTION

DETAIL (2-PANEL)

design of the building. Figs. 38-17 and 38-18 (Page 640).

The doorjamb is the part of the frame which fits inside the masonry opening or rough frame opening. Jambs may be wood or metal. Wood has been the traditional material, but steel and aluminum have gained in popularity and are not uncommon in residential building. The jamb has three parts: the two side jambs and the head jamb across the top. Exterior doorjambs have a stop as part of the jamb. The stop is the portion of the jamb which the face of the door closes against. The jamb is 1 1/8" thick with a 1/2" rabbet serving as a stop.

Wood jambs are manufactured in two standard widths: 5 1/4" for lath and plaster and 4 1/2" for dry wall. Jambs may easily be cut to fit walls of less thickness. If the jamb is not wide enough, strips of wood are nailed on the edges to form an extension.

Jambs may also be custom-made to any size to accommodate various wall thicknesses.

Standard metal jambs are available in the following widths for lath and plaster, concrete block, brick veneer, etc.: 4 3/4", 5 3/4", 6 3/4", and 8 3/4". For dry-wall construction the common widths available are 5 1/2" and 5 5/8".

The sill is the bottom member in the doorframe. It is usually made of oak for wear resistance. When softer wood is used for the sill, a metal nosing and wear strips are included.

The brick mold or outside casings are designed and installed to serve as stops for the screen or combination door, which is 1 1/8" thick. The stops are provided for by

38–15. Combination doors are made in many styles.

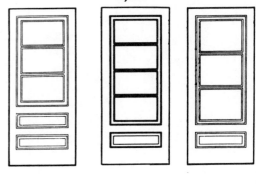

38–14. A variety of muntins are available for different glass sizes.

REGULAR MUNTIN ARRANGEMENTS

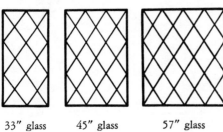

33" glass   45" glass   57" glass

DIAMOND MUNTIN ARRANGEMENTS

33" glass   45" glass   57" glass

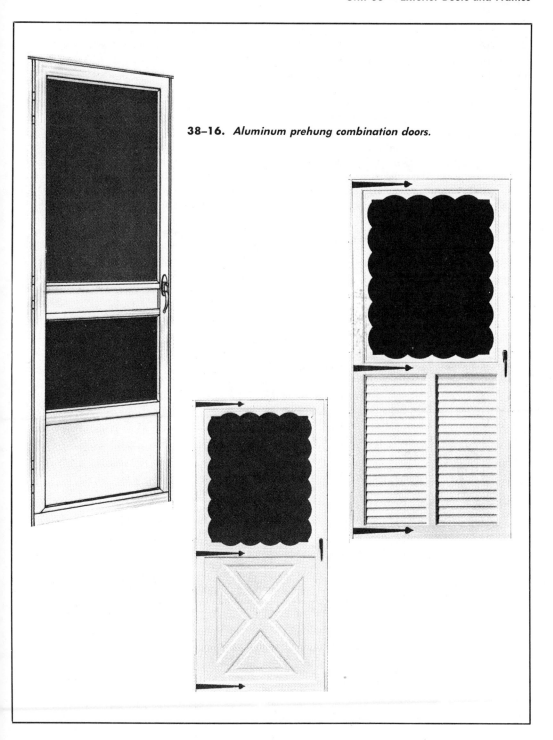

38-16. *Aluminum prehung combination doors.*

the edge of the jamb and the exterior casing thickness. Fig. 38-17.

Doorframes may be purchased knocked down (K.D.) or preassembled with just the exterior casing or brick mold applied. In some cases, they come preassembled with the door hung in the opening. Fig. 38-19 (Page 642). When the doorframe is assembled on the job, nail the side jambs to the head jamb and sill with 10d casing nails. Then nail the casings to the front edges of the jambs with 10d casing nails spaced 16″ on center.

Exterior doors are 1³/₄″ thick and not less than 6′8″ high. The main entrance door is 3′ wide, and the side or rear service door is 2′8″ wide. A hardwood or metal threshold covers the joint between the sill and the finished floor. Fig. 38-20.

Installing the Exterior Doorframe

Before installing the exterior doorframe, prepare the rough opening to receive the frame. (The opening should be somewhat larger each way—3″ wider and 2″ higher—than the size of the door.) The sill should rest firmly on the floor framing, which commonly must be cut out to accommodate the sill. Fig. 38-21. The subfloor, floor joists, and stringer or header joist must be cut to a depth which will place the top of the sill even with the finished floor surface. Fig. 38-22.

Line the rough opening with a strip of 15-pound asphalt felt, 10″ or 12″ wide, as when installing windows. The assembled frame is then set into the opening. Set the sill of the assembled doorframe on the trimmed-out area in the floor framing, tip

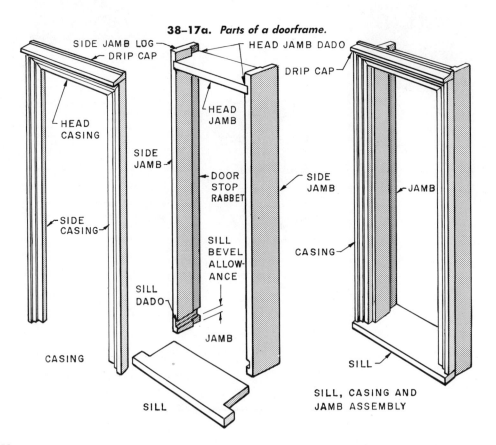

38–17a. *Parts of a doorframe.*

SIDE JAMB LUG
DRIP CAP
HEAD JAMB DADO
DRIP CAP
HEAD CASING
HEAD JAMB
SIDE JAMB
SIDE JAMB
SIDE CASING
DOOR STOP RABBET
SIDE JAMB
JAMB
CASING
SILL BEVEL ALLOWANCE
SILL DADO
JAMB
CASING
SILL
SILL
SILL
SILL, CASING AND JAMB ASSEMBLY

the frame into place, and brace it to keep it from falling out during adjustment. Fig. 38-23.

An outside doorframe is plumbed in the rough opening with wood shingles used as wedges. These are inserted at intervals up the side jambs, between the jambs and the trimmer studs. Check the sill with a level, and wedge it up as necessary. Insert the side jamb wedges. Drive the lower wedges on each side alternately until the space between the side jamb and the trimmer stud is exactly the same on both sides.

Then drive a 16d casing nail through the side casing and into the trimmer studs on each side, near the bottom of the casing, to hold the sill in position. Drive the nails in only partway. Do not drive any nails all the way in until all the nails have been placed and a final check has been made for level and plumb.

Next place the level against one of the side jambs and adjust the remaining wedges on that side until the jamb is perfectly true and plumb. Repeat the same procedure on the other side. Make a final

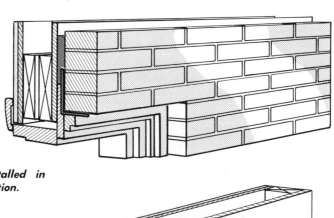

38–17b. Doorframe installed in brick-veneer construction.

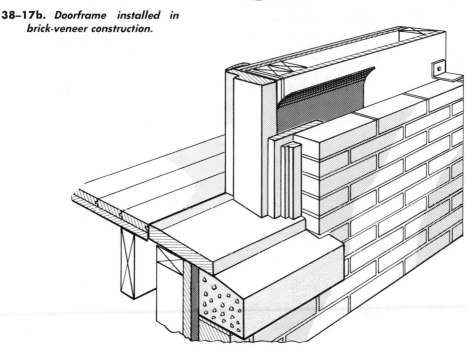

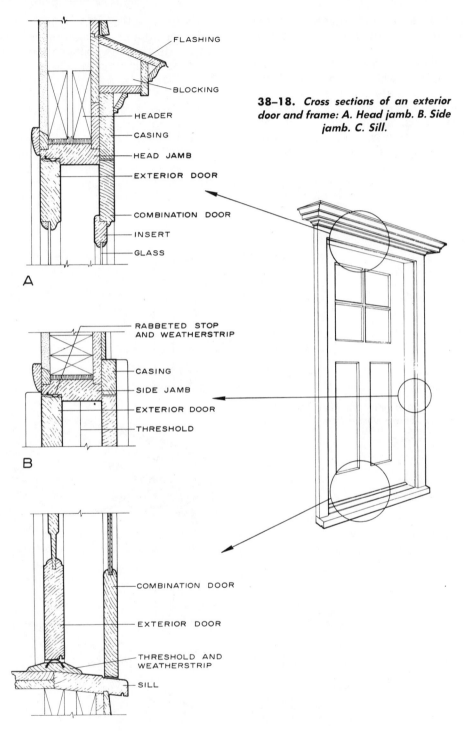

FLASHING

BLOCKING

HEADER

CASING

HEAD JAMB

EXTERIOR DOOR

COMBINATION DOOR

INSERT

GLASS

A

38–18. *Cross sections of an exterior door and frame: A. Head jamb. B. Side jamb. C. Sill.*

RABBETED STOP AND WEATHERSTRIP

CASING

SIDE JAMB

EXTERIOR DOOR

THRESHOLD

B

COMBINATION DOOR

EXTERIOR DOOR

THRESHOLD AND WEATHERSTRIP

SILL

check for level and plumb. Fasten the frame in place with 16d casing nails driven through the casings into the trimmer studs and the door header. Nails are placed ³/₄" from the outer edges of the casings and spaced about 16" on center. Set all nails with a nail set.

After the finish flooring is in place, a hardwood or metal threshold with a plastic weather strip covers the joint between the floor and the sill. Fig. 38-22. Thresholds are installed under exterior doors to close the space allowed for clearance. Weather stripping should be installed around exterior door openings to reduce drafts. Fig. 38-24 (Page 644).

HANDLING THE DOOR AT THE JOB SITE

A door is an important part of the building and has many functions. It guards the building and its possessions, insures privacy, protects against the elements, and lends beauty, refinement, and character to the building. A door is a high-grade precision-made item of cabinetwork and should be treated as such. Proper care and finishing of a door will insure maximum service and satisfaction.

Doors should not be delivered to the building site until after the plaster or concrete is dry, and then the doors should be:
• Stored under cover in a clean, dry, well-ventilated building, not in damp, moist, or freshly plastered areas.
• Stored on edge on a level surface.
• Sealed immediately on the top and bottom edges if they are to be stored at the job site for more than one week.
• Handled with clean gloves; bare hands leave finger marks and soil stains.
• Handled carefully. When moving doors, carry them. Do not drag a door except on the bottom end, and then only if it is protected by scuff strip or skid shoes. Do not drag one door across another.
• Conditioned to the average moisture content of the locality before hanging.

• Finished as soon as the doors are hung in the opening.
• Kept away from abnormal heat, dryness, or humidity. Sudden changes, such as forced heat to dry out a building, should be avoided.
• Straight. Before hanging, warp or bow can usually be eliminated by laying (or piling) the door (or doors) flat under weight. Bow or warp is due to stress forces in the door, usually caused by unequal moisture conditions on the two sides of the door. Improper installation of hinges can also be the cause. When moisture differential is the cause, the door will usually straighten when the moisture equalizes. When improper installation is the cause, hinges should be adjusted.

DETERMINING THE HAND OF A DOOR

A door is designated as having right-hand or left-hand swing. The hand of a door is determined by the location of the hinges when the door is viewed from the outside. For example, if the hinges are on the right when the door is viewed from the outside, the door is considered a right-hand door. Fig. 38-25. In general, the outside of a door is the side from which the hinges are not visible when the door is closed. However, the outside of a closet door is the room side.

FITTING A DOOR

The first step in fitting a door is to determine from the floor plan which edge of the door is the hinge edge and which is the lock edge. Mark both door edges and the corresponding jambs accordingly.

Carefully measure the height of the finished opening on both side jambs and the width of the opening at top and bottom. The finished opening should be perfectly rectangular, but it may not be. Regardless of the shape of the opening, the job is to fit the door accurately to the opening. A well-fitted door, when hung, should conform to the shape of the finished opening, less a

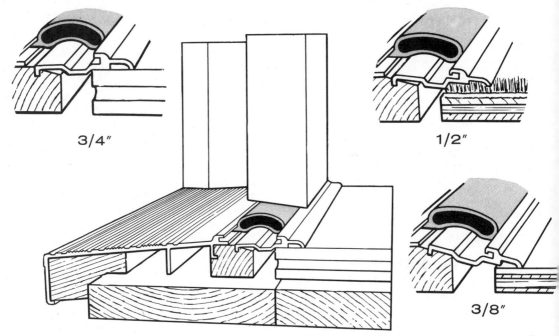

3/4"

1/2"

3/8"

38-19. *Preassembled doorframes offer many conveniences for the worker. This one features a sill which is adjustable to eliminate trimming the floor joists.*

38-20a. *A metal threshold with a vinyl insert.*

38-20b. *Installing a metal threshold.*

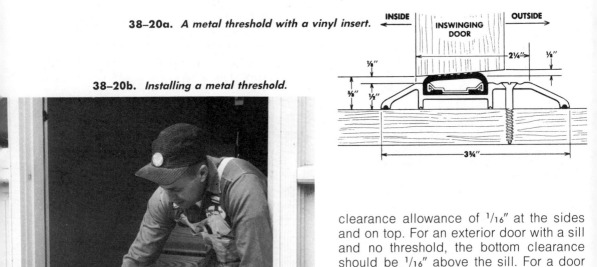

clearance allowance of $1/16''$ at the sides and on top. For an exterior door with a sill and no threshold, the bottom clearance should be $1/16''$ above the sill. For a door with a threshold, the bottom clearance should be $1/8''$ above the threshold. The sill and threshold, if any, should be set in place before the door is hung. Lay out the

642

EXTERIOR DOOR FRAME NEEDS ADDITIONAL SUPPORT

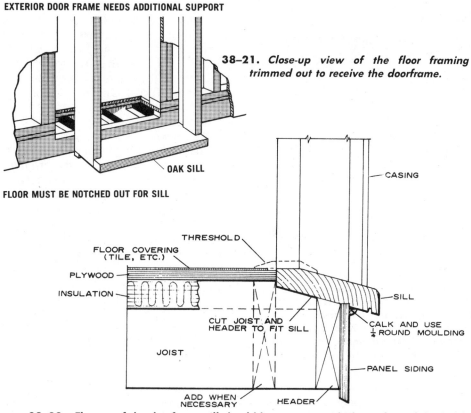

38–21. *Close-up view of the floor framing trimmed out to receive the doorframe.*

OAK SILL

FLOOR MUST BE NOTCHED OUT FOR SILL

CASING

THRESHOLD

FLOOR COVERING
(TILE, ETC.)

PLYWOOD

INSULATION

SILL

CUT JOIST AND
HEADER TO FIT SILL

CALK AND USE
$\frac{1}{4}$ ROUND MOULDING

JOIST

PANEL SIDING

ADD WHEN
NECESSARY

HEADER

38–22. *The top of the doorframe sill should be set even with the surface of the finish floor.*

38–23. *Installing an exterior doorframe.*

measured dimensions of the finished opening, less allowances, on the door.

Check the doorjambs for trueness and transfer any irregularities to the door lines. Plane the door edges to the lines, setting the door in the opening frequently to check the fit. The lock edge of a door must be beveled so that the inside edge will clear

643

Head of Door

Side of Door

Door Bottom

38–24. *There are many kinds of weather stripping available to reduce air infiltration. Shown here are two types: one for the head and side of the door and a second for installation on the door bottom.*

the jamb (at point A in Fig. 38-26) when the door is opened. The bevel required for this clearance is laid out by drawing a line from the point where the hinge pin will be located (B in Fig. 38-26) to the door's other side at the point where it intersects with the door stop (point C). Then place a T bevel on the face of the door and set the blade so that it is parallel to line AC. Fig. 38-26. As shown in the illustration, this can be easily done by placing the blade against the inside edge of the framing square. Plane the edge as necessary, checking frequently with the T bevel to determine the correct angle. When all the planing has been completed, use a piece of sandpaper to form a slight radius on all edges to remove the sharpness.

As an aid in fitting the door, a door jack similar to the one shown in Fig. 38-27 should be constructed. The jack will hold the doors upright for planing edges and for the installation of hardware. Commercially made holders are also available. Fig. 38-28.

38–25. *Determining the hand of a door.*

SINGLE DOOR

<u>HAND OF DOOR</u> MAY BE DETERMINED BY REFERRING TO SKETCHES BELOW. DOOR MUST ALWAYS SWING AWAY FROM POINT VIEWED.

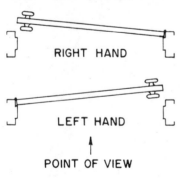

RIGHT HAND

LEFT HAND

↑

POINT OF VIEW

PAIRS OF DOORS

<u>HAND OF DOORS</u> IS DETERMINED BY LOCATION OF ACTIVE LEAF WHEN DOORS SWING AWAY FROM POINT VIEWED.

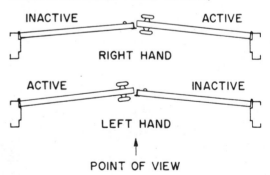

INACTIVE   ACTIVE

RIGHT HAND

ACTIVE   INACTIVE

LEFT HAND

↑

POINT OF VIEW

HANGING A DOOR

The hinge most frequently used for hanging doors on a residential building is the loose-pin butt mortise hinge. Fig. 38-29. This has two rectangular leaves pivoted on a pin which is called a loose pin because it can be removed. The hinge is called a mortise hinge because the leaves are mortised into gains cut in the edge of the door and in the hinge jamb of the doorframe.

After the door has been properly fitted, the first step in hanging it is to lay out the locations of the hinges on the edge of the

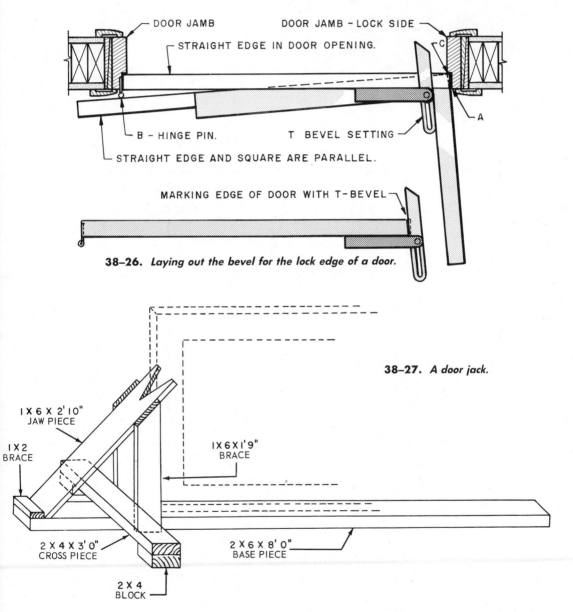

DOOR JAMB   DOOR JAMB – LOCK SIDE

STRAIGHT EDGE IN DOOR OPENING.

C

A

B – HINGE PIN.   T BEVEL SETTING

STRAIGHT EDGE AND SQUARE ARE PARALLEL.

MARKING EDGE OF DOOR WITH T-BEVEL

38–26. Laying out the bevel for the lock edge of a door.

38–27. A door jack.

I X 6 X 2' I0"
JAW PIECE

I X 2
BRACE

I X 6 X I' 9"
BRACE

2 X 4 X 3' 0"
CROSS PIECE

2 X 6 X 8' 0"
BASE PIECE

2 X 4
BLOCK

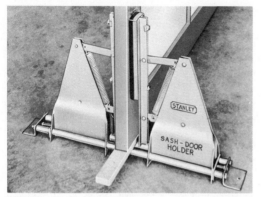

38–28. *A commercially made sash and door holder.*

door and the hinge jamb. Exterior doors usually have three hinges. The following distances may be specified: the vertical distance between the top of the door and the top of the top hinge, and the vertical distance between the top of the finish floor and the bottom of the bottom hinge. If these distances are not specified, the distances customarily used are those shown in Fig. 38-30. The middle hinge is located midway between the other two. The size of a loose-pin butt mortise hinge is designated by the length of a leaf in inches. For an exterior door a 3½" or 4" hinge is recommended.

38–29. *A loose-pin butt mortise hinge.*

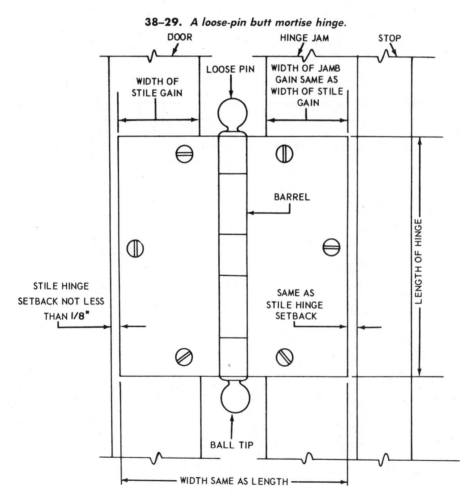

Set the door in the frame and force the hinge edge of the door against the hinge jamb with the wedge marked A in Fig. 38-30. Then insert a 4d finish nail between the top of the door and the head jamb and force the top of the door up against the nail with the wedge marked B in Fig. 38-30. Since a 4d finish nail has a diameter of $^1/_{16}$"

(which is the standard top clearance for a door), the door is now at the correct height.

Measure out the distance from the top of the door to the top of the top hinge and from the floor up to the bottom of the bottom hinge. Mark these locations with a $^1/_2$" chisel or a knife. If a chisel is used, hold it so that the bevel of the chisel is

38–30. *Distances commonly used in laying out hinge locations on the door and the doorjamb.*

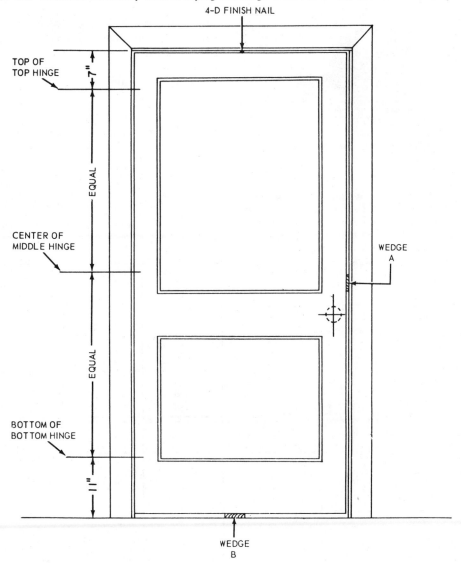

4-D FINISH NAIL

TOP OF TOP HINGE

CENTER OF MIDDLE HINGE

WEDGE A

BOTTOM OF BOTTOM HINGE

WEDGE B

toward the location of the hinge. For example, when marking the bottom hinge, the bevel on the chisel should be held up, and when marking the top hinge, the bevel on the chisel should be held down. Hold the chisel with the cutting edge in a level position so that it is in contact with both the jamb and the edge of the door. Apply pressure and make a small cut into both surfaces to mark the position of the hinge.

38–31. *A hinge butt gauge.*

38–32a. *The door hinge should be set back sufficiently to allow the door to clear the casing when the door is swung wide open. With a $1^3/_4"$ exterior door and $4"$ butt hinges, the maximum clearance is $1"$, as specified on the chart in Fig. 38–32b.*

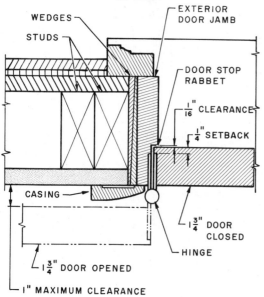

WEDGES
STUDS
EXTERIOR DOOR JAMB
DOOR STOP RABBET
$\frac{1}{16}"$ CLEARANCE
$\frac{1}{4}"$ SETBACK
CASING
$1\frac{3}{4}"$ DOOR CLOSED
HINGE
$1\frac{3}{4}"$ DOOR OPENED
$1"$ MAXIMUM CLEARANCE

When marking for the center hinge, remember that the location line is to the center of the hinge; if a $4"$ hinge is used, measure $2"$ on one side of the location line and mark this point with a chisel. To help avoid mistakes it is best to pencil a small X on the side of the chisel mark where the gain for the hinge will be cut.

Remove the door from the opening. Place the door in a door jack and lay out the outlines of the gains on the edge of the door using a hinge leaf or a hinge butt gauge as a marker. Fig. 38-31. The door-edge hinge setback, shown in Fig. 38-29, should not be less than $1/_8"$. It is usually made about $1/_4"$. Fig. 38-32. Lay out gains of exactly the same size on the hinge jamb.

38–32b. *Trim clearances for wood and metal doors.*

Trim Clearance

The following table gives the clearances for trim of regular stock size butt hinges for wood or hollow metal doors. The clearance is estimated on butt hinges set back $1/_4"$ for doors up to $2^1/_4"$ and $3/_8"$ for doors $2^1/_2"$ to $3"$ in thickness. Where trim presents a specific problem in determining the proper width of the butt hinges for a door, take twice the thickness of the door, plus the thickness of the trim and deduct $1/_2"$ for doors up to $2^1/_4"$ in thickness, and $3/_4"$ for doors $2^1/_2"$ to $3"$ in thickness.

Thickness of Door (Inches)	Size of Butt Hinge (Inches)	Maximum Clearance (Inches)
$1^3/_8$	3 x 3 $3^1/_2$ x $3^1/_2$ 4 x 4	$^3/_4$ $1^1/_4$ $1^3/_4$
$1^9/_{16}$	4 x 4 $4^1/_2$ x $4^1/_2$ 5 x 5 6 x 6	$1^3/_8$ $1^7/_8$ $2^3/_8$ $3^3/_8$
$1^3/_4$	4 x 4 $4^1/_2$ x $4^1/_2$ 5 x 5 6 x 6	1 $1^1/_2$ 2 3
$1^7/_8$	$4^1/_2$ x $4^1/_2$ 5 x 5 6 x 6	$1^1/_4$ $1^3/_4$ $2^3/_4$
2	$4^1/_2$ x $4^1/_2$ 5 x 5 6 x 6	1 $1^1/_2$ $2^1/_2$
$2^1/_4$	5 x 5 6 x 6 6 x 8	1 2 4
$2^1/_2$	5 x 5 6 x 6 6 x 8	$^3/_4$ $1^3/_4$ $3^3/_4$

Chisel out the gains to a depth equal to the thickness of the hinge leaf.

Separate the leaves on the hinges by removing the loose pins. Screw the leaves into the gains on the door and the jamb. Make sure that the leaf in which the pin will be inserted is in the up position when the door is hung in place. Hang the door in place, insert the loose pins, and check the clearances at the side jambs. If the clearance along the hinge jamb is too large (more than 1/16″) and that along the lock jamb is too small (less than 1/16″) extract the pins from the hinges and remove the door. Then remove the hinge leaves from the gains and slightly deepen the gains. If the clearance along the hinge jamb is too small and that along the lock jamb is too large, the gains are too deep. This can be corrected by shimming up the leaves with strips of cardboard placed in the gains under the hinges.

Hinge Butt Routing

A special template is available for hinge butt routing. Fig. 38-33a. The metal template may be adjusted for most common hinge spacings, and it is easily mounted on the door by driving six nails to hold the templates securely on the door. This template guides the router so that the hinge mortises are cut quickly and accurately to size and location. Fig. 38-33b. After the gains or mortises are cut on the door,

the template guide can be transferred to the doorjamb for cutting the hinge mortises to match those on the door. Because the bits leave a radius at the corner of the cut, it is necessary to chisel the corners square for the hinges. It is also possible to purchase hinges designed with round corners. Fig. 38-33c.

INSTALLING A LOCK SET

Lock sets come in many styles, from very simple to ornate. Fig. 38-34. Some are mounted in the center of the door with large decorative plates (escutcheons) behind the knob. Fig. 38-35. The installation instructions for lock sets, particularly the

38–33b. *Hinge butt routing with a door and jamb template.*

38–33c. *Round corner butt hinges save time when installed with the hinge butt router and door and jamb template.*

38–33a. *A door- and jamb-butt template and router accessories.*

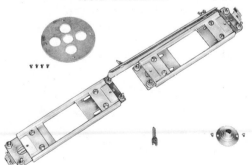

number and size of the holes to be bored in the door, will vary with the manufacturer. Always refer to the instructions which accompany the specific lock set to be installed. The general procedure for installing a lock set is as follows:

1. Open the door to a convenient working position and place wedges under the bottom near the outer edge to hold the door steady.

2. Measure up 36″ from the floor to locate the height of the lock set.

3. Fold and apply the template (which comes with the lock set) to the edge of the door bevel. Fig. 38-36. Mark the center of the door edge and the center of the hole on the door face through the guides on the template. Fig. 38-37. If a boring jig is used, no template is needed. Fig. 38-38.

4. Bore a hole of the recommended diameter in the face of the door. It is recommended that the holes be bored from both sides to prevent splitting. Bore the hole on one side until the point of the bit breaks through and then complete from the other side.

5. Bore a hole of the recommended diameter in the center of the door edge for the latch.

6. Insert the latch in the hole in the door edge. Keep the faceplate parallel to the edge of the door and mark with a sharp

38–34. *Entrance door locks used on residences and on smaller commercial buildings.*

38–35. *A decorative touch is added to the front entrance door by choosing from the wide selection of ornamental escutcheons which are available. The escutcheons not only add beauty to the entrance but also provide protection against finger marks and scratches. When replacing old or damaged locks, the decorative escutcheons may be used to hide unsightly scars and holes.*

pencil around the faceplate. Fig. 38-39. A marking tool may also be used to mark the position of the faceplate. Fig. 38-40.

7. Remove the latch. Chisel out the marked area so that the latch faceplate will be mounted flush with the edge of the door. Fig. 38-41.

8. Install the latch with the curved surface of the latch facing in the direction of the door closing. Insert and tighten the screws. Fig. 38-42.

9. Install the exterior knob by inserting the knob with the spindle into the latch. Make certain that the stems are positioned

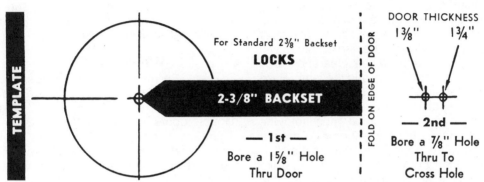

38–36. *A typical template for locating the center of the holes to be bored for a lock set.*

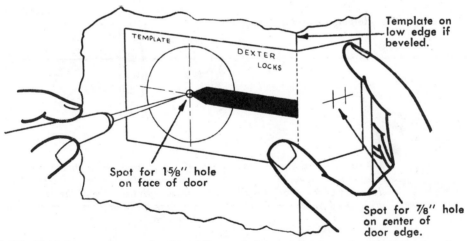

38–37. *Hold the template on the dotted line and place on the door edge. Mark the door through the template with an awl or nail.*

38–38. *When a boring jig is available, it is not necessary to use the template for marking the door previous to boring the holes for the lock set. The jig, when properly adjusted and clamped in position on the door, insures an accurate and rapid boring of the door.*

correctly through the latch holes and pressed flush against the door. Fig. 38-43.

10. Install the interior knob by placing it over the stem and aligning the screw guides with the stems. Push the assembly flush with the door, insert the screws, and tighten until the lock set is firm. Fig. 38-44.

11. Locate the strike on the doorjamb opposite the faceplate of the latch. To locate the strike, place it over the latch in the door. Then carefully close the door against the stops. The strike plate will hang on the latch in the clearance area between the door edge and the jamb. Push the strike plate in against the latch and, with a pencil, mark the top edge of the strike plate on the jamb. Then hold the pencil against the door edge and draw a line down the face of the strike plate.

12. Open the door and hold the strike against the doorjamb just under the line previously marked. Make sure that the line marked on the face of the strike is aligned with the edge of the jamb. Mark around the strike and chisel out the marked area so that the strike will mount flush with the surface of the jamb.

13. Make a clearance hole for the latch bolt by drilling a $^{15}/_{16}''$ hole $^{1}/_{2}''$ deep in the doorjamb on the center line of the screws from top to bottom. Install the strike and tighten the screws. Fig. 38-45.

GARAGE DOORS

There are many types and sizes of garage doors. Fig. 38-46. The standard single door is 9' wide and $6^{1}/_{2}'$ or 7' high. Double doors are usually 16' × $6^{1}/_{2}'$ or 7'. There are many architectural styles available to match the style of the home. Fig. 38-47. To give the door a distinctive cus-

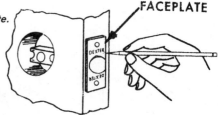

38–39. *Marking around the faceplate.*

FACEPLATE

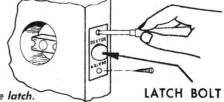

38–40. *When a marking tool is available, it is not necessary to mark around the faceplate as shown in Fig. 38–39. The marking tool is inserted in the hole bored in the edge of the door. It is aligned parallel with the edge of the door and then given a sharp blow with a hammer to outline the area to be chiseled out for the latch face-plate.*

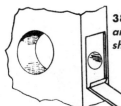

38–41. *Chisel out the marked area. The latch faceplate should mount flush with the edge of the door.*

38–42. *Installing the latch.*

LATCH BOLT

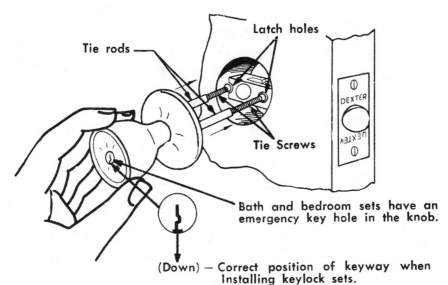

Latch holes

Tie rods

Tie Screws

DEXTER

Bath and bedroom sets have an emergency key hole in the knob.

(Down) — Correct position of keyway when Installing keylock sets.

38–43. *Installing the exterior trim assembly. Tie rods and tie screws must go through the holes in the latch.*

38–44. *Installing the interior knob.*

SCREW GUIDES

SCREW HOLES

38–45. *Installing the strike on the doorjamb.*

Clearance hole for latch bolt

38–46. *Garage doors are available in many sizes.*

Stock Sizes	Standard Sizes	
8' x 7'	8' x 6'6"	15' x 6'6"
9' x 7'	9' x 6'6"	15' x 7'
10' x 7'	10' x 6'6"	16' x 6'6"
16' x 7'	18' x 7'	

38–47. *A variety of garage door styles.*

TWO-CAR GARAGE DOORS

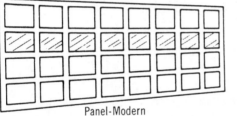

Ranch

Riviera

Ranch-Modern

Contemporary

Panel-Modern

Flush-Modern

SINGLE CAR GARAGE DOORS

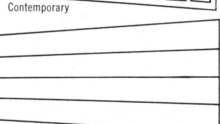

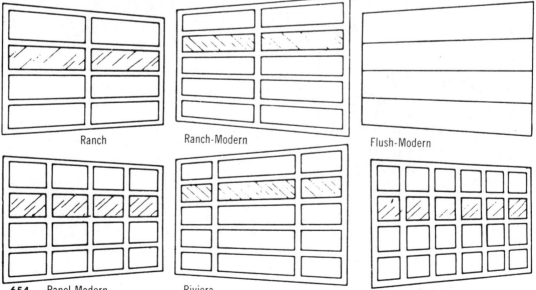

Ranch

Ranch-Modern

Flush-Modern

Panel-Modern

Riviera

Contemporary

tom look, it can be trimmed at any time with easily mounted moldings, rosettes, or monogram plates. Fig. 38-48.

The various types of doors and hardware, with complete instructions for their installation, can be obtained from local suppliers. Figs. 38-49 and 38-50. The three most commonly used doors are the hinged, the overhead swing, and the overhead sectional. Fig. 38-51 (Page 658). Occasionally, folding sliding doors are used.

Hinged doors open outward and are held in position with door holders. These doors are the least expensive and the easiest to install. However, when the door is standing open, it has no protection from rain and snow.

Sliding folding doors are hung from a track above the door. If the track is hung on the outside, the doors are subject to weathering. If the track is hung on the inside, the doors can fold against one another in

several thicknesses, or the track can be curved along the inside wall.

Overhead doors are made in two types: as a single-section (swing) door and as four or five sections hinged together. The swing-up door with the single section operates on a pivot principle with the track mounted on the ceiling and rollers located at the center and top of the door. Fig. 38-52. The sectional overhead door has rollers at each section fitted into a track at the side of the door and the ceiling. It requires more headroom above the opening than the swing door but is by far the most widely used for residential building. Clearance required above the top of the sectional overhead doors is usually about 12″. However, low-headroom brackets are available when such clearance is not possible.

Overhead doors are well protected from rain and wind, and snow and ice offer no

38–48. *Panels for trimming garage doors come in many styles.*

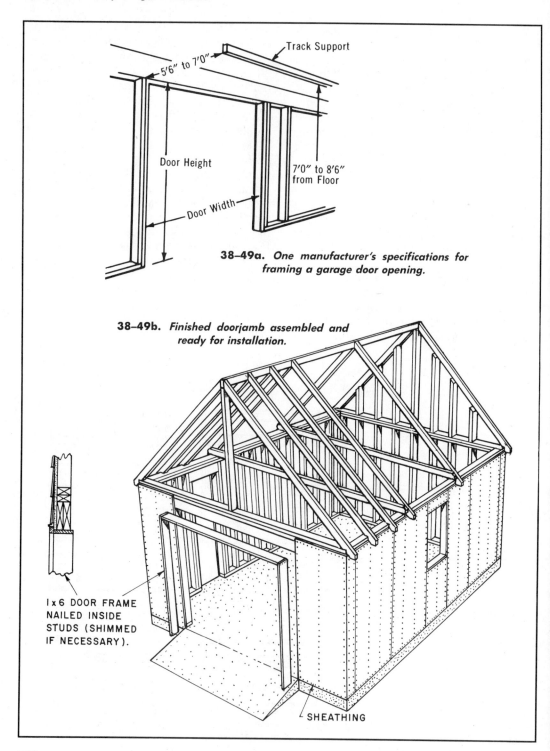

38-49a. One manufacturer's specifications for framing a garage door opening.

Track Support

5'6" to 7'0"

Door Height

7'0" to 8'6" from Floor

Door Width

38-49b. Finished doorjamb assembled and ready for installation.

1 x 6 DOOR FRAME NAILED INSIDE STUDS (SHIMMED IF NECESSARY).

SHEATHING

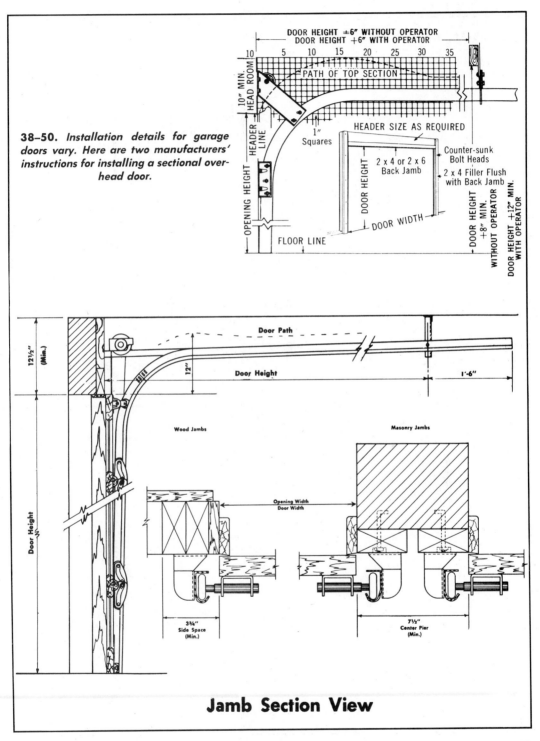

38–50. *Installation details for garage doors vary. Here are two manufacturers' instructions for installing a sectional overhead door.*

DOOR HEIGHT ±6" WITHOUT OPERATOR
DOOR HEIGHT +6" WITH OPERATOR

PATH OF TOP SECTION

10" MIN. HEAD ROOM

HEADER LINE

OPENING HEIGHT

1" Squares

FLOOR LINE

HEADER SIZE AS REQUIRED

DOOR HEIGHT

2 x 4 or 2 x 6 Back Jamb

DOOR WIDTH

Counter-sunk Bolt Heads

2 x 4 Filler Flush with Back Jamb

DOOR HEIGHT +8" MIN. WITHOUT OPERATOR

DOOR HEIGHT +12" MIN. WITH OPERATOR

12½" (Min.)

Door Path

12"

Door Height

1'-6"

Door Height

Wood Jambs

Masonry Jambs

Opening Width
Door Width

3¾" Side Space (Min.)

7½" Center Pier (Min.)

Jamb Section View

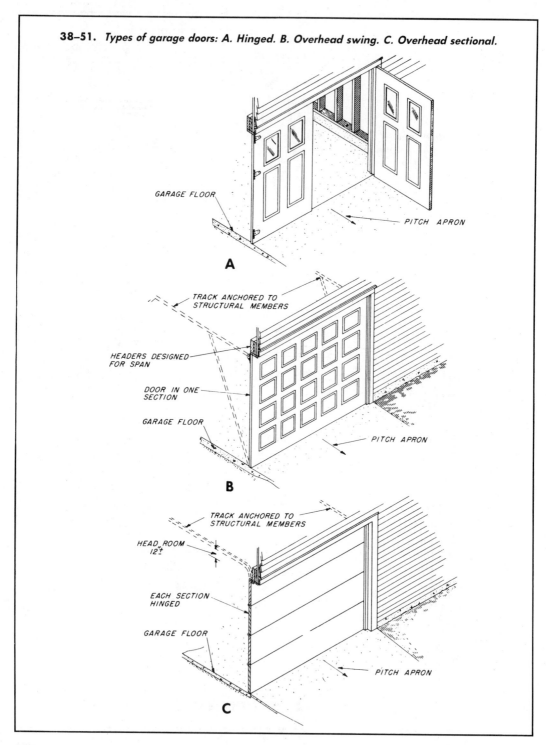

38–51. *Types of garage doors: A. Hinged. B. Overhead swing. C. Overhead sectional.*

A

GARAGE FLOOR

PITCH APRON

TRACK ANCHORED TO
STRUCTURAL MEMBERS

HEADERS DESIGNED
FOR SPAN

DOOR IN ONE
SECTION

GARAGE FLOOR

PITCH APRON

B

TRACK ANCHORED TO
STRUCTURAL MEMBERS

HEAD ROOM
12"±

EACH SECTION
HINGED

GARAGE FLOOR

PITCH APRON

C

particular problem. They are somewhat more difficult to install and more expensive than the hinged doors.

The overhead door has a pair of counterbalance springs mounted on it to help support the weight of the door so that it may be easily opened. These counterbalance springs are of two types: torsion and stretch. Figs. 38-53 and 38-54.

Power door operators are also available. These are electronically controlled by a wall-mounted button within the home or garage or by a portable battery-powered transmitter inside the car. Electric door operators can be installed during construction, or they can be added by the homeowner later. Fig. 38-55.

The bottom edge of a garage door should be scribed and cut to conform to the garage floor. An application of weather stripping is recommended for the bottom rail. It seals any minor irregularities in the floor and acts as a cushion in closing. Fig. 38-56.

The header beam over garage doors should be designed for the snow load which might be imposed on the roof. In wide openings, this may be a steel I-beam or a built-up wood section. For spans of 8'

or 9', two doubled 2 × 10s of high-grade Douglas fir or similar species are commonly used when only snow loads must be considered. If floor loads are also imposed on the header, a steel I-beam or wide-flange beam is usually selected.

ESTIMATING

The cost of the materials for an exterior door and frame depends on the style and trim. An accurate price should be obtained from the local supplier. The installation time required for an exterior doorframe will depend on how ornate the frame is. The conventional exterior frame and brick mold with an oak sill requires approximately two hours to assemble and install. It will require one extra hour to hang the door and half an hour to install the lock set. Combination storm and screen doors require approximately one hour for installation. A 9' × 7' garage door requires approximately 3½ hours and a 16' × 7' garage door about 5½ hours for assembly and installation. Radio-controlled garage door operators take about 2 hours to install.

38–52. *The swing door must be moved outward slightly at the bottom as it is opened.*

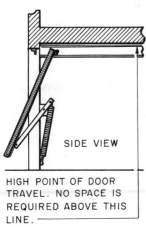

SIDE VIEW

HIGH POINT OF DOOR TRAVEL. NO SPACE IS REQUIRED ABOVE THIS LINE.

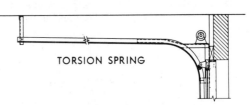

TORSION SPRING

38–53. *This garage door is counterbalanced with a torsion spring.*

38–54. *This garage door is counterbalanced with a stretch spring.*

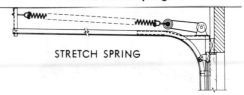

STRETCH SPRING

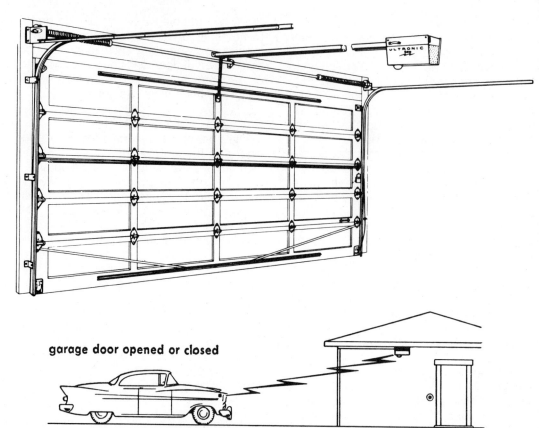

garage door opened or closed

38-55. *An electric door operator will open the garage door from the radio transmitter in the car, a portable hand transmitter, or a wall-mounted button.*

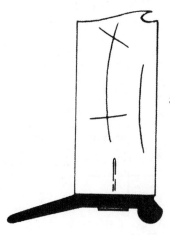

38-56. *Weather stripping should be applied to the bottom rail of a garage door.*

QUESTIONS

1. List five kinds of exterior doors.
2. Name the parts of the doorframe.
3. What is the standard thickness of an exterior door?
4. What is the minimum width door recommended for a main entrance?
5. When the exterior doorframe is set into the opening, how far above the subfloor must the sill project?
6. When a door is fitted to the jamb, why is the lock edge of the door beveled?
7. What are the three types of garage doors most commonly used?

UNIT 39

Exterior Wall Coverings

Today's home builder can select from a wide variety of easy-care materials for the exterior walls of the home. Fig. 39-1. Various materials, shapes, and surface treatments are used to produce over five hundred different wall coverings. An entirely different design effect can be achieved by changing the type of exterior covering on a house. Because it has such a great effect on a home's overall appearance and ease of maintenance, the exterior wall covering should be selected with great care.

39-1. *In this home, brick siding was used on the first level and bevel siding on the second level. The bevel siding is made of cement-asbestos fiber and cellulose fiber. This material has the weathering characteristics of stone and the workability of wood.*

39-2. *Building materials dealers provide their customers with readily accessible displays from which siding selections are made. Each wing of this display holds eight siding samples.*

A wide variety of exterior coverings is possible because changing any one of the following factors can produce a new kind of covering:

- Material used. Wood products used include solid wood and such man-made wood materials as plywood, hardboard, and particle board. Masonry, either solid or veneer, may be of brick, stone, or stucco. Asphalt materials are also used for siding in two forms: as rolled products and as shingles. Common metals used are aluminum and steel, both of which are usually prefinished. Vinyl plastic is also a popular, easy-care exterior wall material.
- Shape and form. Some of the common shapes and forms in which exterior wall covering material is manufactured are bevel and drop siding, vertical tongued and grooved material, large panels, boards and battens, shingles and shakes, and rolled material.
- Surface treatment. Siding can be smooth or rough sawn, plywoods and hardboards can be textured, overlays of

fiber and/or plastics can be added, and materials can be prefinished with paints, enamels, plastics, and other finishes. Most building supply dealers display samples of the wide variety of materials available. Fig. 39-2.

With the exception of solid wood and man-made wood materials, most exterior wall coverings are applied either by brick-layers or masons or by specialty building construction workers. Therefore only those materials most often used in the construction of homes will be considered here. Specific emphasis will be placed on materials commonly installed by the carpenter.

WOOD SIDING

One of the materials most characteristic of the exteriors of North American houses is wood siding. The essential properties required for wood siding are good painting characteristics, easy working qualities, and freedom from warp. These properties are present to a high degree in the cedars, eastern white pine, sugar pine, western white pine, cypress, and redwood. They are present to a good degree in western hemlock, ponderosa pine, spruce, and yellow poplar and to a fair degree in Douglas fir, western larch, and southern yellow pine.

Exterior siding materials should be select grade and should be free from knots, pitch pockets, and waney edges. The moisture content at the time of application should be the same that it would attain in service. This is about 12%, except in the dry southwestern United States, where the moisture content should average about 9%.

Wood siding is made in many shapes and sizes and with various edge treatments. The common types are:
- Bevel.
- Drop.
- Board.

Some types, such as bevel siding, must be installed horizontally. Others, such as board siding, may be installed either horizontally (clapboard) or vertically (board and batten). Fig. 39-3a.

Vertical siding is commonly applied to the gable ends of houses, over entrances, and sometimes on large wall areas. It may consist of plain-surfaced matched boards, patterned matched boards, or square-edge boards covered at the joint with a batten strip. Fig. 39-3a.

Matched vertical siding should preferably be not more than 8″ wide. It should be fastened with two 8d nails not more than 4′ apart. Backer blocks placed between studs provide a good nailing base. The bottom of the boards should be undercut to form a water drip.

Bevel siding. Plain bevel siding is made in nominal 4″, 5″, and 6″ widths with $7/16″$ butts, in 6″, 8″, and 10″ widths with $9/16″$ butts, and in 6″, 8″, 10″, 12″ widths with $11/16″$ butts. Fig. 39-3a. The top edge is $3/16″$ for all sizes. Bevel siding is generally furnished in random lengths varying from 4′ to 16′.

One variation of bevel siding is the Anzac pattern. The Anzac siding pattern was derived from a New Zealand design. This pattern has a relatively thick butt edge which produces a heavy shadow line under each course. On the face of the pattern are two grooves. The deep upper groove acts as a water barrier, and the lower groove is a guideline which aids in aligning the siding properly as it is applied. The back of the siding is machined so that it will lie flat against the studs or sheathing. Fig. 39-3b.

Drop siding. Drop siding is generally $3/4″$ thick, has a flat back, and is made in a variety of patterns with either matched or shiplap edges. Fig. 39-3a. All patterns of drop siding may be applied horizontally. Some patterns may also be applied vertically; for example, at the gable ends of a house.

Drop siding is designed to be applied directly to the studs, and it thereby serves as both sheathing and exterior wall covering. It is widely used in this manner in farm structures, sheds, and garages in all parts of North America, and for houses in mild climates. When drop siding is used over and in contact with other material such as sheathing or sheathing paper, water may work through the joints and be held between the sheathing and the siding. This condition can lead to paint failures and decay. Such conditions are not common when the sidewalls are protected by a good roof overhang. When drop siding is applied vertically, it should also be protected by a wide overhanging roof. Otherwise, water flowing down the face of the siding is led into the joint and held there.

Board siding. Square-edge or clapboard siding made of $25/32″$ board is occasionally selected for architectural effect. In this case wide boards are generally used. Some of this siding is also beveled at the top of the back when used as clapboard siding. This allows the boards to lie rather close to the sheathing, thus providing solid nailing. Fig. 39-3a.

In board and batten siding, when wide square-edged boards are used, they are subject to considerable expansion and contraction because of their width. The batten strips covering the joints should be nailed to only one siding board so that the adjacent board can swell and shrink without splitting the boards or the batten strip. Fig. 39-3a.

INSTALLATION OF WOOD SIDING

Before application, exterior wood siding should be treated with a water repellent. This will improve finish performance no matter what type of finish is used. Fig. 39-4 (Page 666).

Some siding is pretreated (or pre-primed) by the lumber mill. If this has been done, mill instructions accompanying the siding should be followed carefully.

If the siding has not been pretreated, a water repellent is most easily and effectively applied before the siding is put in place. This may be done either by dipping

	Board	Channel Rustic	Drop	Bevel
Patterns	Board on Board Board and Batten Clapboard	(Board and Gap)	T&G Patterns Shiplap Patterns	Plain Rabbeted Edge
	Available surfaced or rough textured		Available in 13 different patterns. Some T&G (as shown), others shiplapped.	Plain Bevel may be used with smooth face exposed or sawn face exposed for textured effect.
Application And Nailing	Recommended 1" minimum overlap. Use 10d Siding nails as shown.	May be applied horizontally or vertically. Has ½" lap and 1¼" channel when installed. Use 8d Siding nails as shown for 6" widths. Wider widths nail twice per bearing.	6d Finish nails for T&G, 8d Siding nails for shiplap.	Recommend 1" minimum overlap on plain bevel siding. Use 6d Siding nails as shown.
Available **Grades** *Most commonly used	No. 1 Common* No. 2 Common* No. 3 Common Or Select Merchantable* Construction* Standard	No. 1 Common* No. 2 Common* No. 3 Common Or C&Btr*, D*, E	No. 1 Common* No. 2 Common* No. 3 Common Or C&Btr*, D*, E	All species except WRC & IWP B&Btr*, C*, D **WRC** Clear-VG-All Heart* A*, B*, Rustic* **IWP** Supreme*, Choice*, Quality
Seasoning	Shipped 15% moisture content or less when specified.	Shipped 15% moisture content or less when specified.	Shipped 15% moisture content or less when specified.	Usually shipped at 12% or less moisture content.

	Bungalow	Dolly Varden	Log Cabin	Tongue & Groove
Patterns	Plain Rabbeted Edge	Rabbeted Edge		Plain
	Thicker and wider than Bevel Siding. Sometimes called "Colonial." Plain bungalow may be used with smooth face exposed for textured effect.	Thicker than Bevel Siding. Rabbeted edge.	1½" at thickest point.	Available in smooth surface or rough surface.
Application And Nailing	Same as for Bevel siding, but use 8d Siding nails.	Same as for Rabbeted Bevel Siding but use 8d Siding nails.	Nail 1½" up from lower edge of piece. Use 10d Casing nails.	Use 6d Finish nails as shown for 6" widths or less. Wider widths, face nail twice per bearing with 8d Siding nails.
Available Grades *Most commonly used	see Bevel Siding Grades	all species except IWP B&Btr*, C*, D **IWP** Supreme* Choice* Quality	No. 1 Common* No. 2 Common* No. 3 Common	No. 1 Common* No. 2 Common* No. 3 Common Or C&Btr*, D*, E
Seasoning	Usually shipped at 12% or less moisture content.	Usually shipped at 12% or less moisture content.	Shipped 15% moisture content or less when specified.	Shipped 15% moisture content or less when specified.

or by brushing the water repellent on the face, back, ends, and edges of each piece. If the siding is stacked to dry after such treatment, stickers (strips of wood) placed between tiers also should be treated with a water repellent. Many boards will be cut to length as the siding is put in place. Freshly cut surfaces also should receive a liberal treatment with a water repellent. Fig. 39-5.

If wood sheathing has been used, the exterior of the building must first be covered with building paper with a 4" lap. The siding may be nailed into the sheathing at 24" intervals. Nails should penetrate at least 1¹/₂" into the studs. Fig. 39-6.

For matched horizontal siding, the weather exposure (that part of the siding which will not be overlapped by another piece of siding) is predetermined by its machined edge treatment. In the applications, each succeeding course is installed up tight against the preceding course.

39–3b. *Anzac siding has a thick butt edge which accentuates the horizontal lines of the siding.*

Since the spacing is predetermined, it is not necessary to lay out the spacing, as it is for plain or bungalow siding.

Laying out the Spacing of Vertical Siding

When laying out board and batten siding, measure the length of the wall on which the siding is to be installed and carefully lay out the spacing of the boards and battens. The spacing will have to be

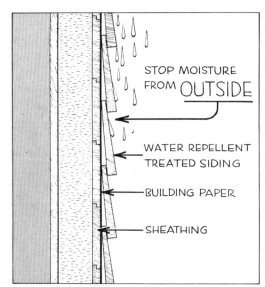

39–4. *To keep moisture out, the siding should be treated with a water repellent. The sheathing should be covered with a water-repellent building paper.*

39–5. *Boards that have been cut to length should have the fresh cut ends treated with a water repellent.*

increased or decreased between the boards so that the widths of the underboards will appear to be the same. The underboards are then cut accurately to length and installed according to this layout. Care should be taken that the boards are plumb as they are installed.

When all the underboards are in place, the battens or the overboards (for board on board siding) are installed. Again, make sure that these members are also plumb. The installation of the overboards is much more critical because they will give the wall its finished appearance.

Laying out the Spacing of Plain Bevel or Bungalow Siding

The spacing for plain bevel or bungalow siding should be carefully laid out before the first board is applied. Siding starts with the bottom course of boards at the foundation. Fig. 39-7. Sometimes the siding is started on a water table, which is a projecting member at the top of the foundation to throw off water. Fig. 39-8. Each succeeding course of bevel siding overlaps the upper edge of the previous course.

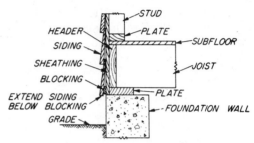

39-7. *The beginning courses of siding at the foundation wall.*

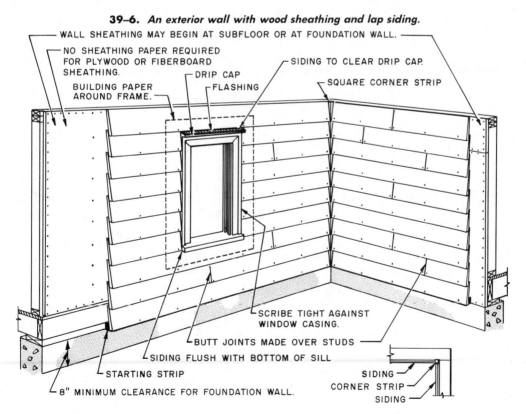

39-6. *An exterior wall with wood sheathing and lap siding.*

667

Determine the number of courses by measuring the distance from at least 1" below the bottom plate to the underside of the soffit and dividing that height by the maximum weather exposure of the siding. Fig. 39-9. To determine the maximum exposure, deduct the minimum overlap, or head lap, from the overall width (dressed dimensions) of the siding. The minimum head lap is 1" for 4" and 6" widths and 1¹/₄" for widths over 6". The dressed dimensions for various sidings can be found in Fig. 39-10.

For example, if a nominal 10" bevel siding is used, the actual or dressed width is 9¹/₄". Fig. 39-10. On 10" plain bevel siding, a minimum overlap of 1¹/₄" is required. Therefore the maximum exposed surface of a 10" piece of siding would be 8" (9¹/₄" − 1¹/₄" = 8"). With a pair of dividers set at 8", make a trial layout on the sidewall beginning at the bottom and "walking off" this exposure dimension. The bottom of the board that passes over the top of the first-floor windows should coincide with the top of the window cap. Fig. 39-11. If the

39–8. *Bevel siding started on a water table (see arrow) at the foundation.*

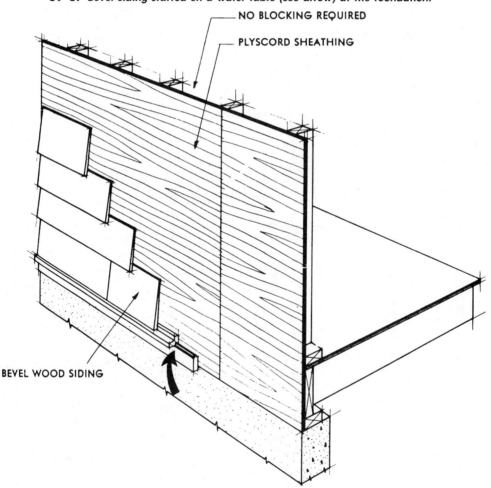

NO BLOCKING REQUIRED

PLYSCORD SHEATHING

BEVEL WOOD SIDING

bottom of this board does not line up, adjust the spacing for each board to something slightly less than 8". (NOTE: Eight inches is the maximum exposure; so do not adjust it to a greater width). Continue to modify the spacing, if at all possible, until the bottom edge of this piece of siding is even with the tops of the windows and doors. The location on the foundation wall for the bottom edge of the first piece of siding may also be adjusted slightly as an aid in making this alignment.

The board spacing should be such that the maximum exposure will not be exceeded. This may mean that the boards will have less than the maximum exposure.

Application of Plain Bevel or Bungalow Siding

The application of the first course of siding determines the level and uniformity of all succeeding courses. To insure the proper application of the first course, proceed as follows:

1. Measure down with a tape from the top of the top plate if possible. If not, measure from the underside of the soffit to a point at least 1" below the bottom plate. This point will usually be on the foundation. Make a mark on the foundation and record this measurement. Fig. 39-12, point A.

2. Repeat Step 1 along the sidewall at about 2' intervals using the same measured length each time. B, Fig. 39-12.

3. Snap a chalk line horizontally along the marks. With bevel, bungalow, or Anzac patterns, nail a furring strip about 3/8" above this line to provide support for the starting course. Fig. 39-13. The bottom edge of the first course of siding will be positioned along the chalk line. Fig. 39-12, line C.

4. Begin the application by placing the butt edge of the siding along the chalk line. Start the first piece of siding at the end of the sidewall and work toward the other end. The first course is nailed to the bottom plate at points just below each

stud. This will mark the nailing locations for the succeeding courses. Fig. 39-14.

5. The amount of overlap is measured along the top of the siding course to be lapped. Snap a chalk line along this mark to locate the butt edge of the second course. Fig. 39-15, line D.

6. The second course is also started at the end of the sidewall. Align the siding horizontally along the chalk line and nail at each studbearing. E, Fig. 39-15. Use only one nail per bearing. Never nail through both courses of siding. F, Fig. 39-15. Tap the nail head flush with the siding surface. The siding courses should fit snug, not tight.

7. Repeat Steps 5 and 6 for the application of successive courses. Make certain that the vertical butt joints between boards are staggered along the sidewall and that they fall on studs, as shown at G in Fig. 39-16 (Page 672).

The siding should be carefully fitted and be in close contact with the adjacent piece. Some carpenters fit the boards so tightly that they have to spring the boards in place. Tight-fitting butt joints are obtained by cutting the closure board of each course approximately 1/16" too long. Bow the piece slightly to get the ends in position, and then snap it into place. Fig. 39-17. This assures a tight joint. Loose-fitting joints allow water to get behind the siding. The water can cause paint deterioration around the joints and also set up

39-9. *Measuring the vertical distance to be covered by the siding.*

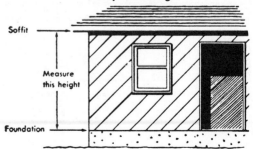

Soffit

Measure this height

Foundation

conditions conducive to decay at the ends of boards.

Siding that passes under a window sill should be cut to fit the groove provided in the bottom of the sill. Fig. 39-18a. Siding

installed over doors and windows should be set on the drip cap. Fig. 39-18b.

8. Trim the last course of siding to fit under the eaves and apply a molding if required.

39–10. *Nominal and dressed dimensions for wood siding.*

Product Description		Nominal Size		Dressed Dimensions		
		Thickness In.	Width In.	Thickness In.	Width In.	Lengths Ft.
Rustic And Drop Siding	(D & M) If ⅜" or ½" T & G specified, same over-all widths apply.	⅝ 1	4 5 6 8 10	9/16 23/32	3⅛ 4⅛ 5⅛ 6⅞ 8⅞	Same
	(Shiplapped, ⅜-in lap)	⅝ 1	4 5 6	9/16 23/32	3 4 5	Same
	(Shiplapped, ½-in. lap)	⅝ 1	4 5 6 8 10 12	9/16 23/32	2⅞ 3⅞ 4⅞ 6⅝ 8⅝ 10⅝	Same
Ceiling And Partition	(S2S & CM)	⅜ ½ ⅝ ¾	3 4 5 6	5/16 7/16 9/16 11/16	2⅛ 3⅛ 4⅛ 5⅛	Same
Bevel Siding Grades	Bevel Siding Western Red Cedar Bevel Siding available in ½", ⅝", ¾" nominal thickness. Corresponding thick edge is 15/32", 9/16" and ¾".	½ 9/16 ⅝ ¾ 1	4 5 6 8 10 12	7/16 butt, 3/16 tip 15/32 butt, 3/16 tip 9/16 butt, 3/16 tip 11/16 butt, 3/16 tip ¾ butt, 3/16 tip	3½ 4½ 5½ 7¼ 9¼ 11¼	Same
	Wide Bevel Siding (Colonial or Bungalow)	¾	8 10 12	11/16 butt, 3/16 tip	7¼ 9¼ 11¼	
Finish And Boards S-Dry	S1S, S2S, S1S2E	⅜ ½ ⅝ ¾ 1 1¼ 1½ 1¾ 2 2½ 3 3½ 4	2 3 4 5 6 7 8 and wider nominal	5/16 7/16 9/16 ⅝ ¾ 1 1¼ 1⅜ 1½ 2 2½ 3 3½	1½ 2½ 3½ 4½ 5½ 6½ ¾ off	3' and longer. In Superior grade, 3% of 3' and 4' and 7% of 5' and 6' are permitted. In Prime grade, 20% of 3' to 6' is permitted.
Factory And Shop Lumber	S2S*	1 (4/4) 1¼ (5/4) 1½ (6/4) 1¾ (7/4) 2 (8/4) 2½ (10/4) 3 (12/4) 4 (16/4)	5 and wider (4" and wider in 4/4 No. 1 Shop and 4/4 No. 2 Shop)	25/32 (4/4) 1 5/32 (5/4) 1 13/32 (6/4) 1 19/32 (7/4) 1 13/16 (8/4) 2⅜ (10/4) 2¾ (12/4) 3¾ (16/4)	(See Rough Sizes Below)	6 ft. and longer in multiples of 1'

*These thicknesses also apply to Tongue & Groove (T&G).
See coverage estimator chart for T&G widths.

Minimum Rough Sizes   Thicknesses and Widths Dry or Unseasoned All Lumber (S1E, S2E, S1S, S2S)
80% of the pieces in a shipment shall be at least ⅛" thicker than the standard surfaced size, the remaining 20% at least 3/32" thicker than the surfaced size. Widths shall be at least ⅛" wider than standard surfaced widths.
When specified to be full sawn, lumber may not be manufactured to a size less than the size specified.

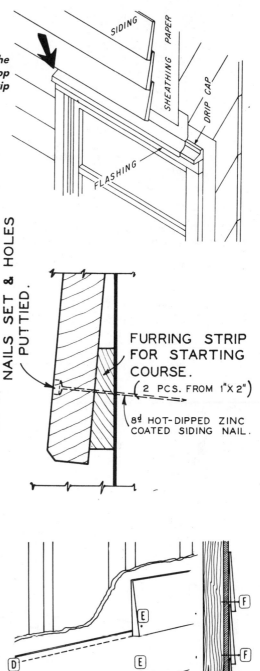

SIDING

SHEATHING PAPER

DRIP CAP

FLASHING

39–11. *Plan the courses of siding so that the bottom edge of the course running across the top of the window will be in alignment with the drip cap.*

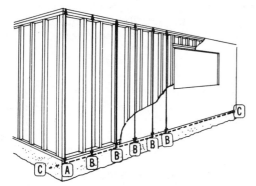

39–12. *Locating the first course of siding on an exterior wall.*

39–13. *The furring strip may be ripped at an angle to provide support along its full width. Notice that two furring strips were ripped from one piece of 1″ × 2″ stock.*

NAILS SET & HOLES PUTTIED.

FURRING STRIP FOR STARTING COURSE.

(2 PCS. FROM 1″X 2″)

8^{d} HOT-DIPPED ZINC COATED SIDING NAIL.

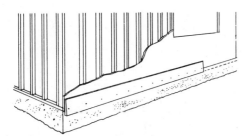

39–14. *Applying the first piece of siding with one nail at each stud location.*

39–15. *Snap a chalk line along the top edge of the siding to locate the butt edge of the succeeding courses.*

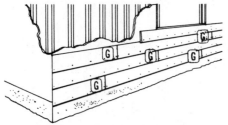

39-16. Apply successive courses, making sure that all the vertical butt joints are staggered and that they fall on studs.

9. Outside corners may be mitered, covered with corner boards, or capped with metal corners. Fig. 39-19. For mitered corners the siding is cut to length and mitered before application. Fig. 39-20. Corner boards are installed before siding application. The siding is then cut to length and butted against the corner boards. If metal corners are used, the siding is cut off even with the outside corner of the building. The metal corners are applied after the siding is in place.

39-17. To fit siding tightly, cut it about 1/16" too long, bow it into position, and then snap it tight.

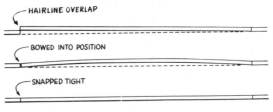

HAIRLINE OVERLAP

BOWED INTO POSITION

SNAPPED TIGHT

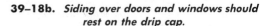

39-18a. Bevel siding under window sills should fit in the groove provided.

SILL

39-18b. Siding over doors and windows should rest on the drip cap.

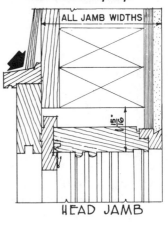

ALL JAMB WIDTHS

HEAD JAMB

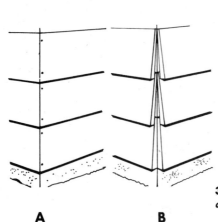

A   B

39-19. Finishing an outside corner: A. Mitered corners. B. Siding installed and ready for metal corners.

Inside corners are cut before application and butted against a square wood corner strip approximately $1^1/8'' \times 1^1/8''$ in size. Fig. 39-21.

Using a Story Pole

You may prepare a story pole when installing horizontal siding to insure accuracy and increase efficiency. Select a straight piece of $1'' \times 2''$ stock for the siding story pole. Place it under the soffit against a dominant wall of the house, usually the front, and mark the total height. Determine

the number of courses and the spacing as described in the section on spacing of bevel siding earlier in this unit. Lay out the spacing on the story pole and check the layout against the building. Fig. 39-22.

Hold the story pole in position up against the soffit. Transfer the marks from the story pole to the house on all corners

39-22. *The story pole should extend from the underside of the soffit to the bottom edge of the first piece of siding as shown from A to B. Hold the story pole in position against the building and lay out the spacing on the story pole as shown at C. Check to be certain that the bottom edge of the piece of siding over the window is even with the top of the window as shown at D.*

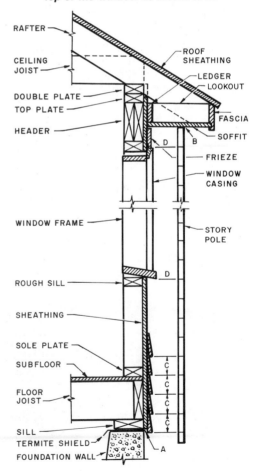

This dimension should equal butt thickness.

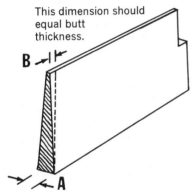

39-20. *Mitering of bevel siding corners must be done carefully to obtain a good joint. To lay out and cut the joint, measure the butt thickness at A. Measure back along the top edge a distance at B equal to the butt thickness shown at A. Then connect these two points as shown by the dotted line. With the saw blade set at about a 47° angle, make the cut beginning at the butt end.*

39-21. *Bevel siding at an inside corner. The bevel siding is butted against a square wood corner strip.*

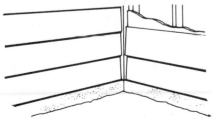

673

and on all window and door casings. Fig. 39-23. Make sure that the bottom marks are clearly visible on the foundation. Snap a chalk line on the bottom marks around the perimeter of the house. Then install the siding as described previously, beginning with Step 4.

Using a Siding Gauge or "Preacher"

A siding gauge, or "preacher," is a small hardwood block used for accurately marking siding pieces to fit between two window casings or a window and a door casing. If corner boards are used at the corners, it may also be used for marking siding between a corner board and a window or door casing. To make a siding gauge, select a piece of $3/8''$ or $1/2''$ hardwood long enough to accommodate the width siding used and proceed as follows:

1. Center the siding on the block of hardwood material. A, Fig. 39-24.

2. Lay out the width of the siding plus $1/4''$ for clearance. B, Fig. 39-24.

3. Lay out the thickness of the siding plus approximately $5/8''$. C. Fig. 39-24.

4. Allow about 1" around all of the inside cuts. D, Fig. 39-24.

5. Cut off the corners as shown at E in Fig. 39-24.

Fig. 39-25 illustrates how to use the preacher for marking siding between a corner board and a window casing. Cut the

end of the siding to fit against the corner board on the spacing mark. A, Fig. 39-25. Align the other end of the siding with the spacing mark on the window casing. B, Fig. 39-25. Place the preacher over the siding and hold it tight against the casing. Holding a pencil against the edge of the preacher, draw a line along the face of the siding. C, Fig. 39-25. Cut the siding to finish length, position it on the marks, and nail it in place.

Siding Nails and Nailing

Good nails and nailing practices are a must for the proper application of wood siding. Nails should be long enough to penetrate into studs (or studs and wood sheathing combined) at least $1^{1}/_{2}''$. When this much penetration is not possible, threaded nails are recommended for increased holding power. Do not nail siding only to composition or pressed fiber sheathing. The nails must penetrate the studs. Nail locations and recommended nail sizes are shown in Fig. 39-3a in the "Application and Nailing" section. However, the following data about nails will be

39-23. *Transferring the marks of the story pole to the house.*

39-24. *Laying out a siding gauge or "preacher".*

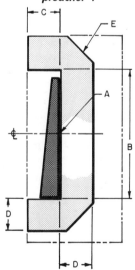

674

very helpful in the selection and use of the right nail for the application.

NAIL REQUIREMENTS

The following characteristics are essential for nails used on wood siding. Such nails:

- Should be rust-resistant, preferably rust proof.
- Must not cause the siding to discolor or stain.
- Should not cause splitting, even when driven near the end or edge of siding.
- Should have adequate strength to avoid the need for predrilling.
- Should be able to be driven easily and rapidly.
- Should not emerge or "pop" at any time after being driven flush with the siding.
- Should not cause an unsightly visible pattern on the sidewall.

Two types of nails which have these characteristics are:
- High tensile strength aluminum nails.
- Galvanized nails.

High tensile strength aluminum nails are corrosion-resistant and will not discolor or deteriorate the wood siding. They are economical when the nail count per pound is considered, although they are somewhat more expensive than the common galvanized nail. Fig. 39-26.

There are two kinds of galvanized nails: the mechanically plated and the hot-dipped. Mechanical plating is an extremely successful process which provides a nail with a uniform coating, giving it outstanding corrosion resistance. With nails that are hot-dipped, the degree of coating protection varies.

NAIL DESIGN

The design of a nail influences the ease with which it can be driven and its holding power. Nail design includes the head, shank, and point. The basic types of nail heads are illustrated in Fig. 39-27. Nail shanks may be smooth or threaded. Nails that are smooth shanked will loosen under

extreme temperature changes. Increased holding power may be obtained by using a ring-threaded or spiral-threaded nail shank. The commonly used nail points include:
- Blunt—reduces splitting.
- Diamond—most widely used.
- Needle—tops in holding but tendency to cause splitting.

For the best possible holding power with the least splitting, a blunt or medium diamond point and a blunt or medium needle point with a ring-threaded shank are recommended.

NAILING RECOMMENDATIONS

When nailing mitered corners or when nailing near the end of a piece, predrill the nail hole or blunt the nail point to avoid splitting the wood. With lapped siding, in order to prevent splitting and to allow expansion clearance, nail just above the lap and not through the tip of the undercourse. Fig. 39-28.

Specific nailing recommendations for standard siding patterns are shown in

39–25. *Using the preacher to mark the length of a piece of siding.*

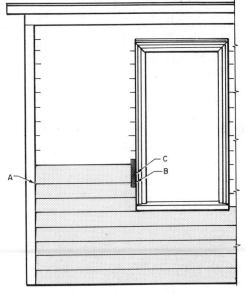

Figs. 39-3a and 39-29. All recommendations refer to nailing at every stud, if siding courses are laid up horizontally. If siding is installed vertically, use 2 × 4 blocking between studs. Fig. 39-30. The blocking should be placed at top and bottom, and intermediately, at not more than 24″ on center.

Bevel and bungalow siding. Face-nail with one nail per bearing only. Use 8d siding nails for ³/₄″ thicknesses; 6d nails for thinner pieces. When applying plain bevel and bungalow siding, drive the nails so that the shank just clears the tip of the preceding course. For rabbeted bevel and bungalow siding, set each course to allow an expansion clearance of ¹/₈″. Drive the nails about 1″ from the lower edge of the course.

Shiplap and rustic siding. Face-nail with two siding nails per bearing for patterns wider than 6″. Space each nail about halfway between the center and the edge of a piece. For narrower courses, one nail per bearing is enough. Drive the nail 1″ from the overlapping edge. Use 8d siding nails for 1″ thicknesses, 6d for thinner pieces.

Tongue-and-groove siding. Siding 4″ or 6″ wide should be blind-nailed through the tongue with 6d finish nails. Use one nail per bearing. For wider patterns, face-nail with two 8d siding nails per bearing.

Board and batten siding. Space the underboards about ¹/₂″ apart and fasten with one 8d siding nail per bearing, driven

39–26. *Comparison of aluminum and hot-dipped galvanized nails.*

Nail Size Specification

Size	Length (Inches)		Siding Nails (Count per lb.)		Approx. lbs. Per 1,000 B.F. of Siding	
	*	**	*	**	*	**
6d	1⁷⁄₈″	2″	566	194	2	6
7d	2¹⁄₈″	2¹⁄₄″	468	172	2¹⁄₂	6¹⁄₂
8d	2³⁄₈″	2¹⁄₂″	319	123	4	9
10d	2⁷⁄₈″	3″	215	103	5¹⁄₂	11

*Aluminum   **Hot-dipped Galv.

39–27. *Nails used for the application of siding.*

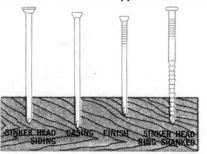

SINKER HEAD SIDING   CASING   FINISH   SINKER HEAD RING SHANKED

39–28. *Correct nailing procedure for lapped siding.*

Rust-resisting nails

1/8″ space allows for seasonal adjustment

Breathing-type building paper

through the center of the piece. Fasten batten strips with one 10d siding nail per bearing, driven through the center of each piece so that the nail shank passes between the underboards. Variations of the board and batten with recommended nailing procedures are shown in Fig. 39-31.

Exterior Corner Treatment

Wood siding is commonly joined at the exterior corners by corner boards, mitered corners, metal corners, or alternately lapped corners. Fig. 39-32 (Page 680). The method of finishing the wood siding at exterior corners is influenced somewhat by overall house design.

Corner boards are used with bevel or drop siding and are generally made of nominal 1" or 1¼" material, depending upon the thickness of the siding. The boards may be plain or molded, depending on the architectural treatment of the house.

The corner boards and the window and door trim may be applied to the sheathing, with the siding fitted tightly against the narrow edge of the corner boards and against the trim. When this method is used, the joints between the siding and the corner boards or trim should be calked or treated with a water repellent. Sometimes corner boards and trim around windows and doors are applied over the siding, a method that minimizes the entrance of water into the ends of the siding. This method works better for panel siding than for bevel siding. Fig. 39-33.

Mitered corners, sometimes used with the thicker patterns, should be cut in a miter box and must fit tightly and smoothly for the full depth of the miter. To maintain a tight fit at the miter, it is important that siding be properly seasoned before delivery and protected from rain when stored at the site. The ends should be set in white lead when the siding is applied, and the exposed faces should be primed immediately after it is applied. Nail mitered ends to the corner posts, not to each other.

Metal corners are made of light-gage metals, such as aluminum or galvanized iron, and are used with bevel siding as a

39–29a. *Suggested nailing methods for typical siding patterns.*

39-29b. *Installing Santa Rosa pattern siding.*

39-30. *To provide backing when applying vertical siding, put 2″ × 4″ blocks between the studs at no more than 24″ on center.*

39-31a. *Red cedar board and batten siding has been installed on the exterior of this house.*

substitute for mitered corners. Fig. 39-33. They can be purchased at most lumberyards. The application of metal corners takes less experience than is required to make good mitered corners or to fit siding to corner boards. Metal corners should be set in white lead paint.

Alternately lapped corners are fitted so that every other piece of siding has the end exposed. Fig. 39-32.

Interior Corner Treatment

Interior corners of siding are butted against a corner strip of nominal 1″ or 1¼″ material, depending upon the thickness of the siding. Fig. 39-33.

Preventing Outside Moisture Problems

Poor construction detailing may enable water to seep into the siding, eventually causing paint or finish deterioration. Poor construction may also result in inadequate insulation, causing discomfort and high heating bills. To avoid these problems take the following precautions at the points where trouble may occur:

CAREFUL FITTING

With any siding pattern, good joints are essential. Accurate cutting of pieces is the only way to insure the proper fit.

Bevel courses should have sufficient lap to prevent wind-driven rain from working up between courses. FHA Minimum Property Standards require a 1″ lap.

Gutter joints and downspouts are other areas which must be carefully fitted.

CALKING

The sealing of all joints helps provide protection against rain, snow, fog, and wind. It is particularly important at the butt

678

39–31b. *Nailing details for various board and batten applications.*

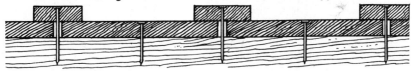

STANDARD BOARD AND BATTEN: One 8d siding nail is driven midway between edges of the underboard, at each bearing. Then apply batten strips and nail with one 10d siding nail at each bearing so that shank passes through space between underboards.

SPECIAL BATTENS: A T-shaped batten or standard batten nailed over a vertical nailing strip, is nailed exactly the same as the standard method; however, in this case an exceptionally good bearing is provided while driving nail through the batten.

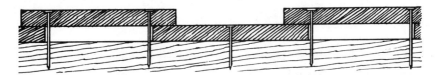

BOARD ON BOARD: Apply underboards first, spacing them to allow 1½-inch overlap by outer boards at both edges. Use standard nailing for underboards, one 8d siding nail per bearing. Outer boards must be nailed twice per bearing to insure proper fastening. Nails, having some free length, do not hold outer boards so rigidly as to cause splitting if there is "movement" from humidity changes. Drive 10d siding nails so that shanks clear edges of underboard approximately ¼-inch. This provides sufficient bearing for nailing, while allowing clearance to enable underboard to expand slightly.

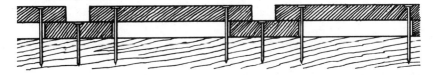

REVERSE BATTEN: Nailing is similar to board on board. Drive one 8d nail per bearing through center of under strip, and two 10d siding nails per bearing through outer boards.

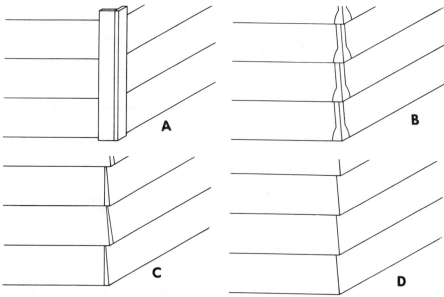

39–32. *Methods of treating bevel siding on an outside corner: A. Corner boards. B. Metal corners. C. Alternately lapped corners. D. Mitered corners.*

39–33. *Lap siding corner details.*

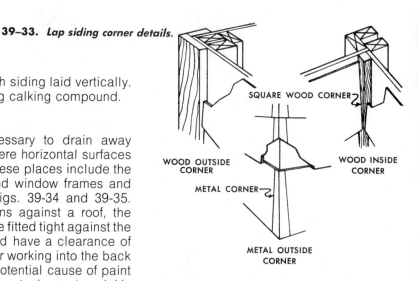

SQUARE WOOD CORNER

WOOD OUTSIDE CORNER

WOOD INSIDE CORNER

METAL CORNER

METAL OUTSIDE CORNER

joints of short length siding laid vertically. Use a nonhardening calking compound.

FLASHING

Flashing is necessary to drain away water at places where horizontal surfaces meet the siding. These places include the areas over door and window frames and around dormers. Figs. 39-34 and 39-35. Where siding returns against a roof, the siding should not be fitted tight against the shingles, but should have a clearance of 2″. Windblown water working into the back of the siding is a potential cause of paint failure. Siding cannot dry out quickly where there is a tight fit. Fig. 39-35.

Flashing should be anchored tightly. It should extend well under the siding and sufficiently over edges and ends of well-sloped water tables to prevent water from running in behind siding or jambs. A bead

of calking should be laid under the flashing to help seal out moisture. Fig. 39-36.

FOUNDATION LINES

The lowest edge of the siding should be at least 8″ above ground level. Fig. 39-37.

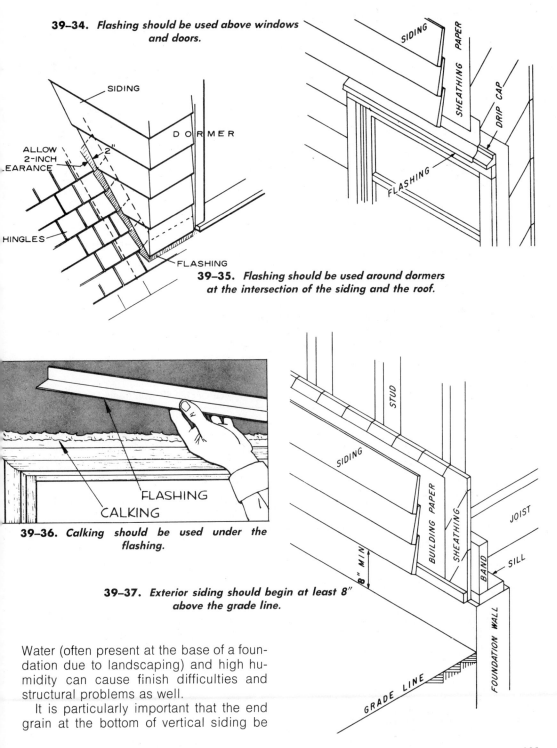

39-34. *Flashing should be used above windows and doors.*

39-35. *Flashing should be used around dormers at the intersection of the siding and the roof.*

39-36. *Calking should be used under the flashing.*

39-37. *Exterior siding should begin at least 8" above the grade line.*

Water (often present at the base of a foundation due to landscaping) and high humidity can cause finish difficulties and structural problems as well.

It is particularly important that the end grain at the bottom of vertical siding be

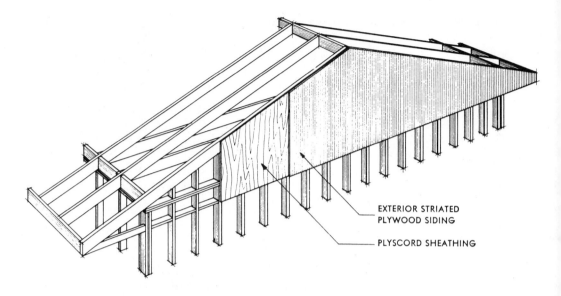

EXTERIOR STRIATED
PLYWOOD SIDING

PLYSCORD SHEATHING

NOTE: PROVIDE ADEQUATE ATTIC VENTILATION THROUGH
GABLE END OR SOFFIT OF ROOF OVERHANG

39–38. *Plywood used as an exterior covering for a gable.*

39–39a. *Plywood paneling used in conjunction with stone veneer.*

T-1-11 on gable blends with frame or masonry walls

T-1-11 used on carport storagewall

horizontal shiplap joint

39–39b. *Plywood panel siding used for the complete exterior of a home. Note the use of the shiplap joint to make a continuous surface when greater lengths are required.*

T-1-11 used as siding on entire house, appropriate in rural or urban settings

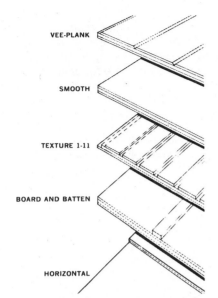

VEE-PLANK

SMOOTH

TEXTURE 1-11

BOARD AND BATTEN

HORIZONTAL

39-40. *Five of many plywood siding styles available.*

39-41. *Applying a texture 1-11 plywood siding panel.*

given a water repellent treatment. The use of a drip cap at the lower edge of the siding will help direct water away from the foundation.

PLYWOOD AND OTHER MAN-MADE MATERIALS
Plywood Siding

Plywood sheets are often used in gable ends, sometimes around windows and porches, and occasionally as overall exterior wall covering. Figs. 39-38 and 39-39. Plywood siding comes in many grades and surface textures, providing almost unlimited freedom of design for all types of construction. Fig. 39-40. For best results with painted surfaces, specify "medium-density overlaid" plywood. This grade has a resin-impregnated fiber surface that is heat-fused to the panel faces. It takes paint well and holds it longer. Sheet siding can be applied directly to studs, thus eliminating the need for sheathing. Fig. 39-41.

Plywood siding is strong. Tests conducted by United States Forests Products Laboratory at Madison, Wisconsin, proved that plywood as thin as 1/4", when nailed directly to studs, provides more than twice the relative rigidity and more than three times the relative strength of 1" × 8" lumber sheathing nailed horizontally to studs.

Plywood siding can be applied horizontally or vertically. The joints can be battens, V-grooves, or flush joints. Sometimes plywood is installed as lap siding. Figs. 39-42 and 39-43 (Page 686).

Plywood panel (sheet) siding and plywood lap siding come in 8' standard and 12', 14', and 16' special lengths. Plywood panel siding also comes in 9' lengths. Fig. 39-44. Lap siding may be 12", 16", or 24" wide. Plywood siding will cover large areas and cut installation time. The wide exposure of the lap siding or the use of panel siding is in keeping with the style of the modern ranch home.

APPLICATION OF PLYWOOD PANEL SIDING

Plywood panel siding is normally installed vertically but may be installed horizontally. All edges of panel siding should be backed with framing members or blocking. Fig. 39-45. To prevent staining of the siding, galvanized, aluminum, or other noncorrosive nails are recommended. Fig. 39-46.

39–42a. *Plywood siding. Consult the chart in Fig. 39–42b for details.*

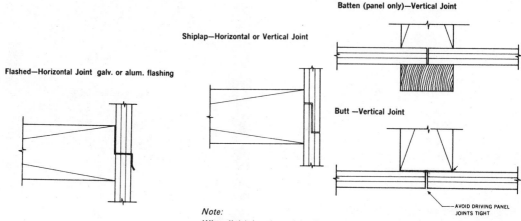

Batten (panel only)—Vertical Joint

Shiplap—Horizontal or Vertical Joint

Flashed—Horizontal Joint galv. or alum. flashing

Butt —Vertical Joint

AVOID DRIVING PANEL JOINTS TIGHT

Note:
When finish is paint, prime all panel edges prior to application. Otherwise treat edges with water repellants or caulk joints. Building paper may be omitted under panel sidings; also under lapped and bevelled siding when plywood sheathing is used. Joints may occur away from studs when plywood or board sheathing is used.

Plywood panel siding

NO DIAGONAL WALL BRACING OR BUILDING PAPER

16" or 24" STUD SPACING

INSULATION AS REQUIRED

AVOID DRIVING PANEL JOINTS TIGHT

EXTERIOR PLYWOOD SIDING

BATTEN

SIDING
(Striated, grooved, embossed, brushed, rough surface, or other)

WHERE PANEL SURFACE WILL BE PAINTED, PAINT ALL PLYWOOD EDGES THOROUGHLY BEFORE INSTALLATION

To apply panel siding follow these suggestions:
• Single wall construction with plywood paneling less than 1/2" thick will permit a maximum stud spacing of 16" on center when no building paper, corner bracing, or sheathing is used. When sheathing is used, the maximum stud spacing may be increased to 24" on center. No building paper or corner bracing is required.
• Nail 6" on center on panel edges and 12" on center at intermediate supports. Fig. 39-42.

APPLICATION OF PLYWOOD LAP SIDING

To apply lap plywood siding, follow these suggestions:
• In single wall construction the maximum stud spacing is 16" on center, and "let-in" corner bracing must be installed to meet FHA requirements. Install building paper between the siding and the studs. Fig. 39-47. The minimum head lap is 2". Wedges are installed at butt ends and at corners. When sheathing is used, corner bracing is not required and the maximum stud spacing is 24" on center. The minimum head lap for 12" widths or less is 1". For widths over 12" use a 1 1/2" head lap.
• Use a 3/8" thick starter strip for the first course. Fig. 39-48 (Page 688).
• Coat the edges of the siding with a primer before application.
• Vertical joints should be staggered. These joints must be centered over studs with a tapered wedge at least 1 5/8" wide behind the joint. Fig. 39-48.
• Use 8d noncorrosive casing or box nails. Insert one nail at each stud on the bottom edge of the siding. At all vertical joints nail 4" on center for siding 12" wide or less. Nail 8" on center for siding 16" wide or more. All nails should be placed 1/4" back from the edge of the plywood. Fig. 39-49. Set and putty all casing nails. Box nails are driven flush.

Hardboard and Particle Board Siding

The hardboard exterior sidings are fabricated for use as lap siding or as large panels up to 16' in length by 4' in width.

39–42b. *Plywood siding details.*

Plywood siding direct to studs (Plywood continuous over two or more spans) / Recommendations apply to all species groups.

Panel Siding	Maximum Stud Spacing c. to c. (inches)	Nail Size & Type[1]	Nail Spacing (inches)	
Minimum Plywood Thickness (inch)			Panel Edges	Intermediate
3/8	16	6d ⎧ non-corrosive	6	12
1/2	24	6d ⎨ siding or casing	6	12
5/8	24	8d ⎪ (galv. or alum.)	6	12
T 1-11 (5/8)	16	8d ⎩	6[2]	12

Lap or Bevel Siding[3]				
Typical Width (inches)	Min. Lap Siding Thickness (inch)	Min. Bevel Butt Thickness (inch)		
12, 16 or 24	3/8	9/16	16	6d ⎧ non-corrosive — One nail per stud along bottom edge — 4" at vertical joint; 8" at studs if siding wider than 12"
	1/2		20	8d ⎨ siding or casing
	5/8		24	8d ⎩ (galv. or alum.)

Notes: (1) Nails through battens must penetrate studs at least 1".
   (2) Use single nail on shiplap edges slant-driven to catch both edges. Can nail to 3/8" from panel edge, but do not set nails. Nails may be set if placed on both sides of joint instead of slant-driven.
   (3) Minimum head-lap 1 1/2".

Look for these typical APA grade-trademarks. See Fig. 5-9, Unit 5, for complete details.

M.D. OVERLAY
GROUP 1 (APA)
EXTERIOR
PS 1-74 — 000 —

Sanded Grades
A-C
GROUP 2 (APA)
EXTERIOR
PS 1-74 — 000 —

Specialty Panels
303 SIDING 16 oc
GROUP 3 (APA)
EXTERIOR
PS 1-74 — 000 —

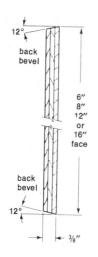

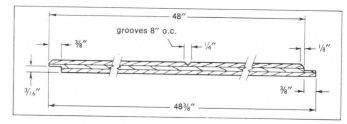

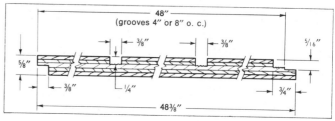

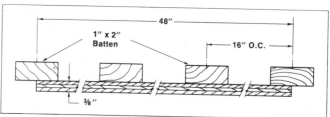

39-43. Details for plywood panel siding.

39-44. Applying plywood lap siding in 16' lengths.

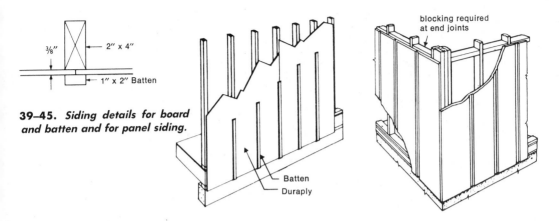

39–45. *Siding details for board and batten and for panel siding.*

3/8″ ← 2″ x 4″
← 1″ x 2″ Batten

Batten
Duraply

blocking required at end joints

39–46. *Framing and nailing schedule for plywood panel siding.*

Framing and Nailing Schedule						
Panel Siding Thickness	5/16″	3/8″	1/2″ grooved	1/2″ flat	5/8″ grooved	5/8″ flat
Single Wall Construction						
Maximum Stud Spacing	—	16″ o.c.	16″ o.c.	24″ o.c.	16″ o.c.	24″ o.c.
Nail Size	—	6d	8d	8d	8d	8d
Over 3/8″ Sheathing						
Maximum Stud Spacing	24″	24″	24″	24″	24″	24″
Nail Size	6d	6d	6d	6d	8d	8d
Approximate Nail Spacing*						
Edges	6″	6″	6″	6″	6″	6″
Intermediate Members	12″	12″	12″	12″	12″	12″

*Use non-corrosive casing, siding, or box nails.

39–47. *Plywood lap siding applied to an exterior wall without sheathing.*

Plywood lap or bevel siding

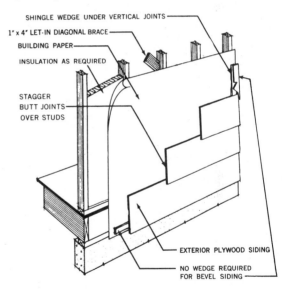

SHINGLE WEDGE UNDER VERTICAL JOINTS
1″ x 4″ LET-IN DIAGONAL BRACE
BUILDING PAPER
INSULATION AS REQUIRED
STAGGER BUTT JOINTS OVER STUDS
EXTERIOR PLYWOOD SIDING
NO WEDGE REQUIRED FOR BEVEL SIDING

They are impregnated with a baked-on tempering compound. This process produces tough, dense, grainless sidings that will not split or splinter and are highly resistant to denting. Panel surfaces are completely free of imperfections.

Hardboard sidings reinforce wall construction, go up quickly, and can be easily worked with both power and regular woodworking tools. Hardboards take paint, enamel, and stain and hold the finish longer than lumber. The sidings are available unprimed or factory primed. Factory-primed siding can be exposed to the weather for as long as 60 days prior to application of the finishing coats.

Particle board can also be used as siding material. It is available in many of the same shapes and surfaces as hardboard or plywood and is applied in the same manner.

APPLICATION OF HARDBOARD SIDING

Hardboard siding may be applied over sheathed walls with studs spaced not more than 24" on center or over unsheathed walls with studs spaced not more than 16" on center. The lowest edge of the siding should be at least 8" above the finished grade level.

When hardboard siding is applied directly to studs or over wood sheathing, moisture-resistant building paper or felt (non-vapor-barrier) should be laid directly under the siding.

As in all frame construction, a vapor barrier must be used in all insulated buildings and in uninsulated buildings located in areas where the average January temperature is below 40° F. The vapor barrier is installed next to the heated wall. The installed vapor barrier must be continuous, with tight joints and with any breaks or tears repaired.

When applying hardboard siding, use rustproof siding nails. Nail only at stud locations and on special members around doors and windows. Fig. 39-50. Nails must be kept back ½" from the ends and edges of the siding pieces. Fig. 39-51.

At inside corners, siding should be butted (with approximately ¹/₁₆" space) against a 1¹/₈" × 1¹/₈" wood corner member. Outside corners may be 1¹/₈" wood corner boards, or metal corners may be used. Fig. 39-52. Calking should be applied wher-

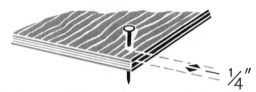

39–49. *Nails for lap siding should be placed ¹/₄" back from the edge of the plywood.*

39–48. *Typical lap siding details.*

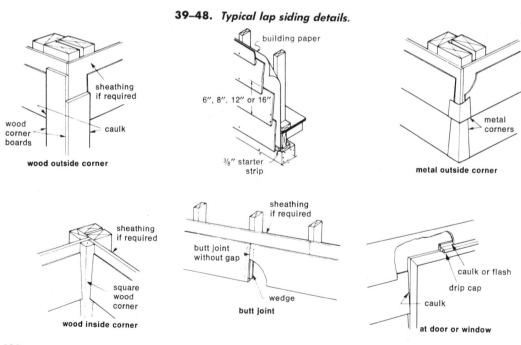

wood outside corner — sheathing if required — wood corner boards — caulk

building paper — 6", 8", 12" or 16" — ³/₈" starter strip

metal outside corner — metal corners

wood inside corner — sheathing if required — square wood corner

butt joint — sheathing if required — butt joint without gap — wedge

at door or window — caulk or flash — drip cap — caulk

39–50. *Hardboard framing and nailing schedule.*

| Panel thickness | Single wall construction | | | Over sheathing | | Approximate nail spacing | | |
	Maximum stud spacing	Nail Size	Maximum stud spacing	Nail Size	Edges	Intermediate members	Distance from edge
7/16″	16″ o.c.	6d	24″	8d	4″	8″	½″

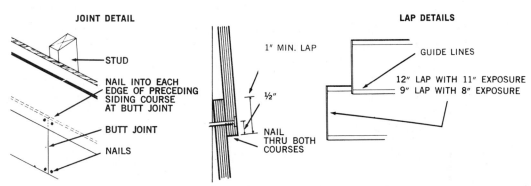

JOINT DETAIL

STUD

NAIL INTO EACH EDGE OF PRECEDING SIDING COURSE AT BUTT JOINT

BUTT JOINT

NAILS

1″ MIN. LAP

½″

NAIL THRU BOTH COURSES

LAP DETAILS

GUIDE LINES

12″ LAP WITH 11″ EXPOSURE
9″ LAP WITH 8″ EXPOSURE

39–51. *Hardboard lap siding application details.*

39–52. *Corner treatment for hardboard lap siding.*

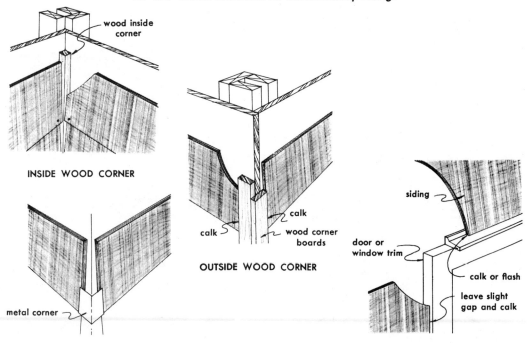

wood inside corner

INSIDE WOOD CORNER

metal corner

OUTSIDE METAL CORNER

calk

calk

wood corner boards

OUTSIDE WOOD CORNER

siding

door or window trim

calk or flash

leave slight gap and calk

DOOR AND WINDOW TREATMENT

39–53. *Hardboard panel siding details.*

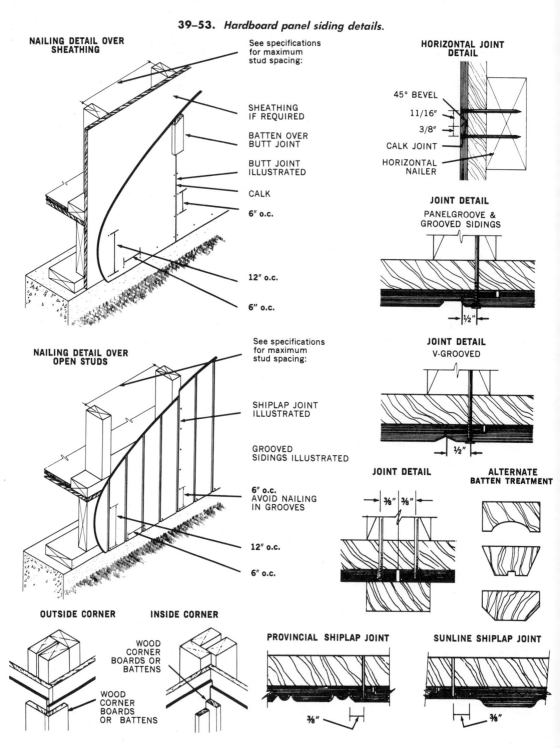

NAILING DETAIL OVER
SHEATHING

See specifications
for maximum
stud spacing:

SHEATHING
IF REQUIRED

BATTEN OVER
BUTT JOINT

BUTT JOINT
ILLUSTRATED

CALK

6" o.c.

12" o.c.

6" o.c.

NAILING DETAIL OVER
OPEN STUDS

See specifications
for maximum
stud spacing:

SHIPLAP JOINT
ILLUSTRATED

GROOVED
SIDINGS ILLUSTRATED

6" o.c.
AVOID NAILING
IN GROOVES

12" o.c.

6" o.c.

OUTSIDE CORNER   INSIDE CORNER

WOOD
CORNER
BOARDS OR
BATTENS

WOOD
CORNER
BOARDS
OR BATTENS

HORIZONTAL JOINT
DETAIL

45° BEVEL

11/16"

3/8"

CALK JOINT

HORIZONTAL
NAILER

JOINT DETAIL
PANELGROOVE &
GROOVED SIDINGS

½"

JOINT DETAIL
V-GROOVED

½"

JOINT DETAIL

3/8" 3/8"

ALTERNATE
BATTEN TREATMENT

PROVINCIAL SHIPLAP JOINT

3/8"

SUNLINE SHIPLAP JOINT

3/8"

ever the siding butts against corner boards, windows, and door casings. Fig. 39-52.

Installing flat panels. Flat panels are installed vertically. All joints and panel edges should fall on the center of framing members. If it is necessary to make a joint with a panel that has been field cut and the shiplap joint removed, use a butt joint. Butter the edges with calking and bring to light contact. Do not force or spring panels into place. Leave a slight space where siding butts against window or door trim and calk. Fig. 39-53.

Installing lap siding. Start the application by fastening a wood starter strip ($3/8'' \times 1^3/8''$) along the bottom edge of the sill. Fig. 39-54. Level and install the first course of siding with the bottom edge at least $1/8''$ below the starter strip. Fasten the first course by nailing $1^1/2''$ from drip edge of siding and $1/2''$ from butt end.

Install subsequent siding courses using a minimum overlap of 1". Fig. 39-51. Butt joints should occur only at stud locations. Factory-primed ends should be used for all vertical butt joints which will not be covered. Adjacent siding pieces should just touch at butt joints, or a $1/16''$ space may be left and filled with a butyl calk. Never force or spring siding into place.

WOOD SHINGLES AND SHAKES

Wood shingles and shakes are widely used for wall coverings, and a large selection is available. Wall shingles come in lengths of 16", 18", and 24". They may be prefinished or finished after installation. Handsplit shakes come in lengths of 18", 24", and 32".

Shingles

Shingles are usually separated into four grades. The first grade is composed of clear shingles, all heart, all edge grain. The second grade consists of shingles with clear butts and allows defects in that part of the shingle that will normally be covered in use. The third grade includes shingles that have defects other than permitted in the second grade. The fourth grade is a utility grade for undercoursing on double-coursed sidewall applications or for interior accent walls.

Shingles are made in random widths. In the No. 1 grade, they vary from 3" to 14", with only a small proportion of the narrow width permitted. Shingles cut uniformly to widths of 4", 5", or 6" are also obtainable. They are known as dimension or rebutted-and-rejointed shingles. These are shingles with edges machine-trimmed so as to be exactly parallel, and with butts retrimmed at precise 90° angles. Dimension shingles are applied with tight-fitting joints to give a strong horizontal line. They are available with the natural "sawed" face or with one face sanded smooth. These shingles may be applied either single or double-coursed.

APPLICATION OF SHINGLES

There are two basic ways to apply shingles: single-course and double-course. In single-coursing, shingles are applied much as in roof construction, but greater weather exposures are permitted. Shingle walls have two layers of shingles at every

39-54. Use a $3/8'' \times 1^3/8''$ starter strip when applying hardboard lap siding.

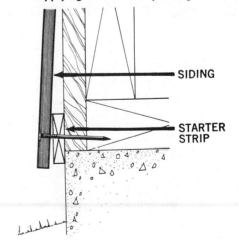

SIDING

STARTER STRIP

39–55. *Application details for wood shingles and shakes: A. Single-course shingles applied over wood sheathing. B. Double-course shingles applied over wood sheathing. C. Double-course shingles applied over non-wood sheathing. Notice the use of wood strips nailed to the studs to provide a good base for nailing the shingles. D. Outside corner formed by alternately overlapped wood shingles. E. Inside corner with the shingles mitered over flashing.*

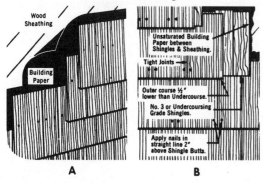

A   B

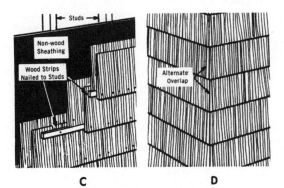

C   D

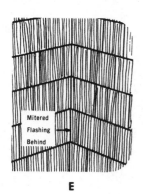

E

point, whereas shingle roofs have three-ply construction. Fig. 39-55 (A). To obtain architectural effect with deep bold shadow lines, shingles are frequently laid in double courses. Fig. 39-55 (B and C). Double-coursing allows for the application of shingles at extended weather exposures over undercoursing-grade shingles which are less expensive. When double-coursed, a shingle wall should be tripled at the foundation line (by using a double underlay). Fig. 39-56. When the wall is single-coursed, the shingles should be doubled at the foundation line. For recommended exposures see Fig. 39-57.

The spacing for the shingle courses is determined the same way as described for bevel siding. When shingles are applied over fiberboard or gypsum sheathing, horizontal 1″ × 4″ nailing strips should first be nailed to the studs. Fig. 39-58. The on-center spacing of these strips should be the same distance as the weather exposure chosen for the shingles, to provide a good base for nailing. Shingles may be staggered for rustic effect. Fig. 39-59.

Shingles should be applied with rust-resistant nails. At least ¼″ space should be allowed between shingles of the same course. It is frequently recommended that no shingle should be laid that is more than 8″ in width. Shingles wider than this should be sawed or split and nailed as two shingles.

For double-coursing, each outer course shingle should be secured with two 5d (1¾″) small head, rust-resistant nails driven about two inches above the butts, ¾″ in from each side. Additional nails should be driven about four inches apart across the face of the shingle. Fig. 39-55 (B and C). Single-coursing involves the same number of nails, but they can be shorter (3d, 1¼″) and should be blind-nailed not more than 1″ above the butt line of the next course. Fig. 39-55 (A). Never drive the nail so hard that its head crushes the wood.

Outside corners should be constructed with an alternate overlap of shingles be-

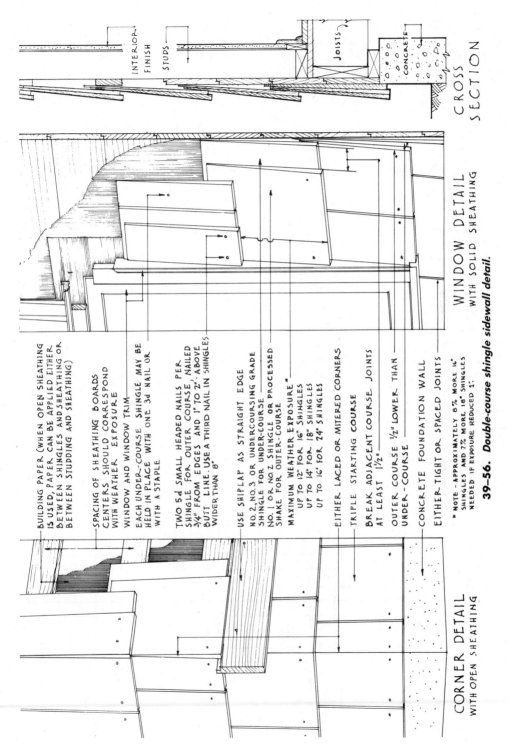

CROSS SECTION

INTERIOR FINISH

STUDS

JOISTS

CONCRETE

WINDOW DETAIL
WITH SOLID SHEATHING

BUILDING PAPER (WHEN OPEN SHEATHING IS USED, PAPER CAN BE APPLIED EITHER BETWEEN SHINGLES AND SHEATHING OR BETWEEN STUDDING AND SHEATHING)

SPACING OF SHEATHING BOARDS CENTERS SHOULD CORRESPOND WITH WEATHER EXPOSURE

WINDOW AND WINDOW TRIM

EACH UNDER-COURSE SHINGLE MAY BE HELD IN PLACE WITH ONE 3d NAIL OR WITH A STAPLE

TWO 5d SMALL HEADED NAILS PER SHINGLE FOR OUTER COURSE, NAILED ¾" FROM EDGES AND 1" TO 2" ABOVE BUTT LINE. USE A THIRD NAIL IN SHINGLES WIDER THAN 8"

USE SHIPLAP AS STRAIGHT EDGE

NO.2, NO.3 OR UNDERCOURSING GRADE SHINGLE FOR UNDER-COURSE

NO.1 OR NO.2 SHINGLE OR PROCESSED SHAKE FOR OUTER-COURSE

MAXIMUM WEATHER EXPOSURE*
UP TO 12" FOR 16" SHINGLES
UP TO 14" FOR 18" SHINGLES
UP TO 16" FOR 24" SHINGLES

EITHER LACED OR MITERED CORNERS

TRIPLE STARTING COURSE

BREAK ADJACENT COURSE JOINTS AT LEAST 1½"

OUTER-COURSE ½" LOWER THAN UNDER-COURSE

CONCRETE FOUNDATION WALL

EITHER TIGHT OR SPACED JOINTS

*NOTE - APPROXIMATELY 8% MORE 16" SHINGLES AND 7% MORE 18" SHINGLES NEEDED IF EXPOSURE REDUCED 1".

39-56. Double-course shingle sidewall detail.

CORNER DETAIL
WITH OPEN SHEATHING

693

tween successive courses. Fig. 39-55 (D). Inside corners may be mitered over a metal flashing. Fig. 39-55 (E). They may also be made by nailing an S4S strip, 1½" or 2" square, in the corner, after which the shingles of each course are jointed to the strip.

Shakes

There are three kinds of shakes:
- Handsplit-and-resawn.
- Tapersplit.
- Straightsplit.

The handsplit-and-resawn shakes have split faces and sawn backs. Cedar logs are first cut into desired lengths. Blanks or boards of proper thickness are split and then run diagonally through a bandsaw to produce two tapered shakes from each blank.

Tapersplit shakes are produced largely by hand, using a sharp-bladed steel froe and a wooden mallet. The natural shingle-like taper is achieved by reversing the block end-for-end with each split.

Straightsplit shakes are produced in the same manner as tapersplit shakes except

that by splitting from the same end of the block the shakes acquire the same thickness throughout. Fig. 39-60.

APPLICATION OF SHAKES

Maximum recommended weather exposure with single-course application is 8½" for 18" shakes, 11½" for 24" shakes, and 15" for 32" shakes. Fig. 39-61. The nailing normally is concealed in single-course applications; that is, the nailing is done at points slightly above (about 1") the butt line of the course to follow.

Double-course application requires an underlay of shakes or regular cedar shingles. With handsplit-resawn or with taper-split shakes, the maximum weather exposure is 14" for 18" shakes and 20" for 24" shakes. If straightsplit shakes are used, the exposure may be 16" for 18" shakes and 22" for 24" shakes. Butt-nailing of shakes is required with double-course application.

Use rust-resistant nails, preferably hot-dipped zinc-coated. The 6d size normally is adequate, but longer nails may be required, depending on the thickness of the shakes and the weather exposure. Do not

39-57. *Wood shingle exposure and coverage chart. The thickness dimension represents the total thickness of a number of shingles. For example, ⁵/2″ means that 5 shingles, measured across the thickest portion, when green, measure 2 full inches.*

LENGTH AND THICKNESS	Approximate coverage of one square (4 bundles) of shingles based on following weather exposures												
16″ x 5/2″	3½″	4″	4½″	5″	5½″	6″	6½″	7″	7½″	8″	8½″	9″	9½″
	70	80	90	100*	110	120	130	140	150𝕀	160	170	180	190
16″ x 5/2″	10″	10½″	11″	11½″	12″	12½″	13″	13½″	14″	14½″	15″	15½″	16″
	200	210	220	230	240†	....	....	....	....	....	....	....	....
18″ x 5/2¼″	3½″	4″	4½″	5″	5½″	6″	6½″	7″	7½″	8″	8½″	9″	9½″
	....	72½	81½	90½	100*	109	118	127	136	145½	154½𝕀	163½	172½
18″ x 5/2¼″	10″	10½″	11″	11½″	12″	12½″	13″	13½″	14″	14½″	15″	15½″	16″
	181½	191	200	209	218	227	236	245½	254½	....	....	....	....
24″ x 4/2″	3½″	4″	4½″	5″	5½″	6″	6½″	7″	7½″	8″	8½″	9″	9½″
	....	....	....	....	....	80	86½	93	100*	106½	113	120	126½
24″ x 4/2″	10″	10½″	11″	11½″	12″	12½″	13″	13½″	14″	14½″	15″	15½″	16″
	133	140	146½	153𝕀	160	166½	173	180	186½	193	200	206½	213†

Notes: * Maximum exposure recommended for roofs.
  𝕀 Maximum exposure recommended for single-coursing on sidewalls.
  † Maximum exposure recommended for double-coursing on sidewalls.

drive nailheads into the shake surface. The methods for constructing the corners when applying shakes are shown in Fig. 39-62.

Finishing Wood Shingles and Shakes

Red cedar shingles and shakes are well equipped by nature to endure without any protective finish or stain. In this state, the wood will eventually weather to a silver or dark gray. The speed of change and final shade depend mainly on atmosphere and climate.

Bleaching agents may be applied, in which case the wood will turn an antique silver gray. So-called natural finishes, which are lightly pigmented and maintain the original appearance of the wood, are available commercially. Stains, whether heavy or semitransparent, are readily "absorbed" by cedar, and it also takes paint well. Quality finishes are strongly recommended because they will prove most economical on a long-term basis.

ASBESTOS-CEMENT SIDING AND SHINGLES

Asbestos-cement siding and shingles come in various sizes and colors. They should be applied in accordance with the manufacturer's directions.

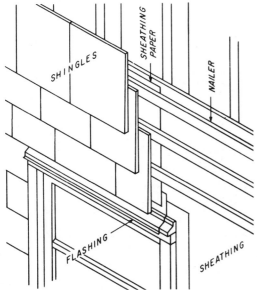

39-58. *The application of wood shingles over non-wood sheathing. Note the use of the nailing strip.*

39-59. *Staggered shingles give a rustic appearance.*

39-60. *Using a froe to split shakes in the 1800's. Today, tapersplit and straightsplit shakes are still produced largely by hand.*

Wood sheathing should be used under asbestos-cement shingles and siding. The siding or shingles are laid over a waterproof paper applied over the sheathing. The courses are laid out in the same manner as for horizontal bevel siding, or a story pole may be used. Noncorroding nails should be used, and care should be taken when driving nails not to crack the shingles. Vertical joints should be flashed with 4" wide strips of saturated felt laid under each joint. Fig. 39-63.

The treatment of corners may be similar to those used for wood siding. However, in most cases, the manufacturers will suggest the type of corner treatment best suited for their products. The corners should be flashed with a wide strip of asphalt paper underlay applied vertically.

STUCCO SIDEWALL FINISH

Stucco when properly used makes a good wall finish. Fig. 39-64. It may be a natural cement color or colored as desired. If stucco is to be applied on houses more than one story high, balloon framing should be used in the outside walls. With platform framing, shrinkage of the joists

and sills of the platform may cause an unsightly bulge or break in the stucco at that point.

Stucco is applied over lath. Three acceptable types of lath are:
• Zinc-coated or galvanized metal, with large openings (1.8 pounds per square yard) or small openings (3.4 pounds per square yard).
• Galvanized woven-wire fabric. This material may be 18-gauge wire with 1" maximum mesh, 17-gauge wire with 1½" maxi-

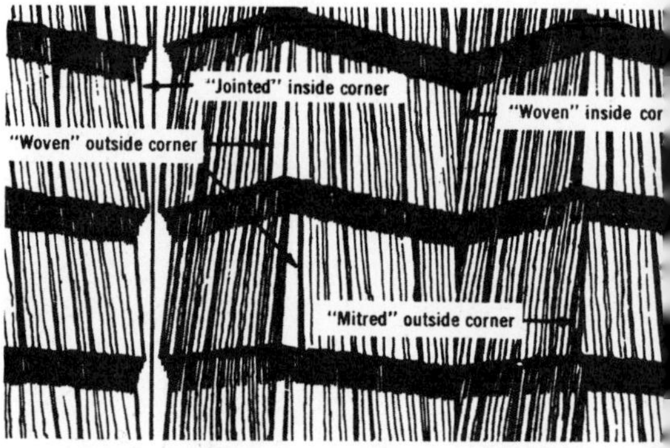

39-62. Wood shake corner treatment.

39-61. Wood shake exposure and coverage chart.

	Approximate sq. ft. coverage of one square of handsplit shakes based on these weather exposures											
	5½"	6½"	7"	7½"	8"	8½"	10"	11½"	13"	14"	15"	16"
18" x ½" to ¾" Handsplit-and-Resawn	55*	65	70	75**	80	85†	. . .	. . .	. . .	140‡	. . .	. . .
18" x ¾" 1¼" Handsplit-and-Resawn	55*	65	70	75**	80	85†	. . .	. . .	. . .	140‡	. . .	. . .
24" x ⅜" Handsplit	. . .	65	70	75***	80	85	100††	115†	. . .	. . .	. . .	. . .
24" x ½" to ¾" Handsplit-and-Resawn	. . .	65	70	75*	80	85	100**	115†	. . .	. . .	. . .	. . .
24" x ¾" to 1¼" Handsplit-and-Resawn	. . .	65	70	75*	80	85	100**	115†	. . .	. . .	. . .	. . .
32" x ¾" to 1¼" Handsplit-and-Resawn	. . .	. . .	. . .	. . .	. . .	. . .	100*	115	130**	140	150†	. . .
24" x ½" to ⅝" Tapersplit	. . .	65	70	75*	80	85	100**	115†	. . .	. . .	. . .	. . .
18" x ⅜" True-Edge Straightsplit	. . .	. . .	. . .	. . .	. . .	. . .	. . .	. . .	. . .	100	106	112‡
18" x ⅜" Straightsplit	65*	75	80	90	95	100†	. . .	. . .	. . .	. . .	. . .	. . .
24" x ⅜" Straightsplit	. . .	65	70	75*	80	85	100	115†	. . .	. . .	. . .	. . .
15" Starter-Finish Course	Use supplementary with shakes applied not over 10" weather exposure.											

Notes:
* Recommended maximum weather exposure for 3-ply roof construction.
** Recommended maximum weather exposure for 2-ply roof construction.
*** Recommended maximum weather exposure for roof pitches of 4/12 to 8/12.
† Recommended maximum weather exposure for single-coursed wall construction.
†† Recommended maximum weather exposure for roof pitches of 8/12 or steeper.
‡ Recommended maximum weather exposure for double-coursed wall construction.

mum mesh, or 16-gauge wire with a 2″ maximum mesh.

• Galvanized welded-wire fabric. This may be made of 16-gauge wire with 2″ × 2″ mesh and waterproof paper backing, or it may be 18-gauge wire with 1″ × 1″ mesh but no paper backing.

The lath should be kept at least 1/4″ away from the sheathing so that the stucco can be forced through the lath and embedded completely. Galvanized furring nails, metal furring strips, or self-furring lath are available for this spacing. Nails should penetrate the wood at least 3/4″. Where

39–63. *Asbestos-cement shingles over plywood sheathing.*

NO BLOCKING REQUIRED

BACKER STRIP AT
EACH VERTICAL JOINT

BUILDING PAPER

PLYSCORD SHEATHING

ASBESTOS-CEMENT
SHINGLES

fiberboard or gypsum sheathing is used, the length of the nail should be such that at least ¾″ penetrates into the wood stud.

Stucco Plaster

The plaster should be one part portland cement, three parts sand, and hydrated lime equal to 10% of the cement by volume. It should be applied in three coats to a total thickness of 1″. The first coat should be forced through the lath and worked so as to embed the lath at all points. Keep fresh stucco shaded and wet for three days. Do not apply stucco when the temperature is below 40° F. It sets very slowly, and there may be a freezing hazard before it has set.

Portland-cement stucco that has been commercially prepared may also be used. It should be applied according to the manufacturer's instructions.

MASONRY VENEER

Brick or stone veneer is often used for part or all of the wall covering over wood frame walls. Fig. 39-65. Although high in initial cost, it is frequently used as a wall covering in residential construction because of the low maintenance. Brick or stone veneer is applied by a skilled mason.

METAL AND PLASTIC SIDING

There are many kinds of metal and plastic sidings that can be applied over sheathing. Metal siding, either aluminum or steel, has a baked-on finish that requires little maintenance or care. Solid vinyl siding in various colors is also available. Standard hand tools such as those shown

39-64. *Stucco applied as an exterior covering over plywood sheathing.*

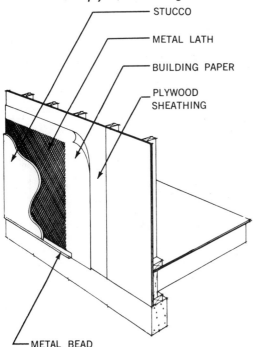

STUCCO

METAL LATH

BUILDING PAPER

PLYWOOD SHEATHING

METAL BEAD

39-65. *Brick veneer as a wall covering over wood-frame construction.*

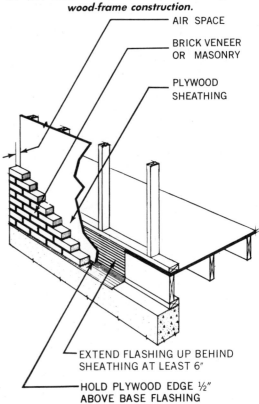

AIR SPACE

BRICK VENEER OR MASONRY

PLYWOOD SHEATHING

EXTEND FLASHING UP BEHIND SHEATHING AT LEAST 6″

HOLD PLYWOOD EDGE ½″ ABOVE BASE FLASHING

in Fig. 39-66 may be used to apply these siding materials.

Application of Metal and Plastic Siding

The interlocking joints of the pieces of siding and the accessory items vary with the manufacturer. Before starting, inspect and plan the job in accordance with the manufacturer's instructions for the material to be applied. Be sure to use the nail recommended to avoid corrosion and stains. Drive the nails firm and snug, but never so tight as to cause waves in the siding.

The general procedure for the application of metal or plastic siding is as follows:

1. Lay out the courses on the sidewalls of the building the same way as for wood bevel siding, or use a story pole. Fig. 39-22.

2. Run a chalk line for the starter strip around the house. Install the starter strip. A, Fig. 39-67. All nails should be driven so that the head is only slightly flush to the material. Do not drive the nails hard enough to bind the material tightly.

3. Nail the inside corner posts 12" on center before the rest of the siding is applied. The bottom of the corner posts should be aligned with the chalk line. The siding later fits into channels on both sides. B, Fig. 39-67.

4. Nail the outside corner posts 12" on center before the siding is applied. C, Fig. 39-67. Here also, siding fits into channels on the sides. Fill the cavity behind a wide corner post with a backer board or wood strips. To close in the open lower end of a wide corner post, cut a piece of "J" channel to the proper length (about 6¹/₈"). Miter it by cutting as shown in Fig. 39-68. Nail it to the wall so that the bottom end of the corner post will fit over it.

5. Attach the door and window trim along the sides of the doors and windows. The gables are also trimmed with this accessory. The same trim is used at the base of a wall intersecting a sloping roof, as on a breezeway. D, Fig. 39-67.

39-66. *Basic tools necessary for the application of metal or plastic siding: A. Chalk line. B. Rule. C. Level. D. Carpenter's square. E. Calking gun. F. Power saw (with a plywood cutting blade). G. Hammer. H. Double-action aviation snips. I. Utility knife. J. Tin snips. K. Crosscut saw.*

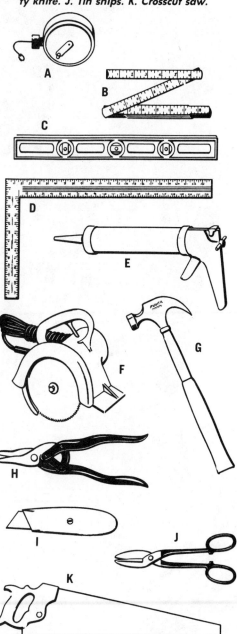

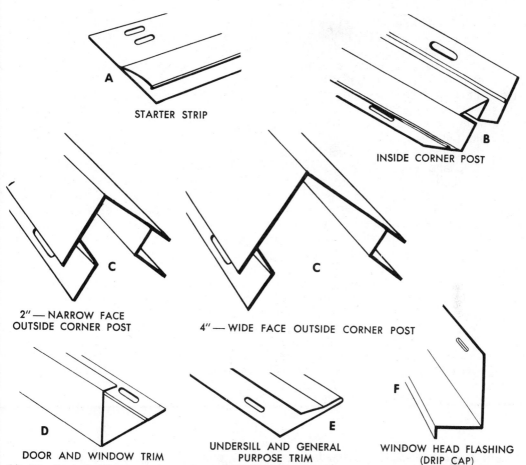

A

STARTER STRIP

B

INSIDE CORNER POST

C

2" — NARROW FACE
OUTSIDE CORNER POST

C

4" — WIDE FACE OUTSIDE CORNER POST

D

DOOR AND WINDOW TRIM

E

UNDERSILL AND GENERAL
PURPOSE TRIM

F

WINDOW HEAD FLASHING
(DRIP CAP)

39–67. *Typical siding accessories. These are for vinyl, but manufacturers of aluminum siding offer similar accessories for the application of their particular product. The picture on page 700 shows how these accessories are used.*

39–68. *Suggested method of bottom closure for the outside corner post shown at C in Fig. 39–67. This piece must be cut and installed before the corner post is nailed in place.*

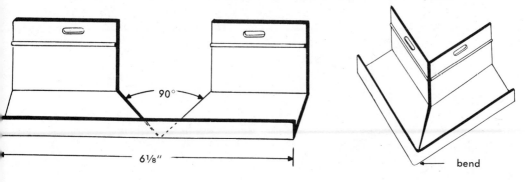

90°

6⅛"

bend

6. Install undersill and general purpose trim under windows and at the tops of walls against soffit moldings, furring where necessary to preserve alignment with the adjacent panels. This may also be used inverted to lock lower cut edges of siding-courses which are located above the level of the starter strip, such as at porch floors and cellar bulkheads. E, Fig. 39-67.

7. Place the first siding panel in the starter strip and lock it securely. Fig. 39-69. Backerboards, if used, are dropped into place. Nail the panel and install succeeding courses similarly. Nails must never be driven so tightly as to cause distortion of the siding.

Allowance should be made for expansion and contraction by leaving a 1/4" space at joints, channels, and corner posts. Field-cut end joints may be made with a sharp knife, hack saw blade, or tin snips. On the end to be overlapped, cuts should be made according to the manufacturer's instructions. For best appearance, overlap the siding away from areas of greatest traffic. Stagger end laps a minimum of 24" and in such a way that one is not directly above another unless separat-

ed by three courses. The end of the uncut panel should lap over the end of the cut panel.

8. Cut and fit the siding as necessary around windows and doors.

9. Attach window head flashing or window and door trim above the windows and doors as application reaches these levels, furring out where necessary for alignment with adjacent panels. See F, Fig. 39-67.

10. Trim the last course of siding to fit under the eaves. Install undersill and general purpose trim, furring where necessary to maintain proper panel angle. Engage the top of the panel with undersill and general purpose trim and lock at lower edge of panel as usual. Nail to secure when necessary.

11. To complete the gables, install windows and door trim above the windows at the gable ends. Cut the siding to the proper angle and install.

12. Calk where required.

13. Finish by washing down the siding to remove fingerprints and soil. Clean up all scrap material around the house.

SHUTTERS AND BLINDS

Shutters, or blinds, are used on today's homes more for architectural effect than for any functional purpose. Fig. 39-70. Historically they were intended as a means of protection from enemy attack or the weather. Shutters are particularly popular with the New England style Cape Cod home, on which they are usually installed adjacent to double-hung windows. Fig. 39-71.

Shutters are made from wood, nylon, vinyl, or aluminum. They are available in a variety of sizes and styles to enhance the architectural effect desired. Figs. 39-72 and 39-73. When shutters are installed for a functional purpose, they are hinged to the window trim or doorjamb and will close and lock over the opening. Shutters used only as decoration are woodscrewed or held with special clips after the siding is applied and finished. Fig. 39-74 (Page 706).

39–69. *Secure the bottom edge of the first piece of siding in the starter strip before nailing the top edge.*

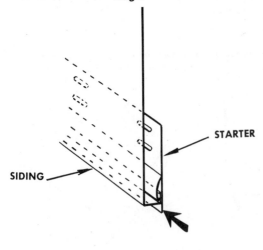

STARTER

SIDING

39-70. *This house has two-panel louvered shutters which add to the architectural effect. They also provide the homeowner with an opportunity to brighten the home with additional color when decorating.*

ESTIMATING
Material

To determine the amount of siding necessary for the exterior of a home, it is first necessary to figure the amount of wall area to be covered. This is done by multiplying the perimeter of the house by the wall height and subtracting window and door openings. For a gable roof, the gable area must also be figured and added to the sidewall area.

WOOD BEVEL SIDING

To figure the amount of wood bevel siding needed to cover the exterior of a house, first determine the area to be covered. Then consider the difference between the nominal and dressed dimensions of the siding material, the amount of lap, and the allowance for trimming and waste.

For example, a house 46' long and 26' wide has a perimeter of 144'. (2 × 46) + (2 × 26) = 144. With 8' sidewalls, the total area to be covered would be 1152 sq. ft.

(144 × 8 = 1152). If the window and door openings equal 250 sq. ft., the total area to be sided then would be 902 sq. ft. (1152 − 250 = 902).

The dressed width of the siding and the amount of lap are accounted for by the *area factor*. To find the area factor, refer to the chart in Fig. 39-75. For example, if 1" × 10" bevel siding is to be used, the chart shows that the area factor is 1.21. Multiply this factor by the area to be covered:

$$902 \times 1.21 = 1091.42.$$

Add five percent of this amount for the trim and waste. Five percent of 1091.42 sq. ft., is 54.57 sq. ft. Added to the original 1091.42 sq. ft., this yields a rounded total of 1146 sq. ft. of siding.

SHINGLES

To determine the number of shingles necessary to cover the exterior sidewall of a house, the area to be covered plus trim and waste allowance must be figured as for bevel siding. For single coursing, one

REVERSIBLE SHUTTER BLIND
BLIND SIDE

REVERSIBLE SHUTTER BLIND
SHUTTER SIDE

39-71. Shutters mounted adjacent to a double-hung window. The shutters from this manufacturer are reversible; they might be used as a blind with louvers or as a shutter with a paneled effect.

39-72. This chart shows a variety of shutter sizes available from one manufacturer. The sizes will vary with the manufacturer and the material from which the shutter or blind is made.

Sizes

1⅛″ thick			
Widths (Pair)			
2′0″			2′10″
2′4″			3′0″
2′6″			3′4″
2′7″			3′8″
2′8″			4′0″
Heights			
1′3″	2′5″	3′7″	5′3″
1′7″	2′7″	3′11″	5′7″
1′9″	2′11″	4′3″	5′11″
1′11″	3′1″	4′7″	6′3″
2′1″	3′3″	4′11″	6′7″
2′3″			6′11″

39-73a. Examples of a few of the many styles and patterns of shutters and blinds which are available.

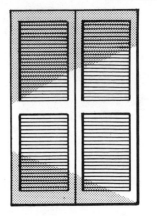

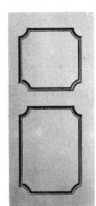

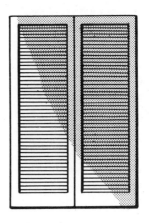

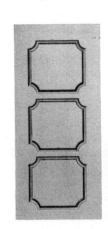

square (4 bundles) of 16″ shingles with a 7½″ exposure will cover 150 sq. ft. of area. Fig. 39-57. Use the previous example of 902 sq. ft. After adding 5 percent to this number for the waste and cutting factor, divide by 150.

$$902 \times 0.05 = 45.1$$
$$902 + 45.1 = 947.1$$

$$\frac{947.1 \text{ (area to be covered)}}{150 \text{ (area covered by 1 square of shingles)}} = 6.31, \text{ or } 6\frac{1}{2} \text{ squares}$$

Since there are 4 bundles in a square, 26 bundles will be required to shingle the house in the example. Shakes would be estimated the same way.

VINYL AND METAL SIDING

Most manufacturers of vinyl and metal siding indicate the number of pieces of a given size siding required to cover one square (100 sq. ft., or a wall area 10′ × 10′). For example, 12 pieces of 8″ siding will cover one square.

Figure the area to be covered (see "Wood Bevel Siding"). Divide this total area by 100 to find the number of squares. Multiply the number of squares by 12 (the number of pieces needed to cover one square). For the house in the previous example, the number of pieces would be calculated as follows:

$$\frac{947.1}{100} = 9.471$$

$$9.471 \times 12 = 113.65, \text{ or } 114 \text{ pieces}$$

NAILS

To determine the number of nails necessary, refer to the chart in Fig. 39-76. For example, bevel siding ¾″ ×10″ (nominal 1″ × 10″) would require ½ (0.5) pound of nails per 100 square feet. Since we have 902 square feet, divide by 100 to find how many hundreds of square feet there are in the area to be covered.

$$\frac{902}{100} = 9.02$$

39–73b. *Two styles of blinds mounted adjacent to an entrance door.*

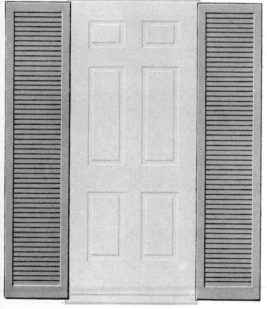

1 Panel

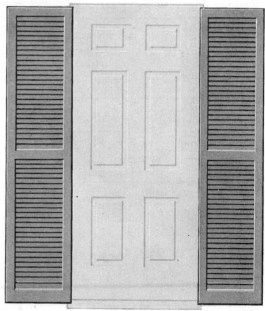

2 Equal Panels

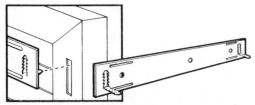

39–74. Shutter mounting bracket. This mounting bracket is fastened to the wall. The projection on the mounting bracket snaps into a plastic insert at the rear of the shutter. With this mounting system, the shutter can be easily removed for cleaning or painting.

Multiply this by the weight of nails required per 100 square feet.

$$9.02 \times 0.5 = 4.51 \text{ (pounds of nails for the siding)}$$

Labor

Again using $3/4'' \times 10''$ bevel siding, refer to Fig. 39-76. The chart indicates that a worker can apply 55 board feet per hour. If the house to be covered contains 1146 square feet, divide this by 55 to get the total number of hours necessary: 20.8.

39–75. Coverage estimator.

The following estimator provides factors for determining the exact amount of material needed for the five basic types of wood paneling. Multiply square footage to be covered by factor (length x width x factor).

	Nominal Size	Width Dress	Width Face	Area Factor*		Nominal Size	Width Dress	Width Face	Area Factor*
Shiplap	1 x 6	5⁷/₁₆	4¹⁵/₁₆	1.22	**Paneling Patterns**	1 x 6	5⁷/₁₆	5¹/₁₆	1.19
	1 x 8	7⅛	8⅝	1.21		1 x 8	7⅛	6¾	1.19
	1 x 10	9⅛	6⅝	1.16		1 x 10	9⅛	8¾	1.14
	1 x 12	11⅛	10⅝	1.13		1 x 12	11⅛	10¾	1.12
Tongue And Groove	1 x 4	3⁷/₁₆	3³/₁₆	1.26	**Bevel Siding**	1 x 4	3½	3½	1.60
	1 x 6	5⁷/₁₆	5³/₁₆	1.16		1 x 6	5½	5½	1.33
	1 x 8	7⅛	6⅞	1.16		1 x 8	7¼	7¼	1.28
	1 x 10	9⅛	8⅞	1.13		1 x 10	9¼	9¼	1.21
	1 x 12	11⅛	10⅞	1.10		1 x 12	11¼	11¼	1.17
S4S	1 x 4	3½	3½	1.14	*Allowance for trim and waste should be added.				
	1 x 6	5½	5½	1.09					
	1 x 8	7¼	7¼	1.10					
	1 x 10	9¼	9¼	1.08					
	1 x 12	11¼	11¼	1.07					

BEVEL SIDING	Nails	Labor
	Per	Board Feet
Size	100 Square Feet	Per Hour
½ x 4	1½ Pounds	30
½ x 5	1½ Pounds	40
½ x 6	1 Pound	45
½ x 8	¾ Pound	50
⅝ x 8	¾ Pound	50
¾ x 8	¾ Pound	50
⅝ x 10	½ Pound	55
¾ x 10	½ Pound	55
¾ x 12	½ Pound	55

39–76. Estimating drop or bevel siding.

DROP SIDING	Nails	Labor
	Per	Board Feet
Size	100 Square Feet	Per Hour
1 x 6	2½ Pounds	50
1 x 8	2 Pounds	55

QUESTIONS

1. What factors are combined to produce the wide variety of exterior wall coverings?

2. What type of wood siding serves as both sheathing and exterior wall coverings?

3. What is the maximum exposure for a piece of 8″ bevel siding?

4. What is the advantage of using a story pole when installing bevel siding?

5. What is the purpose of a "preacher"?

6. List several requirements for nails used on wood siding.

7. List several precautions that should be taken to prevent outside moisture problems.

8. When installing plywood siding, what type of surface should be specified?

9. List several advantages of hardboard siding.

10. What are the two methods of shingle sidewall application?

11. When using shingles as a sidewall covering, under what conditions must nailing strips be used?

12. What are the three kinds of shakes available?

13. When using asbestos-cement siding and shingles, what type of sheathing must be used?

14. When stucco is used as a sidewall finish, what holds it in place?

15. When installing metal and plastic siding, why aren't the nails driven in tightly?

16. When making a lap joint on metal or plastic siding, in which direction should the overlap be made?

17. What is the purpose of shutters and blinds on today's homes?

Section VII

COMPLETING THE INTERIOR

Thermal Insulation and Vapor Barriers

UNIT 40

Insulation is the property of material which slows down the transmission of energy in the form of heat, sound, or electricity. In construction, insulation is usually thought of in relation to heat transmission, although sound is an equally important item to consider. (Sound insulation is discussed in the next unit.) Most building materials—and even the air space between studs—have some heat insulation properties. However, by themselves they are not sufficient for modern homes that must have heating, air conditioning, and other climate control equipment. Insulation materials are products that are used in addition to ordinary building materials for the specific purpose of retarding the passage of heat. Insulation keeps homes warmer in the winter and cooler in the summer.

Better and more efficiently insulated homes are being built by contractors to please quality–conscious buyers. Upgrading of insulation beyond the minimum FHA standards pays good dividends in the form of increased comfort, reduced heating and air conditioning costs, and smaller, less expensive furnaces, cooling equipment, and duct work. Maximum insulation also makes electric heating feasible. In warm climates, the use of insulation with air conditioning is justified because operating costs are reduced and units of smaller capacity are required.

INSULATING MATERIALS

Insulation is manufactured in a variety of forms and types, each with advantages for specific uses. Materials commonly used for insulation may be grouped into the following general classes:

- Flexible (blanket and batt).
- Loose fill.
- Reflective.
- Rigid (structural and nonstructural).
- Miscellaneous types.

Flexible Insulation

Flexible insulation is manufactured in two forms: *blanket* and *batt*. Blanket insulation is furnished in rolls or packages. It comes in widths suited to 16″ and 24″ stud and joist spacing. A, Fig. 40-1. These pieces can be easily cut to length to fit various size openings. Fig. 40-2. Usual thicknesses are 1½, 2, and 3 inches. The body of the blanket is made of felted mats of mineral or vegetable fibers, such as rock wool, glass wool, wood fiber, or cotton. Organic insulations are treated to make them resistant to fire, decay, insects, and vermin. Most blanket insulation is covered with paper or other sheet material with tabs on the sides for fastening to studs or joists. One covering sheet serves as a vapor barrier to resist movement of water vapor and should always face the warm side of the wall. Aluminum foil, asphalt, and plastic-laminated paper are common barrier materials.

Batt insulation is also made of fibrous material preformed to thicknesses of 4″ and 6″ for 16″ and 24″ joist spacing. B, Fig. 40-1. It is supplied with or without a vapor barrier. Fig. 40-3. One friction-type fiber glass batt is supplied without a covering and is designed to remain in place without the normal fastening methods.

Loose Fill Insulation

Loose fill insulation is usually supplied in bags or bales and placed by pouring, blowing, or packing by hand. C, Fig. 40-1. Materials used include rock and glass wool, wood fibers, shredded redwood bark, cork, wood pulp products, vermiculite, sawdust, and shavings.

Fill insulation is best used between first-floor ceiling joists in unheated attics. It is also used in sidewalls of existing houses that were not insulated during construction. Where no vapor barrier was installed during construction, suitable paint coatings, as described later in this unit, should be used for vapor barriers.

Reflective Insulation

Most materials reflect some radiant heat. Radiant heat is heat that flows through air in a direct line from a warm surface to a cooler one. Materials high in reflective properties include aluminum foil, sheet metal with tin coating, and paper products coated with a reflective oxide composition. Such materials can be used in enclosed stud spaces, in attics, and in similar locations to retard heat transfer by radiation.

Reflective insulations are equally effective whether the reflective surface faces the warm or cold side. However, the reflective surface must face an air space at least $3/4''$ deep. Where a reflective surface contacts another material, the reflective properties are lost and the material has little or no insulating value.

Reflective insulation is more effective in preventing summer heat flow through ceilings and walls. It should be considered more for use in the warmer climates than in the North.

Sometimes, reflective insulation of foil is applied to blanket insulation and to the stud-surface side of gypsum lath. The type of reflective insulation shown in D, Fig. 40-1 has air spaces between the reflective surfaces. Metal foil suitably mounted on some supporting base also makes an excellent vapor barrier.

Rigid Insulation

Rigid insulation is manufactured in sheets and other forms. E, Fig. 40-1. The

40–1. *Types of insulation: A. Blanket. B. Batt. C. Loose fill. D. One type of reflective. E. Rigid.*

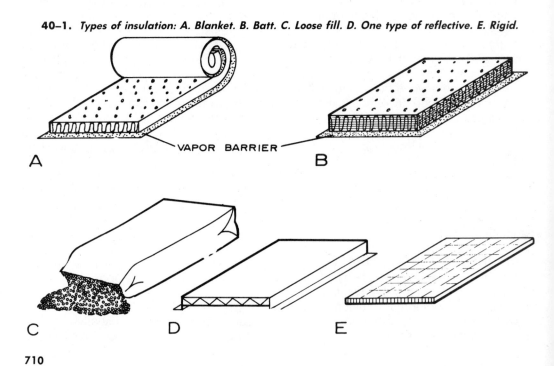

VAPOR BARRIER

A

B

C

D

E

40–2. *To cut insulation, place it on a piece of scrap plywood or 2 × 4, compress the material with one hand, and cut it with a sharp knife. When cutting faced insulation, keep the facing up.*

40–3. *Installing 4" flexible insulation. Note that the vapor barrier is placed so that it faces the warm side of the wall. Sometimes an additional thin sheet of plastic is added to completely seal the wall. This should always be done when insulation without a vapor barrier is used.*

most common types of fiberboard are made from processed wood, sugar cane, or other vegetable products. The insulation may also be made from such inorganic materials as glass fiber.

Rigid insulation may be structural or nonstructural. Structural insulating boards, in densities ranging from 15 to 31 pounds per cubic foot, are used as building boards, roof decking, sheathing, and wallboard, While they have moderately good insulating properties, their primary purpose is structural.

Roof insulation is nonstructural and serves mainly to provide thermal resistance to heat flow. It is called slab or block insulation and is manufactured in rigid units $1/2"$ to $3"$ thick and usually 2' x 4' in size.

In house construction, the most common forms of rigid insulation are sheathing and decorative coverings in sheets or in tile squares. Sheathing board is made in thicknesses of $1/2"$ and $25/32"$. It is coated or impregnated with an asphalt compound to provide water resistance. Sheets are made 2' × 8' for horizontal application and 4' × 8' or longer for vertical application.

Miscellaneous Types

There are several other kinds of insulation. Blanket insulation may be made up of multiple layers of corrugated paper. Sometimes lightweight vermiculite and perlite aggregates are used in plaster to increase its thermal resistance.

Foamed-in-place insulation includes sprayed and foam types. Sprayed insulation is usually inorganic fibrous material blown against a clean surface which has been primed with an adhesive coating. It is often left exposed for acoustical purposes.

Expanded polystyrene and urethane plastic foams may be molded or foamed in place. Urethane insulation may also be applied by spraying. Polystyrene and urethane in board form can be obtained in thicknesses from $1/2"$ to $2"$.

INSULATING VALUES

The thermal properties of most building materials are known, and the rate of heat flow, or coefficient of transmission, for most combinations of construction can be calculated. This coefficient, or *U*-value, is a measure of heat transmission between air on the warm side and air on the cold side of the construction unit. It is defined as the amount of heat (in Btu's) transmitted in 1 hour, through 1 square foot of surface, for each degree Fahrenheit difference in temperature between the inside and outside air. (One Btu, or British thermal unit, is the amount of heat which will raise the temperature of 1 pound of water 1 degree Fahrenheit.)

The insulating value of a wall will vary with different types of construction, with materials used in construction, and with different types and thicknesses of insulation. Comparisons of *U*-values may be used to evaluate different combinations of materials and insulation based on overall heat loss, potential fuel savings, influence on comfort, and installation costs. The amount of insulation required to obtain a desired *U*-value can be determined for any type of construction.

Information regarding the calculated *U*-values for typical constructions with various combinations of insulation may be found in "Thermal Insulation from Wood for Buildings: Effects of Moisture and Its Con-

trol," published by Forest Products Laboratory, Madison, Wisconsin.

The table in Fig. 40-4 provides some comparison of the individual insulating values of various building materials. These are expressed as *k* values, or heat conductivity. Heat conductivity is defined as the amount of heat (in Btu's) that will pass in 1 hour through 1 square foot of material 1 inch thick per 1° F temperature difference between faces of the material. Simply expressed, *k* represents heat loss. The lower this numerical value, the better the insulating qualities.

Building materials are also rated on their resistance or *R* value, which is merely another expression of insulating value. Fig. 40-4. The *R* value is usually expressed as the total resistance of the wall or of a thick insulating blanket or batt, whereas k is the rating per inch of thickness. R = 1/k. Thus, if the *k* value of 1 inch of insulation is 0.25, the resistance, *R*, is 1/0.25, or 4.0. If there are 3 inches of this insulation, the *R* value is 3 × 4.0, or 12.0.

Climate must also be taken into consideration when choosing building materials, insulation, and heating and cooling equipment. The map in Fig. 40-5 shows the average winter low temperatures found in different areas of the United States. Such

40–4.
Thermal Properties Of Various Building Materials Per Inch Of Thickness

Material	Thermal Conductivity K	Thermal Resistance R	Efficiency as as insulator Percent
Wood	0.80	1.25	100.0
Air Space[1]	1.03	0.97	77.6
Cinder Block	3.6	0.28	22.4
Common Brick	5.0	0.20	16.0
Face Brick	9.0	0.11	8.9
Concrete (Sand and Gravel)	12.0	0.08	6.4
Stone (Lime or Sand)	12.5	0.08	6.4
Steel	312.0	0.0032	0.25
Aluminum	1416.0	0.00070	0.06

[1]Thermal properties are for air in a space and apply for air spaces ranging from ¾ to 4 inches in thickness.

data is used in determining the size of heating plant required after calculating heat loss. This information is also useful in figuring the amount of insulation needed for walls, ceilings, and floors.

ACCEPTABLE COMFORT LEVELS

The amount of insulation necessary to provide indoor climate comfort can be determined accurately. The thermal properties of all common building materials are known, and the *U*-value for any combination of construction and insulation can be calculated.

Studies by heating engineers of home heating and air conditioning requirements have resulted in a number of *U*-value design standards recommended for new construction.

These standards are based on the geographical location of the structure and on the cost and type of fuel used.

The most widely used recommendations are in the *All-Weather Comfort Standard*, developed cooperatively by electric power suppliers, equipment makers, and material manufacturers, and in the supplementary performance standards created by the National Mineral Wool Insulation Association. Basically, these standards establish three degrees of comfort which may be attained by varying the amounts of insulation installed in ceilings, walls, and floors of homes.

• The *Maximum Comfort Standards* specify required amounts of insulation for houses in the coldest sections of the country and for energy-saving houses in any location.

• The *Moderate Comfort Standards* are for houses in the midsection of the country.

• The *Minimum Comfort Standards* apply to all homes as the minimum recommendations of many building codes.

40–5. *This map of the United States indicates the lowest temperatures occurring in each zone during an average winter.*

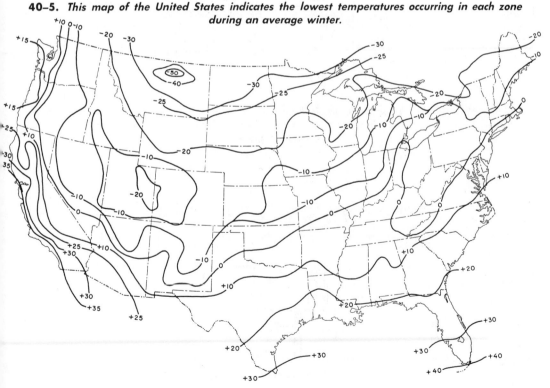

In recent years, the need to conserve energy has caused insulation standards to be raised. The table in Fig. 40-6 shows the currently recommended U-values.

40–6a.
U-Values And Insulation Requirements

Maximum Comfort Standard		
	U-Value	Insulation "R" Number
Ceilings	0.02	R38-42
Walls	0.05	R19
Floors over unheated spaces	0.05	R22
Moderate Comfort Standard		
	U-Value	Insulation "R" Number
Ceilings	0.03	R30-33
Walls	0.05	R19
Floors over unheated spaces	0.05	R19-22
Minimum Comfort Standard		
	U-Value	Insulation "R" Number
Ceilings	0.05	R19
Walls	0.07	R11
Floors over unheated spaces	0.07	R13

on the table are the installed insulation requirements in terms of resistance units, or *R* numbers, which will provide the desired. *U*-value in the building. This system of specifying insulation by total thermal resistance, instead of thickness, stems from the fact that insulating materials vary in density, type of surface, and heat conductivity.

In choosing the best, good, or minimum insulation standards in any climate zone, it should be remembered that comfort and operating economy are dual benefits of insulation. Insulating for maximum comfort automatically provides maximum economy of operation, and reduces initial costs of heating and cooling equipment to a minimum.

WHERE TO INSULATE

To reduce heat loss from the house during cold weather in most climates, all walls, ceilings, roofs, and floors that separate heated from unheated spaces should be insulated. Fig. 40-7.

In houses with unheated crawl spaces, insulation should be placed between the

40-6b. *Minimum insulation standards included in many local codes and recommended by many utilities. These recommendations are the maximum standards for the deep south and the western coastal regions.*

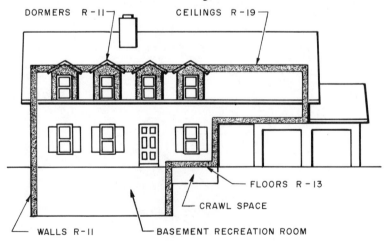

DORMERS R-11

CEILINGS R-19

FLOORS R-13

CRAWL SPACE

WALLS R-11

BASEMENT RECREATION ROOM

floor joists or around the wall perimeter. If flexible insulation is used, it should be well supported between joists by slats and a galvanized wire mesh or by a rigid board. The vapor barrier should be installed toward the subflooring. Fig. 40-8. Press-fit or friction insulation fits tightly between joists and requires only a small amount of support to hold it in place. Reflective insulation is often used for crawl spaces, but only one dead-air space (between the insulation and the subflooring) should be assumed in calculating heat loss when the crawl space is ventilated. A ground cover of roll roofing or plastic film such as polyethylene should be placed on the soil of crawl spaces to decrease the moisture

content of the space as well as of the wood members.

In 1½-story houses, insulation should be placed along all walls, floors, and ceilings that are adjacent to unheated areas. Fig. 40-9. These include stairways, dwarf (knee) walls, and dormers. Provisions should be made for ventilation of the unheated areas.

Where attic space is unheated and a stairway is included, insulation should be installed around the stairway as well as in the first-floor ceiling. Fig. 40-10. The door leading to the attic should be weatherstripped to prevent heat loss. Walls adjoining an unheated garage or porch should also be insulated.

In houses with flat or low-pitched roofs, insulation should be used in the ceiling area with sufficient space allowed above for unobstructed ventilation between the joists. Insulation should be used along the perimeter of houses built on slabs. A vapor barrier should be included under the slab. Fig. 40-11.

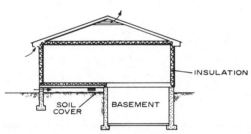

40–7. *A conventional one-story house is insulated in walls, floors over unheated crawl spaces, and ceilings.*

40–8. *Methods of installing insulation between floor joists: A. Wire mesh is stapled to the edges of the joists. B. Pieces of heavy-gauge wire pointed at each end are wedged between the joists to support the insulation.*

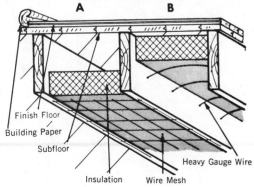

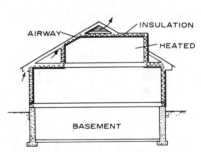

40–9. *Insulating a 1½-story house.*

40–10. *Insulating unheated attic space.*

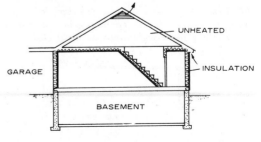

715

In the summer, outside surfaces exposed to the direct rays of the sun may attain temperatures of 50° F or more above shade temperatures and, of course, tend to transfer this heat toward the inside of the house. Insulation in the walls and in attic areas retards the flow of heat, improving summer comfort conditions.

Where air conditioning systems are used, insulation should be placed in all exposed ceilings and walls in the same manner as when insulating against cold-weather heat loss. Shading of glass against direct rays of the sun and the use of insulated glass will aid in reducing the air conditioning load.

Ventilation of attic and roof spaces is an important addition to insulation. Without ventilation, an attic space may become very hot and hold the heat for many hours. (See Unit 53, "Ventilation".) Obviously, more heat will be transmitted through the ceiling when the attic temperature is 150° F than if it is 100° to 120° F. Ventilation methods suggested for protection against cold-weather condensation apply equally well to protection against excessive hot-weather roof temperatures.

The use of storm windows or insulated glass will greatly reduce heat loss. Almost twice as much heat loss occurs through a single glass as through a window glazed with insulated glass or protected by a storm sash. Furthermore, double glass will normally prevent surface condensation and frost from forming on inner glass surfaces in winter. When excessive condensation persists, paint failures or even

decay of the sash rail or other parts can occur.

HOW TO INSTALL INSULATION
Flexible Insulation

Blanket or batt insulation with a vapor barrier should be placed between framing members so that the tabs of the barrier lap the edge of the studs as well as the top and bottom plates. This method is not often popular with the contractor because it is more difficult to apply the dry wall or rock lath (plaster base). However, it assures a minimum amount of vapor loss compared to the loss when tabs are stapled to the sides of the studs. Fig. 40-12. A hand stapler is commonly used to fasten the insulation and the barriers in place. Fig. 40-13.

To protect the head and sole plate as well as the headers over openings, it is good practice to use narrow strips of vapor barrier material along the top and bottom of the wall. A, Fig. 40-14. Ordinarily, these

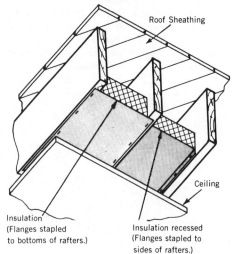

40–11b. Methods of installing flexible insulation between rafters. Regardless of method, it is important that a space be provided between the insulation and the roof sheathing to permit air circulation.

Roof Sheathing

Ceiling

Insulation
(Flanges stapled to bottoms of rafters.)

Insulation recessed (Flanges stapled to sides of rafters.)

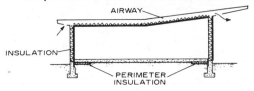

40–11a. When insulating a flat roof, leave an air space for ventilation. For houses built on slabs, use perimeter insulation under the slab.

AIRWAY

INSULATION

PERIMETER INSULATION

areas are not covered too well by the barrier on the blanket or batt.

For insulation without a barrier (press-fit or friction type), a plastic film vapor barrier such as 4-mil polyethylene is commonly

used to envelop the entire exposed wall and ceiling. B, Fig. 40-14. It covers the openings as well as window and door headers and edge studs. This system is one of the best from the standpoint of

40-12. *Methods of installing wall insulation: A. The insulation flange is stapled to the inside edge of the stud. B. The insulation is recessed to provide an air space of at least $^3/_4''$ on each side. C. Insulation on masonry walls is stapled to furring strips which have been nailed to the masonry on 16" centers.*

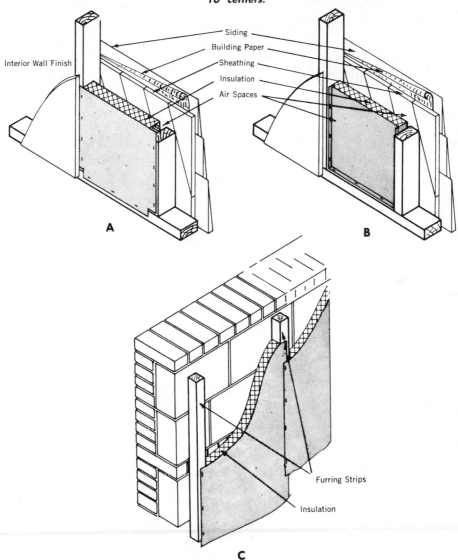

40-13. *Installing batt insulation with a hand stapler following the method shown in B, Fig. 40-12.*

resistance to vapor movement. Furthermore, it does not have the installation inconveniences encountered when tabs of the insulation are stapled over the edges of the studs. After the dry wall is installed or plastering is completed, the film is trimmed around the window and door openings.

Reflective Insulation

Reflective insulation, in single-sheet form with two reflective surfaces, should be placed to divide the space formed by the framing members into two approximately equal spaces. For example, insulation between studs should be placed so as to leave an equal amount of air space on each side of the insulation. Some reflective insulations include air spaces and are furnished with nailing tabs. This type is fastened to the studs in such a way as to provide at least a $3/4''$ space on each side of the reflective surfaces.

Loose Fill Insulation

Loose fill insulation is commonly used in ceiling areas and is poured or blown into place. Fig. 40-15. A vapor barrier should be used on the warm side (the bottom, in case of ceiling joists) before insulation is placed. A leveling board will give a constant insulation thickness. Fig. 40-15.

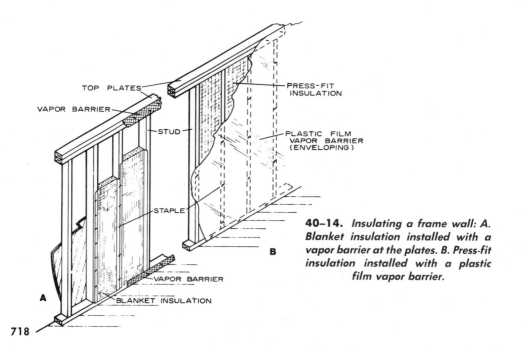

40-14. *Insulating a frame wall: A. Blanket insulation installed with a vapor barrier at the plates. B. Press-fit insulation installed with a plastic film vapor barrier.*

718

Thick batt insulation is also used in ceiling areas. Fig. 40-16. Batt and fill insulation can be combined to obtain the desired thickness. The vapor barrier is placed against the back of the ceiling finish. Ceiling insulation 6" or more thick greatly reduces heat loss in the winter and also provides summertime protection.

Rigid Insulation

Rigid insulation is nailed to sloping rafters through 1" strips of wood. Sheathing is then nailed to the strips, and the shingles are applied. Fig. 40-17. On a roof with wood decking, the insulation is fastened

with nails or adhesive and covered with built-up roofing. Fig. 40-18.

PRECAUTIONS IN INSULATING

Areas over door and window frames and along side and head jambs also require insulation. Because these areas are filled with small sections of insulation, a vapor

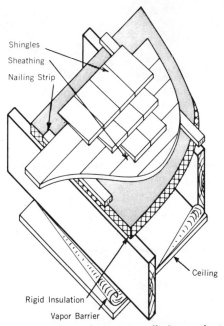

Shingles
Sheathing
Nailing Strip
Ceiling
Rigid Insulation
Vapor Barrier

40–17. *Rigid insulation installed on sloping rafters.*

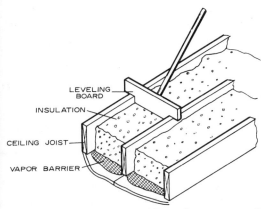

LEVELING BOARD
INSULATION
CEILING JOIST
VAPOR BARRIER

40–15. *Installing loose fill insulation in a ceiling. Note the use of the leveling board.*

40–16. *Loose fill or batt insulation may be installed in a ceiling under an unheated attic.*

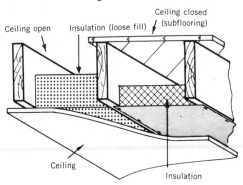

Ceiling open
Insulation (loose fill)
Ceiling closed (subflooring)
Ceiling
Insulation

40–18. *Rigid insulation installed on wood roof decking.*

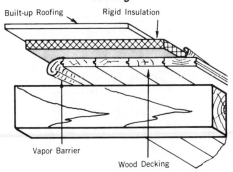

Built-up Roofing
Rigid Insulation
Vapor Barrier
Wood Decking

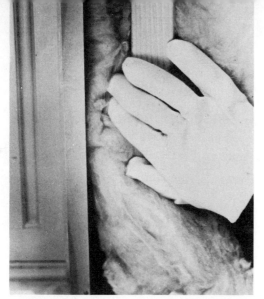

40-19a. *Fill all cracks by forcing pieces of insulation into the voids between windows or door jambs and trimmer studs.*

40-19b. *A vapor barrier should be applied over the insulation around openings.*

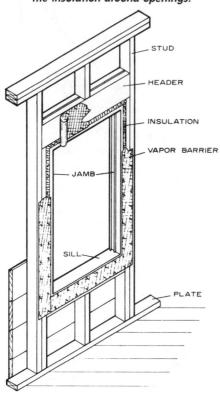

STUD

HEADER

INSULATION

VAPOR BARRIER

JAMB

SILL

PLATE

barrier must be used around the opening as well as over the header above the openings. Fig. 40-19. Enveloping the entire wall eliminates the need for this type of vapor barrier installation.

In 1½-and 2-story houses and in basements, the area at the joist header at outside walls should be insulated and protected with a vapor barrier. Fig. 40-20.

Insulation should be placed behind electrical outlet boxes and other utility connections in exposed walls to minimize condensation. Fig. 40-21a & b.

VAPOR BARRIERS

Some discussion of vapor barriers has been included previously because vapor barriers are usually a part of flexible insulation. However, further information is given in the following paragraphs.

Most building materials are permeable to water vapor. This presents problems because considerable water vapor is generated in a house from cooking, dishwashing, laundering, bathing, humidifiers, and other sources. During cold weather, this vapor may pass through wall and ceiling materials and condense in the wall or attic space. Subsequently, in severe cases, it may damage the exterior paint and interior finish, or even promote decay in structural members. For protection, a material highly resistant to vapor transmission, called a *vapor barrier,* should be used on the warm side of a wall or below the insulation in an attic space.

Among the effective vapor barrier materials are asphalt laminated papers, alumi-

40-20. *A vapor barrier and insulation should be installed in the joist space at outside walls.*

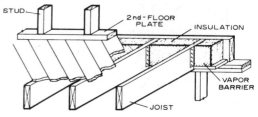

STUD

2nd-FLOOR PLATE

INSULATION

VAPOR BARRIER

JOIST

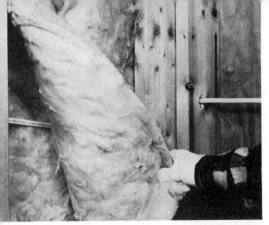

40–21a. *As much insulation as possible should be pushed behind the electrical wiring, outlet boxes, and pipes in exposed walls. Doing so will minimize condensation on cold surfaces.*

40–21b. *Cover these items with additional insulation to completely fill the space between studs.*

num foil, and plastic films. Most blanket and batt insulations are provided with a vapor barrier on one side, some of them with paper-backed aluminum foil. Foil-backed gypsum lath or gypsum boards are also available and serve as excellent vapor barriers.

The effectiveness of vapor barriers is rated by their perm values. (1 perm = 1 grain of vapor transmission per square foot, per hour, for each inch of mercury vapor pressure difference.) Low perm values indicate high resistance to vapor transmission. The perm values of vapor barriers vary, but ordinarily it is good practice to use barriers which have values less than $1/4$ (0.25) perm. Although a value of $1/2$ perm is considered adequate, aging reduces the effectiveness of some materials.

To obtain a positive seal against vapor transmission, wall-height rolls of plastic film vapor barriers should be applied over studs, plates, and window and door headers. Application of the plastic film is discussed on page 717. This system, called "enveloping," is used over insulation having no vapor barrier or to insure excellent protection when used over any type of insulation. The plastic should be fitted tightly around outlet boxes and sealed if necessary. Fig. 40–21c.

A ribbon of sealing compound around an outlet or switch box will minimize vapor

40–21c. *Staple the plastic film vapor barrier to the wood so the insulation is completely sealed.*

loss at this area. Cold-air returns in outside walls should consist of metal ducts to prevent vapor loss and subsequent paint problems.

Paint coatings on plaster may be very effective as vapor barriers if materials are properly chosen and applied. They do not, however, offer protection during construction, and moisture may cause paint blisters on exterior paint before the interior paint can be applied. This is most likely to happen in buildings that are constructed during periods when outdoor temperatures are 25° F or more below inside temperatures. Paint coatings cannot be considered a substitute for the membrane types of vapor barriers. However, they do provide some protection for houses where other types of vapor barriers were not installed during construction.

Of the various types of paint, one coat of aluminum primer followed by two decorative coats of flat wall or lead and oil paint is quite effective. For rough plaster or for buildings in very cold climates, two coats of the aluminum primer may be necessary. A primer and sealer of the pigmented type, followed by decorative finish coats or two coats of rubber-base paint, are also effective in retarding vapor transmission.

No type of vapor barrier can be considered 100% resistive, and some vapor leakage into the wall may be expected. Therefore the flow of vapor to the outside should not be impeded by materials of relatively high vapor resistance on the cold side of the vapor barrier. For example, sheathing paper should be of a type that is waterproof but not highly vapor resistant. This also applies to "permanent" outer coverings or siding. The vapor barrier itself should have a low perm value to prevent the passage of moisture to the cold side of the barrier. This will reduce the danger of condensation on cold surfaces within the wall.

ESTIMATING
Materials

To determine the amount of insulation required for a home, the area to be insulated must first be figured. Using the house plan in Fig. 40-22, round off the outside dimensions of the home to a width of 28' and a length of 52'. The perimeter of the house is 160'. $(2 \times 28) + (2 \times 52) = 160$. If the wall height is 8', the walls will contain 1280 square feet of area $(8 \times 160 = 1280)$. This includes the window and door open-

40–22. *Estimate the amount of insulation required to insulate the outside walls and the ceiling of this house.*

ings, which equal about 150 square feet. Subtract the area of the openings from the total area of 1280 square feet (1280 − 150 = 1130). The total wall area to be insulated is 1130 square feet.

Figure the number of insulating batts required by using the chart in Fig. 40-23. Read down the left column under the size batt to be used, in this case 15″ × 48″. Read across to the number of batts per 100 square feet. The chart indicates 20 batts will cover 100 square feet, but only 95 square feet of wall area (5% waste allowance).

$$\frac{1130 \text{ (total area to be insulated)}}{95 \text{ (sq ft. of wall space covered by 20 batts)}} = \text{approx. } 11.89$$

$$11.89 \times 20 = 238 \text{ batts.}$$

(Since 11.89 is an approximate number, round off the total number of batts to 238 to make certain there will be enough batts.)

The number of staples required to install the insulation batts can also be figured by referring to the chart in Fig. 40-23. For 15″ × 48″ batts, 160 staples are needed for installing 100 square feet of insulation.

$$\frac{1130 \text{ (total area to be insulated)}}{100 \text{ (sq. ft. installed per 160 staples)}} = 11.3$$

$$11.3 \times 160 = 1808 \text{ staples.}$$

The ceiling area must also be insulated. Figure the area of the ceiling by multiplying the width times the length. Calculate the number of insulating batts needed in the same way as described for the walls, using Fig. 40-23. If loose fill insulation is specified, refer to the chart in Fig. 40-24.

40–23. *Estimating insulation batts.*

INSULATION (BATTS)		Material		Staples	Labor
Size	Square Feet	Number of Batts Per 100 Square Feet	Number of Square Feet for 100 Sq. Ft. Wall	per 100 Square Feet	Square Feet per Hour
15 x 24	2.5	40	95	160	80
15 x 48	5.	20	95	160	85
19 x 24	3.7	32	96	160	90
19 x 48	6.33	16	95	160	95
23 x 24	3.84	26	95	160	90
23 x 48	7.67	13	100	160	100

Notes: Glass, Mineral or Rock Wool batts with paper back roll insulation or strip insulation.
     Studding or joist excluded.
     Batts stapled @ 6″ O.C.

40–24. *Estimating loose fill insulation.*

INSULATION (LOOSE FILL)		Material					Labor
		Number of Square Feet Covered per Cubic Foot					Square Feet per Hour
		Density					
Fill Thickness	6 Pounds	7 Pounds	8 Pounds	9 Pounds	10 Pounds		
1″	21.2	18.0	15.9	14.1	13.0		110
2″	10.6	9.1	8.0	7.1	6.4		100
3″	7.1	6.1	5.3	4.7	4.2		90
4″	5.3	4.6	4.0	3.5	3.2		85

Labor

The time required to insulate a house is estimated by dividing the total area to be insulated by the number of square feet insulated in one hour.

For example, the chart in Fig. 40-23 under the column headed "Labor" indi-cates that 85 square feet of 15″ × 48″ insulation batts can be installed each hour. The total amount of time required to install the insulation batts is estimated as follows:

$$\frac{1130 \text{ (total area to be insulated)}}{85 \text{ (sq. ft. insulated per hour)}} = 13.29, \text{ or } 13\frac{1}{3} \text{ hours}$$

QUESTIONS

1. What is insulation?
2. List several types of commercial insulating materials.
3. List some of the materials used for loose fill insulation.
4. What is a Btu?

5. To reduce heat loss, what areas in a structure should be insulated?
6. When installing flexible insulation, to what should the tabs be fastened?
7. When discussing vapor barriers, what is meant by "enveloping"?

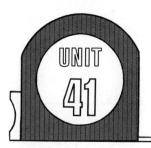

UNIT 41 Sound Insulation

Development of the "quiet" home is becoming more and more important. In the past, sound insulation was more important in apartments, motels, and hotels than in private homes. However, the use of household appliances, television, radio, and stereo systems has increased the noise levels in homes. House designs now often include a family room or "active" living room as well as a "quiet" living room. These rooms should be isolated from the remainder of the house. Sound insulation between the bedroom area and the living area is usually needed, as is isolation of the bathrooms and lavatories. Insulation against outdoor sounds is also desirable. Thus sound control has become a vital part of house design and construction and will be even more important in the coming years.

HOW SOUND TRAVELS

Sound is transmitted by waves. Noises inside a house, such as loud conversation or a barking dog, create sound waves which radiate outward from the source through the air until they strike a wall, floor, or ceiling. These surfaces vibrate as a result of the fluctuating pressure of the sound waves. Because the surface vibrates, it conducts sound to the other side in varying degrees, depending on the construction. Fig. 41-1.

The resistance of a building element, such as a wall, to the passage of airborne sound is rated by its Sound Transmission Class (STC). The higher the number, the better the sound barrier. The approximate effectiveness of walls with varying STC numbers is shown in Fig. 41-2. Most authorities agree that a floor or wall in a multi-occupancy residence should have an STC rating of at least 45, while 50 is considered premium construction. Below 40, privacy and comfort may be impaired because loud speech can be heard as a murmur.

Sound travels readily through the air and also through some materials. When airborne sound strikes a conventional wall, the studs act as sound conductors unless they are separated in some way from the covering material. Electrical switches or convenience outlets placed back to back in a wall readily pass sound. Faulty construction, such as poorly fitted doors, often allows sound to travel through. Thus good construction practices are important in providing sound-resistant walls.

SOUND INSULATION IN WALL CONSTRUCTION

Thick walls of dense materials such as masonry can stop sound. In a wood-frame house, however, an interior masonry wall results in increased costs and in structural problems created by heavy walls. To economically provide a satisfactory sound-resistant wall has been a problem. At one time, sound-resistant frame construction for the home involved much higher costs because it usually meant double walls or suspended ceilings. However, a relatively simple system has been developed using sound-deadening insulating board along with a gypsum board outer covering. This provides good sound control at only slight additional cost. A number of combinations,

41-1. A wall will conduct sound to the other side in varying degrees, depending on its construction.

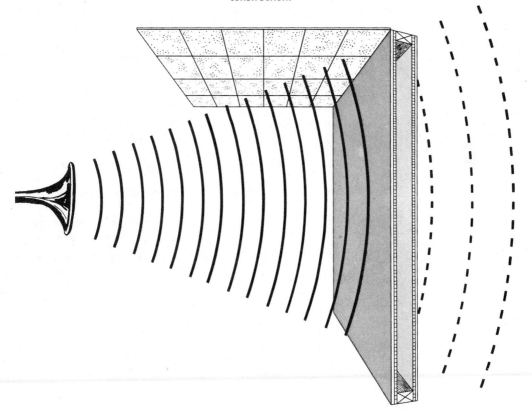

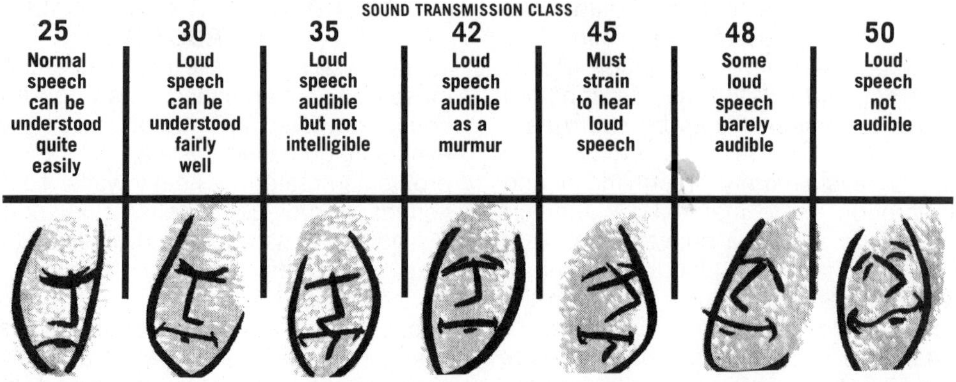

SOUND TRANSMISSION CLASS

25	30	35	42	45	48	50
Normal speech can be understood quite easily	Loud speech can be understood fairly well	Loud speech audible but not intelligible	Loud speech audible as a murmur	Must strain to hear loud speech	Some loud speech barely audible	Loud speech not audible

41–2a. *STC numbers have been adopted by acoustical engineers as a measure of the ability of structural assemblies to reduce airborne noise. The higher the number, the more effective the sound barrier.*

41–2b. *This chart should be used only as a general guide to the partition performance required to meet specific sound control needs. Note that it does not incorporate the effects of interconnecting ducts, wiring, and plumbing.*

Partition Selector Guide

Location	Degree of Privacy	Stc Requirement	Example of Wall Construction Needed
Rural (20 db. background noise level)	High	60 or over	Solid dense masonry min. 12″ thick.
	Moderate	50 to 60	3⅝″ metal studs, 2 layers of ⅝″ fire-stop each side, 1½″ fiber glass insulation in cavity.
	Low	45 to 50	Double solid partition.
Suburban (30 db. background noise level)	High	55 or over	Triple solid partition.
	Moderate	45 to 55	2″ x 4″ wood studs, resilient channel one side, ⅝″ firestop both sides, 1½″ fiber glass insulation in cavity.
	Low	40 to 45	2″ x 4″ wood studs, resilient channel one side, ⅝″ firestop both sides.
City (40 db. background noise level)	High	45 or over	2½″ x 3″ gypsum ribs, 2½″ steel track, ½″ sound deadening board both sides, ⅝″ firestop each side
	Moderate	40 to 45	2″ x 4″ wood studs, 2 layers ⅝″ firestop each side.
	Low	35 to 40	2″ x 4″ wood studs, 2 layers ⅝″ firestop one side, 1 layer ⅝″ firestop other side.

providing different STC ratings, are possible with this system.

As Fig. 41-2a showed, a wall should have an STC rating of 45 or more to provide sufficient resistance to airborne sound.

Gypsum wallboard or lath and plaster are commonly used for partition walls. A and B, Fig. 41-3. However, an STC rating of 45 cannot be obtained with this construction. An 8" concrete block wall has the minimum

41-3. *Sound insulation of single walls.*

WALL DETAIL	DESCRIPTION	STC RATING
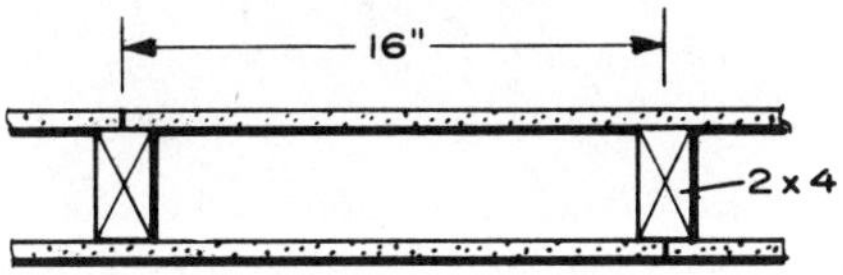 A	½" GYPSUM WALLBOARD	32
	⅝" GYPSUM WALLBOARD	37
B	⅜" GYPSUM LATH (NAILED) PLUS ½" GYPSUM PLASTER WITH WHITECOAT FINISH (EACH SIDE)	39
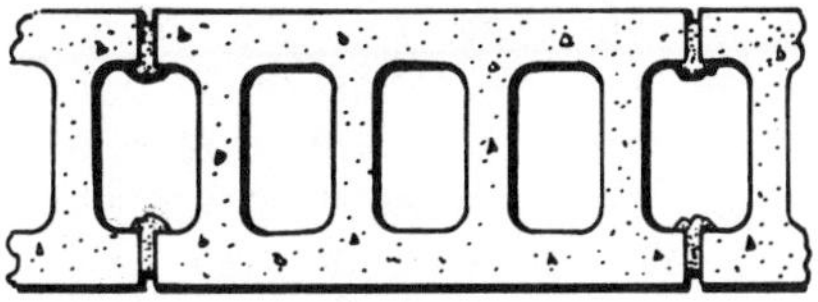 C	8" CONCRETE BLOCK	45
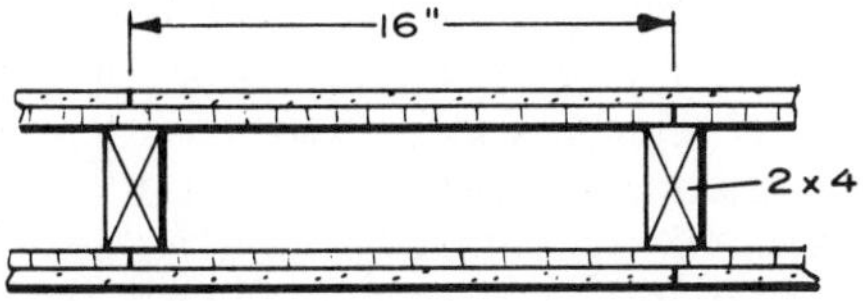 D	½" SOUND DEADENING BOARD (NAILED) ½" GYPSUM WALLBOARD (LAMINATED) (EACH SIDE)	46
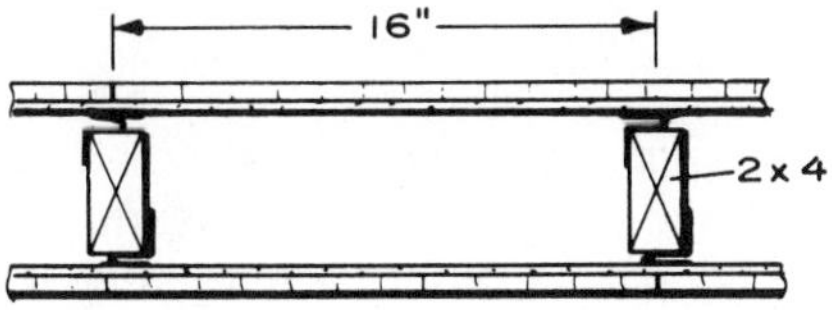 E	RESILIENT CLIPS TO ⅜" GYPSUM BACKER BOARD ½" FIBERBOARD (LAMINATED) (EACH SIDE)	52

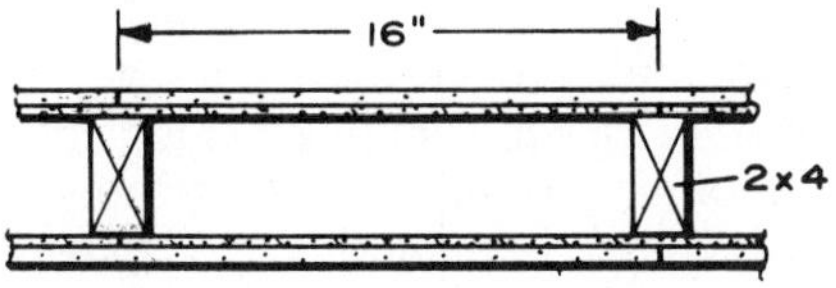

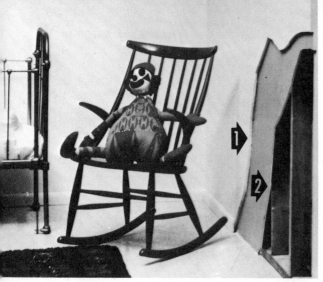

41-4. *Gypsum wallboard (arrow #1) installed over a sound-deadening board (arrow #2) will help assure quiet living.*

rating, but this construction is not always practical in a wood-frame house. C, Fig. 41-3.

In construction of a partition wall, its cost as related to the STC rating should be considered. Good STC ratings can be obtained in a wood-frame wall by using the

41-5. *Sound transmission can be reduced by fastening the interior wall covering to resilient metal furring channels.*

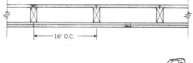

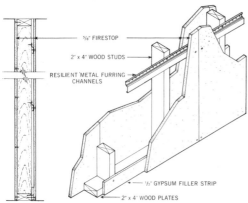

5/8" FIRESTOP

2" x 4" WOOD STUDS

RESILIENT METAL FURRING CHANNELS

1/2" GYPSUM FILLER STRIP

2" x 4" WOOD PLATES

combination of materials shown in D and E of Fig. 41-3. One-half inch of sound-deadening board nailed to the studs, followed by a lamination of 1/2" gypsum wallboard, will provide an STC value of 46 at a relatively low cost. Fig. 41-4. A slightly better rating can be obtained by using 5/8" gypsum wallboard rather than 1/2". A very satisfactory STC rating of 52 can be obtained by using resilient clips to fasten 3/8" gypsum backer boards to the studs, followed by adhesive-laminated 1/2" fiberboard. E, Fig. 41-3. This method further isolates the wall covering from the framing.

A similar isolation system consists of resilient channels nailed horizontally to 2" × 4" studs spaced 16" on center. The channels are spaced 24" apart vertically, and 5/8" gypsum wallboard is screwed to the channels. An STC rating of 44 is thus obtained at a moderately low cost. Fig. 41-5.

A double wall, which may consist of 2" × 6" or wider plate and staggered 2" × 4" studs, is sometimes constructed for sound control. One-half-inch gypsum wallboard on each side of this wall results in an STC value of 45. A, Fig. 41-6. Two layers of 5/8" gypsum wallboard add little, if any, additional sound-transfer resistance. B, Fig.

WALL DETAIL	DESCRIPTION	STC RATING
A ⊢— 16" —⊣ — 2x4	½" GYPSUM WALLBOARD	45
B — 2x4	⅝" GYPSUM WALLBOARD (DOUBLE LAYER EACH SIDE)	45
C 2x4 BETWEEN OR "WOVEN"	½" GYPSUM WALLBOARD 1½" FIBROUS INSULATION	49
D — 2x4	½" SOUND DEADENING BOARD (NAILED) ½" GYPSUM WALLBOARD (LAMINATED)	50

41–6. Sound insulation of double walls.

41-6. However, when 1½" blanket insulation is added to double wall construction, the STC rating increases to 49. C, Fig. 41-6. This insulation may be installed as shown, or placed between studs on a single wall. A single wall with 3½" of insulation is low in cost and will resist sound transfer much better than an open stud space.

The use of ½" sound-deadening board and a lamination of gypsum wallboard in the double wall will result in an STC rating of 50. D, Fig. 41-6. The addition of blanket insulation to this combination will likely provide an even higher value, perhaps 53 or 54. This system, with single-wall con-struction, might also be used to insulate exterior walls against street noises.

SOUND INSULATION IN FLOOR-CEILING CONSTRUCTION

Sound insulation between an upper floor and the ceiling of a lower story involves not only resistance to airborne sounds but also to impact noises. Impact noise results when an object strikes or slides along a wall or floor surface. Footsteps, dropped objects, and furniture being moved all cause impact noise. It may also be caused by the vibration of a dishwasher, food-disposal apparatus, or other equipment. In all instances, the floor is set into vibration by the impact or contact, and sound is radiated from both sides of the floor.

A method of measuring impact noise has been developed and is commonly expressed as the Impact Noise Rating (INR). [INR ratings in some publications are being abandoned in favor of IIC (Impact Insulation Class) ratings.]

The higher the INR, the better the impact sound reduction. For example, an INR of −5 is better than −10, and +5 is better than 0.

INR performance standards for floors are based on criteria established by the FHA. Those criteria range from −8 to +5, depending on location. Fig. 41-7 shows STC and approximate INR (decibel) values for several types of floor construction. A minimum floor assembly with tongued and grooved floor and 3/8" gypsum board ceil-

41-7. STC and INR values in floor-ceiling combinations using 2" × 8" joists.		ESTIMATED VALUES	
DETAIL	DESCRIPTION	STC RATING	APPROX. INR
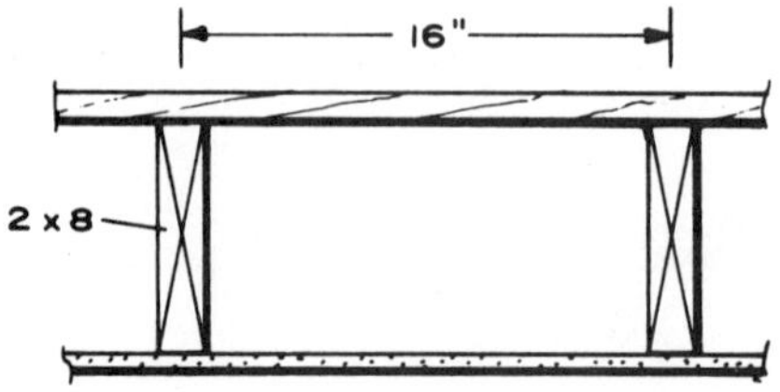 A — 16" — 2 x 8	FLOOR ⅞" T. & G. FLOORING CEILING ⅜" GYPSUM BOARD	30	-18
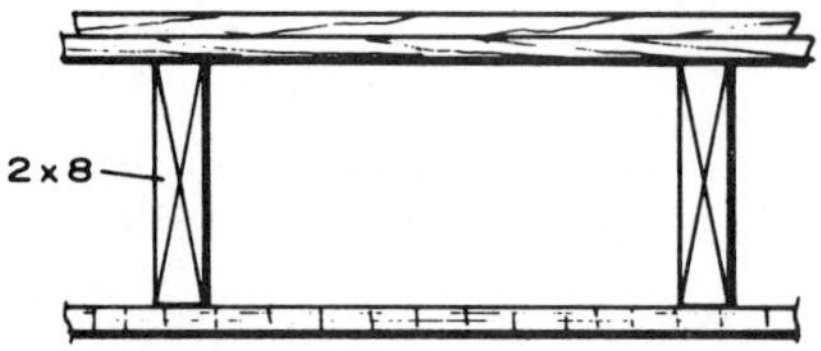 B — 2 x 8	FLOOR ¾" SUBFLOOR ¾" FINISH FLOOR CEILING ¾" FIBERBOARD	42	-12
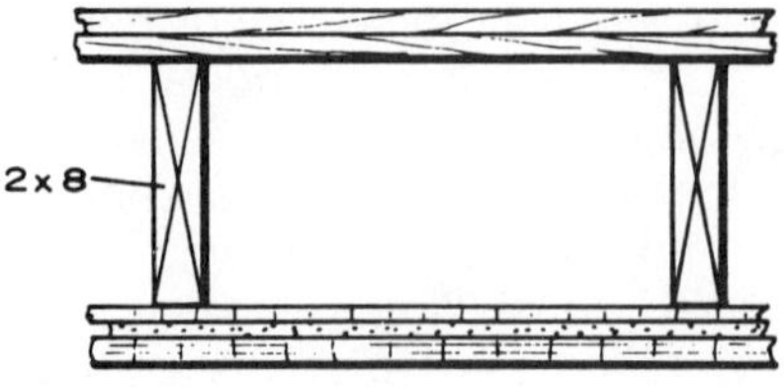 C — 2 x 8	FLOOR ¾" SUBFLOOR ¾" FINISH FLOOR CEILING ½" FIBERBOARD LATH ½" GYPSUM PLASTER ¾" FIBERBOARD	45	-4

ing has an STC value of 30 and an approximate INR value of −18. A, Fig. 41-7. This is improved somewhat by the construction shown in part B, and still further by the combination of materials in part C.

The value of isolating the ceiling joists from a gypsum lath and plaster ceiling by means of spring clips is illustrated in Fig. 41-8, part A. An STC rating of 52 and an approximate INR value of −2 result.

41–8. STC and INR values in floor-ceiling combinations using 2″ × 10″ joists.

DETAIL	DESCRIPTION	ESTIMATED VALUES	
		STC RATING	APPROX. INR

A 2 x 10 16″	FLOOR
	  ¾″ SUBFLOOR
	    (BUILDING PAPER)
	  ¾″ FINISH FLOOR
	CEILING
	  GYPSUM LATH AND
	    SPRING CLIPS
	  ½″ GYPSUM PLASTER
		52	− 2

B 2 x 10	FLOOR
	  ⅝″ PLYWOOD SUBFLOOR
	  ½″ PLYWOOD
	    UNDERLAYMENT
	  ⅛″ VINYL-ASBESTOS
	    TILE
	CEILING
	  ½″ GYPSUM WALLBOARD
		31	−17

C 2 x 10	FLOOR
	  ⅝″ PLYWOOD SUBFLOOR
	  ½″ PLYWOOD
	    UNDERLAYMENT
	  FOAM RUBBER PAD
	  ⅜″ NYLON CARPET
	CEILING
	  ½″ GYPSUM WALLBOARD
		45	+ 5

Floor on sleepers

Weight lbs./sq. ft.	Floor	STC	INR	Test No.
10.9	.075" sheet vinyl on 5/8" T&G plywood underlayment	52	−2	L-224-5&6
11.7	44 oz. Carpet on 40 oz. hair pad on 5/8" T&G Plywood underlayment	52	+27	L-224-7&8
13.0	25/32" wood strip flooring nailed sleepers	53	0	L-224-9&10

Note: Flooring is fastened to sleepers on 1/2" insulation board over a 1/2" plywood subfloor on wood joists with insulation between. 5/8" gypsum board ceiling with taped joints is screwed to resilient channels.

Separate ceiling joists

Weight lbs./sq. ft.	Floor	STC	INR	Test No.
11.0	25/32" hardwood strip flooring on 1/2" plywood subfloor	53	−6	L-224-12&13
10.7	44 oz. Carpet on 40 oz. hair pad on 1 1/8" plywood subfloor (2-4-1 T&G)	51	+29	L-224-14&15

Note: Conventional wood joist floor system with finish floor applied directly on plywood. Separate 2 × 4 ceiling joists, with insulation, and 5/8" gypsum board ceiling with taped joints nailed directly to joists.

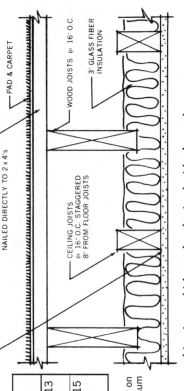

41-9. Two of the sound resistant floor-ceiling combinations which can be obtained with plywood construction.

Foam rubber padding and carpeting improve both the STC and the INR values. The STC rating increases from 31 to 45 and the approximate INR from −17 to +5. B and C, Fig. 41-8. The ratings can probably be further improved by using an isolated ceiling finish with spring clips. The use of sound-deadening board and a lamination of gypsum board for the ceiling would also improve resistance to sound transfer.

An economical construction similar to (but an improvement over) the one shown in part C of Fig. 41-8 has an STC value of 48 and an approximate INR of +18. It consists of the following: (1) a pad and carpet over ⁵⁄₈″ tongued and grooved plywood underlayment, (2) 3″ fiber glass insulating batts between joists, (3) resilient channels spaced 24″ apart across the bottom of the joists, and (4) ⁵⁄₈″ gypsum board screwed to the bottom of the channels and finished with taped joints.

The use of separate floor joists with staggered ceiling joists below provides reasonable values but adds a good deal to construction costs. Separate joists with insulation between and a sound-deadening board between subfloor and finish provide an STC rating of 53 and an approximate INR value of −3. Other combinations are illustrated in Fig. 41-9.

SOUND ABSORPTION

Design of the "quiet" house can include another system of sound insulation, namely, sound absorption. Sound-absorbing materials do not necessarily have resistance to airborne sounds, but they can

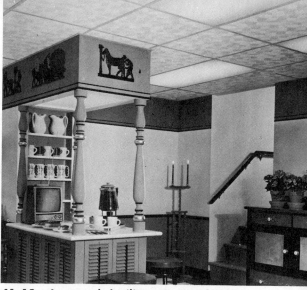

41-10. A suspended-ceiling system with acoustical panel inserts will serve to absorb sound.

minimize the amount of noise by stopping the reflection of sound back into a room. Perhaps the most commonly used sound-absorbing material is acoustic tile or panels. Fig. 41-10. Numerous tiny sound traps on the surfaces may consist of tiny drilled or punched holes, fissured surfaces, or a combination of both. Wood fiber or similar materials are used in the manufacture of the tile and panels, which are usually processed to provide some fire resistance.

Acoustic tile and panels are most often used in the ceiling and other areas, such as above a wall wainscoting, where they are not subject to excessive mechanical damage. Paint or other finishes which fill or cover the tiny holes or fissures for trapping sound will greatly reduce its efficiency.

QUESTIONS

1. How does sound travel?
2. What does STC stand for?
3. What do the higher numbers of STC indicate?
4. Why does an 8″ concrete block wall have a lower STC rating compared to traditional methods of wood-frame wall construction?

5. What is impact noise?
6. What does INR stand for?
7. Which is considered better, and INR of −5 or an INR of +5?
8. What is the most commonly used sound absorbing material in house construction?

UNIT
42

Interior Wall and Ceiling Finish

Before interior wall and ceiling finish is applied, insulation should be in place. Wiring, heating ducts, and other utilities should be roughed in.

Wood, gypsum, wallboard, plywood, and plaster make good finishes for covering the interior framed areas of walls and ceilings. They can serve as a base for paint and other finishes, including wallpaper, or be purchased already finished. The size and thickness of interior finish material should comply with local and national codes. Requirements for interior finish materials used in baths and kitchens normally will be more rigid because of the moisture conditions.

Although there are several types of interior finishes, lath and plaster has been about the most widely used. However, the use of dry-wall materials is increasing steadily. Many builders select dry wall because of the time-saving factor. A plaster finish requires drying time before other interior work can be started, whereas dry wall does not. However, when gypsum dry wall is used as a wall finish, the framing lumber must have a low moisture content to prevent nail pops. These result when framing members dry out, causing the nail head to form small humps on the surface of the gypsum board. It is also very important when applying single-layer gypsum finish that the studs be in alignment. Otherwise, the wall may have a wavy, uneven appearance. Since there are advantages to both plaster and gypsum dry-wall finishes, all factors should be considered along with the initial cost and the future maintenance that may be required.

LATH AND PLASTER

A plaster finish requires a base upon which the plaster is applied. This base, or *lath*, is fastened to the framing members. It must have bonding qualities so that plaster adheres, or is keyed, to it. The most commonly used types of lath are the following:
- Gypsum.
- Insulating fiberboard.
- Metal.

Gypsum Lath

One of the most common types of plaster base used on sidewalls and ceilings is gypsum lath. This lath has paper faces with a gypsum filler. It comes in 16″ × 48″ boards and is applied horizontally across the framing members. Fig. 42-1. For stud or joist spacing of 16″ on center, a ³/₈″ thickness is used. For 24″ spacing, the thickness should be ¹/₂″.

This material can be obtained with a foil back that serves as a vapor barrier. If the foil faces an air space, it also has reflective insulating value. Gypsum lath may also be obtained with perforations, which improve the bond and increase the time the plaster remains intact when exposed to fire. The building codes in some cities require these perforations. A waterproof facing is provided on one type of gypsum board for use as a ceramic tile base when the tile is applied with an adhesive.

INSTALLING GYPSUM LATH

Vertical joints should be made over the center of studs or joists and nailed with 12− or 13−gauge gypsum-lathing nails 1¹/₂″ long with ³/₈″ flat heads. Fig. 42-2. The

42–1. *Installing gypsum board lath.*

42–2. *A gypsum-lathing nail.*

nails should be spaced 5″ on center, or four nails for the 16″ height, and used at each stud or joist crossing. Some manufacturers specify ring-shank nails with a slightly greater spacing. Joints over heads of openings should not occur at the jamb lines. Fig. 42-3. Gypsum lath may also be used in constructions where metal studs are used for framing. The lath is secured to the studs by the use of special clips or tapping screws. Fig. 42-4.

Insulating Fiberboard Lath

Insulating fiberboard lath measuring 16″ × 48″ and ½″ thick is also used as a plaster base. It has greater insulating value than gypsum lath, but horizontal joints must usually be reinforced with metal clips.

INSTALLING INSULATING LATH

Insulating lath is installed much the same as gypsum lath, except that slightly longer blued nails should be used.

Metal Lath

Another type of plaster base is made of sheet metal. The metal is slit and expand-ed in various forms, such as diamond mesh, flat ribbed, and wire lath, to create innumerable openings for the keying of plaster. Fig. 42-5. Metal lath is usually 27″ × 96″ in size and galvanized or painted to resist rusting. Metal lath is usually installed on studs or joists spaced 16″ on center. The minimum weights to be installed on studs or joists spaced 16″ on center are as follows:
- For walls—2.5 lbs. per sq. yd.
- For ceilings—3.4 lbs. per sq. yd. (if rib metal lath is used—2.75 lbs. per sq. yd.)

Metal lath is often used as a plaster base around tub recesses and other bath and kitchen areas. Fig. 42-6. It is also used when ceramic tile is applied over a plaster base. For these uses, the metal lath must be backed with water-resistant sheathing paper placed over the framing.

INSTALLING METAL LATH

Metal lath is applied horizontally over the waterproof backing with side and end joints lapped. It is nailed with No. 11 or No. 12 roofing nails long enough to provide about 1½″ penetration into the framing member or blocking. Fig. 42-7.

Plaster Reinforcing

Because some drying usually takes place in wood framing members after a house is completed, some shrinkage can be expected. This in turn may cause plaster cracks to develop around openings and in corners. To minimize this cracking, expanded metal lath is used in certain key positions over the plaster base as reinforcement. Strips of expanded metal lath about 10″ × 20″ should be placed diago-

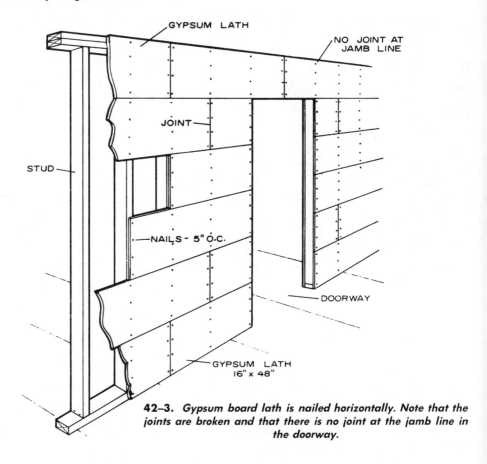

GYPSUM LATH

NO JOINT AT JAMB LINE

JOINT

STUD

NAILS - 5" O.C.

DOORWAY

GYPSUM LATH
16" x 48"

42–3. *Gypsum board lath is nailed horizontally. Note that the joints are broken and that there is no joint at the jamb line in the doorway.*

nally across the upper corners of all window and door openings and tacked in place. Fig. 42-8.

Metal lath should also be installed under flush ceiling beams to prevent plaster cracks. Fig. 42-9. On wood drop beams

extending below the ceiling line, the metal lath is applied with self-furring nails to provide space for keying of the plaster.

Corner beads of expanded metal lath or perforated metal should be installed on all exterior corners. They should be applied

42–4. *Gypsum lath installed on metal studs with special clips and tapping screws.*

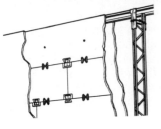

42–5. *Diamond mesh metal lath.*

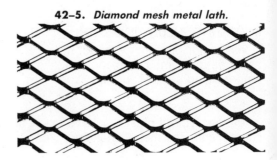

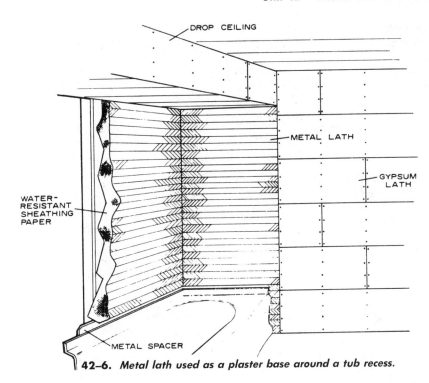

DROP CEILING

METAL LATH

GYPSUM LATH

WATER-RESISTANT SHEATHING PAPER

METAL SPACER

42-6. *Metal lath used as a plaster base around a tub recess.*

plumb and level. The bead acts as a leveling edge when walls are plastered and reinforces the corner against damage, such as from moving furniture. Fig. 42-10.

Inside corners at the intersection of walls and ceilings should also be reinforced. A cornerite of metal lath or wire fabric is tacked lightly in place in these areas. Cornerites provide a key width of 2" to 2¹/₂" at each side for plaster. Fig. 42-11.

Plaster Grounds

Plaster grounds are strips of wood used as guides or strike-off edges when plaster-ing. They are located around window and door openings and at the base of the walls.

42-7. *A No. 11 or No. 12 roofing nail is used to apply metal lath.*

Grounds around interior door openings are often full-width pieces nailed to the door sides over the studs and to the underside of the header. They are 5" in width, which coincides with standard jamb widths for

42-8. *Expanded metal lath is used to help mini-mize plaster cracks.*

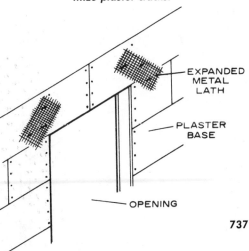

EXPANDED METAL LATH

PLASTER BASE

OPENING

interior walls with a plaster finish. Fig. 42-12. Narrow strip grounds might also be used around these interior openings. Fig. 42-13. These grounds are removed after plaster has dried.

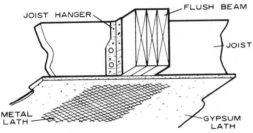

JOIST HANGER

FLUSH BEAM

JOIST

METAL LATH

GYPSUM LATH

42–9. Metal lath used under a flush ceiling beam.

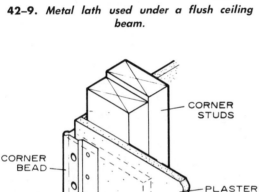

CORNER STUDS

CORNER BEAD

PLASTER BASE

42–10. A corner bead is installed at outside corners to serve as a leveling edge when the plaster is applied.

42–11. A cornerite is installed at inside corners for reinforcement and to minimize plaster cracks.

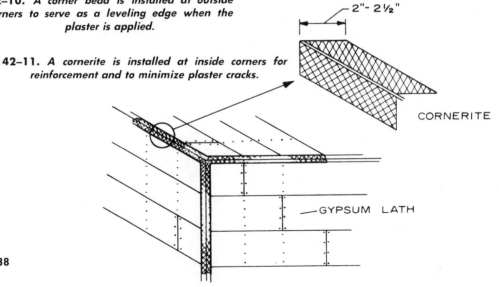

2"- 2½"

CORNERITE

GYPSUM LATH

The frames for window and exterior door openings are normally in place before plaster is applied. Thus the inside edges of the side and head jambs serve as grounds. The edge of the window sill may also be used as a ground, or a narrow ground strip $\frac{7}{8}$" thick and 1" wide may be nailed to the edge of the 2" × 4" sill. The $\frac{7}{8}$" × 1" grounds might also be used around window and door openings. Fig. 42-14. These are normally left in place and are covered by the casing.

A similar narrow ground, or screed, is used at the bottom of a wall for controlling the thickness of the plaster and providing an even surface for the baseboard and molding. Fig. 42-12. These strips are also left in place after the plaster has been applied.

Plaster Materials and Methods of Application

Plaster for interior finishing is made from combinations of sand, lime or prepared plaster, and water. Waterproof finishes for walls are available and should be used in bathrooms, especially in shower and tub recesses when tile is not used, and sometimes on the kitchen wainscot.

Plaster should be applied in three-coat or in two-coat double-up work. The minimum thickness over ³/₈″ gypsum or insulating lath should be about ³/₈″.

Three-coat work is used on metal lath and is usually at least ³/₄″ thick. The first plaster coat over metal lath is called the scratch coat. It is scratched, after a slight set has occurred, to insure a good bond for the second coat. The second coat is called the brown or leveling coat, and leveling is done during its application. The third coat is the finish coat. Fig. 42-15.

Double-up work, combining the scratch and brown coat, is used on gypsum or insulating lath. Leveling and plumbing of the walls and ceilings are done during application.

The final or finish coats are of two general types: the sand-float and the putty finish. In the sand-float finish, lime is mixed with sand, which results in a textured finish. The texture depends on the coarseness of the sand. Putty finish, used without sand, is smooth. This type is common in kitchens and bathrooms where a gloss paint or enamel finish is used, and in other rooms where a smooth finish is desired. Because

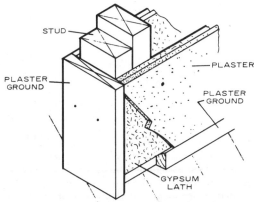

42–12. *A one-piece plaster ground whose width is equal to the finished wall thickness is applied to the trimmer studs and header. Do not drive the nails home. Nails will be pulled and the grounds removed after the plaster is dry.*

42–13. *Narrow strip grounds are sometimes used around interior openings.*

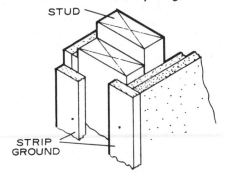

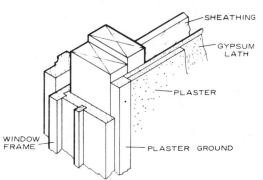

42–14. *When narrow ⁷/₈″ × 1″ plaster grounds are used around a window or door opening, they remain in place and are eventually covered by the casing.*

42–15. *A cross section of plaster on metal lath showing the buildup of the various coats. Notice how the plaster is keyed to the metal lath in the area indicated by the arrow.*

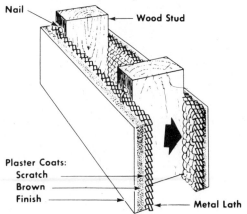

of its durability, keene's cement is often used as a finish plaster in bathrooms.

Plastering should not be done in freezing weather without a source of constant heat. In normal construction, the heating unit is in place before plastering is started.

Insulating plaster, consisting of a vermiculite, perlite, or other aggregate with the plaster mix, may also be used for the finish coat.

DRY WALL

Dry wall is so called because it requires little if any water for application. Gypsum board, plywood, fiberboard, and similar sheet materials, as well as different forms and thicknesses of wood paneling, are classified as dry-wall finishes. Dry-wall materials are versatile, easily applied, decorative, and utilitarian. Fig. 42-16. Most of the several types of paneling can be

used wherever excessive moisture is not a problem. Some are even suitable for moist places like the bathroom and shower stall.

Paneling can be purchased unfinished or prefinished. The unfinished may be stained any color or given a clear finish to preserve its natural color. Some paneling comes with a factory-applied, baked-on plastic finish or a vinyl covering.

When thin sheet material such as gypsum board or plywood is used, the studs and ceiling joists must be in alignment to provide a smooth, even surface. Figs. 42-17 and 42-18. If ceiling joists are uneven, a "strongback" may be used to align the joists. Fig. 42-19.

42–16. *Products offering a variety of textures are available for spray or roller application over gypsum wallboard.*

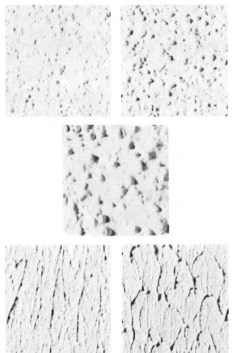

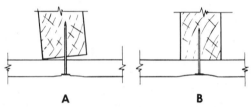

A   **B**

42–17. *A. A framing member has not been properly squared with the plate. This increases the possibility of puncturing the gypsum board paper with the nailhead. There is also the danger of a reverse twisting of the stud as it dries out, in which case the board will be loosely nailed and a "pop" will occur. B. The twisted stud has been squared before the application of the wallboard.*

42–18a. *Improperly aligned studs, joists, or headers will result in the nailheads puncturing the paper or cracking the board.*

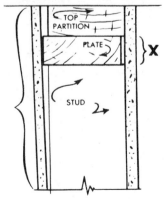

The minimum thicknesses for plywood, fiberboard, and paneling and the required spacing of framing members are shown in Fig. 42-20.

Gypsum Board

Gypsum board is a sheet material made up of gypsum filler faced with paper. These sheets are normally 4' wide and 8' long, but may be obtained in lengths up to 16'. The edges along the length are tapered and, on some types, the ends are tapered also. Tapering allows for a filled and taped joint.

Some gypsum board has a foil back which serves as a vapor barrier on exterior walls. Prefinished gypsum board is also available for single-layer application in new construction. A thickness of 1/2" is recommended. For two-ply laminated applications, two 3/8" thick sheets are used. The maximum spacing of framing members for various thicknesses of gypsum board is shown in Fig. 42-21.

Installing Gypsum Board on Walls

Gypsum board may be applied with nails, screws, or adhesive. Fig. 42-22. The nails must have flat thin heads for flush driving without damage to the surface of the board. Fig. 42-23 (Page 746). Gypsum board 1/2" thick should be applied with a 5d nail (1 5/8" long). For 3/8" thick material,

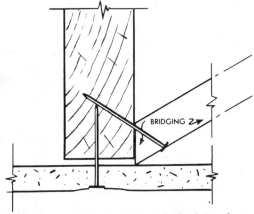

42–18b. *Protrusions, such as the bridging shown here, will puncture the face paper of wallboard. The bridging, which projects beyond the edge of the joists, also prevents the back of the board from being brought into contact with the nailing surface.*

42–20. *Minimum thicknesses for paneling applied to different framing spaces.*

Framing spaced (inches)	Thickness		
	Plywood	Fiberboard	Paneling
16	1/4"	1/2"	3/8"
20	3/8"	3/4"	1/2"
24	3/8"	3/4"	5/6"

42–19. *A strongback used to align the joists.*

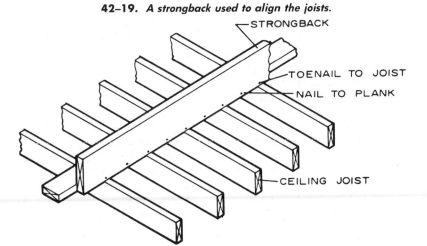

STRONGBACK

TOENAIL TO JOIST

NAIL TO PLANK

CEILING JOIST

42-21. *Maximum framing spacing recommended for various thicknesses of gypsum board.*

Installed long direction of sheet	Minimum thickness	Maximum spacing of supports (on center)	
		Walls	Ceilings
Parallel to framing members	3/8″	16″	
	1/2″	24″	16″
	5/8″	24″	16″
Right angles to framing members	3/8″	16″	16″
	1/2″	24″	24″
	5/8″	24″	24″

use a 4d nail (1³/₈″ long). When ring shank nails are used, a nail about ¹/₈″ shorter will provide adequate holding power. Special screws will help prevent a bulging surface, sometimes referred to as a "nail pop," caused by the drying out of the framing members.

Nail pops are greatly reduced if the moisture content of the framing members is less than 15% when the gypsum board is applied. When framing members have a high moisture content, it is good practice to let them approach moisture equilibrium before application of the gypsum board.

Nails should be spaced 6″ to 8″ on the sidewalls and 5″ to 7″ on the ceiling, with a minimum edge distance of ³/₈″. Nail spacing is the same for horizontal and vertical application. Fig. 42-24 (Page 746).

For studs or joists 16″ on center, screws should be spaced not more than 12″ apart on ceilings and 16″ apart on sidewalls. For studs 24″ on center, screws must not be spaced further than 12″ apart.

Horizontal application is best adapted to rooms in which full-length sheets can be applied because this reduces the number of vertical joints. Any joints necessary should be made at windows or doors. When this is not possible, the end joints should be staggered and centered on the framing members.

Horizontal nailing blocks between studs are not normally required when the studs are 16″ on center and the gypsum board is ³/₈″ or thicker. However, if the spacing is

42-22a. *Applying gypsum wallboard with nails.*

greater or additional support at the joint is required, nailing blocks may be used. Fig. 42-24b.

In single-layer application, the 4′ wide gypsum sheets are installed vertically or horizontally on the walls after the ceiling has been covered. When the sheets are applied vertically, they cover three stud spaces if the studs are spaced 16″ on center and two if the studs are spaced 24″ on center. The edges of the gypsum board should be centered on studs and should make a very light contact with each other.

The laminated, two-ply method of gypsum application is begun by applying an undercourse of ³/₈″ material vertically. To reduce sound transmission between rooms, sound-deadening panels are sometimes used as an undercourse. Fig.

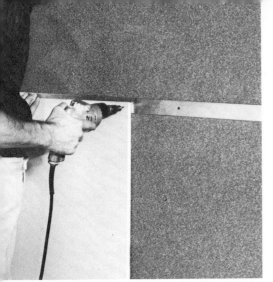

42-22b. *Applying gypsum wallboard with screws.*

42-22c. *Applying gypsum wallboard with an adhesive.*

42-25. The finish ³/₈" sheet is usually in room-size lengths. It is applied horizontally with an adhesive. Be certain to follow the manufacturer's recommendations when applying the adhesive. Nails used in the application of the finish gypsum wallboard should be driven with the head below the surface. The domed head of the hammer will form a small dimple in the wallboard. Fig. 42-26 (Page 748). Do not use a nail set. Care should be taken to avoid breaking the paper face of the gypsum board when nailing.

DOUBLE NAILING SYSTEM

A nail pop is caused by a movement of either the gypsum wallboard or the nail head in relation to the other. These pops may be prevented if the board is held tightly against the framing by the head of the nail at all times. The double nailing system incorporates a second nail in close proximity (2") to the first to insure that the board is nailed tight. The wallboard is first nailed to each framing member with nails spaced approximately 12" on center. This places five nails per 4' width into each framing member. The board is then nailed around its perimeter. The top and bottom are fastened with one nail at each framing member. At the ends of the board, nails are spaced 7" on center. Additional nails are then spaced approximately 2" from each nail on the inner area of the board. Fig. 42-27.

As the second nail in each group of two nails is driven home, the worker can watch for any movement between the board and the head of the first nail driven. Movement indicates that the first nail is not holding the board tight and that it should be given additional blows with the hammer. Always begin nailing at the center of the board and work toward the ends, making sure to hold the gypsum board tight against the framing member.

CUTTING GYPSUM BOARD

Gypsum board may be cut to size by sawing. Another method is to score the finished side with an awl or knife. Fig. 42-28. Snap the board over a straight edge. Fig. 42-29. To complete the separa-

743

42–22d. *Recommended gypsum wallboard fasteners for various applications.*

FASTENER	SPACING	QUANTITY REQUIRED
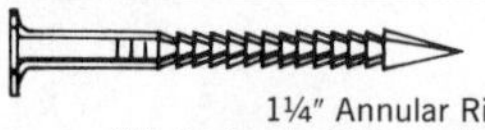 1¼″ Annular Ring Nail—12½ gauge; ¼″ dia. head with a slight taper to a small fillet at shank; bright finish; medium diamond point.	7″ c. to c. on ceilings—8″ c. to c. on walls	5¼ lbs./1000 sq. ft. approx. 325 nails/lb.
 1⅜″ Annular Ring Nail (Specification same as above except for length)	7″ c. to c. on ceilings—8″ c. to c. on walls	5¼ lbs./sq. ft., approx. 321 nails/lb.
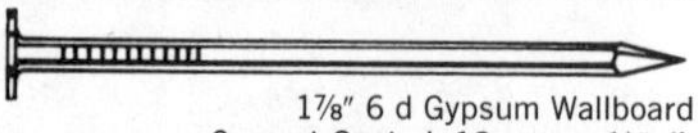 1⅞″ 6 d Gypsum Wallboard Nail— Cement Coated, 13 gauge, ¼″ dia. head	7″ c. to c. on ceilings—8″ c. to c. on walls	6¼ lbs./1000 sq. ft., approx. 275 nails/lb.
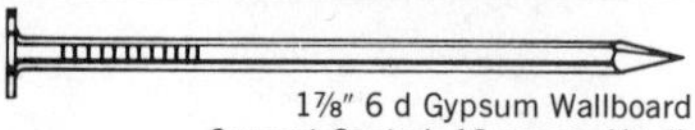 1⅞″ 6 d Gypsum Wallboard Nail— Cement Coated, 13 gauge, ¼″ dia. head	6″ c. to c. on ceilings—7″ c. to c. on walls	6¾ lbs./1000 sq. ft. approx. 278 nails/lb.
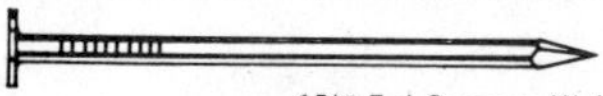 1⅝″ 5 d Gypsum Wallboard Nail— Cement Coated, 13½ gauge, $^{15}/_{64}$″ dia. head	6″ c. to c. on ceilings—7″ c. to c. on walls	5¼ lbs./1000 sq. ft. approx. 366 nails/lb.
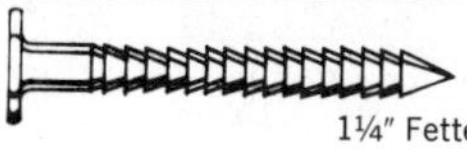 1¼″ Fetter Annular Ring Nail— 11 gauge; $^5/_{16}$″ dia. head	6″ c. to c. on ceilings	6 lbs./1000 sq. ft. approx. 315 nails/lb.
 1⅛″ Matching Color Nail (Steel)	8″ c.to c. on walls	1½ lbs./1000 sq. ft. approx. 1,008 nails/lb.
 1⅞″ Matching Color Nail (Steel)	8″ c. to c. on walls	4½ lbs./1000 sq. ft. approx. 349 nails/lb.
 1⅛″ Matching Color Nail (Brass)	8″ c. to c. on walls	1¾ lbs./1000 sq. ft. approx. 901 nails/lb.
 1¼″ Drywall Screw— Type W	Framing spaced 16″ c. to c. 12″ c. to c. on ceilings—16″ c. to c. on walls Framing spaced 24″ c. to c. 12″ c. to c. on ceilings—12″ c. to c. on walls	Approx. 1000 screws/1000 sq. ft.
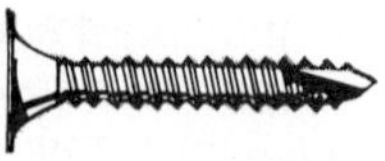 1″ Drywall Screw—Type S	12″ c. to c. on walls and ceilings	Approx. 875 screws/1000 sq. ft.
 1⅝″ Drywall Screw—Type S	16″ c. to c. on walls and ceilings when installed permanently without laminating adhesive; or as required for temporary mechanical attachment while laminating adhesive dries	Varies depending on 2 layer system used.
 1″ Drywall Screw—Type S	12″ c. to c. in field of board and 8″ c. to c. staggered at vertical joints on walls— 12″ c. to c. on ceilings	Approx. 1100 screws/1000 sq. ft.

Attach Gypsum Wallboard with the correct FASTENER

GYPSUM BOARD PRODUCT	TYPE FRAME	APPLICATION CONDITION
½", ⅜" and ¼" Wallboard ½" and ⅜" Gypsum Backing Board	Wood	Direct to framing
⅝" Gypsum Wallboard	Wood	Direct to framing
⅜" and ¼" Gypsum Wallboard	Wood	Over existing surfaces
⅝" Gypsum Wall Board with special core for increased fire resistance	Wood	Direct to wood frame
½" Gypsum Wallboard with special core for increased fire resistance	Wood	Direct to framing
⅝" Gypsum Wallboard ⅝" Gypsum Backing Board with special core for increased fire resistance	Steel Nailing Channel	Direct to framing
Woodgrained Gypsum Wallboard Vinyl-coated Woodgrained Gypsum Wallboard	Wood	Direct to framing
	Wood	Over existing surfaces
Vinyl-Coated Gypsum Wallboard	Wood	Direct to framing
⅝", ½" and ⅜" Gypsum Wallboard ½" and ⅜" Gypsum Backing Board	Wood	Direct to framing
⅝", ½" and ⅜" Gypsum Wallboard ½" and ⅜" Gypsum Backing Board	Metal Stud and/or Drywall Furring Channel	Direct to framing
⅝", ½" and ⅜" Gypsum Wallboard	Metal Stud and/or Drywall Furring Channel	Permanent or temporary face layer attachment with double wall installation
⅝" Gypsum Wallboard	Metal Stud and/or Drywall Furring Channel	Direct to framing

NAIL ATTACHMENT

Nails for the attachment of single layer ⅜, ½, and ⅝-inch gypsum board comply with the "Performance Standards for Nails for Application of Gypsum Wallboard," as adopted and published by the Gypsum Association. Other nails may be used provided they have immediate and delayed holding power with penetration into wood frame not exceeding ⅞-inch; tear-through resistance and dimpling characteristics; and meet the dimensional requirements of nails so described under this performance standard.

SCREW ATTACHMENT

Designed to provide positive mechanical attachment of single layer Gypsum Wallboard.

tion, score the back of the board with a sharp knife and snap the board forward toward the face side for a clean, straight joint break. Fig. 42-30.

Small cutouts for electrical outlets and other openings are made with a keyhole saw or a saber saw. To mark the location of the electrical outlet cutouts, hold the panel in place against the wall. With a wood block to protect the board, tap around the outlet with a hammer. An indentation of the outlet box will be made on the back of the board to use as a guide for cutting.

When notching gypsum board for door or window openings, make two saw cuts to the correct depth. The final cut is made by scoring and snapping, the same as when

42–23. *The gypsum wallboard nail has a thin head and is ring shanked for greater holding power.*

cutting the sheet to size. Fig. 42-31. The cut edges of gypsum board may be smoothed with #2 sandpaper wrapped around a wood block. Fig. 42-32 (Page 750).

JOINT TREATMENT

The joints between the panels are made smooth by applying joint cement, perforated tape, and additional coats of joint cement, and then sanding the surface level with the wall surface. Fig. 42-33. Joint cement can be purchased in either premixed or powder form. The powder is mixed with water to a soft putty consistency that can be easily applied.

Use a 5″ wide spackling knife or a mechanical applicator to fill the joints with the cement. Fig. 42-34. Press the tape into the recess with a wide, flat knife until the joint cement is forced through the perforations in the tape. Fig. 42-35. Next cover the tape with additional cement, feathering the outer edges. Fig. 42-36. After the cement

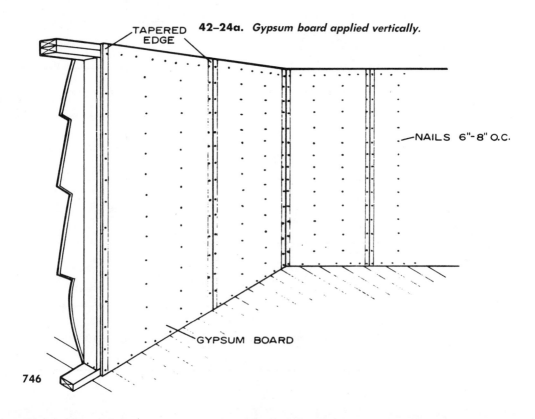

TAPERED EDGE **42–24a.** *Gypsum board applied vertically.*

NAILS 6″-8″ O.C.

GYPSUM BOARD

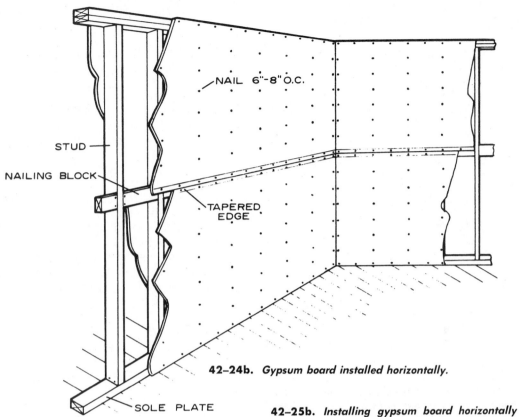

NAIL 6"-8" O.C.

STUD

NAILING BLOCK

TAPERED EDGE

42–24b. *Gypsum board installed horizontally.*

SOLE PLATE

42–25a. *Applying a sound-deadening board to the studs as an undercourse for gypsum board. This special-density fiberboard helps to reduce sound transmission between rooms.*

42–25b. *Installing gypsum board horizontally over the sound-deadening panels, which were applied vertically. Openings in the sound-deadening board for electrical outlets, heating vents, etc, must be calked carefully, since even a small hole will destroy the sound-deadening effectiveness of the wall.*

747

has dried, sand the joint lightly and then apply a second coat, again feathering the edges. Sometimes a steel trowel is used to apply the second coat. For best results, a third coat is applied and the edges are feathered beyond the second coat. After

the joint cement is completely dry, sand the joint smooth and even with the wall surface.

To tape interior corners, fold the tape down the center to form a right angle. Fig. 42-37. Apply the cement in the corner and press the tape in place. Then finish the corner with joint cement and sand smooth when dry. Apply a second coat if necessary and sand it smooth, flat, and even with the wall surface. The same procedure is followed for interior corners between a wall and ceiling, or a molding of some type is installed. Fig. 42-38. To hide hammer indentations, fill them with joint cement and sand them smooth when they are dry. Usually a second coat is necessary. Figs. 42-39 and 42-40.

Temperature and humidity have a direct effect on the drying time of joint treatment products. Very little can be done to alter temperature and humidity under job conditions. However, care should be taken to note the average differences in drying time under different atmospheric conditions so that problems may be minimized. Joint treatment products must be thoroughly dry before successive coats and/or final deco-

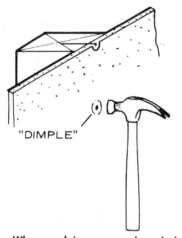

"DIMPLE"

42–26. *When applying gypsum board, the final hammer blows should make a slight dimple on the face of the board around the nailhead. Avoid a heavy blow that would break the face paper or crush the gypsum core.*

42–27. *The double nailing system for installing gypsum wallboard. Note that the nails are spaced 7″ on center at the panel ends.*

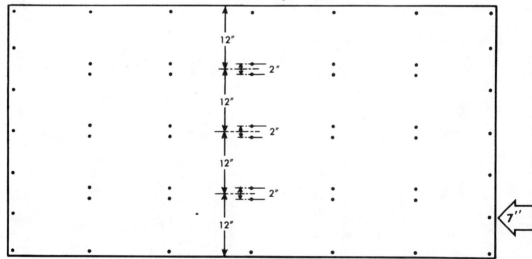

rations are applied. In all cases, a well-ventilated area assists in proper drying of these materials. The chart in Fig. 42-41 indicates the average drying periods for joint treatment products under different temperature and humidity conditions.

Installing Gypsum Board on the Ceiling

Gypsum board applied to the ceiling is nailed to ceiling joists or to the bottom chord of a truss. Nails are spaced 5" to 7"

apart and dimpled in the same manner as when applying gypsum board to sidewalls. During installation, the gypsum board can be held in place with one or two braces about 1" longer than the height of the ceiling. Fig. 42-42 (Page 752). Joints should be staggered and centered over framing members. Fig. 42-43. Joint treatment is the same for the ceiling as for the walls.

Plywood Paneling

Plywood paneling usually comes in sheets that are 4' wide by 8' long. However, other lengths are available. Some types are narrower than 4' for greater ease of handling or to create interesting patterns. Plywood is available in a number of

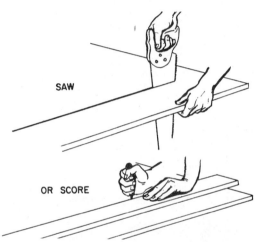

SAW

OR SCORE

42-28. *Gypsum wallboard may be sawed, or it may be scored and snapped.*

42-29. *Break the gypsum core by placing the scored line of the sheet over the edge of the table or bench and snapping down. Hold the sheet firmly on the table and support the cutoff with your other hand.*

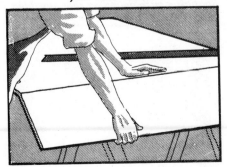

42-30. *Complete the separation by scoring the back paper of the board with a sharp knife. Then snap the board forward toward the face side.*

42-31. *Making a cutout for a door or window. Make two saw cuts to the correct depth, then score and snap.*

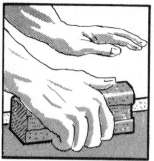

42-32. Sanding the rough edge of gypsum board.

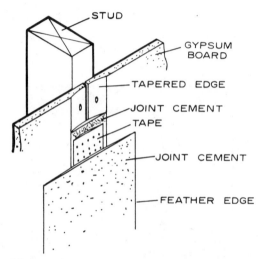

STUD

GYPSUM BOARD

TAPERED EDGE

JOINT CEMENT

TAPE

JOINT CEMENT

FEATHER EDGE

42-33. The tapered edge of the gypsum board is filled with joint cement and taped. Additional joint cement is then applied and feathered out to provide a smooth surface.

42-34a. Applying joint cement with a wide spackling knife.

42-34b. Applying joint compound with a mechanical applicator.

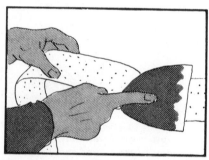

42-35. Press the perforated tape into the cement, forcing the excess cement from under the tape.

42-36. Spread a thin coat of cement over the tape. If necessary, follow with a second and third coat after each preceding coat has dried.

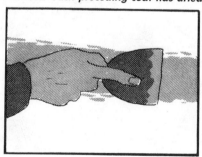

species. Because of its beauty, warmth, and ease of maintenance, plywood paneling may be used for accent walls or to cover entire room wall areas. Fig. 42-44. Plywood panels can be applied either vertically or horizontally, as long as solid backing is provided at all edges. For framing spaced 16" on center, plywood $\frac{1}{4}$" thick is considered minimum. For framing 20" or 24" on center, $\frac{3}{8}$" plywood is the minimum thickness. Sometimes the wall is first covered with $\frac{3}{8}$" or $\frac{1}{2}$" gypsum board, and the plywood is then applied to the surface of the gypsum board.

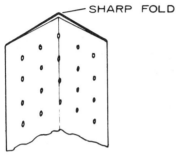

42-37. *Fold the perforated tape down the center to form a right angle when taping interior corners.*

42-39. *Filling the nail dimples with joint cement. Repeat with second and third coats of cement, if necessary, and sand smooth.*

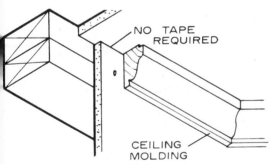

42-38. *When a molding is installed between the wall and the ceiling, it is not necessary to tape the joint.*

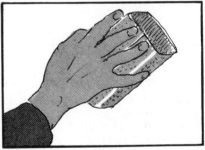

42-40. *Sanding the joints and nail dimples after the joint cement is completely dried. This is the last step in preparing the gypsum board for a decorative treatment.*

42-41. *Drying time for joint cements.*

Relative Humidity	0	20%	40%	50%	60%	70%	80%	90%	98%
Temp. ° F. 40	28 H	34 H	44 H	2 D	2½ D	3½ D	4½ D	9 D	37 D
60	13 H	16 H	20 H	24 H	29 H	38 H	2½ D	4½ D	18 D
80	6 H	8 H	10 H	12 H	13½ H	19½ H	27 H	49 H	9 D
100	3 H	4 H	5 H	6 H	8 H	10 H	14 H	26 H	5 D

H = hours   D = days   (24 hours)

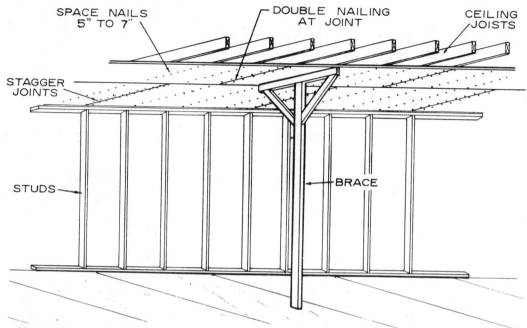

SPACE NAILS
5" TO 7"

DOUBLE NAILING
AT JOINT

CEILING
JOISTS

STAGGER
JOINTS

STUDS

BRACE

42–42. *Gypsum board ceiling installation details. A brace is used to hold the material in position for nailing.*

Installing Plywood Paneling

On exterior masonry walls above or below grade, be sure the wall is properly waterproofed before the studding or furring is applied. Where extreme humidity may cause condensation on the inside of the exterior masonry wall, apply a vapor barrier, paper, or film to prevent moisture penetration to the panel. Fig. 42-45.

Furring strips must be used on masonry and plaster walls. When paneling over an existing wall that is not masonry or plaster, sand any uneven or rough spots to obtain a flush surface. If sanding will not remove severe unevenness, furring strips should be used. Paneling may be applied directly to furring strips or studs. However, for additional strength, fire resistance, and sound deadening, 3/8" or 1/2" gypsum wall board is recommended as a backing behind plywood paneling. Plywood sheathing 5/16" thick is also ideal for application

to the studs as a backing under finish paneling.

APPLYING FURRING STRIPS TO MASONRY WALLS

On masonry walls, apply furring strips horizontally every 16". Fig. 42-46. Allow a clearance of at least 1/4" between the top furring strip and the ceiling and between the bottom furring strip and the floor. The top furring strip is nailed to the bottom edge of the ceiling joists or to a nailing block. Fig. 42-47 (Page 756). Insert vertical strips every 48" to support the panel edges. Figs. 42-46 and 42-48. The furring strips are attached to the masonry walls with masonry nails, screws, nails driven into shields or wood dowels, nail anchors, adhesive anchors, or bolt anchors. Fig. 42-49. Uneven furring strips can be leveled by placing shims in the low spots and driving a nail through the furring strip and the shims to hold them in place. Fig. 42-50.

42–43. *Installing gypsum board on the ceiling. Note the staggering of the end joints.*

APPLYING FURRING STRIPS TO PLASTER WALLS

If the paneling is to be applied on a plaster wall, the furring strips should be nailed horizontally to the studs, starting at the floor line and continuing up the wall every 16″. Nail vertical strips every 48″ to support the panel edges. The furring strips are shimmed as necessary with wood shingles to obtain a flush surface.

FURRING STRIPS FOR SPECIAL APPLICATIONS

For walls over 8′ high, additional furring strips are nailed horizontally with the center of one of the strips 8′ from the floor and another at the ceiling. Fig. 42-5l.

When wainscoting is installed, nail an extra strip horizontally at the wainscot height. This is usually about 32″, since three 32″ pieces can be cut from a full 8′ panel (96″). Fig. 42-52 (Page 758).

LAYING OUT THE JOB

Set up the panels around the room to plan their sequence. Arrange them so that the natural color variations form a pleasing pattern. Fig. 42-53. For most interiors, it is practical to start paneling from one corner

42–44a. *The indentations of age can be felt as well as seen in this antiqued plywood paneling used as an accent wall.*

42–44b. *A pecan plywood paneling with darkly stained grooves was used to accent this living room wall.*

and work around the room. After deciding in which corner to start the paneling, stack the panels in the correct sequence so that as the job proceeds, the panels may be removed from the stack in the proper order.

INSTALLING PANELS

Accurately measure the height of the wall in several places. The panels should be cut so that they have a ¼" clearance at top and bottom. If the location of the studs is not visible, locate them and mark the center of each stud with a chalk line. Also lightly mark the stud center locations on the floor and ceiling to serve as a guide when nailing each panel in position. Fig. 42-54 (Page 759). If the panel has grooves, these will usually be spaced to line up with the studs. Fig. 42-55. (Page 760).

Place the first panel in position and butt it to the adjacent wall in the corner. Make

sure the panel is perfectly plumb and the outer edge is directly over the chalk line which marks the center of the stud. Fig. 42-56. If this edge does not fall directly on the stud, cut the other edge of the panel so that it will. In most cases, the corner where the paneling begins is irregular. There may be a fireplace, concrete blocks, or uneven plaster. Scribe the panel with a small compass to insure a perfect fit. Position the panel at the proper height by setting it on a block and shimming it with a shingle to allow for ¼" clearance at the top and bottom. Fig. 42-57. When the panel is set perfectly plumb and at the correct height, set the compass for the amount to be cut off and scribe the line as shown in Fig. 42-58. Cut the panel along this line to fit the irregular wall.

Set the panel back in place against the wall to which it has been scribed and

42–45a. *Plywood paneling applied to furring strips on a basement wall. A waterproof coating has been applied to the cement. Insulation with a vapor barrier has been installed between the furring strips.*

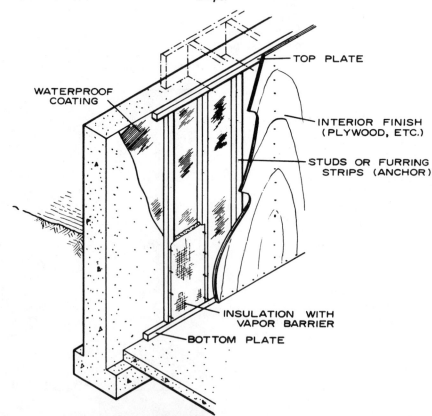

WATERPROOF COATING

TOP PLATE

INTERIOR FINISH (PLYWOOD, ETC.)

STUDS OR FURRING STRIPS (ANCHOR)

INSULATION WITH VAPOR BARRIER

BOTTOM PLATE

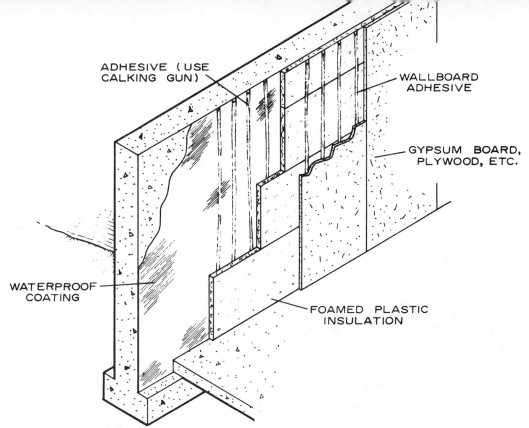

42–45b. *Basement paneling applied to a foamed plastic insulation with wallboard adhesive.*

42–46. *Correct placement of furring strips in preparation for paneling.*

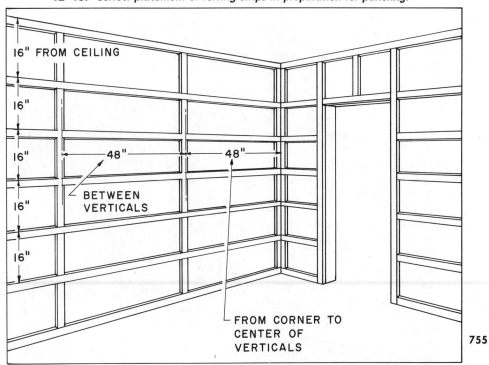

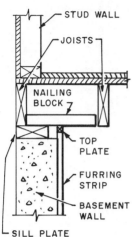

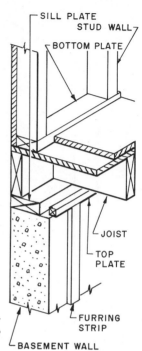

42–47a. *When the wall to be furred runs at right angles to the joists, nail the top plate (top furring strip) to the underside of the joists.*

42–47b. *When the wall to be furred runs parallel to the joists, install a nailing block to which the top plate can be nailed.*

42–48. *Furring strips must be installed so that all panel edges are supported. Attach the furring strips securely to the masonry wall.*

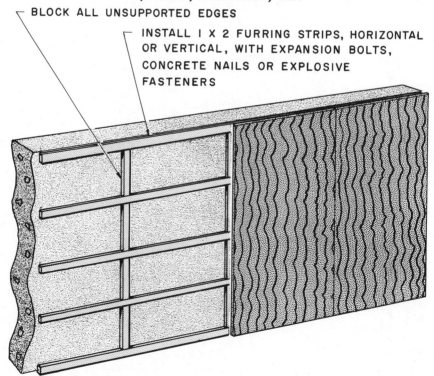

BLOCK ALL UNSUPPORTED EDGES

INSTALL 1 X 2 FURRING STRIPS, HORIZONTAL OR VERTICAL, WITH EXPANSION BOLTS, CONCRETE NAILS OR EXPLOSIVE FASTENERS

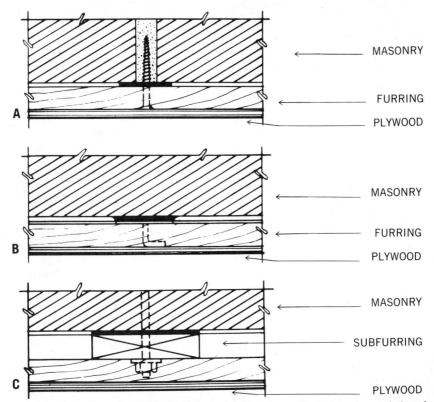

42–49. *Methods of attaching furring strips to a masonry wall: A. Insert wood dowels in the masonry to which the furring strips can be wood screwed or nailed. B. Use nail anchors or adhesive anchors. C. Bolt anchors may be used for attaching a 1″ × 3″ subfurring to the wall, then attaching the 1″ × 2″ furring to the 1″ × 3″ as if to studs.*

42–50. *Wood shingles are used as shims behind furring.*

42–51a. *A wall over 8′ high must have a horizontal furring strip positioned so that the top of an 8′ panel will be aligned with the center of the furring strip at that point.*

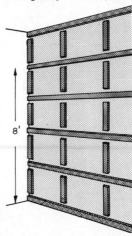

42-51b. *Installing paneling on a wall over 8' high.*

42-51c. *The finished wall.*

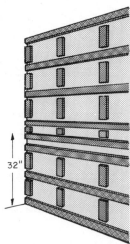

42-52. *An extra furring strip is nailed horizontally at the wainscot height.*

block the panel up as previously to the correct height. Fasten the panel to the wall. It may be applied with nails or adhesive; this will be discussed later. Butt the second panel to the first. With the first panel properly positioned on the studs, the edges of the remaining 4' panels will also land on stud centers, assuming that the stud spacing is uniform across the wall surface. Fig. 42-59. When paneling is being applied over a backer board with adhesive, the panel edges do not need to meet on a stud.

FASTENING PANELS WITH NAILS

For paneling applied directly to studs, use 4d finish nails spaced 6" along the panel edges and 12" elsewhere in the grooves. Fig. 42-60. If the paneling is applied over furring, use 2d nails spaced 8" apart along the panel edges and 16" apart elsewhere. Fig. 42-61. For paneling applied over backer board, use 6d finish nails spaced 4" along the panel edges and 6" elsewhere. Fig. 42-62 (Page 762). Countersink the nails 1/32" below the surface. These holes can be filled later with a putty stick to match the color of the panel. Colored nails which blend with the wood

758

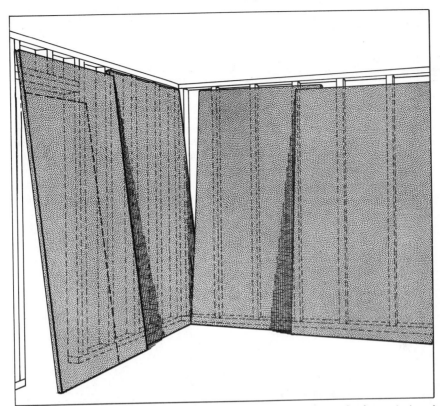

42-53. *Arrange the panels around the room to show the best pattern of color variations in both daylight and artificial light.*

42-54. *Snapping a chalk line on the wall to indicate the center of each wall stud.*

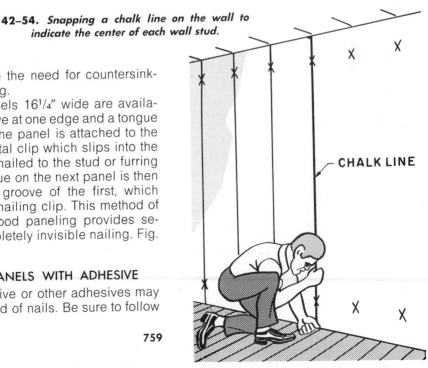

CHALK LINE

finish eliminate the need for countersinking and puttying.

Plywood panels 16¼" wide are available with a groove at one edge and a tongue on the other. The panel is attached to the wall with a metal clip which slips into the groove and is nailed to the stud or furring strip. The tongue on the next panel is then placed in the groove of the first, which covers up the nailing clip. This method of applying plywood paneling provides secure and completely invisible nailing. Fig. 42-63.

FASTENING PANELS WITH ADHESIVE

Panel adhesive or other adhesives may be used instead of nails. Be sure to follow

GROOVE SPACING

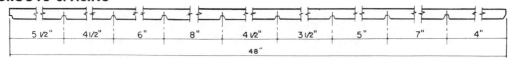

5 1/2"	4 1/2"	6"	8"	4 1/2"	3 1/2"	5"	7"	4"

48"

42–55. *A typical groove spacing for plywood paneling. The groove locations appear to be randomly spaced. However, when the dimensions are added together, the grooves fall on 16" and 24" centers. In this way, the panel can be nailed through the grooves to the studs.*

the manufacturer's instructions. After the panels are properly cut and fitted, the adhesive is applied to the studs, furring strips, or backing. Fig. 42-64 (Page 764). Apply the adhesive in continuous 1/8" wide beads or in intermittent beads 3" apart to all stud or furring strip surfaces. Apply a continuous 1/8" wide bead at the corners and around cutouts. Position the panel and press it firmly against the adhesive. Place

three or four finishing nails across the top of the panel to hold it in place.

Pull the bottom of the panel 8" to 10" away from the wall, allowing the nails at the top to serve as a hinge. Hold the panel

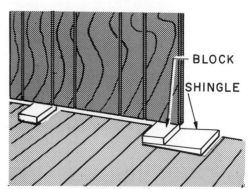

42–57. *Use a block and a shingle to position the panel at the correct height.*

42–58. *Scribing the panel edge to fit the adjacent wall.*

42–56. *Set the first panel in position, making certain that the edge is plumb.*

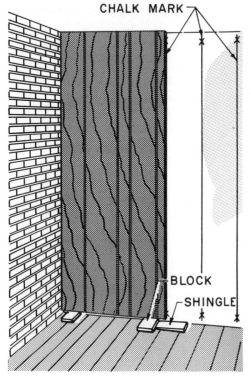

42–59. *The panel edges must meet on the center of a stud or furring strip.*

42–60c. *Inspecting a plywood panel installation. This paneling has been applied directly to the studs and is ready for door installation and trimming.*

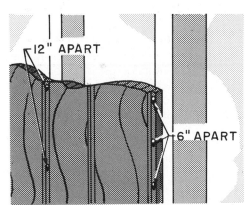

42–60a. *Nail spacing for plywood panels applied directly to the studs.*

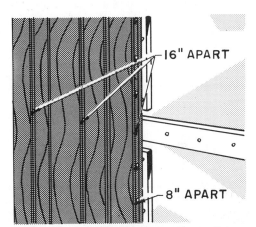

42–61. *Nail spacing for plywood paneling applied over furring.*

42–60b. *Nailing recommendations for interior plywood paneling.*

Interior plywood paneling/Recommendations apply to all species groups.

Plywood Thickness (inch)	Max. Support Spacing (inches)	Nail Size & Type	Nail Spacing (inches)	
			Panel Edges	Intermediate
¼	16[1]	4d casing or finish	6	12
⅜	24	6d casing or finish	6	12

Notes: (1) Can be 20″ if face grain of paneling is across supports.

out with a spacer block for 8 to 10 minutes to allow the adhesive to dry. Fig. 42-65. Remove the spacer block and reposition the panel to fit perfectly. Place a clean block of soft wood against the panel and tap the block with a hammer or rubber mallet to obtain a full surface contact. Fig. 42-66.

If a panel is not flush with the stud or furring strip surfaces, small finishing nails may be needed to hold the panel in position until the adhesive acquires full strength.

CUTTING PLYWOOD PANELS

When cutting panels with a crosscut saw or table saw, cut with the face side up. If an electric hand saw or saber saw is used, cut with the face side down. Never use a rip saw, since this will tear the veneer on the edge of the panel.

When cutting the panel for outlets, locate the opening by chalking the edges of the outlet box and carefully fitting the panel loosely over the chalked box. Strike the face of the panel sharply several times with the heel of your hand to transfer the box outline to the back of the panel. Drill pilot holes in the corners from the back side; then cut out the outlet hole from the front side with a keyhole saw. Fig. 42-67. If a saber saw is used, a plunge cut can be

made to eliminate the drilling of the pilot holes at the corners.

MOLDINGS

There are several styles of wood and metal moldings for wood paneling. Pine moldings are sometimes used. They can be stained to harmonize with the prefinished paneling. Prefinished moldings to match the panel finish are also available.

Accurate measurements are essential for a good, professional-looking molding job. Measure along the ceiling line for the cove or crown molding. Measure along the floor for the exact length of the base and shoe molding. Do not assume that the ceiling and floor are the same length. Wood moldings should be scribed, mitered at 45°, or coped as described in Unit 47, "Interior Trim." Construction details showing the use of various metal and wood moldings at corners, doors, windows, floors, and ceilings are illustrated in Fig. 42-68 (Page 766).

Hardboard and Fiberboard

Hardboard and fiberboard are applied in the same manner as plywood. Hardboard should be at least $1/4''$ thick when applied over open framing spaced 16" on center. It should be at least $7/16''$ thick for framing spaced 24" on center. When $1/8''$ hardboard is used, a rigid backing of some type is required. Fiberboard in tongue-and-groove planks or sheet form must be $1/2''$ thick when framing members are spaced 16" on center and $3/4''$ thick for 24" spacing. For best results, vertical furring should be used with hardboard paneling. However, horizontal furring can be applied 16" on center over studs spaced 48" on center. Fig. 42-69 (Page 768).

The paneling may be nailed or applied with an adhesive. Fig. 42-70. Nails should penetrate into the studs at least $3/4''$ and should be spaced 4" on center at all joints and along the edges. At all intermediate supports, nail 8" on center. Nails around the perimeter of the panel should be $1/4''$

42–62. *Nail spacing for plywood paneling applied over backer board.*

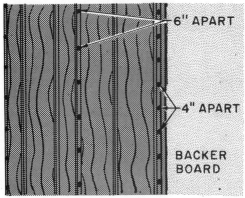

6" APART

4" APART

BACKER BOARD

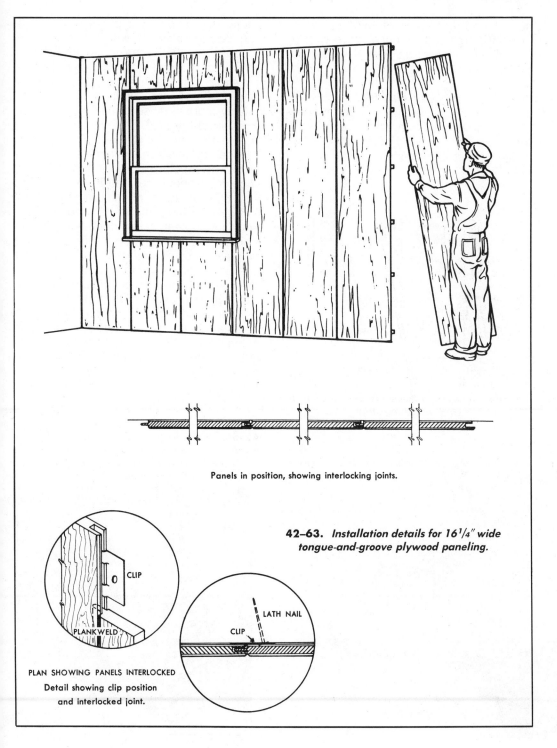

Panels in position, showing interlocking joints.

42–63. *Installation details for 16¼″ wide tongue-and-groove plywood paneling.*

CLIP

PLANKWELD

LATH NAIL

CLIP

PLAN SHOWING PANELS INTERLOCKED
Detail showing clip position
and interlocked joint.

from the edge. Fig. 42-71. Wood or metal moldings are used as trim and are applied in the same manner as over plywood. Fig. 42-72 (Page 770).

Wood Paneling

Many kinds of wood are made into paneling. For example, a rustic or informal look can be obtained with knotty pine, white pocket Douglas fir, sound wormy chestnut, and pecky cypress. The panels can be cut plain or with a tongue and groove. Fig. 42-73. These may be finished natural or stained and varnished. Wood paneling may be used to cover one or more walls or partial walls of a room.

Only thoroughly seasoned wood paneling should be used. The moisture content should be near the average it reaches in service, about 8% in most areas. However, in the dry southwestern United States, it should be about 6%, and in the southern and coastal areas of the country, about 11%. Allow the material to reach correct moisture content by storing it in the area in which it will be installed in such a way that air may circulate around all surfaces of the boards. Wood paneling on the inside of an exterior wall should be installed over a vapor barrier and insulation.

Wood paneling should not be too wide; a nominal 8" is recommended for most parts of the country. Boards wider than 8" should not be used except when they have a long tongue or matched edge. The boards may be applied horizontally or vertically.

42–64a. *Applying adhesive to the edges of the studs or furring strips.*

42–64b. *Applying adhesive to backer board.*

42–65. *After the panel has been nailed at the top for hinge action, pull the panel out from the wall and block it in this position to allow the adhesive to partially set.*

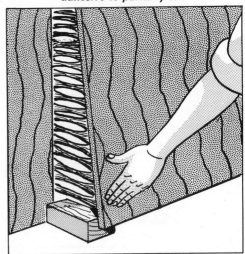

INSTALLATION OF VERTICAL WOOD PANELING

For paneling that is to be installed vertically, adequate blocking should be installed between the studs to provide nailing support. The blocking should not be more than 24" o.c. Fig. 42-74 (Page 771).

A common practice when installing wood paneling is to nail a 1" × 8" board at the floor line. The 1" × 4" baseboard is then face-nailed to the 1" × 8" board Fig. 42-75 (Page 772). The ends of the vertical paneling will rest on the top edge of the 1" × 4" base. This is a much cleaner application than resting the paneling ends on the floor and applying the base to the face of the paneling.

Plumb and scribe the first piece of paneling to the wall. Undercut the edge about 5° to insure a snug fit against the wall. Fig. 42-76. Blind-nail all paneling in place using 5d or 6d casing or finishing nails. Fig. 42-77. Continue to install the pieces of paneling, checking for plumb periodically. If necessary, adjust slightly on each added panel until the pieces are again in plumb. The tongue and groove is used for the adjustment. Fig. 42-78.

42–66. *Tapping the panel with a soft wood block to press it firmly into place. You may put a cloth under the block to further protect the paneling.*

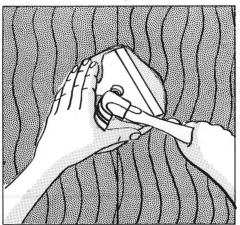

On the last piece of paneling to be installed on a wall, the edge that is to fit into the corner should be scribed and undercut at about 5° angle. The groove of the panel can then be slipped over the tongue of the preceding piece and the panel snapped into place. Fig. 42-79. Apply a cove or a crown molding at the

42–67a. *Cutting an opening for an electrical outlet box.*

42–67b. *Installing a plywood panel that has an opening cut for an outlet box.*

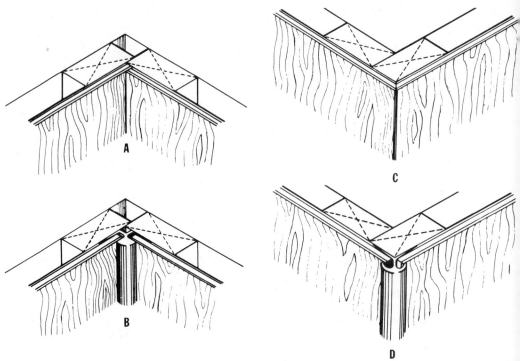

42–68a. *Corner details: A. An inside corner with the first panel butted into the corner and the second panel scribed to the face of the first. B. An inside corner trimmed with a veneer-faced aluminum molding. C. An outside corner mitered. D. An outside corner trimmed with a veneer-faced aluminum molding.*

42–68b. *Joint details between panels: A. Shallow V-joint. B. Wide joint using $1/4'' \times 2\,1/2''$ furring strips of matching or contrasting paneling. If a prefinished furring strip is not used, the strip should be finished before the panels are installed. C. Veneer-faced aluminum molding installed as a divider strip between panels.*

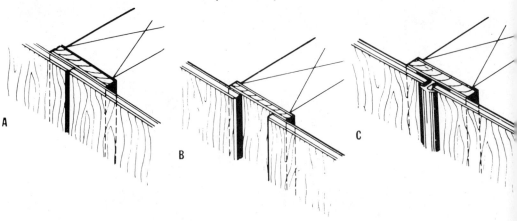

A

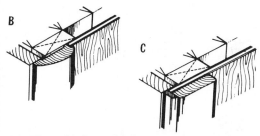

B

C

42–68c. *Window and door trimming details: A. Casing installed over paneling on furring strips. B. Rabbeted casing installed over plywood on furring strips. C. Quarter-round molding installed at the jamb to cover the joint between the plywood and the square-edge casing.*

42–68d. *Base installation details: A. This method is used for installing wainscoting when it is desirable to gain a few inches of wall height. The panel is held up off the floor, and a piece of thicker furring is attached to the wall at the floor. The base is then nailed at the top and bottom to the two furring strips. B. This method is frequently used in remodeling. The walls are furred out, and the thickness of the old base is used as the bottom furring strip. The new base is then nailed to the face of the plywood paneling. This same method may be employed for new construction by using a furring strip at the floor line.*

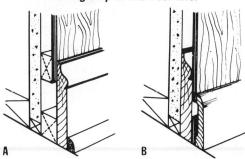

A   B

42–68e. *Plywood installation details at the ceiling: A. Cove lighting framed and covered with plywood paneling. Note the use of the veneer-faced aluminum cap and inside and outside corner moldings. B. Crown molding. C. A strip of prefinished paneling cut from leftover pieces and scribed to the ceiling with a quarter-round attached at the bottom edge. D. A strip of prefinished paneling ripped at 45° and installed at the ceiling line.*

A

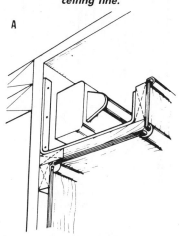

B

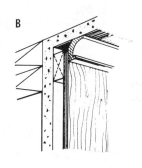

C   D

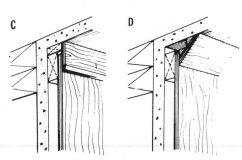

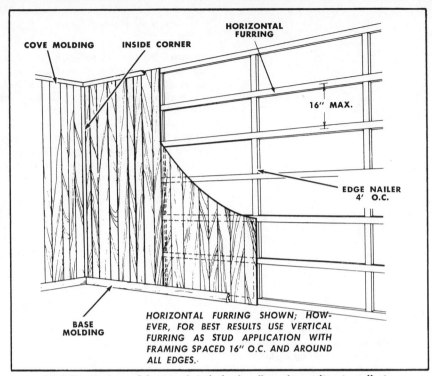

42–69. *Horizontal furring details for hardboard paneling installation.*

42–70a. *Installation details for hardboard applied with an adhesive.*

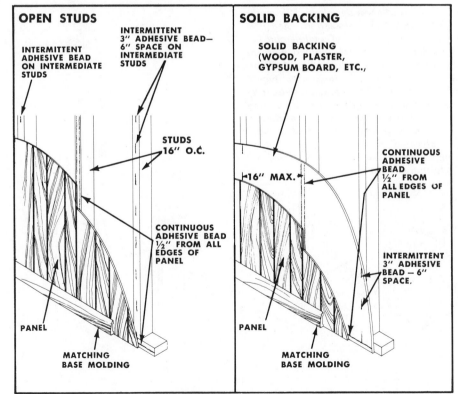

ceiling and wall intersection and a base shoe at the floor. Fig. 42-80 (Page 773). If necessary, install quarter-round trim in the corners.

INSTALLATION OF HORIZONTAL WOOD PANELING

Horizontal paneling, while not as common as vertical paneling, has some advan-

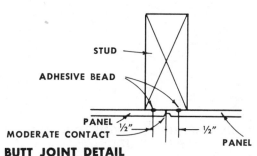

STUD

ADHESIVE BEAD

PANEL ½"
MODERATE CONTACT — ½"
PANEL

BUTT JOINT DETAIL

42–70b. *Butt joint installation detail for hardboard applied with an adhesive.*

tages. This method of application requires fewer pieces to shape and install, and it is therefore much faster to apply. It also gives the room the appearance of being longer or larger with a lower ceiling.

Apply vertical nailing strips 18" on center as shown in Fig. 42-81. Horizontal paneling may also be nailed directly to the studs. Fig. 42-82 (Page 774). Begin the paneling at the floor line, making certain that the first piece is installed level. Undercut the ends slightly to provide a tight joint at the inside corners. Outside corners should be mitered. Fig. 42-83. Blind-nail all the paneling as described for the vertical application, checking periodically to make certain the paneling remains level.

If no molding is to be used at the ceiling, undercut the last panel edge at a 5° angle to insure a snug fit against the ceiling line. When desired, apply cove or crown moldings at the ceiling and wall joints and

42–71. *Hardboard nailing details.*

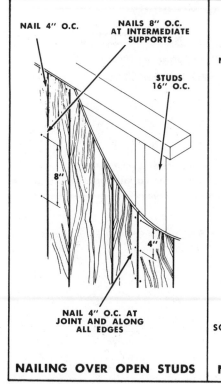

NAIL 4" O.C.

NAILS 8" O.C. AT INTERMEDIATE SUPPORTS

STUDS 16" O.C.

8"

4"

NAIL 4" O.C. AT JOINT AND ALONG ALL EDGES

NAILING OVER OPEN STUDS

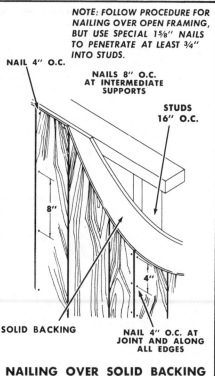

NOTE: FOLLOW PROCEDURE FOR NAILING OVER OPEN FRAMING, BUT USE SPECIAL 1⅝" NAILS TO PENETRATE AT LEAST ¾" INTO STUDS.

NAIL 4" O.C.

NAILS 8" O.C. AT INTERMEDIATE SUPPORTS

STUDS 16" O.C.

8"

4"

SOLID BACKING

NAIL 4" O.C. AT JOINT AND ALONG ALL EDGES

NAILING OVER SOLID BACKING

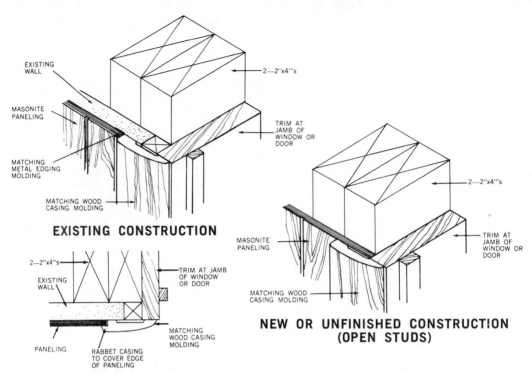

EXISTING
WALL

MASONITE
PANELING

MATCHING
METAL EDGING
MOLDING

MATCHING WOOD
CASING MOLDING

2—2"x4"'s

TRIM AT
JAMB OF
WINDOW OR
DOOR

EXISTING CONSTRUCTION

2—2"x4"'s

EXISTING
WALL

TRIM AT JAMB
OF WINDOW
OR DOOR

PANELING

RABBET CASING
TO COVER EDGE
OF PANELING

MATCHING
WOOD CASING
MOLDING

MASONITE
PANELING

2—2"x4"'s

TRIM AT
JAMB OF
WINDOW OR
DOOR

MATCHING WOOD
CASING MOLDING

NEW OR UNFINISHED CONSTRUCTION
(OPEN STUDS)

OPTIONAL DETAIL   **42–72a.** *Window and door trim details.*

TRIM DETAILS
MATCHING METAL MOLDINGS

Outside Corner   Edging   Inside Corner   Division

MATCHING VINYL-CLAD WOOD MOLDINGS

Shoe   Stop   Casing   Base   Outside Corner   Inside Corner   Cove

42–72b. *Metal and wood moldings in different styles for various applications are made to match prefinished paneling.*

42–73a. *Six popular tongue-and-groove paneling patterns. Most retail lumberyards carry two or three of these patterns in stock.*

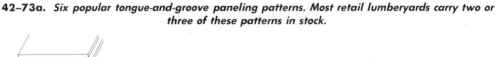

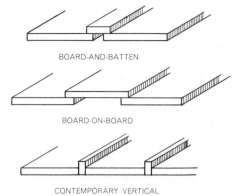

BOARD-AND-BATTEN

BOARD-ON-BOARD

CONTEMPORARY VERTICAL

42–73b. *Installation patterns for plain lumber paneling.*

42–73c. *An attractive board-on-board paneling installation.*

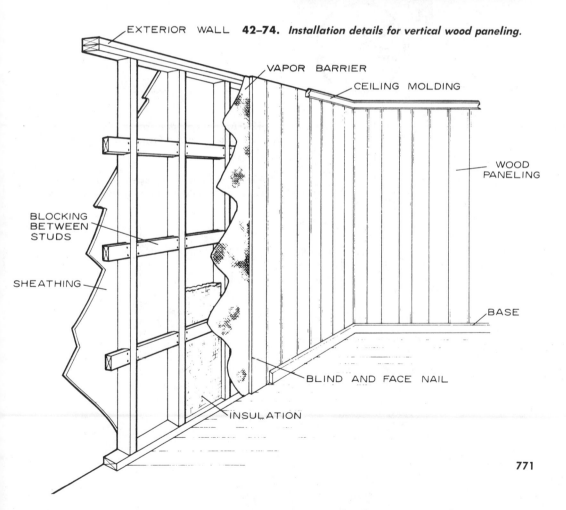

EXTERIOR WALL **42–74.** *Installation details for vertical wood paneling.*

VAPOR BARRIER

CEILING MOLDING

WOOD PANELING

BLOCKING BETWEEN STUDS

SHEATHING

BASE

BLIND AND FACE NAIL

INSULATION

771

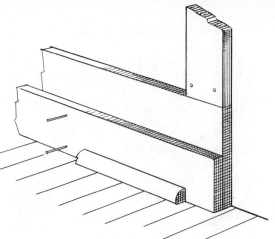

42–75. *The baseboard is nailed to a 1" × 8" furring strip at the floorline.*

apply base shoe at the floor. If necessary, quarter-round trim may be installed in the corners.

INSTALLING WOOD PANELING IN A HERRINGBONE PATTERN

The herringbone style of application is very interesting, but it is also the most

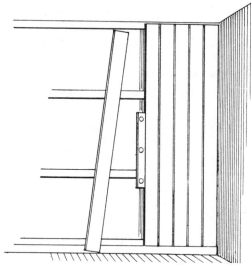

42–78. *Periodically check the paneling for plumb as the installation progresses.*

42–79. *Scribe and undercut the last piece of paneling to insure a tight fit against the adjacent wall.*

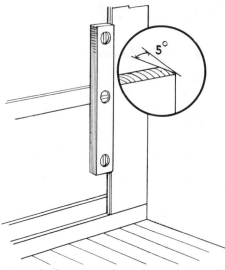

42–76. *The first piece of paneling to be installed is scribed in a plumb position to the adjacent wall and undercut about 5° to provide a tight joint in the corner.*

42–77. *Blind-nailing details for lumber paneling.*

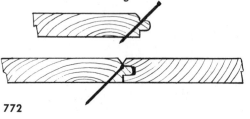

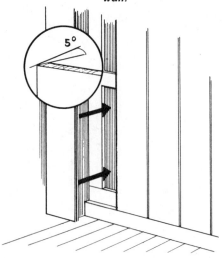

demanding in craftsmanship. Fig. 42-84. Apply vertical nailing strips so that the space between them is evenly divided. For example, if the wall is 12′ long, space the strips 18″ on center as shown in Fig. 42-85. When laying out, begin the actual measurement from the adjacent wall next to the end strip and measure to the center of the other strips. Make sure that each strip is plumb.

Install the baseboard as shown in Fig. 42-86. Then, using a long level, draw a plumb line at the center of every other nailing strip as shown at a, b, and c in Fig. 42-85. For a 12′ wall, these lines should be as close as possible to 36″ apart. Saw two pieces of paneling in the shape of a triangle with the tongue on the long edge, as shown in Fig. 42-87. Install them with the vertical joint on one of the plumb lines. Extreme care should be taken to make the paneling butt even with the baseboard. A molding strip will be applied later at the vertical joint.

In the same manner, measure and cut to length the second pieces to be installed. Install each paneling piece all the way across the wall, building toward the top. Use the play in the tongue and groove to keep the panels aligned at the "V" (along the vertical joint). After all the pieces have been applied, install the cove or crown molding at the ceiling. Next apply the molding at the butt joint of the paneling.

This molding should stop at the baseboard and butt into the cove or crown molding at the ceiling. Fig. 42-88. A cove or quarter-round may be used at the corners and a base shoe at the floor if necessary.

Bathroom Wall Covering

When a complete, prefabricated shower or combination shower and tub stall is installed in the bathroom instead of a tub, special wall finishes are not required. Fig. 42-89 (Page 776). When tubs are used, however, some type of waterproof wall covering is normally required around it to protect the wall. There are several types of finish, including coated hardboard paneling and various ceramic, plastic, and similar tiles. Fig. 42-90.

Plastic-surfaced hardboard materials are applied in sheet form and are fastened with an adhesive or nails as described under "Installing Gypsum Board" or "Hardboard" earlier in this unit. The method of application depends on the nature of the material. Moldings are placed on inside corners, on tub edges at the joints, and as end caps. Fig. 42-91. Several types of calking sealants are also available which will provide excellent results.

Ceramic, plastic, and metal tile is installed over water-resistant gypsum board. The adhesive is spread with a serrated trowel, and the tiles are pressed into place.

42-80. *Apply cove or crown molding at the intersection of wall and ceiling and a base shoe at the floor.*

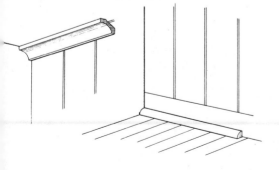

42-81. *When applying wood paneling horizontally, apply the nailing strips vertically 18″ on center. Begin the measurement at the wall on the end strips.*

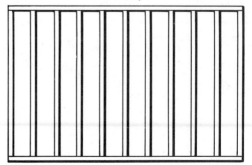

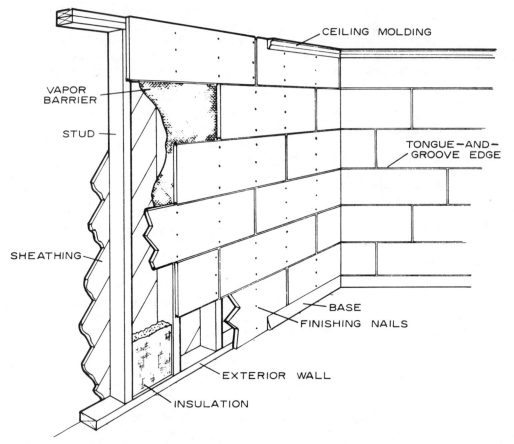

CEILING MOLDING

VAPOR BARRIER

STUD

TONGUE—AND— GROOVE EDGE

SHEATHING

BASE

FINISHING NAILS

EXTERIOR WALL

INSULATION

42–82. *Wood paneling applied horizontally to studs. Face-nailing is used because of the wide paneling stock.*

42–83. *Wood paneling applied horizontally is mitered at outside corners. In this installation, instead of a baseboard, a "reveal" is shown.*

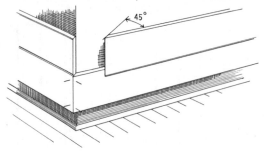

45°

42–84. *Wood paneling applied in a herringbone style.*

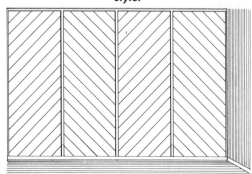

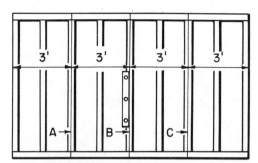

42–85. Locating the furring strips on a 12' wall for a herringbone application.

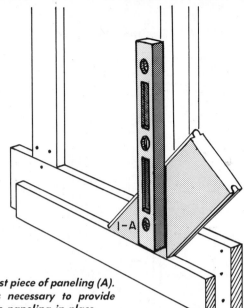

42–86. Laying out the first piece of paneling (A). Baseboard installation is necessary to provide backing for nailing the paneling in place.

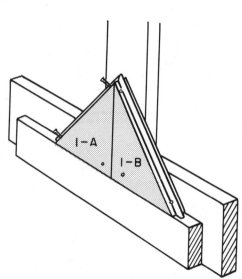

42–87. Lay out and cut the second piece (1B) before nailing the first piece (1A) in place.

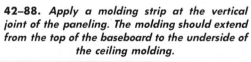

42–88. Apply a molding strip at the vertical joint of the paneling. The molding should extend from the top of the baseboard to the underside of the ceiling molding.

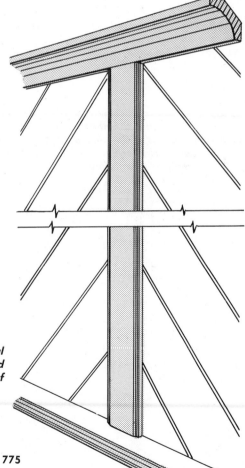

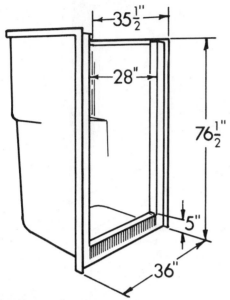

42–89a. *A shower cove bathing module made from plastic reinforced with fiber glass. Bathing modules are available in white and in several colors.*

42–89b. *A bath and shower bathing module. This module includes a bathtub and walled alcove. The tub has a flat rim for seating or to accommodate a shower door. The unit is plastic reinforced with fiber glass.*

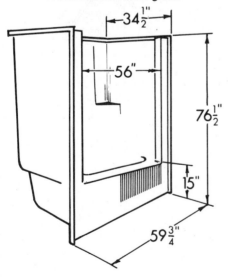

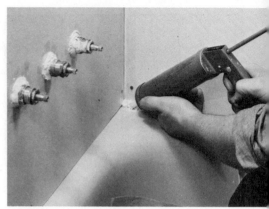

42–90. *Moisture-resistant gypsum wallboard installed in a tub alcove. Corners and fittings must be calked.*

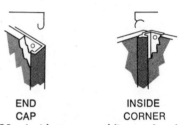

END
CAP

INSIDE
CORNER

42–91. *Inside corner molding and end cap.*

A grout cement is inserted in the joints of the tile after the adhesive has set. The plastic, metal, or ceramic type of wall covering around the tub area is usually installed by subcontractors specializing in this craft.

ESTIMATING
Determining the Area to Be Covered

The area of the walls and ceiling will need to be known in order to estimate labor costs and to determine material needs for gypsum lath, plaster, wood paneling, and some wall finishes. To calculate the area to be covered, multiply the perimeter of the room by the ceiling height and subtract all openings such as doors, windows, and fireplaces. To calculate the ceiling area, multiply the length of the room by its width.

42–92. Wall areas can be determined by using this chart. The left-hand column shows ceiling heights. The top row represents wall lengths. For example, a wall 8' high and 16' long has an area of 128 sq. ft.

Wall Areas

Ceiling Height	6	8	10	12	14	16	18	20	22	24	26	28	30
7'6"	45	60	75	90	105	120	135	150	165	180	195	210	225
8'0"	48	64	80	96	112	128	144	160	176	192	208	224	240
8'6"	51	68	85	102	119	136	153	170	187	204	221	238	255
9'0"	54	72	90	108	126	144	162	180	198	216	234	252	270
9'6"	57	76	95	114	133	152	171	190	209	228	247	266	285
10'0"	60	80	100	120	140	160	180	200	220	240	260	280	300

Note: Square Feet of Area per Wall Length.

The chart in Fig. 42-92 provides a quick means of figuring wall area. This chart tells the square feet of area per wall length. To find the total area to be covered, add up the areas of all the walls. An additional chart of wall and ceiling areas is found in Fig. 48-34, p.966.

Determining the Amount of Sheet Paneling

Figure the perimeter of the room to be paneled. Convert the perimeter into the number of panels needed by using the conversion table in Fig. 42-93. For example, the room in Fig. 42-94 has a perimeter of 66'. The conversion table indicates that 17 panels are required. Should the perimeter fall between figures, use the next higher number.

If the ceiling height is more than 8' (the standard length of a panel), determine the additional height. For example, if the room in Fig. 42-94 has a 10' ceiling, 2' of additional height are required. Use 10' panels if they are available, or cut 2' pieces from an 8' panel. Four 2' pieces can be cut from each 8' panel. Since 17 panels are required to go around the room, 4¼ (or, rounded off, 5) additional panels will be required. (17 ÷ 4 = 4¼.) This makes a total of 22 panels.

Deduct for areas such as doors, windows, and fireplaces. For estimating purposes, deduct ⅔ of a panel for a door and

42–93. Find the perimeter of the room in the left-hand column. Follow across to the right-hand column headed "Panels Needed" to determine the number of panels for a room with a given perimeter. For example, a room with a perimeter of 60' would require 15 panels.

Conversion Table	
Perimeter	**Panels needed**
20'	5
24'	6
28'	7
32'	8
60'	15
64'	16
66'	17
68'	17
72'	18
92'	23

½ for a window or fireplace. Subtract the total deductions from the number of panels originally determined and round off the remainder to the next higher number. Our example for an 8' ceiling showed 17 panels minus 2 ⅚ panels for doors, windows, and a fireplace, or a net of 14 ⅙ panels. Fig. 42-96. Therefore it will be necessary to order 15 panels. For a room with a 10' ceiling, an additional 4¼ panels are necessary. A total of 18 5/12 panels are needed. (14 ⅙ + 4 ¼ = 18 5/12.) Round off this number to 19. This same system can be used for estimating all standard 4' wide wall covering materials.

Estimating Wood Paneling

To determine the amount of wood paneling required to panel a room, first figure the wall area to be covered. Using the room in Fig. 42-94 as an example, with a perimeter of 66' and a ceiling height of 8', the wall area is 528 sq. ft. Subtract the window, door, and fireplace areas. Assuming that these areas total 112 sq. ft., subtract this from the wall area. A total of 416 sq. ft. is to be covered by wood paneling $(528 - 112 = 416)$.

If tongue-and-groove paneling is to be installed, multiply the total area to be covered by the area factor. Fig. 42-95. Add an allowance for trim and waste, usually 5%. The area factor for tongue-and-groove 1" × 8" paneling is 1.16. The room in our example contains 416 sq. ft. Multiplying 416 × 1.16 equals 482.56, or 483 board feet of paneling. To this is added a trim and waste factor of 5%, or 24 more board feet. (483 × 0.05 = 24.15, or 24). The total amount required is then 507 board feet (483 + 24 = 507). Multiply this figure by the cost per

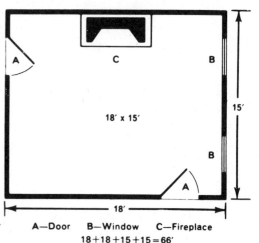

42–94. *The perimeter of a room is calculated by adding the lengths of the four walls together.*

A—Door   B—Window   C—Fireplace

18+18+15+15 = 66'

42–95. *For most installations, an allowance of 5% will be adequate for trim and waste. Sometimes, rather than add 5%, the area of the doors and windows is not subtracted but is used as a trim and waste allowance.*

Coverage Estimator

		Nominal Size	Width Dress	Width Face	Area Factor*			Nominal Size	Width Dress	Width Face	Area Factor*
Shiplap		1 x 6	5⁷/₁₆	4¹⁵/₁₆	1.22			1 x 6	5⁷/₁₆	5¹/₁₆	1.19
		1 x 8	7⅛	6⅝	1.21	**Paneling**		1 x 8	7⅛	6¾	1.19
		1 x 10	9⅛	8⅝	1.16	**Patterns**		1 x 10	9⅛	8¾	1.14
		1 x 12	11⅛	10⅝	1.13			1 x 12	11⅛	10¾	1.12
		1 x 4	3⁷/₁₆	3³/₁₆	1.26			1 x 4	3½	3½	1.60
Tongue		1 x 6	5⁷/₁₆	5³/₁₆	1.16			1 x 6	5½	5½	1.33
And		1 x 8	7⅛	6⅞	1.16	**Bevel**		1 x 8	7¼	7¼	1.28
Groove		1 x 10	9⅛	8⅞	1.13	**Siding**		1 x 10	9¼	9¼	1.21
		1 x 12	11⅛	10⅞	1.10			1 x 12	11¼	11¼	1.17
		1 x 4	3½	3½	1.14						
		1 x 6	5½	5½	1.09						
S4S		1 x 8	7¼	7¼	1.10						
		1 x 10	9¼	9¼	1.08						
		1 x 12	11¼	11¼	1.07						

*Allowance for trim and waste should be added.

42–96. *Standard wallboard sizes and the amount of adhesive or nails required for installation per 100 square feet.*

WALL BOARDS		Material		Adhesive or Nails
Material	Size	Fastened by		Per 100 Sq. Ft.
Gypsum Board..........	48″ x 96″	Nailing to Studs		5 Pounds
Plank T&G Board	8″ to 12″ x 96″	Nailing to Studs		2 Pounds
Tempered Tileboard	48″ x 48″	Nailing to Studs		1 Pound
Tempered Tileboard	48″ x 48″	Adhesive to Walls		1.5 Gallons
Plywood Panels	48″ x 96″	Nailing to Studs or Wall		1.25 Pounds
Rock Lath	16″ x 48″	Nailing to Studs		5 Pounds
Perforated Hardboard ...	48″ x 96″	Nailing to Studs		4 Pounds

board foot to determine the total cost of the paneling to be installed.

Estimating Lath and Plaster

Gypsum lath is packaged in bundles of eight 24″ × 48″ pieces. A standard lath bundle therefore contains 64 square feet. To determine the number of bundles required, divide 64 into the total area to be covered. For example, suppose that the wall and ceiling area of the room in Fig. 42-94 is to be finished with lath and plaster. The total wall and ceiling area equals 686 square feet (416 + 270 = 686). Divide the number of square feet to be covered by the number of square feet in a bundle of gypsum lath (64) for a total of 10.7, or 11 bundles of gypsum lath.

A plasterer calculates the cost of a job by the number of square yards to be covered. Convert the square feet in the room to square yards by dividing by 9 (1 square yard equals 9 square feet). In our example, 686 ÷ 9 equals approximately 76 square yards.

Estimating Adhesive and Nails

The amount of adhesive or nails required to install various wall finishes can be determined by using the chart in Fig. 42-96. To figure the amount of nails required for installing the gypsum lath in the previous example, refer to the information on rock lath. Five pounds are required to install 100 square feet of rock lath. Since approximately 700 square feet (686 actual square feet) of rock lath are to be installed, 35 pounds of nails are required.

Labor

The labor for the application of wall and ceiling finishes can be determined by referring to the chart in Fig. 42-97. For example, if wood paneling is to be installed in a

42–97. *Labor estimates for installing wall and ceiling finishes.*

Estimating Labor

Wall Finishes	
Wood paneling	
4′ high—7′ high	2.2 hrs. per 100 sq. ft.
8′ high	2.5 hrs. per 100 sq. ft.
Gypsum wallboard	
with joint system	2.5 hrs. per 100 sq. ft.
wood grained	2.0 hrs. per 100 sq. ft.
Insulating plank	2.5 hrs. per 100 sq. ft.
Insulating wallboard	2.0 hrs. per 100 sq. ft.
Standard hardboard	2.0 hrs. per 100 sq. ft.
Asbestos wallboard	2.0 hrs. per 100 sq. ft.
⅛″ perf. hardboard	2.0 hrs. per 100 sq. ft.
¼″ plywood	2.0 hrs. per 100 sq. ft.
Ceiling Finishes	
Gypsum wallboard	
with joint system	3.0 hrs. per 100 sq. ft.
Ceiling tile,	
12′x12′—16′x32′	2.5 hrs. per 100 sq. ft.

room with an 8′ ceiling, the chart indicates that 2½ hours are required to install 100 square feet. In our wood paneling example, 416 square feet of wall space was to be covered. To calculate the number of hours required to install this material, divide 416 by 100 to determine how many hundreds of square feet are to be covered. Multiply by 2.5 (hours required per 100 square feet). The product is 10.4 hours (4.16 × 2.5 = 10.4). Multiply this by the cost of labor per hour to determine the total cost for the installation of the wood paneling.

QUESTIONS

1. List several materials that are commonly used for interior wall and ceiling finish.

2. What materials are most frequently used as a plaster base?

3. Why is plaster reinforced at certain key positions?

4. What material is used for plaster reinforcing?

5. What are plaster grounds?

6. What are the ingredients of plaster?

7. What two plaster coats are combined when they are applied on gypsum or insulating lath?

8. When plastering, what is the difference between a sand-float and a putty finish?

9. What materials are classified as dry-wall finish?

10. When applying gypsum board dry wall, what is meant by double nailing?

11. Describe how gypsum board is held in place on a ceiling during application.

12. When plywood is used as a wainscot, why is it usually about 32″ high?

13. Describe how an outlet box cutout is located on a plywood panel.

14. What is the recommended nominal width for wood paneling used in most parts of the country?

15. Which method of wood paneling installation will give a room the appearance of being longer or larger and having a lower ceiling?

16. List some waterproof materials that are commonly used for a bathroom wall covering.

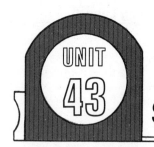

Ceiling Tile and Suspended Ceilings

UNIT 43

In addition to lath and plaster and the sheet materials discussed in Unit 42, insulating board or ceiling tile makes a good ceiling. Fig. 43-1. In most cases, this type of ceiling finish has excellent acoustic and insulating qualities. It is also fire retardant. Ceiling tile may be installed in several ways, depending on the type of ceiling or roof construction. When a flat surface is present, such as between beams of a beamed ceiling in a low sloped roof, the tiles are fastened with adhesive. When the tile is edgematched, it can be stapled in place. Another method is to apply a suspended ceiling with small metal or wood hangers which form supports for drop-in

43-1a. *A suspended ceiling system with 2′ × 4′ drop-in panels. The embossed swirl design gives the effect of a notch-troweled plaster ceiling.*

43-1b. *An acoustic tile ceiling with a vinyl-acrylic finish that is resistant to staining. This tile absorbs over 60% of all the sound waves striking its surface.*

panels. The most common method of installing ceiling tile is with 1″ × 3″ wood furring strips nailed across the ceiling joists or roof trusses. Fig. 43-2a.

CEILING TILE

Standard ceiling tiles and acoustical tiles are fiberboard products made from natural wood or cane fibers. They are designed for decoration and sound insulation and are used in new construction as well as remodeling. The tiles are factory predecorated, requiring no painting or other finishing. Many designs, colors, and patterns are available in either the standard or acoustical type. Surface characteristics vary from smooth to various textured and sculptured effects. In acoustical tiles, surface openings provide for sound absorption. These openings may be holes drilled or punched in various patterns, or they may be slots, striations, or fissures.

The most popular sizes of ceiling tile are 12″ × 12″ and 12″ × 24″. The 12″ × 24″ size is available with or without center scoring to represent two tiles. Ceiling tiles come in a nominal ¹/₂″ thickness with an interlocking tongue-and-groove joint which provides for self-leveling and concealed attachment. Some tiles are available with beveled or curved edges. Fig. 43-2b. They

also come in 16″ × 16″ and 16″ × 32″ sizes in nominal thicknesses from ¹/₂″ to ³/₄″.

Ceiling tile may be cemented to gypsum wallboard if the wallboard is at least ³/₈″ thick and nailed on not more than 16″ centers. Ceiling tile may also be applied directly to a plaster ceiling, provided the plaster is solid and level. If the existing surface in either case does not meet these minimum requirements, furring strips should be installed to keep the new ceiling level and prevent future trouble.

Determining Room Layout

The following steps are taken for figuring the ceiling tile layout of a rectangular room. If an L-shaped room is to be tiled, divide the room into two rectangles or squares and figure accordingly.

A ceiling has a better appearance if border tiles (those adjacent to the walls) are the same width on opposite sides of the room. To find the border tile width for the long walls of a room, follow these steps:

1. Measure one of the short walls in the room.

2. If this measurement is not an exact number of feet, add 12″ to the inches left over.

3. Divide this total by two.

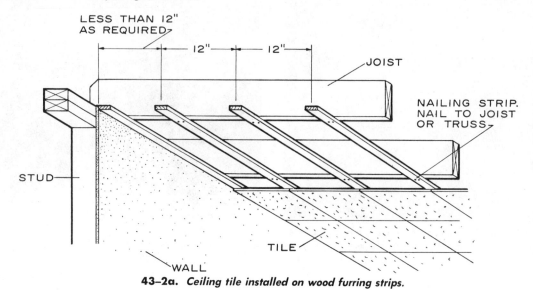

43-2a. *Ceiling tile installed on wood furring strips.*

This will give you the width of the border tile for the long walls.
Example:

    Short wall = 10' 8"
    Extra inches: 8"
    Add: 12"
    Divide: 2)20"
    Border tile for long wall: 10"

To figure border tile for the short walls, follow the same procedure:

1. Measure one of the long walls in the room.

2. If this measurement is not an exact number of feet, add 12" to the inches left over.

43-2b. *Ceiling tile joint details.*

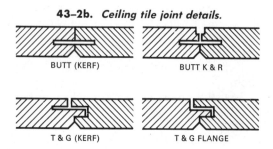

BUTT (KERF)   BUTT K & R

T & G (KERF)   T & G FLANGE

3. Divide this total by two.
This will give you the width of the border tile for the short walls.
Example:

    Long wall = 12' 4"
    Extra inches: 4"
    Add: 12"
    Divide: 2)16"
    Border tile for short wall: 8"

When using 16" × 16" tile, convert the wall measurement into inches. Divide this measurement by 16. Treat the extra inches the same way as with the 12" tile in the example above, but add 16" (instead of 12") and divide by 2.

Installing Furring Strips

Furring strips can be nailed directly to wood joists or through an existing ceiling into the joists to provide a solid base for stapling the tiles. The first two furring strips must be carefully placed so that the border tiles are properly aligned. Place the first furring strip flush against the wall at right angles to the ceiling joists. Nail the strip into place with two 8d nails at each joist location. Fig. 43-3.

43-3. *The first furring strip is nailed flush against a wall at right angles to the joists. Use two 8d nails at each joist crossing.*

43-4. *The position of the second furring strip depends upon the width of the border tile to be used. All strips thereafter are installed 12" on center.*

If the joists are concealed by an existing ceiling, locate and mark the position of each joist before nailing the furring strips in place. Generally, joists are spaced 16" on center and are perpendicular to the long wall. If you are unable to determine the direction of the joist by tapping on the existing ceiling, check the floor above the joists. If the finished floor is wood, it will be nailed across the joists.

The second furring strip should be placed parallel to the first at the border width distance from the wall. In the example, the border width measurement is 10". Add 1/2" for the stapling flange, which overlaps the finished bevel edge of the tile. Thus the second furring strip should be placed so that the center is 10½" from the wall.

After installing the second strip, work across the ceiling, nailing the rest of the furring strips in place 12" on center from each other. Fig. 43-4. Because the border tile size was determined earlier, the next to last strip will be positioned automatically—in this case 10½" from the wall. The last strip should be nailed flush against the wall in the same manner as the first strip. Fig. 43-5.

If using 16" × 16" tile, nail the first strip flush against the wall. Nail the second strip at the appropriate border width distance from the wall and then nail remaining strips 16" on center.

Make certain that the furring strips are level by checking them with a straightedge. Correct unevenness by driving thin wood shims between the strips and the joists. Fig. 43-6.

Installing Special Furring Strips

If pipes or wire cables are located below ceiling joists or project less than 1½" below the existing ceiling structure, install double furring strips over the entire ceiling to avoid the projections. The first course should be spaced 24" to 32" on center and fastened directly to the joists. The second course is then applied perpendicular to the first, appropriately spaced to receive the ceiling tile. Fig. 43-7.

Pipes or ducts that project more than 1½" below the ceiling joists should be boxed in with furring strips before the ceiling is installed. After the ceiling tile has been installed, wood molding can be used to provide a finish detail for inside and outside edges.

Ceiling Layout

Construct two reference lines before beginning the tile application. These will

783

align the first rows of tile installed in both directions. To make certain these lines are accurate, follow these steps:

1. Partially drive a nail into both ends of the second furring strip, using the appro-priate border tile measurement. In the ex-ample, the border tile measurement from the long wall is 10″. Thus the nails should be positioned at each end of the second furring strip, 10½″ from the long wall. The

43–5. The layout for ceiling tile on a 10′ 8″ × 12′ 4″ ceiling.

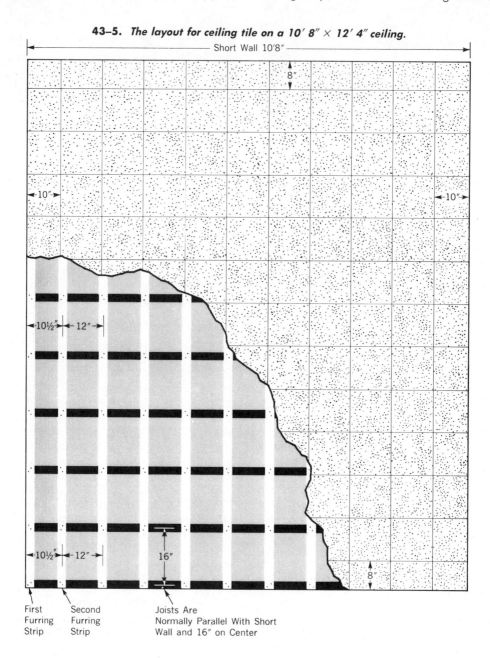

784

stapling flange of the tile, when installed, will line up with the exact center of the furring strip. Fig. 43-8.

2. Stretch a chalk line tightly between the two nails and snap it.

To be assured of proper tile alignment, the second reference line must be constructed exactly at a right angle to the first. This is done by constructing a 3' × 4' × 5' right triangle, using the first reference line as a base.

1. Locate point A on the first reference line using the short wall border tile measurement. Fig. 43-8.

2. From point A measure in exactly 3' along the reference line to locate point B.

3. Drive small nails into the furring strips at points A and B.

4. Starting with point A measure off exactly 4' across the furring strips. Mark a small arc on the furring strip at this point.

5. From point B measure exactly 5' toward the first arc, marking the point of intersection as point C.

6. Snap a chalk line through points A and C across all furring strips. This second reference line is perpendicular to the first. Fig. 43-9.

Cutting Tile

Measure and cut each of the border tiles individually. Remove the tongue edge and leave the wide flanges for stapling. *Include the face and the flange of each tile in your measurement.* Cut the tile face up with a coping saw or a very sharp fiberboard knife.

Cut the first tile so that it fits into the corner. In the example, the first tile would be cut 8½" by 10½". Fig. 43-10. After aligning the flanges with reference lines AB and AD, staple this tile into position. Fig. 43-11.

Cut a second tile so that one of the tongue edges fits into the corner tile and the stapling flange falls directly on line AB. Fig. 43-11

Cut a third tile so that the tongue edge will fit into the corner tile and the stapling

43–6. Level the furring strips with pieces of wood shingle. Check the strips with a carpenter's level or pull a line taut from end to end.

43–7. When cables or pipes are mounted below the joists, install a double layer of furring strips over the entire ceiling to eliminate the projection. The first layer of strips is nailed on about 24" to 32" centers parallel to the pipes.

flange will fall directly on line AC. Fig. 43-11.

Work across the ceiling, installing about two tiles at a time along the borders and filling in between with full-sized tile. When you reach the opposite wall, measure each border tile individually. Fasten the tile by stapling into remaining flange and face-nailing into furring along the wall. Finish off the ceiling with a crown or cove molding.

43–8. *A furring strip layout for a ceiling measuring 10' 8" × 12' 4".*

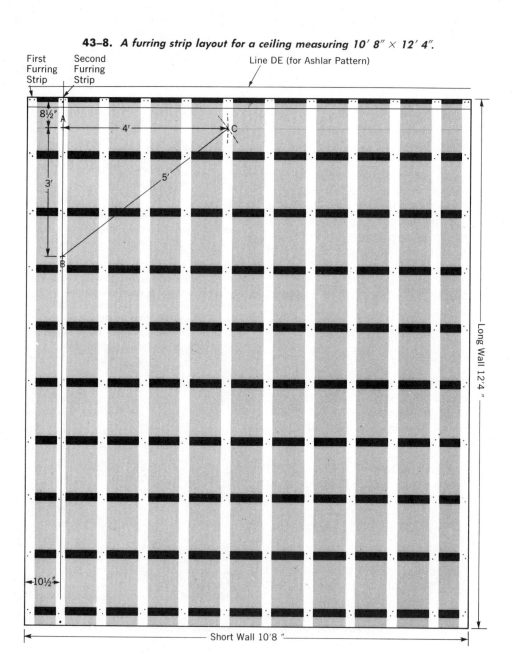

Fastening Tile

STAPLING

Using $^{9}/_{16}''$ staples, fasten each tile in place with three staples in the flange parallel to the reference line AB and one staple in the flange on the furring strip closest to the wall. Fig. 43-11 and 43-12. If you are using $16'' \times 16''$ tile, put four staples in the flange parallel to the reference line. Fig. 43-13.

When sliding the tiles into position, be sure they join snugly but *do not force them tightly together*. Fig. 43-14.

CEMENTING

Brush-on ceiling cement can be used only when installing tiles that have a tongue-and-groove joint detail. Acoustic cement should be used for butt-edge tile.

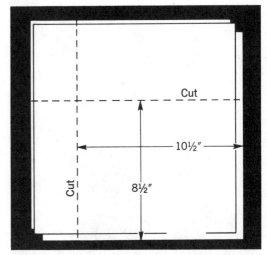

43–10. *Laying out the first tile so that it fits into the corner. Be sure to leave a flange (tongue). The tile will be nailed or stapled in place through the flange.*

43–9. *Snap a chalk line across the furring strips to indicate the location of the short wall border tile.*

43–11. *Installation details for the first border tile along each wall.*

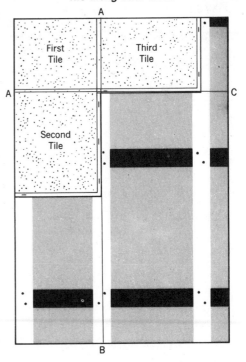

43-12. Stapling ceiling tile. When finishing the border on the opposite side of the room, remove the stapling flanges from the tiles so that the tiles will fit against the walls. These tiles are face-nailed into the furring along the wall.

43-13. Stapling details: A. Use four staples to fasten each 12″ × 12″ tile in place; three staples in one flange, one in the other. B. Use five staples for each 16″ × 16″ tile; put four staples in one flange and one in the other. C. For 12″ × 24″ tile use six staples for each tile; one in the 12″ flange, five in the 24″ flange.

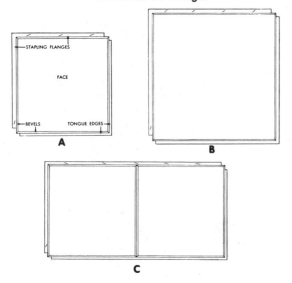

Before starting to cement, plan the room layout. Then lay out the ceiling and cut the tile as described earlier. The reference lines can be snapped directly on the existing ceiling surface.

For 12″ × 12″ tile, put five daubs of the cement on the back of each tile with a paintbrush. Place one daub on each corner of the tile and one in the center. Each spot of cement should be about $1\frac{1}{2}$″ across and at least the thickness of a nickel. Keep the daubs of cement away from the edges of the tile to allow them room to spread when the tile is installed. Fig. 43-15.

Slip the first border tile into position and press it tightly against the ceiling. Fasten it in place with two $\frac{3}{8}$″ or $\frac{5}{16}$″ staples in each flange. The staples are needed to hold the tile in place long enough for the cement to set up.

Follow a similar procedure for 16″ × 16″ tile, but use nine daubs of cement. Place these daubs evenly over the back of the tile. Put three staples in each flange of the 16″ × 16″ tile. Fig. 43-15.

The Ashlar Pattern

The ashlar pattern of installation creates an interesting design effect by staggering the tile in "brickwork" fashion. Fig. 43-16. To install a ceiling in the ashlar pattern, proceed as follows:

43-14. The tongue-and-groove joint is designed to give each tile room to expand and contract as the indoor climate changes. When sliding tiles into position, be sure to join them snugly, but do not force them tightly together.

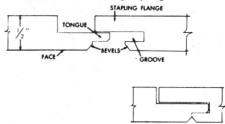

1. Make chalk lines for the long and short walls as described earlier. After making a chalk line for the short wall (this is the AC line), snap a third chalk line (DE) 6″ closer to the wall. Fig. 43-8.

2. Cut the first tile to fit into the corner where the first and second chalk lines cross. One of the stapling flanges should fall directly along the AB line and the other along the AC line. Staple into position.

3. Remove one of the tongue edges from the second tile so that the cut side will fit flush against the long wall and one of the stapling flanges falls on the AB line. The tile's remaining tongue edge should fit into the groove of the corner tile. Fig. 43-16.

4. Cut a third tile so that it will fit flush against the short wall. One of the stapling flanges should fall directly on line DE, and its remaining tongue edge should fit into the corner tile. Fig. 43-16.

5. Work across the ceiling, installing about two tiles at a time along the borders

as indicated in Fig. 43-16. Fill in between them with full-sized tiles.

6. When you reach the opposite wall, measure each border tile individually. Fasten the tile in place by stapling into the remaining flange.

7. Finish off the ceiling with a crown or cove molding.

Ceiling Light Fixtures for Use with Ceiling Tile

Several styles of ceiling lights are made to go with ceiling tile. Fixtures can be standard, chandelier, or fluorescent. Fig. 43-17. Also available is a lighting fixture

43–15. *Adhesive application details: A. Put five daubs of cement on a 12″ × 12″ tile. B. Put nine daubs of cement on a 16″ × 16″ tile.*

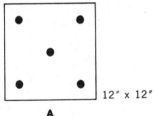

12″ x 12″

A

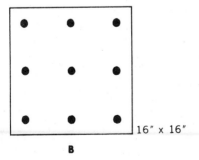

16″ x 16″

B

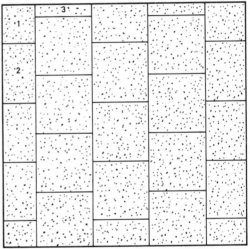

43–16. *Ceiling tile installed using the ashlar pattern.*

43–17. *A fluorescent lighting fixture installed directly under the ceiling tile.*

43-18. *Installing a recessed fixture in place of a 12″ × 12″ tile.*

43-19. *The recessed fixture installed, complete with lens frame.*

measuring 12″ × 12″, the same size as a standard ceiling tile. It can be easily inserted in place of a tile during the ceiling installation. An adapter plate is attached to the furring strips with four wood screws. Fig. 43-18. This adapter plate fits into the tongue and groove of the tiles that surround it and serves as a base for the fixture's other parts. After the junction box is installed and wired, the reflector dome is snapped into place. The lens frame is then attached to the adapter plate and pushed into place. Fig. 43-19.

SUSPENDED CEILINGS

Suspended ceilings consist of panels held in place by a grid system at a desired distance from the existing ceiling structure. The panels are made of fiber glass or plastic and are 24″ × 48″ or smaller. The grid system which supports these panels includes main runners, cross tees, and wall molding. Main runners are usually 12′ long and are spaced 2′ or 4′ on center. Cross tees are installed at right angles to the main runners. There are many types of suspended ceiling grid components. Two kinds are shown in Fig. 43-20.

The suspended ceiling reduces noise in two ways. It absorbs a large amount of the noise striking its surface. Also, because of its suspension, it does not transmit the sound vibrations into the framing above as readily as materials applied directly to framing.

A suspended ceiling is easily and quickly installed. It conveniently covers up bare joists, exposed pipes, and wiring, and it may be used to lower a high ceiling. Accessibility to unsightly valves, switches, and controls hidden by the suspended ceiling is no problem because the panel in question can merely be slid to one side.

The first step when installing a suspended ceiling is to determine the ceiling height. Sometimes, such as for a basement ceiling, it is desirable that the ceiling be as high as possible to provide maximum headroom. Care should be taken, however, to keep the top edges of the grid system at least 2″ to 2½″ below the bottom of the framing. This space is necessary for the insertion of the ceiling panels after the grid system is in place.

Mark or snap a level chalk line on each wall of the room, ¾″ (the width of the wall molding) above the intended height. This permits the wall molding to be installed below the chalk line and eliminates any undesirable marks on the wall below the ceiling level after installation is completed. To insure a level ceiling, check these lines carefully with a carpenter's level.

Slide Lock

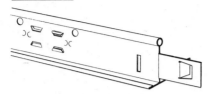

A. Main runner with
   splicer attached

B. Cross tees—2-foot—4-foot

C. Wall molding

Custom Grid

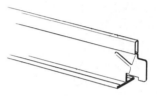

A. Main runner with
   splicer attached

B. Cross tees—2-foot—4-foot

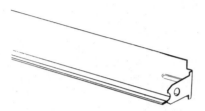

C. Wall molding

43–20. *Two of the many types of suspended ceiling grid components.*

Determining Room Layout

First determine the direction of the main runners. In most cases, the main runners should be installed perpendicular to the ceiling joists and parallel to the long wall.

Determine where the first main runner will be placed. To locate the distance of the first main runner from the long wall, proceed as follows:

1. Measure the length of one of the short walls and convert to inches.

2. Divide this figure by 48″.

3. Take the number of inches left over (if any), and add 48″ to it.

4. Divide this figure by 2. You now have the distance that the first main runner should be placed from the long wall. This also will be the size of the long wall border panel.
Example:

Short wall = 10′4″ = 124″
Divide: 48)124 = 2, remainder 28
Add: 28″ + 48″ = 76″
Divide: 2)76 = 38″, The distance of
                     main runner and
                     long wall border
                     panel size

NOTE: When using 24″ × 24″ ceiling panels, divide by 24″ and add 24″ instead of 48″.

Measure out this exact distance (in the example, 38″) on both short walls from the long wall. Fasten reference string A at the ceiling height line and stretch it between the two points. The first main runner will eventually be placed along this string. Fig. 43-21.

To determine the short wall border panel size and the location of cross tees, follow these steps:

1. Measure the length of one of the long walls and convert to inches.

2. Divide this figure by 24″.

3. Take the number of inches left over (if any), and add 24″ to it.

4. Divide this figure by 2.

Example:

Long wall = 18′8″ = 224″
Divide: $24\overline{)224}$ = 9, remainder 8
Add: 8″ + 24″ = 32″
Divide: $2\overline{)32}$ = 16″ Distance of first cross tee and short wall border panel size

Measure out this exact distance (in the example, 16″) on both long walls from the short walls. Fasten a second reference string, B, at the ceiling height line between these two points. Fig. 43-21. The first row of cross tees will be installed in line with this string.

Since walls are seldom perfectly straight, it is imperative that the second reference string be perpendicular to the first. When stretching this string, make certain that it is exactly at a 90° angle.

Installing Wall Molding

Fasten the metal molding to the walls, making certain that the top of the molding is in line with the level chalk mark. Fig. 43-22a. If the molding cannot be nailed directly to the wall, hang a suspended main runner in place of the regular wall molding. For inside corners, lap one piece of molding over the other. Outside corners are formed by mitering the two wall moldings or by overlapping. Fig. 43-22b. If the molding is to be nailed to a cinderblock wall, use concrete stud nails. Drive the nail between the mortar joint and the edge of the cinder block.

Installing Main Runners

To insure that the short wall border panels are of equal size, the main runners must be accurately cut. To cut a main runner, follow these steps.

1. Subtract the short wall border measurement from 24″. (In the example this measurement was 16″; 16″ from 24″ equals 8″.)

2. Add 6″ to remaining inches (6″ + 8″ = 14″).

3. Cut the main runner by this amount. (In the example, the main runner is cut 14″ from the end.) A, Fig. 43-23. NOTE: Be sure that a cross tee tab falls directly above reference string B. Figs. 43-21 and 43-23.

The main runners are suspended from the joists by hanger wires. To find the location of the first hanger wire, rest the cut end of the runner on the wall molding and directly above reference string A. See B, Fig. 43-23. Directly above any hole near the uncut end of the main runner fasten the first hanger wire to the existing ceiling structure. Run the wire through the hole in the main runner, but do not attach it permanently. Install the other hanger wires for this runner at 4′ intervals.

Attach wires to the ceiling structure at each main runner location. Place them in line with the wires attached to the first main runner. Figs. 43-23 and 43-24 (Page 796).

To insure that all main runners will be perfectly level, follow these steps:

1. Stretch a string across all hanger wires in the direction of the main runner. Attach each reference string so that it is in line with the holes in the runner to which the hanger wires will be attached. Fig.

43–21. *Layout details for a 10′ 4″ × 18′ 8″ suspended ceiling using 24″ × 48″ panels.*

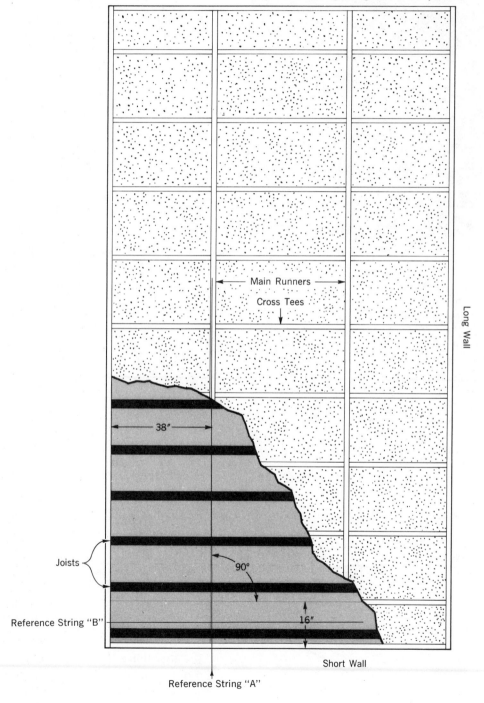

43-22a. *Nailing the wall molding in place.*

43-23, string C. For the system shown in Fig. 43-23, these holes should be located $1^{1}/_{8}$" above the bottom edge of the runners. Thus string C is $1^{1}/_{8}$" above string A. This string must be kept very tight.

2. Make sharp 90° bends where the wires come across the string. The best method is to clamp pliers horizontally to the wire so that the bottom of the pliers is at string level. With the other hand, firmly bend the wire tightly against the bottom of the pliers.

Install the remaining runners that go perpendicular to the joists. Measure each

43-22b. *Wall molding corner details.*

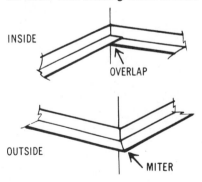

new main runner individually. After cutting a main runner, two of the holes on the runner should be directly above reference string B. Insert a hanger wire into one of these holes. Let one end of the main runner rest on the wall molding and support the other end with a hanger wire.

If the grid is uneven, fasten additional hanger wires as needed to the holes in the main runners. Bend the wire at the top of the hole, insert it through the hole, and wrap any excess wire around itself. Fig. 43-25. More than one runner may be needed to reach the opposite wall. For longer lengths, connect the main runners by inserting the end tabs into precut holes and bending them over.

Attaching Cross Tees to Main Runners

The first set of cross tees will be installed directly above reference string B. Figs. 43-21 and 43-23. Follow these steps:

1. Measure the distance from the long wall to reference string A. (In the example, this is 38".)

2. Cut the cross tee to this length and install it.

3. Complete the installation of cross tees above reference string B. Working in the same direction, install the balance of the cross tees in rows across the room. Fig. 43-26. The tees may be 2' or 4' apart.

Depending on the spacing of the main runners, the cross tees may be 24" or 48" long. If the cross tees are 48" long and 24" apart, and 24" × 24" ceiling panels are to be installed, attach 24" cross tees between the 48" cross tees, parallel to the main runners.

Cutting the Panels

Measure and cut each of the border panels individually. A panel is cut face up, with a coping saw or a very sharp fiberboard knife.

Installing Ceiling Panels

Drop-in panels are installed by resting these units on the flanges of cross tees and main runners. Fig. 43-27. Exercise care when handling ceiling panels to avoid marring the surface. Handle the panels by the edges, keeping the fingers off the finished side of the board as much as possible.

Ceiling Light Fixtures for Use with Suspended Ceilings

Standard ceiling fixtures and chandeliers may be used with a suspended ceiling. Recessed lighting can also be conveniently installed at any point. Simply install any one of several styles of translucent panels in place of a ceiling tile. Fig. 43-28.

Fluorescent lighting fixtures can be suspended from the wood framing between the floor joists. Special mounting brackets can also be obtained for attaching the fixture directly to the suspended grid. Fig. 43-29. Some fluorescent fixtures are designed to fit flush against the ceiling surface. Fig. 43-30. Vaulted ceiling lighting modules which fit the grid system are also available. Fig. 43-31.

ESTIMATING
Ceiling Tile

To figure the amount of 12″ × 12″ (1 sq. ft.) ceiling tile necessary, first calculate the ceiling area in square feet. Round off the room dimensions to the next foot and multiply the length times the width. Add the length and width of the room to the product. It is necessary to add the length and width of the room because of the border tile. For example, the room in Fig. 43-5 is 10′8″ × 12′4″, which is rounded off to 11′ × 13′. The area is 143 sq. ft. Therefore the room would require 143 ceiling tiles plus 13 tiles for the 12′4″ dimension and 11 tiles for the 10′8″ dimension. If there are any offsets or irregular cuts to be made, additional tiles should be allowed for waste and trim.

Labor

To calculate the amount of time necessary to install ceiling tile, refer to the chart in Fig. 42-96 of Unit 42, "Interior Wall and Ceiling Finish." The ceiling in the example contained 143 square feet. According to the chart, it will take 2.5 hours to install 100 square feet. To determine the total amount of time required for the ceiling in the example, divide by 100 to find how many hundreds of square feet are in the ceiling area.

$$\frac{143 \text{ (area of ceiling)}}{100} = 143 \text{ (hundreds of sq. ft. in ceiling)}$$

43–23. *Laying out the main runner for the room layout shown in Fig. 43–21. A. Fourteen inches should be cut from the end of the main runner. This end will rest on the wall molding. B. Suspend the main runner from the joists with wire. Use a string pulled taut to align the main runner. Make certain that the cross tee tab connection is located directly above the reference string B. This is essential so that a cross tee will be properly located to provide support for the border tile.*

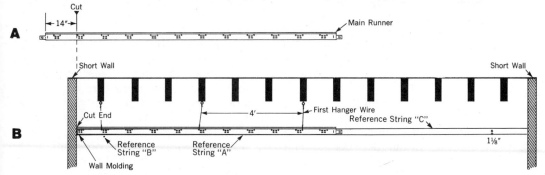

43-24. *The hanger wires are attached to the joists at 4' intervals.*

43-26. *Installing the cross tees between the main runners.*

43-25. *Fastening the main runner of the metal framework to the hanger wires.*

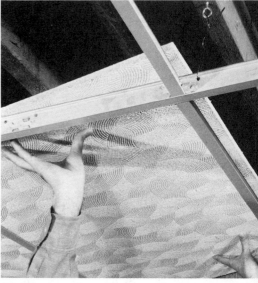

43-27. *Lay the ceiling panels into the grid formed by the main runners and the cross tees.*

Multiply this figure by the installation time required per 100 sq. ft. of ceiling tile.

$$1.43 \times 2.5 = 3.575, \text{ or } 3.6 \text{ hours}$$

Suspended Ceilings

To determine the materials necessary to install a suspended ceiling, measure the room carefully. Draw the room outline on a piece of graph paper, with each square representing 2' × 2'. Fig. 43-32. Select the desired grid pattern either for 2' × 4' or 2' × 2' panels. Draw in the main runners perpendicular to the joists and space them either 2' or 4' apart. Be sure to keep the border panels equal on both sides and as large as possible. Draw in the cross tees so that the border panels at the room ends are equal and as large as possible.

For a 2' × 2' pattern, the main runners may be spaced 4' on center with 4' cross

43-28. Translucent panels used with recessed lighting fixtures.

43-30. A fluorescent lighting fixture hung from the grid system. Special mounting brackets permit the fluorescent lighting fixture to be installed anywhere on the grid system of a suspended ceiling.

43-31. A vaulted lighting system for suspended ceilings.

43-29. The fluorescent fixture shown here is mounted on the suspended ceiling grid system.

tees spaced 2' apart. A 2' cross tee is then inserted between the 4' cross tees. Another pattern for 2' × 2' panels would be to space the main runners 2' apart and to use 2' cross tees also spaced 2' on center. Fig. 43-33.

For 2' × 4' panels, the main runners may be spaced 4' on center, with the cross tees spaced 2' on center. Another method is to space the cross tees 4' on center and the main runners 2' on center.

The wall angle is available in 12' lengths. Measure the perimeter of the room and divide by 12 to determine the number of pieces of wall angle. Add one piece for any portion of a piece.

The main tees (main runners) are also available in 12′ lengths. Determine the number of 12′ main tees required from the layout. If main tees longer than 12′ are required, they may be extended by splicing two or more tees together.

Count the number of 2′ cross tees in the layout. When determining the number of 2′ cross tees for border panels, no more than two border tees may be cut from any one 2′ cross tee. If 4′ cross tees are required, count the number shown in the layout. In determining 4′ cross tees for border panels, no more than two border tees may be cut from any one 4′ cross tee.

Count the number of lighting panels required. The remaining panels are the ceiling panels needed. The border panels are counted as whole panels.

LABOR

The labor for installing a suspended ceiling depends a great deal upon the complexity of the room, or the number of inside and outside corners and special

cutting. As a general rule, it can be estimated that a worker can lay out, install the grid system, and apply the panels for approximately 25 sq. ft. of area per hour. The lay out in Fig. 43-33 shows a ceiling 15′ × 21′, or 315 sq. ft. The installation time is figured by dividing the ceiling area by 25.

$$\frac{315 \text{ (area of ceiling in sq. ft.)}}{25 \text{ (sq. ft. installed in 1 hr.)}} = 12^{15}/_{25}, \text{ or about } 12^{1}/_{2} \text{ hrs.}$$

43–33. Two possible suspended ceiling layouts for the same room. The top layout has the main tees 2′ apart. In the second layout, the main tees are 4′ apart.

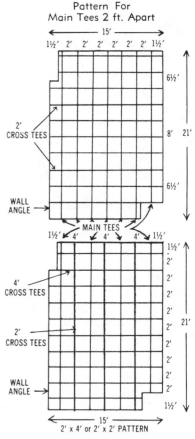

Pattern For
Main Tees 2 ft. Apart

Pattern For Main Tees 4 ft. Apart

43–32. Use a piece of graph paper, with each block representing a 2′ × 2′ area, for laying out a suspended ceiling system.

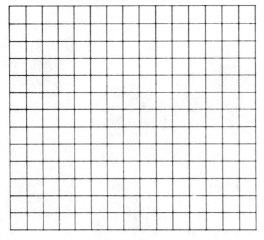

QUESTIONS

1. What are some of the advantages of ceiling tile and suspended ceilings?

2. When laying out a room for ceiling tile, why is it important that the border tile on opposite sides of the room be the same width?

3. Describe briefly the procedure for determing the width of the border tile for the ceiling.

4. Why is the distance between the wall and the first furring strip different from the distance between centers on all other furring strips?

5. What tools may be used for cutting the tile?

6. What size staples should be used when installing ceiling tile?

7. Describe the ashlar pattern.

8. Why is the chalk line snapped above the location of the wall moldings for suspended ceilings?

9. Calculate the width of the border tile for a ceiling using 24″ × 24″ ceiling panels. The room measures 14′6″ × 23′4″.

10. List some of the lighting systems that can be used with a suspended ceiling.

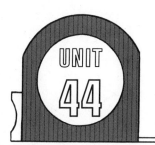

Finish Flooring

Finish flooring is the final wearing surface applied to a floor. Many materials are used as finish flooring, each one having properties suited to a particular usage. Durability and ease of cleaning are essential in all cases. Specific service requirements may call for special properties such as resistance to hard wear, comfort to users, and attractive appearance.

There is a wide selection of flooring materials. Hardwoods are available as strip flooring in a variety of widths and thicknesses and as random-width planks, parquetry, and block flooring. Other materials include plain and inlaid linoleum, cork, asphalt, vinyl, rubber, plastic vinyl, ceramic tile, and wall-to-wall carpeting. When a finish floor other than wood is to be installed on a subfloor of wood boards, an underlayment is required. This may be plywood, hardboard, or particle board.

HARDWOOD STRIP FLOORING

Strip flooring is the most widely used type of hardwood flooring. Practically all species are produced in this form. As the name implies, it consists of flooring pieces cut in narrow strips of varying thicknesses. The thinner strips are used chiefly in remodeling when new floor is laid over old and it is not desirable to reduce room height by using thick, more expensive flooring.

The most common hardwoods for strip flooring are oak, maple, beech, birch, and pecan. Oak, the most plentiful, is the most popular by far. It constitutes about 80% of the residential hardwood flooring in the United States.

Despite its extensive usage, strip flooring should not be considered commonplace. Its popularity is due to its high quality. No two hardwood floors are exactly

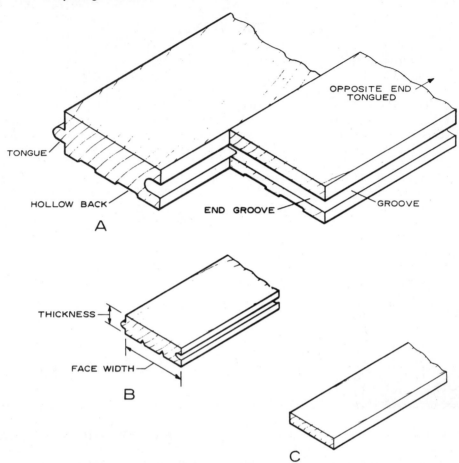

TONGUE

HOLLOW BACK

OPPOSITE END
TONGUED

END GROOVE

GROOVE

A

THICKNESS

FACE WIDTH

B

C

44–1. *Strip flooring. A. Side- and end-matched. B. Side-matched. C. Square-edged.*

alike. Each has individuality of character and beauty of grain. Most floors of this type are composed of strips of uniform width. Interesting patterns are formed by use of stock selected for variations in color or other natural irregularities. Attractive designs also may be achieved with strips of random widths. Most hardwood strip flooring today is tongued and grooved at the factory so that each piece joins the next one snugly when laid. Fig. 44-1.

Sizes

Strip flooring of oak, maple, beech, birch, and pecan is manufactured in a variety of sizes, ranging in width from 1" to 3½" and in thickness from $5/16$" to $33/32$". The standard thicknesses for tongued and grooved strip flooring are $1/2$", $3/8$", and $25/32$". Since strips of these thicknesses are the most commonly chosen for homes, they are produced in greatest volume and are available at lower cost than special thicknesses. The $25/32$" strips, used the most, are manufactured in four widths: 1½", 2", 2¼', and 3¼'. Most popular is the 2¼".

Square-edge oak strip flooring is $5/16$" thick and comes in widths of 1", 1⅛", 1½", and 2". Square-edge maple, beech, birch, and pecan strip flooring comes in two thicknesses: $25/32$" and $33/32$". Widths of 2½" and 3½" are available in each thickness.

The length of the strips in a bundle of flooring varies, but average lengths are specified for each grade. Some strips may be as long as 16 feet.

Grading

Through two major trade associations, the principal American producers of hardwood strip flooring have adopted uniform grading rules and regulations for commercial practice. Approved by the Bureau of Standards of the U.S. Department of Commerce, these rules and regulations are enforced rigidly, in part by the organizations themselves. As a result, dealers and consumers are assured of high quality flooring, and the industry is protected against sharp practices. The two organizations are the National Oak Flooring Manufacturers' Association, with headquarters at Memphis, Tennessee, and the Maple Flooring Manufacturers' Association, Chicago, Illinois. Every bundle of flooring produced by a member of either association is identified as to grade and is guaranteed to meet all established specifications. Usually the manufacturer's name and a mill mark of identification are stated on each bundle.

The hardwood flooring grades in Canada are identical to those for the United States.

There are no official grading rules for plank, parquet, and block flooring. Generally the different grades correspond to those of strip flooring. Hardwood strip flooring grades are based principally on appearance. Since all regular grades have adequate strength, durability, and resistance to wear, these qualities are not factors. Chiefly considered are such characteristics as knots, streaks, pin wormholes and, in some cases, sapwood and variations in color. Slight imperfections in processing also are factors.

OAK

Oak is classified into two grades of quarter-sawed stock and four grades plain-sawed. In descending order the quarter-sawed grades are: Clear and Select. Plain-sawed grades are: Clear, Select, No. 1 Common, and No. 2 Common.

In Clear Grade Oak the amount of sapwood is limited. Otherwise, variations in color are disregarded in grading. Red oak and white oak ordinarily are separated, but that does not affect their grading. In most cases the average length of strip flooring pieces is greater in the higher grades.

MAPLE, BEECH, BIRCH

Rules governing the grading of maple, beech, and birch are virtually identical for all three species. Neither sapwood nor varying natural color is considered a defect in standard grades. These standard grades are first, second, and third, with first being the highest. Fig. 44-2. Each of these grades also is available in a special grade selected for uniformity of color.

PECAN

Pecan is processed in six standard grades. Two of the grades specify all heartwood, and one specifies bright sapwood. Otherwise color variation is not considered.

STORAGE AND DELIVERY

Certified hardwood flooring is kiln-dried at the factory to a low moisture content. However, moisture content later equalizes itself to the moisture conditions in the area where the flooring is used. The flooring must be protected from the elements during storage and delivery to guard against excessive shrinkage or expansion, which may cause cracks or buckling after the floor has been laid. Manufacturers who ship hardwood flooring in closed boxcars recommend the following precautions in handling:

• Do not unload, truck, or transfer hardwood flooring in rain or snow. Cover it with tarpaulin if the atmosphere is foggy or damp.

• Flooring should be stored in airy, well-ventilated buildings, preferably with

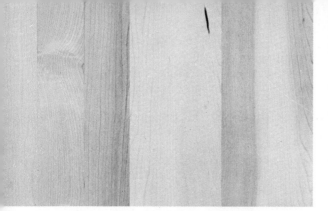

44–2a. Typical first grade MFMA (Maple Flooring Manufacturers' Association) maple flooring.

44–2c. Typical third grade MFMA maple flooring.

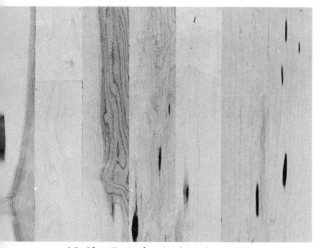

44–2b. Typical second grade MFMA maple flooring.

weathertight windows that will admit sunlight.

• Do not pile flooring on storage floors that are less than 18″ from the ground and which do not have good air circulation underneath.

• Do not store or lay flooring in a cold or damp building. Wait until the plaster and concrete work have dried thoroughly and all but the final woodwork and trim have been installed.

• Especially in winter construction, the building in which the flooring is to be used should first be heated to 70° F. Then the flooring should be stored in the building at least four or five days before being laid.

This permits the flooring to reach a moisture content equivalent to that of the building.

INSTALLATION

Installation of the finish flooring should be the last construction operation in a house. Fig. 44-3. All plumbing, electrical wiring, and plastering should be completed before the application of the finish floor is begun. Only the final interior trim work should remain.

Before laying the flooring, the bundles should be opened and the flooring spread out and exposed to warm, dry air for at least 24 hours, preferably 48 hours. Moisture content of the flooring should be 6% for the dry Southwest, 10% for southern and coastal areas, and 7% for the remainder of North America.

Preparing the Subfloor

Just before installation of the finish flooring is to begin, the subfloors should be examined carefully and any defects corrected. Raised nails, for instance, should be driven down and loose or warped boards replaced. The subfloors should be swept thoroughly and scraped if necessary to remove all plaster, mortar, or other foreign materials. These precautions must be taken if the finish flooring is to be laid properly.

The last operation before actual installation of the finish flooring should be the application of a good quality asphalt-

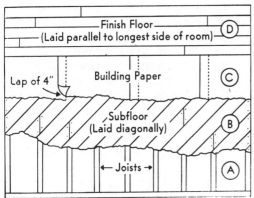

44-3. Cutaway view of floor, showing details of construction: A. Joist. B. Subfloor. C. Building paper. D. Finish floor.

44-4. Laying asphalt-coated building paper over the subfloor in preparation for the finish floor. NOTE: The building paper will cover up the nails in the subfloor which indicate the location of the floor joists. It is therefore advisable to mark the location of the floor joists on the walls before laying the building paper. Later chalk lines may be snapped on the building paper to indicate the location of the floor joists below for nailing the strip flooring.

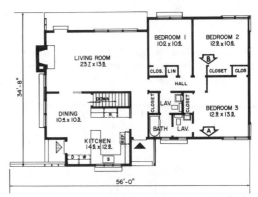

44-5. The first piece of strip flooring should be laid at the wall indicated by arrow A above. At arrow B a piece of strip flooring is reversed as shown in Fig. 44-6. This will permit blind-nailing into the closet area.

coated building paper over the subflooring. Fig. 44-4. The paper will protect the finish floor and the interior of the home from dust, cold, and moisture which might seep through the floor seams. In the area directly over the heating plant, it is advisable to lay double weight building paper or standard insulating board. This will protect the finish floor from excessive heat which might shrink the boards.

Where to Start

Before the actual laying of strip flooring begins, the floor plans of the house should be studied. Careful consideration must be given to the area of the house in which the installation will begin. Strip flooring should be laid so that there is an uninterrupted flow from one room through the hall to the other rooms. Plan the job to eliminate having to align in the hall the courses of strip flooring from two rooms.

For example, in Fig. 44-5 if the courses from bedrooms 1 and 2 are laid through the doorways and into the hall, it is very unlikely that they will line up. Therefore it is best to start the first strip of flooring against the wall indicated by arrow A in bedroom 3. Fig. 44-5. Work across the bedroom and through the door into the hall. Continue along the hall wall into the living room.

Work the courses across the hall and the first few courses of the living room together. As the courses approach the rear wall of the hall, work through the doorways into bedrooms 1 and 2, and complete the laying of the bedroom floors. Then return to the living room and finish laying the living room floor, working toward the rear of the house.

In working in this manner the closets in bedrooms 1 and 2 will require placing the grooves of two strips together, with a spline between the grooves. The tongues can then be blind-nailed on each piece. This will permit the courses of strip flooring in the closet to be blind-nailed as well as the pieces which will go across the bedroom floor toward the back wall. Fig. 44-6.

In a split-level, as shown in Fig. 44-7, the first strip of flooring is laid against the back wall of bedroom 1. The floor is laid through the door, across the hall, and into the two front bedrooms. The living room is done separately, and should be started at its front wall and laid in the direction of the arrow across the floor toward the dining room. If hardwood flooring is to be laid in the dining room, continue through the archway and to the rear wall of the house. If another type of flooring is to be used in the dining room, the strip flooring should terminate at the wall of the dining room. Fig. 44-7, line A.

When different floor coverings are laid in adjoining rooms and a doorway connects the two rooms, the hardwood flooring should terminate under the center of the door when the door is closed. The second type of flooring should begin at this point. In this way, when the door is closed, only the flooring of the room in which the individual is standing may be seen. When finish floors of different materials come together in an archway between two rooms, the hardwood floor is usually laid through the archway and even with the wall line of the adjacent room.

Arrangement of Flooring Pieces

In most houses strip flooring presents the most attractive appearance when laid lengthwise of the room's longest dimension. In some cases, however, it is considered acceptable to lay the flooring cross-

44–6. *A spline is inserted between the two pieces of flooring at the wall line indicated by arrow B in Fig. 44–5.*

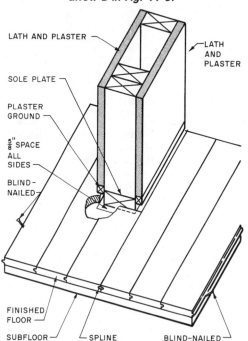

LATH AND PLASTER

LATH AND PLASTER

SOLE PLATE

PLASTER GROUND

$\frac{5}{8}''$ SPACE ALL SIDES

BLIND-NAILED

FINISHED FLOOR

SUBFLOOR

SPLINE

BLIND-NAILED

44–7. *In a split-level home, laying strip flooring is simplified. Each level is done independently, which eliminates many of the problems of lining up courses in several rooms.*

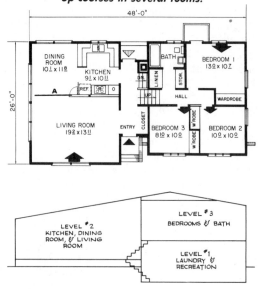

wise. This is true only if the rooms are sufficiently wide. Wood-strip flooring is usually laid at right angles to the direction of the joists under the largest room on that story. Fig. 44-8. Since interior thresholds are omitted today, the flooring should run continuously between adjoining rooms.

Proper placing of the strips of flooring calls for the use of the shorter lengths in closets and in the general floor area. The longer pieces should be used at entrances and doorways and for starting and finishing in a room. This arrangement results in maximum attractiveness of the floor as a whole. Care should be taken to stagger the end joints of the flooring pieces so that several of them are not grouped closely together. Fig. 44-9.

Importance of Adequate Nailing

Adequate nailing is absolutely essential in the finish floor as well as the subfloor. Insufficient nailing may result in loose or

44–9. *Laying out the strip flooring in preparation for cutting the end pieces so that the joints are staggered.*

44–8. *Strip flooring should be laid at right angles to the floor joists.*

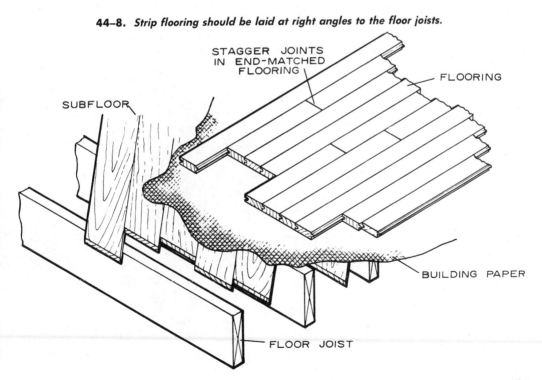

STAGGER JOINTS
IN END-MATCHED
FLOORING

FLOORING

SUBFLOOR

BUILDING PAPER

FLOOR JOIST

44–10a. *Nailing schedule for wood flooring.*

Nail Schedule

A

B

C

44–10b. *Flooring nails: A. Barbed. B. Screw. C. Cut steel.*

Tongued and Grooved Flooring Must Always Be Blind-Nailed, Square-Edge Flooring Face-Nailed		
Size Flooring	Type and Size of Nails	Spacing
(Tongued & Grooved) $^{25}/_{32}$ x 3¼	7d or 8d screw type, cut steel nails or 2″ barbed fasteners*	10-12 in. apart
(Tongued & Grooved) $^{25}/_{32}$ x 2¼	Same as above	Same as above
(Tongued & Grooved) $^{25}/_{32}$ x 1½	Same as above	Same as above
(Tongued & Grooved) ½ x 2, ½ x 1½	5d screw, cut or wire nail. Or, 1½″ barbed fasteners*	8-10 in. apart
Following flooring must be laid on wood subfloor		
(Tongued & Grooved) ⅜ x 2, ⅜ x 1½	4d bright casing, wire, cut, screw nail or 1¼″ barbed fasteners*	6-8 in. apart
(Square-Edge) $^{5}/_{16}$ x 2, $^{5}/_{16}$ x 1½	1-in. 15-gauge fully barbed flooring brad, preferably cement coated.	2 nails every 7 in.

*If steel wire flooring nails are used they should be 8d, preferably cement coated. Newly developed machine-driven barbed fasteners, used as recommended by the manufacturer, are acceptable.

the room between two nails placed about 8″ from a side wall. Many walls are not perfectly true. By lining up the first courses of flooring at a uniform distance from the string rather than from the wall itself, a straighter course is assured. Fig. 44-11.

Place a long piece of flooring with the groove edge about ½″ to ⅝″ from a side wall and the groove end nearest an end wall. The ½″ to ⅝″ space is for expansion. It will be hidden by shoe molding. Face-nail the flooring piece near the end. Fig. 44-12. Measure as you nail toward the other end, maintaining the same fixed dis-

44–11. *To line up the first course of flooring, stretch a string or snap a chalk line about 8″ from the wall. The first course is face-nailed.*

squeaky flooring, one of the most annoying deficiencies a house may develop. The building paper should be chalklined at the joists as a guide in nailing the strip flooring. Tongue-and-groove flooring should be blind-nailed. The nails should be driven at an angle of about 45° or 50° at the point where the tongue leaves the shoulder. Square-edge flooring should be face-nailed. It is recommended that nail heads be countersunk. Fig. 44-10 includes information about the types of nails and the spacing commonly used.

Laying Strip Flooring over Subflooring

Cover the subfloor with building paper, preferably 15-pound, asphalt-saturated felt. Fig. 44-4. Stretch a string the length of

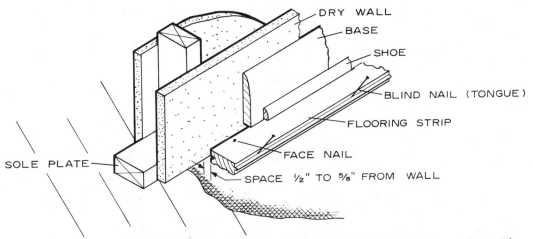

DRY WALL
BASE
SHOE
BLIND NAIL (TONGUE)
FLOORING STRIP
FACE NAIL
SPACE ½" TO ⅝" FROM WALL
SOLE PLATE

44-12. *Cross section of a wall showing the first piece of strip flooring nailed in place. NOTE: With lath and plaster walls, sometimes the ground is kept up off the subfloor about ⁷/₈", and the edge of the first piece of strip flooring is set about even with the wall line. The flooring is then allowed to expand under the ground.*

tance from the guide string. Fig. 44-11. Do likewise in lining up succeeding pieces in the course. Drive one nail at each joist crossing, or every 10" to 12" if the joists run parallel to the flooring.

Observe nailing recommendations of the flooring manufacturer. For example, with flooring ²⁵/₃₂" thick and 1¹/₂" or more wide, 7d or 8d screw nails or cut steel nails are best. If steel wire flooring nails are used, they should be 8d, preferably cement coated. Fig. 44-10.

After face-nailing the first course, toenail the pieces through the tongue edge, following the same spacing. Fig. 44-13. Drive the nail at the point where the tongue of the flooring leaves the shoulder, at an angle of about 45° or 50°. Fig. 44-14. Each nail should be driven down to the point where another blow or two might cause the hammer to damage the edge of the strip. Use a nail set to drive the nail the rest of the way home. Fig. 44-15. The best nailing procedure is to stand on the strip, with toes in line with the outer edge, and strike the nail from a stooping position which will bring the hammer head square against the nail.

44-13. *Succeeding courses are toenailed into the flooring at the point where the tongue leaves the shoulder.*

44–14. *Strip flooring is nailed in place with the nail driven at about a 50° angle to the floor.*

44–15. *When setting nails in strip flooring, many workers place the nail set on top of the nail as shown here. The nail set is struck with a sharp blow against its side. This sets the nail and at the same time tends to drive the piece of flooring up tightly against the piece previously laid.*

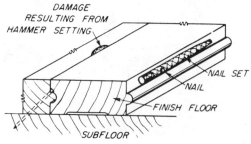

Fit the groove edges of pieces in each succeeding course with the tongues of those in the preceding course. Fit the groove end of each piece in a course with the tongue end of the previous pieces.

When a piece of flooring cannot be readily found to fit the remaining space in a course, cut one to size. Lay a piece down in reversed position from that in which it will be nailed. Draw a line at the point where it should be sawed. Fig. 44-16. Be sure the piece is reversed for this marking so that the tongue end will be cut off. The groove end is needed for joining with the tongue end of the previous piece.

44–16. *When fitting a piece of flooring to the remaining space in a course, place it in reverse position for marking so that the tongue end is cut off. The groove end is needed for joining with the tongue of the previous piece.*

44–17. *When installing strip flooring, it is customary for two people to work together in the room. One person nails the flooring in place, while the second works ahead laying out pieces, staggering joints, and cutting the end pieces to length. Notice in this picture that the man laying the flooring is using a nailing machine rather than nailing by hand as shown in Fig. 44–13.*

44–18. *After nailing three or four courses, place a piece of scrap flooring along the tongue edge of the flooring already laid. Drive the flooring up tight with two or three sharp hammer blows.*

Stagger the joints of neighboring pieces so that they will not be grouped closely. Fig. 44-17. A joint should not be closer than 6" to another in a previous course. To provide for this, arrange several pieces in their approximate positions in succeeding courses before they are nailed. After nailing three or four courses, place a piece of scrap flooring at intervals against the tongue edges of pieces in the last course. Strike the scrap piece a couple of good hammer blows. This drives the nailed flooring pieces up tightly. Fig. 44-18.

To fit flooring around a jutting door frame, place it flush against the frame. Fig. 44-19. Measure the gap between the face of the previous piece and the groove edge of the piece to be nailed. Where the jutting begins, draw a straight line on the flooring to the same distance as the width of the gap. Fig. 44-19. Do the same on the other side of the door frame. Draw a straight line connecting the ends of these lines. Cut the flooring along the lines.

44–19. *The flooring should be fitted around a doorframe or other projection. Place the strip tight against the frame, measure the gap between the face edge of the previous piece and the groove edge of the piece to be nailed. Mark this distance, and the width of the frame, on the flooring and notch it out.*

44–20. *When laying the last few courses in the room, there will not be enough space to swing a hammer for toenailing. Therefore these courses are face-nailed. Place the last few courses in position, and pull the flooring up tightly when nailing. When prying the pieces with the crowbar, put something behind the crowbar to protect the plaster. A piece of paper is used here. However, a scrap piece of flooring or other piece of wood is better, particularly if it is a little longer and will span between the studs in the wall so that the plaster is not punctured.*

On reaching the opposite side of the room, you will find there is no space between the wall and the flooring to permit toenailing of the last two courses. Therefore just fit the pieces in. Face-nail the last few courses, at the same time pulling the flooring up tightly by exerting pressure against it with a chisel or crowbar driven into the subflooring. Fig. 44-20. Protect molding with cardboard or a piece of scrap lumber. After the last course, if the remaining space is too large to be covered by the shoe molding, rip pieces of flooring to the proper width and insert them in the spaces. Face-nail these strips. If they are very narrow, drill holes for the nails to prevent splitting.

Laying Strip Flooring over Concrete

Wood subflooring ordinarily is omitted in homes built on concrete slabs. In such a

dwelling the slab serves the major function of subflooring because it offers a strong, solid support for the finish floors and provides a working surface for the building mechanics. Omission of wood subflooring in a slab home has a definite cost advantage, since it permits savings both in material and labor. An adequate nailing surface for the finish floors must be provided, of course. The method described here is economical and results in finish floors of ample strength and resiliency. Relatively

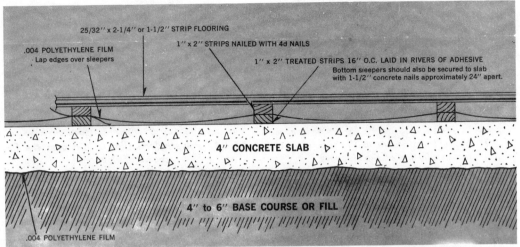

25/32" x 2-1/4" or 1-1/2" STRIP FLOORING

1" x 2" STRIPS NAILED WITH 4d NAILS

.004 POLYETHYLENE FILM
· Lap edges over sleepers

1" x 2" TREATED STRIPS 16" O.C. LAID IN RIVERS OF ADHESIVE
Bottom sleepers should also be secured to slab
with 1-1/2" concrete nails approximately 24" apart.

4" CONCRETE SLAB

4" to 6" BASE COURSE OR FILL

.004 POLYETHYLENE FILM

44–21. *A cross section of hardwood strip flooring installed over a concrete slab.*

new, it has gained widespread popularity in recent years. Fig. 44-21.

1. Sweep the slab clean and prime the surface. After the primer is dry, snap chalk lines 16" apart at right angles to the direction the flooring will run. Cover the lines with rivers of adhesive applied to a width of about 2". The adhesive can be an asphalt mastic designed for bonding wood to concrete or a suitable adhesive of another type designed for the same purpose. Fig. 44-22. (If heating is in the slab, the adhesive should be resistant to heat.)

2. The bottom sleepers should be 1" × 2" strips treated with wood preservative. Imbed the strips in the adhesive and also secure them to the slab with $1\frac{1}{2}$" concrete nails, approximately 24" apart. Fig. 44-23. The sleepers should be of random lengths and laid end to end with slight spaces between the ends, not butted tightly together.

3. After all bottom sleepers have been installed, 0.004 polyethylene film should be laid over the strips. Fig. 44-24. Join the polyethylene sheets by lapping edges over sleepers.

4. The second course of 1 × 2s (which do not have to be preservative treated)

would be nailed with 4d nails, 16" to 24" apart. Nail through the top sleeper and the polyethylene into the bottom sleeper. Fig. 44-25.

5. Install the flooring at right angles to the sleepers by blind-nailing to each sleeper, driving at an angle of approximately 50°. Fig. 44-26. Nails should be the threaded or screw type, cut steel, or barbed. Two adjoining flooring courses should not have joints on the same sleeper. Each strip should bear on at least one sleeper. Provide a minimum of ½" clearance between flooring and wall to allow for expansion. Fig. 44-27.

FINISHING
Sanding

Although certified unfinished hardwood flooring is smoothly surfaced at the factory, scratches and other slight marks caused by handling usually show after the floor has been laid. These may be removed by sanding or scraping. The work should be done by a specialist in that line. An electrically operated sanding machine generally is used for this work, since sanding or scraping by hand is laborious and time-consuming. Fig. 44-28. For fine floors,

44-22. *Applying adhesive to the concrete slab for installation of the sleepers.*

44-24. *Polyethylene film is laid over the first course of sleepers.*

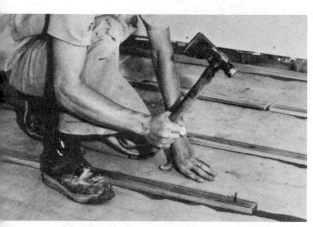

44-23. *The bottom sleepers are placed over the ribbons of adhesive and nailed in place with concrete nails.*

44-25. *The second course of 1" × 2" sleepers is nailed with 4d nails through the polyethylene film and into the bottom sleepers.*

most manufacturers advise at least four sandings, starting with No. 2 sandpaper and graduating down to No. 1/2, No. 0, and No. 00. A final buffing with No. 00 or No. 000 sandpaper assures an even smoother surface. Many authorities say that best results are obtained when the final traverse is made by hand.

Stain

It is important that the first coat of stain or other finish be applied the same day as the last sanding. Otherwise the wood grain will have risen, and the finish consequently will be slightly rough. Stain is not used if the finish is to retain the natural color of the wood. If stain is used, it is applied first, before wood filler or other finishes. It should be put on evenly, preferably with a brush 3" or 4" wide.

Wood Filler

Paste wood filler customarily is used to fill the minute surface crevices in oak and other hardwoods with large pores. It gives the floor the perfectly smooth surface required for a lustrous appearance. Filler is applied after stains and sometimes after floor seals but always before other finishing materials. It should be allowed to dry

44-26. *The strip flooring is laid over the sleepers at right angles.*

44-28a. *Although flooring is sanded at the mill, additional sanding is required after the floor is laid for a good finishing job.*

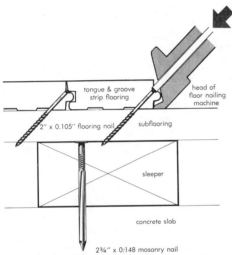

44-27. *This drawing shows correct nailing procedure when tongue-and-groove strip flooring is installed over subflooring and a single sleeper is used over a concrete slab.*

44-28b. *A small power sander called an edger is used for sanding floors near walls, in closets, and in other small areas where the large floor machine is inefficient.*

24 hours before the next operation is begun. Wood filler may be colorless, or it may contain pigment to bring out the grain of the wood. For residential oak flooring, wood filler is always recommended.

Types of Finish

A finish for hardwood floors ideally should have the following qualities:

- Attractive appearance.
- Durability.
- Ease of maintenance.

- Capacity for being retouched in worn spots without revealing a patched appearance.

A finish applied to high grade flooring of considerable natural beauty should also be transparent in order to accentuate that beauty. The three principal types of finishes are floor seal, varnish, and shellac. Lacquer also is used occasionally.

FLOOR SEAL

Floor seal, a relatively new material, is being used on an increasingly large scale

44-29. There are several finishing materials available for hardwood floors. However, most hardwood flooring producers recommend a floor seal. Floor seal is a tough wear-resistant finish that can be mopped on easily with a long-handled lamb's wool or sponge applicator or a clean string mop. The floor should first be swept and dusted.

44-30. After a few square feet of floor seal has been applied and allowed to penetrate, the excess seal should be wiped off with clean cloths. With some brands of seal, however, the excess is rubbed into the wood. The room should be well ventilated during all finishing operations.

for residential as well as heavy duty flooring. It differs from other finishes in this important respect: rather than forming a surface coating, it penetrates the wood fibers, sealing them together. In effect it becomes a part of the wood itself. It wears only as the wood wears, does not chip or scratch, and is practically immune to ordinary stains and spots. While it does not provide as shiny an appearance as other finishes, it has the advantage of being easily retouched. Worn spots may be refinished without presenting a patched appearance. Floor seals are available either with or without color.

It is difficult to give specific directions for applying floor seal because directions of different manufacturers vary widely. Generally it is applied across the grain first, then smoothed out in the direction of the grain. A wide brush, a squeegee, or a wool applicator may be used. Fig. 44-29. After a period of 15 minutes to 2 hours, depending on specific directions of the

manufacturer, the excess seal should be wiped off with clean cloths or a rubber squeegee. Fig. 44-30.

For best results the floor then should be buffed with No. 2 steel wool. An electric buffer makes this task relatively simple. Fig. 44-31. If a power buffer is not available, a sanding machine equipped with steel wool pads may be used, or the buffing may be done by hand. Although one application of seal sometimes is sufficient, a second coat frequently is recommended for new floors or floors just sanded. Floor seal is a complete finish in itself. However, it may also be used as a base for a surface finish such as varnish.

VARNISH

Varnish presents a glossy appearance and is quite durable. It is fairly resistant to stains and spots, but shows scratches. It is difficult to patch worn spots without leaving lines of demarcation between old and new varnish. New types of varnish dry in eight hours or less. Like other types of finish, varnish is satisfactory if properly waxed and otherwise maintained.

Precise directions for application of varnish usually are stated on containers. Varnish made especially for floors is much preferred. So-called all-purpose varnish ordinarily is not so durable when used on floors. As a rule three coats are required when varnish is applied to bare wood. Two coats usually are adequate when wood filler has been used or a coat of shellac has been applied. Cleanliness of both floor and applicator is essential for smooth finish. Drying action is hastened when room temperature is at least 70° F and plenty of ventilation is provided.

SHELLAC

One of the chief reasons shellac is so widely used is because it dries quickly. Workers, starting with floors in the front of a house and moving toward the rear, may begin applying the second coat by the time they have finished the first. Shellac spots rather readily if water or other liquids remain on it long. It is transparent and has a high gloss. It does not darken with age as quickly as varnish.

Shellac to be used on floors should be fresh or at least stored in a glass container. If it remains too long in a metal container, it may accumulate salts of iron which will discolor oak and other hardwoods containing tannin. Shellac should not be mixed with cheaper resins, but before use should be thinned with 188-proof No. 1 denatured alcohol. The recommended proportion is 1 quart of thinner to a gallon of 5-pound cut shellac. A wide brush that covers three boards of strip flooring is the most effective and convenient size. Strokes should be long and even with laps joined smoothly.

The first coat on bare wood will dry in 15 to 20 minutes. After drying, the floor should be rubbed lightly with steel wool or sandpaper, then swept clean. A second coat should be applied and allowed to dry for 2 to 3 hours. The floor should be rubbed

44-31. *After the floor seal has dried completely, buff the floor lightly with a pad of fine steel wool on the polishing machine. Wax may also be polished with the machine pictured here by detaching the steel wool pad and attaching a brush provided with the machine.*

again with steel wool or sandpaper and swept. A third coat then is applied. If necessary, the floor may be walked on in about 3 hours, but preferably it should remain out of service overnight.

LACQUER

Lacquer is a glossy finish with about the same durability as varnish. Because it dries so rapidly, it requires considerable skill in application. Worn spots may be retouched with fairly good results, since a new coat of lacquer dissolves the original coat.

If possible, lacquer should be applied with a spray gun. The first coat or sealer should be sanded with a 150 or 180 grit aluminum oxide or garnet abrasive paper. Additional top coats are then applied. Unlike other materials, lacquer does not have to be sanded between coats unless some imperfection occurs and the spot has to be sanded smooth.

Wax

All hardwood floors should be waxed after the finish has dried thoroughly. In some cases two or three coats are recommended for best results. Wax not only imparts a lustrous sheen to a floor, but forms a protective film that prevents dirt from penetrating the wood pores. When wax becomes dirty, it is easily removed and new wax applied.

Hardwood floor wax is available in paste or liquid form. The liquid is known as rubbing wax. Considered about equal in performance, both forms are applied in much the same manner. Usually the wax is mopped on with a cloth, then polished after an interval of 15 to 30 minutes with a soft cloth, a weighted floor brush, or an electric polisher. The latter is preferred, for it eliminates a great deal of labor and does the job equally well, if not better. Some electric polishers apply the wax and polish it in the same operation. Power-driven polishing machines, as well as sanding and buffing machines helpful in the earlier finishing steps, can usually be rented.

Prefinished Hardwood Flooring

When hardwood flooring is purchased in unfinished form, sanding and other finishing operations are performed after the floor has been installed. Some manufacturers produce flooring which is completely prefinished at the factory. It is ready for use immediately after being laid. All species and types of hardwood flooring may not be readily available in this form. However, a manufacturer equipped to make factory finished flooring usually can furnish it in any species and type ordinarily produced unfinished.

OTHER TYPES OF HARDWOOD FLOORING
Plank Flooring

One of the oldest types of hardwood floors, plank flooring, dates back to the handicraft era. In its crude early form it was widely used in medieval Europe, and it later became a popular flooring for American colonial homes. Its use has been increasing in homes, clubs, and other buildings where an atmosphere of rugged informality is desired. Colonial plank flooring derived much of its charm from its rough effects and interesting irregularities, which were unintentional. This charm is retained in modern plank flooring even though it is now a precision-made product.

Oak is preferred for plank floors. Production of planks ordinarily is confined to that species, but they can also be obtained in solid walnut and East Indian teak, as well as in various veneers. The planks usually come in random widths. Generally the pieces are tongued and grooved, with square or matched ends, but sometimes they are produced with square edges and ends. Frequently the edges of planks are beveled to reproduce the effect of the large cracks which characterize early hand-hewn plank floors. The wood pegs by which the old plank floors were fastened also are simulated. This is done by gluing wood plugs in holes on top of the countersunk screws which actually fasten the planks to the subfloor.

Pattern or Parquet Flooring

Pattern floors, also known as parquet and design floors, appeared as early as the 14th century in Europe. While today, as then, they are the most elaborate and expensive type of hardwood flooring, they have been simplified to a great extent. Early pattern floors presented bizarre effects which would be incompatible with present architectural styles. Today's parquetry uses squares, rectangles, and herringbone patterns to achieve an almost infinite variety of effects. Fig. 44-32. Literally hundreds of designs are available, many featuring various species or different shades of the same species. Most parquet flooring is of oak, although it also is produced in maple, beech, birch, walnut, mahogany, East Indian teak, and ebonized wood. The latter consists of dyed white maple or holly. Parquetry is manufactured in short lengths of individual pieces. Each piece must be cut to exact dimensions so that it will match perfectly the dimensions of another piece, or multiples thereof. Customarily tongued and grooved and end-matched, the pieces are laid separately, either by nailing or setting in mastic.

Block Flooring

Block flooring is really a form of parquetry, since it constitutes a definite pattern. It differs from conventional parquetry in that the pieces are assembled into square or rectangular blocks at the factory. They are held together by various means on the back, sides, or ends with matching tongue-and-groove edges. Fig. 44-33. These prefabricated blocks are laid as units, either in mastic or by nailing.

Installing Plank, Block, and Parquet Flooring

Each piece in plank flooring, in addition to being blind-nailed, should be face-nailed or screwed about every 30 inches. The ends of each piece also should be fastened to the subfloor with countersunk screws. Wood plugs glued in the plank conceal screwheads and lend a decorative effect.

Parquet flooring, including prefabricated blocks, may be nailed over a wood subfloor or old finish floor, or it may be laid

in mastic over concrete. Preferably the subfloor should cross the joists at right angles rather than on a diagonal. When the subfloor is installed before the type of finish flooring has been decided upon, it is best to lay it at a 60° angle. This "compromise" will make it adaptable to any type of flooring.

LINOLEUM AND TILE FLOORS

Use $\frac{1}{4}''$ underlayment for linoleum, asphalt tile, or vinyl tile floors. It may be laid directly on $\frac{24}{32}''$ tongue-and-groove wood flooring strips, maximum width $3\frac{1}{4}''$, with a joist spacing of 16'' on center. Where these finish floors are used for the floor covering in one room and wood floors are used in adjacent room over a subfloor of common level, a suitable base floor is required for

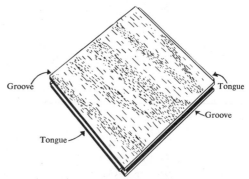

44-33a. *Wood block flooring. Note the tongue and groove locations.*

44-33b. *Laminated block flooring.*

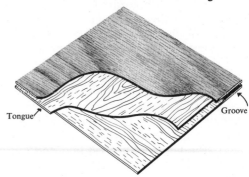

44-32. *A few examples of pattern or parquet floors.*

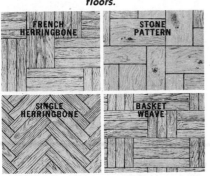

44–34. *Plywood is frequently used as an under-layment for a tile floor.*

the nonwood finish flooring. This base floor may also be tongue-and-groove flooring or plywood. The thickness of the base floor plus the thickness of the nonwood finish floor should equal the thickness of the wood finish floors in adjacent rooms so that the floors will be at the same level.

Linoleum

Linoleum is manufactured in thicknesses ranging from $1/16''$ to $1/4''$ and is generally 6 feet wide. It is made in various grades, in plain colors, inlaid, or embossed. Linoleum may be laid on wood or plywood base floors, but not on concrete slabs on the ground. Since linoleum follows the contour of the base floor over which it is laid, it is essential that the base be uniform and level. When wood floors are used as a base, they should be sanded smooth and be level and dry. When plywood base floors are used, the sheets should be carefully joined together. After the base floor is correctly prepared, the adhesive is applied. The linoleum is then laid and thoroughly rolled to insure complete adhesion to the floor.

Asphalt and Vinyl Tile

Asphalt tile is widely used as a covering over concrete slabs and is sometimes used over an underlayment. It is the least costly of the commonly used floor-covering materials. This tile is about $1/8''$ thick and $9'' \times 9''$ or $12'' \times 12''$. Most types of asphalt tile are damaged by grease and oil and for that reason are not recommended for use in kitchens.

It is important that the subfloor or base be suitably prepared. Otherwise, the finish floor will not give satisfactory performance. Most manufacturers provide directions on the preparation of the base and recommend the type of adhesive that is best for their product. The tile should be laid according to the manufacturer's directions. When a wax finish is recommended, the wax should have a water base. Vinyl tile can be used in the same manner as asphalt tile. It has the advantage of being impervious to grease and oil.

Rubber Tile

Rubber tile flooring is resilient, noise-absorbing, waterproof, and highly wear resistant. It may be applied over wood subfloors or concrete floor slabs, except slabs on ground. The finish may be plain or marbleized in various designs, with the colors running throughout the body of the tile. Rubber tile is made in square shapes ranging in size from $4'' \times 4''$ to $18'' \times 18''$. It also comes in rectangular shapes ranging from $9'' \times 18''$ to $9''$ to $36''$. Thickness is from $1/8''$ to $3/16''$. The tile is generally laid in a waterproof rubber cement and thoroughly rolled.

Wood subfloors for rubber tile should be above grade. If the subfloor is plywood, rubber tile may be laid directly on the wood surface. Make sure that plywood joints do not coincide with rubber tile joints, but come midway between the tile joints. If tile is laid on a solid wood subfloor, the floor should first be sanded smooth and sealed. A layer of 15- to 30-pound saturated lining felt should then be bonded to the surface. Joints in the tile should not coincide with subfloor joints. Otherwise, expansion or contraction of the subfloor will also affect the rubber tile. The tile should be installed following the manufacturer's recommendations for both method and materials.

Installing a Tile Floor

Tile can be laid directly over wood flooring or over a plywood floor. Fig. 44-34. If a hardwood floor is to be used in one room and the adjacent room is to have a tile floor, a base is required for the tile so that the floors will be exactly the same height. In other words, the thickness of the base floor plus the thickness of the tile should equal the thickness of the finish floor in an adjacent room. Fig. 44-35. Some kinds of tile require an underfelt that must be applied over the floor before the tiles are installed. Figs. 44-36 through 44-58 and the accompanying text describe how to install 9″ × 9″ vinyl-asbestos tile. The same general procedure applies to other types and sizes of tile.

• Prepare the surface on which the tile will be laid. The appearance of the finished floor will depend a great deal on the condition of the subfloor. Make sure the floor is smooth and completely free of wax, paint, varnish, grease, or oil. Holes or cracks in concrete subfloors should be filled with crack filler. Plane down high

44–36. *Planing down high spots.*

spots and renail loose boards of wood floors. Fig. 44-36.

• If a wood subfloor is only a single layer or if it is a double-layer floor and the boards are in bad condition, the old floor should be covered with an underlayment. Fig. 44-37.

• Allow a little less than $1/32″$, or the thickness of a paper matchbook, between each panel of underlayment or plywood to allow for expansion. Stagger the joints as shown. Fig. 44-38.

44–35. *The underlayment, in this case plywood, is usually thinner than the hardwood floor to compensate for the thickness of the tile or linoleum.*

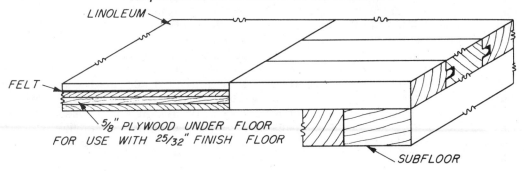

LINOLEUM

FELT

⅝″ PLYWOOD UNDER FLOOR FOR USE WITH 25/32″ FINISH FLOOR

SUBFLOOR

44-37. *Covering the old floor with underlayment.*

44-39. *Nailing the underlayment.*

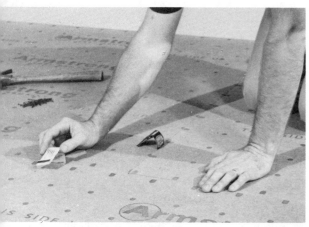

44-38. *Allow a distance equal to the thickness of a matchbook between the panels of underlayment.*

44-40. *Using a nailing machine to install underlayment.*

● Nail the underlayment or plywood with coated or ring-grooved nails at least every four inches along all edges and over the entire face of the panels. Fig. 44-39. Most manufacturers of underlayment provide a nailing guide in the form of lines or dots stenciled on the surface to indicate the nail locations.

● The underlayment or plywood may be nailed with a nailing machine to speed up the operation. Fig. 44-40.

● Find the center points of the two ends walls of the room. Connect these points with a chalk line and snap a line down the middle of the room. Fig. 44-41.

● Locate the midpoint of this line. Use a framing square to draw a perpendicular line from this point. Fig. 44-42. If a square is not available, a tile may be used with one edge on the center line to establish the perpendicular.

● The center lines should be at an exact 90° angle (right angle) to each other. This can be assured by constructing a 3′ × 4′ × 5′ right triangle from the two lines and a diagonal. Fig. 44-43. In a large room, a 6′ × 8′ × 10′ triangle can be used.

● Along the perpendicular line, strike a chalk line which extends to both of the side walls. Along the chalk lines, lay one test

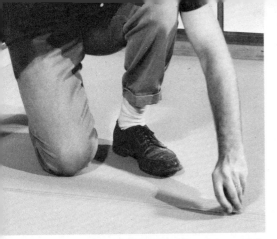

44-41. *Snapping a chalk line down the middle of the room.*

44-42. *Using a framing square to draw a perpendicular from the centerline.*

row of uncemented tile from the center point to one side wall and to one end wall. Fig. 44-44.

● The uncemented tiles are placed along the line with the edge of the tiles exactly on the line. The rows may be shifted by moving one tile at a time until the border space at the end of each of the rows is equal. Count the tiles in each row. If there is an even number of tiles, those at the center of each row (tiles A, B, and C) should meet exactly on the center lines. Fig. 44-45.

● If there is an odd number of tiles, the center lines on the floor will bisect the tiles. Fig. 44-46.

● Border tiles should not be less than half a tile wide. Measure the distance between the wall and the last tile. Fig. 44-47. If the distance is less than 2″ or more than 8″, move the center line parallel to that wall 4½″ closer to the wall.

● Locate the new center line with the chalk line and snap it. Fig. 44-48. Moving the center line closer to the wall creates wider borders. For example, if the borders are 1½″ wide, moving the center line 4½″ closer to one wall will take away the 1½″ border plus 3″ from the next tile. The border tile on that side will then be 6″ wide. On the opposite side, 4½″ will be gained. The border will thus become 6″ wide (4½″ + 1½″). Since the line is moved 4½″, half the size of one tile, the border tile remains uniform on both sides.

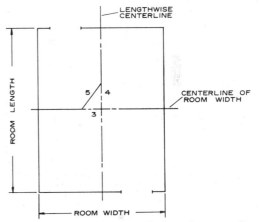

44-43. *The two lines must be exactly perpendicular.*

● Choose the correct adhesive for the tile. Spread a thin coat over one-fourth of the room. Do not cover the chalk lines. The adhesive may be trowled or brushed on, depending on the type of tile to be laid and the adhesive used. Fig. 44-49.

● Allow the adhesive to dry about 15 minutes. Then test it for proper tackiness by touching lightly with the thumb. Fig. 44-50. It should feel tacky but not stick to the thumb. If it sticks to the thumb, allow more drying time.

● For an even number of tiles, start at one inside corner and lay the first tile exactly in line with the marked center lines. For an

44-44. *Placing test rows of uncemented tile along the chalk lines.*

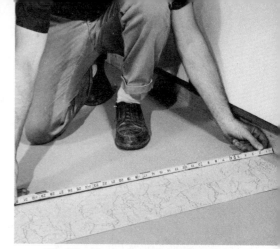

44-47. *Measuring the border space.*

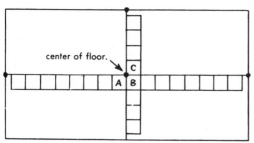

44-45. *Layout for an even number of tiles.*

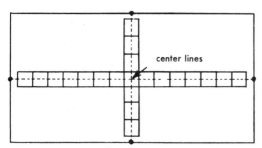

44-46. *Layout for an odd number of tiles.*

44-48. *Relocating the center line.*

odd number of tiles, center the first tile over the intersection of the center lines. The second tile can be laid adjacent to the first on one side. The third tile is laid adjacent to the first on the other side. Continue laying tiles along the center line

and filling in between until the entire section is covered. Fig. 44-51. The remaining three sections can be covered in the same way. Do not slide tiles into position. Some kinds of tile require only pressing in place; others should be rolled after installation for better adherence.

● To fit border tiles, place a loose tile (A) exactly over the last tile in the row. Take another tile (B) and place it against the wall, overlapping tile A. Mark tile A with a pencil along the edge of B. Fig. 44-52.

● With household shears, tin-snips, or a knife, cut tile A along the pencil mark. Fig. 44-53.

● The cut portion of tile A will fit exactly into the border space. Fig. 44-54. Place the

44–49. *Spreading the adhesive.*

44–50. *Testing the adhesive for proper tackiness.*

hand-cut edge against the wall. Repeat this procedure until the border area is completely covered. A clearance of $1/8''$ to $1/4''$ should be allowed at all sides for expansion. This space is covered with a cove base of the same resilient material as the tile or with a standard wood base. Wood base is usually lower in cost than the resilient cove base, but installation costs are somewhat greater. When installing vinyl tile, each quarter of the room should be rolled with a linoleum roller or other smooth roller as it is completed.

• To cut around pipes or other obstructions in a room, first make a paper pattern to fit the space exactly. Then trace the outline onto the tile and cut along the outline. Some tile can be cut with scissors. Fig. 44-55.

• Vinyl-asbestos feature strips in solid colors can be used to create an unusual effect. Several colors can be combined to provide a room with individuality and an interesting complement to the room's decor. Spread the adhesive, allow it to dry, and lay the feature strips in place. Fig. 44-56.

• Vinyl cove base can be installed along the walls of the room. After cutting the proper lengths to fit into place, apply the adhesive to the back of the vinyl cove base and press the material against the wall. Fig. 44-57. This completes the floor installation. Fig. 44-58.

44–51a. *Installing the tiles.*

ESTIMATING

Regardless of the type of flooring to be installed, it is first necessary to determine the number of square feet of floor space to be covered. The number of square feet is figured by multiplying the length of the room times the width of the room. Any offsets or closets may be figured separately by multiplying their length times their width and adding the product to the total number of square feet in the main room. For example, if the room is 10' by 12', the room contains 120 sq. ft. of floor area (10 ×

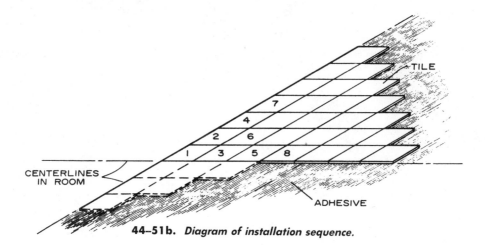

TILE

7

4

2 6

1 3 5 8

CENTERLINES
IN ROOM

ADHESIVE

44–51b. *Diagram of installation sequence.*

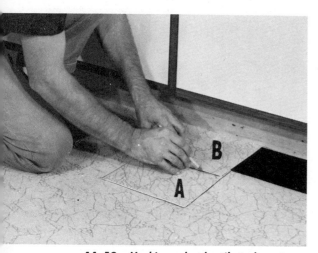

44–52. Marking a border tile to be cut.

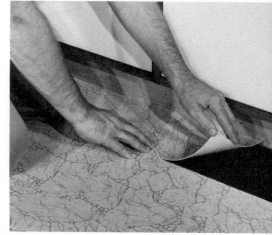

44–54. Installing border tile.

44–53. Cutting the border tile to the correct size.

44–55. Fitting tile around a pipe.

824

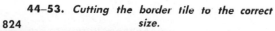

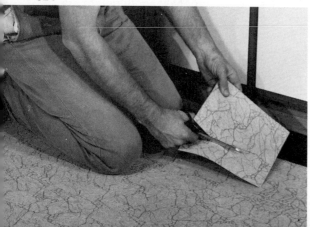

12 = 120). If the closet is 2' wide by 8' long, the closet contains a total of 16 sq. ft. The main room and closet contain 136 sq. ft. of floor area (120 + 16 = 136).

Strip Flooring

To determine the amount of strip flooring needed, refer to the chart in Fig. 44-59. Under the section entitled strip flooring read across on the line for the size flooring to be used. If $^{25}/_{32}$″ × $2^{1}/_{4}$″ strip flooring is selected, it will take 138.3 board feet to lay 100 square feet of floor area. In the sample room described earlier, there are 136 square feet of floor space; 188.09 board feet of strip flooring will be required.

$$\frac{136 \text{ (sq. ft. of floor space)}}{100} = 1.36 \text{ (floor area in hundreds of sq. ft.)}$$

$$1.36 \times 138.3 \text{ (board ft. per 100 sq. ft.)} = 188.09 \text{ board ft.}$$

MATERIAL COST

To figure the cost of the material, multiply the cost per board foot times the total number of board feet required for the job.

44–56. *Installing vinyl-asbestos strips.*

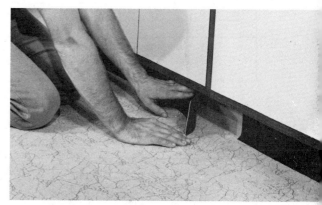

44–57. *Applying vinyl cove base.*

44–58. *A basement recreation room. Vinyl-asbestos floor tiles, feature strips in three different colors, and a vinyl cove base create an attractive custom-designed floor.*

825

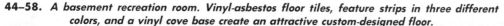

The chart in Fig. 44-59 also has information for determining the number of nails necessary. Read across on the line for the size strip flooring used to the heading "Nails." The chart shows it will take 3 pounds of nails to lay 100 square feet of $2^1/_4$" strip flooring. The floor in the example contains 136 square feet and will require a little over 4 pounds of nails.

$$\frac{136 \text{ (sq. ft. of floor area)}}{100} = 1.36 \text{ (floor area in hundreds of sq. ft.)}$$

$$1.36 \times 3 \text{ (lbs. of nails per 100 sq. ft.)} = 4.08 \text{ lbs.}$$

Multiply the cost per pound times the number of pounds required to find the cost of the nails.

44–59. *These charts can be used to figure materials and labor required for various types of flooring.*

WOOD BLOCK FLOORING	Material			Labor	
	Block	Adhesive	Nails	Per 100 Square Feet	
Size	Per 100 Sq. Ft.	Per 100 Sq. Ft.	Lbs. per 100 Sq. Ft.	Nailed	Adhesive
8 x 8	225	1 Gallon	4.0 Pounds	6.5 Hours	3.5 Hours
9 x 9	178	1 Gallon	3.5 Pounds	5.0 Hours	2.2 Hours
12 x 12	100	1 Gallon	2.8 Pounds	3.5 Hours	1.8 Hours

RESILIENT FLOORING	Labor						Material
	Hours per 100 Square Feet						Mastic
Type	4 x 4	6 x 6	9 x 9	6 x 12	12 x 12	9 x 18	Per 100 Square Feet
Rubber Tile	5.0	3.3	2.2	2.2	1.8	1.3	0.75 Gallon
Asphalt Tile		3.3	2.0	2.0	1.3	1.3	0.75 Gallon
Linotile		7.0	6.4		5.7		1.50 Gallon
Plastic Tile			2.4				1.35 Gallon
Cork Tile		6.0	3.5		2.5		1.50 Gallon

STRIP FLOORING	Material		Nails	Labor		
	Board Feet per 100 Square Feet	1000 Board Feet will lay Square Feet	Per 100 Square Feet	Hours per 100 Square Feet		
Size				Laying	Sanding	Finishing
$^{25}/_{32}$ x $1^1/_2$	155.0	645.0	3.7 Pounds	3.7 Hours		
$^{25}/_{32}$ x 2	142.5	701.8	3.0 Pounds	3.4 Hours		
$^{25}/_{32}$ x $2^1/_4$	138.3	723.0	3.0 Pounds	3.0 Hours	Average	Average
$^{25}/_{32}$ x $3^1/_4$	129.0	775.2	2.3 Pounds	2.6 Hours	1.3	2.6
$^3/_8$ x $1^1/_2$	138.3	723.0	3.7 Pounds	3.7 Hours		
$^3/_8$ x 2	130.0	769.0	3.0 Pounds	3.4 Hours	Hours	Hours
$^1/_2$ x $1^1/_2$	138.3	723.0	3.7 Pounds	3.7 Hours		
$^1/_2$ x 2	130.0	769.2	3.0 Pounds	3.4 Hours		

LABOR COST

To figure the labor for laying the floor in the example, refer to the column headed "Labor" on the table in Fig. 44-59. The table shows that a worker can lay 100 square feet of $^{25}/_{32}"\times 2^1/_4"$ strip flooring in 3 hours. By using the table, you can determine that the room in the example will require 4.08 hours to lay.

$$\frac{136 \text{ (sq. ft. of floor area)}}{100} = 1.36 \text{ (floor area in hundreds of sq. ft.)}$$

$$1.36 \times 3 \text{ (hrs. of labor per 100 sq. ft.)} = 4.08 \text{ hrs.}$$

Multiply this figure by the hourly rate to determine total labor cost. If it is desirable to determine the cost for sanding and finishing, this may be figured in the same way.

Other Types of Flooring

The charts in Fig. 44-59 for resilient flooring and wood block flooring may be used to determine the amount of material and the cost for labor once the total number of square feet in the room has been determined.

The chart in Fig. 44-60 may be used for estimating the necessary number of floor

44–60. *Tile flooring estimating chart.*

Estimating Floor Tile

Square Feet	Number of Tiles			
	9" x 9"	12" x 12"	6" x 6"	9" x 18"
1	2	1	4	1
2	4	2	8	2
3	6	3	12	3
4	8	4	16	4
5	9	5	20	5
6	11	6	24	6
7	13	7	28	7
8	15	8	32	8
9	16	9	36	8
10	18	10	40	9
20	36	20	80	18
30	54	30	120	27
40	72	40	160	36
50	89	50	200	45
60	107	60	240	54
70	125	70	280	63
80	143	80	320	72
90	160	90	360	80
100	178	100	400	90
200	356	200	800	178
300	534	300	1200	267
400	712	400	1600	356
500	890	500	2000	445
Labor per 100 Sq. Ft.	2 Hours	1.3 Hours	3.3 Hours	1.3 Hours

Flooring Adhesives

Type and Uses	Approximate Coverage in Square Feet
Primer—For treating on-or-below-grade concrete subfloors before installing asphalt tile	250 to 350
Asphalt Cement—For installing asphalt tile over primed concrete subfloors in direct contact with the ground	200
Emulsion Adhesive—Adhesive used for installing asphalt tile over lining felt	130 to 150
Lining Paste—For cementing lining felt to wood subfloors	160
Floor and Wall Size—Used to prime chalky or dusty suspended concrete subfloors before installing resilient tiles other than asphalt	
Waterproof Cement—Recommended for installing linoleum tile, rubber, and cork tile over any type of suspended subfloor in areas where surface moisture is a problem	130 to 150

Tile Waste Allowances

1 to	50 sq. ft.....................................	14%
50 to	100 sq. ft.....................................	10%
100 to	200 sq. ft.....................................	8%
200 to	300 sq. ft.....................................	7%
300 to	1000 sq. ft....................................	5%
Over 1000 sq. ft.....................................		3%

To find the number of tiles needed for an area not shown on the chart, such as the number of 9" x 9" tile needed for 850 sq. ft., add the number of tiles for 50 sq. ft. to the number of tiles needed for 800 sq. ft. The result will then be 1513 to which must be added 5% for waste (see table). Total 1589.

44-61. *Standard sizes, counts, and weights of wood flooring.*

Standard Sizes, Counts And Weights

OAK			
Nominal	Actual	Counted	Wts. M Ft.
Tongued And Grooved-End Matched			
$25/32$ x 3¼ in.	$25/32$ x 3¼ in.	1 x 4 in.	2300 lbs.
$25/32$ x 2¼ in.	$25/32$ x 2¼ in.	1 x 3 in.	2100 lbs.
$25/32$ x 2 in.	$25/32$ x 2 in.	1 x 2¾ in.	2000 lbs.
$25/32$ x 1½ in.	$25/32$ x 1½ in.	1 x 2¼ in.	1900 lbs.
⅜ x 2 in.	$11/32$ x 2 in.	1 x 2½ in.	1000 lbs.
⅜ x 1½ in.	$11/32$ x 1½ in.	1 x 2 in.	1000 lbs.
½ x 2 in.	$15/32$ x 2 in.	1 x 2½ in.	1350 lbs.
½ x 1½ in.	$15/32$ x 1½ in.	1 x 2 in.	1300 lbs.
Square Edge			
$5/16$ x 2 in.	$5/16$ x 2 in.	face count	1200 lbs.
$5/16$ x 1½ in.	$5/16$ x 1½ in.	face count	1200 lbs.
BEECH, BIRCH, HARD MAPLE AND PECAN			
Nominal	Actual	Counted	Wts. M Ft.
Tongued And Grooved-End Matched			
$25/32$ x 3¼ in.	$25/32$ x 3¼ in.	1 x 4 in.	2300 lbs.
$25/32$ x 2¼ in.	$25/32$ x 2¼ in.	1 x 3 in.	2100 lbs.
$25/32$ x 2 in.	$25/32$ x 2 in.	1 x 2¾ in.	2000 lbs.
$25/32$ x 1½ in.	$25/32$ x 1½ in.	1 x 2¼ in.	1900 lbs.
⅜ x 2 in.	$11/32$ x 2 in.	1 x 2½ in.	1000 lbs.
⅜ x 1½ in.	$11/32$ x 1½ in.	1 x 2 in.	1000 lbs.
½ x 2 in.	$15/32$ x 2 in.	1 x 2½ in.	1350 lbs.
½ x 1½ in.	$15/32$ x 1½ in.	1 x 2 in.	1300 lbs.
Special Thicknesses (T and G, End Matched)			
$17/16$ x 3¼ in.	$33/32$ x 3¼ in.	1¼ x 4 in.	2400 lbs.
$17/16$ x 2¼ in.	$33/32$ x 2¼ in.	1¼ x 3 in.	2250 lbs.
$17/16$ x 2 in.	$33/32$ x 2 in.	1¼ x 2¾ in.	2250 lbs.
Jointed Flooring—i.e., Square Edge			
$25/32$ x 2½ in.	$25/32$ x 2½ in.	1 x 3¼ in.	2250 lbs.
$25/32$ x 3¼ in.	$25/32$ x 3¼ in.	1 x 4 in.	2400 lbs.
$25/32$ x 3½ in.	$25/32$ x 3½ in.	1 x 4¼ in.	2500 lbs.
$17/16$ x 2½ in.	$33/32$ x 2½ in.	1¼ x 3¼ in.	2500 lbs.
$17/16$ x 3½ in.	$33/32$ x 3½ in.	1¼ x 4¼ in.	2600 lbs.

"Nominal size" is the term ordinarily used in referring to hardwood flooring sizes. In all cases the nominal and the actual widths are the same. In some instances, however, the nominal thickness is $1/32$-inch greater. Nominal sizes probably have their origin in the sizes in which flooring actually was made years ago. Their continued use is a matter of convenience in the trade.

"Actual size" is the thickness and face width to which the flooring measures after it has been processed at the mill. It does not include the width of the tongue in side-matched flooring.

"Counted size" is the size used in determining the board feet in a shipment. In arriving at the counted size, pieces which are 1 inch or less in thickness are considered to be 1-inch flooring. Allowance is made over and above the actual face width for flooring which is tongued and grooved at the sides. In $25/32$-inch flooring and thicker, the extra allowance is ¾-inch, while in other sizes the extra allowance is ½-inch. Thus a piece of side-matched flooring which measures $25/32$-inch x 2¼ inches in actual or face width is referred to as 3 inches in counted width.

tiles of various sizes, the amount of flooring adhesive needed, and the waste allowances for laying tile in various size rooms.

The two charts shown in Figs. 44-61 and 44-62 also contain information pertaining to wood flooring and synthetic flooring. These two charts will provide additional helpful information when estimating and ordering finish flooring materials.

44-62. *Weights and gauges of various kinds of synthetic flooring.*

Weights And Gauges: Floorings

Material	Approximate Thickness, Inches	Finished Gauge, Inches	Average Net Wt. per Sq. Ft. in Lbs.	Roll Width in Feet
Asphalt Tile	⅛	0.125	1.16	
	$3/16$	0.187	1.75	
Asphalt Tile	⅛	0.125	1.17	
(Greaseproof)	$3/16$	0.188	1.74	
Conductive Asphalt Tile (Regular	⅛	0.125	0.97	
& Greaseproof types)	$3/16$	0.187	1.45	
Felt—Lining Felt	$1/25$	0.040	1.40	3
Industrial Asphalt Tile	⅛	0.125	0.90	
	$3/16$	0.187	1.35	
Rubber Tile	⅛	0.125	1.24	
Vinyl Tile..............	$3/16$	0.187	1.86	
	$3/32$	0.0925	0.93	
Linoleum				
Battleship	⅛	0.125	0.83	6
Heavy Gauge				
Embossed Inlaid				
Standard Gauge	$3/32$	0.0925	0.60	6
Jaspe				
Heavy Gauge	⅛	0.125	0.83	6
Standard Gauge	$3/32$	0.0925	0.65	6
Marbleized				
Heavy Gauge	⅛	0.125	0.92	6
Standard Gauge	$3/32$	0.0925	0.65	6
Light Gauge	$1/16$	0.070	0.46	6
Plain				
Heavy Gauge	⅛	0.125	0.83	6
Standard Gauge	$3/32$	0.0925	0.60	6
Straight Line Inlaid,				
Standard Gauge	$3/32$	0.0925	0.62	6
Light Gauge	$1/16$	0.070	0.46	6

*The weights and gauges in this table are manufacturing standards. Slight variations will occur, but for practical purposes, these figures are substantially correct.

QUESTIONS

1. What is finish flooring?

2. What are some of the finish flooring materials?

3. What are the most common hardwoods used for wood strip flooring?

4. What wood species is used most often for wood strip flooring?

5. What is the most popular thickness and width of strip flooring?

6. Who enforces the grading rules and regulations for hardwood flooring?

7. What are the grades of oak strip flooring?

8. What is the last operation before actual installation of finish flooring?

9. What factors must be taken into consideration when deciding what area of the house to begin the installation?

10. When two different floor coverings are used in adjoining rooms, they should terminate under the center of the door when the door is closed. Why?

11. What kind and size of nails are recommended for laying strip flooring?

12. How is strip flooring installed over concrete?

13. What are some other types of hardwood flooring besides strip flooring?

14. What thickness of underlayment is used for linoleum floors?

Stairs

There are two general types of stairs: principal and service. The principal stairs are designed to provide ease and comfort and are often made a feature of house design. The service stairs lead to the basement or attic. They are usually somewhat steeper and constructed of less expensive materials.

Stairs may be built on the job or assembled from units built in a mill. All parts for a finish stairway can be purchased from a lumberyard as stock mill items. Stairways may have a straight, continuous run with or without an intermediate platform. They may also consist of two or more runs at angles to each other. Usually there is a platform at the angle. The turn may also be made by radiating treads called winders.

Fig. 45-1 shows the stair patterns most often found in homes. Winders are not often used because they are not as safe as platforms. Fig. 45-2. The stairway for most homes is a straight, continuous run, although a stairway with a landing or platform is sometimes used to conserve space. Fig. 45-3. Details for stair building are shown in the stairwell section of most house plans. Fig. 45-4 (Page 833).

There are many different kinds of stairs, but all have two main parts in common: the treads people walk on and the stringers which support the treads. The simplest stairway has a pair of straight-edged stringers and a series of plank treads. Fig. 45-5. It is called a cleat stairway because the treads are supported by cleats nailed

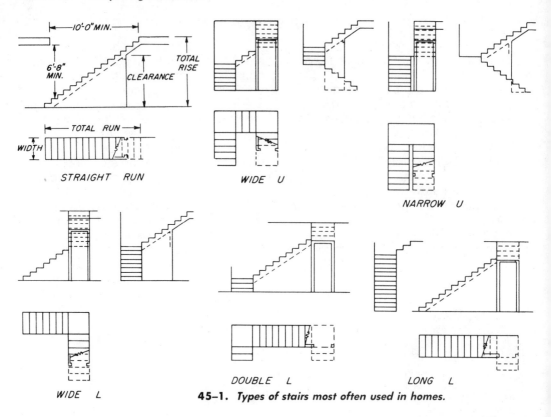

45-1. Types of stairs most often used in homes.

45-2. A stairway with winders. At the inner corner where all the winders meet, there is very little if any tread to support one's foot. This makes winders rather dangerous, and they are not often used.

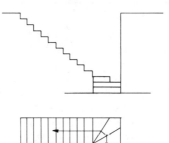

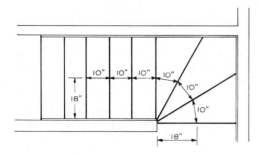

to the stringers. A complete stairway includes two or more sawtooth-edged stringers, a series of treads, and a series of risers. Fig. 45-6 (Page 834). The stringers shown in Fig. 45-6 are cut out of solid pieces (usually 2″ × 12″), and are therefore called cutout or sawed stringers. In some stairways the treads and risers are supported on triangular stair blocks nailed to the upper edges of straight-edged stringers. Fig. 45-7.

830

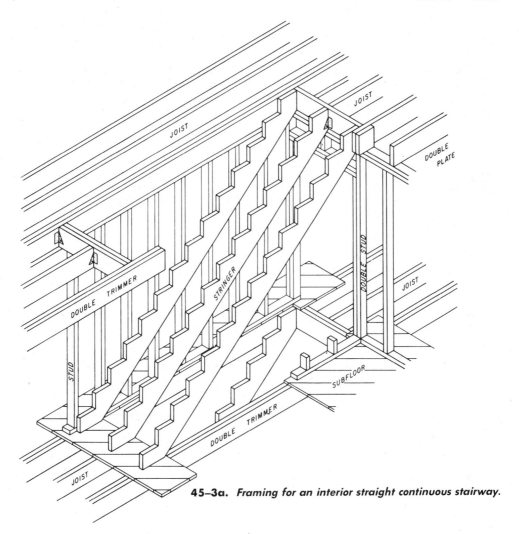

45-3a. *Framing for an interior straight continuous stairway.*

COMMON STAIR PARTS AND TERMS

String or *stringer*, sometimes called a *carriage*, or *horse*. One of the inclined sides of a stair which support the treads and risers. Open (plain) stringers can be either rough or finish stock and are cut to follow the lines of the treads and risers. Closed stringers have parallel sides, with the risers and treads housed into them. The term also applies to any similar member, whether a support or not, such as finish stock placed outside the carriage on open stairs and next to the walls on closed

stairs. Figs. 45-5, 45-6, 45-7, 45-8 (Page 836), 45-9, and 45-10 show various kinds of stringers.

Riser. The vertical face of one step. Fig. 45-11.

Tread. The horizontal face of one step. Fig. 45-11.

Winders. Radiating or wedge-shaped treads at turns of stairs. Fig. 45-12 (Page 838).

Nosing. The projection of tread beyond the face of the riser. Fig. 45-11.

Railing. The protection on the open side of a run of stairs. Fig. 45-11.

Newel. The main post of the railing at the start of the stairs and the stiffening post at angles or platforms. Fig. 45-11.

Handrail. The top finishing piece on the railing to be grasped by the hand when going up or down the stairs. Fig. 45-11.

Balusters. The vertical members supporting the handrail on open stairs. Fig. 45-11. For closed stairs where there is no railing, the handrail is attached to the wall with brackets.

Platform. The intermediate area between two parts of a flight of stairs. Fig. 45-13 (Page 838), arrow 1.

Landing. The floor at the top or bottom of each story where the flight of stairs ends or begins. Fig. 45-13, arrow 2.

Total rise. The total vertical distance from one floor to the next. Fig. 45-1.

Total run. The total horizontal length of the stairs. Fig. 45-1.

FRAMING A STAIRWELL

When large openings are made in floors, such as for stairwells, one or more joists must be cut. The location of openings in the floor determines the method of framing. When the length of the stairway is parallel

45-3b. *Framing for an interior stairway with a platform.*

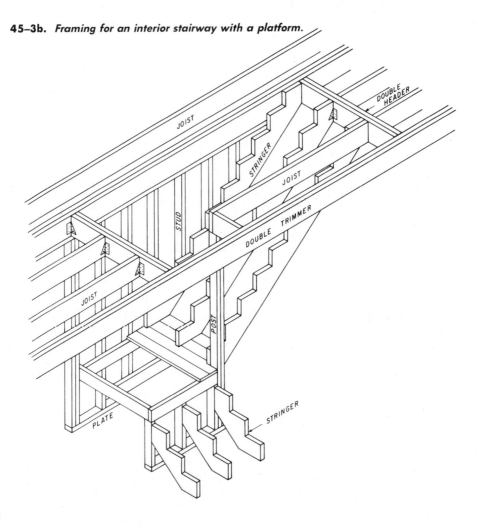

to the joists, the opening is framed as shown in Fig. 45-14. When the stairway is arranged so that the opening is perpendicular to the length of the joists, the framing should follow the details shown in Fig. 45-15 (Page 839). Fig. 45-16 (Page 840) shows typical framing for a stair landing. Nailing and framing of stairwell openings should comply with the principles explained in Unit 24, "Floor Framing."

DESIGNING A STAIRWAY

Stairways are designed, arranged, and installed for safety, adequate headroom, and space for the passage of furniture. There are three important considerations when designing the stairway:
- The stair width.
- The headroom.

- The relationship between the height of the riser and the width of the tread.

Stair Width

Staircases must be wide enough to allow two people to pass comfortably on the

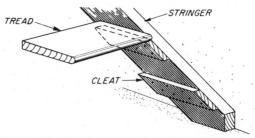

45–5. *The simplest type of stairs, a cleat stairway.*

45–4. *A stairwell section drawing for a set of house plans. Note that the architect has indicated the riser to be 7⁷/₈″ and the tread to be 9″.*

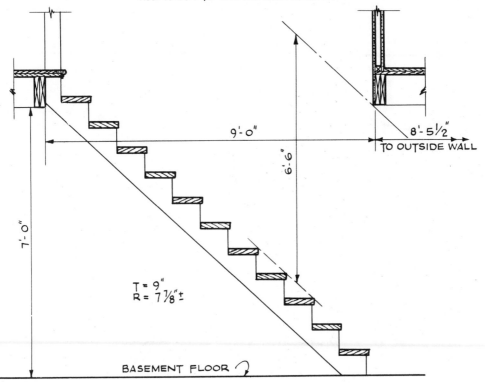

$T = 9''$
$R = 7\frac{1}{8}'' \pm$

BASEMENT FLOOR

stairs and to permit furniture to be carried up or down. The minimum width is 3 feet; however, 3½ feet is better. Fig. 45-17 (Page 840).

The width needed for the passage of furniture varies. Stairs which are open on one side, including open-well stairs, are best for moving large pieces of furniture. The furniture can usually be raised up over the handrails and newel posts.

Headroom

Headroom is the distance or clearance above a stair. It is measured from the outside corner of the tread and riser to the lowest point of the soffit or ceiling directly overhead. If two or more flights of stairs are arranged one above the other in the same stairwell (for example, cellar stairs under, or attic stairs over, the main staircase), getting enough headroom beneath the upper stair is a planning problem and must be carefully considered. Although the minimum required by the FHA is 6' 8", this is usually not enough for the main stairway. It has been found from studies of the dimensions of the average man or woman that headroom may vary with the steepness of

45–6. _Parts of a stairway._

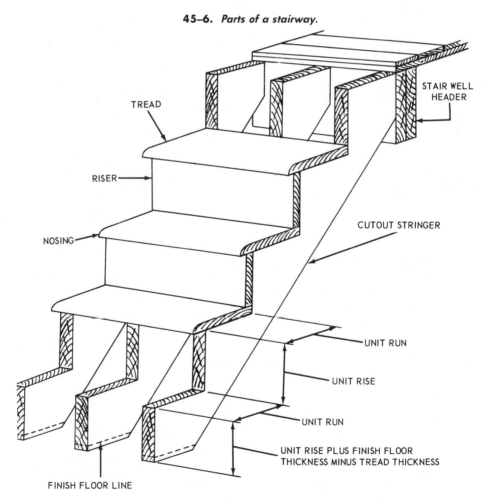

TREAD

RISER

NOSING

STAIR WELL HEADER

CUTOUT STRINGER

UNIT RUN

UNIT RISE

UNIT RUN

UNIT RISE PLUS FINISH FLOOR THICKNESS MINUS TREAD THICKNESS

FINISH FLOOR LINE

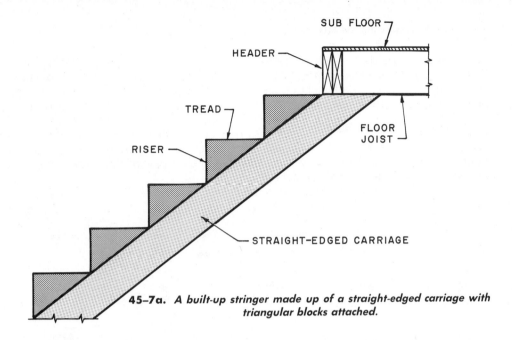

45–7a. *A built-up stringer made up of a straight-edged carriage with triangular blocks attached.*

45–7b. *If additional support is needed for a wide stairway, a third stringer is used in the center. The triangular blocks may be nailed to the edge or to the face of the straight stringer (carriage).*

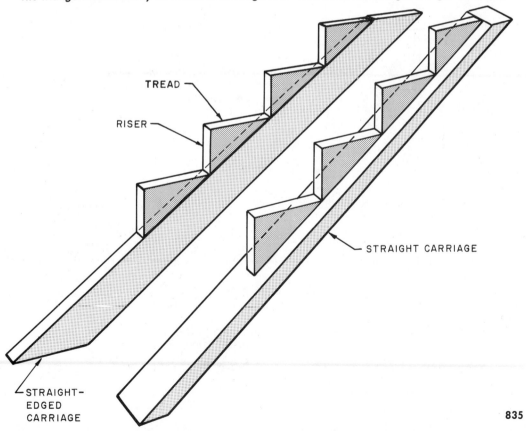

the stairs, but should generally be between 7' 4" and 7' 7". This allows for the arm to be swung up over the head without hitting anything. Fig. 45-18 (Page 841).

Riser and Tread Relationship

It is very important that the stairway be built with the proper rise and run. The relationship between the height of the riser and the width of the tread determines the ease with which the stairs may be ascended or descended. If the combination of run and rise is too great, the steps are tiring. There is a strain on the leg muscles and the heart. If the combination is too short, the foot may kick the riser at each step. An attempt to shorten one's stride may be tiring. If one of the three rules described

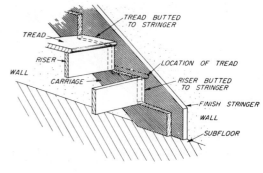

45–8. *A finished wall stringer and carriage.*

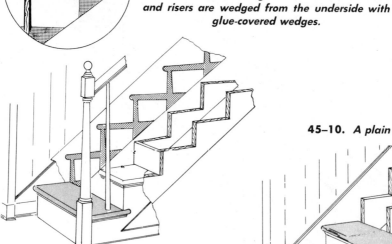

45–9. *A housed stringer. Note that the treads and risers are wedged from the underside with glue-covered wedges.*

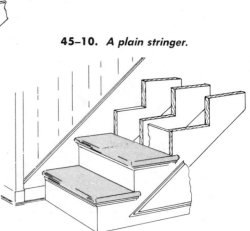

45–10. *A plain stringer.*

45–11. *The parts of this finished stairway can be purchased from a lumber-yard as stock mill items.*

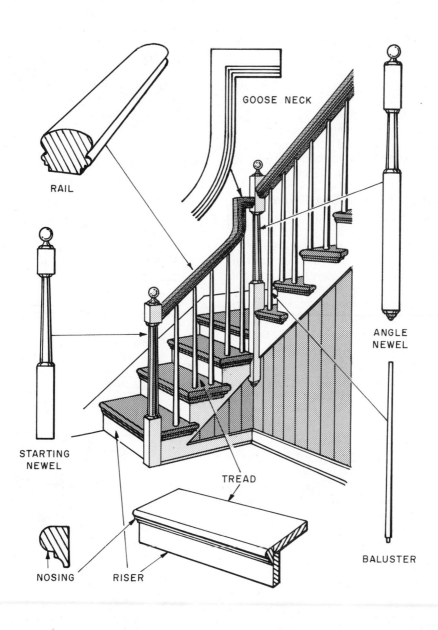

RAIL

GOOSE NECK

ANGLE NEWEL

STARTING NEWEL

TREAD

NOSING

RISER

BALUSTER

below is followed, the stairway will be easy and safe to ascend and descend. Figs. 45-19 and 45-20.

• The sum of two risers and one tread should be between 24″ and 25″. Acceptable therefore would be a riser 7″ to 7¹/₂″ and tread 10″ to 11″. (Example: 7¹/₂″ + 7¹/₂″ + 10″.)

• The sum of one riser and one tread should equal between 17″ and 18″. (Example: 7¹/₂″ + 10″.)

• The product obtained by multiplying the height of the riser by the width of the tread should be between 70″ and 75″. (Example: 7″ × 10″.) For the main staircase in a house, risers should not be higher than 7⁵/₈″ or less than 7″, combined with a tread width of 10″ to 11″ (not including nosing).

LAYING OUT A STAIRWAY

The first step in stairway layout is to determine the *unit rise* and *unit run* per step. Fig. 45-21. The unit rise (height of one riser) is calculated on the basis of the total rise of the stairway, and the fact that the unit rise for stairs should be about 7″.

The total rise is the vertical distance between the lower finish floor level and the upper finish floor level. It is given on the

elevations and wall sections. Fig. 45-22 (Page 842). The actual distance, however, may vary slightly from the specified distance. Measure the actual distance. If both the lower and the upper floor are to be covered with finish flooring of the same thickness, the vertical distance between the subfloor levels will be the same as the vertical distance between the finish floor levels. However, you may be measuring up from a finish floor (a concrete basement floor, for example) to a floor which will be covered with finish flooring. In this case, you must add the thickness of the upper

45-13. *The platform is shown at arrow 1, the landing at arrow 2.*

45-12. *A stairway with winders, shown at arrow 1.*

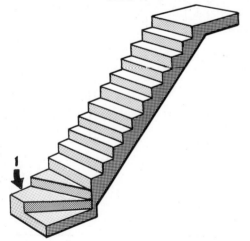

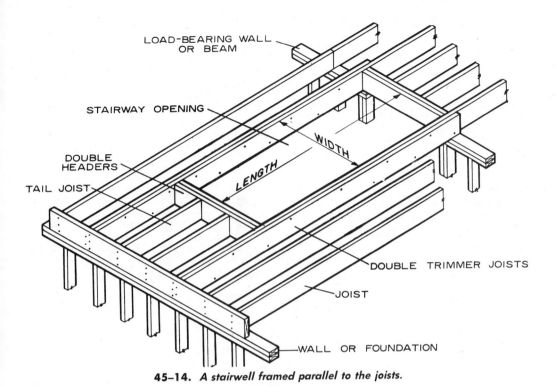

45-14. *A stairwell framed parallel to the joists.*

Labels on figure 45-14:
LOAD-BEARING WALL OR BEAM
STAIRWAY OPENING
WIDTH
LENGTH
DOUBLE HEADERS
TAIL JOIST
DOUBLE TRIMMER JOISTS
JOIST
WALL OR FOUNDATION

45-15. *A stairwell framed perpendicular to the joists.*

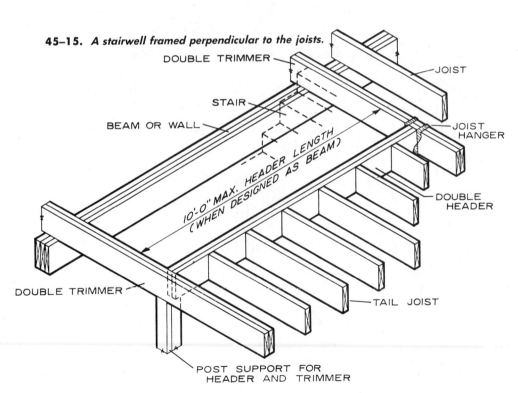

Labels on figure 45-15:
DOUBLE TRIMMER
STAIR
BEAM OR WALL
JOIST
JOIST HANGER
10'-0" MAX. HEADER LENGTH (WHEN DESIGNED AS BEAM)
DOUBLE HEADER
DOUBLE TRIMMER
TAIL JOIST
POST SUPPORT FOR HEADER AND TRIMMER

floor finish flooring to the vertical distance between the lower finish floor and the upper subfloor.

Let's assume that the total rise, or vertical distance between finish floors, is 8'11". The unit rise can be determined from the total rise as follows:

- The total rise is 107" (8' 11" = 107").
- A permissible unit rise is anything near 7". Divide 107" by 7".
- The result (disregarding any fraction) is the total number of risers in the stairway. In this case, the total is 15 risers (107 ÷ 7 = 15).
- To get the unit rise, divide the total rise by the number of risers. The unit rise in this case is 7⅛" (107 ÷ 15 = 7.13, or 7⅛).

The unit run is equal to the width of a tread, less the width of the nosing. Fig. 45-21. It is calculated on the basis of the unit rise and a rule, discussed earlier, that the sum of one riser and one tread should equal between 17" and 18". Subtract the

45–17. *The stair width should be adequate. Three feet is acceptable, but wider stairs are preferred.*

45–16. *Framing for a landing.*

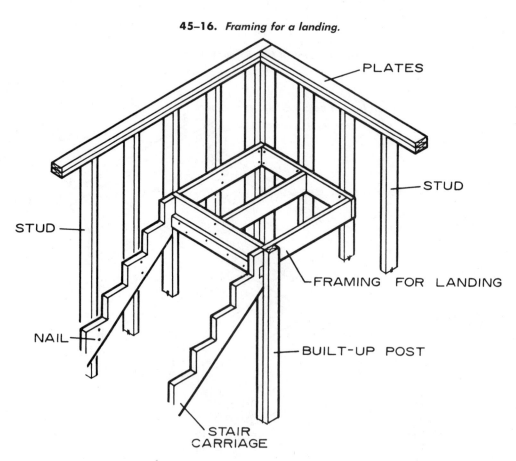

STUD

PLATES

STUD

STUD

FRAMING FOR LANDING

NAIL

BUILT-UP POST

STAIR CARRIAGE

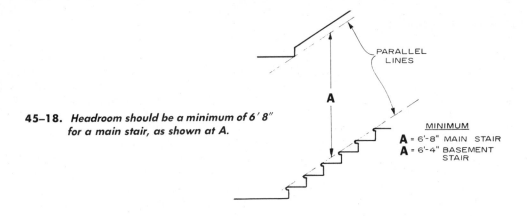

45-18. *Headroom should be a minimum of 6' 8" for a main stair, as shown at A.*

PARALLEL
LINES

A

MINIMUM
A = 6'-8" MAIN STAIR
A = 6'-4" BASEMENT
STAIR

45-19. *A stairway should be built with the proper rise and run: A. Stairway with a tread too wide and the riser too short. B. Stairway with a tread too narrow and the riser too high. C. Stairway with a correct tread width (10") and the correct riser height (7¹/₂").*

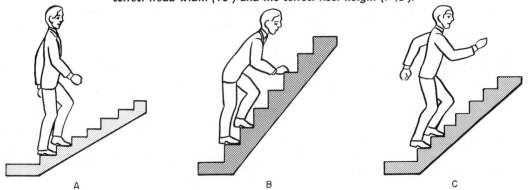

A

B

C

45-20. *Riser and tread relationships.*

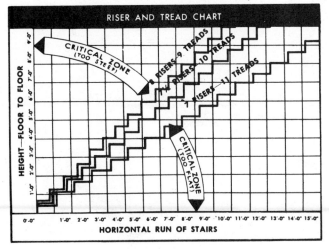

RISER AND TREAD CHART

HEIGHT—FLOOR TO FLOOR

CRITICAL ZONE (TOO STEEP)

8 RISERS-9 TREADS
10 RISERS-10 TREADS
11 RISERS-11 TREADS

CRITICAL ZONE (TOO FLAT)

HORIZONTAL RUN OF STAIRS

45-21. *The unit run (arrow A) is the distance from the face of one riser to the face of the next riser. It does not include the nosing. The unit rise (arrow B) is the distance from the top of one tread to the top of the next tread.*

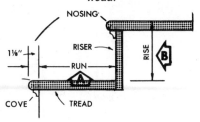

NOSING

RISER

1⅛"

RUN

RISE

B

COVE

TREAD

841

unit rise from this sum. Let's assume that the sum of one riser and one tread should be 17^1/$_2$". If the unit rise is 7^1/$_8$", the unit run will be 10^3/$_8$" (17^1/$_2$" − 7^1/$_8$" = 10^3/$_8$").

You now have all the information you need to lay out, cut, and install a cutout stringer except the total run of the stairway. The total run is equal to the unit run times the number of treads in the stairway. The total number of treads depends on the manner in which the upper end of the stairway is anchored to the upper landing. Three common types of anchorage are shown in Fig. 45-23.

In A, Fig. 45-23, there is a complete tread at the top of the stairway. This means that the number of treads in the stairway is the same as the number of risers. If there are 15 risers and the unit run is 10^3/$_8$", the total run of the stairway is 12' 11^5/$_8$" (15 × 10^3/$_8$ = 155^5/$_8$", or 12' 11^5/$_8$").

In B, Fig. 45-23, there is only part of a tread at the top of the stairway. In this case, the number of complete treads is one less than the number of risers, or 14. The total run of the stairway is 14 × 10^3/$_8$, plus the run of the partial tread at the top. This run

may be shown in detail. If not, you will have to estimate it as closely as possible. Let's assume it's about 7". The total run, then, is 12' 8^1/$_4$" (14 × 10^3/$_8$ = 145^1/$_4$"; 145^1/$_4$" + 7" = 152^1/$_4$", or 12' 8^1/$_4$").

In C, Fig. 45-23, there is no tread at the top of the stairway. The upper finish flooring serves as a top tread. In this case the number of treads is one less than the number of risers, or 14. The total run is 12' 1^1/$_4$" (14 × 10^3/$_8$ = 145^1/$_4$", or 12' 1^1/$_4$").

After you have calculated the total run of the stairway, drop a plumb bob from the stairwell header to the floor below. Measure off along the floor the total run, starting at the plumb bob. You have now located the anchoring point for the lower end of the stairway. Some standard stair layouts can be found in the chart in Fig. 45-24.

STRINGERS

The treads and risers are supported by stringers, or carriages, that are solidly fixed in place, level and true, upon the framework of the building. The stringers may be cut or routed to fit the outline of the treads and risers. A third stringer should

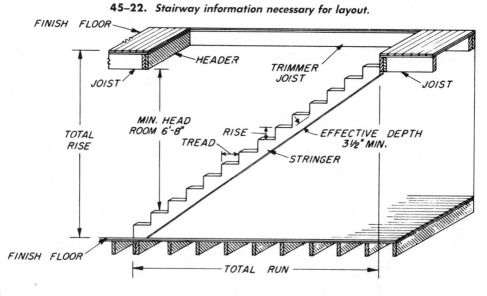

45–22. *Stairway information necessary for layout.*

FINISH FLOOR

HEADER

TRIMMER JOIST

JOIST

JOIST

TOTAL RISE

MIN. HEAD ROOM 6'-8"

RISE

TREAD

RISE

EFFECTIVE DEPTH 3½" MIN.

STRINGER

FINISH FLOOR

TOTAL RUN

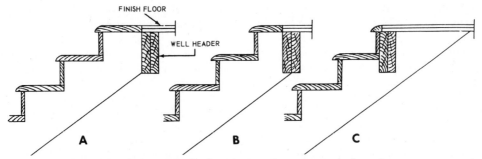

FINISH FLOOR

WELL HEADER

A   B   C

45–23. Three methods of anchoring the upper end of a stringer.

45–24. Layout dimensions for some standard stairways.

Stair With Landing

This type is easier to climb, safer, and reduces the length of space required. The landing provides a resting point and a logical place to have a right angle turn. Landing near bottom with quarter-turn is basis of calling this type "dog-legged" or "platform" stairs.

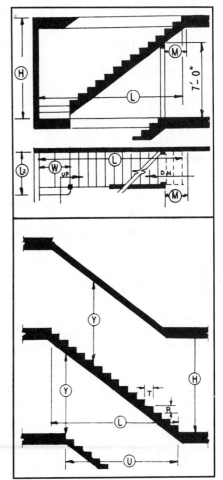

Height Floor to Floor H	Number of Risers	Height of Risers R	Width of Tread T	Run Number of Risers	Run L	Run Number of Risers	Run L2
8'0"	13	7⅜" +	10"	11	8'4" + W	2	0'10" + W
8'6"	14	7⁵/₁₆" −	10"	12	9'2" + W	2	0'10" + W
9'0"	15	7³/₁₆" +	10"	13	10'0" + W	2	0'10" + W
9'6"	16	7⅛"	10"	14	10'10" + W	2	0'10" + W

Straight Stairs

Simplest and least costly; requires a long hallway which may sometimes be a disadvantage. May have walls on both sides (closed string) or may have open balustrade on one side (open string).

Height Floor to Floor H	Number of Risers	Height of Risers R	Width of Treads T	Total Run L	Minimum Head Rm. Y	Well Opening U
8'0"	12	8"	9"	8'3"	6'6"	8'1"
	13	7⅜" +	9½"	9'6"	6'6"	9'2½"
	13	7⅜" +	10"	10'0"	6'6"	9'8½"
8'6"	13	7⅞" −	9"	9'0"	6'6"	8'3"
	14	7⁵/₁₆" −	9½"	10'3½"	6'6"	9'4"
	14	7⁵/₁₆" −	10"	10'10"	6'6"	9'10"
9'0"	14	7¹¹/₁₆" +	9"	9'9"	6'6"	8'5"
	15	7³/₁₆" +	9½"	11'1"	6'6"	9'6½"
	15	7³/₁₆" +	10"	11'8"	6'6"	9'11½"
9'6"	15	7⅝" −	9"	10'6"	6'6"	8'6½"
	16	7⅛"	9½"	11'10½"	6'6"	9'7"
	16	7⅛"	10"	12'6"	6'6"	10'1"

Note: Dimensions shown under well opening "U" are based on 6'6" minimum headroom. If headroom is increased well opening also increases.

be installed in the middle of the stairs when the treads are less than 1⅛″ thick or the stairs are more than 2′ 6″ wide.

In some cases rough stringers with rough treads nailed across them are used during the construction period. These are installed for the convenience of the workers until wall finish is applied. When the wall finish is completed, the temporary stairway is removed and finished stairs, which usually have been made in a mill, are erected or built in place.

For a housed stringer, the wall stringer is routed out to the exact profile of the tread, riser, and nosing, with sufficient space at the back to take the wedges. Fig. 45-25. The top of the riser is rabbeted to fit into a groove in the bottom front of the tread. The back of the tread is rabbeted into a groove in the bottom of the next riser. The wall stringer is spiked to the inside of the wall. The treads and risers are fitted together and forced into the wall stringer housing, where they are set tight by driving and gluing wood wedges behind them. Fig. 45-26. The wall stringer thus shows above the profiles of the treads and risers as a finish against the wall. It is often made continuous with the baseboard of the upper and lower landing.

If the outside stringer (stair carriage) is an open stringer, it is cut out to fit risers and treads and nailed against the finish stringer. The edges of the risers are mitered with the corresponding edges of the stringer. The nosing of the tread is returned upon its outside edge along the face of the stringer. Fig. 45-27. Another method would be to butt the stringer to the riser and cover the joint with an inexpensive stair bracket.

Fig. 45-28 shows a finish stringer nailed in position on the wall and a rough carriage nailed in place against the stringer. If there are walls on both sides of the staircase, the other stringer and carriage are nailed in the same way. The risers are nailed to the riser cuts of the carriage on each side and butted against the stringers.

The treads are nailed to the tread cuts of the carriage and butted against the stringers. This is the least expensive of the types described and perhaps the best type of construction to use when the treads and risers are to be nailed to the carriages.

Another method of fitting the treads and risers to wall stringers is shown in Fig. 45-29. The finish stringers are laid out with the same rise and run as the stair carriages, but they are cut out in reverse. The risers are butted and nailed from the back to the riser cuts of the wall stringers, and the assembled stringers and risers are laid over the carriage. The treads are butted to the stringers and nailed to the risers. Sometimes the treads are allowed to run underneath the tread cut of the stringer. This makes it necessary to notch the tread at the nosing to fit around the finish stringer. Fig. 45-29.

Laying Out a Cutout Stringer

Cutout stringers for main stairways are usually made from 2″ × 12″ stock, To lay out a cutout stringer, you must first determine how long a piece of stock you will need. Using the example from "Laying Out a Stairway," let's assume that the method of upper-end anchorage is the one shown in C, Fig. 45-23. In that case the total run of the stairway is 12′ 1¼″. The total rise in the example is 8′ 11″. On the framing square twelfth scale, measure the distance between a little over 12¹⁄₁₂″ on the blade and 8¹¹⁄₁₂″ on the tongue. You'll find that it comes to just about 15″. Therefore you'll need a piece of stock at least 15′ long. It is better to allow about 4′ more for waste and for the part that extends beyond the header at the upper end. Select or cut a piece about 19′ long. Proceed to lay out the stringer from the lower end as follows:

Set the framing square to the unit run and unit rise as shown in Fig. 45-30. Draw the line AB along the blade and the line BC along the tongue. AB indicates the first tread, BC the second riser. Reverse the

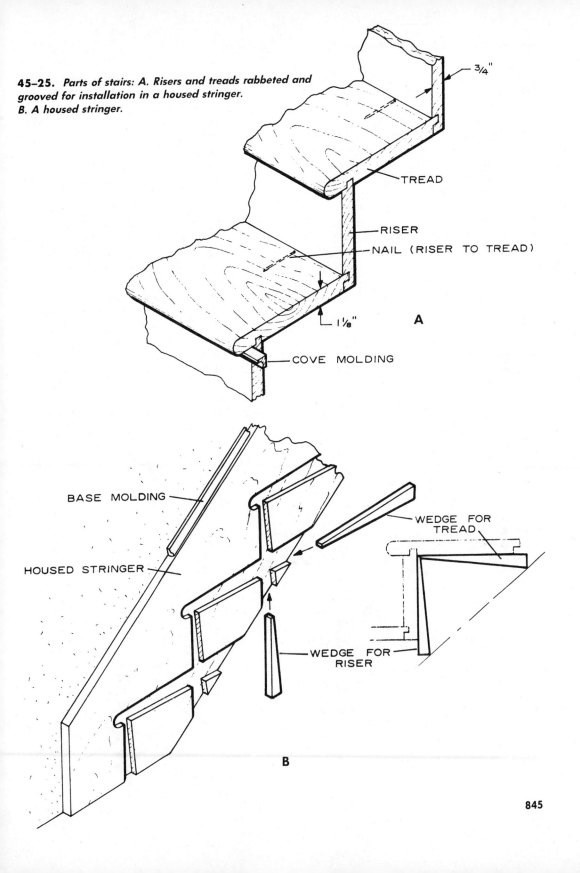

45-25. *Parts of stairs: A. Risers and treads rabbeted and grooved for installation in a housed stringer.*
B. A housed stringer.

3/4"

TREAD

RISER

NAIL (RISER TO TREAD)

1⅛"

A

COVE MOLDING

BASE MOLDING

HOUSED STRINGER

WEDGE FOR TREAD

WEDGE FOR RISER

B

845

square and draw line AD from A, perpendicular to AB, and equal in length to the unit rise.

Line AD indicates the first riser. The first riser has to be shortened, a process which is called dropping the stringer. Fig. 45-31. As you can see, in the completed stairway the unit rise is measured from the top of one tread to the top of the next. Let's assume that the bottom of the stairway is to be anchored on a finished floor, such as a concrete basement floor. If AD were cut to

the unit rise, when the first tread was put on, the height of the first step would be the unit rise plus the thickness of the tread. To make the height of the first step equal to the unit rise, you must shorten AD by the thickness of a tread.

If the bottom of the stringer is to be anchored on a subfloor to which finish flooring will be applied, shorten AD by the thickness of a tread less the thickness of the finish flooring. When you have shortened AD as required, proceed to step off the unit run and unit rise as many times as the stairway has treads—in this case, 14.

Finish the layout at the upper end as shown in Fig. 45-32. You are going to anchor the upper end by the method shown in C, Fig. 45-23, in which the stringer fits around the well header and extends beyond it to end level with the upper edges of the floor joists. Lay out the line AB, which indicates the last of the treads. Lay out the dotted line BC, which indicates the face of the well header. Extend BC down to

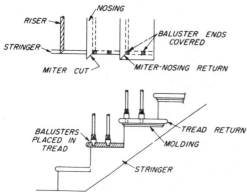

45–26. *A method of joining treads to risers.*

45–27. *An open stringer with the risers mitered to the stringer. The balusters are set into the treads and trimmed with a nosing return and a molding.*

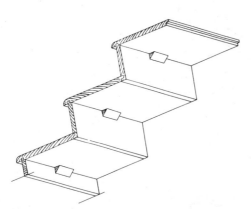

45–28. *A finish stringer and a rough carriage nailed in place.*

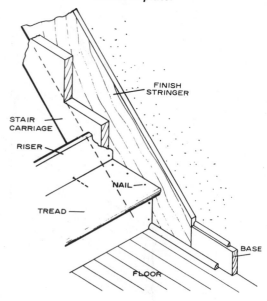

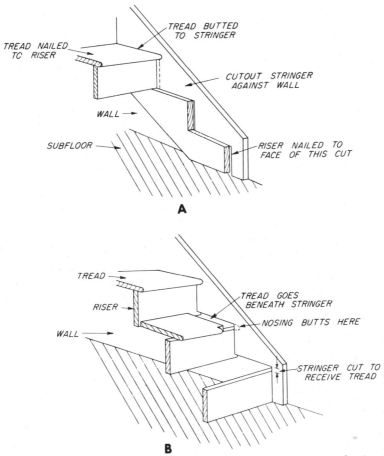

TREAD BUTTED
TO STRINGER

TREAD NAILED
TO RISER

CUTOUT STRINGER
AGAINST WALL

WALL →

SUBFLOOR →

RISER NAILED TO
FACE OF THIS CUT

A

TREAD →

RISER →

WALL →

TREAD GOES
BENEATH STRINGER

NOSING BUTTS HERE

STRINGER CUT TO
RECEIVE TREAD

B

45–29. *In this method the stringer is cut out in reverse of the carriage. In A, the riser fits between the stringer and the carriage. The treads butt against the stringer. In B, both the tread and riser are fitted between the stringer and carriage, with the tread nosing notched and butted to the stringer.*

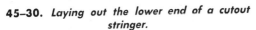

45–30. *Laying out the lower end of a cutout stringer.*

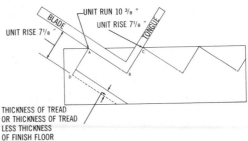

UNIT RUN 10 3/8 "

BLADE

UNIT RISE 7 1/8 "

UNIT RISE 7 1/8 "

TONGUE

THICKNESS OF TREAD
OR THICKNESS OF TREAD
LESS THICKNESS
OF FINISH FLOOR

D, so that BC plus BD will equal the depth of the header plus the thickness of the flooring.

To make the stringer fit close up under the lower edge of the header, you must shorten BD by the amount the stringer was dropped, as was shown in Fig. 45-31. Draw EF equal in length to the thickness of the header. From F draw FG equal in length to the depth of the header, and from G draw GH. When the stringer is set in place, the edge indicated by GH will lie close up under the subflooring, level with the upper edges of the joists.

Carefully cut out the first stringer, set it in position, and check it. Then use this as a pattern for cutting one or two more.

Installing Stringers

Methods used in framing stairways and securing stringers vary in different areas of the country. Regardless of method, the object of the stair builder should be the installation of a structurally strong, safe stairway. A few suggested ways of securing stringers are shown in Fig. 45-33. In A, the upper rough stringer is notched to fit the stairwell header. The stringers are hung by means of a metal supporting strap. B, Fig. 45-33, shows the stringer notched out for the stairwell header and supported by a ledger strip. This method will slightly reduce the headroom underneath.

The method shown in C, Fig. 45-33, requires a larger well "opening" and is not used too often, yet it offers the full bearing of the rough stringer against the stairwell header. The support is a ledger strip. In D, Fig. 45-33, a piece of plywood sheathing is used as a bearing surface and ledger. The stringers are secured by nailing from the back. This method would apply most often

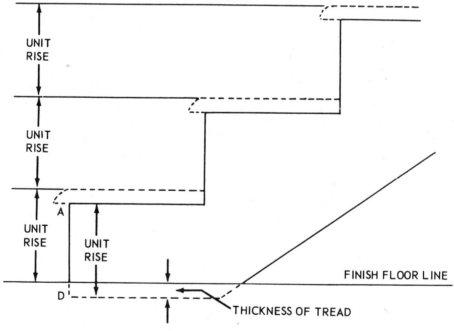

45–31. *Dropping the stringer to compensate for the thickness of the first tread keeps the unit rise uniform.*

UNIT RISE

UNIT RISE

UNIT RISE

A

UNIT RISE

UNIT RISE

FINISH FLOOR LINE

D

THICKNESS OF TREAD

at a platform where the headers are usually of less depth than the floor joists. It also affords full headroom underneath.

In E, Fig. 45-33, the stair carriages are framed to a header or trimmer with metal brackets. Another method of installing stringers is to cut the top stair tread deeper to permit the stringer to pass under the header. Then attach the stringer to a special framing member between the floor joists as shown in F, Fig. 45-33. Note also in this illustration that the stringer has been cut off at the bottom to allow for the thickness of the first tread.

Erecting the Stairway

The rough stringers are the first stairway members erected—except when a side of the stairway butts against a wall, in which case the wall (finish) stringer must be nailed on first. Temporarily nail the rough stringers in position. Check each stringer for plumb by holding the carpenter's level vertically against a riser cut. Then check the stringers for levelness with each other by setting the carpenter's level across the stringers on the tread cuts.

A stringer which lies against a trimmer joist should be spiked to the joist with at least three 16d nails. A stringer which is installed as shown at C in Fig. 45-33 and is not adjacent to a joist (a center stringer in a three-stringer stairway, for example) should be toenailed to the well header with 10d nails, three to each side of the stringer. The bottom of a stringer which is anchored on subflooring should be toenailed with 10d nails, four to each side if possible, driven into the subflooring and if possible into a joist below.

After the stringers are mounted to the wall and treads and risers cut to length, nail the bottom riser to each stringer with two 6d, 8d, or 10d nails, depending on the thickness of the stock.

The first tread, if $1^{1}/_{16}$" thick, is then nailed to each stringer with two 10d finish nails and to the riser below with at least two 10d finish nails. Proceed up the stair in this same manner. If $1^{5}/_{8}$" thick treads are used, a 12d finish nail may be required. Use three nails at each stringer, but eliminate nailing to the riser below. All finish nails should be set.

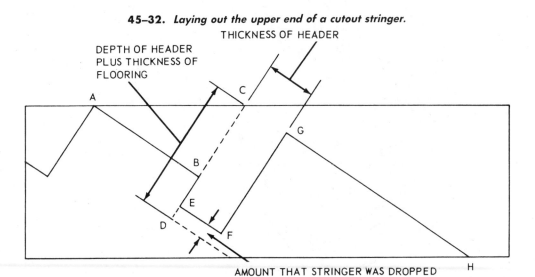

45–32. *Laying out the upper end of a cutout stringer.*

THICKNESS OF HEADER

DEPTH OF HEADER
PLUS THICKNESS OF
FLOORING

A

C

G

B

E

D

F

AMOUNT THAT STRINGER WAS DROPPED   H

45–33. *Methods of securing stringers.*

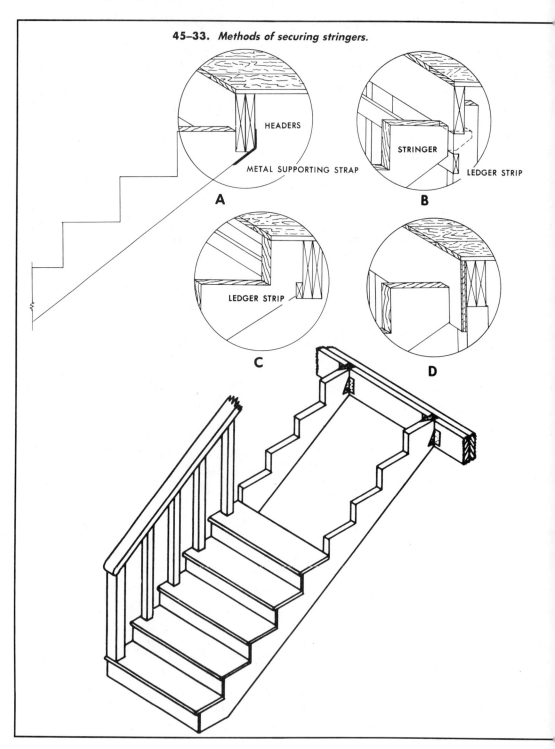

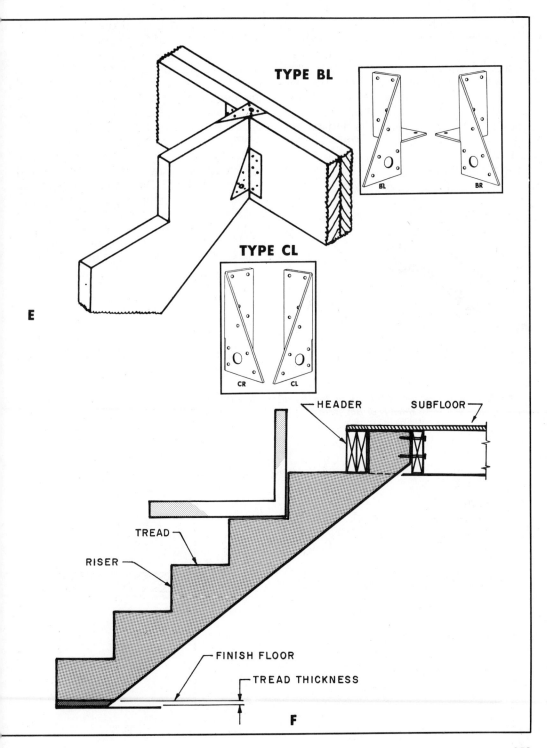

TYPE BL

BL   BR

TYPE CL

CR   CL

E

HEADER   SUBFLOOR

TREAD

RISER

FINISH FLOOR

TREAD THICKNESS

F

RAILING

All stairways should have a handrail extending from one floor to the next. For closed stairways the rail is attached to the wall with suitable metal brackets. The rail should be set 30" above the tread at the riser line and 34" above the floor on a landing. Fig. 45-34. Handrails and balusters are used for open stairs and for open spaces around stairs. The handrails end against newel posts.

Stairs should be laid out so that stock parts may be used for newels, rails, balusters, and goosenecks. Fig. 45-11. These parts may be very plain or elaborate, but they should be in keeping with the style of the house. The balusters are doweled or dovetailed into the treads and in some cases are covered by a return nosing. Fig. 45-35. For the dovetail method, a strip called a nosing return is cut off the end of the tread, as shown in the upper (plan) view of Fig. 45-35. Dovetails are shaped on the lower ends of the balusters, and dovetail recesses of corresponding size are cut in the end section of the tread. The dovetails on the balusters are glued into the recesses in the tread, and the nosing return is then nailed back in place.

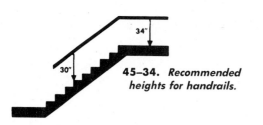

45–34. Recommended heights for handrails.

45–35. Balusters are attached to the treads with either dowels or dovetails.

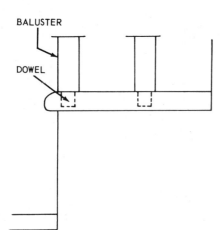

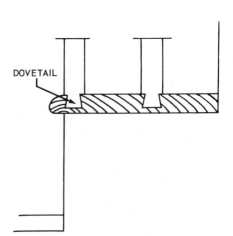

Newel posts should be firmly anchored. Where half-newels are attached to a wall, blocking should be provided at the time the wall is framed.

BASEMENT STAIRS

Basement stairs may be built either with or without riser boards. Cutout stringers are probably the most widely used supports for the treads, but in some cases the stringer is not cut out and the tread is fastened to the stringers by cleats. Fig. 45-36. The lower end of a basement stairway is usually anchored against a kicker plate which has been bolted to the concrete. Fig. 45-37.

Laying Out and Framing Cleat Stairway

A cleat stairway is inexpensive to build, it does not require risers, and the treads are usually made of softwood. First determine the total rise and run. Divide the rise by 7. If this does not result in even spacing adjust the divisor until equal spacings are obtained. Try to keep this spacing between $6^1/_2''$ and $7^1/_2''$.

For the first tread position use the determined riser height *minus the thickness of the tread* and measure up the stringer at 90° from the bottom cut to establish point A. Fig. 45-38. Set the T bevel to the angle formed by the front edge of the stringer and the bottom cut as shown at B in Fig. 45–38. Slide the T bevel up the stringer until the tongue of the T bevel is at point A. Fig. 45-38. Mark a line across the stringer at this point.

Measure up from line A using the riser height and establish point C. Position the T bevel on the new point and mark another line across the stringer. Continue this operation until all tread positions are located. These lines locate the bottom of the tread and the top edge of the cleat.

Cut the cleats from 1" × 2" stock and nail the cleats in position below the line. Place the stringers in the stairwell and fasten them in place. Cut the treads to length. Starting with the bottom tread, work up, placing each tread in position and nailing securely.

EXTERIOR STAIRS

Proportioning of risers and treads in laying out porch steps or approaches to terraces should be as carefully considered as the design of interior stairways. Similar riser-to-tread ratios can be used. However, the riser used in principal exterior steps should be between 6" and 7". The need for a good support or foundation for outside steps is often overlooked. If wood steps are used, the bottom step should be concrete. Fig. 45-39. If the steps are located over backfill or disturbed ground, the foundation should be carried down to undisturbed ground.

METAL SPIRAL STAIRWAYS

The standardized spiral stairway system provides unlimited versatility in design. These stairs are adaptable for use in all types of buildings, from a modest cottage to the most elegant residence, and are suitable for interior or exterior installation. Fig. 45-40 (Page 856).

Standardized construction provides a light, simple, strong, and durable stairway, meeting code requirements for design and carrying capacities. The basic stair structure is steel. To this can be added various types of treads, handrails, and railings.

The stairways are simple to install. Fig. 45-41. No welding is necessary. Handrails are attached to balusters with adjustable brackets. The stairs are shipped knocked down for easy handling at the job site. All of the parts are matched and marked, and the manufacturer furnishes complete shop drawings and installation instructions.

DISAPPEARING STAIRS

When attics are used primarily for storage and space for a fixed stairway is not

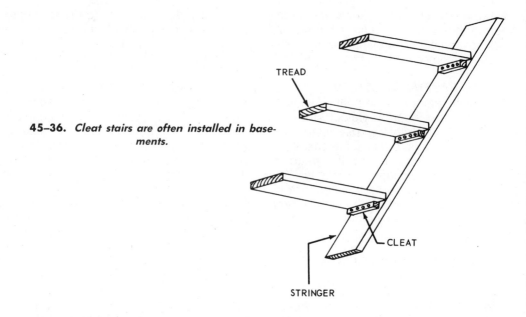

TREAD

CLEAT

STRINGER

45–36. *Cleat stairs are often installed in basements.*

45–37. *Basement stairs: A. Carriage details. B. Ledger for carriage. C. Kicker plate. The lower end of the stringer should be anchored against a kicker plate which has been bolted to the concrete floor.*

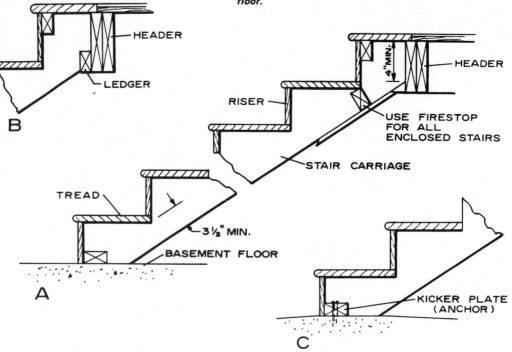

HEADER

LEDGER

B

RISER

HEADER

4" MIN.

USE FIRESTOP
FOR ALL
ENCLOSED STAIRS

STAIR CARRIAGE

TREAD

3½" MIN.

BASEMENT FLOOR

A

KICKER PLATE
(ANCHOR)

C

available, hinged or disappearing stairs are often used. Disappearing stairways may be purchased ready to install. They operate through an opening in the ceiling and swing up into the attic space, out of the way, when not in use. Fig. 45-42. Where such stairs are installed, the attic floor should be designed for regular floor loading.

ESTIMATING
Materials

To estimate the materials for a stairway, either precut or cut on-the job, a complete material bill must first be made. The cost of the items can then be totaled to determine the material cost for the stairway.

45–38. *Laying out a cleat stairway. Note that the distance from A to C is the same as the distance from C to D and is equal to the riser height. The distance between the floor and line A, however, is less than the riser height to allow for the thickness of the first tread.*

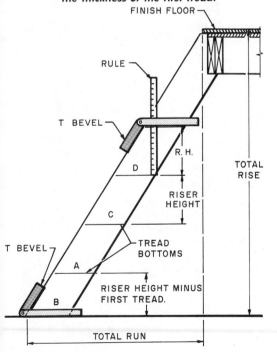

Labor

Labor costs can be only roughly estimated because of the many variables which will affect the construction time. For example, for an open stairway less than 12' long and 42" wide, construction time is estimated at 8³/₄ hours. This includes rough-cutting the stringers and framing and installing the stringers, treads, and risers. Add about 3 more hours if there is a turn in the stairway involving a platform or landing. If a handrail is to be installed, add 1 more hour; for an open stairway, add 2¹/₂ hours for the installation of the newel posts, rails, and balusters.

For assembling a precut stairway no longer than 12' and less than 42" wide, 6 hours are required. If this stairway has a turn including a platform, add about 3 more hours. A handrail takes 1 hour, and for an open stairway with newel posts, rails, and balusters, add 2¹/₂ more hours. A folding stairway which has been prefabricated will take approximately 2 hours to install.

It must be understood that these are very rough approximations which depend on the species of wood, the style of the stairway, and the worker's experience.

45–39. *Exterior steps of wood should have a bottom step of concrete. The stringer should be secured to the bottom step against a kick plate which has been bolted to the concrete.*

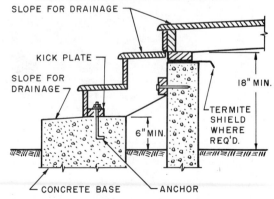

45–40a. A simple metal spiral stairway with carpeted treads.

45–40b. Note the use of wood on the balusters and the wood treads on this spiral stairway.

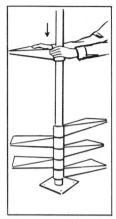

1. Place center column in position and slide on treads.

2. Attach rail standards.

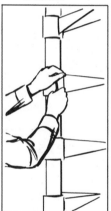

3. Fasten treads to center column.

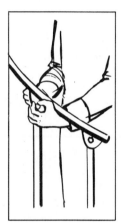

4. Connect handrail to rail standards with adjustable brackets.

45–41. *Installing a metal spiral stairway.*

45–42. *Disappearing stairs. When the stairs are in the stored position, the plywood door is barely noticeable.*

QUESTIONS

1. What are the two general types of stairs?

2. What are the two main parts of a stairway?

3. What is the difference between a landing and a platform?

4. When designing a stairway, what are the three important considerations?

5. The total rise of a stairway is the vertical distance between what two points?

(Questions continued on page 858)

6. When laying out a stringer for a basement stairway, what is subtracted from the bottom of the stringer?

7. After the stair stringers are installed and the other parts are cut to length, what is the next piece to be installed?

UNIT 46

Cabinets and Built-Ins

Kitchen cabinets, wardrobes, linen closets, china cases, and similar pieces of millwork are installed at the same time as the interior trim. This work is usually done after the finish floor is in place. The kitchen normally contains more millwork than all other rooms combined, in the form of wall and base cabinets, broom closets, and other storage centers. Kitchen design concepts have changed radically in recent years. Today's kitchen is open, creating a feeling of warmth and homeyness. It is often combined with the family room to create a center for casual everyday living or informal entertaining. Sometimes the kitchen is linked to a patio or terrace for indoor-outdoor enjoyment. Most important, today's kitchens are more beautiful, functional, efficient, and convenient. Fig. 46-1.

KITCHEN FLOOR PLANS

There are four basic layouts commonly used in kitchen design:
- U-shape.
- L-shape.
- Parallel wall, or Pullman.
- Sidewall.

The U-shape kitchen, with the sink at the bottom of the U and the range and refrigerator on opposite sides, is very efficient. Fig. 46-2, A. The L-shape, also very popular, has the sink and range on one leg and the refrigerator on the other. Sometimes the dining space is located in the opposite

corner. Fig. 46-2, B. The parallel wall, or Pullman, kitchen plan is often used where there is limited space. It can be quite efficient with proper arrangement of sink, range, and refrigerator. Fig. 46-2, C. The sidewall kitchen is usually preferred for small apartments. The cabinets, sink, range, and refrigerator are all located along one wall. With this design, counter space is somewhat limited when kitchens are small. Fig. 46-2, D.

KITCHEN WORK CENTERS

When planning a kitchen layout, three work centers must be kept in mind: the food preparation center, the cooking center, and the clean-up center. Fig. 46-3. All equipment, storage space, and surface work areas for each activity should be located in their respective work centers. The food preparation center should be planned around the refrigerator and food storage area. Fig. 46-4. Ideally, the clean-up center (sink and dishwasher) should be located between the food preparation center and the cooking center. 46-5 and 46-6. Whenever possible, counter space should be continuous between work centers.

Arrange the work centers in a logical sequence, with the flow of materials in the same direction as when preparing a typical meal. Most people work from right to left. Therefore, whenever possible, the work centers and equipment are arranged

46–1. *An attractive, well-planned kitchen with adequate storage for complete modern living.*

46–2. *Four of the most popular kitchen layouts: A. U-shape. B. L-shape. C. Parallel wall. D. Side-wall.*

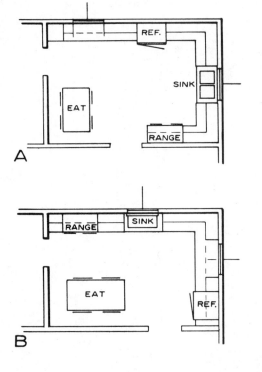

accordingly. A work triangle is developed among the work centers. Fig. 46-7. The three sides of the triangle should add up to at least 12'. The ideal total is 15' to 20'.

Ample storage in a kitchen is a necessity. A good rule of thumb is to allow 6 sq. ft. of storage cabinet shelf space for every member of the family, then add 12 sq. ft. of storage for staple items used in the kitchen. Of course, this rule can be used only when the size of the family is known. Normally the base and wall cabinet needs are determined by the size of the home. Figs. 46-8 and 46-9.

KITCHEN CABINET DIMENSIONS

Base and wall units should be designed and installed to a standard height and depth. There are also standard clearances for wall cabinets over appliances and work centers. Fig. 46-10. While the limits for counter height range from 30" to 38", the standard height is 36". Wall cabinets vary in height. Depending on the type of installation at the counter, they may be anywhere from 12" to 33" high. The usual height is 30". The 12" wall cabinets are

46–3. *The three basic work centers: A. Food preparation. B. Cleanup. C. Cooking and serving. The minimum counter space for each area is shown by the arrows.*

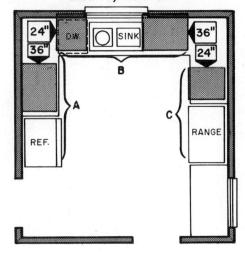

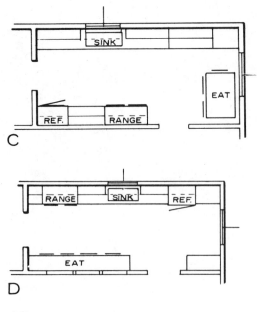

usually placed over refrigerators. When a range or sink is located under them, wall cabinets should not be more than 21″ high. The tops of wall cabinets should all be at the same height. Wall cabinets may be hung free or under a 12″ soffit (drop ceiling) or storage cabinet. Figs. 46-11 (Page 864) and 46-12. The important thing is to place the cabinets, counters, and shelves at heights designed for working efficiency, convenience, and comfort. Fig. 46-13.

ELECTRICITY IN THE KITCHEN

Good lighting is important. A light source for each work center should be provided, as well as general overall lighting operated by a conveniently located switch. Adequate wiring must also be pro-

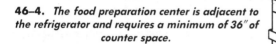

46-4. *The food preparation center is adjacent to the refrigerator and requires a minimum of 36″ of counter space.*

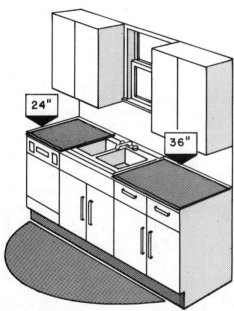

46-5. *The clean-up center includes the dishwasher and the sink and should provide at least 60″ of counter space. If a dishwasher is not provided, the counter space to the left of the sink should be increased to 30″, making a total counter top requirement of 66″.*

46-6. *The cooking and serving center requires a minimum of 24″ of counter space.*

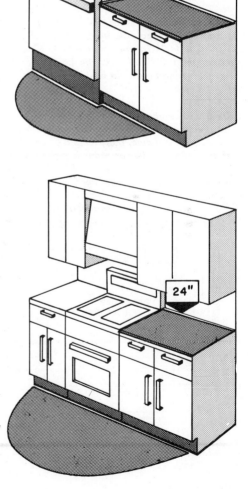

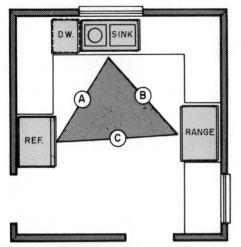

46-7. The work triangle has the major appliance for one of the work centers at each point. The distance at A should be from 4' to 7'; at B, 4' to 6'; at C, 4' to 9'. The minimum total is 12' and the maximum is 22'. Ideally, the perimeter of the triangle should be 15' to 20'.

46-8. To determine the number of base cabinets required, measure the base-cabinet frontage. For a 1,000 sq. ft. home, there should be a minimum of 6', although 10' or more is preferred. For a home of 1,000 to 1,400 sq. ft., 8' of base cabinet is minimum and 12' or more is preferred. For homes over 1,400 sq. ft., the minimum is 10', and 14' or more is preferred.

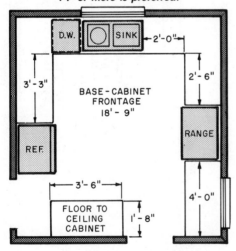

vided, with separate circuits for the electric range, water heater, dishwasher, and any other appliances which require heavy-duty circuits. Each work center should have at least one double convenience outlet. These should not all be on one circuit, and they must conform to local codes. Proper ventilation should also be provided in the form of an exhaust fan and hood to control cooking odors and grease. Fig. 46-14. The electrical work is usually handled by an electrical contractor. The electrical contractor's work is coordinated with the builder's.

KITCHEN CABINET CONSTRUCTION

There are three basic ways in which kitchen cabinets are obtained. A cabinet-maker may build the cabinets on the job site, piece by piece, from the plans supplied by the architect. The second method

46-9. The number of wall cabinets required is determined by measuring the wall-cabinet frontage. For a home with 1,000 sq. ft., a minimum of 6' of wall cabinet frontage is required; 10' or more is preferred. For a home of 1,000 to 1,400 sq. ft., the minimum is 8' and 12' or more is preferred. For a home over 1,400 sq. ft., 10' is the minimum and 14' or more is preferred.

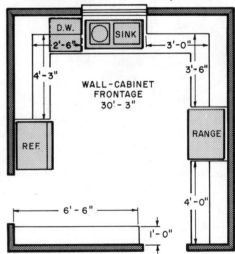

is to purchase knocked-down kitchen cabinets which have been produced in a mill. The cabinetmaker assembles the parts on the job site and installs the cabinets as a unit. The third method is to purchase pre-built kitchen cabinets that have been mass produced in shops or factories. More information about how cabinets are built and construction details are available in *Cabinetmaking and Millwork* by John L. Feirer. Construction details for one method of job-site kitchen cabinet construction are shown in Fig. 46-15 (Page 866).

Planning a Kitchen with Factory-built Cabinets

When factory-built cabinets are to be installed in a kitchen, first draw a scale outline of the kitchen. A piece of graph paper is ideal for this. Fig. 46-16. Measure the kitchen carefully, taking the dimen-

46–10a. *Minimum and standard dimensions for kitchen cabinets.*

46–10b. *Typical kitchen cabinet dimensions.*

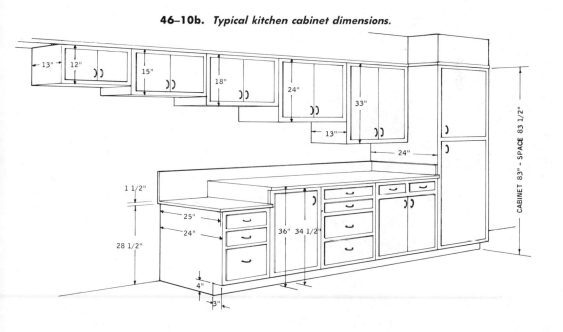

46–11. *These wall cabinets are hung free with the tops aligned.*

sions 36″ from the floor. This is the height at which the counter tops must fit snugly to the wall. Draw the outline of the kitchen on paper. If graph paper is to be used, a scale of ¹/₂″ = 1′ will work conveniently. Indicate window and door openings (including trim), as well as other obstructions, insets, and wall surface details. Fig. 46-17 (Page 868).

After the layout has been drawn to scale, locate the appliances and utilities such as the sink, range, oven, refrigerator, and dishwasher. From a catalog, select and draw in the corner base cabinets, then fill in the empty spaces with the proper size base cabinets. Always select the base cabinets first. Begin the planning of the wall cabinets by drawing in the proper cabinet over the refrigerator and range,

46–12. *These wall cabinets are installed under a soffit (drop ceiling).*

46–13. *Cabinets should be designed so that the height of the counters and shelves is convenient and comfortable. For example, counter tops should be 34″ to 36″ high. The most comfortable reach is from 30″ to 60″ above the floor, as shown by the arrow.*

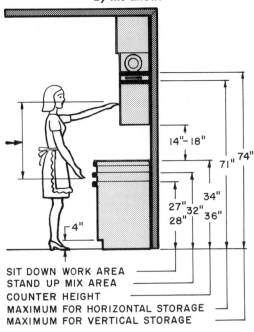

SIT DOWN WORK AREA
STAND UP MIX AREA
COUNTER HEIGHT
MAXIMUM FOR HORIZONTAL STORAGE
MAXIMUM FOR VERTICAL STORAGE

taking care to match each appliance with an appropriate cabinet of the same width. Next, select the corner wall cabinets. Finally, fill in the empty spaces with the proper size wall cabinets. Figs. 46-18 and 46-19 (Pages 869-873).

Ordering Factory-Built Cabinets

Refer to the catalog and be sure to use the correct stock number listed with each illustration. Most manufacturer's stock numbers have a reference to the size and type of cabinet. Stock numbers for the cabinets shown in Fig. 46-19 are one example. The beginning letters indicate the cabinet type, and the first two numbers indicate the cabinet width. The second pair of numbers, if any, indicates the cabinet height. On single-door base cabinets, always indicate whether the door is to be hinged on the right or the left side. Also provide the manufacturer with the size of the sink and the opening needed for built-ins such as the oven, dishwasher, and refrigerator. Be sure also to include the style of the cabinet and the finish desired. Fig. 46-20 (Page 874).

Installing Factory-Built Cabinets

Factory-built cabinets may be installed in either of two ways. Some cabinetmakers prefer to install the base cabinets first and then build platforms which rest on top of the base cabinets. The wall cabinets are set onto the platforms and held in place for installation. Other cabinetmakers prefer to install the wall cabinets first. This is sometimes more convenient because it allows the workers to stand up close to the wall when working on the wall cabinets, rather than having to reach over and climb onto the base cabinets during installation. With both procedures, the same precautions must be taken when installing the cabinets to insure that they are properly aligned and installed plumb and level.

46–14a. *This cooking surface has an exhaust hood and fan to help eliminate cooking odors and grease.*

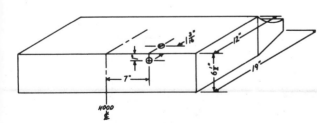

46–14b. *When installing a range hood, the specifications for installation are available from the manufacturer. The drawing here was taken from a typical manufacturer's catalog.*

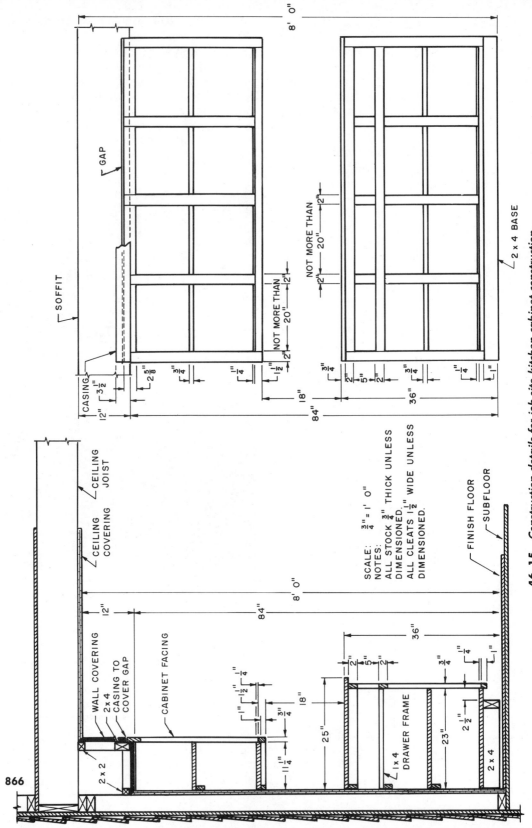

46-15. Construction details for job-site kitchen cabinet construction.

WALL CABINETS

It is imperative that the wall cabinet be mounted securely and that it will easily bear normal heavy loads. The wall cabinets are attached to the wall with wood screws or toggle bolts through the upper and lower back rail, through the wall covering, and into each stud or, in some instances, just through the wall covering. Fig. 46-21 (Page 876). The wood screws should be at least a #9 or #10 round head and long enough to go through the ¾" back rail and the wall covering and extend at least 1" into the studding. A minimum of four screws should be used for each cabinet. Never rely on nails for hanging wall cabinets, or on screws driven only into plaster. These will not hold securely.

Locate and mark the wall studs on the wall. Then mark the location of the bottom

46–16. *Graph paper, similar to that shown here, is usually provided in the catalog of factory-built cabinets as an aid to planning.*

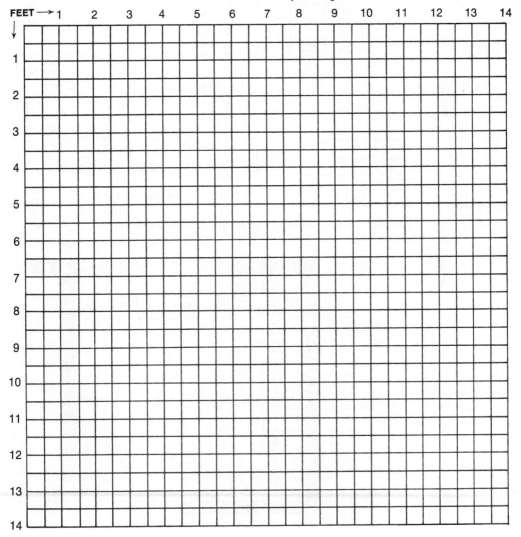

of the wall cabinet on the wall. The most common distance between a counter top and the bottom of an upper cabinet is 18" to accommodate taller electrical appliances.

Align the tops of the wall cabinets 7' from the floor. Start the installation from the corner and work toward windows and doorways. Place each of the cabinets in position and brace them so that the cabinets are held securely. Fig. 46-22 (Page 876). Make certain the cabinet is against the wall and up tight against the soffit. If there is no soffit, be sure the tops of the cabinets are properly aligned.

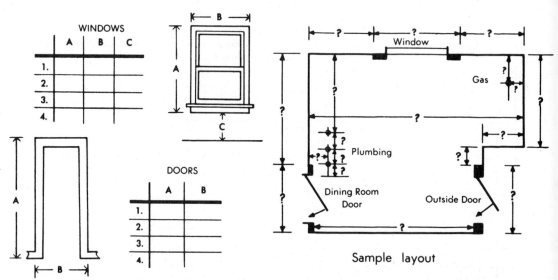

46–17. *When drawing a room to scale, be sure to include all of the necessary information.*

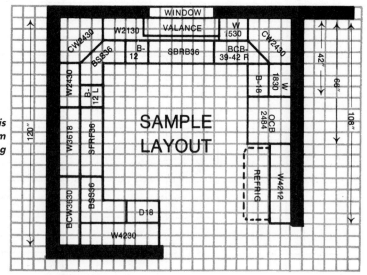

46–18. *The cabinets for this kitchen layout were taken from the manufacturer's catalog shown in Fig. 46–19.*

46–19. *A manufacturer's catalog of kitchen cabinets showing stock numbers and cabinet dimensions available.*

WALL CABINETS

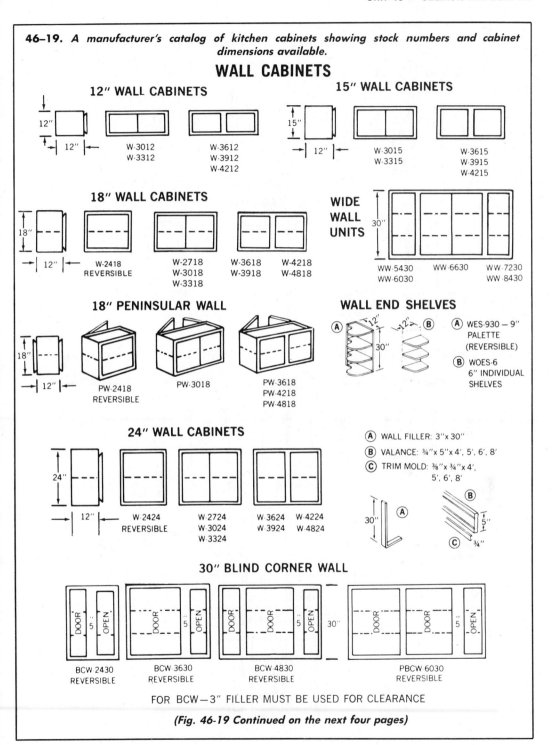

12" WALL CABINETS

12"
12"

W-3012
W-3312

W-3612
W-3912
W-4212

15" WALL CABINETS

15"
12"

W-3015
W-3315

W-3615
W-3915
W-4215

18" WALL CABINETS

18"
12"

W-2418
REVERSIBLE

W-2718
W-3018
W-3318

W-3618
W-3918

W-4218
W-4818

WIDE WALL UNITS

30"

WW-5430
WW-6030

WW-6630

WW-7230
WW-8430

18" PENINSULAR WALL

18"
12"

PW-2418
REVERSIBLE

PW-3018

PW-3618
PW-4218
PW-4818

WALL END SHELVES

Ⓐ 12" 30" 12" Ⓑ

Ⓐ WES-930 – 9"
PALETTE
(REVERSIBLE)

Ⓑ WOES-6
6" INDIVIDUAL
SHELVES

24" WALL CABINETS

24"
12"

W-2424
REVERSIBLE

W-2724
W-3024
W-3324

W-3624
W-3924

W-4224
W-4824

Ⓐ WALL FILLER: 3"x 30"

Ⓑ VALANCE: ¾"x 5"x 4', 5', 6', 8'

Ⓒ TRIM MOLD: ⅜"x ¾"x 4', 5', 6', 8'

30" Ⓐ

Ⓑ 5"

Ⓒ ¾"

30" BLIND CORNER WALL

DOOR 5" OPEN

DOOR 5" OPEN

DOOR DOOR 5" OPEN 30"

DOOR DOOR 5" OPEN

BCW 2430
REVERSIBLE

BCW-3630
REVERSIBLE

BCW-4830
REVERSIBLE

PBCW-6030
REVERSIBLE

FOR BCW–3" FILLER MUST BE USED FOR CLEARANCE

(Fig. 46-19 Continued on the next four pages)

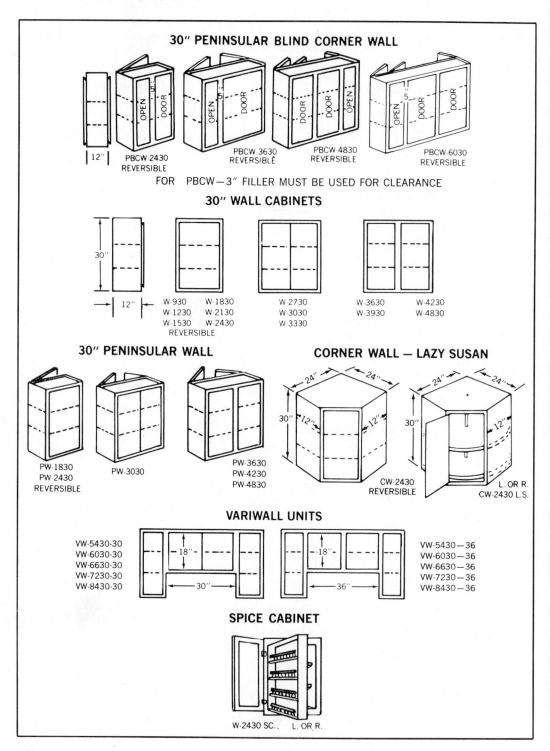

30" PENINSULAR BLIND CORNER WALL

OPEN — 5" — DOOR
PBCW-2430 REVERSIBLE

12"

OPEN — 5" — DOOR
PBCW 3630 REVERSIBLE

DOOR DOOR OPEN
PBCW-4830 REVERSIBLE

OPEN — 5" — DOOR — DOOR
PBCW-6030 REVERSIBLE

FOR   PBCW—3" FILLER MUST BE USED FOR CLEARANCE

30" WALL CABINETS

30"

12"

W-930	W-1830
W-1230	W-2130
W-1530	W-2430
	REVERSIBLE

W-2730
W-3030
W-3330

W-3630
W-3930

W-4230
W-4830

30" PENINSULAR WALL

PW-1830
PW-2430
REVERSIBLE

PW-3030

PW-3630
PW-4230
PW-4830

CORNER WALL – LAZY SUSAN

24"   24"
30" 12"   12"
CW-2430 REVERSIBLE

24"   24"
30"   12"
L. OR R.
CW-2430 L.S.

VARIWALL UNITS

VW-5430-30
VW-6030-30
VW-6630-30
VW-7230-30
VW-8430-30

18"   30"

18"   36"

VW-5430—36
VW-6030—36
VW-6630—36
VW-7230—36
VW-8430—36

SPICE CABINET

W-2430 SC..   L. OR R.

BASE CABINETS

BASE CABINETS

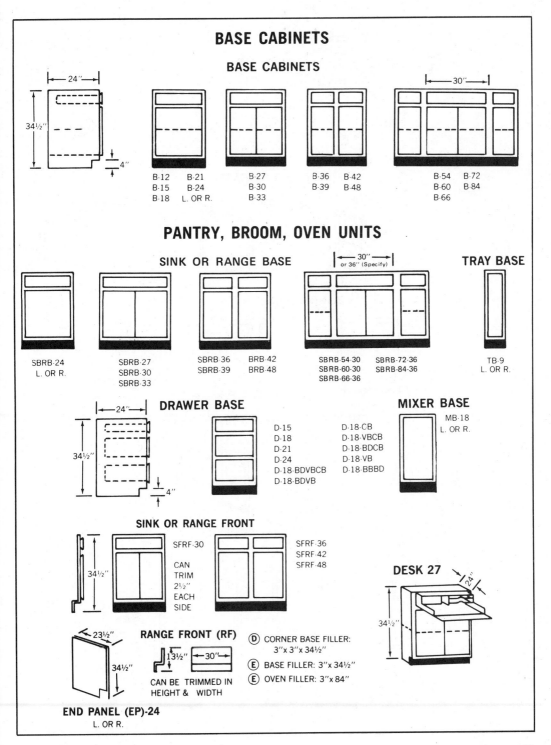

24″

34½″

4″

30″

B-12 B-21
B-15 B-24
B-18 L. OR R.

B-27
B-30
B-33

B-36 B-42
B-39 B-48

B-54 B-72
B-60 B-84
B-66

PANTRY, BROOM, OVEN UNITS

SINK OR RANGE BASE

30″
or 36″ (Specify)

TRAY BASE

SBRB-24
L. OR R.

SBRB-27
SBRB-30
SBRB-33

SBRB-36 BRB-42
SBRB-39 BRB-48

SBRB-54-30 SBRB-72-36
SBRB-60-30 SBRB-84-36
SBRB-66-36

TB-9
L. OR R.

DRAWER BASE

24″

34½″

4″

D-15
D-18
D-21
D-24
D-18-BDVBCB
D-18-BDVB

D-18-CB
D-18-VBCB
D-18-BDCB
D-18-VB
D-18-BBBD

MIXER BASE

MB-18
L. OR R.

SINK OR RANGE FRONT

34½″

SFRF-30

CAN
TRIM
2½″
EACH
SIDE

SFRF-36
SFRF-42
SFRF-48

DESK 27

24″

34½″

RANGE FRONT (RF)

23½″

34½″

13½″ 30″

CAN BE TRIMMED IN
HEIGHT & WIDTH

(D) CORNER BASE FILLER:
3″x 3″x 34½″

(E) BASE FILLER: 3″x 34½″

(E) OVEN FILLER: 3″x 84″

END PANEL (EP)-24
L. OR R.

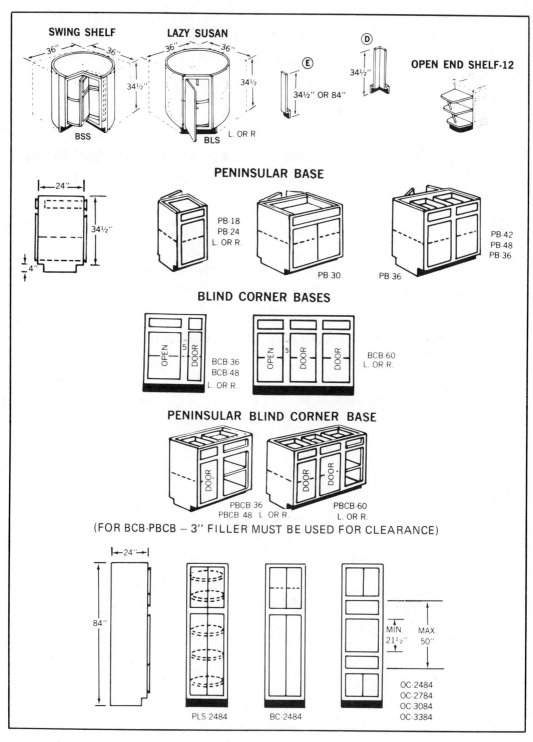

SWING SHELF

36" 36"

34½"

BSS

LAZY SUSAN

36" 36"

34½

BLS L. OR R.

E

34½" OR 84"

D

34½"

OPEN END SHELF-12

24"

34½"

4"

PENINSULAR BASE

PB-18
PB-24
L. OR R.

PB-30

PB-36

PB-42
PB-48
PB-36

BLIND CORNER BASES

OPEN 5" DOOR

BCB-36
BCB-48
L. OR R.

OPEN 5" DOOR DOOR

BCB-60
L. OR R.

PENINSULAR BLIND CORNER BASE

DOOR

PBCB 36
PBCB 48 L. OR R.

DOOR DOOR

PBCB-60
L. OR R.

(FOR BCB-PBCB — 3" FILLER MUST BE USED FOR CLEARANCE)

24"

84"

MIN.
21½"

MAX.
50"

PLS-2484

BC-2484

OC-2484
OC-2784
OC-3084
OC-3384

BATHROOM VANITIES

VANITY

VANITY BASE FILLER

END PANEL

7"

V-18 V-24 V-30

VSTD-27"

30½"

30½"

20¾"

L. OR R.

VBF-1   V-30-EP

VANITY BASE BOWL

21"

30½"

VBB-24   L. OR R.

VBB-27   VBB-30

VBB-36   VBB-48

VANITY BASE DOOR

VANITY BASE DRAWER

VANITY BASE BOWL FRONT

30½"

VBD-12   VBD-24   L. OR R.

VD-15   VD-18   VD-24

VBBF-27"

LEGEND

B—Base Cabinet	**LS**—Lazy Susan	**SFRF**—Sink or Range Front
BB—Bread Board (for BD-18 only)	**MB**—Mixer Base	**TB**—Tray Base
BC—Broom Cabinet	**OC**—Oven Cabinet	**VB**—Vegetable Bin
BCB—Blind Corner Base	**PW**—Peninsular Wall Cabinet	**V**—Vanity
BCW—Blind Corner Wall Cabinet	**PB**—Peninsular Base	**VBB**—Vanity Base Bowl
BD—Bread Drawer	**PBCB**—Peninsular Blind Corner Base	**VBBF**—Vanity Base Bowl Front
BLS—Base Lazy Susan		**VBD**—Vanity Base Door
BOES—Base Open End Shelf	**PBCW**—Peninsular Blind Corner Wall Cabinet	**VBF**—Vanity Base Filler
BSS—Base Swing Shelf		**VBD**—Vanity Base Drawer
CB—Chop Block (for D-18-CB only)	**PLS**—Pantry Lazy Susan	**VSTD**—Vanity Single Tray Drawer
	RB—Range Base	**VW**—Vari Wall
CW—Corner Wall Cabinet	**RFP**—Range Front Panel	**W**—Wall Cabinet
D—Drawer Base	**SBRB**—Sink or Range Base	**WES**—Wall End Shelf
EP—End Panel	**SC**—Spice Cabinet	**WOES**—Wall Open End Shelf
	SF—Sink Front	**WW**—Wide Wall

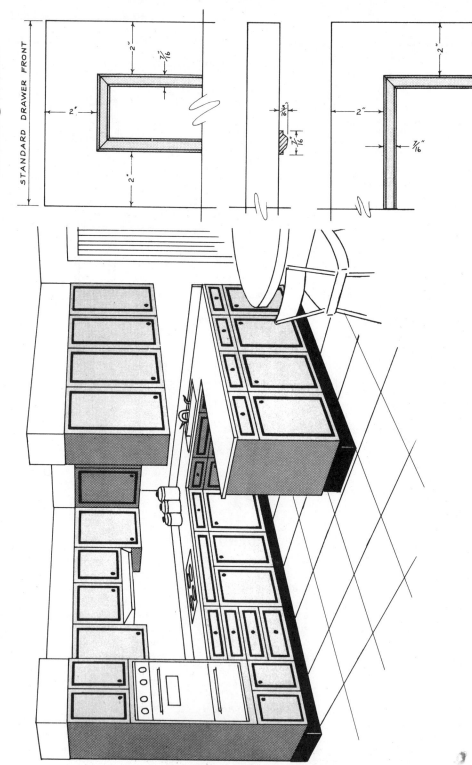

2"

7/16"

2"

2"

2"

3/16"

7/16"

2"

2"

7/16"

46–20a. The style of many cabinets can be changed by the application of various wood moldings to the drawers and door fronts.

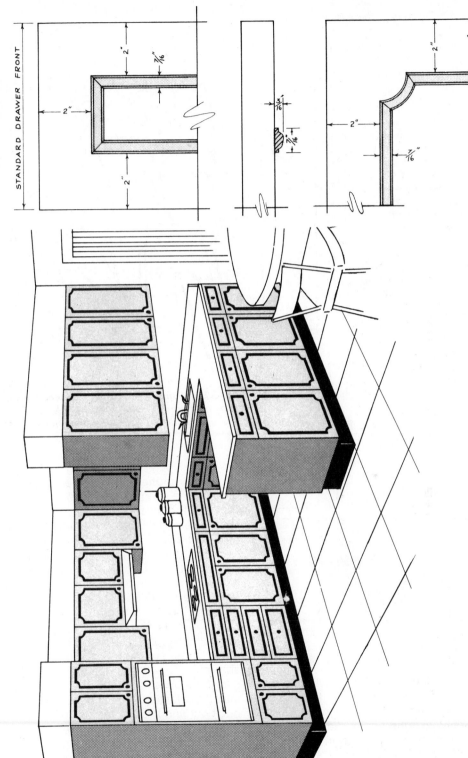

STANDARD DRAWER FRONT

2"

2"

2"

7/16"

2"

3/16"

7/16"

2"

2"

7/16"

46–20b. Notice the different look of these cabinets compared to those shown in Fig. 46–20a. It was done by changing the style of the molding on the door and drawer fronts.

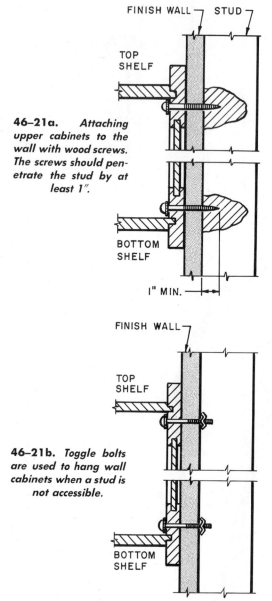

FINISH WALL ⌐ STUD ⌐

TOP
SHELF

46–21a. *Attaching upper cabinets to the wall with wood screws. The screws should penetrate the stud by at least 1".*

BOTTOM
SHELF

1" MIN.

FINISH WALL ⌐

TOP
SHELF

46–21b. *Toggle bolts are used to hang wall cabinets when a stud is not accessible.*

BOTTOM
SHELF

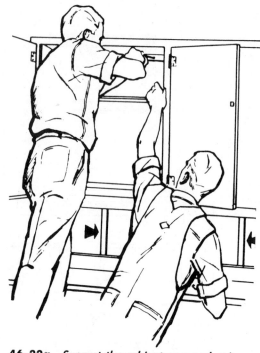

46–22a. Support the cabinet on wood strips as shown at arrows. One worker holds the front of the unit while the second fastens the cabinet to the studs.

By measuring, determine where the back cabinet rails cross over the exact center of the studs. Drill through the cabinet rail only with the shank diameter drill (the screw should slip through this hole freely). With the cabinet in position and properly aligned, drill the pilot hole through the shank hole and the wall covering into the stud. Drill the pilot hole to a depth which will accommodate the length of the screw.

For proper installation, each wall cabinet should be fastened with at least two screws, toggle bolts, or hollow wall fasteners in the upper rail and two in the lower rail. Fig. 46-23. Should the cabinet span two studs, as shown in Fig. 46-24a, it is logical to secure at these points. However, if the cabinet spans only one stud, as shown in Fig. 46-24b, fasten wood screws into the stud and $3/16'' \times 3\frac{1}{2}''$ toggle bolts into the plaster. Drill a hole through the back rail and the plaster just large enough for a slip fit of the folded leaves of the toggle. Then use a washer large enough to cover the resulting hole.

46–23b. Using a hollow wall anchor to install wall cabinets.

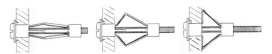

46–23c. As the anchor is inserted in the opening and the screw is tightened, the legs of the fastener expand and are drawn up tightly against the inside of the wall. The advantage of this system is that once the anchor has been set, the screw may be removed as often as necessary.

46–24a. When the cabinet spans two studs, use wood screws.

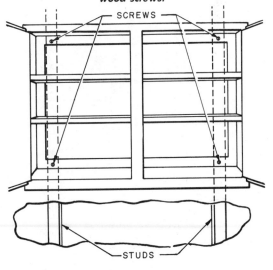

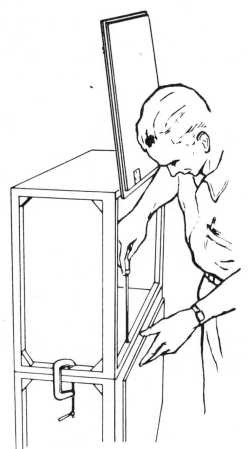

46–22b. When two or more narrow wall cabinets are placed side by side, it is better to fasten them together on the floor and mount them on the wall as one unit. Make certain that the joining faces of the stiles are flush and that the tops and bottoms of the cabinets are aligned before drilling.

46–23a. The holes which are drilled to receive toggle bolts must be drilled large enough to accept the wings. When the installation is complete, the wings form a perfect 90° angle with the wall, allowing the entire length of the wings to come in contact with the interior wall surface.

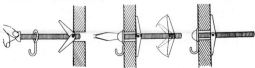

46–24b. When the cabinet spans only one stud, a toggle bolt or hollow wall fastener is needed to install the wall cabinet securely.

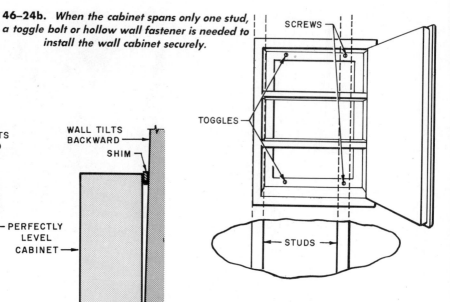

SCREWS

TOGGLES

STUDS

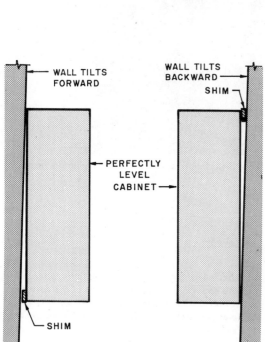

WALL TILTS FORWARD

WALL TILTS BACKWARD

SHIM

PERFECTLY LEVEL CABINET

SHIM

46–25. If scribing the cabinet to the wall is not possible, place wood shims behind the wall cabinet so that it can be pulled up tightly against the wall in a plumb and level position.

46–26. Install the corner cabinets first. When installing a revolving corner base, nail or screw 1" × 2" strips to the wall as shown at the arrows. The strips provide a level platform for the counter top to rest on.

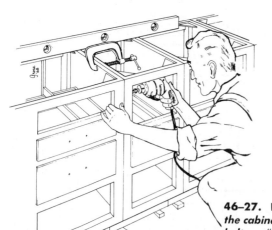

46–27. When more than one cabinet is installed, the cabinets should be fastened together with 1/4" bolts or #10 wood screws to unify the construction.

Uneven walls sometimes make it difficult to obtain proper alignment and get a snug fit. If the walls are out of plumb, the cabinet should be shimmed. Fig. 46-25. If material for scribing is provided on the back of the cabinet, the cabinet should be held in place, scribed, and cut to fit the irregular wall surface. (Scribing is discussed on page 754.) When a high spot in the plaster occurs at the center of the cabinet, take care when securing the cabinet to the wall that stress is not placed on the front framing members of the cabinet. Such stress may cause a checking and breaking of the glued joints. This can be avoided by shimming out the ends of the cabinets equal to the high spot in the plaster.

If the cabinets are to be fastened together through the ends for additional strength and for a tight fit between the facings of the cabinets, use bolts or #10 wood screws of sufficient length. Once again, caution is needed when drawing the cabinets together because of the resulting strain on the facing joints. However, if the cabinets are properly aligned to the wall and fit snugly at the junction between the cabinets, there should be no problem.

BASE CABINETS

Before installing the base cabinets, clean the walls. On remodeling jobs, remove any moldings such as quarter rounds and baseboards. Set the base cabinets loosely into their respective positions and check the measurements with the plans. Begin setting the cabinets at the corners or side wall. Fig. 46-26. If they fit snugly to the wall at the back and are level, they may be fastened together with one or two evenly spaced #10 wood screws or 1/4" bolts through the panels just behind the front stiles. Fig. 46-27.

If a sink front is included in the assembly, the opening should be carefully checked and the facing under the sink should be properly fitted between the adjoining base cabinets. Secure this facing

on the inside with wood screws through a 3/4" × 3/4" wood cleat or angle irons.

Sometimes the floor and/or wall is uneven and the cabinets do not set plumb and level and tightly against the wall. In that case, they must be shimmed or scribed and trimmed. Fig. 46-28. To trim a cabinet to fit the floor, scribe it or measure the distance A and mark the end panel as shown at B in Fig. 46-28. Cut the end panel along this line to fit the floor. Fig. 46-29. When the wall is uneven, the rear of the cabinet may be scribed to fit the wall if the end panel is exposed. If the end panel is not exposed, the cabinet may be shimmed

46–28. *To make the base cabinet set plumb and level, it can be shimmed as shown at A or scribed as shown at B. If the cabinet is scribed, it should be cut off as shown in Fig. 46–29.*

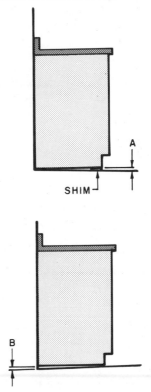

46–29. Cutting off the scribed base to fit the floor.

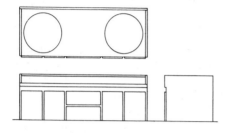

Dimensions of model shown are 60" wide, 22" deep, 31" high.

46–30a. A wall-hung, factory-built bathroom cabinet of contemporary design.

46–30b. A factory-built bathroom cabinet of Italian Provincial design.

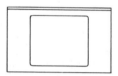

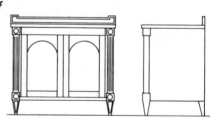

Dimensions of model shown are 36" wide, 22" deep, 31" high.

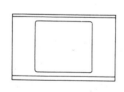

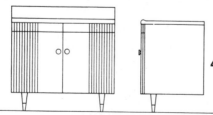

46–30c. A factory-built bathroom cabinet of contemporary design.

Dimensions of model shown are 36" wide, 22" deep, 31" high.

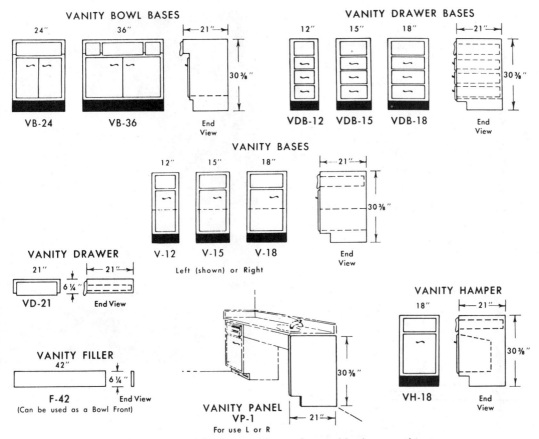

46–31a. *A typical selection of manufactured bathroom cabinets.*

and securely wood screwed through the wall covering and into the stud.

BATHROOM CABINETS

Bathroom cabinets, like kitchen cabinets, are obtained in three basic ways. They may be constructed on the job site; purchased knocked-down, then assembled and installed as a unit; or purchased prebuilt and ready to install from factories which mass-produce cabinets. Factory-built bathroom cabinets are installed in the same manner as described earlier in the discussion of kitchen cabinets. Fig. 46-30. Bathroom cabinets differ from kitchen cabinets slightly in size. They are usually 30"

high and 21" deep. The drawer opening is usually made 4" deep, rather than 5" as in the case of the kitchen cabinet. Fig. 46-31. See *Cabinetmaking and Millwork* by John L. Feirer for additional information and construction details.

COUNTER TOPS

Counter tops for kitchen cabinets, bathroom vanities, built-in desks, and room dividers are often covered with plastic laminates. Therefore the cabinetmaker or finish carpenter should have some knowledge of their use and methods of installation. Some common trade names for plastic laminates are Textolite®, Formica®, and

46–31b. *The layout of an attractive bathroom with factory-built cabinets installed.*

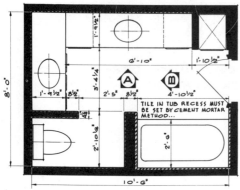

PLAN

46–31c. *Factory-built cabinets in a large bathroom with a dual lavatory installation.*

ELEVATION "A"

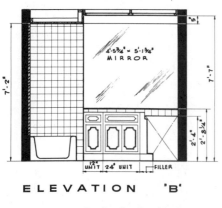

ELEVATION "B"

SCALE

Panelyte®. They are available in widths of 24″, 30″, 36″, 48″, and 60″. The common lengths are 60″, 72″, 84″, 96″, 120″, and 144″. Most manufacturers provide enough extra material in the stock sizes for one or two saw cuts to be made. For example, two 12″ widths can usually be cut from a 24″ width. Plastic laminates are sold by the square foot, and the standard grade for counter tops is $1/16$″ thick.

A plastic laminate is quite thin and brittle: so it must be fastened or adhered to a core. Common core materials are plywood, particle board, and hardboard. These should be $3/4$″ thick with no defects or voids in the surface. For most job site applications, the laminate is adhered to the core with a contact cement. Other adhesives such as casein, polyvinyl, urea, and other slow setting glues can be used, but the laminate must be clamped to the core long enough to allow the glue to cure. Therefore a neoprene-base adhesive in either water or a flammable solvent (toluol and naphtha) is most popular for job-site installation.

Cutting the Laminate to Size

Plastic laminates may be sawed, routed, filed, drilled, and otherwise worked and fitted. The tools used may be hand- or power-operated. Since decorative laminate dulls tools more quickly than wood, the tools must be sharpened often. Dull tools may cause chipping. Whenever possible, carbide-tipped cutting tools should be used. Plastic laminate may also be sized by cutting it with special shears or by scoring the plastic with a scratch awl. When using the scoring method, score a cutting line deeply on the decorative side of the plastic with a scratch awl. Place a straightedge along the scored line. Bend evenly and upward only. Fig. 46-32. Do not try to tear the laminate. This method will not give as clean an edge as those methods previously mentioned. Therefore allow at least a half-inch oversize on all surface dimensions so that the edge may be trimmed back even with the edge of the core stock.

Measure the area to be covered. Allow 1/4" oversize on all edges unless the scored method of cutting is used. Cut the laminate to size. If there are any seams to be made, dress the edges smooth and straight until they fit together perfectly. Undercut the two edges slightly to allow a closer fit at the surface. These edges may be dressed with a block plane and touched up with a fine abrasive paper or a file.

Preparation of the Surface for Installation

For best results, all materials should be at room temperature (70° F) or higher before installation. The core should have a smooth, sound surface. Make certain the surface is clean, dry, and free of oil,

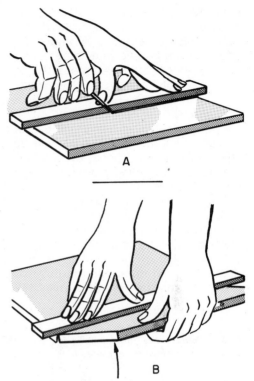

46–32. *Cutting plastic laminate: A. Score the face side of the laminate, using a straightedge as a guide. B. With the straightedge placed along the scribed line, bend the laminate up to break it off.*

grease, or wax. Fill holes and cracks with a spackling compound and then sand smooth and even with the surface. Dust all surfaces.

Applying the Adhesive

The adhesive is applied to both the back of the plastic laminate and the top of the core. (NOTE: If the edge trim is plastic laminate, apply the edge trim first. Coat and bond the top surface after the edge trim is bonded and trimmed.) Stir the adhesive thoroughly from the bottom of the can and pour it onto the core surface and the back of the laminate. Fig. 46-33. Spread the adhesive evenly, using a roller or a spreader with a notched edge. Fig. 46-34. If a spreader is used, the notched edge should be held at a 45° angle to the surface of the plastic laminate. Hold the spreader at 90° to the surface of the core material. A brush may also be used to apply adhesive and is recommended when adhesive is applied to vertical surfaces or edges, or whenever the use of a spreader is impractical. With a brush, both surfaces will need at least two coats. Be sure to allow the adhesive adequate drying time between coats.

To make certain that enough adhesive has been applied, look across the surface into the light after the adhesive is completely dry. There should be a glossy film. Dull spots after drying indicate that more adhesive should be applied to these areas. Let the adhesive dry according to the

label directions on the can. Test dryness of the adhesive by pressing a piece of wrapping paper lightly against it and pulling it away. Fig. 46-35. If the adhesive sticks to the paper, more drying time is needed. If, for any reason, bonding of the plastic laminate to the core is not completed within two hours, the adhesive should be reactivated by applying another thin coat to both the back of the laminate and the core.

Bonding and Finishing

If the edge trim is plastic laminate, install it first. If the edge trim is metal, install it after the top surface has been bonded in place. Fig. 46-36. The plastic laminate edge trim is carefully positioned so that its bottom edge is flush with the bottom edge of the core. The top edge of the trim will

46–33. *Pouring the contact adhesive onto the surface to be covered.*

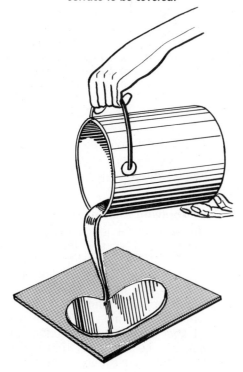

46–34a. *Spreading contact adhesive with a roller.*

46–34b. *Spreading contact adhesive with a notched spreader.*

extend about ¼″ or more above the top of the core surface. Press the edge trim in place by sliding a soft wood block along it. Tap the wood block with a hammer to complete the bond. Fig. 46-37. Carefully trim off excess material and file until the edge trim is flush with the top of the core surface. Fig. 46-38. Be sure the edge trim is bonded and completely trimmed before coating the top surface.

Apply adhesive to the top surface. *Use extreme care in aligning the plastic laminate with the core because bonding is immediate upon contact.* Place several dowel rods or strips of scrap plastic laminate with the smooth side down on top of

the core. Fig. 46-39. Align the plastic laminate with the core so that an equal amount of laminate hangs over all edges. Now gently slip the center piece of plastic scrap or dowel rod from beneath the plastic laminate, leaving the others in place. The two adhesive surfaces will come in contact with each other. Press to make the bond. Remove the other scraps one at a

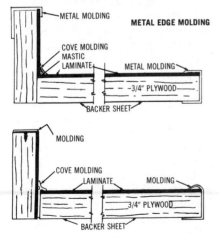

46-35. Testing contact adhesive to make sure it is completely dry.

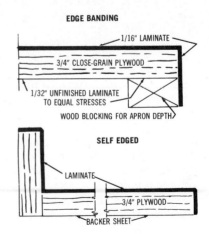

46-37. Complete the bond between the plastic laminate and the core by sliding a soft wood block along the trim, tapping it with a hammer.

46-36. Cross sections of job-site fabricated counter tops. They can be trimmed with metal molding or with plastic laminate which has been cut from the same material used on the surface.

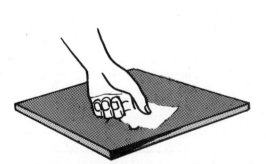

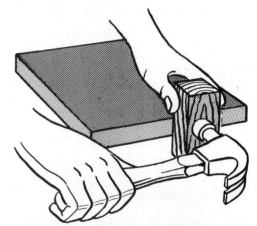

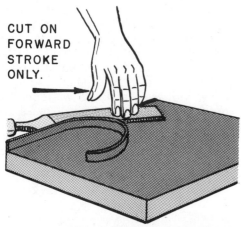

CUT ON
FORWARD
STROKE
ONLY.

46-38. *Carefully file the excess material until the edge trim is flush with the top surface. Be sure to cut on the forward stroke only, to avoid chipping the edges of the laminate.*

46-39. *Use scraps of plastic laminate to support the surface sheet of plastic laminate while aligning it with the core for bonding.*

time, working from the center toward the ends. Be careful not to jar the work as you move the scraps because the two pieces will bond immediately.

With a small roller, roll the surface from the center toward the edges in all direc-

46-40. *With a small roller, roll the surface of the plastic laminate from the center toward the edges in all directions to complete the bond.*

tions. Fig. 46-40. If a roller is not available, use a block of soft wood. Place this on the center and work toward the edges, tapping sharply with a hammer. Tap or roll the entire surface to insure a complete bond. Use a hand plane or plastic laminate trimmer to remove the excess plastic laminate which overhangs the edges of the core. File the edges smooth with a flat mill file at a 20° or 30° angle to the edge. Fig. 46-41. Use long, smooth downward strokes. If using metal or wood edge trim, file the laminate flush with the edge of the core. Then file a 45° bevel on the edge of the laminate to allow for expansion under the metal or wood edge trim. To butt seams, cut the seam as previously described. Apply one piece at a time along the predetermined seam line marked on the core surface. Position and adhere the laminate accurately to insure that the seam will be neat.

To insure a good plastic laminate job be sure that:
● The core and the back of the laminate are completely covered with glossy adhesive films.
● Both films are dry.
● Both adhesive films are brought into firm contact by rolling or tapping to complete the bond.

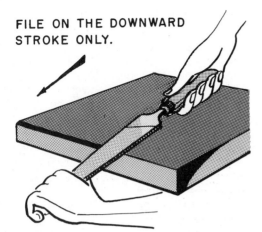

FILE ON THE DOWNWARD STROKE ONLY.

46–41. *File the edges of the surface laminate smooth on the downward stroke only, to avoid lifting and tearing the edge of the surface material.*

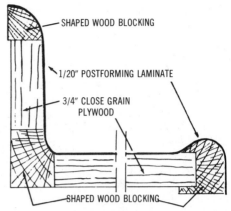

SHAPED WOOD BLOCKING

1/20" POSTFORMING LAMINATE

3/4" CLOSE GRAIN PLYWOOD

SHAPED WOOD BLOCKING

46–42. *A cross section of a preformed plastic laminate counter top.*

Preformed Plastic Laminate Counter Tops

Preformed counter tops, ready for installation, can be bought in slabs or custom made. Fig. 46-42. The preformed slab counter top is generally available in 8', 10', and 12' lengths. The choice of patterns and colors is somewhat limited. For an exposed end of the slab, preformed end pieces are available.

The custom-made counter top is available in most patterns and colors of plastic laminate. The counter top is made to the exact measurements. For an L- or U-shaped kitchen, the custom counter top is made up as a single unit whenever possible. If it is impossible to get an assembled counter into the kitchen for installation, it can be made in sections and held together at the corners with draw bolts.

INSTALLING A PREFORMED COUNTER TOP

Set the counter on the cabinets so that the front face of the counter (nearest the user) is 2^1/$_8$" from the desired final position. (This dimension may vary with the make of laminate trimmer.) For example, if

the counter is to overhang the cabinet by 1", then the top should be pulled out 3^1/$_8$". Clamp the counter securely in position. Use a laminate trimmer to fit the counter top to the wall. Fig. 46-43. Place the trimmer on the top edge of the backsplash and move the tool from left to right while keeping the wall guide bracket in contact with the wall. This will trim the rear edge of the plastic laminate. At the same time, the contour of the wall is duplicated on the trimmed edge of the laminate. This trimming operation will create a gap 2^1/$_8$" wide between the wall and the trimmed edge so that, when the clamps are removed from the counter top, the counter top can be pushed back snug against the wall.

If a trimmer is not available, the top edge of the backsplash on the preformed top will have to be hand-filed to fit the contour of the wall. After the top is secured to the cabinets, apply a small bead of caulking compound at the joint between the wall and the back top edge of the backsplash.

The counter top is fastened in place by driving wood screws through cleats on the underside of the cabinet up into the bottom of the counter top. Be careful to drill the correct size pilot hole and take special pains not to drill through the top. Be sure to use the correct length wood screw so that it does not penetrate the plastic laminate when the counter top is pulled down snug against the top of the cabinets.

Installing the Sink

Lay out the sink opening by positioning the sink rim upright on the counter top. Carefully trace around the outside bottom edge of the rim. Allow an additional 2" of space between the cutout and cabinet partitions or other parts to provide room for attaching the clips. Fig. 46-44. Cut out the opening with a saber saw or an electric hand circular saw. Use a saber saw for the corners.

The sink may be held in place with a one-piece, preformed sink frame. Another method is to fasten the sink with clips which attach to the bottom of the rim or, in the case of the self-rim type, to the bottom of the sink. Fig. 46-45. Apply calking to the underside of the sink frame molding. Fig. 46-46. When installing a self-rim sink, place a bead of calking around the underside of the sink at the edge so that, when the sink is pulled down against the top, it will seal properly.

LINEN CLOSETS

Built-in storage areas, such as a linen closet, reduce the amount of furniture required. A linen closet may be simply a series of shelves behind a flush or panel door, or it may consist of an open cabinet with doors and drawers built directly into a corner or wall area. Linen closets are usually located in hallways near the bedroom and bath. Adjustable shelves are very practical because of the large variety of sizes stored. If the shelves are not adjustable and are supported by cleats which are nailed to the wall, do not fasten the shelves to the cleats. The shelves can then be removed for convenience when painting and decorating. See *Cabinetmaking and Millwork* by John L. Feirer for additional information and construction details.

CHINA CASES

Another millwork item often found in a formal or traditional dining room is the china cabinet. It is usually designed to fit into a corner of the room. Sometimes a pair of cabinets is installed in two corners of the room. A corner cabinet often has glazed

46–43a. *A plastic laminate trimmer.*

46–43b. *Using a laminate trimmer to trim the end piece on a preformed counter top.*

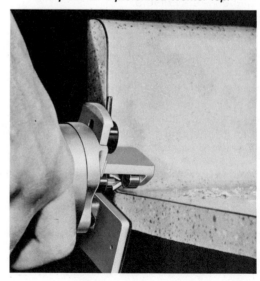

doors above and single or double paneled doors below. Fig. 46-47. Generally it is 7' or more high and about 3' wide across the front, with a drop ceiling above. Shelves are installed in both the upper and lower cabinet.

China cases or storage shelves in dining rooms of contemporary homes may be built in place by the contractor, or prefab cabinets can be installed. A row of cabinets or shelves installed between dining room and kitchen may act as a room divid-

er and serve as a storage area for both rooms. Factory-built cabinets can be fitted with accessories which are available from the manufacturer. These cabinets and accessories are combined to form free-standing or built-in pieces such as bars, hutches, sideboards and barbeques. Many different arrangements can be assembled for use throughout the entire home. Fig. 46-48 (Pages 890-896). For additional in-

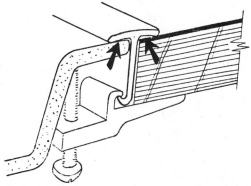

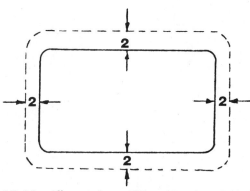

46–44. *Allow at least 2″ between the edge of the cutout and any cabinet partitions or other cabinet parts to provide room for attaching the clips used to install the sink.*

46–46. *Apply calking under the molding on both sides, as shown at the arrows. Use a screwdriver or special wrench to draw up the clamp screws. Wipe off the excess calking compound that squeezes out.*

46–45. *A clip used to fasten the sink. Two of these clamps should be placed at each corner, with the remaining clamps at equidistant points around the sink.*

46–47. *A built-in corner cabinet.*

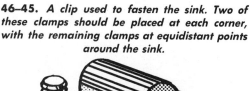

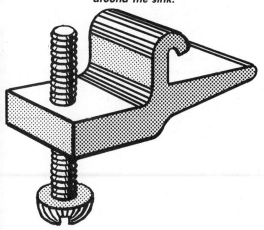

SPECIFICATIONS

46—48a. A typical manufacturer's catalog showing cabinets which can be used in conjunction with the accessory items shown in Fig. 46—48b.

WALL CABINETS

All cabinet ends are finished gumwood. If same finish as face is desired use maple end panels.

12-inch high

12" — 12½"

33" 3312 36" 3612 33" 3315 36" 3615

15-inch high

15" — 12½"

18-inch high

18" — 12½"

24" P2424 24" 2418 27" 2718 30" 3018 33" 3318 36" 3618

39" 3918 42" 4218 48" 4818

2418 thru 4218 — 1 Adjustable Shelf; 4818 — 1 Fixed Shelf.

24-inch high peninsula

24" — 12½"

24" P2424 27" P2724 30" P3024 36" P3624 42" P4224 48" P4824

P2424 thru P4224—1 Adjustable Shelf; P4824—1 Fixed Shelf.

corner

30" — 12½" 35"

C3530 C4230 C4830

C3530 and C4230—2 adjustable shelves; C4830 and 2330—2 fixed shelves.
No holes drilled for pull; apply pull after installation.

23" 23" 23" 30" 30"

2330 2330SR
3 Adjustable Spin Shelves

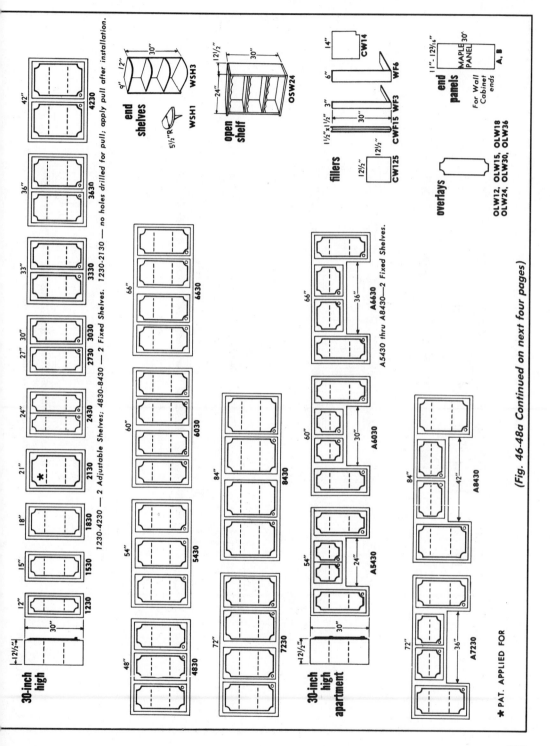

(Fig. 46-48a Continued on next four pages)

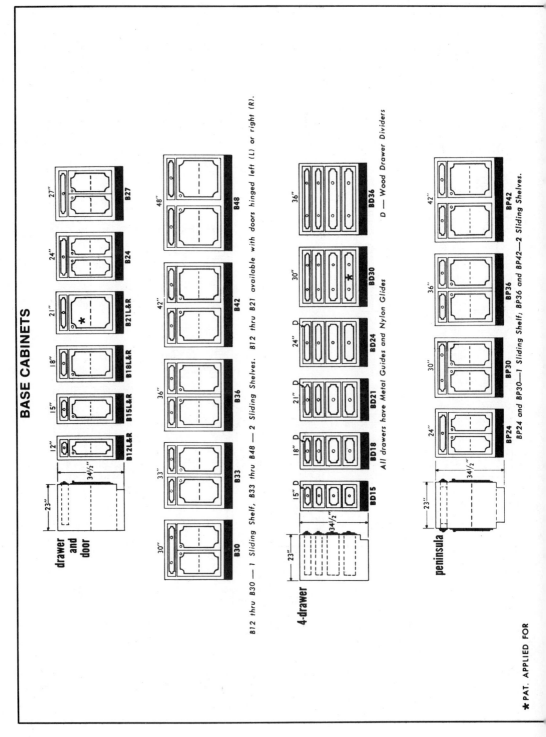

BASE CABINETS

drawer and door

23″ · 34½″

12″ — B12L&R
15″ — B15L&R
18″ — B18L&R
21″ — B21L&R ✱
24″ — B24
27″ — B27

30″ — B30
33″ — B33
36″ — B36
42″ — B42
48″ — B48

B12 thru B30 — 1 Sliding Shelf. B33 thru B48 — 2 Sliding Shelves. B12 thru B21 available with doors hinged left (L) or right (R).

4-drawer

23″ · 34½″

15″ D — BD15
18″ D — BD18
21″ D — BD21
24″ D — BD24
30″ — BD30 ✱
36″ — BD36

All drawers have Metal Guides and Nylon Glides   D — Wood Drawer Dividers

peninsula

23″ · 34½″

24″ — BP24
30″ — BP30
36″ — BP36
42″ — BP42

BP24 and BP30 — 1 Sliding Shelf; BP36 and BP42 — 2 Sliding Shelves.

✱ **PAT. APPLIED FOR**

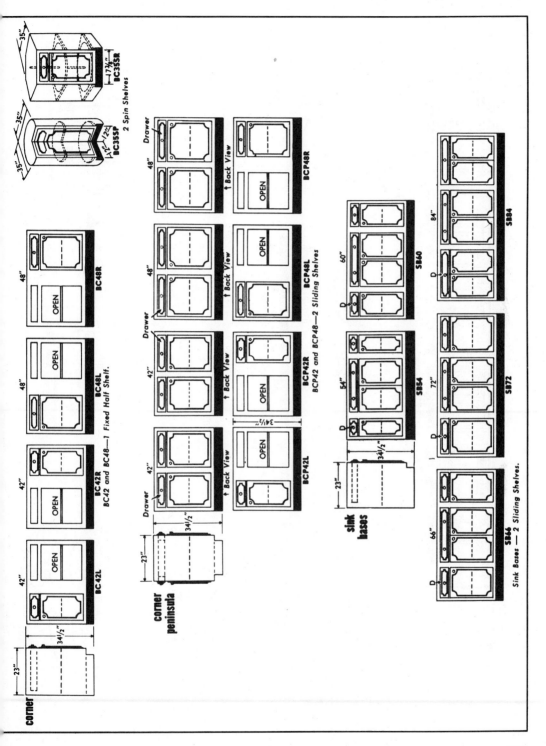

35"

17¾"
BC35SR
2 Spin Shelves

35"
1¼"/2"
BC35SP

corner

Drawer 48"   ↑ Back View   48"
BCP48R

BC48R
48"

Drawer 48"   ↑ Back View   48"   OPEN
BCP48L
BCP42 and BCP48—2 Sliding Shelves

BC48L
BC42 and BC48—1 Fixed Half Shelf.
48"   OPEN

Drawer 42"   ↑ Back View   42"   OPEN
BCP42R

BC42R
42"   OPEN

Drawer 42"   ↑ Back View   42"   OPEN
34½"
BCP42L

BC42L
42"   OPEN

corner peninsula

34½"
23"

84"
SB84

60"
D
SB60

54"
34½"
SB54

72"
D
SB72

sink bases

23"

66"
D
SB66
Sink Bases — 2 Sliding Shelves.

corner
34½"
23"

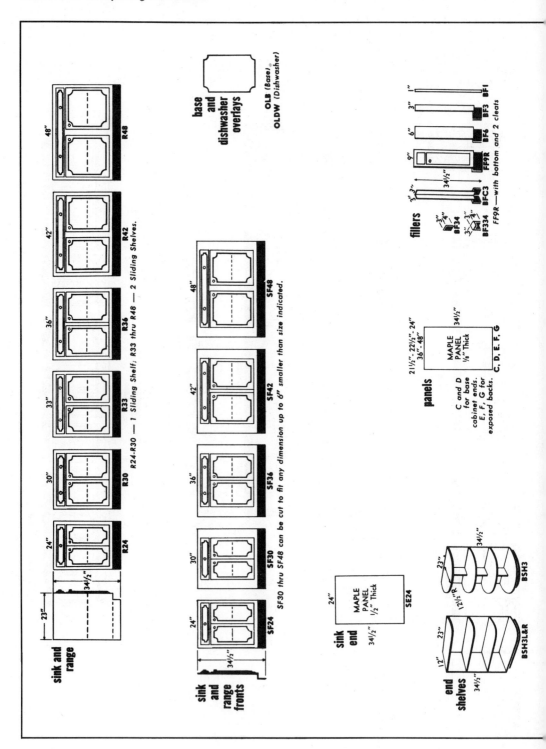

sink and range

R24 — 24"
R30 — 30"
R33 — 33"
R36 — 36"
R42 — 42"
R48 — 48"

R24-R30 — 1 Sliding Shelf; R33 thru R48 — 2 Sliding Shelves.

sink and range fronts

SF24 — 24"
SF30 — 30"
SF36 — 36"
SF42 — 42"
SF48 — 48"

SF30 thru SF48 can be cut to fit any dimension up to 6" smaller than size indicated.

base and dishwasher overlays

OLB (Base)
OLDW (Dishwasher)

sink end

MAPLE PANEL ½" Thick

SE24

panels

MAPLE PANEL ⅛" Thick

C and D for base cabinet ends. E, F, G for exposed backs.

C, D, E, F, G

fillers

BF1 — 1"
BF3 — 3"
BF4 — 6"
FF9R — 9"
BFC3 — 3"-3"
BF34
BF334

FF9R—with bottom and 2 cleats

end shelves

BSH3L&R

BSH3

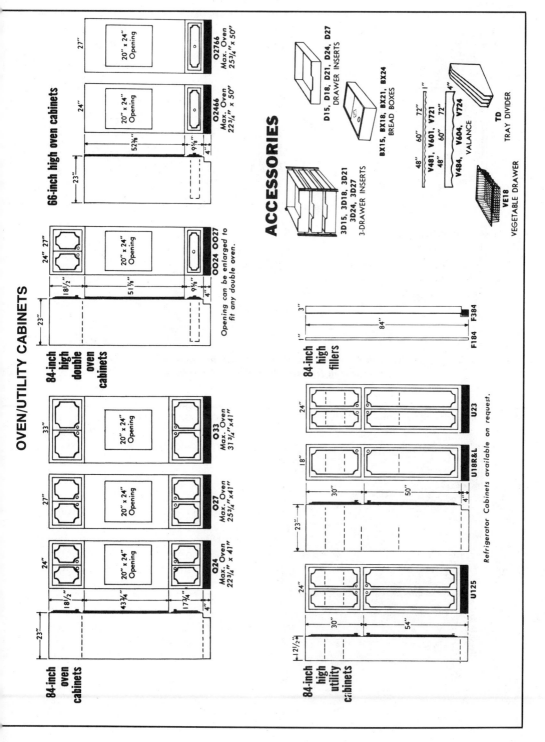

OVEN/UTILITY CABINETS

84-inch oven cabinets

O24 Max. Oven 22¾" x 41"

O27 Max. Oven 25¾" x 41"

O33 Max. Oven 31¾" x 41"

84-inch high double oven cabinets

OO24 OO27 Opening can be enlarged to fit any double oven.

66-inch high oven cabinets

O2466 Max. Oven 22¾" x 50"

O2766 Max. Oven 25¾" x 50"

20" x 24" Opening

ACCESSORIES

D15, D18, D21, D24, D27 DRAWER INSERTS

BX15, BX18, BX21, BX24 BREAD BOXES

3D15, 3D18, 3D21 3D24, 3D27 3-DRAWER INSERTS

V481, V601, V721 V484, V604, V724 VALANCE

TD TRAY DIVIDER

VE18 VEGETABLE DRAWER

84-inch high fillers

F184   F384

84-inch high utility cabinets

U125

U18R&L

U23

Refrigerator Cabinets available on request.

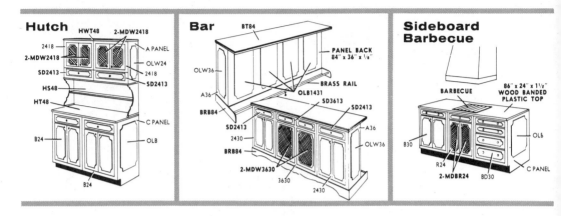

Hutch

HWT48 2-MDW2418
2418 A PANEL
2-MDW2418 OLW24
SD2413 2418
HS48 SD2413
HT48
C PANEL
B24 OLB
B24

Bar

BT84
PANEL BACK
84" x 36" x ¹/₈"
OLW36
BRASS RAIL
A36 OLB1431
BRB84 SD3613
SD2413 SD2413
2430 A36
BRB84 OLW36
2-MDW3630
3630
2430

Sideboard Barbecue

BARBECUE
86" x 24" x 1¹/₂"
WOOD BANDED
PLASTIC TOP
B30 OLB
R24
2-MDBR24 BD30
C PANEL

Furniture Groupings Accessories Specifications

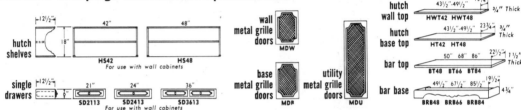

hutch shelves
12¹/₂"
18"
42" 48"
HS42 HS48
For use with wall cabinets

single drawers
12¹/₂"
21" 24" 36"
SD2113 SD2413 SD3613
For use with wall cabinets

wall metal grille doors
MDW

base metal grille doors
MDB

utility metal grille doors
MDU

hutch wall top
43¹/₂"-49¹/₂" 13¹/₄"
HWT42 HWT48 ³/₄" Thick

hutch base top
43¹/₂"-49¹/₂" 23³/₄"
HT42 HT48 ³/₄" Thick

bar top
50" 68" 86" 22¹/₂"
BT48 BT66 BT84 1¹/₂" Thick

bar base
49¹/₂" 67¹/₂" 85¹/₂" 19¹/₄"
BRB48 BRB66 BRB84 4³/₄"

46–48b. *Details of accessories to be used with standard manufactured base cabinets for developing free-standing pieces of furniture.*

46–48c. *Manufactured cabinets installed in a dining area.*

formation on designing and building built-in cabinets, see *Cabinetmaking and Millwork* by John L. Feirer.

ESTIMATING
Materials

Cabinets are of three types: prefab, K.D. (knocked-down), and built on the job site. The prefab cabinet costs are found by selecting cabinets for the building, listing them with the identifying stock number, and pricing them according to current price lists. Knocked-down cabinets would be priced in much the same manner. Job-site built cabinets will require a complete bill of materials with the pricing for each item. The estimate must include all items required for a complete installation, such as counter tops, hardware, drawers (if they are special drawers), glass, shelves, any

46–49a. *Approximate installation times for factory-built cabinets.*

Type of Cabinet	Approximate Installation Time (Hours)
Base cabinet containing one door and one drawer	½
Base cabinet containing two doors and two drawers	¾
Base corner cabinet	1
Broom closet	1
Drawer cabinet—4 drawers	½
Oven cabinet	1¼
Sink cabinet	1½
Wall cabinet—two doors (refrigerator cabinet)	½
Wall cabinet—two doors (standard height)	½
China case—corner unit with 36″ front	2
Bathroom vanity—up to 84″ long	2

46–49b. *Approximate installation times for preformed counter tops.*

Counter Tops	Approximate Installation Time (Hours)
Preformed plastic laminate counter top	¾ (per linear foot)
25″-wide plastic laminate counter top with a 4″ back splash and self edge.	1 (per linear foot)
Preformed plastic laminate mitered corner (L- or U-shaped kitchen)	1
End cap on a preformed plastic laminate top	⅓

The list would include core material, plastic laminate, adhesive, moldings, and any other items that might be required for a complete installation.

Labor

The time required to install prefab cabinets will vary with the complexity of the kitchen layout and the variety of cabinets to be installed. The approximate labor cost can be determined by totaling the installation time for the selected cabinets and multiplying this by the hourly rate. The approximate times for installing various cabinets and counter tops are listed in Fig. 46-49.

special grill work, and any other items needed.

Counter tops are priced according to type. The preformed slab counter top is sold by the linear foot, and current prices would be available from a local dealer. Counter tops to be built on the job site would require that materials be itemized.

QUESTIONS

1. What are the four basic layouts commonly used in kitchen design?

2. List the kitchen work centers.

3. What is the standard height of the kitchen cabinet base unit?

4. When measuring along the kitchen wall for the installation of factory-built cabinets, how high off the floor are the dimensions taken?

5. Why are the dimensions taken at some point above the floor?

6. When installing factory-built base cabinets in a kitchen, which cabinets are installed first?

7. What is the most widely used material for kitchen cabinet counter top surfaces?

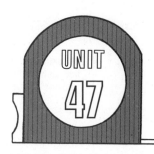

Interior Trim

Wood or plastic molding and trim add individuality to home design. Figs. 47-1 and 47-2. Sometimes special patterns of millwork are required, but it is possible to create unique effects with standard patterns. Fig. 47-3. The standard patterns are readily available from local lumber dealers, allowing the development of many interesting design with maximum economy. Wood or plastic moldings can be used with wallpaper or fabric, as accent walls, or to create interesting shadow and highlight effects on flat surfaces. Figs. 47-4, 47-5 (Page 903), and 47-6 (Page 906). They may be painted to blend or contrast, or they may be stained to match or accent natural wood grains. Wood and plastic moldings in the home are most often used for trimming door and window frames, for base moldings, and, some-

47-2. With the addition of molding, a plain room achieves definite character and depth. Moldings soften harsh corners, highlight room features, and, as can be seen in this stairway, emphasize direction.

47-1. Moldings may be installed on painted walls or with wallpaper. These moldings can be painted, stained, or even antiqued.

47–3. *A variety of standard wood molding styles.*

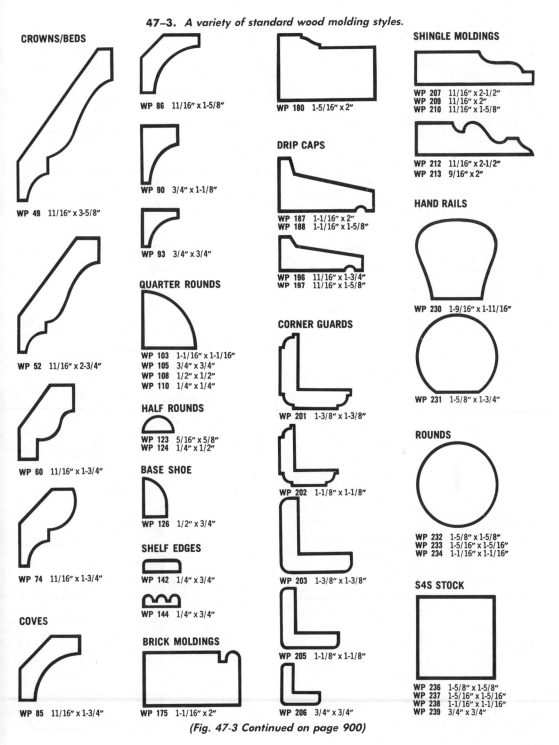

CROWNS/BEDS

WP 49   11/16" x 3-5/8"

WP 52   11/16" x 2-3/4"

WP 60   11/16" x 1-3/4"

WP 74   11/16" x 1-3/4"

COVES

WP 85   11/16" x 1-3/4"

WP 86   11/16" x 1-5/8"

WP 90   3/4" x 1-1/8"

WP 93   3/4" x 3/4"

QUARTER ROUNDS

WP 103   1-1/16" x 1-1/16"
WP 105   3/4" x 3/4"
WP 108   1/2" x 1/2"
WP 110   1/4" x 1/4"

HALF ROUNDS

WP 123   5/16" x 5/8"
WP 124   1/4" x 1/2"

BASE SHOE

WP 126   1/2" x 3/4"

SHELF EDGES

WP 142   1/4" x 3/4"

WP 144   1/4" x 3/4"

BRICK MOLDINGS

WP 175   1-1/16" x 2"

WP 180   1-5/16" x 2"

DRIP CAPS

WP 187   1-1/16" x 2"
WP 188   1-1/16" x 1-5/8"

WP 196   11/16" x 1-3/4"
WP 197   11/16" x 1-5/8"

CORNER GUARDS

WP 201   1-3/8" x 1-3/8"

WP 202   1-1/8" x 1-1/8"

WP 203   1-3/8" x 1-3/8"

WP 205   1-1/8" x 1-1/8"

WP 206   3/4" x 3/4"

SHINGLE MOLDINGS

WP 207   11/16" x 2-1/2"
WP 209   11/16" x 2"
WP 210   11/16" x 1-5/8"

WP 212   11/16" x 2-1/2"
WP 213   9/16" x 2"

HAND RAILS

WP 230   1-9/16" x 1-11/16"

WP 231   1-5/8" x 1-3/4"

ROUNDS

WP 232   1-5/8" x 1-5/8"
WP 233   1-5/16" x 1-5/16"
WP 234   1-1/16" x 1-1/16"

S4S STOCK

WP 236   1-5/8" x 1-5/8"
WP 237   1-5/16" x 1-5/16"
WP 238   1-1/16" x 1-1/16"
WP 239   3/4" x 3/4"

(Fig. 47-3 Continued on page 900)

(Fig. 47-3 Continued from page 899)

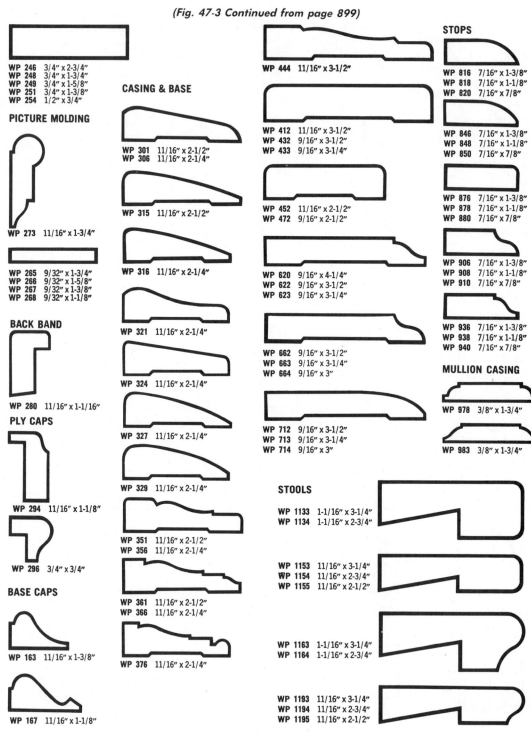

WP 246   3/4" x 2-3/4"
WP 248   3/4" x 1-3/4"
WP 249   3/4" x 1-5/8"
WP 251   3/4" x 1-3/8"
WP 254   1/2" x 3/4"

PICTURE MOLDING

WP 273   11/16" x 1-3/4"

WP 265   9/32" x 1-3/4"
WP 266   9/32" x 1-5/8"
WP 267   9/32" x 1-3/8"
WP 268   9/32" x 1-1/8"

BACK BAND

WP 280   11/16" x 1-1/16"

PLY CAPS

WP 294   11/16" x 1-1/8"

WP 296   3/4" x 3/4"

BASE CAPS

WP 163   11/16" x 1-3/8"

WP 167   11/16" x 1-1/8"

CASING & BASE

WP 301   11/16" x 2-1/2"
WP 306   11/16" x 2-1/4"

WP 315   11/16" x 2-1/2"

WP 316   11/16" x 2-1/4"

WP 321   11/16" x 2-1/4"

WP 324   11/16" x 2-1/4"

WP 327   11/16" x 2-1/4"

WP 329   11/16" x 2-1/4"

WP 351   11/16" x 2-1/2"
WP 356   11/16" x 2-1/4"

WP 361   11/16" x 2-1/2"
WP 366   11/16" x 2-1/4"

WP 376   11/16" x 2-1/4"

WP 444   11/16" x 3-1/2"

WP 412   11/16" x 3-1/2"
WP 432   9/16" x 3-1/2"
WP 433   9/16" x 3-1/4"

WP 452   11/16" x 2-1/2"
WP 472   9/16" x 2-1/2"

WP 620   9/16" x 4-1/4"
WP 622   9/16" x 3-1/2"
WP 623   9/16" x 3-1/4"

WP 662   9/16" x 3-1/2"
WP 663   9/16" x 3-1/4"
WP 664   9/16" x 3"

WP 712   9/16" x 3-1/2"
WP 713   9/16" x 3-1/4"
WP 714   9/16" x 3"

STOOLS

WP 1133   1-1/16" x 3-1/4"
WP 1134   1-1/16" x 2-3/4"

WP 1153   11/16" x 3-1/4"
WP 1154   11/16" x 2-3/4"
WP 1155   11/16" x 2-1/2"

WP 1163   1-1/16" x 3-1/4"
WP 1164   1-1/16" x 2-3/4"

WP 1193   11/16" x 3-1/4"
WP 1194   11/16" x 2-3/4"
WP 1195   11/16" x 2-1/2"

STOPS

WP 816   7/16" x 1-3/8"
WP 818   7/16" x 1-1/8"
WP 820   7/16" x 7/8"

WP 846   7/16" x 1-3/8"
WP 848   7/16" x 1-1/8"
WP 850   7/16" x 7/8"

WP 876   7/16" x 1-3/8"
WP 878   7/16" x 1-1/8"
WP 880   7/16" x 7/8"

WP 906   7/16" x 1-3/8"
WP 908   7/16" x 1-1/8"
WP 910   7/16" x 7/8"

WP 936   7/16" x 1-3/8"
WP 938   7/16" x 1-1/8"
WP 940   7/16" x 7/8"

MULLION CASING

WP 978   3/8" x 1-3/4"

WP 983   3/8" x 1-3/4"

47–4a. *Procedure and bill of materials for traditional wall design shown in Fig. 47–4b.*

TRADITIONAL WALL (8′ high; 12′ long)

Base
Saw and nail 12′ base as shown in Fig. 47-4b, Part 2.

Wall
At approximately 16″ on center (on studs), glue up vertical pieces of screen stock WP-246 which have been sawn to approximate 7′ lengths. Always measure wall for exact length, subtracting the width of the base and top. Fig. 47-4b, Part 1.

Top
Saw and nail 12′ top section. Fig. 47-4b, Part 3.

Finish
Measure, cut, and apply wallpaper, burlap, grass cloth or other decorative wall covering. Measure, cut, and miter WP-163 cap molding to fit. Fig. 47-4b, Part 3.

Materials List

1 piece	1″ x 8″ stock finish lumber	12′ long
1 piece	WP-131 base shoe	12′ long
1 piece	WP-664 base	12′ long
9 pieces	WP-246 screen stock	7′ (approx.)
1 piece	WP-93 cove molding	12′ long
16 pieces	WP-163 cap molding	13″ long (approx.)
16 pieces	WP-163 cap molding	7′ long
8 sections	Wallpaper, burlap, grass cloth, etc.	1′ x 7′ (approx.)
1 lb.	5d finishing nails	
½ lb.	No. 18 x 1¼″ brads	
½ pint	White woodworking glue	
As needed	Paint or stain	

times, for ceiling and wall moldings. Fig. 47-7 (Page 909).

Interior doors, trim, and other millwork may be painted or given a natural finish with stain, varnish, or other nonpigmented material. The paint or natural finish desired for the woodwork in various rooms often determines the species of wood to be used. Woodwork to be painted should be smooth, close-grained, and free from pitch streaks. Some species having these qualities in a high degree include ponderosa pine, northern white pine, redwood, and spruce. When hardness and resistance to hard usage are additional requirements, species such as birch or yellow poplar may be used.

When the finish is to be natural, a pleasing figure, hardness, and uniform color are usually desirable. Species with these qualities include ash, birch, cherry, maple, oak, and walnut. Some require staining for best appearance.

The recommended moisture content for interior trim varies from 6 to 11 percent, depending on climate. The average moisture content for various parts of the United States is shown in Fig. 47-8 (Page 910).

INTERIOR DOOR AND WINDOW TRIM

After interior wall covering has been applied and the finish floor laid, all floor and wall surfaces should be scraped clean and free of any irregularities. Mark the

47–4b. *Wood moldings used with wallpaper or fabric: 1. A wall enriched with moldings in a traditional design. 2. Base detail. 3. Ceiling detail.*

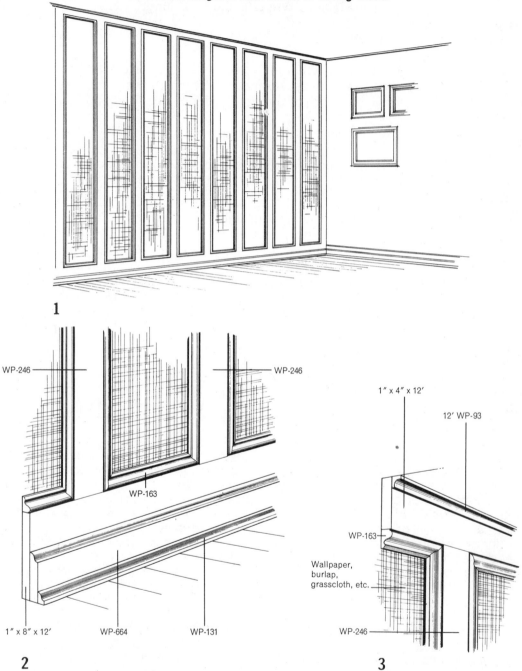

1

2

WP-246

WP-246

WP-163

1″ x 8″ x 12′

WP-664

WP-131

3

1″ x 4″ x 12′

12′ WP-93

WP-163

Wallpaper,
burlap,
grasscloth, etc.

WP-246

47–5a. *Procedure and bill of materials for constructing the Spanish wall shown in Fig. 47–5b.*

SPANISH WALL (8' high; 12' long)

It will be helpful to think of each section of this wall as a picture frame. Therefore the base consists of 24 frames one foot square; the upper section consists of six frames measuring 24" x 63". Care must be taken to insure exactness in dimensions.

Base

The base is applied in two parts: backing (furring) and baseboard. Fig. 47-5b, Part 2.

Wall

● In a jig (shown in Fig. 47-5c) make 24 one-foot square (outside dimensions) "picture frames" of WP-93 cove molding.

● Glue miters. Use masking tape at the corners while mitered joints are drying.

● In a jig, make 6 "picture frames" 24" wide and 63" high (outside dimensions) using WP-93 cove molding. Height given is approximate. Always measure the wall first for the exact height.

● Glue miters and tape.

● Starting at the base and working up, glue the one-foot square sections to the wall and to each other. Nail a few brads in each section to secure frame to wall while it dries. Remove nails when glue is dry and fill holes, or countersink the nails.

● Glue the six 24" x 63" sections to the wall and to each other, again securing them to wall with nails while drying.

● Nail 12' section of WP-252 screen stock to wall at top of frames.

● Nail 12' piece of crown molding, WP-49, to ceiling and back piece. Fig. 47-5b, Part 3.

● Make 24 rosettes as shown in Parts 4 and 5 of Fig. 47-5b. Center in position and glue to wall as shown in Part 1. A template will simplify centering the rosettes. Cut a piece of plywood to fit into the 12" frames. Then cut a 6½"-square hole in the center and use it as a guide.

● Before finishing, sand off excess glue.

Materials List

96 pieces	WP-93 cove molding	12" long
12 pieces	WP-93 cove molding	24" long
12 pieces	WP-93 cove molding	63" long (approx.)
1 piece	WP-624 base molding	12' long
1 piece	¾" x 3" furring strip	12' long
1 piece	WP-49 crown molding	12' long
1 piece	WP-252 screen stock	12' long
96 pieces	WP-624 base molding	6½" long
96 pieces	WP-163 cap molding	2¾" long
½ pint	Woodworking glue	
1 lb.	5d finishing nails	
1 lb.	No. 18 x 1¼" brads	
As needed	Masking tape	
As needed	Paint or stain	

47–5b. *Moldings and woodwork are used to give this wall a Spanish motif: 1. Spanish wall and divider. 2. Wall base detail. 3. Wall, ceiling detail. 4. Cutting miters in the molding to make the rosettes. 5. Rosette detail.*

1   Spanish Wall & Divider

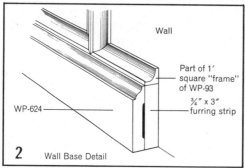

Wall

Part of 1'
square "frame"
of WP-93

¾" x 3"
furring strip

WP-624

2   Wall Base Detail

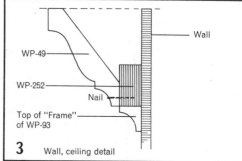

Wall

WP-49

WP-252

Nail

Top of "Frame"
of WP-93

3   Wall, ceiling detail

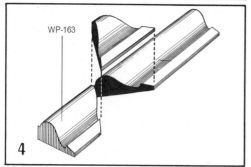

WP-163

4

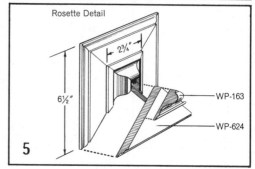

Rosette Detail

2¾"

6½"

WP-163

WP-624

5

location of all wall studs lightly on the floor or wall. Usually the marks are placed on the floor because the interior wall covering may be a finished surface.

Door and window frames are usually trimmed first to allow other trim such as base moldings or wall moldings to be properly fitted between the door and window casings. Cabinets, built-in bookcases, fireplace mantels, and other millwork items are also installed at this time.

Doorframes

Rough openings in the stud walls for interior doors are usually framed out to be 3″ more than the door height and 2¹/₂″ more than the door width. This provides room for plumbing and leveling the frame in the opening. Interior doorframes are made up of two side jambs and a head jamb and include stop moldings which the door closes against. One-piece jambs are the most common. Fig. 47-9 (Page 910). They may be obtained in 5¹/₄″ widths for plaster walls and 4¹/₂″ widths for walls with ¹/₂″ dry-wall finish. Two- and three-piece adjustable jambs are also available. Fig. 47-10. Their chief advantage is in being adaptable to different wall thicknesses.

Some manufacturers produce interior doorframes with the door fitted and prehung, ready for installation. Application of the casing completes the job. When used with two- or three-piece jambs, casings are installed at the factory. Figs. 47-11 and 47-12 (Page 914).

DOORFRAME INSTALLATION

When the frames and doors are not assembled and prefitted, the side jambs should be nailed through the notch into the head jamb with three 7d or 8d coated nails. Fig. 47-9. Cut a spreader to a length exactly equal to the distance between the jambs at the head jamb. Fig. 47-13 (Page 916).

The assembled frame is fastened in the rough opening by shingle wedges placed between the side jamb and the stud. Fig.

47-14. One jamb, usually the hinge jamb, is plumbed using four or five sets of shingle wedges for the height of the frame. Two 8d finishing nails are installed at each wedged area, one driven so the the door-stop will cover it. Fig. 47-14.

Place the spreader in position at the floor line. Fig. 47-13. Fasten the opposite side jamb in place with shingle wedges and finishing nails, using the first jamb as a guide in keeping a uniform width. This can be done by using a second precut spreader as a gauge, checking several points, or by carefully measuring at various points along the height of the doorframe between the side jambs.

Door Trim

Door trim, or *casing*, is nailed around interior door openings and is also used to finish the room side of windows and exterior doorframes. The most commonly used casings vary in width from 2¹/₄″ to 3¹/₂″, depending on the style. Fig. 47-3. Thicknesses vary from ¹/₂″ to ³/₄″, although ¹¹/₁₆″ is standard in many of the narrow-line patterns. Two of the more common patterns are shown in Fig. 47-15.

DOOR TRIM INSTALLATION

Casings are nailed to both the jamb and the framing studs or header, allowing about a ³/₁₆″ edge distance from the face of

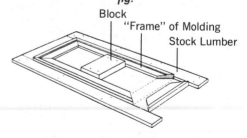

47–5c. Making a jig. The inside dimensions of the jig equal the outside dimensions of the frame. The jig consists of stock pieces of lumber nailed to any flat, nailable surface. Blocks can be used (as shown) to straighten moldings against side of jig.

Block

"Frame" of Molding

Stock Lumber

47–6a. *Procedure and bill of materials for constructing the room divider shown in Fig. 47–5b.*

ROOM DIVIDER (Approximately 4′ wide and 8′ high)

● Think of the base as a sandwich, as shown in the cutaway view. Fig. 47-6b, Part 5. A sheet of plywood ¼″ x 2′3″ x 4′0″ is sandwiched between two 4′ long pieces of ¾″ x 3″ furring and two 4′1″ sections of WP-624 base molding, mitered at the outside ends.

● Saw and nail an 8′ piece of WP-248 screen stock to the end of the base. Fig. 47-6b, Part 3.

● In a jig (Fig. 47-5c) make 16 one-foot square "frames" of WP-93 cove molding (8 for each side). Glue to the plywood and to each other on both sides, using brads to hold in place while drying. Fig. 47-6b, Part 3.

● In a jig, make 24 (12 for each side) frames of WP-93 cove molding, approximately 4″ x 63″. Measure on-the-spot for exact length.

● Clear a large floor area and cover with tarpaulin or paper. Glue and nail 12 of the frames together. Nail and glue 4 pieces of 4′ long WP-226 or 267 lattice (one at the top, one at the bottom, and the other two equally spaced) to back of glued frames. Fig. 47-6b, Part 2.

● Glue remaining 12 frames together and attach to other section.

● Attach the 24 assembled frames to the top of the base section.

● Glue and nail 4′ strip of WP-248 screen stock to top of divider as shown in Part 4 of Fig. 47-6b.

● Glue and nail WP-49 crown molding to top of divider. Fig. 47-6b, Part 4.

● Miter and cope wedges of crown molding to fit end. Fig. 47-6b, Part 2.

● Glue two 8′ sections of WP-93 cove molding together. Glue and nail to the WP-248 screen stock piece at end of divider. Fig. 47-6b, Part 2. Cope to fit over WP-49 crown molding.

● Miter and cope wedges of WP-624 base molding to fit base. Fig. 47-6b, Part 3.

the jamb. Fig. 47-14. Finish or casing nails either 6d or 7d, depending on the thickness of the casing, are used to nail into the stud. Arrow 1, Fig. 47-14. Fourpenny or 5d finishing nails or 1½″ brads are used to fasten the thinner edge of the casing to the jamb. Arrow 2, Fig. 47-14. With hardwood, it is advisable to predrill to prevent splitting. Nails in the casing are located in pairs and spaced about 16″ apart along the full height of the opening and at the head jamb. Fig. 47-14.

Casing with a molded shape must have mitered corner joints. A, Fig. 47-16 (Page 917). When casing is square-edged, a butt joint may be made at the junction of the side and head casing. B, Fig. 47-16. If the moisture content of the casing is well above the recommended amount, a mitered joint may open slightly at the outer edge as the material dries. This can be minimized by installing a small glued spline at the corner of the mitered joint. Actually, use of a spline joint under any moisture condition in considered good practice. Some prefitted jamb, door, and casing units are provided with splined joints. Nailing into the joint after drilling will aid in retaining a close fit. Fig. 47-16.

The door opening is now complete except for fitting and securing the hardware and nailing the stops in proper position.

● Secure two or three ¾" blocks (tapered to fit inside crown molding) to ceiling with molly bolts.
● Nail divider to ceiling blocks. Fig. 47-6b, Part 4.
● Make 16 rosettes (8 for each side) from WP-624 base molding and WP-163 cap molding as shown in Parts 4 and 5 of Fig. 47-5b, Spanish wall instructions.
● Within each one-foot square frame, center the 16 completed rosettes and glue to plywood.
● Before finishing, sand off excess glue.

Materials List

64 pieces	WP-93 cove molding	12" long
48 pieces	WP-93 cove molding	4" long
48 pieces	WP-93 cove molding	63" long (approx.)
2 pieces	WP-624 base molding	4'6" long
2 pieces	WP-93 cove molding	8' long
1 piece	WP-248 screen stock	8' long
2 pieces	WP-49 crown molding	4'6" long
4 pieces	WP-266 or WP-267 lattice	4' long
1 piece	Plywood	¼" x 2'3" x 4'0"
1 piece	WP-248 screen stock	4' long
2 pieces	¾" x 3" furring	4' long
64 pieces	WP-624 base molding	6½" long
64 pieces	WP-163 cap molding	2¾" long
2 or 3	Wood blocks	¾" thick
1 lb.	5d finishing nails	
1 lb.	No. 18 x 1¼" brads	
1 pint	White woodworking glue	
As needed	Molly bolts	

Interior Doors

As in exterior door styles, the two general interior types are the flush and the panel. Bifold and sliding door units might be flush or louvered. Most standard interior doors are 1⅜" thick. Folding and sliding doors are usually 1⅛" thick.

The flush interior door is usually made up of a hollow core of light framework faced with thin plywood or hardboard. A, Fig. 47-17. Plywood-faced flush doors may be obtained in gum, birch, oak, mahogany, and woods of other species, most of which are suitable for natural finish. Nonselected grades are usually painted, as are hardboard-faced doors.

The panel door consists of solid stiles (vertical side members), rails (cross-pieces), and panels of various types. The five-cross panel and the Colonial panel doors are perhaps the most common of this style. B and C, Fig. 47-17. The louvered door is also popular and is commonly used for closets because it provides some ventilation. D, Fig. 47-17. Sliding or folding doors are installed in openings for wardrobes. These are usually flush or louvered. E, Fig. 47-17.

Common minimum widths for single interior doors are as follows:
● Bedrooms and other habitable rooms— 2'6".

47–6b. *Details for the room divider shown in Fig. 47–5b: 1. Panel details. 2. Top detail. 3. Base detail. 4. Top cutaway. 5. Base cutaway.*

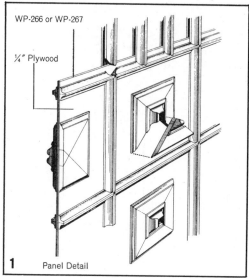

WP-266 or WP-267

¼″ Plywood

1   Panel Detail

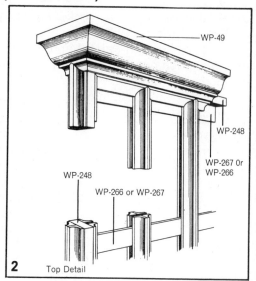

WP-49

WP-248

WP-267 Or WP-266

WP-248

WP-266 or WP-267

2   Top Detail

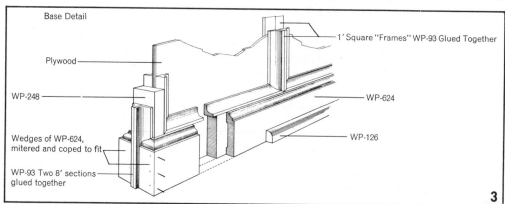

Base Detail

Plywood

WP-248

Wedges of WP-624, mitered and coped to fit

WP-93 Two 8′ sections glued together

1′ Square "Frames" WP-93 Glued Together

WP-624

WP-126

3

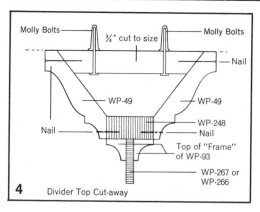

Molly Bolts

¾″ cut to size

Molly Bolts

Nail

WP-49

WP-49

WP-248

Nail

Nail

Top of "Frame" of WP-93

WP-267 or WP-266

4   Divider Top Cut-away

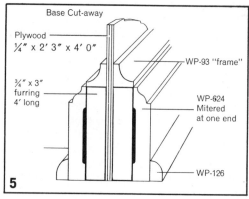

Base Cut-away

Plywood

¼″ x 2′ 3″ x 4′ 0″

¾″ x 3″ furring 4′ long

WP-93 "frame"

WP-624 Mitered at one end

WP-126

5

47–7a. Wood moldings added to an ordinary room divider will give a room a new focal point and added interest.

47–7c. Wood moldings are spaced evenly in square patterns on this wall. The molding at the ceiling caps the formal effect.

47–7b. A plain wall becomes interesting and decorative with the addition of wood moldings.

47–7d. Molding can also be used to emphasize wall hangings.

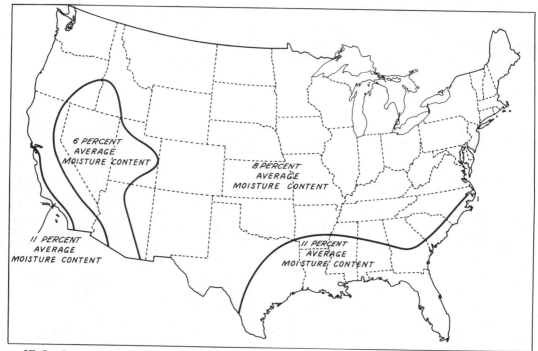

47–8. Recommended average moisture content for interior wood trim in various parts of the United States. In Canada the recommended moisture contents are as follows: Vancouver, 11%; Saskatoon, 7%; Ottawa, 8%; Halifax, 9%. (These cities represent four major geographical areas.)

47–9. Interior doorframe details.

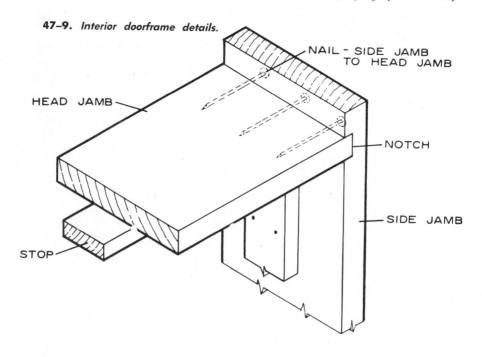

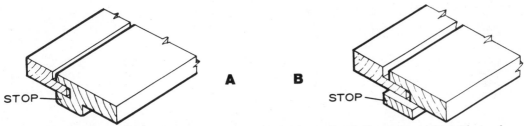

47–10. *Adjustable doorjambs: A. Two-piece adjustable jamb. B. Three-piece adjustable jamb. Two- and three-piece adjustable doorjambs adapt to various wall thicknesses.*

47–11a. *This prehung door has a two-piece jamb with pins for alignment behind the doorstop. The jamb can be adjusted to slight variations in wall thickness. However, when ordering a prehung door unit, be sure to specify the wall thickness. The manufacturer of this unit recommends a 2″ allowance in width and height over the nominal door size for the rough stud opening. The height is figured from the finished floor.*

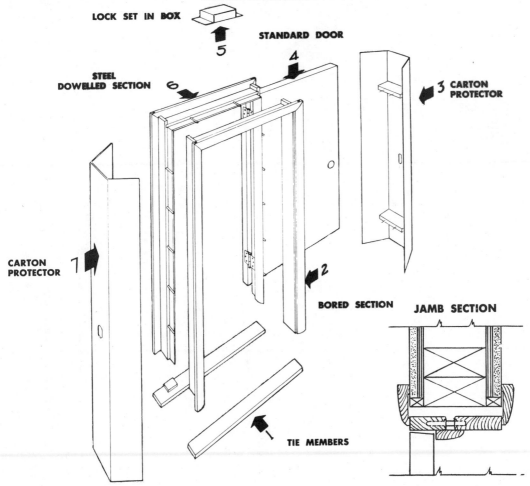

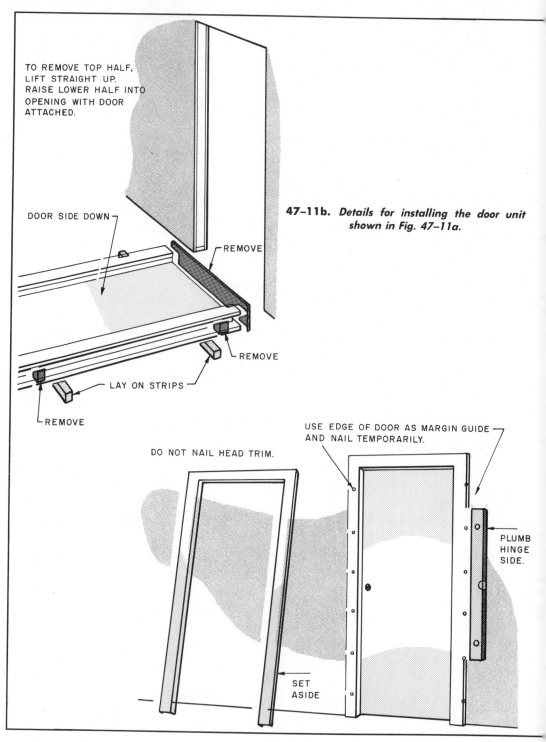

TO REMOVE TOP HALF,
LIFT STRAIGHT UP.
RAISE LOWER HALF INTO
OPENING WITH DOOR
ATTACHED.

DOOR SIDE DOWN

REMOVE

47–11b. *Details for installing the door unit shown in Fig. 47–11a.*

REMOVE

LAY ON STRIPS

REMOVE

USE EDGE OF DOOR AS MARGIN GUIDE
AND NAIL TEMPORARILY.

DO NOT NAIL HEAD TRIM.

PLUMB
HINGE
SIDE.

SET
ASIDE

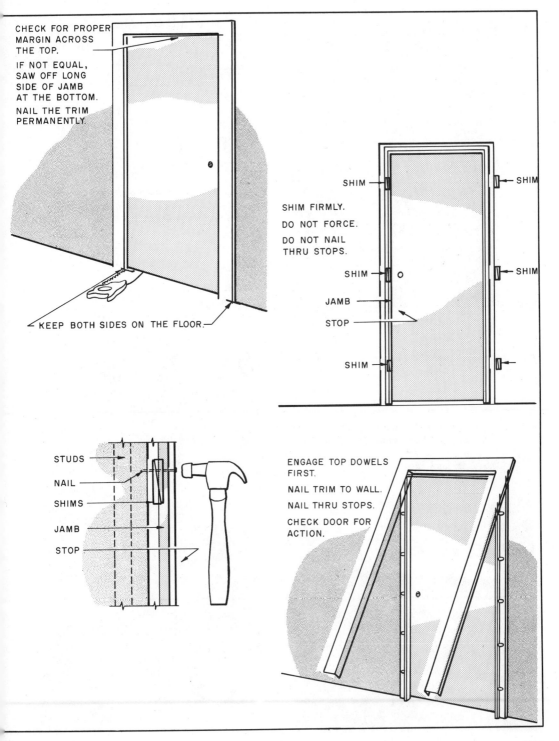

CHECK FOR PROPER MARGIN ACROSS THE TOP.

IF NOT EQUAL, SAW OFF LONG SIDE OF JAMB AT THE BOTTOM.

NAIL THE TRIM PERMANENTLY.

KEEP BOTH SIDES ON THE FLOOR.

SHIM

SHIM

SHIM FIRMLY.

DO NOT FORCE.

DO NOT NAIL THRU STOPS.

SHIM

SHIM

JAMB

STOP

SHIM

STUDS

NAIL

SHIMS

JAMB

STOP

ENGAGE TOP DOWELS FIRST.

NAIL TRIM TO WALL.

NAIL THRU STOPS.

CHECK DOOR FOR ACTION.

47–12a. *Some prehung doors are prefinished.*

47–12b. *The doors are delivered to the job site individually packaged in cartons for protection.*

JAMB DETAIL

HINGE

10 MIL
VINYL FILM

2¼"

4¼" TO 5⅜"
VARIABLE

Jamb widths from 3¼" to 7½" available
on special order.

47–12c. *The prefinished doors shown here have an adjustable two-piece split jamb which will accommodate walls from 4¹/₄″ to 5³/₈″ thick.*

47–12d. *The side of the doorjamb to which the door is hinged is removed from the carton and installed in the rough opening.*

47-12e. *Use wood shingles to shim the jambs when plumbing and leveling the unit in the opening.*

47-12f. *After the unit has been properly aligned in the opening, face-nail through the door casing to hold the unit in position.*

47-12g. *Install the casing and second half of the split jamb on the other side of the opening and nail in position.*

- Bathrooms—2′4″.
- Small closets and linen closets—2′.
These sizes can vary a great deal. Sliding doors or folding door units, used for wardrobes, may be 6′ or more in width. However, in most cases, the jamb, stop, and casing parts are still used to frame and finish the opening. The standard interior door height for first floors in 6′8″. Doors on upper floors are sometimes 6′6″.

Hinged doors should open or swing in the direction of natural entry, against a blank wall whenever possible. They should not be obstructed by other swinging doors.

Doors should never be hinged to swing into a hallway. The door swing is designated as either right or left hand. A right-hand door is one in which the latch is on the right when a person faces a closed door on the hinge side. A left-hand door is one in which the latch is on the left when a person faces a closed door on the hinge side. Fig. 47-18.

INTERIOR DOOR INSTALLATION

Interior doors are normally hung with two 3½″ × 3½″ loose-pin butt hinges. The door is fitted into the opening with the clearances shown in Fig. 47-19. The clearance and location of hinges, lock set, and

47–13. *Cut a spreader equal to the distance (X) between the side jambs just below the head jamb. Place the spreader at the floorline to hold the side jambs parallel.*

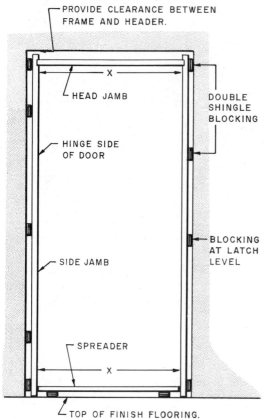

PROVIDE CLEARANCE BETWEEN FRAME AND HEADER.

HEAD JAMB

DOUBLE SHINGLE BLOCKING

HINGE SIDE OF DOOR

SIDE JAMB

BLOCKING AT LATCH LEVEL

SPREADER

X

TOP OF FINISH FLOORING.

47–14. *Doorframe and trim installation details. Use a 6d or 7d finish nail at arrow 1 to nail through the casing into the wall stud. At arrow 2, use a 4d or 5d finish nail to fasten the casing to the jamb.*

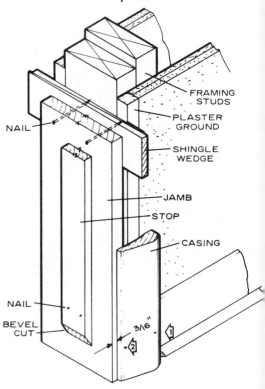

NAIL

FRAMING STUDS

PLASTER GROUND

SHINGLE WEDGE

JAMB

STOP

CASING

NAIL

BEVEL CUT

3/16″

doorknob may vary somewhat, but the dimensions in Fig. 47-19 are generally accepted and conform to most millwork standards. The edge of the lock stile should be beveled slightly to permit the door to clear the jamb when swung open. See "Fitting a Door" in Unit 38. If the door is to swing across heavy carpeting, the bottom clearance should be increased.

When fitting doors, the stops are usually temporarily nailed in place until the door has been hung. Stops for doors in single-piece jambs are generally $7/16''$ thick and may be $3/4''$ to $2 1/4''$ wide. They are installed

with mitered joints at the junction of the side and head jambs. A 45° bevel cut at the bottom of the stop, about 1" to $1 1/2''$ above the finished floor line, will eliminate a dirt pocket and make cleaning or refinishing the floor easier. Fig. 47-14. Review

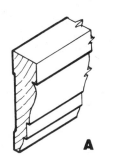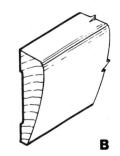

A　　　　　　　**B**

47–15. *Two common casings used for interior trim: A. Colonial. B. Ranch casing.*

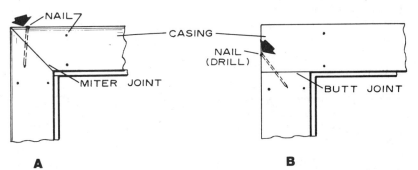

A　　　　　　　　　　　**B**

47–16. *A. Molded casing must have a mitered joint at the corner. B. Square-edge casing may be joined with a butt joint. In both cases, the joints may be reinforced by nailing at the arrows.*

47–17. *Interior doors: A. Flush. B. Panel (5-cross). C. Panel (Colonial). D. Louvered. E. Folding (louvered).*

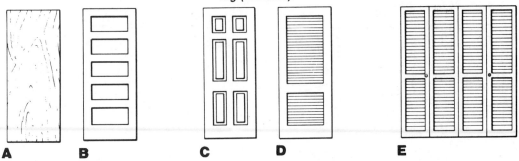

A　　　　**B**　　　　　　**C**　　　**D**　　　　　**E**

917

"Fitting a Door" and "Hanging a Door," Unit 38.

Door Hardware

Hardware for doors is made in a number of finishes, with brass, bronze, and nickel perhaps the most common. There are three kinds of door sets:
- Entry locks for exterior doors.
- Bathroom sets (inside lock control with safety slot for opening from the outside).
- Passage set (without lock). Fig. 47-20.

HINGES

Use two or three hinges for interior doors. Three hinges reduce the possibility of warpage and are useful on doors that lead to unheated attics or on wider and heavier doors.

Loose-pin butt hinges should be used. They must be the proper size for the door they will support. For 1¾" interior doors, choose 3½" butts. After the door is fitted to

the framed opening, with the proper clearances, hinge halves are fitted to the door. They are routed into the door edge with a back distance of about ³/₁₆". Fig. 47-21.

LOCKS

Types of door locks differ with regard to installation, cost, and the amount of labor required to set them. Follow the installation instructions supplied with the lock set. Some types require drilling of the edge and face of the door and routing of the edge to accommodate the lock set and faceplate. A, Fig. 47-22. A more common bored type is much easier to install. It requires only one hole drilled in the edge and one in the face of the door. B, Fig. 47-22. Boring jigs and faceplate markers

47–18. Determining the hand of a door. When standing on the hinge side of the door, with the door closed, the knob is on the left for a left-hand door. For a right-hand door, the knob is on the right.

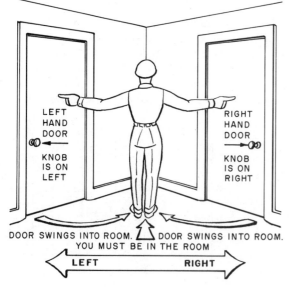

DOOR SWINGS INTO ROOM. DOOR SWINGS INTO ROOM.
YOU MUST BE IN THE ROOM
LEFT   RIGHT

47–19. Door clearances.

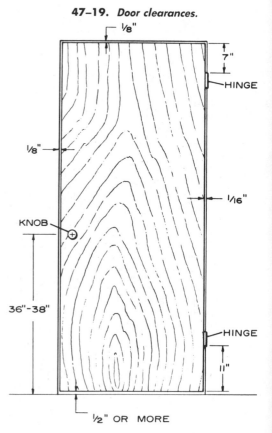

are available to provide accurate installation. Locks should be installed so that the doorknob is 36″ to 38″ above the floor line. Most sets come with paper templates marking the location of the lock and size of the holes to be drilled. See "Installing a Lock Set," Unit 38.

STRIKE PLATE

The strike plate, which is routed into the doorjamb, holds the door in place by contact with the latch. Fig. 47-23. When the door is latched, its face should be flush with the edge of the jamb. Review "Installing a Lock Set," Unit 38.

47–22. *Installation of lock set: A. Mortise lock. B. Bored lock.*

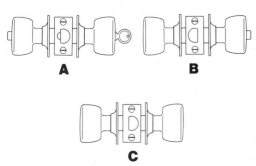

A   **B**

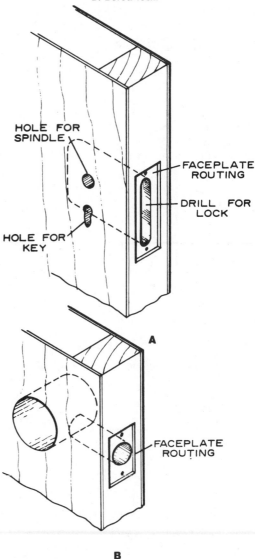

HOLE FOR SPINDLE

FACEPLATE ROUTING

DRILL FOR LOCK

HOLE FOR KEY

A

FACEPLATE ROUTING

B

47–20. *Door sets: A. Exterior lock. Push button locks the outside knob, and a key is needed to unlock it. B. Bathroom or privacy door lock. Push button locks the outside knob. An emergency key may be used to unlock the door from the outside. C. Passage door lock. Both knobs are always free.*

47–21. *Installation of the door hinge.*

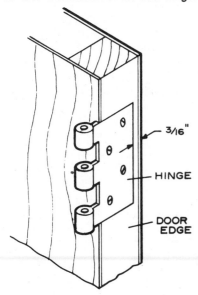

3/16″

HINGE

DOOR EDGE

Doorstops

The stops which have been set temporarily during fitting of the door and installation of the hardware may now be permanently nailed in place. Finish nails or brads, 1½" long, should be used. The stop at the lock side should be nailed first, setting it tight against the door face when the door is latched. Space the nails 16" apart in pairs. Fig. 47-23.

The stop behind the hinge side is nailed next. A ¹⁄₃₂" clearance from the door face should be allowed in order to prevent scraping as the door is opened. Fig. 47-24. The head-jamb stop is then nailed in place. Remember that when door and trim are painted, some of the clearances will be taken up.

Sliding Doors

The bypass sliding door is designed for closets and storage walls. It requires no open swinging area, thus permitting a more effective and varied arrangement of furniture. No valuable floor space is lost due to door swing. Full access to the storage area is obtained by sliding the doors right or left. Fig. 47-25.

Sliding doors are usually installed in a standard doorframe. The track is mounted below the head jamb and then hidden from view with a piece of trim. Fig. 47-26. Standard size interior doors, 1³⁄₈" thick, 6'8" or 7'0" high, and any width, may be used. Most sliding door hardware will also adapt to ¾" or 1⅛" door thicknesses. Fig. 47-27. The door rollers are adjustable so that the door may be plumbed and aligned with the opening. Fig. 47-28. The doors are guided at the bottom by a small guide which is mounted on the floor where the doors overlap at the center of the opening. Fig. 47-29. Rough opening sizes differ slightly from one manufacturer to another. Be sure to consult the specifications provided with each particular door unit. Fig. 47-30.

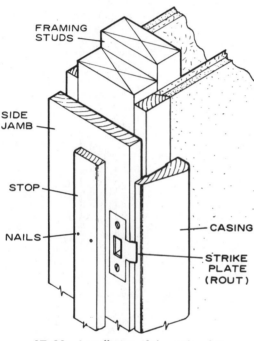

47–23. *Installation of the strike plate.*

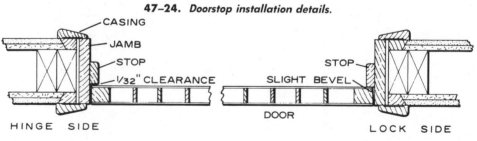

47–24. *Doorstop installation details.*

HINGE SIDE   LOCK SIDE

PLAN VIEW

Bifold Doors

Bifold doors may be used to enclose a closet area, storage wall, or laundry area. The doors may be wood, metal, or coated with plastic. They come in a large variety of styles to match the architecture of the home. Fig. 47-31 (Page 924).

The bifold unit has the advantage of opening up so that the entire opening is exposed at one time. With sliding bypass doors, the entire opening is accessible, but only half is exposed at one time. Fig. 47-32.

The doors are available in 6'8", 7'6", and 8'0" heights and in widths of 3', 4', 5', and 6'. These are four-panel units and, if desired, the tracks may be cut in half and a two-panel unit installed. For example, two panels of the 3'0" size could be used for a 1'6" linen closet opening.

The bifold door is installed in a conventional doorframe. The frame may be trimmed with door casing to match the trim in the remainder of the house or, if desired, the jamb may be finished the same as the walls. Fig. 47-33 (Page 926). The rough opening is framed in the same way as for the conventional swinging door. The finish opening size, however, may vary with the manufacturer. Fig. 47-34.

To install a bifold door, install the top track first. Fasten the lower track to the floor, directly under the top track. Install the doors by inserting the bottom pivot into the bottom track socket. Insert the upper pivot into the top track socket. Adjust the panels to the opening by adjusting the track sockets. To make the tops of the panels even, raise or lower the panels by adjusting the lower pivot pin. Fig. 47-35.

Folding Doors

A folding door may be used as a room divider or to close off a laundry area,

47-26. Hanging the sliding door in the track. Note that the track is concealed behind a piece of trim mounted below the head jamb.

47-25. Sliding bypass doors used on a storage wall.

47–27. *Adjustable sliding door hardware: A. The bracket, when attached to the roller, is turned away from the roller. The nail may be inserted in either of two positions, making the roller adaptable to either 1¹/₈″ or 1³/₈″ door thicknesses. B.When it is in position for a ³/₄″ door thickness, the bracket is turned so that it is installed almost directly under the roller.*

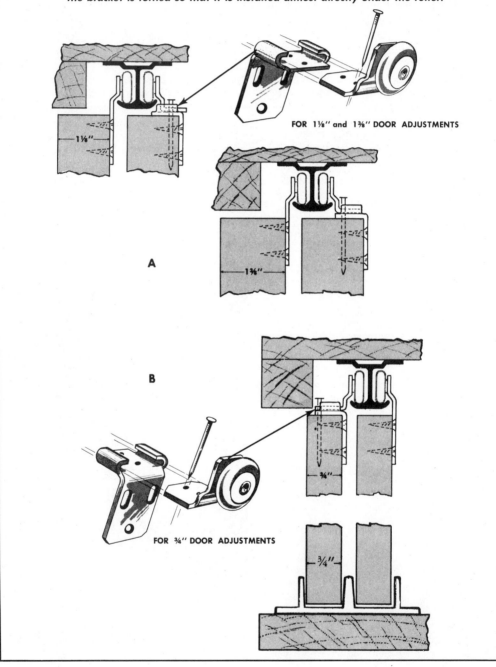

FOR 1⅛″ and 1⅜″ DOOR ADJUSTMENTS

A

B

FOR ¾″ DOOR ADJUSTMENTS

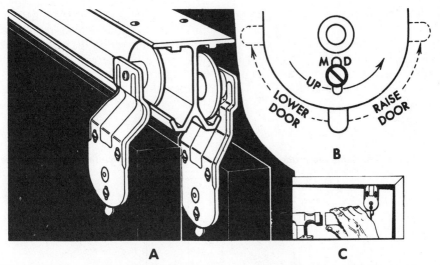

47-28. *Sliding door hardware is designed to permit aligning the doors with the opening without repositioning the hangers: A. The doors are mounted on the hangers and installed on the track. B. Details for adjusting the hangers when aligning the door. C. Tapping the adjustment with a block of wood and a hammer to make the alignment.*

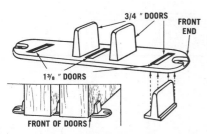

47-29. *The door guide on the floor is also adjustable. The plastic guides may be inserted in different slots of the metal plate to accommodate various door thicknesses.*

47-30. *A typical specification chart for sliding door units. Be sure to consult the manufacturer's specifications for each door to be installed.*

Specifications

Front Opening Size	For 2 Doors	Rough Stud Opening
3′0″ x 6′8″	1′6″ x 6′8″	3′1½″ x 7′0″
4′0″ x 6′8″	2′0″ x 6′8″	4′1½″ x 7′0″
5′0″ x 6′8″	2′6″ x 6′8″	5′1½″ x 7′0″
5′4″ x 6′8″	2′8″ x 6′8″	5′5½″ x 7′0″
6′0″ x 6′8″	3′0″ x 6′8″	6′1½″ x 7′0″

closet, or storage wall. Figs. 47-36 (Page 927) and 47-37. Folding doors are made from wood, reinforced vinyl, or plastic-coated wood. Wood folding doors, when closed, look like paneling. Folding doors made of a metal framework and covered with fabric or vinyl-coated materials are also available. These are called accordion-fold doors.

Folding doors are both convenient and attractive. They fold right inside their own

doorway so that full advantage can be taken of every square foot of living space. The doors are hung on nylon rollers that glide smoothly in an aluminum track. The track at the top is concealed with beveled matching wood molding installed on each side. There is no exposed hardware to detract from the beauty of the door. Fig. 47-36.

Folding doors come from the factory already assembled. They are shipped complete in a package containing the door, hardware, latch fittings, and installation instructions. Standard or stock doors are available 6'8" high and 2'4", 2'8", 3', or 4' wide.

The folding door is installed in a standard doorframe. It may be trimmed in a conventional manner, or plaster jambs may be used. Fig. 47-38 (Page 928). A plaster channel is also available. This channel is installed before plastering, and the track is mounted after the plaster is applied, eliminating the need for the wood head molding normally used. Fig. 47-39. Folding doors may also be installed to fit into a wall cavity when opened. The cavity must be at least 7½" wide. Fig. 47-40.

47–31a. *Bifold doors are made in several styles to match the architecture of the home.*

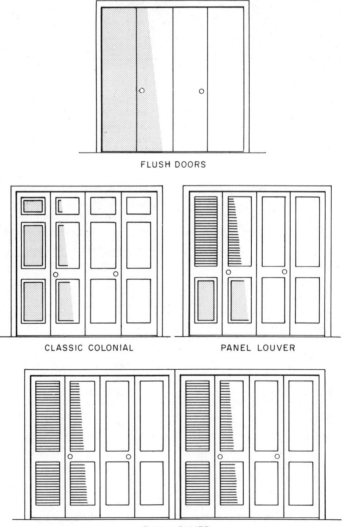

FLUSH DOORS

CLASSIC COLONIAL

PANEL LOUVER

FULL LOUVER

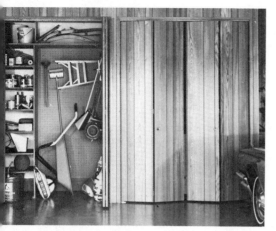

47–31b. *Wood bifold doors used to conceal a storage area.*

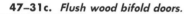

47–31c. *Flush wood bifold doors.*

47–31d. *Wood louvered bifold doors used on a storage cabinet in a dining area.*

47–32. *A bifold door unit opens up to expose the entire opening.*

FULL OPENING

SAVES
FLOOR SPACE

Sliding Pocket Doors

The pocket type sliding door, in which the door slides into the wall, is often installed in places where the door is seldom closed. When the door stands open, it is out of the way. It is also convenient where there is minimum space for clearance of a swinging door. Fig. 47-41 (Page 930).

The unit can be bought complete with hardware, door, and trim. Universal sliding door hardware that will fit all door sizes can also be purchased. Fig. 47-42. Standard widths are 2'0", 2'4", 2'6", 2'8", and 3'0". Any style of door with a thickness of 1³⁄₈" can be installed in the pocket to match the style of the other doors in the home.

When the opening for the door pocket unit is roughed in, the manufacturer should be consulted for specifications. The rough opening is usually 6'11¹⁄₂" or 7' high and twice the door width plus 2" or 2¹⁄₂". The wall header above the pocket must be adequate to support any weight on the wall and the structural frame of the building so that there is no weight on the pocket. Fig. 47-43.

The sliding door pocket frame comes complete and ready for installation into the rough wall opening. Install the pocket in the opening with 8d finish nails through shingle wedges at the head and the side jambs to level and plumb the unit. Fig.

47-34. A typical manufacturer's specifications for bifold doors.

Two-Panel Units*

Size	Door Width	Opening Width	Finished Heights		
			6'8" Units	7'6" Units	8' Units
1'6"	17¹⁄₂"	18¹⁄₂"	6'8¾"	7'5¼"	7'11¼"
2'0"	23¹⁄₂"	24¹⁄₂"	6'8¾"	7'5¼"	7'11¼"
2'6"	29¹⁄₂"	30¹⁄₂"	6'8¾"	7'5¼"	7'11¼"
3'0"	35¹⁄₂"	36¹⁄₂"	6'8¾"	7'5¼"	7'11¼"

Four-Panel Units

Size	Door Width	Opening Width	Finished Heights		
			6'8" Units	7'6" Units	8' Units
3'0"	35"	36"	6'8¾"	7'5¼"	7'11¼"
4'0"	47"	48"	6'8¾"	7'5¼"	7'11¼"
5'0"	59"	60"	6'8¾"	7'5¼"	7'11¼"
6'0"	71"	72"	6'8¾"	7'5¼"	7'11¼"

*Note: 2-panel units may be made on the job by cutting 4-panel unit tracks in half.

47-33. Various header construction and trim details for a bifold door.

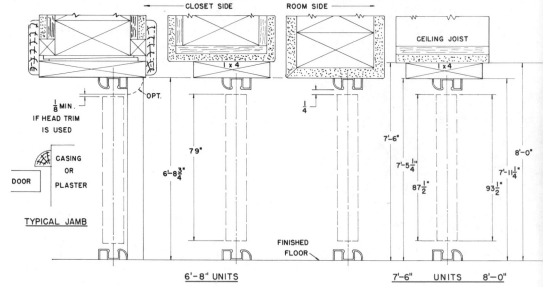

47–35. *Installation details for a bifold door unit. The track at the top is hidden by the matching wood molding.*

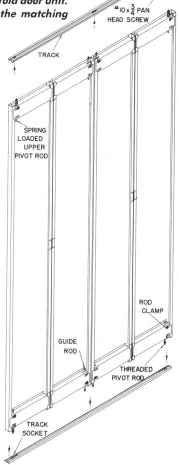

#10 x ¾ PAN HEAD SCREW

TRACK

SPRING LOADED UPPER PIVOT ROD

ROD CLAMP

GUIDE ROD

THREADED PIVOT ROD

TRACK SOCKET

47–36. *Wood folding doors installed on a closet.*

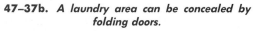

47–37a. *Wood folding doors used in an apartment to divide a large room.*

47–37b. *A laundry area can be concealed by folding doors.*

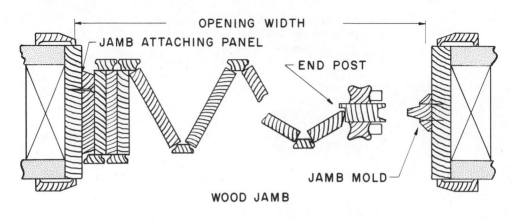

OPENING WIDTH

JAMB ATTACHING PANEL

END POST

JAMB MOLD

WOOD JAMB

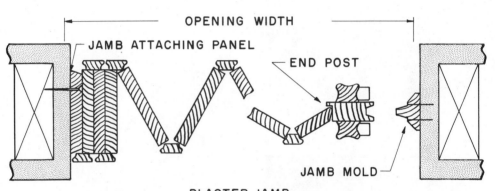

OPENING WIDTH

JAMB ATTACHING PANEL

END POST

JAMB MOLD

PLASTER JAMB

47–38. *Jamb sections showing the door installed with a wood jamb or a plaster jamb.*

47–39. *Head sections of folding doors showing the track installation details for a channel mounted on wood, plaster, or recessed in plaster.*

SURFACE MOUNTED ON PLASTER　　　**RECESSED IN PLASTER**　　　**SURFACE MOUNTED ON WOOD**

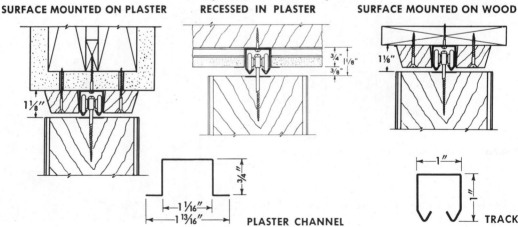

PLASTER CHANNEL

TRACK

47–40. *A. Hinged recessed door. The door can be stacked into the recessed area and the panel closed to cover the end of the door. B. Sliding recessed panel. When the door is in use, the panel travels forward and covers the opening of the recessed area.*

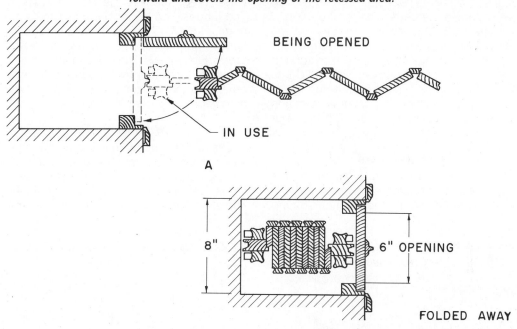

BEING OPENED

IN USE

A

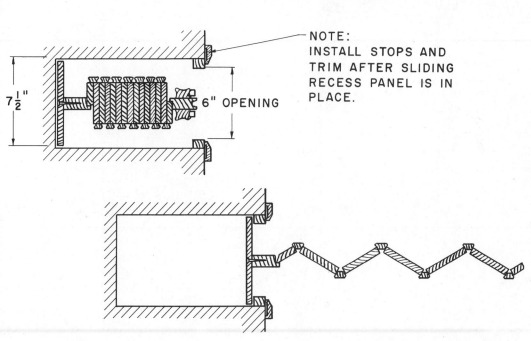

8"   6" OPENING

FOLDED AWAY

7½"   6" OPENING

NOTE:
INSTALL STOPS AND
TRIM AFTER SLIDING
RECESS PANEL IS IN
PLACE.

47-44. After the wall covering has been applied, hang the door and install the stops. Fig. 47-45. Special care should be taken to use the correct nails when applying the wall covering. If the nails are too long and project through the frame slats into the opening, they may either scratch the surface of the door when it is slid into the pocket or prohibit it from entering the pocket. The same care in nail selection should be taken when installing the door casing and base molding.

Cafe Doors

The cafe door will add charm to the home and is adaptable to many uses. It may be installed in the kitchen, dining area, family room, or recreation room. Cafe doors swing from either direction and always freely return to the closed position. The hinges also permit the doors to be left open. Fig. 47-46. These doors are normally hung in pairs and come in a large variety of sizes. Fig. 47-47. They are installed with the door tops slightly below eye level to enable someone using the door to see anyone coming from the other side.

Window Trim

Casing for window trim should be of the same pattern as that selected for door casing. Other trim parts consist of sash stops, the stool, and the apron. Fig. 47-48. The stool is the horizontal trim member that

47-41a. *Sliding pocket door unit without wall covering.*

47-41b. *Sliding pocket door unit after wall covering is applied.*

laps the window sill and extends beyond the casing. The apron serves as a finish member below the stool. There are two common methods of installing wood window trim:

• With a stool and apron. Fig. 47-49 (Page 934).

• With complete casing trim. Fig. 47-50. Metal casing is sometimes used in place of wood trim around the window opening. Fig. 47-51.

WINDOW STOOL INSTALLATION

The stool is normally the first piece of window trim to be installed. It is notched out between the jambs so that the forward edge contacts the lower sash rail. Fig. 47-52.

The upper part of Fig. 47-52 shows a plan view of a stool in place. The lower

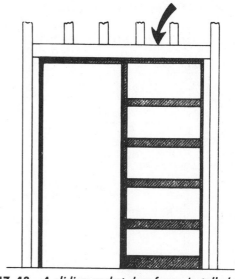

47–43. *A sliding pocket door frame installed in the rough opening. The wall header (arrow) must span the entire opening, which includes the pocket and the doorway.*

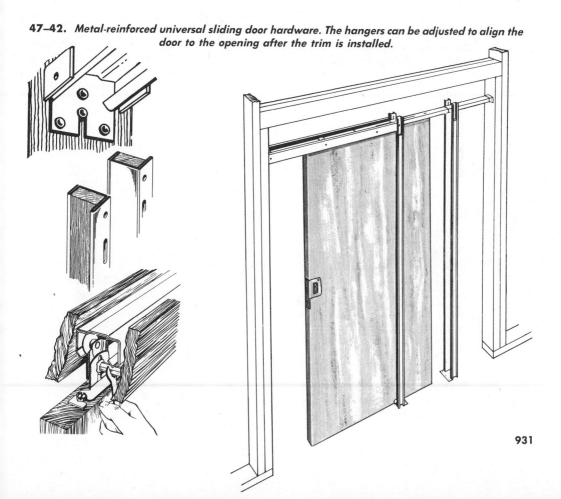

47–42. *Metal-reinforced universal sliding door hardware. The hangers can be adjusted to align the door to the opening after the trim is installed.*

931

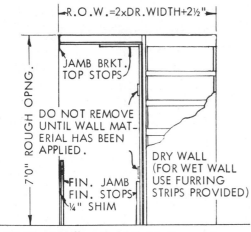

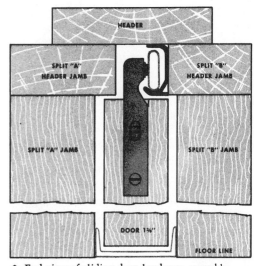

47–44. Installing the pocket in the rough opening. Use shingle wedges at the head and side jambs to level and plumb the unit. Care should be taken to keep the jambs straight.

◆ End view of sliding door hardware assembly

◆ Top view of jambs showing heavy metal reinforcing

47–45. Sliding door pocket frames are available for wet wall ("A" jamb) and dry wall ("B" jamb).

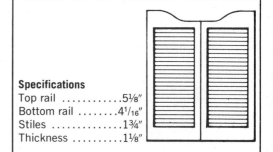

47–46. The hinges on a cafe door will permit the door to stand in the open position or freely return to the closed position.

Specifications
Top rail ....5⅛"
Bottom rail ....4¹/₁₆"
Stiles ....1¾"
Thickness ....1⅛"

47–47a. One manufacturer's specifications of standard cafe door sizes.

Size Of Pair	Size Of Pair
2'6" x 3'0"	2'8" x 4'0"
2'6" x 3'6"	3'0" x 3'0"
2'6" x 4'0"	3'0" x 3'6"
2'8" x 3'0"	3'0" x 4'0"
2'8" x 3'6"	

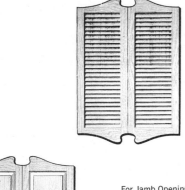

For Jamb Opening Widths
2'6", 2'8" and 3'0"

Door Sizes
1⅛" thick
2'5⅛"x3'7½"
2'7⅛"x3'7½"
2'11⅛"x3'7½"

47-47b. *Another manufacturer's specifications for standard cafe door sizes showing different door styles.*

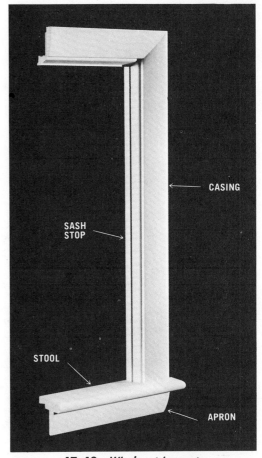

47-48. *Window trim parts.*

view shows the stool laid out and cut, ready for installation. The distance A, the overall length of the stool, is equal to: the width of the finished opening, plus twice the width of the side casing, plus twice the amount that each end of the stool extends beyond the outer edge of the side casing. Distance B is equal to the width of the finished opening.

Distance C is equal to the horizontal distance measured along the face of the side jamb between the inside edge of the side jamb and the inside face of the lower sash. An allowance of about ¹/₃₂" should be deducted for clearance between the sash and the stool. Lay out this width from the outside edge of the stool. If the stool is held in place (centered) against the inside edge of the window jamb, the distance between the front edge of the stool and the lower sash, dimension C, can be set on a

scriber. The distance along the wall can then be scribed onto the top of the stool. Cut out the notch at each corner of the stool on the layout lines.

The stool is blind-nailed at the ends with 8d finish nails so that the casing at the sides will cover the nailheads. With hardwood, predrilling is usually required to prevent splitting. The stool should also be nailed at the center to the sill, and to the apron when it is installed. Toenailing may be substituted for face-nailing to the sill. Fig. 47-49.

933

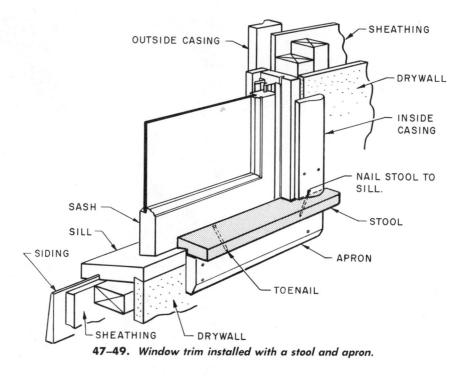

47–49. *Window trim installed with a stool and apron.*

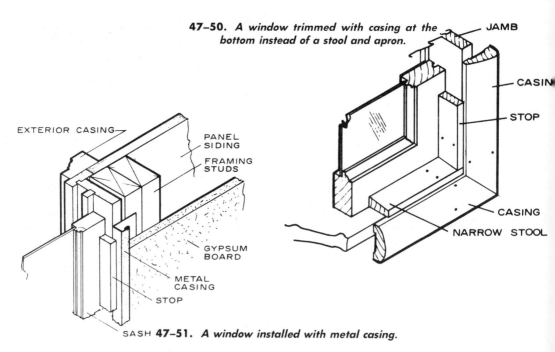

47–50. *A window trimmed with casing at the bottom instead of a stool and apron.*

47–51. *A window installed with metal casing.*

WINDOW CASING INSTALLATION

The casing is applied after the stool is installed. It is nailed as described for the doorframes (Fig. 47-14) except that on some window types the inner edge is flush with the inner face of the jambs so that the stop covers the joint. Fig. 47-49. When stops are to be installed, they are fitted like the interior door stops and placed against the lower sash so that the sash can slide freely. A 4d casing nail or 1½" brad should be used. Place nails in pairs spaced about 12" apart. When the window is made with channel type weather stripping, it includes full-width metal subjambs in which the upper and lower sash slide, replacing the parting strip. In this case, the stops are located against the subjambs instead of the sash to provide a small amount of pressure.

When metal casing is used as trim around window openings, it is applied to the sill as well as at the sides and head of the frame. Consequently, the jambs and sill of the frame are not as deep as when wood casing is used. The stops are also narrower by the thickness of the wall cov-ering. The metal casing is installed flush with the inside edge of the window jamb. Fig. 47-51. This type of trim is installed at the same time as the wall covering.

WINDOW APRON INSTALLATION

Cut the apron to a length equal to the outer width of the casing line. The ends of the apron should be cut with a coping saw to appear as though they had been returned, or a return may be cut and nailed in place. Fig. 47-53. The apron is attached to the framing sill below with 8d finish nails. Fig. 47-49.

When casing is used instead of a stool and apron to finish the bottom of the window frame, a narrow stool or stop may be needed on some window types. Miter the side casings at the bottom corners and apply the bottom casing in the same way as the side and head casings of the window. Fig. 47-50. When this method is used to trim a window, the casings can be cut to finished length with miters on each end. The four pieces of mitered casing are then laid face down on a clean, smooth surface and fastened together with ¼" × 5" corru-

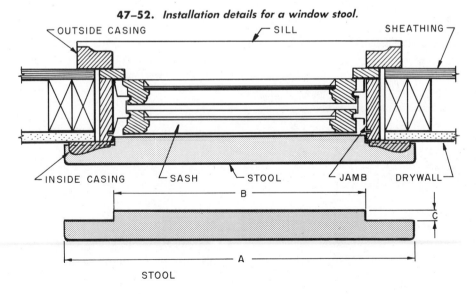

47-52. *Installation details for a window stool.*

OUTSIDE CASING — SILL — SHEATHING

INSIDE CASING — SASH — STOOL — JAMB — DRYWALL

B

C

A

STOOL

gated fasteners from the back. The assembled casing, much like a picture frame, is then nailed as a unit to the window jambs and studs as described earlier.

Other types of windows, such as the awning, hopper, or casement, are trimmed about the same as the double-hung window. Casings of the same types are used for trimming these units.

Interior Shutters

Movable interior shutters were popular in the great mansions of New Orleans and in many of the finer homes of America from 1700 to the early part of the 19th century. The use of shutters has once again become popular. Fig. 47-54. They are found in the traditional interior, the provincial (country style) setting, and in the modern home, studio, or office. Shutters offer better control of light, air, visibility, and privacy than most other types of window treatment. They also make ideal foldaway partitions or room separators that distinguish one living area from another.

47–53. *The ends of the window apron should be coped to match the profile of the molding or mitered with a return nailed in place.*

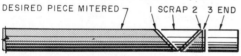

DESIRED PIECE MITERED   I SCRAP 2 3 END

THE TOP VIEW WITH THREE NECESSARY CUTS.

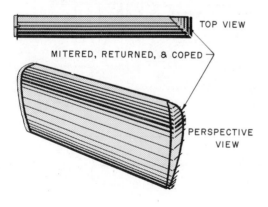

TOP VIEW

MITERED, RETURNED, & COPED →

PERSPECTIVE VIEW

To determine the size of the louver to be installed in a window, measure the width between the inside edges of the jamb or casing, depending on where you wish to hinge the shutters. Measure the height of the opening from the top of the sill to the inside edge of the top jamb or casing. If a cross rail is desired, it will be located at the approximate center of the shutter, referred to as the break. Fig. 47-55.

INSTALLING MOVABLE INTERIOR SHUTTERS

When installing movable shutters on a double-hung window, a hinge strip replaces the sash stop. Fig. 47-56. The strip should be 3/4" wide and have a depth equal to the distance from the front of the jamb to the sash. Mount the shutter to these strips with 3/4" flush door hinges.

On casings that are reasonably flat, the shutter may be applied directly with a 3/4" flush door hinge. The hinge should be set no more than 1" back from the inside edge of the casing. Fig. 47-57.

Another method of hanging shutters to casings that are reasonably flat is shown in Fig. 47-58. In this installation the hinges are applied in reverse so that the shutter will not cover the casing.

When there are no obstructions in back of the shutter, loose-pin butt hinges may be applied directly to the jamb. Fig. 47-59.

If the window jamb is other than wood and the wall is not cased, or if the casing has been removed, apply a hanging strip to the wall and use either loose-pin butt hinges or 3/4" flush door hinges. Fig. 47-60.

When a casing is at least 3/4" thick, the shutter may be applied directly to the casing with either type hinge. Fig. 47-61.

On French doors or casement windows, install a framing strip, 3/4" wide × 1" deep, to the sash. Fig. 47-62. The shutter may then be installed with 3/4" flush door hinges. Shutter and frame should overlap each side and top and bottom outside the glass area by 3/4" to leave room for the hinges and operating clearance for louvers.

BASE, CEILING, AND WALL MOLDING

Installing Base Molding

Base molding is usually about the last trim to be installed. It must be installed after all the doors are trimmed and the cabinets are in place. Also, it usually butts wall openings such as warm and cold air registers. The base molding serves as a finish between the finished wall and floor. It is made in a number of sizes and shapes. Fig. 47-3.

Base molding may have several parts. Two-piece base consists of a baseboard topped with a small base cap. A, Fig. 47-63. When the wall covering is not straight and true, the small base cap mold-

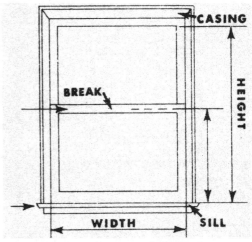

47–55. *Measuring a window for shutters.*

47–54. *Louvered shutters may be used throughout the home. They allow light and air to pass through, yet assure complete privacy and freedom from drafts.*

BATHROOM

KITCHEN

BEDROOM

ARCHED WINDOW

LIVING ROOM

BAR

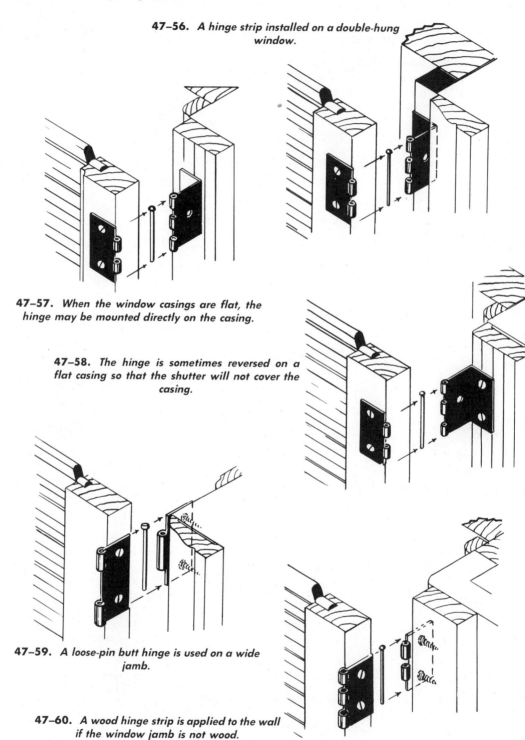

47–56. *A hinge strip installed on a double-hung window.*

47–57. *When the window casings are flat, the hinge may be mounted directly on the casing.*

47–58. *The hinge is sometimes reversed on a flat casing so that the shutter will not cover the casing.*

47–59. *A loose-pin butt hinge is used on a wide jamb.*

47–60. *A wood hinge strip is applied to the wall if the window jamb is not wood.*

ing will conform more closely to the variations than will a wider base alone. A common size for two-piece base is $5/8'' \times 3^1/4''$ or wider. One-piece base varies in size from $7/16'' \times 2^1/4''$ to $1/2'' \times 3^1/4''$ and wider.

Most baseboards are finished with a base shoe, $1/2'' \times 3/4''$ in size. B, and C, Fig. 47-63. Base molding without the shoe is sometimes placed at the wall-floor junction, expecially when carpeting is installed. Although a baseboard is desirable at the wall-floor junction to serve as a protective "bumper," wood trim is sometimes eliminated entirely.

Square-edged baseboards should be installed with a butt joint at inside corners and a miter joint at outside corners. Molded baseboards are also mitered at outside corners but they are coped at inside corners. Fig. 47-64. These methods of joining are necessary to provide tight joints because the walls at corner baseboard locations may not be perfectly vertical. When cutting molding to fit between walls, always cut it a little long so that the molding can be bowed slightly and sprung into place. When it is necessary to use more than one length of molding along a wall, join the pieces on a wall stud with a lap miter joint. Fig. 47-65. The baseboard is secured to each stud with two 8d finishing nails.

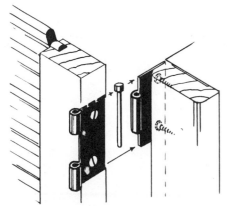

47–61. On a thick casing, the hinge may be applied directly to the edge.

47–62. Louvers applied to French doors or casement windows must be furred out to provide sufficient operating clearance for hardware.

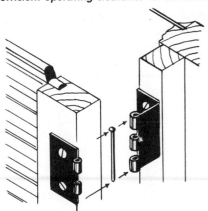

47–63. Base moldings: A. Two-piece baseboard with a square-edged base and a base cap. B. Narrow ranch base. C. Wide ranch base.

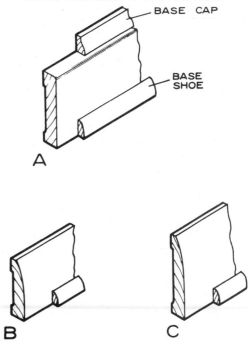

BASE CAP

BASE SHOE

A

B   C

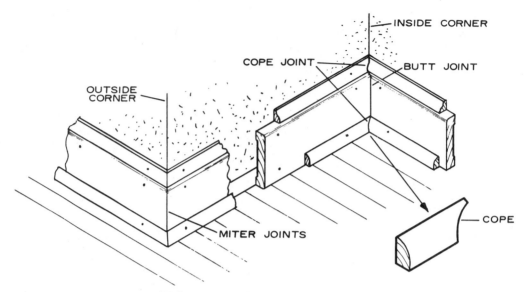

47-64. *Base molding installation details.*

47-65. *When several pieces of molding are needed to provide sufficient length, they should be joined with a lap miter joint on a wall stud. Use two 8d finish nails. The bottom nail should be close enough to the floorline to be covered by the base shoe molding.*

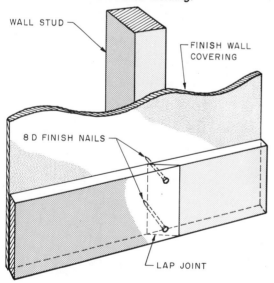

When the face of the base shoe projects beyond the face of the molding to which it butts, the end of the base shoe should be returned onto itself. The return is cut at 45° on the end of the shoe molding. This return will eliminate dirt pockets and give a better appearance. Fig. 47-66. The base shoe should be nailed into the subfloor and not into the baseboard itself. Thus, if there is a small amount of joint shrinkage, no opening will occur under the shoe. Fig. 47-67.

SCRIBING A BUTT JOINT

To butt join a piece of square-edged baseboard to another piece already in place at an inside corner, set the piece to be joined in position on the floor. Bring the end against or near the face of the piece already installed and scribe a line parallel to the installed piece on the face of the piece to be joined. Fig. 47-68. Be careful to hold the legs of the scriber at right angles to the reference surface (the baseboard which has already been installed). This will insure a parallel line. Follow the same

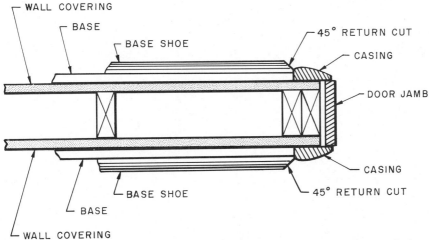

WALL COVERING

BASE

BASE SHOE

45° RETURN CUT

CASING

DOOR JAMB

CASING

BASE SHOE

45° RETURN CUT

BASE

WALL COVERING

47–66. *When the face of the base shoe projects beyond the face of the molding which it abuts, a 45° return cut should be made on the base shoe.*

procedure when putting ends of moldings against the side of door casings or wall registers.

MITERING A JOINT

When miter joining an outside corner, set a marker piece of baseboard across the wall corner. A, Fig. 47-69. Mark a line on the floor along the edge of the piece. Set the piece to be mitered in place. Mark the place where the wall corner intersects the top edge and the point where the mark on the floor intersects the bottom edge. B, Fig. 47-69. Draw a 45° line across the top edge from the point marked at arrow 1 and draw a 45° line across the bottom edge from the point shown at arrow 2 in Fig. 47-69. Connect these lines with a line drawn across the face of the board and cut along the line.

COPING A JOINT

Inside corner joints between molding trim members are usually made by cutting the end of one member to fit against the face of the other. Shaping the end of the butting member to fit the face of the other

47–67. *Base shoe installation details. The nail is driven through the base shoe and the finished floor into the subfloor.*

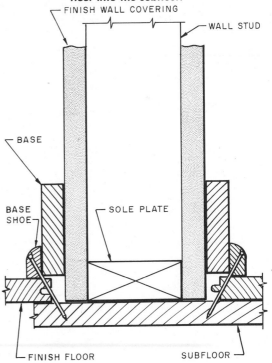

FINISH WALL COVERING

WALL STUD

BASE

BASE SHOE

SOLE PLATE

FINISH FLOOR

SUBFLOOR

member is called coping. Fig. 47-64. To cope a molding, miter the end at 45° the same way as if the molding were to have a plain mitered inside corner joint. A, Fig. 47-70. Then set the coping saw on the line at the top of the miter cut. Hold the saw at 90° to the back of the molding and saw along the face contour line created by the 45° miter cut. B and C, Fig. 47-70. The end profile of the coped member will match the face of the other member. D, Fig. 47-70. The result is a good, tight joint. It will not open up when the molding is nailed in place, and it is not likely to open up as the wood shrinks after installation.

BASE MOLDING INSTALLATION SEQUENCE

When applying base moldings in a room, the installation sequence should be carefully planned before starting the job to save time in making necessary and difficult cuts. The drawing in Fig. 47-71 shows the outline of a room with one door. Here is one suggested sequence for installing the molding: Cut and fit a piece of molding to go along the wall marked 1, scribing each end to fit walls 2 and 3. Install molding on the walls marked 2 and 3. The molding for

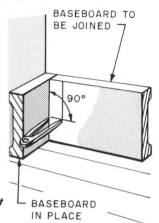

47–68. *When an inside corner is butt-joined, it must be scribed to insure a good tight joint.*

47–69. *Laying out a miter joint at an outside corner.*

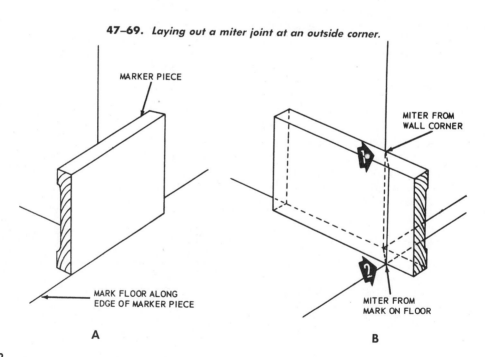

A

B

these walls should have one end coped to fit the molding on wall 1. Cope the joint first; then measure and mark the piece and cut it to finish length. Cope one end of the moldings for walls 4 and 5 to fit against the face of the moldings for walls 2 and 3. Cut the other end to fit against the door casing.

Another approach to trimming this same room (Fig. 47-71) is to begin at the right and work around the room in a counter-

clockwise direction. The first piece is cut to fit along wall 5 between the door casing and the end wall. The piece of molding on wall 2 is coped on one end to fit against the molding on wall 5. It is then measured, marked, and cut to length so that the other end fits against wall 1. The first end of wall molding 1 is coped to fit against piece 2, then measured, marked, and cut to length. This procedure is continued for walls 3 and 4.

A right-handed person will find in working around the room counterclockwise that the coping of the molding will be much easier. This is true for most molding shapes because the sawing is started at the top of the molding, where it is lighter and where most moldings are shaped and have a narrow edge. This narrow edge is weak and, if the cut is started at the bottom, the molding may break as the cut is finished.

Ceiling Molding

Moldings are sometimes used at the junction of wall and ceiling for architectural effect or to terminate dry wall or wood paneling. Fig. 47-72. A cut-back edge at the top of the molding will partially conceal any unevenness of the plaster and make

47-70. *Coping a joint: A. Make a 45° miter cut. B. Set the coping saw at 90° to the back edge. C. Make the cut, following the contour line created by the 45° cut. D. The coped end of the molding will fit tightly against the face of the other member.*

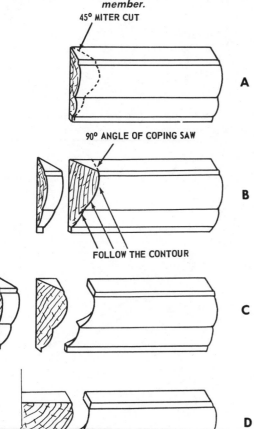

45° MITER CUT

A

90° ANGLE OF COPING SAW

B

FOLLOW THE CONTOUR

C

D

47-71. *An outline of a room with one door. Can you figure the best sequence for cutting and installing base molding in this room?*

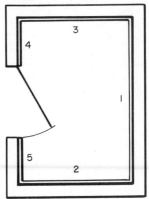

painting easier when molding and ceiling are different colors. B, Fig. 47-72. For gypsum dry wall construction a small simple molding might be preferred. C, Fig. 47-72.

Cut and fit ceiling molding in the same way as described for base moldings to insure tight joints and retain a good fit if there are minor moisture changes. To secure the molding, a finish nail should be driven through the molding and into the upper wall plates. For large moldings, if possible, also drive a nail through the molding into each ceiling joist.

Wall Molding

Wall molding, sometimes called dado molding or chair rail, consists of strips of molding which run along the walls at 3′ to 4′ above the floor. This type of molding serves to protect the finished wall from the backs of chairs. It may also serve as trim between two different wall finishes. For example, a wall may be painted below the molding and wallpapered above. Regular casing, base, band, and cap moldings may be used as wall moldings, allowing a wide range of choices with an almost unlimited selection of thicknesses and widths. Wall moldings are installed in the same manner as base moldings.

Applying Molding to a Flush Door

The style of a flush door can be varied by the application of different moldings. Fig. 47-73a. A Spanish style door is shown in Part 1 of Fig. 47-73b. This door is trimmed with two molding patterns used in conjunction with a screen stock. Make a jig with the inside dimensions equal to the outside dimensions of the frame (12″ × 10″). Cut the molding to length with mitered ends and assemble them in the jig. Fig. 47-5c.

Other moldings applied to a flush door can give it a contemporary or a traditional style. Parts 2 and 3, Fig. 47-73b. Silhou-

47–72. Ceiling moldings: A. Installation details at an inside corner. B. Crown molding with a cut-back edge. C. A small crown molding.

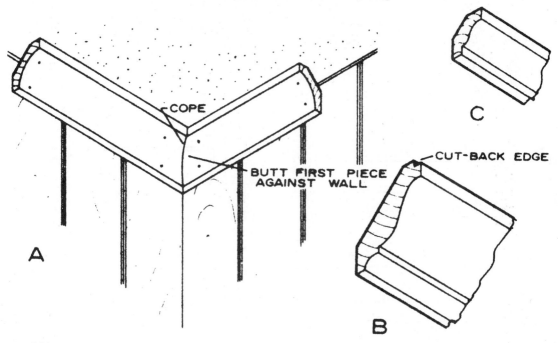

COPE

BUTT FIRST PIECE AGAINST WALL

CUT-BACK EDGE

A

B

C

47-73a. *Materials list and procedure for creating different style doors by applying moldings to a flush door.*

FLUSH DOORS

Materials lists and dimensions are for flush door 3' x 6'8". Use waterproof woodworking glue.

Spanish Door

Two molding patterns (WP-126 and WP-93) and screen stock (WP-242) are used together as shown in Part 1, Fig. 47-73b to make "frames." Make jig with inside dimensions 12" wide and 10" high. Cut and miter 10 frames. Glue frames to door as shown in Part 1, Fig. 47-73b. Sand off excess glue before finishing.

Materials Needed (Allow For Waste)

20 pieces	WP-126 base shoe	13" long
20 pieces	WP-126 base shoe	11" long
20 pieces	WP-93 cove molding	8" long
20 pieces	WP-93 cove molding	6" long
20 pieces	WP-242 screen stock	11½" long
20 pieces	WP-242 screen stock	10" long
As needed	White woodworking glue, waterproof	
½ lb.	No. 18 x 1¼" brads	

Traditional Door

Cut, miter, and glue "frames" from picture molding. Part 2, Fig. 47-73b. Make two frames 12" x 24"; one frame 28" x 11"; and two frames 12" x 25". Glue to door as shown in Part 2, Fig. 47-73b. Sand off excess glue before finishing. Note: There are many other molding patterns equally suitable for this project.

Materials Needed (Allow For Waste)

8 pieces	WP-271 picture molding	12" long
4 pieces	WP-271 picture molding	24" long
4 pieces	WP-271 picture molding	25" long
2 pieces	WP-271 picture molding	28" long
2 pieces	WP-271 picture molding	11" long
As needed	White woodworking glue, waterproof	
½ lb.	No. 18 x 1¼" brads	

Contemporary Door

Cut 16 strips of WP-93 cove molding to 6' lengths. Glue together as shown in Part 3 of Fig. 47-73b. Glue the eight resulting strips to door. Part 3, Fig. 47-73b. Sand off excess glue before finishing.

Materials Needed (Allow For Waste)

16 pieces	WP-93 cove molding	6' long
As needed	White woodworking glue, waterproof	
½ lb.	No. 18 x 1¼" brads	

47–73b. *Details for applying molding to a flush door: 1. Spanish style. 2. Traditional. 3. Contemporary.*

Spanish Style

Traditional Style

WP-242
WP-93
WP-126
Nail and Glue

1

Glue
WP-271

2

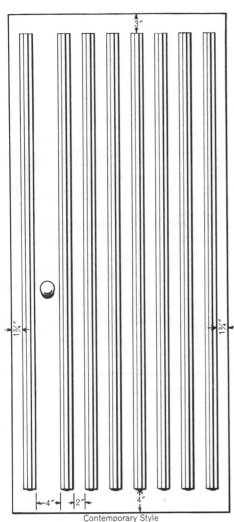

3″

1¾″

1¾″

←4″→ →2″← 4″

Contemporary Style

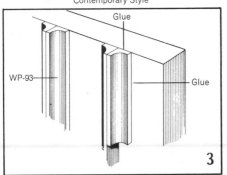

Glue

WP-93

Glue

3

ettes of the various molding styles and their pattern numbers are shown in Fig. 47-74. Review the sections on cutting and coping moldings in this unit before cutting the moldings to size for the applications shown here or applications of your own design.

Trimming a Clothes Closet

Base and shoe moldings of the same pattern used in the adjoining room are installed at the floorline of clothes closets. These moldings are cut, fit, and installed in the same manner as the base moldings in the room.

A piece of wood molding or a piece of 1″ × 3″ clear stock may be used as a hook strip and a shelf support in which the closet pole is installed. Many times a piece of wood base turned upside down (with the square edge up) is used as a hook strip. Locate the height of the hook strip, usually about 66″ above the finished floor line. Fig. 47-75. Level a line at this point on each wall of the closet. Cut and fit the molding in the same manner as the base molding.

Lay out and bore holes for the clothes pole. On one side, make a cut down from the top to remove the stock above the hole. Point A, Fig. 47-75. This will allow the pole to be inserted into the other hole and then slipped into the slot. Install the molding as you would base molding.

Cut the shelf to length and set it on top of the hook strip. The shelf is usually not nailed in order to make cleaning and painting easier.

Metal adjustable shelving is available prefinished in a variety of colors and wood grains. Also available are wall plates and brackets for mounting and a rod which adjusts to various lengths and fits into the wall plates. Fig. 47-76. Another accessory for clothes closets is a clothes hanger bar which attaches to the front edge of a wood shelf. This unit, which will reinforce and trim the front edge of the shelf, comes in varying lengths up to 14′. Fig. 47-77.

ESTIMATING
Materials

The cost of interior trim such as door-frames, doors, and moldings varies a great deal with wood species and styles. For example, pine casing used as trim for door and window frames may cost only half as much as some hardwood trim. However, some types of interior panel doors made of pine may cost twice as much as a mahogany flush door. Choice of materials should be based both on cost and utility. Such details are covered in the building plans and specifications.

Labor

To figure the total labor time, multiply the estimated labor rate by the number of

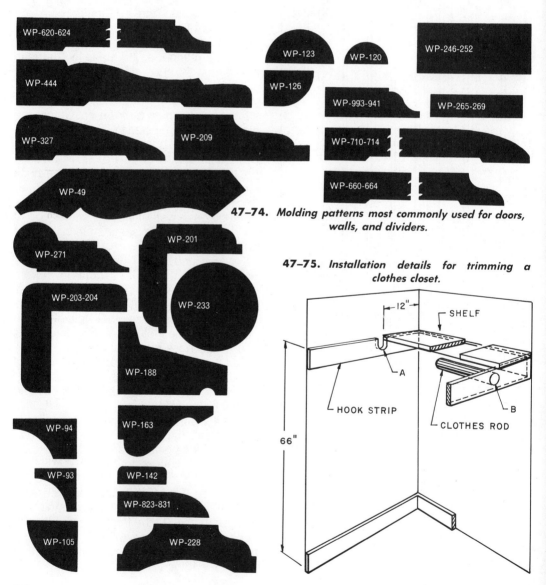

47-74. *Molding patterns most commonly used for doors, walls, and dividers.*

47-75. *Installation details for trimming a clothes closet.*

windows, doors, or the total amount of molding to be installed. Fig. 47-78. This will give the total number of hours required for interior trimming. To obtain the total labor cost, multiply the total number of hours by the labor cost per hour.

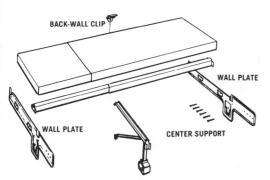

47-76. *Prefinished adjustable shelving, wall plates, brackets, and clothes rods made from metal are available for closet installations.*

47-77. *A metal clothes hanger bar which attaches to the front edge of the closet shelf.*

For example, if ceiling molding were to be installed in a room 22′ × 15′, the total amount of molding to be installed would be 74′ (22′ + 22′ for the side walls and 15′ + 15′ for the end walls). The total labor cost is then figured as follows:

$$\frac{74 \text{ (total amount of ceiling molding to be installed)}}{40 \text{ (linear feet of ceiling molding installed per hour)}} = 1.85 \text{ (hours to install the ceiling molding)}$$

Multiply this by the rate per hour to obtain the total labor cost.

47-78. *Interior trim labor estimates.*

Interior Trim Labor Estimates

Item	Estimated Labor Rate
Window and casement units (includes window installation and casement trim)	2.8 hrs. each
Interior doors (including jambs and trim)	4 hrs. each
Prehung Doors	½ hr. each
Shutters	1 hr. each
Ceiling and wall molding	40 linear ft. per hr.
Base (softwood—add 25% for hardwood)	
One-piece base	20 linear ft. per hr.
Two-piece base	15 linear ft. per hr.

QUESTIONS

1. What are the most common uses of wood or plastic moldings in the home?

2. What work must be completed before trimming interior doors and windows in a home?

3. What is the main advantage of a two- or three-piece doorjamb?

4. What is the standard thickness of interior doors?

5. What is the standard height of an interior door?

6. What is the advantage of a bifold door over a sliding door for a closet installation?

7. What are the two common methods of installing window trim?

8. What is the first piece of window trim to be installed?

9. How is the length of the window apron determined?

10. Why is base molding about the last trim to be installed?

11. What is the standard height for a clothes closet shelf?

Section VIII

SPECIAL CONSTRUCTION ACTIVITIES

Special Construction Techniques

Some important construction techniques are not commonly used in the standard house. Some of these techniques meet regional needs, such as the need for special construction along seacoasts. Others are useful primarily for cluster housing, such as apartments. Still others are too new to be in common use yet.

COASTAL HIGH-HAZARD CONSTRUCTION

Building in coastal high-hazard areas involves special problems. Houses must be protected against tides, storm surges, wind-generated waves, and high winds. Special care must be taken to assure that houses have sound foundations. Sand is the most common soil in most coastal areas. It does not provide a good base for building unless special techniques are used. In some areas, clay is under the sand. Generally, clay soils are a better base for building than sandy soils. Before deciding on what kind of construction to use, it is a good idea to take boring samples.

Foundation Systems

Several types of foundations are suitable for elevated residential housing in coastal high-hazard areas. Wood pile is the most common foundation. Fig. 48-1. A pile is a round or rectangular post driven into the soil. Wood piles must always be

treated to minimize decay and damage from fungus and marine borers. Concrete or steel piles can also be used, but these are much more expensive.

The effectiveness of a pile foundation depends, in part, on the method of inserting the pile into the ground. The best way is to use a pile driver. A less desirable method is a technique known as *jetting*. Jetting involves shooting a high-pressure stream of water through a pipe advanced alongside the pile. The water blows a hole in the sand into which the pile is driven until the required depth is reached. If precast concrete piles or steel piles are used for the foundation, a regular pile driver is required.

48-1. *For protection against flooding, houses may be built on wood piles.*

951

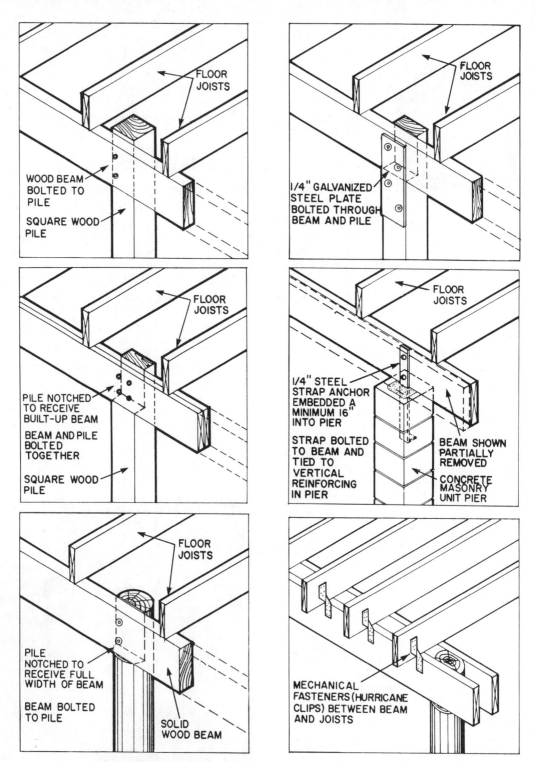

48-2. *Floor framing may be anchored to piles in a variety of ways.*

Another foundation consists of wood posts resting on some kind of spread footing. Wood posts are satisfactory for areas where there is less danger of wind and flooding. In some areas, it is common to use reinforced masonry piers for elevated residential structures. A good place for pier construction is far enough away from the beach so that flood waters will not greatly affect them. Bracing of the foundation piers can be very effective in minimizing storm damage.

Floors

Once the required elevation is obtained, the framing must be attached. Primary floor beams spanning the supports should be parallel to the flow of water. If joists are used, cross bridging is necessary for added support. Subflooring must be waterproof plywood or either 1" × 4" or 1" × 6" boards placed diagonally.

Walls

The most commonly used studs are 2 × 4s, 16 inches on center. However, 2 × 6s will permit 50 percent thicker insulation for cold weather areas. Wood frame walls should be braced to resist wind forces. Plywood is the normal wall sheathing used for exterior walls to add structural strength.

Anchorage

Anchorage is one of the most critical building requirements in coastal areas. Metal anchors or clips should be installed to attach all parts together wherever possible. These mechanical fasteners greatly strengthen the total building. Some methods for anchoring the foundation and floor framing are shown in Fig. 48-2.

CLUSTER HOUSING

Cluster housing includes apartments, condominiums, cooperatives, town houses, and other types of row housing. Fig. 48-3. If wood is the primary building material, construction of these units is similar to that of single homes, with two major exceptions—fire control and sound control.

48-3. *A great many people live in some type of cluster housing.*

Fire Control

Building codes for cluster housing generally require that the walls and floors between apartments be built to resist fire for a specific time. The walls between apartments, for example, may be required to have a fire resistance rating of one hour. The fire resistance rating depends on framing techniques and on the materials used in the wall.

Fire stops are one important way of resisting the spread of flames. Fire stops block the flow of air upward inside walls and horizontally between enclosed floor joists. By blocking air flow, they slow down the spread of flames. Structural elements,

such as top and bottom plates, function as fire blocks. Fig. 48-4. Nonstructural fire stops may also be nailed between studs or between floor joists.

Normal wood-and-panel assembly will meet the required fire resistance rating if the proper building materials are used. Each material and assembly is rated in tests performed by the Underwriters' Laboratories (UL). For example, a wall built of one layer of ⅝" type X gypsum board nailed to wood studs on 16" centers has a fire resistance rating of one hour. (Type X

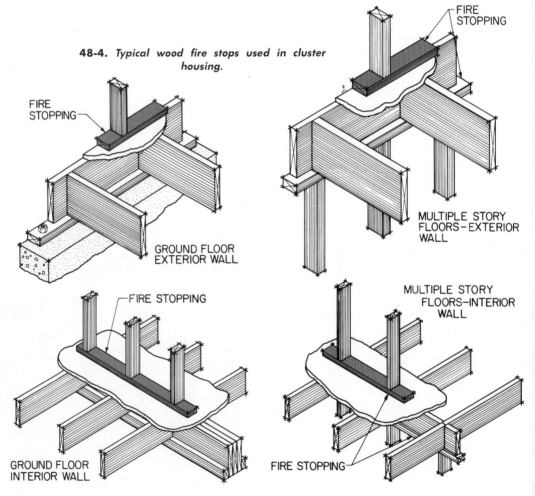

48-4. *Typical wood fire stops used in cluster housing.*

FIRE STOPPING

FIRE STOPPING

GROUND FLOOR EXTERIOR WALL

MULTIPLE STORY FLOORS – EXTERIOR WALL

FIRE STOPPING

MULTIPLE STORY FLOORS–INTERIOR WALL

GROUND FLOOR INTERIOR WALL

FIRE STOPPING

48-5. *A typical wall and floor design to provide for an adequate fire resistance rating.*

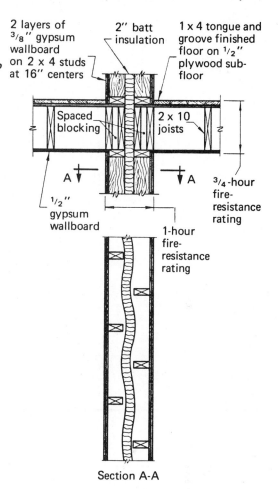

2 layers of ³/₈" gypsum wallboard on 2 x 4 studs at 16" centers

2" batt insulation

1 x 4 tongue and groove finished floor on ¹/₂" plywood sub-floor

Spaced blocking

2 x 10 joists

A

A

³/₄-hour fire-resistance rating

¹/₂" gypsum wallboard

1-hour fire-resistance rating

Section A-A

gypsum board is a gypsum board with a special fire resistant core.) The fire resistance rating of the wall could be increased by using a double thickness of gypsum board, by using a mineral wool insulation in the wall, and by many other means. Fig. 48-5 shows a typical fire-rated assembly. The *UL Fire Resistance Directory* lists the ratings of many materials and assemblies. It is the architect's responsibility to design buildings using these UL ratings to meet the local building codes.

Sound Control

In a well-built apartment building, sound should not travel easily from one unit to another. There are several methods for retarding the movement of sound.

INCREASING MASS

Heavier materials block sound better than light materials. Therefore, adding another layer of gypsum wall board will make a wall more soundproof. As a general rule, every doubling of the weight of the wall increases sound transmission loss by an additional 5 to 6 decibels.

BREAKING VIBRATION PATHS

Another efficient way of controlling sound transmission is to break the vibration path. This can be done by staggering the studs. The wall of one unit is then attached to one set of studs, and the wall of the other unit to another set of studs. Another approach is to use metal studs. They are more resilient than wood studs and reduce the transmission vibration between one wall and the other. With wood studs, metal clips can be placed between the gypsum board and the studs to break the vibration path.

ABSORBING SOUND IN THE WALL

Sound transmission can be decreased by filling the wall cavity with sound-absorbing material such as building insulation. Insulation is a good way to absorb sound without adding significant weight or mass to the partition.

ABSORBING SOUND IN THE UNIT

Another method of reducing sound transmission is to add sound-absorbing material in the unit itself. For example, rugs could be placed on the floor, or sound-proof ceiling tile can be added to the ceiling.

OTHER METHODS OF SOUND CONTROL

Some other ways to reduce the passage of sound include:

● Installing an air seal around the perimeter of the wall to create a proper acoustical seal.

● Reducing the window area by using fewer and smaller windows.

● Installing electrical items (including telephone bells and other wiring) on a well-insulated interior wall—never on a party wall.

● Insulating the plumbing from the adjacent structure so that the pipes will not carry noise. Fig. 48-6.

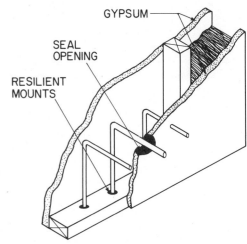

48-6. *Isolating pipes from surrounding structures with resilient mounts will reduce sound transmission.*

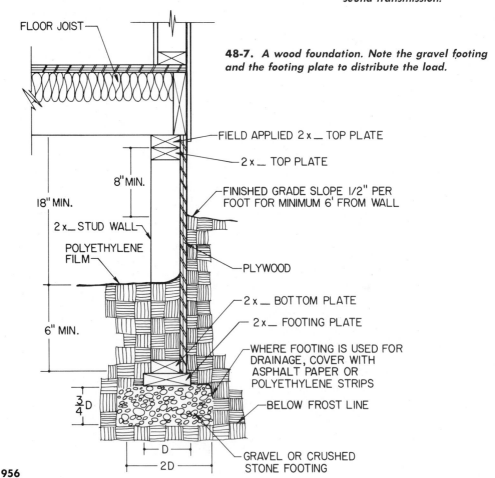

48-7. *A wood foundation. Note the gravel footing and the footing plate to distribute the load.*

• Installing furnace ducts so that noise cannot be transmitted through the heating and ventilating system.

PERMANENT WOOD FOUNDATION

The permanent wood foundation system provides a way to build a foundation easily and quickly. It can be installed in less than a day by a small crew. Wet and freezing weather does not delay construction. Components can be fabricated in a factory, so the cost is low. Once in place, wood frame basement walls are easy to finish. Board or paneling can be nailed directly to the studs. The walls will not crack, and the basement will not smell damp or musty. The cost of construction is less than a comparable foundation of poured concrete or concrete block.

Materials

All lumber and plywood in the foundation must be pressure-treated to the standards of the American Wood Preservers Institute for full protection against termites and decay. The fasteners must be silicon,

bronze, copper, or stainless steel. Hot-dipped, zinc-coated stell nails meet certain construction requirements.

Procedure

The first step in building a wood foundation is to provide footings. A layer of gravel, coarse sand, or crushed stone is required. A wood footing plate on top of this distributes the load. Fig. 48-7. The wall panels fabricated in a factory consist of the footing plate, bottom and top plates, studs, and plywood. Fig. 48-8. The footing plate is set in on one end of the bottom plate and extends out over the other end. Thus, the bottom plate of each panel overlaps the footing plate of an adjoining panel. The design also permits the interlocking of panels at the corners. The panels are set in place by a screw and the corners are solidly fitted together. The poured concrete subfloor conists of a layer of concrete that covers the treated sill and keeps the wall studs from moving.

If the exterior is to be brick veneered, a wood knee wall will provide a base for the

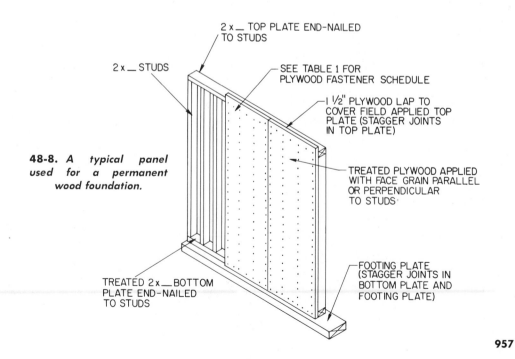

2 x __ TOP PLATE END-NAILED TO STUDS

2 x __ STUDS

SEE TABLE 1 FOR PLYWOOD FASTENER SCHEDULE

1 1/2" PLYWOOD LAP TO COVER FIELD APPLIED TOP PLATE (STAGGER JOINTS IN TOP PLATE)

48-8. A typical panel used for a permanent wood foundation.

TREATED PLYWOOD APPLIED WITH FACE GRAIN PARALLEL OR PERPENDICULAR TO STUDS

TREATED 2 x __ BOTTOM PLATE END-NAILED TO STUDS

FOOTING PLATE (STAGGER JOINTS IN BOTTOM PLATE AND FOOTING PLATE)

48-9. *This is how the permanent wood foundation looks when assembled.*

brick. The knee wall must rest on an oversize treated sill. With an adequate crew and prefabricated panels, the foundation can be assembled in less than two hours. Fig. 48-9. Detailed directions and technical information concerning the permanent wood foundation can be obtained from the National Forest Products Association.

FLOOR TRUSSES

Modern construction increasingly makes use of prefabricated trusses. Floor trusses, like roof trusses, are lightweight and easy to handle. They can be installed rapidly on the job. They provide many advantages for both builders and homeowners:
- Longer floor span.
- Wider spacing.
- Elimination of support beams in many cases.
- Elimination of bridging.
- Less lumber to be cut on the job.
- Less waste material on the job.
- Increased subfloor nailing area.
- Bottom chords to which ceiling material can be attached.
- Great flexibility in placing partitions.
- Reduction of floor squeaks.
- Reduction of sound transmission.

Floor trusses can be grouped in two basic types, open and closed. The open type allows duct work, electrical wires, and pipes to run through the trusses. They therefore increase the height clearance in the basement. Closed trusses are similar in appearance to floor joists. There are three common types of open trusses. The *wood web system* consists of upper and lower chords of 2 × 4s joined with metal fasteners to web members of 2 × 4s. Fig. 48-10. The *metal web truss* consists of 2 × 4s joined with specially designed commercial metal webs. Fig. 48-11. The *pipe web truss* consists of 2 × 4s or 2 × 6s joined together by specially designed pipes. Fig. 48-12.

A common closed truss is the *plywood I-beam* truss, which consists of wood top and bottom chords with a plywood center. Fig. 48-13. The *wood/metal I-beam* truss consists of top and bottom wood chords with a corrugated metal center. Floor trusses must be built with machine stress rated (MSR) lumber selected to meet the requirements for a specific type of truss.

48-10. *A wood web system for a floor truss.*

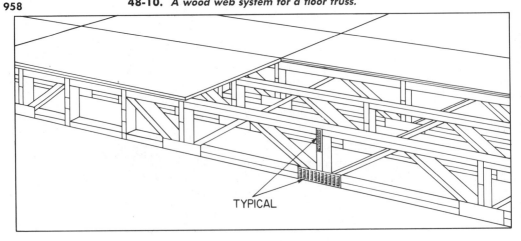

TYPICAL

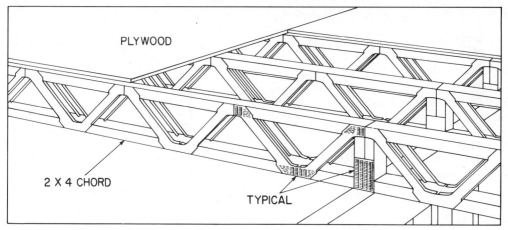

48-11. *A metal web truss.*

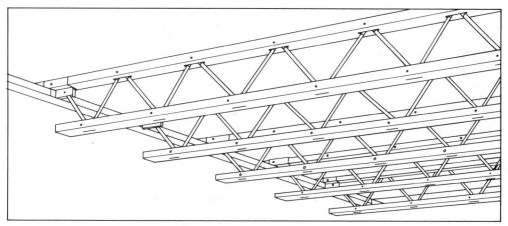

48-12. *A pipe web truss.*

48-13. *A plywood I-Beam.*

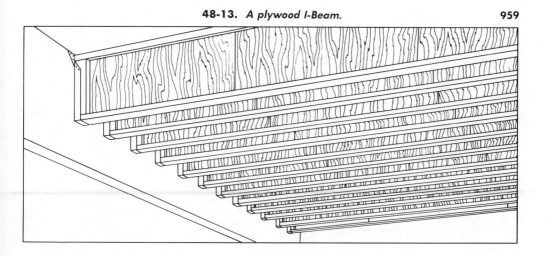

TRUSS-FRAMED SYSTEM

The truss-framed system (TFS) of construction consists of a roof truss, floor truss, and wall studs fastened together in a single rigid unit. Fig. 48-14. The frames are placed 24" on center to form the skeleton of the building. This method saves about 30 percent of the lumber used in the traditionally built home, where wall studs are placed 16" on center. The truss frames are manufactured in a factory and then shipped to the construction site. Frames for the entire house can be erected in less than three hours by a small crew. Fig. 48-15. The house can then be covered with wall sheathing and roofing to form a shell so that the completely enclosed structure is ready in one day. This is important during cold weather because it enables electricians, plumbers, and other contractors to finish the house under shelter.

Once the trusses are erected and the sheathing and roofing installed, the interior can be completed in a variety of designs. Since the inside is one large open space, different room arrangements are possible

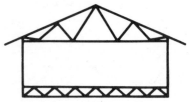

48-14. A conventional truss frame.

with the same exterior shell. It is easy to install the wiring and plumbing in the truss-framed house. Strength is another advantage of the truss-framed system. This system is used in hurricane country where disaster-resistant features are important. The system greatly increases the probability that the home will withstand hurrican forces. The truss-frame system can be used for any house with rectangular subdivisions, including L-shaped, U-shaped, and H-shaped designs. The system can be used for both single story and two-story construction. The roof may be of any type found on conventional homes.

48-15. A truss frame being raised.

QUESTIONS

1. Describe two methods of installing piles for a home in a coastal region.

2. Why might it sometimes be preferable to use 2 × 6s instead of 2 × 4s as studs for exterior walls?

3. In what two ways does the construction of cluster housing differ from single home construction?

4. How does type X gypsum board differ from regular gypsum board?

5. What features of an apartment building help assure that sound does not travel easily from one unit to another?

6. What are some of the advantages of a permanent wood foundation?

7. What are some of the advantages of floor trusses?

8. How do open and closed floor trusses differ?

9. What are some of the advantages of using a truss-frame system?

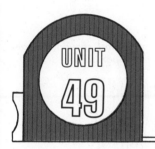

Avoiding Problems of Moisture

In recent years there have been remarkable changes in the methods and materials used in house construction. The average house has grown tighter and provides more livability in less space. Ceilings are lower. Basements are often omitted and thus utilities such as furnaces and laundry equipment are installed on the first floor. Exterior design also has changed. Many homes no longer have any overhang at eaves and gable ends to protect the walls.

Today people make greater use of water and of appliances which discharge moisture into living areas. Daily activities such as bathing and cooking add gallons of water vapor to the air. The use of sheet materials for sheathing, plaster base, and interior wall finish has reduced air exchange through walls. Insulation in walls, roofs, and under floors has helped to seal houses more tightly. Weather-stripped doors and windows and the modern window frame, while increasing comfort within the home, also trap air and moisture inside the house.

Condensed water vapor, in the form of free water or ice, often collects behind the siding of a building. This excess moisture may absorb chemicals from the wood and then stain the siding as it runs out over the surface. In some cases, the siding is thoroughly wetted. There is a loss of paint adhesion, and water-filled blisters form under the paint film.

MOISTURE CONTROL
Walls

Protection of exterior walls against moisture from both inside and outside sources is very important. There are a number of effective methods for preventing condensation. Described here are control measures and construction features which curb moisture.

A vapor barrier will protect outside walls from moisture coming from within the home. Fig. 49-1. Such barriers should be carefully applied to provide a complete envelope, preventing water vapor from entering enclosed wall spaces where condensation may occur. The goal of vapor protection is to make the warm side of the wall as vapor-tight as possible, and the cold side permeable enough to permit passage of water vapor to the outside.

Dull-surfaced asphalt- or tar-saturated sheet products, commonly used as sheathing papers, are water-repellent but not vapor-proof. They adequately serve as cold-side, vapor-resistant materials.

Good warm-side barriers are available in many forms. Some insulation is made with a vapor barrier on one face. Also, many interior wall-facing materials used by builders are backed with vapor barriers. If not present in one of these forms, however, a separate barrier is needed. Sheet metals, metal foils, asphalt-laminated papers, and foil laminates are highly vapor-resistant and perform well as vapor barriers. Polyethylene films and certain other plastic sheet materials also are satisfactory.

To prevent accidental punctures, the warm-side vapor barrier should be installed after heat ducts, plumbing, and wiring are in place. Wherever possible, barrier-faced material should be compressed and tucked behind conduits, piping, and ducts. Where there are outlet boxes, electrical receptacles, and wall projections, place the insulation behind the obstruction. Carefully cut and fit the

vapor barrier to insure maximum protection. Nail or staple the barrier over the edges of the studs. To avoid holes or gaps, apply plenty of fasteners, spaced no more than 6″ apart. The entire stud space, from floor to ceiling, should be covered in this manner.

In homes built without vapor barriers, good vapor protection can be achieved by applying two coats of a low-permeability paint system to inside walls and ceilings. However, such measures do not provide complete protection. Paints of this type are intended to supplement the action of vapor barriers in new homes.

Ceilings

Vapor barriers in ceilings are considered necessary only where winter temperatures commonly fall below -20°F or where roof slopes are very low. In these cases, vapor

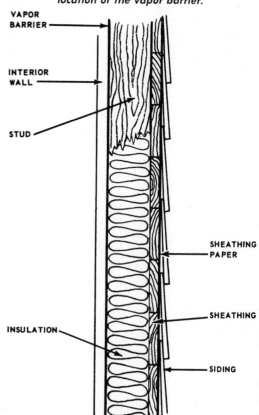

49-1. *Cross section of a wall showing proper location of the vapor barrier.*

VAPOR BARRIER

INTERIOR WALL

STUD

SHEATHING PAPER

SHEATHING

INSULATION

SIDING

barriers should be installed in ceilings under attics and flat roofs in the same manner as in walls.

Crawl Spaces

Except in very dry climates, the soil in crawl spaces should be covered with a layer of vapor-resistant durable material. Asphalt-saturated felt roll roofing or polyethylene film may be used. The ground surface should be leveled, and the cover material turned at walls and piers and lapped at least two inches. This material need not be sealed.

Basement Floors

If a concrete slab is to be placed on the ground, a vapor barrier should be installed directly under the slab. Otherwise moisture from the ground will move up through the slab and into the basement. The barrier must be strong enough to resist puncturing when the slab is poured. It must also be a type that will not deteriorate with age.

Basement Walls

Unit-masonry basement walls should be thoroughly damp-proofed by applying mortar and asphalt coating to the outside surfaces.

VENTILATION

Good ventilation controls condensation in buildings. In the winter, fresh, dry air from outside replaces moisture-laden air within the house. In this way, high vapor pressures which produce condensation are reduced.

Attic Spaces

In attic spaces and under flat roofs, condensation can occur in the same manner as in walls. It has become standard practice for builders to install louvered openings in the gable ends of houses to provide ventilation through the attic spaces. When these openings function properly, there is generally little condensation.

However, the ventilation must be adequate. Too little ventilation is seldom better than no ventilation at all. Fig. 49-2.

For gable roofs, where the slope is greater than 3 in 12, screened louvers are usually installed on opposite sides near the ridge. The total area of the openings should be $1/300$ of the ceiling area below the roof. In other words, 1 sq. ft. of louvers must be provided for every 300 sq. ft. of ceiling area. For a house with 1500 sq. ft. of ceiling area, at least 5 sq. ft. of ventilator area should be provided ($1/300 \times 1500 = 5$). When a ¾" slot is provided beneath the eaves, the ventilating area may be reduced to $1/900$. This would be $1\frac{2}{3}$ sq. ft. of ventilating area for the example.

For hip roofs, it is customary to provide a ¾" or larger slot beneath the eaves and a sheet-metal ventilator near the peak. In this case the net area of the inlet should be $1/900$ and that of the outlet should be $1/1600$ of the ceiling area.

For flat roofs, blocking and bridging should be arranged to prevent interference with movement of air. Such roofs may be ventilated along overhanging eaves. The net area of openings should equal $1/250$ of the ceiling area.

Crawl Spaces

All crawl spaces under houses without basements, and such unexcavated spaces as porches, breezeways, and patios should be ventilated by openings in the foundation walls. The spaces should be kept open all year and provided with access panels so that they may be easily inspected. The vent openings should have a net area of not less than 2 sq. ft. for each 100 linear feet of exterior wall, plus ⅓ sq. ft. for each 100 sq. ft. of crawl space.

Openings should be arranged to provide cross ventilation. They should be covered with corrosion-resistant wire mesh, not less than ¼" nor more than ½" in any dimension. No unventilated, inaccessible spaces should be permitted.

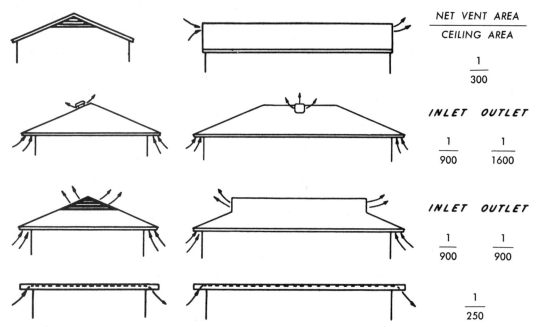

	NET VENT AREA
	CEILING AREA
	$\dfrac{1}{300}$

| INLET | OUTLET |
| $\dfrac{1}{900}$ | $\dfrac{1}{1600}$ |

| INLET | OUTLET |
| $\dfrac{1}{900}$ | $\dfrac{1}{900}$ |

| | $\dfrac{1}{250}$ |

49-2a. *The relationship of net vent area to ceiling area in various roof styles.*

Vent Protection	Increase in Total Area of Vent Opening Per Cent
¼-inch mesh	0
No. 8 mesh	25
No. 16 mesh	100
Louver + 1-inch mesh	100
Louver + No. 8 mesh	225
Louver + No. 16 mesh	300

49-2b. *Percent of increase required for a vent opening when vent protection material is used.*

Louvers and Screened Ventilators

Where crawl space, attic, or soffit ventilators are protected by louvers and/or screens, the recommended total area of the vent openings increases. In the example of the house with 1500 sq. ft. of ceiling area, 5 sq. ft. of vent area are needed. If a louver with #16 mesh screen is used, however, a 300% increase in the total area of the vent opening is needed. The louvers and the small mesh of the screen restrict air flow. The house in the example would therefore require 20 sq. ft. of vent area. 5 (sq. ft. of normal vent area) × 3.00 = 15 (sq. ft. of increase) 15 + 5 = 20 (sq. ft. of vent area).

GENERAL RECOMMENDATIONS FOR GOOD CONSTRUCTION

In addition to vapor barriers and effective ventilation of houses, there are other ways to protect houses from moisture under all exposure conditons.

Site Drainage

In all cases, the building site should be graded to provide positive drainage away from foundation walls. Fig. 49-3. In determining the height of the foundation, it is important to take into consideration the proposed height of the finished grade in order to assure proper clearance for wood members resting on top of the foundation.

Building Drainage

All exposed wood surfaces. as well as adjoining areas, should be pitched to pro-

964

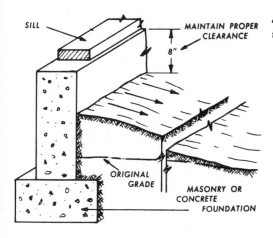

49-3. *The grade should be at least 8" below the top of the foundation wall and must slope away from the structure.*

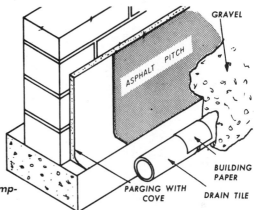

49-4. *The foundation wall should be damp-proofed and proper drainage provided.*

vide rapid water runoff. Construction details which tend to trap moisture in end-grain joints should be avoided.

Good foundation drainage and damp-proofing of walls, below grade, will prevent undesirable moisture conditions from developing in crawl spaces and basements. Fig. 49-4.

Poured concrete walls are moisture-proofed by applying two coats of asphalt or tar from footing to finish grade. A masonry wall requires a ½" layer of portland-cement mortar applied from footing to finish grade. A cove is left at the junction of the footing and wall. The mortar should then be mopped with asphalt or tar. Footing drains should be installed around foundations enclosing any living space below grade. Drain tile may also be provided in very damp crawl spaces. Fig. 49-5.

For slab-on-ground construction and for basement floors, earth should be covered

with not less than 4" of coarse gravel or crushed stone. A membrane is placed over this before pouring the slab. Where basements contain living space, this membrane should be a vapor barrier.

To protect exterior walls and roofs against the entrance of water, flashing is needed in the following places:

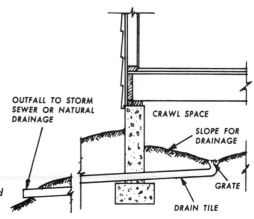

49-5. *Proper drainage of the crawl space should be provided.*

Paint Requirements

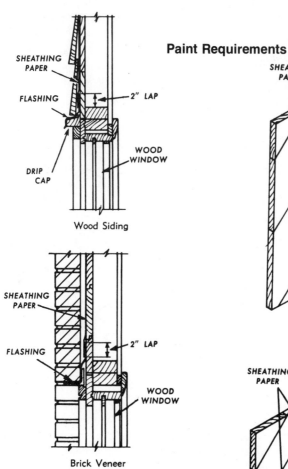

Wood Siding

SHEATHING PAPER

FLASHING

DRIP CAP

2" LAP

WOOD WINDOW

Brick Veneer

SHEATHING PAPER

FLASHING

2" LAP

WOOD WINDOW

49-6. *Openings in the exterior walls must be flashed.*

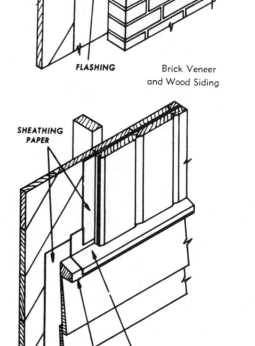

SHEATHING PAPER

FLASHING

Brick Veneer and Wood Siding

SHEATHING PAPER

FLASHING

DRIP CAP

Horizontal and Vertical Wood Siding

49-8. *Flashing must be installed where there are changes in the exterior surface material.*

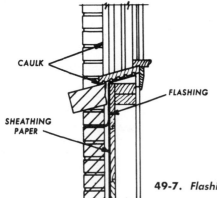

CAULK

FLASHING

SHEATHING PAPER

49-7. *Flashing under the window in an exterior wall.*

• Heads of all openings in exterior walls, unless protected by overhanging eaves. Fig. 49-6.

• Under window openings where masonry veneer is used. Fig. 49-7.

• Where changes in exterior finish materials occur. Fig. 49-8.

• Between concrete porch or patio floors and wood finish, where such floors are above the top of the foundation. Fig. 49-9.

• At the junction of roof deck and wood siding, at roof valleys, and around chimneys and vent stacks. Fig. 49-10.

Exterior wood siding should extend at least one inch below the top of the foundation to provide a drip line protecting the sill from rainwater. Fig. 49-11.

Where the exterior finish is masonry veneer, a 1″ space should be allowed between the sheathing and the veneer. Weep holes and flashing should be provided at the base of the veneer wall to permit drainage.

On roofs, shingles should project at least 1½″ beyond the roof sheathing to provide a drip line. When asphalt shingles are used, a starting course of wood shingles or a metal drip edge is desirable.

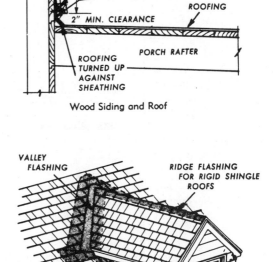

Wood Siding and Roof

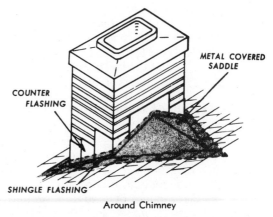

At Valley and Dormer

49-9. *Flashing must be installed between the exterior wall and a concrete porch or patio floor.*

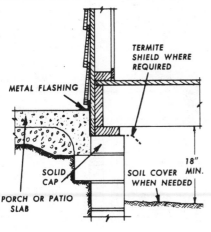

49-10. *Flashing at various points on a roof.*

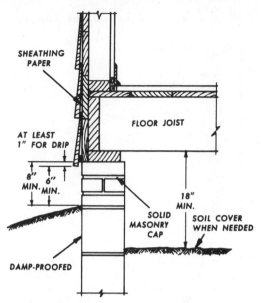

FLOOR JOIST

AT LEAST
1" FOR DRIP

8"
MIN.

6"
MIN.

18"
MIN.

SOLID
MASONRY
CAP

SOIL COVER
WHEN NEEDED

DAMP-PROOFED

49-11. *Wood siding should extend at least 1"
below the top of the foundation.*

Well-drained gutters and downspouts are needed, unless eave projection is 24" or greater. Roof runoff should be drained away from the structure.

In northern areas, melting snow may result in ice dams on roof overhang and gutters. The ice dams may cause water to work back under shingles and drop into attic spaces and walls. Protection is provided by applying 55-pound roll roofing, extending it over the roof sheathing and behind the gutter. Fig. 49-12.

Gutters should be placed so that the lip is even with the extended shingle line. When ice is likely to form, gutters should be kept ¾" away from the fascia by blocks spaced at 16" to 24" intervals.

Separation of Wood from Ground

Wood structural components may be protected from the effects of ground moisture by following these accepted building practices:
• In crawl spaces and other unexcavated spaces, clearance between the ground and the bottom of wood joists, or structural planks without joists, should be at least

18". Between the bottom of wood girders or wood posts and the ground, the clearance should be at least 12".
• Wood sills, which rest on concrete or masonry exterior walls, should be at least 8" above the exposed earth.
• Wood siding and trim should be at least 6" above exposed earth on the exterior of a structure.
• Structural portions of exterior wood stairs, such as stringers and posts, should be at least 6" above the finished grade.
• In porches, breezeways, and patios, all supporting floor framing such as beams, headers, and posts should be at least 12" above the ground. Floor joists should be at least 18" above the ground. Posts which rest on wood, concrete, or masonry floors should be supported on pedestals extending at least 2" above the floor, or at least 6" above exposed earth.
• Ends of main structural members which are exposed to weather and which support roofs or floors should rest on foundations which provide a clearance of a least 12" above the ground or 6" above concrete.

49-12. *Flashing at the eave should extend at least 6" beyond the inside face of the wall. The flashing also extends behind the gutter.*

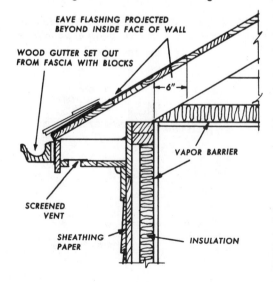

EAVE FLASHING PROJECTED
BEYOND INSIDE FACE OF WALL

WOOD GUTTER SET OUT
FROM FASCIA WITH BLOCKS

6"

VAPOR BARRIER

SCREENED
VENT

SHEATHING
PAPER

INSULATION

• Shutters, window boxes, and other decorative attachments should be separated from exterior siding to avoid trapping rainwater.

Exterior Siding and Covering

Correct application of siding and covering materials helps to insure their freedom of movement with changing weather conditions. It minimizes the possibility of cracking, buckling, and resultant water leakage. The construction practices described in this section will thus help assure that exterior paint will give satisfactory appearance and performance.

Weathertight walls are provided by sheathing covered on the outside with asphalt-saturated felt or with other impregnated paper having the same water-repellent properties. Sheathing paper must not be of a type which would act as a vapor barrier.

Beginning at the bottom of the wall, the felt should be lapped 4" at horizontal joints and 6" at vertical joints. Strips of sheathing paper, about 6" wide, should be installed behind all exterior trim and around all openings.

WOOD SIDING

Many types and patterns of wood siding are available. Names vary with the locality and include bevel, bungalow, Colonial, rustic, shiplap, and many others.

All siding and exterior trim should be thoroughly dry when installed. It must be free of dampness or condensation when painted. Most siding has been water-repellent treated or paint primed on the back or on both surfaces. Such applications protect the siding from any later effects of excess moisture. Preprimed siding has a complete prime coat of paint on the siding face as well as a protective primer or treatment on the reverse side. Factory priming provides a superior base for additional coats of quality paint. Excellent service can be expected of these pre-primed products.

For painted houses, where entrance of rainwater is suspected, water-repellent preservative should be applied along the butt edge of siding boards before repainting.

Where wood sheathing is used, siding may be nailed at 24" intervals with corrosion-resistant nails, usually galvanized steel or aluminum. Where other types of sheathing are used, nails should be driven through the sheathing into the studs. The nail length to be used will vary with thickness of siding and type of sheathing.

Bevel siding and square-edge boards should be lapped and applied horizontally. Nails should be driven just above the lap to permit movement caused by changes in moisture conditions. Fig. 49-13.

49-13. *Proper nailing of various styles of siding.*

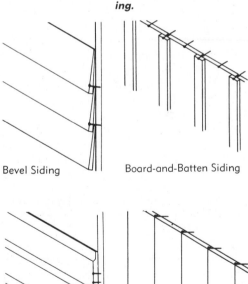

Bevel Siding

Board-and-Batten Siding

Drop Siding

Tongued and Grooved Board Siding

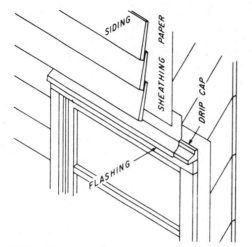

49-14. *The bottom of a siding course should coincide with the tops of door and window trim.*

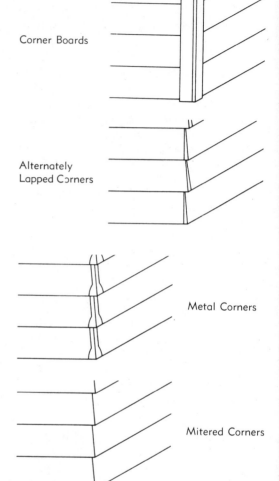

Corner Boards

Alternately
Lapped Corners

Metal Corners

Mitered Corners

49-15. *Methods of finishing outside corners of bevel siding.*

Space siding so that the bottom of a piece coincides with the top of the trim over door and window openings. Fig. 49-14. A liberal coating of water repellent on the end surfaces is good practice.

Corner treatment of siding depends upon the overall house design. It may involve corner boards, mitered corners, metal corners, or alternately lapped corners. Fig. 49-15.

Siding is frequently applied vertically. Where tongued and grooved boards are used, the siding should be blind-nailed to wood sheathing at 24" intervals.

Square-edge boards used with battens are spaced about ½" apart and nailed only at the center. The batten is attached by one nail driven through the center so that it passes between the boards. This arrangement permits movement with change in moisture conditions. Where sheathing other than wood is used, blocking should be provided between studs to permit nailing of the vertical siding.

Protection may be provided against rainwater entering behind the batten by knifing white lead paste on the back of the batten

49-16.

Recommended Weather Exposures For Wood Shingles

Shingle Length	Maximum Exposure	
	Single Course	Double Course
16"	7½"	12"
18"	8½"	14"
24"	11½"	16"

49-17. *Install 1″ × 3″ nailer strips over nonwood sheathing.*

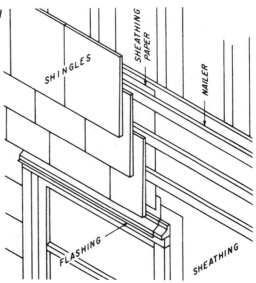

strips before nailing in place. Brushing a paintable water-repellent preservative into each batten-siding joint after installation serves the same purpose.

WOOD SHINGLES

No. 1 grade shingles, other than shakes, when used as exterior wall covering, should have the weather exposures shown in Fig. 49-16.

If shingles are installed in double courses, the butt of the exposed shingle should extend about ½″ below the undercourse in order to produce a shadow line.

Shingles should be nailed with corrosion-resistant nails that are long enough to penetrate the sheathing. Use two nails for shingles up to 8″ wide and three nails for wider shingles.

For single coursing, nails should be driven approximately 1″ above the butt line of the succeeding course. For double coursing, the undercourse should be attached to the sheathing with 3d nails or staples. The outer course should be attached with small-headed nails driven about 2″ above the butts and ¾″ from edges.

Apply 1″ × 3″ horizontal nailing strips over nonwood sheathing. The strips should be spaced to correspond to the weather exposure of the shingles. Fig. 49-17.

QUESTIONS

1. What is the purpose of a vapor barrier?

2. What will control condensation in buildings?

3. What is used to protect the exterior walls and roof against the entrance of water?

4. How are the wood structural components of a building protected from the effects of ground moisture?

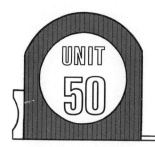

Remodeling and Renovation

An older house can sometimes be a good buy for someone who wants to fix it up. Some houses, however, need extensive and expensive repairs to put them in good shape. Time can take its toll on even the best built houses, and older houses did not have all the benefits of modern building technology. It is not easy to judge the condition of an older house. A first glance is often deceiving. Sometimes a home that looks run-down may be in better condition than one that has just been patched up and painted. To determine whether it is cost-effective to renovate an older home, a thorough inspection must be conducted. Fig. 50-1.

50-1.
Inspection Checklist

House Exterior

☐ I. Site plan
  A. House orientation
    1. North-South
    2. East-West
  B. Trees and shrubs
    1. Provide windbreak
    2. Provide shade
    3. Allow south sunlight
  C. House location
    1. On hill
    2. Protected by trees
    3. In open plain

☐ II. House appearance
  A. Yard
    1. Clean, well kept
    2. In need of landscaping
    3. Unkept, messy
  B. House
    1. Well-maintained appearance
    2. Needs paint
    3. Needs roof
    4. Broken windows

☐ III. Electrical
  A. Service from pole
    1. 2-wire 110V/120V
    2. 3-wire 220V/240V

☐ IV. Siding
  A. Type
    1. Wood
    2. Stucco
    3. Masonry
    4. Aluminum or vinyl
    5. Other
  B. Condition
    1. Satisfactory
    2. Repairs needed

☐ V. Roof
  A. Type
    1. Asphalt shingle
    2. Metal
    3. Rolled asphalt
    4. Other
  B. Condition
    1. Satisfactory
    2. Repairs needed

  C. Chimney
    1. Satisfactory condition
    2. Repairs needed
  D. Flashing
    1. Satisfactory condition
    2. Repairs needed
  E. Eaves and downspouts
    1. Type
    2. Condition

☐ VI. Exterior foundation
  A. Type
    1. Stone
    2. Stone capped with mortar
    3. Cement block
    4. Other
  B. Condition
    1. Satisfactory
    2. Repairs needed

(Continued on next page)

Inspection Checklist (continued)

House Interior

☐ I. Basement

   A. Floor
      1. Stone
      2. Cement
      3. Dirt
   B. Walls
      1. Stone
      2. Block
      3. Cement
   C. Condition
      1. Wet or damp
      2. Dry—no sign of
         moisture
   D. Joists, sillplate, and
      header
      1. Good condition,
         straight joists, no
         cracks
      2. Rotting wood, in-
         sect infestation,
         sagging wood
   E. Heating System
      1. Type
      2. Appearance
      3. Original or new
      4. Type of fuel

☐ II. Insulation

   A. Type
   B. Location (how many?)
      1. Wall
      2. Ceiling
      3. Floor

☐ III. Electrical

   A. Service entrance
      1. 60 amperes
      2. 100 amperes
      3. 200 amperes
      4. Fuse box
      5. Circuit breakers
   B. Branch circuits (how
      many?)
      1. Area lighting cir-
         cuits
      2. Appliance circuits
      3. Electric range/dryer
        /hot water heater
        circuit

☐ IV. Plumbing

   A. Sewage system
      1. City sewer
      2. Septic tank
      3. Other
   B. Septic tank (if appro-
      priate)
      1. Concrete
      2. Metal
      3. Size (500, 800,
         1000 gal.)
   C. Supply lines
      1. City water
      2. Well (dug or
         drilled)
      3. Flow rate

   D. Water lines
      1. Copper
      2. Galvanized iron
      3. PVC plastic
      4. Other
   E. Hot water tank
      1. Gas
      2. Electric
      3. Fuel oil
      4. Age and condition
      5. Size
   F. Well pump
      1. Type
      2. Condition

☐ V. Room layout

   A. Traffic flow
      1. Satisfactory
      2. Changes needed
   B. Kitchen
      1. Modern/convenient
      2. Changes needed
   C. Interior Walls
      1. Type (plaster, pan-
         eled, sheetrock)
      2. Condition

GENERAL INSPECTION

If you know what to look for, you can conduct much of the inspection yourself. Be sure to wear old clothes so you can check the attic and crawl spaces. Equip yourself with flashlight, pocketknife, pen, notepaper, and tape measure. As you inspect the house, remember that old homes are often not very energy efficient. Look for any flaws that waste heat. Fig. 50-2.

Building Site

Before entering, inspect the house from the outside. Walk around the house several times, checking the surrounding site, exterior walls, roof, windows, and doors. Locate the north and south exposures of the house. Ideally, the south side should have the most window space. The north side should be protected by a windbreak such as vegetation, a garage, or a hill. Trees on

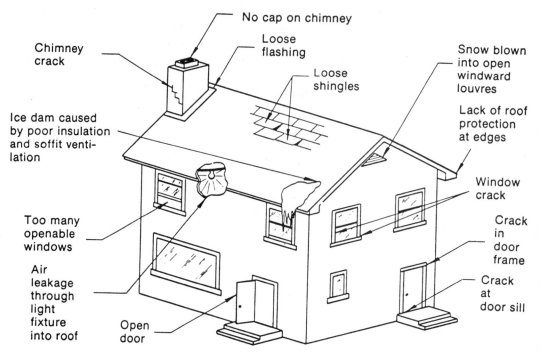

No cap on chimney

Loose
flashing

Chimney
crack

Loose
shingles

Snow blown
into open
windward
louvres

Ice dam caused
by poor insulation
and soffit venti-
lation

Lack of roof
protection
at edges

Window
crack

Too many
openable
windows

Crack
in
door
frame

Air
leakage
through
light
fixture
into roof

Open
door

Crack
at
door sill

50-2. *An older home often needs insulation and other repairs to make it more energy efficient.*

the south side should be deciduous, providing shade in the summer and allowing sunlight to enter in the winter. Trees on the north side should be evergreen, providing a windbreak during the winter.

Exterior Walls

Inspect exterior siding to see if it needs to be replaced or if it has been replaced. Houses with wood siding should be checked for peeling paint. Peeling paint or bare wood in random areas around a house indicates a moisture problem coming from inside the house. Paint peeling away from another layer of paint indicates that different kinds of paint have been used on the same surface. In either case, the old paint must be completely removed or new siding must be attached.

Check the wood trim around the eaves and soffit to make sure there is no dry rot due to ice damage, leaky eaves troughs,

insects, or mildew. In older homes, it takes much longer to repair and paint the trim than the main body of the house.

Aluminum or vinyl siding is often put on an older home over the original wood or asbestos siding. Sometimes this is done to cover a paint-peeling problem caused by excessive moisture inside the house. Moisture finds its way into the wood siding and forces the paint off as it escapes. Sealing the wood siding with metal or vinyl siding captures this moisture in the wall spaces. The trapped moisture eventually causes dry rot or insulation damage. For this reason, unvented aluminum, vinyl, or composition siding should not be installed on a house that has inadequate vapor barriers.

Roof

Foundation settling in an older home will often show up in a swayback or leaning roof ridge. Fig. 50-3. If the roof ridge is not

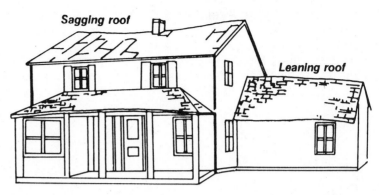

50-3. *The roof often indicates serious problems in a house's foundation.*

straight, make a note to check the basement walls, support columns, and joists. Roof shingles must lie flat and intact. Old roof shingles often curl and lose their granular surface. They can also curl under and break. Sometimes too many asphalt shingles have been placed over each other. There should not be more than two depths of shingles on a roof. Check the roof for large amounts of black tar; this indicates there have been roof leaks. Sometimes these roof leaks will cause the roof boards to rot. Check the chimney for loose mortar or broken bricks, and make sure that the metal flashing is in good condition.

Windows and Doors

Check to see that doors and windows are snug and well-fitted. Make sure also that they have storm doors and windows. Poorly fitted doors and windows, and those without storms, are a major cause of energy loss. Entrance doors should also be protected by an overhang. Wood frame windows are best for energy savings. Aluminum frames can cause considerable heat loss in the winter. Be sure to check the window operation by raising and lowering double-hung windows and by cranking open casement windows.

Basement

Check the basement or crawl space, floors, and walls for damp spots. Musti-ness indicates that moisture may be finding its way into the living space. Flooding in a basement will eventually undermine the foundation. Cracks on the inside of a foundation wall often indicate that the wall is being pushed in from the outside. Check the structural members for termites and dry rot. Also, make sure the floors are level.

Attic Vents

There should be at least two gable vents. If there is little or not attic ventilation, check the attic ceiling and rafters for condensation stains or dark colors. Look carefully for any signs of leaks. Check also to see what kinds of insulation have been installed in the house.

Interior Walls and Ceilings

Most older homes have plastered walls. These should be checked for cracks. Drywall or gypsum board is the surface material used instead of plaster in buildings today. Examine the walls and ceilings carefully for signs of leaks.

Check the insulation in the walls and determine if any must be added.

Common residential construction techniques years ago involved post-and-beam and balloon framing. The old post-and-beam construction is the most difficult to insulate. Balloon frame houses are not common any more.

Professional Assistance

You may need some professional help in making your inspection. A commercial termite inspection will often uncover damage that the untrained person will overlook. If there has been extensive damage due to termites, marine borers, or dry rot, repairs to the house may be very costly.

The heating system, plumbing, and electrical wiring should be looked at close-

50-4a. *This house, typical of the 1950s, will be remodeled without adding floor space.*

50-4b. *The same house—with a stunning, contemporary exterior.*

50-4c. *The old kitchen.*

50-4d. *The remodeled kitchen.*

ly by someone who knows these systems. Problems with these systems often develop in older houses. They too can mean costly repairs.

Plans for Remodeling

After a house has been examined and found fit for remodeling, plans for improvement can be made. The type of improvements to be made depend on the condition of the house and the needs of the owner. A serious problem that needs repair—such as a leaky roof—will probably deserve first attention. Work that will make the home more energy efficient should be considered. The homeowner may also want to make changes that will provide more space, add modern conveniences, or improve the appearance of the house.

Types of Remodeling

There are two general types of remodeling. One involves making changes to the exterior or interior without adding to the size of the structure. Fig. 50-4. The other involves making additions that increase the size of the structure. Fig. 50-5.

Most projects fall into the first category. A few typical specific remodeling projects are listed below:
• Changing the layout. It may be important to alter the interior layout of the home. Fig. 50-6. Large rooms can be made by removing walls between small rooms. A large room may be divided into smaller rooms by adding a wall. Doors can be relocated to change traffic patterns.
• Replacing windows. Old windows may be replaced with new ones to make the house weather tight. The number, size, or arrangement of windows can be changed to make the house more attractive.
• Adding closets. If a house does not have enough closet space, the homeowner may need to add some.
• Remodeling the kitchen. Appliances such as a modern range and a dishwasher can be built into an old kitchen. Old, worn-out cabinets and counters may need to be replaced. So may the sinks and plumbing. Kitchen remodeling in an older house is often necessary and often costly. (The same is true in the case of bathroom remodeling.)

50-5a. *The foundation and rough floor for an addition to a house.*

50-5b. *Working on the rough exterior.*

50-5c. *The completed addition.*

● Adding a room within an existing structure. An attic, basement, or garage may have space for an additional room. For a room in the attic, it may be necessary to add a dormer. Sometimes a garage can be converted into a room, particularly if the garage is attached to the house. An unfinished basement is one of the best places

for expansion. The homeowner should check local building codes, however, to make sure the minimum ceiling height does not exceed the height of the basement. A new entrance to the basement can improve its appearance. A possible problem with a room in the basement is the lack of natural light.

● Adding a porch or a deck. Many older homes have no porch or deck. Adding one can improve the appearance of a house and make it more comfortable as well. Fig. 50-7.

THE VALUE OF REMODELING

Before remodeling a house, the homeowner should consider the cost of the project. That figure should be compared to the amount that the project will add to the resale value of the house. Generally speaking, the more that can be recovered in resale, the better the investment.

● Adding a room. Adding a room can be a better investment than buying a larger house. Adding a room can allow the homeowner to maintain a low mortgage rate. An attractive addition designed in the same style as the house can increase its resale value by quite a lot. Generally speaking, 55 to 60 percent of the cost of a room addition can be recaptured in the resale of

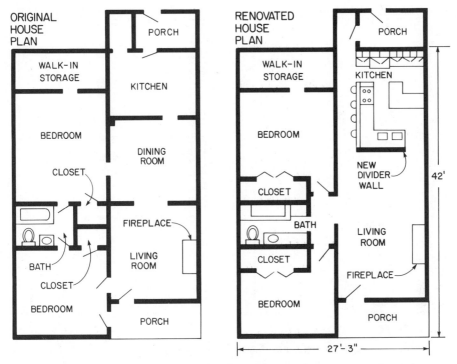

50-6. *A house's floor plan can be changed to meet a new owner's needs.*

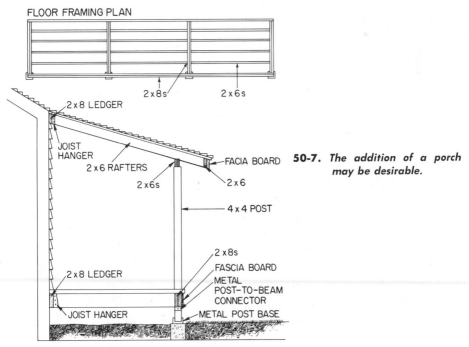

50-7. *The addition of a porch may be desirable.*

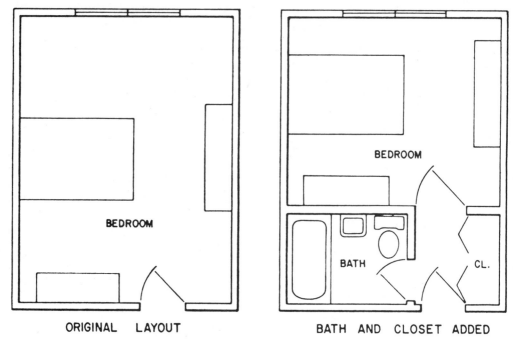

ORIGINAL   LAYOUT   BATH   AND   CLOSET   ADDED

50-8. *A private bath can add to parents' comfort.*

the house. A room addition is a good investment, provided it does not make the house much larger than other homes in the area. If a house is much larger than those around it, its owner will have a hard time selling it for its true value.

• Kitchen Remodeling. A major remodeling would include new cabinets, counter tops, flooring, appliances, and decorating. Since the kitchen can be a major concern of the homemaker, remodeling it can add greatly to the value of the house. As much as 75 to 80 percent of remodeling costs can be recovered in resale. Remodeling the kitchen, therefore, is a good investment, even if the home is sold soon afterwards.

• Bathroom Remodeling. This involves new walls, floors, and fixtures. The remodeled bath will not only beautify but also update a home. However, in the short term, only about 30 percent of the cost will be recaptured in resale.

• Adding a Full Bath. A full bath includes a tub and shower, toilet, vanity, sink, and medicine cabinet. An extra bath is an important addition, particularly when there are three bedrooms and only one existing bath. It is a very strong selling feature. From 80 to 100 percent of the cost can be recaptured when selling the home. Fig. 50-8.

• Windows, and Doors. New windows and doors can improve the insulation of a home and reduce heating and air conditioning costs. They can also increase the security of the home. This improvement is an excellent investment for someone planning to keep a home. However, only 25 to 30 percent of the cost will be recovered if the house is sold in a very short time.

• Siding. Siding usually includes added insulation plus aluminum or plastic siding. Siding will improve the appearance of a house, make it more maintenance free, and increase its energy efficiency. This is a

good investment even in the short term since up to 70 percent can be recovered.
● Insulation. Adding insulation to levels of R-42 in the attic and R-19 in the walls will improve energy efficiency. Such a project will pay for itself in three to five years. It is a good investment for anyone planning to live in the home for four to five years or more. However, in the shorter term, only about 50 percent will be recovered in the sale price.
● Patio and Deck. A patio or a deck can be added to the house to improve its appearance and livability. They are of greater value in climates where they can be used more often. They can be good investments in the sun belt states. In colder climates, only about 40 to 50 percent of the cost can be recaptured.

ADDING A ROOM

Three steps are necessary when adding a room to an existing house.
1. Do the necessary research to make sure the addition is practical.
2. Do the design to make sure that the addition is going to blend with the existing structure.
3. Do the actual construction.

Research

Check the building codes to be sure the addition will meet the site requirements for your property. The building code will tell you how close the house can be to the property line and how high the building can be. Check the utility lines. They must be available to the addition, and the new construction must not be built over them.

Consider your family's needs carefully during this initial stage. Be sure that the proposed addition will meet all the needs for which it is intended.

Consider the layout of the existing house when planning a new room. In general, when adding to a home, dens should be near the living room or master bedroom. Family rooms should be placed by the

kitchen or dining area. Bedrooms should be off by themselves, not close to the family room. Avoid awkward floor plans, such as a bedroom that opens onto a kitchen or dining area.

It is also important to check structural aspects of your home such as the foundation, framing, roof, and drainage system. You will need to consider such factors in your design plan. After all of these steps are taken, you are ready to do the designing of the addition.

Design

If a set of plans for the house itself is available, they should be used as the basis for the room addition. Determine the scale of the existing plan; then use squared paper that will fit the same scale. Attach the squared paper to the existing plans where the addition is to take place. Draw to scale a floor plan of the addition, indicating the location of windows, doors, and all other structural details. Once the floor plan is complete, you must design the roof line and the elevation. The roof line should be influenced by the current roof line of your house. The shape (hip, gable, shed) and the pitch of the roof should either duplicate the present roof or harmonize with it. Fig. 50-9. The elevations will show the windows, doors, and other external features. Determine whether the house will be built with the conventional framing pattern or with panelized wall construction. Finally, using the completed design, develop a bill of materials.

Construction

The first step after the plans are approved is to get the necessary building permit. Then the foundation should be planned. It may be slab-on-ground, crawl space, or full basement. The masonry and construction work is done in much the same way as in building a new home. Fig. 50-10.

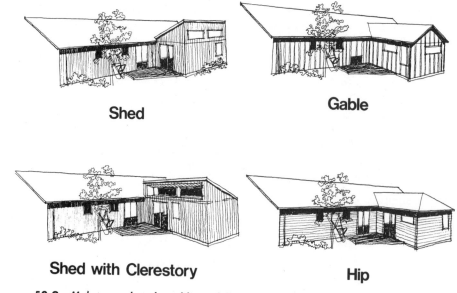

Shed

Gable

Shed with Clerestory

Hip

50-9. *Make sure that the addition follows the same roof line as the home itself.*

50-10. *This addition has one wall up even before construction starts. The extension of the existing roof can be done with conventional 2 × 6 rafters. The glass wall opening to the deck makes the project ideal for indoor/outdoor living.*

REMODELING TECHNIQUES

Before doing any remodeling, make a thorough review of the structure. Make sure that you have carefully planned for the work to be done and that the new materials needed will be available. For example, before you begin to remove a wall, be sure to check the location of utilities such as heating, wiring, and plumbing. You could run into a problem that would require moving those lines before remodeling could proceed.

Before beginning any remodeling, review the units in this book on new construction. For windows and doors to be replaced, reread Units, 37, 38, and 47.

Demolition Process

The major difference between new construction and remodeling is the demolition. In demolition, the construction process is reversed. Materials are removed, one piece at a time, until the rebuilding can begin. The general procedure is to remove first all hardware, then the trim, then the millwork, such as doors, windows, and flooring, and finally the wall covering and structural parts. If utilities must be removed and replaced, other subcontractors will become involved.

Be sure to consider whether the old materials can be reused. If any original material, such as the trim, can be used again, then demolition must be done with special care. In removing trim around a window or door, for example, start at one corner to lift the trim away from the wall. When the trim is loose, tap it back with a soft-faced mallet. This will usually expose the finishing nails. These can be then removed one at a time with a hammer or heavy pliers. By using this technique, you will not crack the trim as you pull it away from the frame.

Replacing Windows

Doors and windows are often replaced to increase the energy efficiency of a house and improve its appearance. Before you begin, measure carefully the width and height of the opening of the door or window. Find out if you can obtain replacements that will fit into the existing opening.

REMOVING OLD WINDOWS

The basic procedures for removing all types of windows are about the same. Since double-hung windows are most common, they will be used as an illustration. Fig. 50-11. Run a razor blade around the casing (trim) to make a clean break from the paint or wallpaper. Remove the inside stops around the window with a pry bar or wide chisel. Remove the interior casing around the window. Raise the lower sash and remove the apron and stool. Now check carefully to see how the window unit is held in place. In older windows, the jambs are nailed directly into the studs. In newer types that were installed as a total unit, the window is held in place by nails driven into the outside casing. If the older type is being removed, drive the nails through the jamb using a nail set and hammer. Sometimes it may be necessary to chisel the wood around the head and pull the nail out with a hammer of pliers. It may also be necessary to use a thin hacksaw blade to cut thin nails between the jamb and studs. If the nails are in the exterior casing, drive them through with a nail set and hammer or remove them from the outside. Now remove the window unit from the outside of the house.

INSTALLING A WINDOW

Select the correct size and kind of window. Often a different style is selected. Make sure the unit will fit the existing opening. Install the window unit from the outside, and nail it in place. Figs. 50-12 and 13. Calk around the outside of the unit. Fig. 50-14. From the inside, place insulation around the window between the jambs and the rough opening. Then add the inside trim (casing).

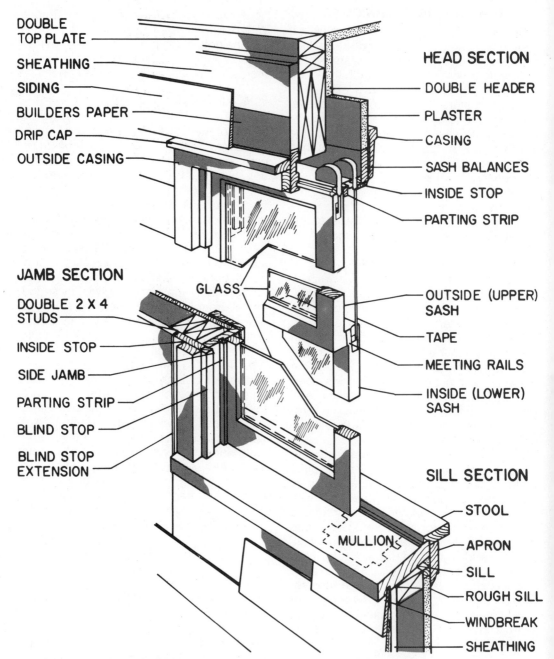

DOUBLE TOP PLATE
SHEATHING
SIDING
BUILDERS PAPER
DRIP CAP
OUTSIDE CASING

HEAD SECTION
DOUBLE HEADER
PLASTER
CASING
SASH BALANCES
INSIDE STOP
PARTING STRIP

JAMB SECTION
DOUBLE 2 X 4 STUDS
INSIDE STOP
SIDE JAMB
PARTING STRIP
BLIND STOP
BLIND STOP EXTENSION

GLASS

OUTSIDE (UPPER) SASH
TAPE
MEETING RAILS
INSIDE (LOWER) SASH

SILL SECTION
STOOL
APRON
SILL
ROUGH SILL
WINDBREAK
SHEATHING

MULLION

50-11. *Parts of a double-hung window. There are two sash that slide up and down in the grooves of the frame. The sash can be opened from either the top or the bottom. In some types, the sash can be removed for cleaning. Balances at the sides support the sash and make them easier to raise.*

50-12. *Installing a new window unit.*

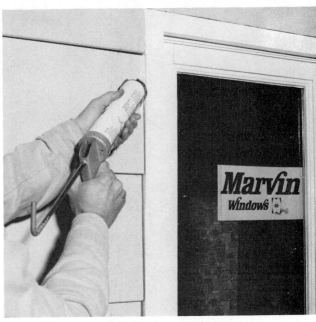

50-14. *Calking seals against the weather.*

50-13. *The unit is held in place with nails.*

Replacing Doors

There are two common ways of replacing old doors. You can replace the door only using the existing door frame, or you can replace the entire door and frame with a prehung door. Some prehung doors are designed to fit into the existing door frame.

FITTING AND HANGING A WOOD DOOR

1. Remove the old door.
2. Check jambs and stop; they should be square and plumb. Before trimming a door to proper height, measure the inside of the jamb from floor to header. Do not go by the measurements of the door being replaced, as they may not be accurate. Measure both the height and width in several places, as the jamb may be out of square. Allow room at the bottom for the sill or carpet. No more than ¾″ should be trimmed from the width with a maximum of ⅜″ from each side.

50-15. *Cutting the gains for hinges.*

3. Use a fine tooth saw for trimming the door to proper height, across the grain of the stiles. A wood plane can be used for minor trimming to width; otherwise a circular saw can be used to achieve the desired bevel. The best rule is to allow approximately 3/16" clearance under perfectly dry conditions. (Weatherstripping may require more trimming, depending upon the type used, but do not trim for weatherstripping until the door is hung and operational.)

4. Use three hinges on doors less than 7' high, four on doors over 7'. Position the top hinge 7" from the top and the bottom hinge 11" from the bottom. Use a sharp wood chisel (or a special tool made for this purpose) to cut insets for the hinges in the door stile and door jamb. Fig. 50-15. Check frequently that the hinges are installed in a straight line to prevent binding. Hang the door on its hinges and make final trim adjustments for a precise fit before installing the lockset.

50-16. *Marking the location for the lockset.*

5. When installing a lockset, follow the manufacturer's instructions and use the template provided for drilling and cutting holes. Fig. 50-16. Bore from one side of the door until the point of the hole saw protrudes from the opposite side. Then reverse the procedure and complete boring the holes from the other side. Fig. 50-17. Bore the holes from the edge of the door. Fig. 50-18. Use a sharp wood chisel to cut away wood. Leave at least 1" of solid wood in the stile behind the lockset to maintain the strength of the door. Take care in installing the striker plate accurately.

INSTALLING A PREHUNG DOOR

Prehung door units come in many sizes and styles. Indoor units are all wood construction. Exterior door units may have a wood frame with a steel door that has an insulated core. When selecting the unit, consider its type, size, and design. Consider also the swing of the door and the kind of hardware needed. To install a prehung door, remove the casing, the old door, and the door frame. (Some units are

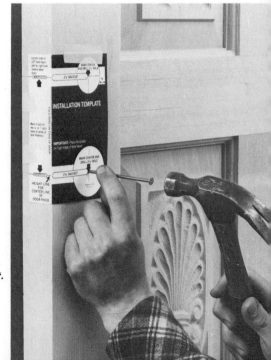

50-17. Boring holes for the lockset.

designed to fit into the old door frame, so check for this before buying the door.) Replace with a prehung door following the instructions in Unit 47.

Installing a New Door in an Existing Wall

Before beginning the job, check the following:
- Will it be an interior or exterior door?
- Will the door be installed in a bearing or non-bearing wall? All outside walls are bearing walls and one or more inside walls may be also. Usually if only one or two studs need to be removed, it will not be necessary to shore up the wall. However, shoring will always be necessary if a double-hung door is installed.
- What size will the door be? Check Units 36 and 47 for door sizes. The door opening is determined by the overall size of the door frame and the location of the studs. Usually the smallest rough opening must be 4" wider and 2" to 4" higher than the door frame size.
- Where are the utilities such as wiring, plumbing, or heating ducts? Avoid cutting a door where plumbing and heating ducts might be located. A single electrical wire running through the area can be moved; an electrical outlet will present a greater problem.
- Where are the studs in the wall? You can locate the studs in several different ways. The simplest method is to use a magnetic or electronic stud finder. You can also check carefully for taped wall joints or nails in the trim to help locate the studs. Still another method is to measure either 16" or 24" from the corner of the room and tap the surface with the handle of a hammer until you hear a solid sound. Whenever possible, the door opening should start from one side of a stud so that not more than two studs will have to be removed to install a single door.

50-18. Boring holes from the edge of the door.

CUTTING THE OPENING FOR THE DOOR

1. Carefully mark the outline of the rough opening on one wall. If an outside door is to be installed, start the job on the inside wall. Check the lines with a plumb and level.

2. Cut an inspection opening near the middle of the outline. An 8" × 8" opening is about the right size. Mark this opening and drill a starting hole at each corner. Use a wallboard saw to cut the drywall at an angle so that the opening can be closed, if necessary, by replacing the orignal drywall or cutting a new piece to fit. If the wall is insulated, remove as much as possible.

3. Inspect the interior by shining a light into the opening to observe the location of the studs or any electrical wires. Another method is to use a stiff wire (a straightened wire coat hanger will do) to "fish" around the opening to locate the studs and any electric wire. If necessary, move the rough opening to the right or left a few inches so that one side of the rough opening will be next to a stud.

4. Drill starting holes in the upper corners and just above the floor trim for the rough opening. IF THERE IS ANY DANGER THAT AN ELECTRICAL WIRE RUNS THROUGH THIS AREA, TURN OFF THE APPROPRIATE CIRCUIT.

5. Cut the drywall with a drywall saw, a portable saber saw, or a reciprocating saw. If electricity is needed, use a long extension cord plugged into a live circuit. IT IS A GOOD IDEA TO USE A METAL-CUTTING SAW BLADE ON POWER SAWS. Cut along a straight line from corner to corner. If you hit a stud, tip the power saw slightly. When the cut is complete, remove the drywall in large sections, exposing the studs. Use a hand saw to cut the section of the floor trim and sole plate, including the base board and shoe. Check to see if an electric wire runs through the opening. When necessary, this wire can be spliced and run over the door frame to complete the circuit.

6. Now cut open the opposite wall in a similar manner.

7. Cut off and remove parts of the studs. This is best done by first cutting through the center of each stud. Then cut and remove each section of the studs for the opening.

8. To install an exterior door, the finished floor and subfloor must be notched out. Sometimes the joists must also be notched so that the sill will be even with the finished floor. Fig. 38-22.

9. Rough in the opening and hang the door. (See Units 38 and 47.)

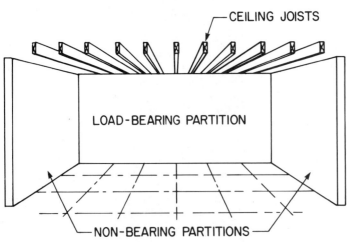

CEILING JOISTS

LOAD-BEARING PARTITION

NON-BEARING PARTITIONS

50-19. Load bearing and non-bearing partitions. A second floor load may place a load on any partition.

50-20. *The center partition will be load-bearing. Note how the joists are spliced over the girder.*

Changing Partitions

Often rooms are not the right size and it may be necessary to remove some partitions. This is not difficult if the partition is a non-bearing wall and if plumbing, electrical, and heating utilities are not involved.

All outside walls and usually the center interior partition parallel to the long side of the structure are bearing walls. To dtermine whether a wall is load-bearing, check the direction of the floor and ceiling joists. Partitions parallel to the joists are usually non-bearing walls. Fig. 50-19. However, the wall may support a second floor load, so that must be checked. In most house construction, when the second floor joists are perpendicular to the partition, the joists require support. The walls, therefore, are load bearing. An exception occurs when trusses are used instead of the traditional roof framing. If the trusses span the total width of the building, then all interior partitions are non-load bearing. If joists are spliced over a girder, then the wall below and often the wall above are bearing walls. Fig. 50-20. The same is true for ceiling joists.

To take out a bearing wall, carefully remove the trim around any doors and along the flooring and ceiling. Remove the hard-

ware and then the door. Remove the door frame. Then remove the drywall from both sides of the wall.

The correct way to remove the structural wall depends on how the wall was assembled. If the studs are nailed through the top plate and toenailed through the sole plate, use a pry bar or nail remover to take out the lower nails. Then pull the stud out from the top plate. If the wall was preassembled before erection using a nailing machine, the best approach is not to try to save the studs. Cut across the middle of each stud and remove each piece, one at a time. However, if a reciprocating saw with a metal cutting blade is available, cut the studs just below the top plate and just above the sole plate and remove the major portion of the studs, which then can be reused.

After the studs are removed, pry loose the sole plate and remove the top plate. IF THE ENDS ARE TIED TO THE SIDE WALLS, CUT THEM FLUSH WITH THE WALL. After the wall is removed, the ceiling, walls, and floor will require repair where the partition intersected them.

REMOVING PART OR ALL OF A BEARING WALL

To change partitions or to cut a large opening to a new room addition, it is often necessary to remove part or all of a bearing wall. Any load bearing wall must be shored up (temporarily supported) before it can be removed. If it is an outside wall, then shoring is needed only on the inside. If the bearing wall is on the inside, then shoring is needed on both sides of the wall. If a bearing wall is to be removed from the second floor, it is a good practice to start the shoring at the basement level. Before deciding on the shoring to use, review these definitions:

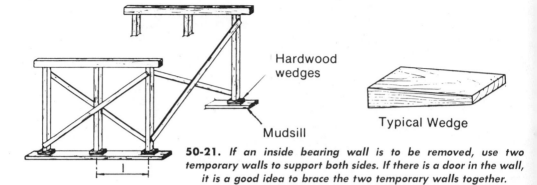

Hardwood wedges

Mudsill

Typical Wedge

50-21. *If an inside bearing wall is to be removed, use two temporary walls to support both sides. If there is a door in the wall, it is a good idea to brace the two temporary walls together.*

• Brace or bracing is a piece of wood or other material that helps to hold the total unit in place and resists weight or pressure.
• Shore is a prop placed against or beneath an object to support it.
• Shore head is a horizontal, heavy wood beam placed across a shore or shores.
• Shoring is a temporary support that is used for many purposes such as the removal of a bearing wall.
• Beam bottom or mudsill is a heavy wood beam or plank of wood which supports the shore or shores.
• Stringer is a heavy wood beam placed across the top of several shores as a horizontal support.
• Wedge is a tapered piece of wood or metal used to tighten the shoring.
There are several kinds of shoring that can be used:
• A temporary wall (called falsework) built ½" to 1" less than the ceiling height. Wedges are used to "block in place" the temporary wall so that it will give firm support to the ceiling. Fig. 50-21.
• A T-shore using wedges. Fig. 50-22.
• Jack posts (house jacks) between the shore head and the beam bottom or mudsill. Fig. 50-23.
• A patented shore device (adjustable) instead of house jacks. Fig. 50-24.
• Steel posts between the shore head and the beam bottom or mudsill.

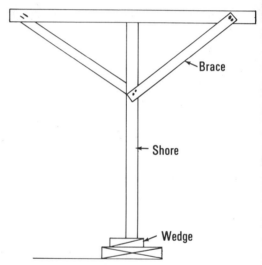

Brace

Shore

Wedge

50-22. *A T-shore can be use for cutting a small opening or for supporting large sheets of drywall or panel that are attached to the ceiling after the wall is removed.*

Erect the shoring. Make sure the shoring is supported by several joists. At times, such as when cutting an opening in the outside end wall of the house, the shoring may be parallel to the joists. In such cases, place heavy plywood on the floor so that the weight is distributed over several joists. To protect the finished floor and ceiling, place scraps of rugs or towels over and under the shoring. Locate the shoring

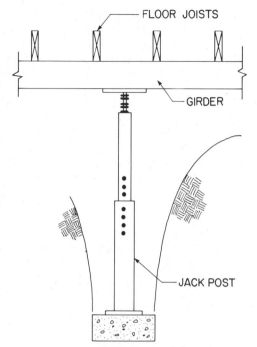

FLOOR JOISTS

GIRDER

JACK POST

50-23. *Jack posts can be used for temporary shoring. In an old house they can also be used to prop up a sagging floor.*

about 2' away from the wall so that you can work between the shoring and wall. Use wedges or adjust the shoring to bring the unit to full height for solid support.

Once the shoring is in place, remove the wall as you would a non-bearing wall. Frame in the opening using a header (either of solid lumber, built-up lumber, or a box beam) supported by trimmer studs.

REPAIRS

There are hundreds of major and minor repair jobs that need to be done to keep a housing unit in good condition. Smaller repairs can be done by the home owner using one of the many books on home repair. Some major repairs, such as those described here, can be done by a general carpenter. Other major repairs and improvements are best left to skilled specialists. Such jobs include reroofing, installing

new siding (usually of metal or plastic), replacing gutters and downspouts, and installing insulation.

Repairing Roof Leaks

When a leak develops, it must be repaired immediately. Even a small leak will cause damage such as discoloration of the ceiling and walls or stains on the finish flooring. If the leak is left unattended, the ceiling and wall will become so damp that parts or all of both will need to be replaced.

LOCATING A ROOF LEAK

It is very difficult to locate the point of leakage from the wet spot that appears on the ceiling. This is because the water will follow the rafters or trusses before dripping down on the ceiling or wall. If the attic is heavily insulated, the leakage can continue for some time before it appears on the inside of the house. It is also very difficult to locate the hole from the top of the roof.

If the rafters or trusses are exposed, the best method is to locate the holes from the

50-24. *Patented shores can be purchased or rented for remodeling.*

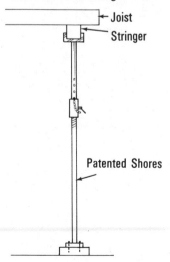

Joist

Stringer

Patented Shores

inside of the attic on a very bright day. Even a small hole will be visible. Its location should be marked by pushing a small nail or wire through to the roof surface. Leaks can also be detected during a rain storm using a bright flashlight to check the attic. First, watch where the water starts to drip on the ceiling and then work back to where the leak starts. The rafter or truss will be wet up to the place where the leak begins.

REPAIRING FLAT ROOFS

Make the repairs following the directions and the detailed steps shown in Fig. 50-25.

• Loose Felt Edges. Using a brush, clean out any dirt that may have blown under the loose felt (Fig. 1). Then, using large-head roofing nails, nail the loose felt in place. Start nailing away from the felt edge and work toward the edge, to prevent making a blister in the felt. Place the nails 1" to 1½" apart (Fig. 2). After the felt is nailed in place, cover the patched area with asphalt cement. Make sure the cement extends 1" to 1½" beyond the repair area (Fig. 3).

• Blisters in the Felt. Using a knife, cut the blister (Fig. 4). Then proceed as if you were repairing a crack, using the instructions below.

• Cracks in the Roofing. Clean out the crack and the area around the crack. Using the brush or putty knife, place a thin layer of asphalt cement over the crack. The cement should completely cover the cleaned area around the crack (Fig. 5). Cut a piece of roofing felt a little larger than the crack. Place it over the cement and press it firmly in place. Nail the edges on the felt piece, spacing the nails 1" to 1½" apart (Fig. 6). Spread another layer of cement over the felt. Make sure the cement extends 1" to 1½" beyond all edges of the felt piece (Fig. 7).

• Deteriorated or Damaged Roofing. This condition is harder to repair. Never build up or cover the old or damaged roofing with a series of felt layers. This may change the drainage pattern of the roof and create more problems. Cut out the damaged roofing in a rectangle and remove it (Fig. 8). Clean the surfaces in and around the cut-out area. Now cut pieces of felt to fit the cut-out area neatly. The number of felt strips placed should equal the number of layers of felt removed. Cut the top patch 2" to 3" larger so that it will overlap the cut-out area on all sides. Spread a thin layer of asphalt cement over the cut-out area, and press the first felt strip firmly into place (Fig. 10). Now spread a thin layer of cement over that strip. Then place and firmly press down a second felt strip. Continue placing strips this way until the cut-out area is level with the original roofing. After the cut-out area has been built up to its original level, spread a thin layer of cement over the cut-out area. Spread the cement so that it extends 3" to 4" beyond all edges of the area (Fig. 11). Place the oversized felt strip over the cement and press it firmly in place. Using broad-head roofing nails, nail the strip along all four edges (Fig. 6). Then cover the strip with cement, making sure that the cement extends 1" to 1½" beyond all edges of the strip (Fig. 7).

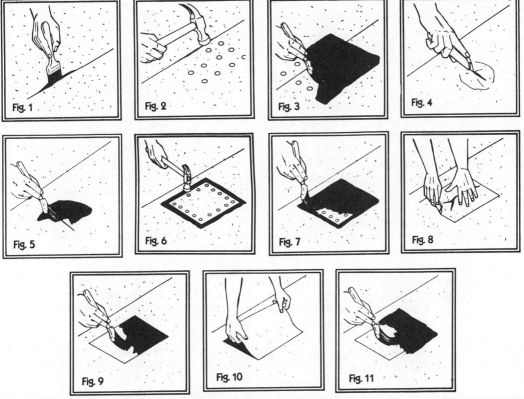

50-25. *Repairing a flat roof.*

REPAIRING SHINGLE ROOFS

Make the repairs following the directions and the detailed steps shown in Fig. 50-26.

• Wood Shingles. If the shingle is cracked, it is better to repair the crack rather than replace the shingle. If the crack is small (¼" or less), pull out loose splinters so that only the large, solid pieces remain. Check the roofing material under the shingles to determine where the nails should go. Sometimes shingles are nailed to wood slats spaced 4" or 5" apart (Fig. 1). Sometimes they are nailed to wood sheathing. After the loose splinters are removed, butt the solid pieces tightly together and nail the split shingle together with galvanized roofing nails (Fig. 1). Do not drive the heads of the nails into the shingle and damage its surface. Cover the crack fully with asphalt roofing cement. Apply a dab of cement over the nailheads (Fig. 3).

If the crack is wide, add a sheet metal patch. To do this, drive a square piece of sheet metal up under the cracked shingle (Fig. 4). Make sure that the top of the sheet metal goes beyond the upper edge of the crack. Now complete the job as described above for the small crack.

If shingles are damaged beyond repair, replace them. This can be more tricky than repairing a cracked shingle. Using a screwdriver or chisel, cut the damaged shingle into smaller pieces that can be removed by pulling with your fingers (Fig. 5). Remove the damaged shingle. Using a hacksaw blade, cut the nails off flush with the wood slats or sheathing (Fig. 6). Since shingles overlap, you may have to pry up the shingle above to get at all the nails.

Take care not to crack the good shingle. Measure the empty space and cut a replacement shingle to fit the space. Using a block of wood and hammer, drive the replacement shingle into place (Fig. 7). Nail the new shingle in place with galvanized roofing nails (Fig. 8). Apply a dab of asphalt cement to cover the nailheads.

• Asphalt Shingles. Locate the damaged area and examine the condition. If the shingle does not need replacing, simply raise the damaged or torn shingle and apply an ample amount of asphalt cement to the underside (Fig. 9). Press the shingle firmly into place and nail it down with broad-headed, galvanized or aluminum roofing nails. Fig. 35-31. (Always remember to apply asphalt cement to the nailheads.)

If the shingle needs replacing, select a strip the same as the piece to be replaced. Your asphalt roofing will usually come in shingle strips (Fig. 10). Some roofing is in single, separate shingles (Fig. 11). Raise the shingles above the damaged one. Pull the nails from the damaged shingles with the claw hammer (Fig. 12). If nails cannot be reached with a hammer, cut them off with a hacksaw blade. Remove the damaged shingle, and slip the new shingle into place (Fig. 13). Nail the new shingle in place with roofing nails, placing two nails in each tab (Fig. 14). The shingle should be "blind nailed." That is, the nails should be covered by the upper shingles when they are lowered into place. Apply a dab of asphalt cement over the nailheads and lower the upper shingles into place.

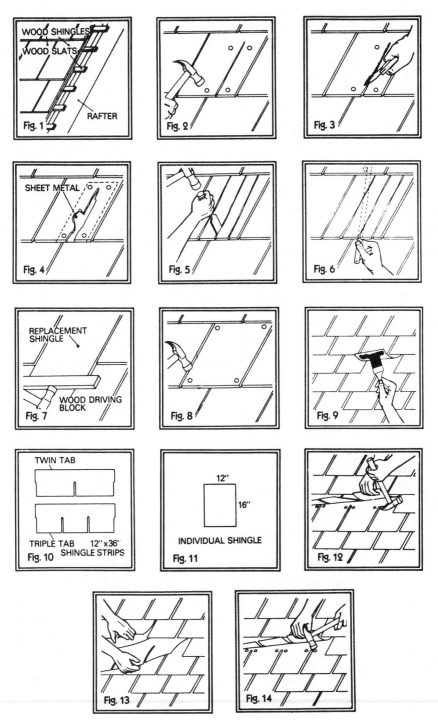

50-26. *Repairing a shingled roof.*

50-27. *Repairing siding.*

Repairing Siding

Make the repairs following the directions and the detailed steps shown in Fig. 50-27.

● Warped Boards (or Wood Shingles). Use screws, rather than nails, to straighten a warped board. First, drill guide holes for the screws into the thicker portion of the board (Fig. 1). Then drill the larger holes to countersink the screws. Pull the warped board into line by tightening the screws into the sheathing (Fig. 2). Cover the head of each screw with putty.

● Split Boards (or Wood Shingles). First, cut a piece of building paper to slip underneath the split board or shingle. Make it wide enough to fit between the in-place nails. Butt the two halves of the split shingle tightly together. Then nail both halves into place with galvanized or aluminum roofing nails. Fig. 39-27. Countersink the nailheads and cover them with putty.

● Damaged Wood Shingles. First, using the chisel and hammer, splinter the shingle into small, slender pieces (Fig. 3). Carefully remove the splintered pieces so as not to damage the remaining shingles. Pull the exposed nails with a claw hammer. Examine the building paper underneath and patch any tears or cuts with asphalt cement (Fig. 4). Slip the new shingle into position (Fig. 5). Nail the shingle in place with galvanized or aluminum shingle nails (Fig. 6).

● Damaged Wood Siding. Instead of replacing the entire siding board, it is easier to cut out the damaged portion. Using the square, mark the board for cut lines (Fig. 7). Pry up the bottom edge of the board and insert wedges underneath (Fig. 8). Using the saw, cut out the damaged portion of the siding (Fig. 9). Make the cut carefully. Don't damage siding boards above or below. Splinter the damaged portion into smaller pieces, using the hammer and chisel. Remove the pieces with a pry bar or chisel. Remove the remaining nails with the claw hammer.

Examine the building paper underneath. Patch any tears or cuts with asphalt cement. (Use asphalt cement sparingly, as too much will prevent "breathing" of the exterior.)

Measure the damaged board opening, mark the saw cut lines, and cut the replacement board to fit the opening. Slip the new board into position and drive it into place with the hammer. Hammer against a small wood block to avoid damaging the board (Fig. 10). Nail the board in place with galvanized siding nails, using the existing nailing pattern.

• Damaged Asbestos Shingles. Remove the damaged shingle by simply shattering it with the hammer. If the shingle is not brittle enough to shatter, splinter it into pieces. Remove the shingle pieces and the exposed nails. Drill the nail holes in the new shingle at its lower edge. Position the holes as they were in the old shingle. Patch any tears or cuts in the exposed building paper with asphalt cement (Fig. 4). Slip the new shingle into position (Fig. 5) and nail it in place with galvanized or aluminum shingle nails (Fig. 6).

QUESTIONS

1. What features of a house should you examine during an inspection to see if the house is worth remodeling?

2. What does a leaning roof ridge often indicate about a house?

3. In what three areas of a home can living space often be added without increasing the size of the structure?

4. What two types of remodeling generally prove to be the best investments when a house is resold?

5. What are the three general steps in adding a room to a house?

6. What is the major difference between remodeling and new construction?

7. Why are old doors and windows often replaced?

8. What are two different ways for replacing an old door?

9. Is it necessary to use shoring when installing a single door in an existing wall?

10. How do you determine the difference between a bearing and a non-bearing wall?

11. Briefly explain the process of removing a non-bearing wall.

12. Name three devices that can be used to shore up a bearing wall.

13. Describe two methods of finding a leak in a roof.

14. Explain how to repair wood siding.

Section IX

RELATED ACTIVITIES

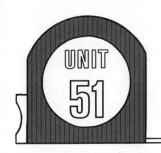

UNIT 51

Chimneys and Fireplaces

Proper construction of chimneys and fireplaces is essential for safe, efficient operation. Local codes vary slightly, so be sure to consult and follow them. It is recommended that chimneys and fireplaces be designed and the construction supervised by someone experienced in this field of work. The chimney and the fireplace must be carefully built in order to be free of fire hazards, and it is desirable to have them in harmony with the architectural style of the house. Fig. 51-1.

CHIMNEYS

All fireplaces and fuel-burning equipment such as stoves and furnaces require some type of chimney. Fig. 51-2. Chimneys are generally constructed of masonry units supported on a suitable foundation. Lightweight, prefabricated chimneys that do not require masonry protection or concrete foundations are now accepted for certain uses by fire underwriters. Make certain, however, that they are approved and listed by Underwriters' Laboratories, Inc. The chimney must be designed and built so that it produces sufficient draft to supply an adequate quantity of fresh air to the fire and to expel smoke and harmful gases given off by the fire or fuel-burning equipment.

The greater the difference in temperature between chimney gases and outside atmosphere, the better the draft. Thus an interior chimney will have better draft because the masonry retains heat longer.

Chimney Construction

FLUE SIZE

The flue is the passage in the chimney through which the air, gases, and smoke travel. Proper construction of the flue is important. Its size (area), height, shape, tightness, and smoothness determine the effectiveness of the chimney in producing adequate draft and in expelling smoke and gases.

Follow the manufacturer's specifications for fuel-burning equipment. These specifications usually include the requirements for equipment supplied by the manufacturer.

CHIMNEY HEIGHT

The height of the chimney is also an important factor in providing sufficient draft. The height of a chimney above the roofline usually depends upon its location in relation to the ridge. The top of the extending flue liners should not be less than 3' above flat roofs and at least 2' above a roof ridge or any raised part of a roof within 10' of the chimney. Fig. 51-3. A hood should be provided if a chimney cannot be built high enough above a ridge to prevent trouble from irregular air currents caused by wind being deflected from the roof. The open ends of the hood should be parallel to the ridge. C, Fig. 51-4 (Page 1002).

CHIMNEY FOUNDATIONS

The chimney is usually the heaviest part of a building. It must rest on a solid foundation to prevent uneven settling of the building.

Concrete footings must be designed to distribute the load over an area wide enough to avoid exceeding the safe load-bearing capacity of the soil. The footings should extend at least 6" beyond the chimney on all sides and should be 8" thick for

999

51-1. *Three variations of contemporary fireplace styling.*

one-story houses and 12″ thick for two-story houses having basements.

Chimneys in frame buildings should usually be built from the ground up. They may also rest on the building foundation or basement walls if the walls are of solid masonry 12″ thick and have adequate footings. For houses with basements, the footings for the walls and fireplace are usually poured together and at the same elevation. If there is no basement, footings for an exterior chimney are poured on solid ground below the frostline.

MORTAR

Brickwork around chimney flues and fireplaces should be laid with cement mortar. It is more resistant to the action of heat and flue gases than lime mortar.

A good mortar to use in setting flue linings and all chimney masonry, except firebrick, consists of 1 part portland cement, 1 part hydrated lime (or slaked-lime putty), and 6 parts clean sand, measured

1000

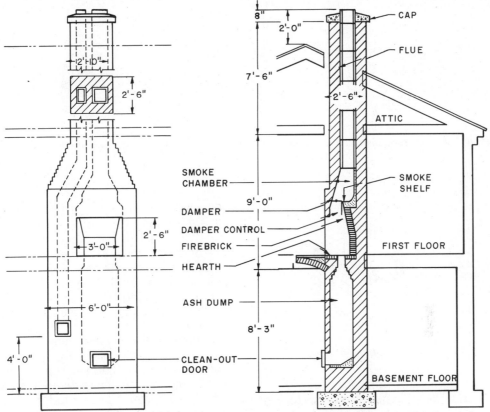

Dimension labels on diagram: 2'-10", 2'-6", 3'-0", 2'-6", 6'-0", 4'-0", 8", 2'-0", 7'-6", 2'-6", 9'-0", 8'-3"

SMOKE CHAMBER

DAMPER

DAMPER CONTROL

FIREBRICK

HEARTH

ASH DUMP

CLEAN-OUT DOOR

CAP

FLUE

ATTIC

SMOKE SHELF

FIRST FLOOR

BASEMENT FLOOR

51–2. *This chimney is designed to serve the house-heating unit and one fireplace.*

by volume. Firebrick should be laid with fireclay.

FLUE LINING

The size of the chimney depends on the number of flues, the presence of a fireplace, and the design of the house. The house design may include a room-wide brick or stone fireplace wall which extends through the roof. While only two or three flues may be required for heating units and fireplaces, several "false" flues may be added at the top for appearance. Each fireplace should have a separate flue and, for best performance, flues should be separated by a 4" wide brick spacer (wythe) between them. Fig. 51-5.

Chimneys are sometimes built without flue lining to reduce cost, but those with lined flues are safer and more efficient. Lined flues are recommended for brick

chimneys. When the flue is not lined, mortar and bricks directly exposed to the action of flue gases disintegrate. This disintegration plus that caused by temperature changes can open cracks in the masonry, which will reduce the draft and increase the fire hazard.

51–3. *The top of the flue lining should be at least 2' above the ridge line.*

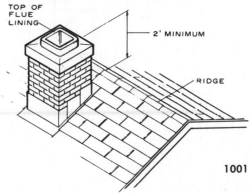

TOP OF FLUE LINING

2' MINIMUM

RIDGE

1001

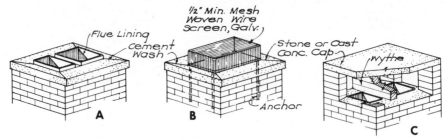

51-4. *Chimney top construction: A. Standard chimney top. B. Spark arrester or bird screen. C. Hood to keep out the rain.*

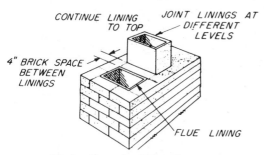

51-5. *Flues should be 4" apart.*

51-6. *Flue lining should extend at least 8" below the smoke pipe thimble. Note also the soot pocket and the cast-iron cleanout access door.*

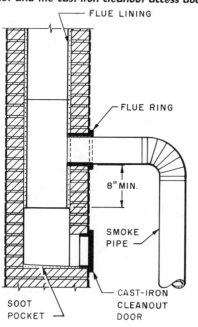

Flue lining can be omitted if the chimney walls are made of reinforced concrete at least 6" thick or of unreinforced concrete or brick at least 8" thick. However, the cost of the extra brick or masonry and the labor involved are most likely greater than the cost of flue lining. Furthermore, a well-installed flue lining will result in a safer chimney.

Rectangular fireclay flue linings or round vitrified (glazed) tile can normally be used in all chimneys. Vitrified tile or a stainless steel lining is usually required for gas-burning equipment. Local codes outline specific requirements.

Rectangular flue lining is made in 2' lengths and in sizes of 8" × 8", 8" × 12", 12" × 12", 12" × 16", and up to 20" × 20". Wall thicknesses of the flue linings vary with the size of the flue. The smaller sizes have a 5/8" thick wall. The larger sizes vary from 3/4" to 1 3/8" in thickness.

Vitrified tiles, 8" in diameter, are most commonly used for the flues of heating units, although larger sizes are also available. This tile has a bell joint.

Each length of lining should be set in cement mortar with the joint struck smooth on the inside. The brick should then be laid around it. If the lining is slipped down after several courses of brick have been laid, the joints cannot be filled and leakage will occur. In masonry chimneys with walls less than 8" thick, there should be space between the lining and the chimney walls. This space should not be filled with mortar. Use only enough mortar to make good joints and to hold the flue lining in position.

Unless it rests on solid masonry at the bottom of the flue, the lower section of lining must be supported on at least three sides by brick courses projecting to the inside surface of the lining. The lining should extend to a point at least 8" below the smoke pipe thimble. Fig. 51-6. In fireplaces, the flue liner should start at the top of the throat and extend to the top of the chimney.

Flues should be as nearly vertical as possible. If a change in direction is necessary, the angle should never exceed 45°. An angle of 30° or less is better because sharp turns set up eddies (irregular air currents) which affect the motion of smoke and gases. Where a flue does change directions, the lining joints should be made tight by mitering or cutting equally the ends of the adjoining sections. Fig. 51-7. Cut the lining before it is built into the chimney. If cut after, it may break and fall out of place. To cut the lining, stuff a sack of damp sand into it and then tap a sharp chisel with a light hammer along the desired line of cut.

When laying lining and brick, draw a tight-fitting bag of straw up the flue as the work progresses to catch material that might fall and block the flue.

METAL FLUES

Gas-fired house heaters and built-in unit heaters can be connected to metal flues instead of to a masonry chimney. Fig. 51-8. The flues should be made of corrosion-resistant metal not lighter than 20 gauge. They should be properly insulated with asbestos or other fireproof material that complies with the recommendations of Underwriters' Laboratories, Inc. The flues must extend through the roof.

CHIMNEY WALLS

Walls of chimneys that have lined flues and are not more than 30' high should be at least 4" thick if made of brick or reinforced concrete and at least 12" thick if made of stone. A minimum thickness of 8" is recommended for the exterior wall of a chimney exposed to the weather.

51-7. *Offsetting a chimney. For structural safety, the amount of offset must be limited. The centerline of the upper flue (XY) should not fall beyond the center of the lower-flue wall. A. Start to offset the left wall of an unlined flue two brick courses higher than the right wall so that the area of the sloping section will not be reduced after plastering. B. Method of cutting a flue lining to make a tight joint.*

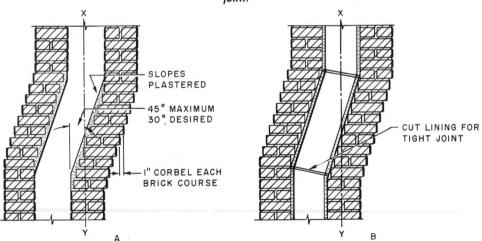

The flue sizes are made to conform to the width and length of a brick so that full-length bricks can be used to enclose the flue lining. Thus an 8″ × 8″ flue lining (about 8½″ × 8½″ in outside dimensions) with the minimum 4″ thickness of surrounding masonry will use six standard bricks for each course. A, Fig. 51-9. An 8″ × 12″ flue lining (8½″ × 13″ in outside dimen-

sions) will be enclosed by seven bricks at each course. B, Fig. 51-9. A 12″ × 12″ flue (13″ × 13″ in outside dimensions) will be enclosed by eight bricks, and so on. C, Fig. 51-9.

Brick chimneys that extend up through the roof may sway enough in heavy winds to open up mortar joints at the roof line. Openings to the flue at that point are dangerous because sparks from the flue may start fires in the woodwork or roofing. A good practice is to make the upper walls 8″ thick by starting to offset the bricks at least 6″ below the underside of roof joists or rafters. Fig. 51-10.

Chimneys may contain more than one flue. Building codes generally require a separate flue for each fireplace, furnace, or boiler. If a chimney contains three or more lined flues, each group of two flues must be separated from the other flues by brick divisions, or wythes, at least 3¾″ thick. Fig. 51-11. Two flues grouped together without a dividing wall should have the lining joints staggered at least 7″, and the joints must be completely filled with mortar. If a chimney contains two or more unlined flues, the flues must be separated by a well-bonded wythe at least 8″ thick.

SOOT POCKET AND CLEANOUT

A soot pocket and cleanout are recommended for each flue. Fig. 51-6. Deep soot pockets permit the accumulation of too much soot, which may catch fire. Therefore the pocket should be only deep enough to permit the installation of a cleanout door below the smoke pipe connection. The lower part of the chimney from the bottom of the soot pocket to the base of the chimney is filled with solid masonry.

The cleanout door should be made of cast iron. It should fit snugly and be kept tightly closed to keep air out. A cleanout should serve only one flue. If two or more flues are connected to the same cleanout, air drawn from one to another will affect the draft in all the flues.

51–8. Metal flues are frequently used for factory-built fireplaces.

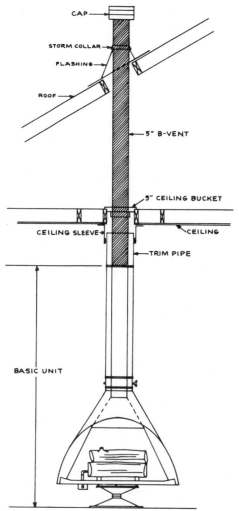

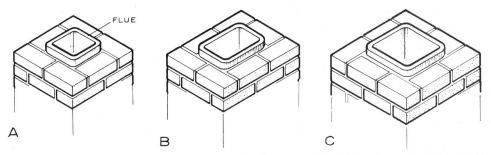

51-9. *Brick and flue combinations: A. 8″ × 8″ flue lining. B. 8″ × 12″ flue lining. C. 12″ × 12″ flue lining.*

SMOKE PIPE

No range, stove, fireplace, or other equipment should be connected to the flue for the central heating unit. In fact, as stated previously each unit should be connected to a separate flue. If there are two or more connections to the same flue, fires may occur from sparks passing into one flue opening and out through another flue opening.

Smoke pipes connect a stove, furnace, or metal fireplace to a flue. They must be correctly installed and connected to the chimney for safe operation. A smoke pipe should enter the chimney horizontally and should not extend into the flue. Fig. 51-6. The hole in the chimney wall should be lined with fireclay, or metal thimbles should be tightly built into the masonry. Metal thimbles or flue rings are available in diameters of 6″, 7″, 8″, 10″, and 12″, and in lengths of 4½″, 6″, 9″, and 12″. To make an airtight connection where the pipe enters the wall, install a closely fitting collar and apply boiler putty, good cement mortar, or stiff clay.

A smoke pipe should never be closer than 9″ to woodwork or other combustible material. If it is less than 18″ from woodwork or other combustible material, cover at least that half of the pipe nearest the woodwork with fire-resistant material. Commercial fireproof pipe covering is available.

51-10. *Corbeling a chimney (offsetting the bricks) to provide 8″ walls for the section above the roofline.*

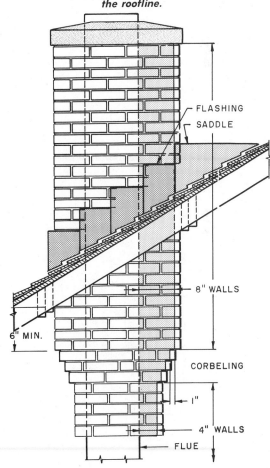

If a smoke pipe must pass through a wood partition, the woodwork has to be protected. Either cut an opening in the partition and insert a galvanized-iron, double-wall ventilating shield at least 12"

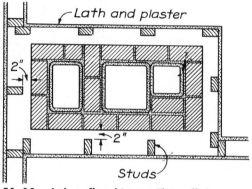

51–11. *A three-flue chimney. The walls between the flues are bonded by staggering the joints of successive courses. Wood framing should be at least 2″ from the masonry.*

larger than the pipe, or install at least 4" of brickwork or other incombustible material around the pipe. Fig. 51-12.

Smoke pipes should never pass through floors, closets, or concealed spaces or enter the chimney in the attic.

INSULATION

No wood should be in contact with the chimney. Leave a 2" space between the chimney walls and all wooden beams or joists. Fig. 51-11. Solid masonry walls 8" thick can be within ½" of the chimney masonry.

Fill the space between wall and floor framing with porous, nonmetallic, incombustible material, such as loose cinders. Fig. 51-13. Do not use brickwork, mortar, or concrete. Place the filling before the floor is laid. It not only forms a firestop but also prevents the accumulation of shavings or other combustible material. Flooring and subflooring can be laid within ¾" of the masonry.

51–12. *A wood partition must be protected when a smoke pipe passes through it.*

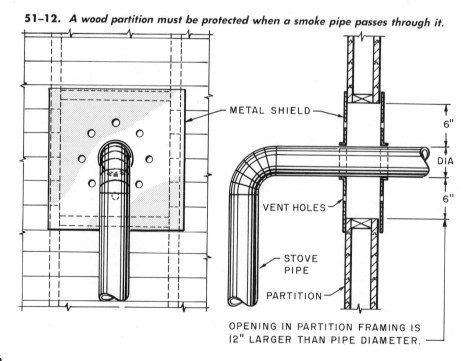

A coat of cement plaster should be applied to chimney walls that will be surrounded by wood partitions or other combustible materials. Wood studding, furring, or lathing should be at least 2" from chimney walls. Plaster can be applied directly to the masonry, or to metal lath laid over the masonry. However, this is not recommended because settlement of the chimney may crack the plaster.

Sometimes baseboards are fastened to plaster that is in direct contact with the chimney wall. In that case, install a layer of fireproof material, such as asbestos, at least 1/8" thick between the baseboard and the plaster. Fig. 51-13.

ROOF CONNECTION

Where the chimney passes through the roof, a 2" clearance between the wood framing and the masonry is required for fire protection. This clearance will also permit expansion due to temperature changes, settlement, and slight movement during heavy winds.

Chimneys must be flashed and counterflashed to make the junction with the roof watertight. Fig. 51-14. When the chimney is located on the slope of a roof, a saddle (sometimes called a cricket) is built high enough to shed water around the chimney. Fig. 51-15. (See Unit 32, p. 474.) Corrosion-resistant metal, such as copper, aluminum, zinc, or lead, should be used for flashing. Galvanized or tinned sheet steel will require occasional painting.

Chimney Top Construction

To prevent moisture from entering between the brick and flue lining, a concrete cap is usually poured over the top course of brick. Precast or stone caps with a cement wash are also used. The flue lining extends at least 4" above the cap or top course of brick and is surrounded by at least 2" of cement mortar. The mortar is finished with a straight or concave slope to direct air currents upward at the top of the flue and to drain water from the top of the chimney. Fig. 51-4A and 51-16.

Hoods are used to keep rain out of chimneys and to prevent downdraft due to nearby buildings, trees, or other objects. Common types are the arched brick hood and the flat stone or cast concrete cap. If the hood covers more than one flue, it should be divided by wythes so that each flue has a separate section. The area of the hood opening for each flue must be larger than the area of the flue. C, Fig. 51-4.

Spark arresters are recommended when burning fuels that give off sparks, such as sawdust, or when burning paper or other

51–13. *Wood floor joists and the baseboard must be insulated at a chimney.*

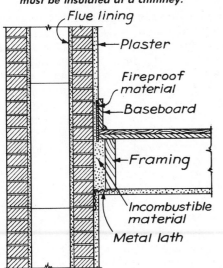

Flue lining
Plaster
Fireproof material
Baseboard
Framing
Incombustible material
Metal lath

51–14. *Flashing at a chimney located on a ridge.*

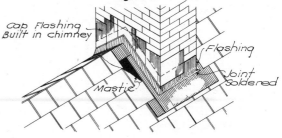

Cap Flashing Built in chimney
Flashing
Joint Soldered
Mastic

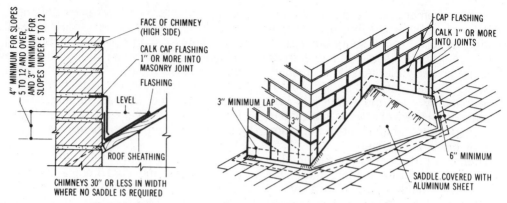

51–15. *Flashing a chimney that projects through a sloping roof.*

trash. B, Fig. 51-4. Spark arresters may be required when chimneys are on or near combustible roofs, woodland, lumber, or other combustible material. They are not recommended when burning soft coal because they may become plugged with soot.

Spark arresters do not entirely eliminate the discharge of sparks. However, if properly built and installed, they greatly reduce the hazard. They should be made of rust-resistant material and should have screen openings not larger than $5/8''$ nor smaller than $5/16''$. They should completely enclose the flue discharge area and must be securely fastened to the top of the chimney.

Prefabricated Chimneys

Many new types of lightweight chimneys that require no masonry protection nor

51–16. *A chimney cap.*

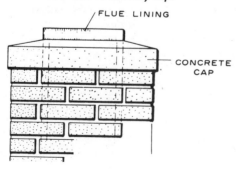

concrete footings are available. If a heating system is such that this type of chimney can be used, the following precautions should be observed.

● Make sure the model has been tested and listed by the Underwriters' Laboratories, Inc.

● Install the chimney in strict accordance with the manufacturer's instructions.

● Make sure the unit conforms to the local building code.

FIREPLACES

A fireplace is a luxury except in mild climates or in locations where other heating systems are not available. Since an ordinary fireplace has an efficiency of only about 10%, its value as a heating unit is low compared to its decorative value and to the cheerful and homelike atmosphere it creates. The heating efficiency of a fireplace can be materially increased by the use of a factory-made metal unit that is incorporated in the fireplace structure. This unit allows air to be heated and circulated throughout a room, separate from the direct heat of the fire.

Design

Varied fireplace designs are possible. Figs. 51-17 and 51-18. A fireplace should harmonize in detail and proportion with the room in which it is located, but safety and utility should not be sacrificed for appearance. A fireplace should not be located near doors.

Fireplace openings are usually made from 2′ to 6′ wide. The kind of fuel to be used will determine the width. For example, if cordwood (4′ long) is to be cut in half and burned, an opening 30″ wide is desirable. However, if coal is to be burned, a narrower opening can be used.

The height of the opening can range from 18″ for an opening 2′ wide to 28″ for one that is 6′ wide. The higher the opening, the more chance of a smoky fireplace.

In general, the wider the opening, the greater the depth. A shallow opening throws out relatively more heat than a deep one, but holds smaller pieces of wood. The choice, then, is between a deeper opening that holds larger, longer-burning logs and a shallower one that takes smaller pieces of wood but throws out more heat. In small fireplaces, a depth of 12″ may permit good draft, but a minimum depth of 16″ is recommended to lessen the danger of firebrands falling out on the floor. Suitable screens should be placed in front of all fireplaces to minimize the danger from brands and sparks.

Because of the reduced flue height, second-floor fireplaces are usually made smaller than first-floor ones.

51–17. A fireplace installed and trimmed becomes an attractive and functional part of the home.

Construction

Fireplace construction is basically the same regardless of design. The construction of a typical fireplace and recommended dimensions for essential parts or areas of fireplaces of various sizes are shown in Fig. 51-19 (Page 1012).

FOOTINGS

Foundation and footing construction for chimneys with fireplaces is similar to that for chimneys without fireplaces. (Page 999.) Be sure the footings rest on good firm soil below the frostline.

HEARTH

The hearth consists of two parts: the front, or finish, hearth and the back hearth, under the fire. Because the back hearth must withstand intense heat, it is built of heat-resistant materials, such as firebrick. The front hearth is simply a precaution against flying sparks. While it must be noncumbustible, it need not resist intense prolonged heat. The hearth should project at least 16″ from the front of the fireplace and should be 24″ wider than the fireplace opening. (That is, the hearth should be 12″ wider than the fireplace on each side.)

The hearth can be flush with the floor so that sweepings can be brushed into the fireplace, or it can be raised. Raising the hearth to various levels and extending its length as desired is presently common practice, especially in contemporary design. If there is a basement, a convenient

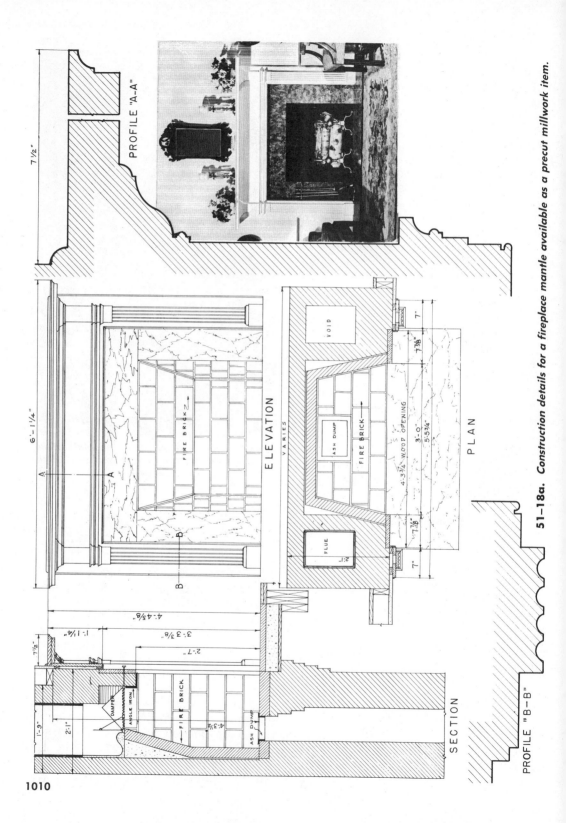

PROFILE "A-A"

PROFILE "B-B"

ELEVATION

PLAN

SECTION

VARIES

VOID

FLUE

ASH DUMP

FIRE BRICK

FIRE BRICK

FIRE BRICK

DAMPER

ANGLE IRON

ASH DUMP

7½"

6'-1¼"

4'-4⅝"

3'-3⅜"

2'-7"

1'-1¼"

7½"

1'-9"

2'-1"

4'-3½"

4'-3½"

4'-3¾" WOOD OPENING

3'-0"

5'-5¾"

7"

7⅞"

7⅞"

7"

2:1

2:1

51-18a. *Construction details for a fireplace mantle available as a precut millwork item.*

1010

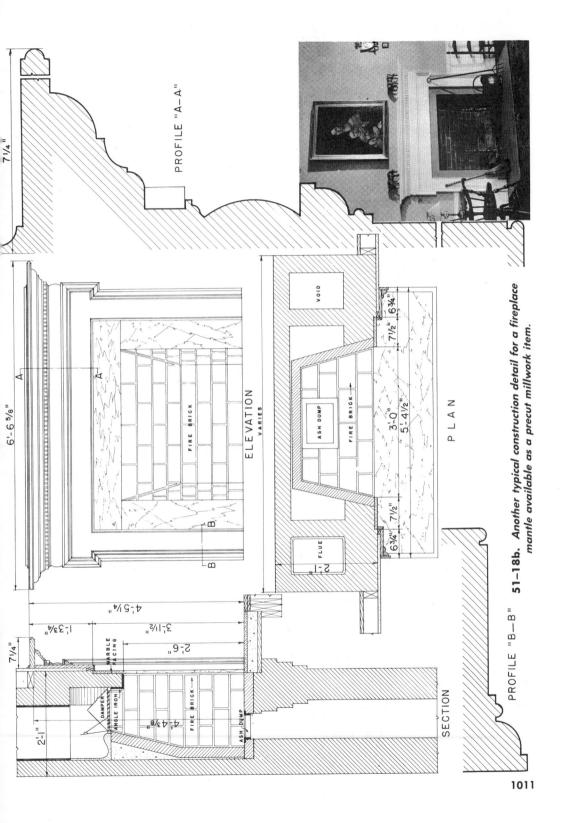

PROFILE "A-A"

7¼"

6'-6⅝"

FIRE BRICK

A

A

B

B

ELEVATION

VARIES

VOID

ASH DUMP

FIRE BRICK

FLUE

2'-1"

6¾"

7½"

7½"

6¾"

3'-0"

5'-4½"

PLAN

7¼"

1'-3¾"

4'-5¼"

3'-11½"

MARBLE FACING

2'-6"

DAMPER

ANGLE IRON

FIRE BRICK

4'-4⅜"

ASH DUMP

2'-1"

SECTION

PROFILE "B-B"

51-18b. *Another typical construction detail for a fireplace mantle available as a precut millwork item.*

51–19. *Fireplace construction details.*
Recommended Dimensions For Fireplaces And Size of Flue Lining Required

Size of fireplace opening						Size of flue lining required	
Width w Inches	Height h Inches	Depth d Inches	Minimum width of back wall c Inches	Height of vertical back wall a Inches	Height of inclined back wall b Inches	Standard rectangular (outside dimensions) Inches	Standard round (inside diameter) Inches
24	24	16–18	14	14	16	8½ x 8½	10
28	24	16–18	14	14	16	8½ x 8½	10
30	28–30	16–18	16	14	18	8½ x 13	10
36	28–30	16–18	22	14	18	8½ x 13	12
42	28–32	16–18	28	14	18	13 x 13	12
48	32	18–20	32	14	24	13 x 13	15
54	36	18–20	36	14	28	13 x 18	15
60	36	18–20	44	14	28	13 x 18	15
54	40	20–22	36	17	29	13 x 18	15
60	40	20–22	42	17	30	18 x 18	18
66	40	20–22	44	17	30	18 x 18	18
72	40	22–28	51	17	30	18 x 18	18

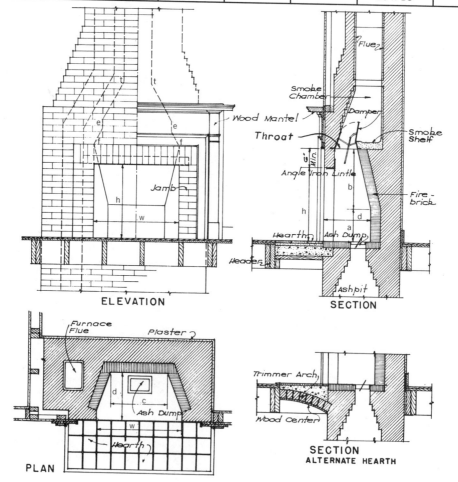

ELEVATION

SECTION

PLAN

SECTION
ALTERNATE HEARTH

ash dump can be built under the back of the hearth. Fig. 51-20.

In buildings with wooden floors, the hearth in front of the fireplace should be supported by masonry trimmer arches or other fire-resistant construction. Fig. 51-21. Wood centering under the arches used during construction of the hearth and hearth extension should be removed when construction is completed. Fig. 51-22 shows the recommended method of installing floor framing around the hearth.

WALLS

Building codes generally require that the back and sides of fireplaces be constructed of solid masonry or reinforced concrete at least 8″ thick and that they be lined with firebrick or other approved noncombustible material not less than 2″ thick or with steel lining not less than 1/4″ thick. Such lining may be omitted when the walls are of solid masonry or reinforced concrete at least 12″ thick.

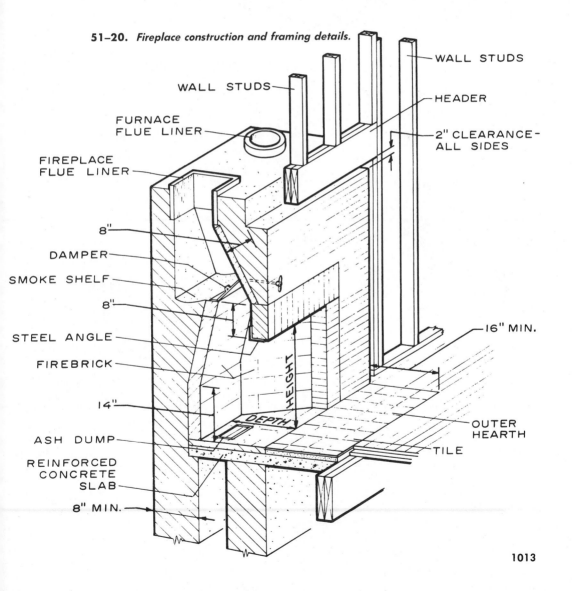

51-20. *Fireplace construction and framing details.*

WALL STUDS

WALL STUDS

HEADER

FURNACE
FLUE LINER

2″ CLEARANCE-
ALL SIDES

FIREPLACE
FLUE LINER

8″

DAMPER

SMOKE SHELF

8″

16″ MIN.

STEEL ANGLE

FIREBRICK

HEIGHT

14″

DEPTH

OUTER
HEARTH

ASH DUMP

TILE

REINFORCED
CONCRETE
SLAB

8″ MIN.

JAMBS

The jambs of the fireplace should be wide enough to provide stability and to present a pleasing appearance. For a fireplace opening 3' wide or less, the jambs can be 12" wide if a wood mantel will be used or 16" wide if they will be of exposed masonry. For wider fireplace openings, or if the fireplace is in a large room, the jambs should be proportionately wider. Fireplace jambs are frequently faced with ornamental brick or tile. Fig. 51-19.

No woodwork should be placed within 6" of the fireplace opening. Woodwork above and projecting more than 1½" from the fireplace opening should be placed at least 12" above the top of the fireplace opening. The mantel height above the opening can also be figured by adding 6" to the width of the mantel. Fig. 51-23.

LINTELS

A lintel must be installed across the top of the fireplace opening to support the masonry. For fireplace openings 4' wide or less, ½" × 3" flat steel bars, 3½" × 3½" × ¼" angle irons, or specially designed

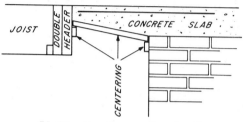

51–21. *Hearth centering detail.*

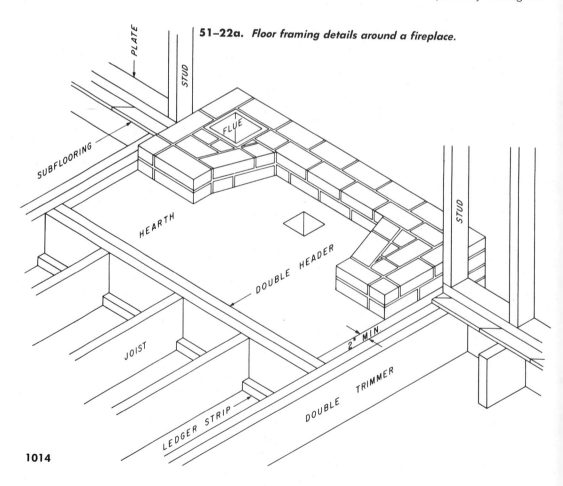

51–22a. *Floor framing details around a fireplace.*

damper frames may be used. Wider openings will require heavier lintels.

If a masonry arch is used over the opening, the fireplace jambs must be heavy enough to resist the thrust of the arch.

THROAT

Proper construction of the throat area is essential for a satisfactory fireplace. Fig. 51-19, f-f. The sides of the fireplace must be vertical up to the throat, which should be 6″ to 8″ or more above the bottom of the lintel. The area of the throat must be not

51-22b. *A fireplace opening framed in a house without basement. Note the reinforcing rod which will be used to tie the exterior fireplace masonry to the building.*

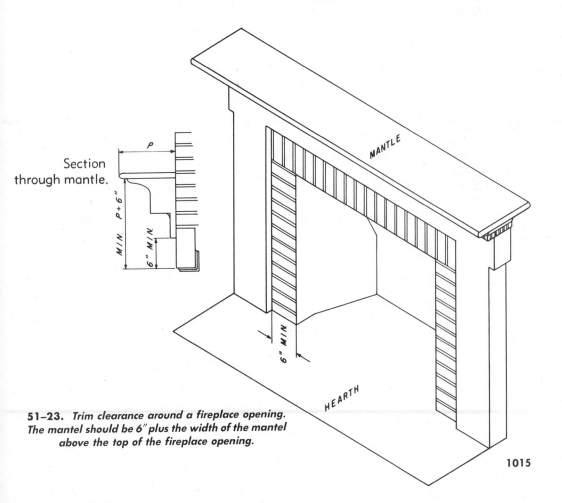

Section through mantle.

51-23. *Trim clearance around a fireplace opening. The mantel should be 6″ plus the width of the mantel above the top of the fireplace opening.*

1015

less than that of the flue. The length must be equal to the width of the fireplace opening, and the width of the throat will depend on the width of the damper frame (if a damper is installed). Five inches above the throat (at e-e in Fig. 51-19) the sidewalls should start sloping inward to meet the flue (at t-t in Fig. 51-19).

DAMPER

A damper consists of a cast iron frame with a hinged lid that opens or closes to vary the throat opening. Dampers are not always installed, but they are recommended, especially in cold climates.

With a well-designed, properly installed damper, you can:

• Regulate the draft.

• Close the fireplace flue to prevent loss of heat from the room when there is no fire in the fireplace. In the summer, the flue should be closed to prevent loss of cool air from the air-conditioning system.

• Adjust the throat opening according to the type of fire and thus reduce heat loss. For example, a roaring pine fire may require a full throat opening, but a slow-burning hardwood log fire may require an opening of only 1″ or 2″. Closing the damper to the opening will reduce loss of heat up the chimney.

• Close or partially close the flue to prevent loss of heat from the main heating system. When air heated by a furnace goes up a chimney, an excessive amount of fuel may be wasted.

• Close the flue in the summer to prevent insects from entering the house through the chimney

Dampers of various designs are on the market. Some support the masonry over fireplace openings, thus replacing ordinary lintels. It is important that the full damper opening equal the area of the flue.

SMOKE SHELF AND CHAMBER

A smoke shelf prevents downdraft. Fig. 51-19. It is made by setting the brickwork at the top of the throat back to the line of the flue wall for the full length of the throat. Depth of the shelf may be 6″ to 12″ or more, depending on the depth of the fireplace. The smoke shelf is concave to retain any slight amount of rain that may enter.

The smoke chamber is the area from the top of the throat (e-e in Fig. 51-19) to the bottom of the flue (t-t in Fig. 51-19). As stated under "Throat" the sidewalls should slope inward to meet the flue. The smoke shelf and the smoke chamber walls should be plastered with cement mortar at least ½″ thick.

FIREPLACE FLUE

Proper proportion between the area of the fireplace opening, area of the flue, and height of the flue is essential for satisfactory operation of the fireplace. The area of a lined flue 22′ high should be at least $\frac{1}{12}$ of the area of the fireplace opening. The area of an unlined flue or a flue less than 22′ high should be $\frac{1}{10}$ of the area of the fireplace opening.

The table in Fig. 51-19 lists dimensions of fireplace openings and indicates the size of flue lining required. Flue-construction principles given under "Chimneys" apply also to fireplace flues.

Fireplaces with two or more openings require much larger flues than the conventional fireplace. Fig. 51-24. For example, a fireplace with two open adjacent faces would require a 12″ × 16″ flue for an opening 34″ wide by 20″ deep by 30″ high. A, Fig. 51-24. Local building regulations usually cover the proper sizes for these types of fireplaces.

Modified Fireplaces

Modified fireplaces are manufactured fireplace units. They are made of heavy metal and are designed to be set in place and concealed by the usual brickwork or other construction. Fig. 51-25. They contain all the essential fireplace parts: firebox, damper, throat, and smoke shelf and

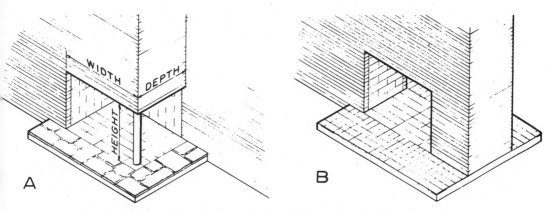

51-24. *Dual-opening fireplaces: A. Adjacent opening. B. Through fireplace.*

chamber. In the completed installation, only grilles show. Fig. 51-25a.

These fireplace units are available in a variety of styles for different room locations. Fig. 51-26. Multilevel installations are also possible through the use of chimney offsets, elbows, and various types of roof terminations. Figs. 51-27 (Page 1020) and 51-28.

Modified fireplaces offer two advantages:
• The correctly designed and proportioned firebox provides a ready-made form for the masonry. This reduces the chance of faulty construction and assures a smokeless fireplace.
• When properly installed, the better-designed units heat more efficiently than ordinary fireplaces. They circulate heat into the cold corners of rooms and can deliver heated air through ducts to upper or adjoining rooms.

Even a well-designed modified fireplace unit will not operate properly if the chimney is inadequate. Therefore, when a conventional masonry chimney is used, proper chimney construction is as important for a modified unit as it is for ordinary fireplaces.

The modified fireplace system is easy to handle. Factory-built components are

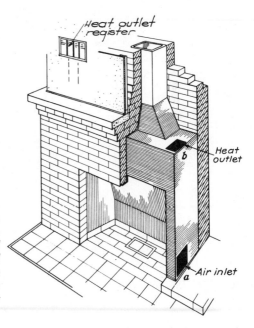

*51-25a. **Modified fireplace.** Air is drawn through inlet "a" from the room being heated. The air is heated upon contact with the metal and discharged through outlet "b". The inlets and outlets are connected to registers which may be located at the front, as shown, at the ends of the fireplace, or even in an adjacent or second-story room.*

quickly installed for a complete operating fireplace, hearth to chimney top. Fig. 51-29 (Pages 1021-1024) illustrates the installation of a modified fireplace system.

Prefabricated Fireplaces and Chimneys

Prefabricated fireplace and chimney units, with all the parts needed for a complete fireplace-to-chimney installation, are also available. Figs. 51-30 and 51-31 (Pages 1024-1026).

Such units offer these features:
- Wide selection of styles, shapes, and colors.
- Pretested design that is highly efficient in operation.
- Easy and versatile installation. They can be installed freestanding or flush against a wall in practically any part of a house.
- Light weight.
- Lower cost than comparable masonry units.

The basic part of the prefabricated fire-

place is a specially insulated metal firebox shell. Since it is light in weight, the fireplace can be set directly on the floor without the heavy footing required for masonry fireplaces.

Prefabricated chimneys can be used for furnaces, heaters, and incinerators as well as for prefabricated fireplaces. The chimneys are tested and approved by Underwriters' Laboratories, Inc., and other nationally recognized testing laboratories. They are rapidly being accepted for use by building codes in many cities.

Outdoor Fireplaces

Outdoor fireplaces range from simple makeshift units to elaborate structures designed to harmonize with and enhance the look of the house and the landscape.

Built-in features, such as ovens, grills, storage compartments, sinks, and benches add to the appearance and convenience of fireplaces. Many homes now include backyard picnic and cooking facilities. Fig. 51-32.

51–25b. *Manufacturer's details of two modified fireplace units.*

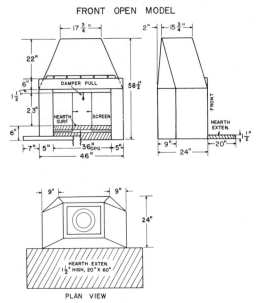

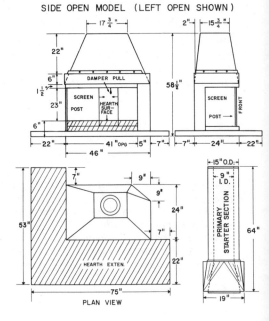

51-25c. *A finished modified fireplace with contemporary styling. This unit is of the type shown at the right in Fig. 51-25b.*

51-25d. *A finished modified fireplace with traditional styling. This unit is of the type shown at the left in Fig. 51-25b.*

51-26. *The modified fireplace system can be built into or out from any wall, used as part of a room divider, or placed into any corner.*

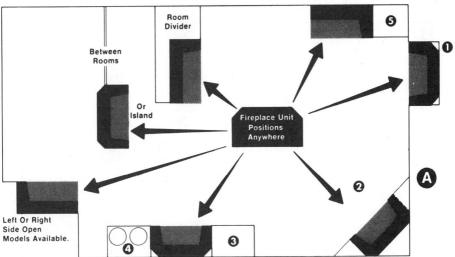

1. In Outside Chase, Breezeway, Garage, Or in Unused Attic If On Upper Floor.
2. Position In Corner And Save Additional Floor Space.
3. Built-In Bookcases Or Storage Space.
4. Venting From Other Units.
5. Log Storage, Etc.

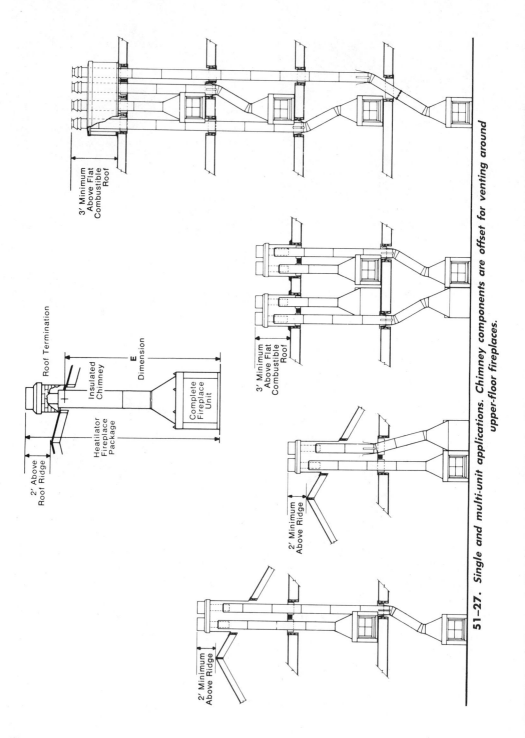

51-27. Single and multi-unit applications. Chimney components are offset for venting around upper-floor fireplaces.

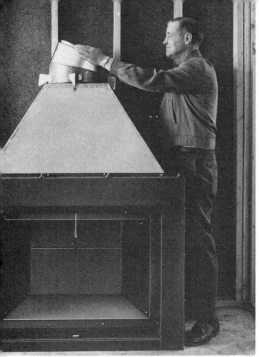

51-28c. *Installing the offset return. This is secured to the wood members of the framed opening. These members must then carry the weight of the vertical chimney.*

51-28a. *For installations requiring an inclined chimney similar to the one shown in Fig. 51-27, install the offset starter first. It allows for a 15° or 30° incline of the chimney to left, right, or rear of the fireplace.*

51-29a. *Framing completely ready for the fireplace installation. This system requires no special support foundation or masonry.*

51-28b. *Installing the starter section.*

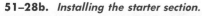

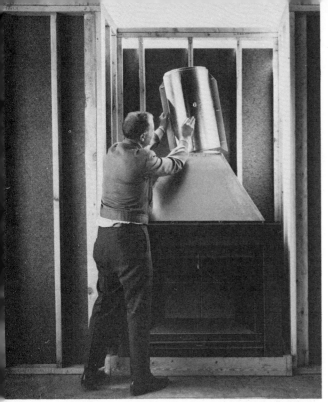

51–29b. *The fireplace is set in position and the 2′ starter section is installed.*

51–29c. *The insulated chimney is completed with 2′ or 3′ sections to the roof termination. Dimension "E" in Fig. 51–27.*

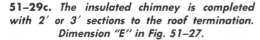

51–29d. *The housing panels for the chimney unit are snapped together.*

51–29e. *Installing the 45° braces across the corners of the chimney unit.*

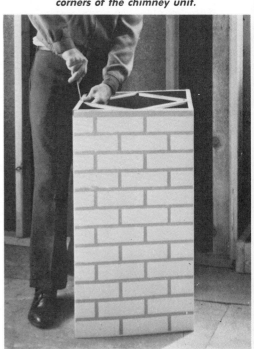

51-29f. *The chimney panels are cut to fit the roof pitch.*

51-29g. *Installing the flashing and securing the chimney panels to the flashing.*

51-29h. *The chimney terminal cap is slipped into position.*

51-29i. *The corners of the cap are pressed down onto the housing top and locked into position.*

51-29j. *A fireplace framed with headers ready for the application of the interior wall covering.*

51-30. *Prefabricated fireplaces are an attractive addition to a room. They do not require the heavy footing needed for masonry fireplaces.*

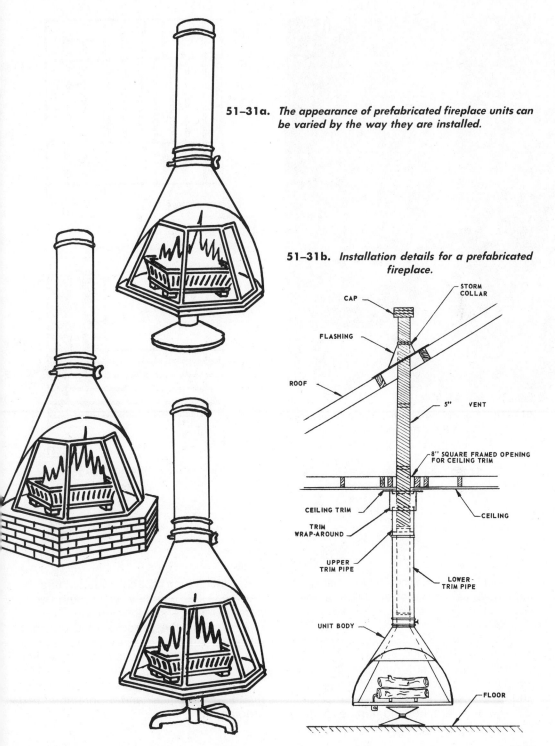

51–31a. *The appearance of prefabricated fireplace units can be varied by the way they are installed.*

51–31b. *Installation details for a prefabricated fireplace.*

CAP

STORM COLLAR

FLASHING

ROOF

5" VENT

8" SQUARE FRAMED OPENING FOR CEILING TRIM

CEILING TRIM

CEILING

TRIM WRAP-AROUND

UPPER TRIM PIPE

LOWER TRIM PIPE

UNIT BODY

FLOOR

51–31c. *A prefabricated fireplace unit installed in a colonial setting.*

51–32. *An outdoor fireplace, such as this one, can be attractive and convenient.*

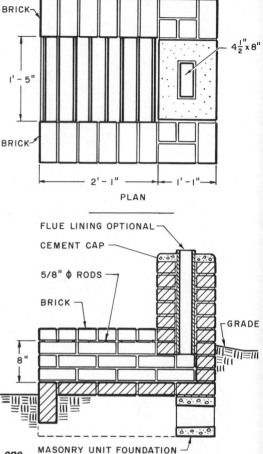

BRICK

$4\frac{1}{2}" \times 8"$

1' – 5"

BRICK

2' – 1"   1' – 1"

PLAN

FLUE LINING OPTIONAL

CEMENT CAP

5/8" ⌀ RODS

BRICK

GRADE

8"

MASONRY UNIT FOUNDATION

SECTION

QUESTIONS

1. Why will a chimney at the interior of the house have a better draft than a chimney built on an exterior wall?

2. What is a flue?

3. What are two important factors that affect the draft of a chimney?

4. Why is cement mortar rather than lime mortar used around chimney flues and fireplaces?

5. What determines the number of flues contained in a chimney?

6. How much space should there be between the chimney walls and wood framing members of the house?

7. What is the purpose of a concrete cap on a chimney?

8. What is the purpose of a smoke shelf in a fireplace?

9. What is the difference between a modified fireplace and a prefabricated fireplace?

Protection Against
Decay and Termites

Wood used under conditions where it will always be dry, or even where it is wetted briefly and rapidly dried, will not decay. However, all wood and wood products in construction use are subject to decay if kept wet for long periods under temperature conditions favorable to the growth of decay organisms. Most of the wood used in a house is not subjected to such conditions. There are places where water can work into the structure, but such places can be protected by proper design and construction, by use of suitable materials, and in some cases by using treated material.

Wood is also subject to attack by termites and some other insects. Termites can be grouped into two main classes: subterranean and dry-wood. Subterranean termites are found in the northernmost states and in Hawaii. Serious damage is confined to scattered, localized areas of infestation. Fig. 52-1. Buildings may be fully protected against subterranean ter-

52–1. *Line A is the northern limit of damage by subterranean termites. Line B is the northern limit of damage by dry-wood termites.*

mites by employing fairly inexpensive protection measures during construction.

Since 1966 the Formosan subterranean termite has been discovered in several locations in the southern United States. It is a serious pest because its colonies contain large numbers of the worker caste and cause damage rapidly. Though presently in localized areas, it could spread. Controls are similar to those for other subterranean species.

Dry-wood termites are found principally in Florida, southern California, the Gulf Coast states, and in Hawaii. They are more difficult to control, but the damage is less serious than that caused by subterranean termites.

Wood has proved itself through the years to be an excellent building material. Damage from decay and termites has been small in proportion to the total value of wood in homes, but it has been a troublesome problem to many homeowners. With changes in design and use of new building materials, it is important to note the basic safeguards which protect buildings against both decay and termites.

DECAY

Wood decay is caused by certain fungi that can utilize wood for food. These fungi, like the higher plants, require air, warmth, food, and moisture for growth. Early stages of decay caused by these fungi may be accompanied by a discoloration of the wood. Paint also may become discolored where the underlying wood is rotting. Advanced decay is easily recognized because the wood has by then undergone definite changes in properties and appearance. In advanced stages of building decay, the affected wood generally is brown and crumbly, but sometimes may be rather white and spongy. These changes may not be apparent on the surface. The loss of sound wood inside, however, is often indicated by sunken areas on the surface or by a hollow sound when the wood is tapped with a hammer. Where the surrounding atmosphere is very damp, the decay fungus may grow out on the surface, appearing as white or brownish growths in patches or strands or in special cases as vinelike structures.

Fungi grow most rapidly at temperatures of about 70° to 85° F. Elevated temperatures such as those used in kiln-drying of lumber kill fungi. Low temperatures, even far below zero, merely cause them to remain dormant.

Moisture requirements of fungi are within definite limits. Wood-destroying fungi cannot grow in dry wood. A moisture content of 20% or less is safe. Moisture contents greater than this are practically never reached in wood that is sheltered against rain and protected, if necessary, against wetting by condensation or fog. Decay can be permanently stopped by simply taking measures to dry out the infected wood and to keep it dry. Brown crumbly decay, in the dry condition, is sometimes called *dry rot*, but this is a misnomer. Such wood must be damp if rotting is to occur.

The presence of mold or of fungus stains should serve as a warning that conditions are or have been suitable for decay fungi. Heavily molded or stained lumber, therefore, should be examined for evidence of decay. Such discolored wood is not entirely satisfactory for exterior millwork because it has greater water absorptiveness than bright wood.

The natural decay resistance of all common native species of wood lies in the heartwood. The sapwood of all species, when untreated, has low resistance to decay and usually has a short life under decay-producing conditions. Of the wood species commonly used in house construction, the heartwood of bald cypress, redwood, and the cedars is classified as being highest in decay resistance. All-heartwood, quality lumber is becoming more and more difficult to obtain, however, as increasing amounts of timber are cut from the smaller trees of second-growth stands. In general, when substantial decay resistance is needed in loadbearing members that are difficult and expensive to replace, preservative-treated wood is recommended.

Safeguards against Decay

Except for special cases of wetting by condensation or fog, a dry piece of wood, when placed off the ground under a tight roof with wide overhang, will stay dry and never decay. This principle of "umbrella protection," when applied to houses of proper design and construction, is a good precaution. The use of dry lumber in designs that will keep the wood dry is the simplest way to avoid decay in buildings.

LUMBER

Construction lumber that is green or partially seasoned may be infected with one or more of the staining, molding, or decay fungi and should be avoided. Such wood may contribute to serious decay in both the substructure and exterior parts of buildings. If wet lumber must be used, or if wetting occurs during construction, the wood should not be fully enclosed or painted until thoroughly dried.

DESIGN DETAILS

Untreated wood should not come in contact with the soil. Foundation walls should have a clearance of at least 8″ above the exterior finish grade. Floor construction should have a clearance of 18″ or more from the bottom of the joists to the ground in basementless spaces. Fig. 52-2. The foundation should be accessible at all points for inspection. Porches that prevent access should be isolated from the soil by concrete or from the building itself by metal barriers or aprons. Fig. 52-3. Steps and stair carriages, posts, wallplates, and sills should be insulated from the ground with concrete or masonry. Figs. 52-4 and 52-5. Sill plates and other wood in contact with concrete near the ground should be protected by a moistureproof membrane, such as heavy roll roofing or

6-mil polyethylene. Girder and joist openings in masonry walls should be big enough to assure an air space around the ends of these members. Fig. 52-6.

Surfaces like steps, porches, door and window frames, roofs, and other projections should be sloped to promote runoff of water. Fig. 52-7 (Page 1032). Noncorroding flashing should be used around chimneys, windows, doors, or other places where water might seep in. Figs. 52-8 through 52-11. Roofs with considerable overhang give added protection to the siding and other parts of the house. Gutters and downspouts should be placed and maintained to divert water away from the building. Porch columns and screen rails should be shimmed above the floor to allow quick drying, or posts should slightly overhang raised concrete bases. Fig. 52-12 (Page 1034).

WOOD PRESERVATIVES

Exterior steps, rails, and porch floors exposed to rain need protection from de-

52-2. *Clearance between the wood members of a structure and the soil.*

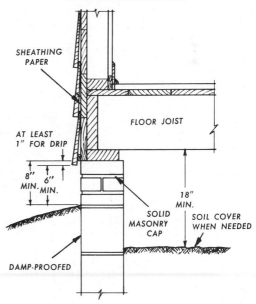

SHEATHING PAPER

FLOOR JOIST

AT LEAST 1″ FOR DRIP

8″ MIN. 6″ MIN.

18″ MIN.

SOLID MASONRY CAP

SOIL COVER WHEN NEEDED

DAMP-PROOFED

cay, particularly in warm, damp climates. Wood may be treated by the pressure method, in which it is impregnated with toxic chemicals at elevated pressures and temperatures. Pressure treatment of the wood provides a high degree of protection against decay and termite attack. One of the following classes of preservatives is commonly used:

- Creosote and creosote solutions.
- Oilborne preservatives.
- Waterborne preservatives.
- Water-repellent preservatives.

Where the chance of decay is relatively small or where pressure-treated wood is not readily obtainable, on-the-job application of water-repellent preservatives by brushing, dipping, or soaking has been found worthwhile. Preservatives used for non-pressure-treated wood are of two types:

- Water-repellent preservative—a solution of light petroleum solvent containing water-repellent materials and a minimum of 5% by weight of pentachlorophenol, meeting the standards of the National Woodwork Manufacturers' Association. This treatment can be painted.
- Oilborne preservative—a solution of heavier petroleum solvent containing 5%

52–3. *Metal flashing is used to protect the wood where the porch slab is connected. Note also the termite shield between the foundation wall and sill, and the vapor barrier in the crawl space.*

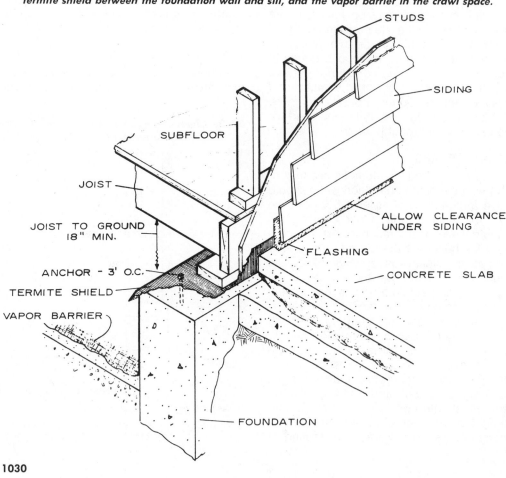

STUDS

SIDING

SUBFLOOR

JOIST

ALLOW CLEARANCE UNDER SIDING

JOIST TO GROUND 18" MIN.

FLASHING

ANCHOR - 3' O.C.

TERMITE SHIELD

CONCRETE SLAB

VAPOR BARRIER

FOUNDATION

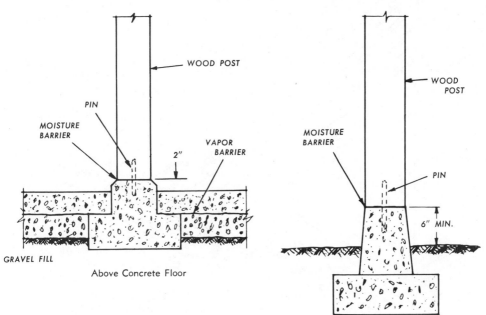

WOOD POST

PIN

MOISTURE
BARRIER

VAPOR
BARRIER

2"

GRAVEL FILL

Above Concrete Floor

WOOD
POST

MOISTURE
BARRIER

PIN

6" MIN.

Above Earth Floor

52–4. Wood posts should be insulated from the ground with concrete or masonry.

by weight of pentachlorophenol meeting the American Wood-Preservers' Association standard P9. This treatment is not recommended for painting.

The wood should be dried, cut to final dimensions, and then brushed, dipped, or soaked. Soaking is the best of these non-pressure methods, and the ends of the

boards should be soaked for at least 3 minutes. It is important to protect the end grain of wood at joints, for this area absorbs water easily and is the most common infection point. The edges of porch flooring

52–6. Girder and joist openings in masonry walls should be large enough to allow air circulation around the ends of these members.

52–5. Exterior wood steps should be at least 6" off the ground and supported on a concrete footing.

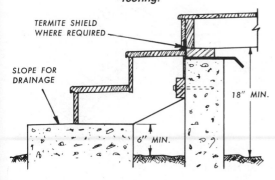

TERMITE SHIELD
WHERE REQUIRED

SLOPE FOR
DRAINAGE

18" MIN.

6" MIN.

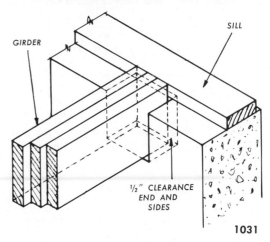

GIRDER

SILL

½" CLEARANCE
END AND
SIDES

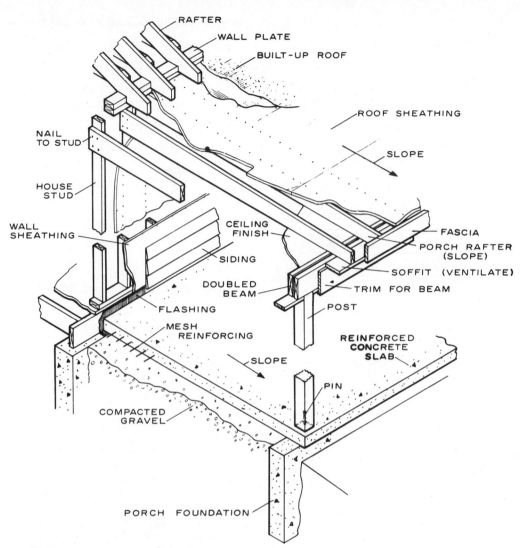

52–7. *Roofs and porches should slope away from the building to promote runoff of water.*

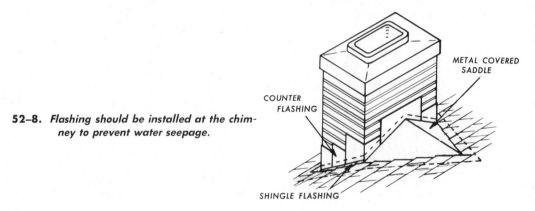

52–8. *Flashing should be installed at the chimney to prevent water seepage.*

52–9. *Flashing should also be installed over windows and doors to prevent water seepage.*

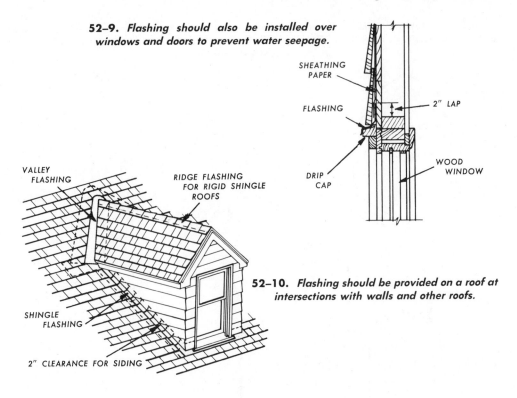

SHEATHING PAPER

FLASHING

2" LAP

DRIP CAP

WOOD WINDOW

VALLEY FLASHING

RIDGE FLASHING FOR RIGID SHINGLE ROOFS

SHINGLE FLASHING

2" CLEARANCE FOR SIDING

52–10. *Flashing should be provided on a roof at intersections with walls and other roofs.*

52–11. *There should be flashing at a break in an exterior wall.*

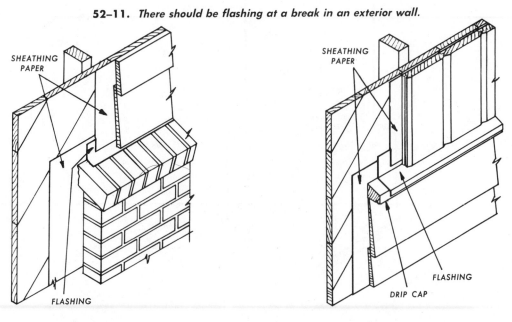

SHEATHING PAPER

FLASHING

Brick Veneer and Wood Siding

SHEATHING PAPER

FLASHING

DRIP CAP

Horizontal and Vertical Wood Siding

52–12. *Wood post details: A. Cased post. B. Post pinned to a concrete floor. Note the use of the galvanized washer as a spacer. C. Flashing at the base of the post. Note that the base molding is not set down tightly on the flashing.*

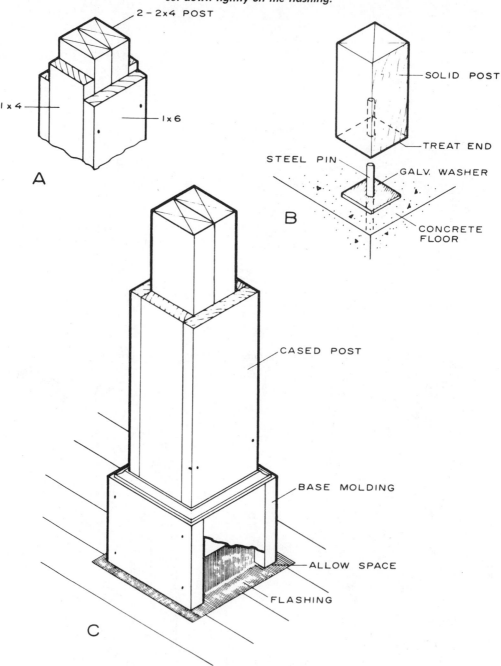

should be coated with thick white lead or other durable coating as the porch is laid.

WATER VAPOR FROM THE SOIL

Crawl spaces of houses built on poorly-drained sites may be subjected to high humidity. During the winter when the sills and outer joists are cold, moisture condenses on them and, in time, the wood absorbs so much moisture that it is susceptible to attack by fungi. Unless this moisture dries out before temperatures favorable for fungus growth are reached, considerable decay may result. However, this decay may progress so slowly that no weakening of the wood becomes apparent for a few years. Placing a layer of 45-pound or heavier roll roofing or a 6-mil sheet of polyethylene over the soil to keep the vapor from getting into the crawl space would prevent such decay. This might be recommended for all sites where, during the cold months, the soil is wet enough to be compressed in the hand. Fig. 52-3.

If the floor is uninsulated, there is an advantage in closing the foundation vents during the coldest months from the standpoint of fuel savings. However, unless the crawl space is used as a heat plenum chamber, insulation is usually located between floor joists. The vents could then remain open. Crawl-space vents can be very small when soil covers are used; only 10% of the area required without covers. (See Unit 53, "Ventilation.")

WATER FROM HOUSEHOLD ACTIVITIES

Water vapor is given off during cooking, washing, and other household activities. This vapor can pass through walls and ceilings during very cold weather and condense on sheathing, studs, and rafters, causing condensation problems. A vapor barrier of an approved type is needed on the warm side of walls. (See Unit 50.) It is also important that the attic space be ventilated as discussed in Unit 53, "Ventilation."

Leaking pipes should be fixed immediately to prevent damage to the house, as well as to guard against possible decay.

WATER SUPPLIED BY THE FUNGUS ITSELF

In the warmer coastal areas principally, some substructure decay is caused by a fungus that provides its own needed moisture by conducting it through a vinelike structure from moist ground to the wood. The total damage caused by this water-conducting fungus is not large, but in individual instances it tends to be unusually severe. Preventive and remedial measures depend on getting the soil dry and avoiding untreated wood "bridges," such as posts, between ground and sills or beams.

TERMITES

Subterranean Termites

Subterranean termites are the most destructive of the insects that infest wood in houses. The chance of infestation is great enough to justify preventive measures in the design and construction of buildings in areas where termites are common.

Subterranean termites are common throughout the southern two-thirds of the United States and in Hawaii except in mountainous and extremely dry areas.

Subterranean termites thrive in moist, warm soil containing an abundant supply of food in the form of wood or other cellulosic material. In their search for additional food (wood), they build earthlike shelter tubes over foundation walls, in cracks in the walls, or on pipes or supports leading from the soil to the house. These flattened tubes are from $1/4''$ to $1/2''$ or more in width. They serve to protect the termites in their travels between food and shelter.

Since subterranean termites eat the interior of the wood, they may cause much damage before they are discovered. They honeycomb the wood with tunnels that are separated by thin layers of sound wood.

Decay fungi, on the other hand, soften the wood and eventually cause it to shrink, crack, and crumble without producing anything like these continuous tunnels. When both decay fungi and subterranean termites are present in the same wood, even the layers between the termite tunnels will be softened.

Dry-Wood Termites

Dry-wood termites fly directly to the wood and bore into it instead of building tunnels from the ground as do the subterranean termites. Dry-wood termites are common in the tropics. They have also been found in the United States in a narrow strip along the Atlantic Coast from Cape Henry, Virginia, to the Florida Keys, and westward along the coast of the Gulf of Mexico to the Pacific Coast as far as northern California. Fig. 52-1. They may infest structural timber and other woodwork in buildings, and also furniture, particularly where the surface is not adequately protected by paint or other finishes.

Dry-wood termites cut across the grain of the wood and excavate broad pockets, or chambers. These chambers are connected by tunnels about the diameter of the termite's body. Dry-wood termites destroy both springwood and the usually harder summerwood, whereas subterranean termites principally attack springwood. Dry-wood termites remain hidden in the wood and are seldom seen, except when they make dispersal flights.

Safeguards against Termites

The best time to provide protection against termites is during the planning and construction of the building. Remove all woody debris, like stumps and discarded form boards, from the soil at the building site before and after construction. Steps should also be taken to keep the soil under the house as dry as possible.

The foundation should be made impervious to subterranean termites to prevent them from crawling up through hidden cracks to the wood in the building above. Fig. 52-13 and 52-14. Properly reinforced concrete makes the best foundation. Unit-construction walls or piers capped with at least 4" of reinforced concrete are also satisfactory.

The heartwood of foundation-grade redwood, particularly when painted, is more resistant to attack than most other native commercial species. No wood member of the house structure should be in contact with the soil.

The best protection against subterranean termites is treatment of the soil near the foundation or under an entire slab founda-

52–13. *Termite shields: Note in (a) the installation of the termite shield on the pipe which goes up through the flooring.*

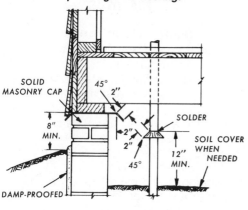

(a) At Exterior Wall

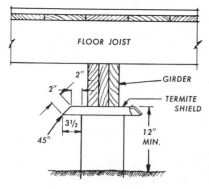

(b) Over Interior Pier

tion. The effective soil treatments are water emulsions of aldrin (0.5%), chlordane (1.0%), dieldrin (0.5%), and heptachlor (0.5%). The rate of application is 4 gallons per 10 linear feet at the edge and along expansion joints of slabs or along a foundation. For brick or hollow-block foundations, the rate is 4 gallons per 10 linear feet for each foot of depth to the footing. One to 1½ gallons of emulsion per 10 square feet of surface area is recommended for overall treatment before pouring concrete slab foundations. Fig. 52-15. Any wood used in such places as wall extensions, decorative fences, and gates should be pressure-treated with a good preservative.

In regions where dry-wood termites are present, the following measures should be taken to prevent damage.

• All lumber, particularly secondhand material, should be carefully inspected before use. If infected, discard the piece.

• All doors, windows (especially attic windows), and other ventilation openings should be screened with metal wire with not less than 20 meshes to the inch.

• Preservative treatment can be used to prevent attack in construction timber and lumber.

• Several coats of house paint will provide considerable protection to exterior woodwork in buildings. All cracks, crevices, and joints between exterior wood members should be filled with a mastic calking or plastic wood before painting.

Handling Pesticides

Pesticides used improperly can be injurious to people, animals, and plants. Follow the directions and heed all precautions on the labels.

Store pesticides in original containers, out of reach of children and pets and away from food.

Apply pesticides selectively and carefully. Do not apply a pesticide when there is danger of drift to other areas. Avoid prolonged inhalation of a pesticide spray or dust. When applying a pesticide, it is advisable that you be fully clothed.

After handling a pesticide, do not eat, drink, or smoke until you have washed. In case a pesticide is swallowed or gets in the eyes, follow the first aid treatment

52–14. *Termite barriers on concrete slab-on-ground construction: A. Termite shield at the exterior wall. B. The use of a coal-tar pitch filler at the joint between the concrete floor and the cement block wall.*

A Termite Shield

B Filler at Joints

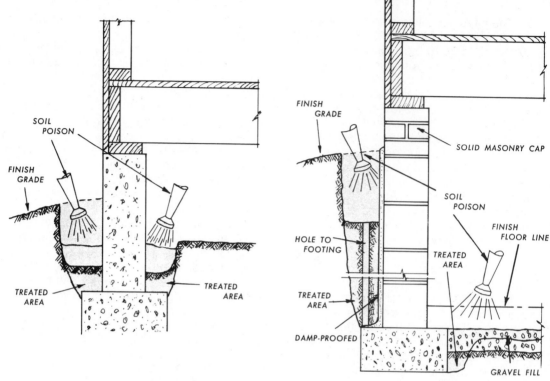

SOIL
POISON

FINISH
GRADE

TREATED
AREA

TREATED
AREA

FINISH
GRADE

SOLID MASONRY CAP

SOIL
POISON

FINISH
FLOOR LINE

HOLE TO
FOOTING

TREATED
AREA

TREATED
AREA

DAMP-PROOFED

GRAVEL FILL

Poured Concrete Wall and Crawl Space

Unit Masonry Wall and Basement

52–15. *Chemical treatment of the soil for termite protection.*

given on the label, and get prompt medical attention. If the pesticide is spilled on skin or clothing, remove clothing immediately and wash skin thoroughly.

Dispose of empty pesticide containers by wrapping them in several layers of newspaper and placing them in the trash can.

It is difficult to remove all traces of a herbicide (weed killer) from equipment.

Therefore, to prevent injury to desirable plants, do not use the same equipment for insecticides and fungicides that you use for a herbicide.

NOTE: Registrations of pesticides are under constant review by the U. S. Department of Agriculture. Use only pesticides that bear the USDA registration number and carry directions for home and garden use.

QUESTIONS

1. What causes decay in wood?
2. What are the two main classes of termites?
3. Below what moisture content is wood safe from decay?
4. What is the simplest way to avoid decay in buildings?

5. List several ways water may get into a structure.
6. Which of the two classes of termites is more destructive?
7. When is the best time to provide protection against termites?

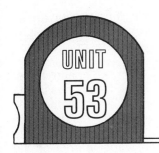

Ventilation

During cold weather, condensation of water vapor may occur in attic spaces and under flat roofs. Even when vapor barriers have been installed, some vapor will work into spaces around pipes and other inadequately protected areas and some through the vapor barrier itself. Although the amount may be unimportant if equally distributed, it may be sufficiently concentrated in cold spots to cause damage. While wood shingle and wood shake roofs do not resist vapor movement, such roofings as asphalt shingles and built-up roofs are highly resistant. The most practical method of removing the moisture is by adequately ventilating the roof spaces.

In addition to water condensation, a warm attic that is inadequately ventilated and insulated may cause formation of ice dams at the cornice. After a heavy snowfall, heat causes the snow on the roof to melt. Water running down the roof freezes on the colder surface of the cornice, often forming an ice dam at the gutter. The ice dam may cause water to back up at the eaves and into the wall and ceiling. Similar dams often form in roof valleys. (See "Roof Coverings," Fig. 35-5.) Ventilation provides part of the answer to this problem. With a well-insulated ceiling and adequate ventilation, attic temperatures are low and melting of snow over the attic space will be greatly reduced.

Tight construction (including storm windows and storm doors) and the use of humidifiers have also created potential moisture problems which must be resolved by adequate ventilation as well as the proper use of vapor barriers.

In hot weather, ventilation of attic and roof spaces allows hot air to escape, thereby lowering the temperature in these spaces. Insulation should be put between the ceiling joists located below the attic or roof space to further retard heat flow into the rooms below and to improve comfort conditions. Room temperature and heat in the attic and roof spaces can be quickly reduced with a ventilating fan. The fan is located so as to draw the cool evening air into the house and expel it through the roof ventilators. Fig. 53-1.

It is common practice to install louvered openings in the end walls of gable roofs for ventilation. Air movement through such

53–1. Ventilating fans are ideal for pulling the cool evening air in to reduce interior temperature quickly. A. The fan may be mounted horizontally for a low roof installation. B. The fan may be mounted vertically with a suction box installation.

A

B

Ventilation Guide

Length Feet	Width (In Feet)											
	20	22	24	26	28	30	32	34	36	38	40	42
20	192	211	230	250	269	288	307	326	346	365	384	403
22	211	232	253	275	296	317	338	359	380	401	422	444
24	230	253	276	300	323	346	369	392	415	438	461	484
26	250	275	300	324	349	374	399	424	449	474	499	524
28	269	296	323	349	376	403	430	457	484	511	538	564
30	288	317	346	374	403	432	461	490	518	547	576	605
32	307	338	369	399	430	461	492	522	553	584	614	645
34	326	359	392	424	457	490	522	555	588	620	653	685
36	346	380	415	449	484	518	553	588	622	657	691	726
38	365	401	438	474	511	547	584	620	657	693	730	766
40	384	422	461	499	538	576	614	653	691	730	768	806
42	403	444	484	524	564	605	645	685	726	766	806	847
44	422	465	507	549	591	634	676	718	760	803	845	887
46	442	486	530	574	618	662	707	751	795	839	883	927
48	461	507	553	599	645	691	737	783	829	876	922	968
50	480	528	576	624	672	720	768	816	864	912	960	1008

53–2. *Use this chart to figure the square inches of ventilation required to provide $1/300$ of the ceiling area. Find the ceiling length in the left-hand column. Read across to the column headed by the ceiling width. The number shown equals $1/300$ of the ceiling area in square inches. If there is no vapor barrier and the roof slope is less than 2 in 12, multiply this number by 2.*

openings depends primarily on wind direction and velocity. Little movement can be expected unless there is wind and one or more openings face the wind. More positive air movement can be obtained by providing openings in the soffit areas of the roof overhang in addition to openings at the gable ends or ridge.

Hip-roof houses are best ventilated by inlet ventilators in the soffit area and by outlet ventilators along the ridge. The differences in temperature between the attic and the outside will then create an air movement independent of the wind, and also a more positive movement when there is wind.

Where there is a crawl space under house or porch, ventilation is necessary to remove water vapor rising from the soil. Such vapor may otherwise condense on the wood below the floor and facilitate decay. A permanent vapor barrier on the soil of the crawl space greatly reduces the amount of ventilating area required.

AREA OF VENTILATORS

These are the FHA minimum requirements for ventilation of attics and foundations.

Attic Area

Provide cross ventilation for all spaces between roof and top-floor ceiling with corrosion-resistant 8-mesh screened louvers as follows:
- Roofs with slopes 2 in 12 or greater: $1/300$ of the horizontal projection of the roof area over each space. One-half the required ventilation shall be in the upper part of the ventilated space as near the high point of the roof as practicable. The chart in Fig. 53-2 can be used to find the square inches in $1/300$ of the ceiling area.
- Roofs with slopes less than 2 in 12: $1/150$ of the horizontal projection area over each space unless a complete continuous vapor barrier is provided. If the vapor barrier is provided, the requirement is $1/300$.

Foundation

Provide cross ventilation by corrosion-resistant screened vents, 8 mesh per inch, as follows: 2 sq. ft. per 100 linear feet of foundation plus $1/300$ of the ground area. Minimum—4 vents located at corners.

ROOF VENTILATION

Types of ventilators and minimum recommended sizes have been established for various types of roofs. Fig. 53-3, 53-4, and 53-5. The minimum area for attic or

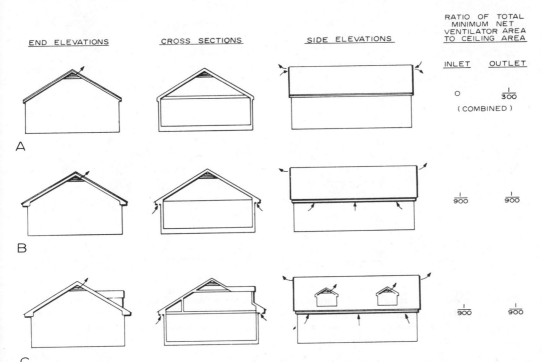

53–3. *Ventilating areas of gable roofs: A. End wall louver outlets. B. End wall louver outlets with soffit area inlets. C. End wall louver and dormer outlets with eave inlets.*

53–4. *Hip roof ventilation: A. Air inlet beneath the eaves and outlet vent near the ridge. B. Air inlet openings beneath the eaves and outlets in the small gable of a Dutch hip roof.*

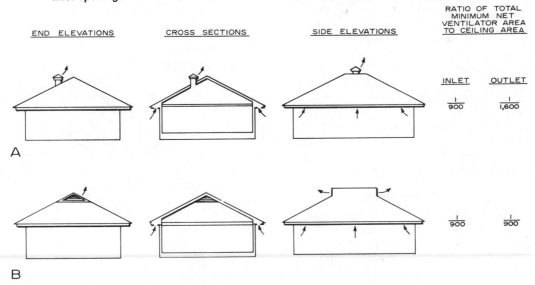

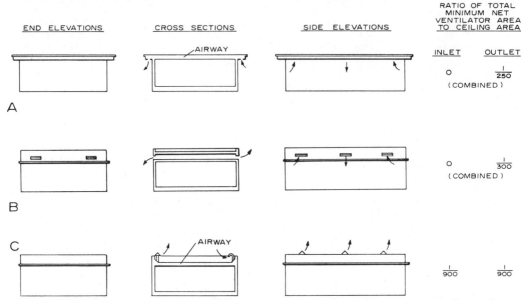

END ELEVATIONS	CROSS SECTIONS	SIDE ELEVATIONS	RATIO OF TOTAL MINIMUM NET VENTILATOR AREA TO CEILING AREA	
			INLET	OUTLET
	AIRWAY		O	$\frac{1}{250}$
A			(COMBINED)	
B			O	$\frac{1}{300}$
			(COMBINED)	
C	AIRWAY		$\frac{1}{900}$	$\frac{1}{900}$

53–5. *Flat roof ventilating area: A. Ventilator openings under the overhanging eaves where ceiling and roof joists are combined. B. Ventilating a roof with a parapet where roof and ceiling joists are separate. C. Ventilating a roof with a parapet where roof and ceiling joists are combined.*

roof-space ventilators is based on the projected ceiling area of the rooms below. The ratios given in Figs. 53-3, — 53-5 are of *net* ventilator area to ceiling area. The actual ventilator area must be increased to allow for any restrictions such as louvers and wire cloth or screen. The screen area should be double the specified net area.

To obtain extra screen area without adding to the area of the vent, use a frame of required size to hold the screen away from the ventilator opening. Use as coarse a screen as conditions permit, not smaller than No. 16, because lint and dirt tend to clog fine-mesh screens. Screens should be installed in such a way that paint brushes will not easily contact the screen and close the mesh with paint.

Gable Roofs

Louvered openings are generally provided in the end walls of gable roofs. They should be as close to the ridge as possible. The net area for the openings should be $1/300$ of the ceiling area. A, Fig. 53-3. For

example, if the ceiling area equals 1,200 square feet, the minimum total net area of the ventilators should be 4 square feet.

As previously explained, more positive air movement can be obtained if additional

53–6. *Installing a continuous soffit vent.*

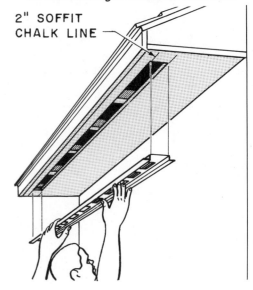

2" SOFFIT
CHALK LINE

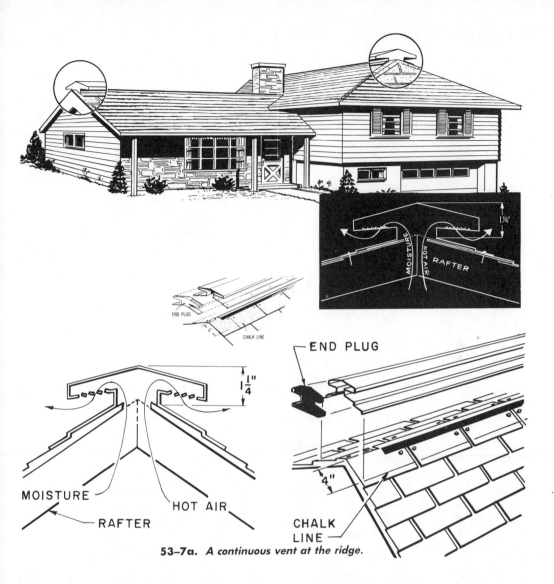

53-7a. *A continuous vent at the ridge.*

openings are provided in the soffit area. The minimum ventilation areas for this method are shown in B, Fig. 53-3.

Where there are rooms in the attic with sloping ceilings, the insulation should follow the roof slope and be so placed that there is a free opening of at least 1½″ between the roof boards and the insulation for air movement. C, Fig. 53-3.

Hip Roofs

Hip roofs should have air inlet openings in the soffit area of the eaves and outlet openings at or near the peak. For minimum net areas of openings see A, Fig. 53-4. The most efficient type of inlet openings is the continuous slot, which should provide a free opening of not less than ³/₄″. Fig. 53-6. The air outlet opening near the peak can be a continuous vent, a cupola, or several smaller roof ventilators located near the ridge. Fig. 53-7. The smaller roof ventilators can be located below the peak on the rear slope of the roof so that they will not be visible from the front of the house. Gabled extensions of a hip-roof house are some-

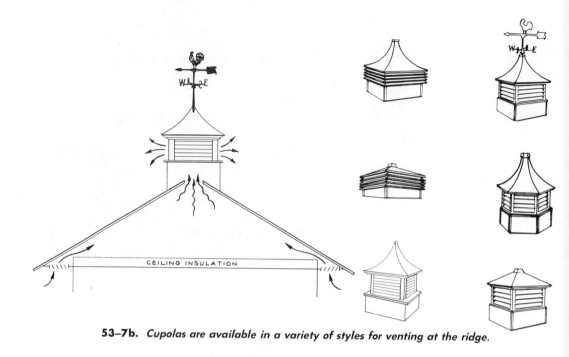

53–7b. *Cupolas are available in a variety of styles for venting at the ridge.*

53–7c. *Pitched roof ventilators are installed below the ridge on the rear slope of the roof.*

SLIDE FLANGE OVER LOWEST SHINGLE AND UNDER UPPER SHINGLES.

times used to provide efficient outlet ventilators. B, Fig. 53-4.

Flat Roofs

A greater ratio of ventilating area is required in some types of flat roofs than in pitched roofs because the air movement is less positive and is dependent upon wind. It is important that there be a clear open space above the ceiling insulation and below the roof sheathing for free air movement from inlet to outlet openings. Solid blocking should not be used for bridging or for bracing over bearing partitions if its use prevents air circulation.

Perhaps the most common type of flat or low-pitched roof is one in which the rafters extend beyond the wall, forming an overhang. A, Fig. 53-5. When soffits are used, this area can contain the combined inlet-outlet ventilators, preferably a continuous slot. When single ventilators are used, they

53-8. *Two types of attic ventilators installed in the end of a gable roof.*

should be distributed evenly along the overhang.

A parapet-type wall and flat roof combination may be constructed with the ceiling joists separate from the roof joists or combined. When members are separate, the space between can be used for an airway. B, Fig. 53-5. Inlet and outlet vents are then located as shown, or a series of outlet stack vents can be used along the center-line of the roof in combination with the inlet vents. When ceiling joists and flat rafters are combined in parapet construction, vents may be located as shown in C, Fig. 53-5. Wall inlet ventilators combined with center stack outlet vents are another option in this type of roof.

OUTLET VENTILATORS

Various styles of gable-end ventilators are available ready for installation. Fig. 53-8. Many are made with metal louvers and frames, while others may be made of wood to fit the house design more closely. However, the most important considerations are to have sufficient net ventilating area and to locate ventilators as close to the ridge as possible without affecting house appearance.

One of the types commonly used fits the slope of the roof and is located near the ridge. A, Fig. 53-9. It can be made of wood or metal. In metal it is often adjustable to conform to the roof slope. Fig. 53-10. A wood ventilator of this type is enclosed in a frame and placed in the rough opening much as a window frame. B, Fig. 53-9. Other forms of gable-end ventilators which might be used are also shown in Fig. 53-9.

Another system of attic ventilation can be used on houses with a wide roof overhang at the gable end. It consists of a series of small vents or a continuous slot located on the underside of the soffit areas. F, Fig. 53-9. Several large openings located near the ridge might also be used. This system is especially desirable on low-pitched roofs where standard wall ventilators may not be suitable.

It is important that the roof framing at the wall line does not block off ventilation spaces to the attic area. Ventilation space might be provided by the use of a "ladder" frame extension. A flat nailing block used at the wall line will provide airways into the attic. Fig. 53-11. This can also be adapted to narrower rake sections, providing ventilating areas to the attic.

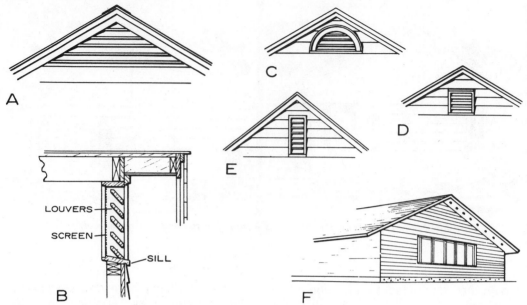

53-9. *A variety of outlet ventilators: A. Triangular. B. Typical ventilator cross section. C. Half-circle. D. Square. E. Vertical. F. Soffit.*

INLET VENTILATORS

Small, well-distributed ventilators or a continuous slot in the soffit provide inlet ventilation. These small louvered and screened vents can be obtained in most lumberyards or hardware stores and are simple to install.

Only small sections need to be cut out of the soffit. These can be sawed out before the soffit is applied. It is more desirable to use a number of smaller well-distributed ventilators than several large ones. A, Fig. 53-12. Any blocking which might be required between rafters at the wall line should be installed so as to provide an airway into the attic area.

A continuous screened slot is often desirable. It should be located near the outer edge of the soffit near the fascia. B, Fig. 53-12. Locating the slot in this area will minimize the chances of snow entering. This type may also be used on the extension of flat roofs.

CRAWL-SPACE VENTILATION AND SOIL COVER

The crawl space below the floor of a basementless house and under porches should be ventilated and protected from

53-10. *Installing a multi-pitch gable louver.*

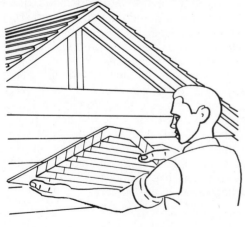

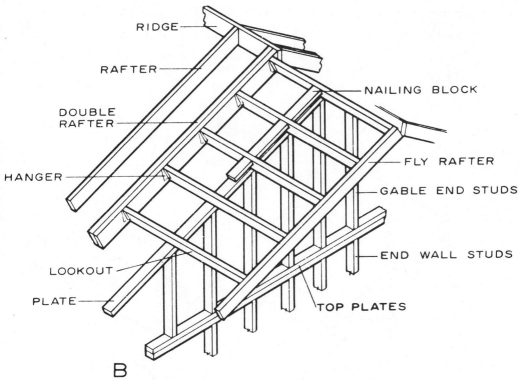

RIDGE

RAFTER

DOUBLE
RAFTER

HANGER

LOOKOUT

PLATE

NAILING BLOCK

FLY RAFTER

GABLE END STUDS

END WALL STUDS

TOP PLATES

B

53–11. *The nailing blocks are laid flat on the plate to allow airways into the attic.*

ground moisture by the use of a soil cover. Fig. 53-13. The soil cover should be a vapor barrier such as plastic film, roll roofing, or asphalt-laminated paper. Such protection will minimize the effect of ground moisture on the wood framing members. High moisture content and humidity encourage staining and decay of untreated members.

Where there is a partial basement open to a crawl-space area, no wall vents are required if there is some type of operable window. The use of a soil cover in the crawl space is still important, however.

For crawl spaces with no basement area, provide at least four foundation wall vents near corners of the building. The total free (net) area of the ventilators should be equal to $1/160$ of the ground area when no soil cover is used. Thus, for a ground area

of 1,200 square feet, a total net ventilating area of about 8 square feet is required, or 2 square feet for each of four ventilators. A greater number of smaller ventilators having the same net ratio would also provide satisfactory ventilation.

When a vapor-barrier ground cover is used, the required ventilating area is $1/1600$ of the ground area. For the 1,200-square-foot house, this would be 0.75 square foot. This area should be divided between two small ventilators located on opposite sides of the crawl space. Vents should be covered with a corrosion-resistant screen of No. 8 mesh. Fig. 53-13.

The use of a ground cover is recommended under all conditions. It not only protects wood framing members from ground moisture but also allows the use of small, inconspicuous ventilators.

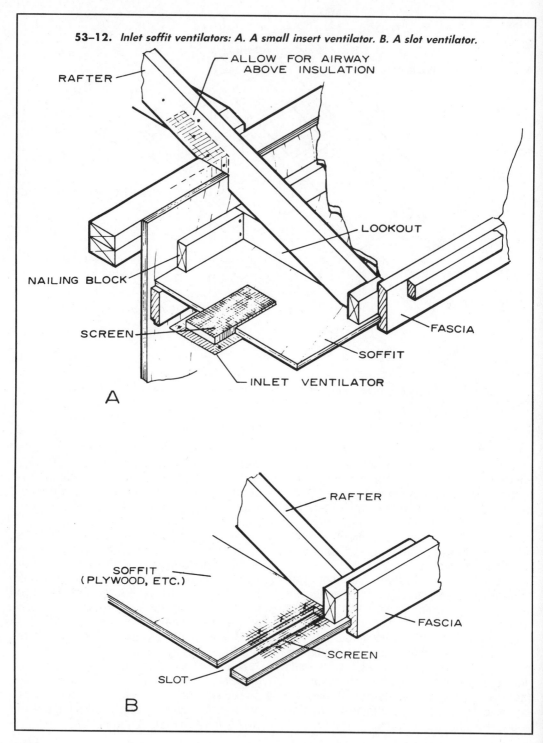

53–12. *Inlet soffit ventilators: A. A small insert ventilator. B. A slot ventilator.*

ALLOW FOR AIRWAY ABOVE INSULATION

RAFTER

LOOKOUT

NAILING BLOCK

SCREEN

FASCIA

SOFFIT

INLET VENTILATOR

A

RAFTER

SOFFIT (PLYWOOD, ETC.)

FASCIA

SCREEN

SLOT

B

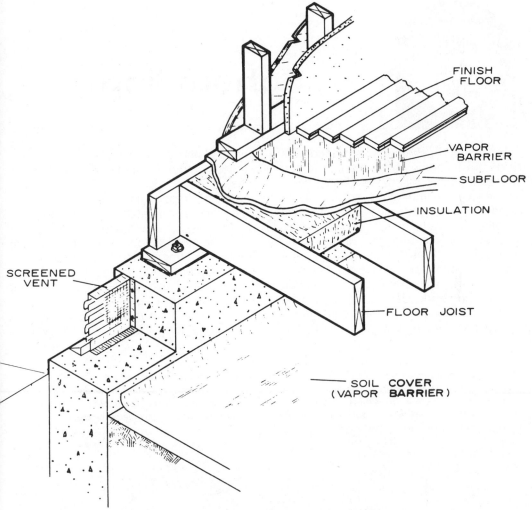

FINISH
FLOOR

VAPOR
BARRIER

SUBFLOOR

INSULATION

SCREENED
VENT

FLOOR   JOIST

SOIL   COVER
(VAPOR   **BARRIER**)

53–13. *A ventilated crawl space with vapor barrier.*

QUESTIONS

1. What is the most practical method of removing the moisture in a home?

2. What is meant by an ice dam?

3. What is the minimum net ventilator area required for a gable-roof home with an area of 1950 square feet?

4. Why is a greater ratio of ventilating area required in flat roofs than in pitched roofs?

5. Why is the use of ground cover recommended in crawl-space areas?

UNIT 54

Scheduling

Two types of scheduling are needed in building a home: job scheduling and material scheduling. It is the responsibility of the general contractor to set up these schedules and make sure they function smoothly. Naturally the size of the building project will have a great deal to do with the complexity of the scheduling. Figs. 54-1 and 54-2. A contractor who is building only a few houses with perhaps two or three workers often works part-time as part of the crew on the job site. The rest of the time is spent in coordinating the delivery of material and the work of the subcontractors. Larger contractors spend all their time on these matters.

It is very important that each subcontractor's work is done at the correct time so as not to delay the overall construction progress. The more subcontractors used by the general contractor, the fewer will be the general subcontractor's responsibilities in coordinating material scheduling, since subcontractors arrange for their own materials. One of the most common faults of small general contractors is that they try to

54-1. *In a building project of this size a special crew is responsible for each phase of construction. For example, one crew puts in the footings and the foundation wall; a second crew does the floor framing; another crew, the wall framing, etc. Supervisors who are directly responsible to the general contractor coordinate material deliveries and scheduling of the various work crews.*

do too many specialized jobs themselves such as tile setting or concrete work. Most small general contractors are former carpenters; so they should concentrate on this aspect of the building project.

The general contractor should always subcontract those jobs that require special equipment and skill. Such tradesmen as plumbers, heating and air-conditioning specialists, and electricians have the knowledge and skill to do their work efficiently. In many areas of the country the amount of specialized work of this type that the general contractor can do is limited by local building codes and trade union regulations. Most of the following jobs normally are subcontracted: surveying, excavating, concrete and masonry work, plumbing, electrical work, termite control, climate control, heating, air-conditioning, sheet-metal work, interior wall finishing (except wood wall paneling), painting and decorating, ceramic tile, floor covering, and landscaping.

MATERIAL SCHEDULING

Material scheduling must be coordinated by the lumber supply dealer who works with the general contractor. Since each general contractor operates in a slightly different manner, the lumber supplier must adapt the material schedule to the contractor's wishes. Naturally, the material deliveries vary depending on the type and size of the project, the number of people working on it, and the time set for completing the project. Fig. 54-3.

Generally, material deliveries are made as follows:
• The first load includes all items needed to complete the house up to and including the subfloor.
• The second load includes wall framing and ceiling joists.
• Third load—roof framing materials. If roof trusses are used, these will be shipped to the site on a special truck. This truck sometimes has a crane to lift trusses into position.

54-2. Many large homes are built by contractors with a work force of only two or three people.

54-3. General contractors who work on a building project such as that shown in Fig. 54-1 buy large quantities of materials from a single source in order to obtain a better price. These materials are then delivered to a central receiving area for redistribution at the job site.

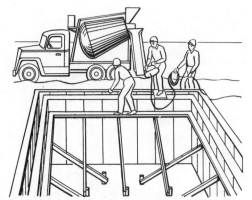

54–4. *Pouring a concrete foundation wall. Note that the forms are well braced to keep them aligned while the concrete is being placed.*

• Fourth load—exterior doors, windows, exterior trim, and siding as necessary. After the building is enclosed with doors and windows and can be locked up, the interior wall finish is applied. If the walls are plastered, adequate drying time must be allowed before additional material shipments are made.

• Fifth load—hardwood flooring and underlay materials.

• Sixth load—interior doors, trim, and built-in cabinet materials.

Materials to be delivered to the job site are stacked on the truck in the sequence in which they are to be used. In this way, after the materials are unloaded, those materials needed first will be on top of the pile.

It is the lumber supplier's responsibility to make frequent trips to the job site to keep track of the construction progress. The supplier must check on any missing materials and send the additional items needed to complete a job. Often some materials are left over, and these can be returned to the supplier for credit. The supplier keeps a running tally of the cost of materials shipped to the job site as well as any credits for returns. The general contractor is expected to pay for these materials on some time schedule, such as every week or month.

JOB SCHEDULING

In order to keep the building progressing at a smooth rate, the general contractor also has the responsibility of scheduling the jobs. Whenever it is necessary to have a subcontractor come in, the general contractor must arrange the time. Careful scheduling can minimize the frustrations of undue delays caused by subcontractors whose work needs to be done before other progress can be made.

The following is a list of steps in house construction. It is the general contractor's responsibility to see that these steps are carried out.

1. The job site is surveyed and the abstract brought up to date so that application for title insurance can be made.

2. A building permit is obtained from proper authorities so that work can begin.

3. The excavator brings in power equipment and strips the topsoil away, piling it in one corner of the lot for future use. If the building will have a basement, it is excavated at this time.

4. The electrical company is contacted to arrange for a power pole to be set in place on the building site and a hookup made. The electricity is needed for operating power tools.

5. On some job sites the plumber makes the temporary water hookup, which must be coordinated with the city utilities. Sometimes a power-driven water pump for a well is used. In existing neighborhoods, water can be obtained from a neighbor. In this case, the permanent hookup for water to the building is not made until the foundation walls have been installed.

6. Footings and foundation walls are installed by the concrete and masonry subcontractor. Fig. 54-4.

7. In areas that require it, a termite control specialist will treat the soil at the base of the foundation wall and the footings.

8. Pipelines for plumbing are installed in the subsoil by the plumbing contractor.

9. At this point the general contractor

must make certain decisions concerning how the foundation walls are to be braced to permit backfilling. Backfilling cannot be done unless the foundation walls are braced or unless the building is framed to provide plenty of weight to the foundation. Even when the walls and roof have been framed, the foundation must be braced to some extent before backfilling. Whether to backfill immediately after the foundation is in or to wait until after the framing is completed is the general contractor's choice.

If backfilling is delayed until framing is completed, the workers have the inconvenience of working around a foundation with a large excavation. This is unsafe and can lead to injuries because the carpenters normally have to use planking to carry materials over the excavation into the building. On the other hand, if backfilling is done before construction, extra time and costs are involved in bracing the foundation securely so that it does not cave in. Before any backfilling can be completed, the exterior walls of the foundation must be moisture-proofed.

10. The carpenters can now do the floor, wall, and roof framing. Fig. 54-5.

11. Chimneys and fireplaces are built by the masonry contractor after the rough framing is completed.

12. At this point in construction, a variety of activities may be carried out simultaneously, or at least in rapid succession. These include plumbing, heating, and electrical work. All of these mechanical subcontractors must work in two stages; namely, the "rough-in" and the "finish" work. Figs. 54-6–54-13 (Pages 1054-1058).

For example, when the rough framing is complete, the electrician comes in to do the rough wiring, including installing the outlet boxes and feeding all the wires through the framing. This is the rough-in portion of the work. Later after the interior walls are all completed, the electrician comes back to install the switches, outlets, and fixtures. This is also true of the plumb-

54–5. When the foundation is finished, the framing begins.

er and the heating and air-conditioning workers. (The plumber will install bathtubs with the rough-in, since tubs are a built-in feature of the house.)

13. While the mechanical subcontractors are doing the rough-in work, the carpenters install exterior doors, windows, and special framing. Fig. 54-14.

14. After all rough-in work is done, insulation is installed in the walls and as necessary in the ceiling. Fig. 54-15.

15. Most interiors of homes are finished either with gypsum dry wall or with plaster. If plastering is specified, this should be done immediately. At the same time, carpenters can work on the exterior of the building installing siding, exterior trim, and the garage door.

Normally, plaster is applied in two stages, the rough coat and the finish coat. Plenty of time must be allowed for drying between stages and after the finish coat. Often a week or ten days must be allowed

54–6a. *Boring holes in joists for the electric rough-in.*

54–6b. *A floor plan showing the electric details. The electrician uses this plan to locate the electric outlets. To pull the wire between the various outlets, holes must be bored in the framing members as can be seen in Fig. 54–6a. Each mechanical subcontractor works from a plan such as this.*

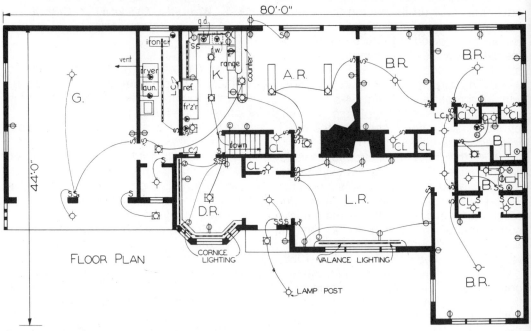

80'·0"

44'·0"

FLOOR PLAN

G.

K.

A.R.

B.R.

B.R.

D.R.

L.R.

B.

CORNICE LIGHTING

VALANCE LIGHTING

LAMP POST

B.R.

ELECTRIC LEGEND

LIGHTING OUTLET	SPECIAL PURPOSE OUTLET	S SWITCH OUTLET
DUPLEX CONVENIENCE OUTLET	FAN OUTLET	NITE-LITE
RANGE OUTLET	CLOCK OUTLET	FLUORESCENT VALANCE OR CORNICE LIGHTING

before proceeding with any interior work. If gypsum dry wall is installed, the drying period is much shorter since the only wet application is taping the joints and covering nailheads. Fig. 54-16 (Page 1059).

16. If the house has a basement, the concrete floor is poured after the rough plumbing is in but before the finish interior work is done, since concrete, too, must dry out thoroughly. The garage floor is put in anytime after the backfill is completed. Often this is done at the same time as the basement floor. The concrete is delivered to the site by special trucks. Fig. 54-17 (Page 1060). Concrete driveways and sidewalks, however, are installed at the very last stages of the building construction, after finish grading.

17. At this stage the carpenters are ready to do the interior finishing, provided the plaster and concrete are thoroughly dry. If lumber is delivered and stored in a house where there is high humidity due to wet plaster and concrete, the wood absorbs the moisture and swells. Later it will dry out and show large cracks.

18. In completing the interior, the first job is to install underlayment, then flooring, wood paneling, and finally the finish flooring. The interior frames, doors, and cabinets are next installed. Finally the interior moldings are applied, including the base, shoe, ceiling, and window trim. Fig. 54-18.

19. As carpenters are working on the inside of the house, painters can be finishing the exterior. The ideal arrangement is for the painters to work closely behind the carpenters on the exterior so that the wood is properly sealed. If exterior trim has been preprimed at the factory, this schedule is not so critical.

20. While the carpenters are completing the interior of the house, the exterior grading can be done and all flat concrete work such as sidewalks and driveways can be installed. Fig. 54-19.

21. The final step in completing the exterior of the house is the landscaping.

22. When the carpenters have completed the interior of the house, the painting can be done.

23. After the paint is dry, the wall tile, floor tile, and linoleum are installed. Fig. 54-20.

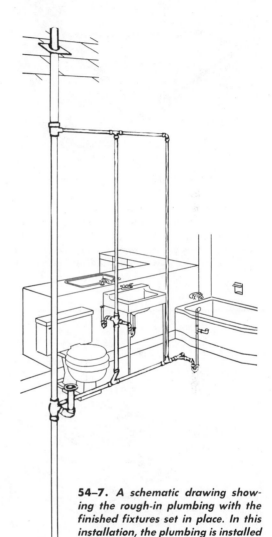

54-7. *A schematic drawing showing the rough-in plumbing with the finished fixtures set in place. In this installation, the plumbing is installed back to back on a partition wall. The kitchen sink is on one side with the bathroom fixtures on the other side.*

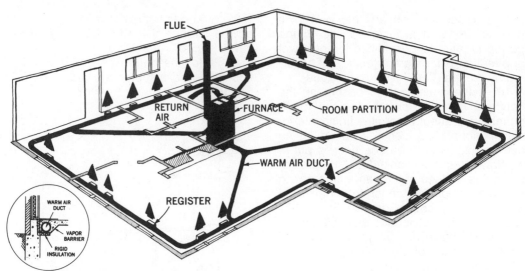

54–8. *Perimeter loop heating systems are often used in basementless houses built on a concrete slab.*

54–9. *Forced warm air systems are the most popular heating systems. Most installations have a cold-air return in each room (except the bathroom and the kitchen). When the basement is heated, additional ducts should deliver hot air near the basement floor along the outside walls.*

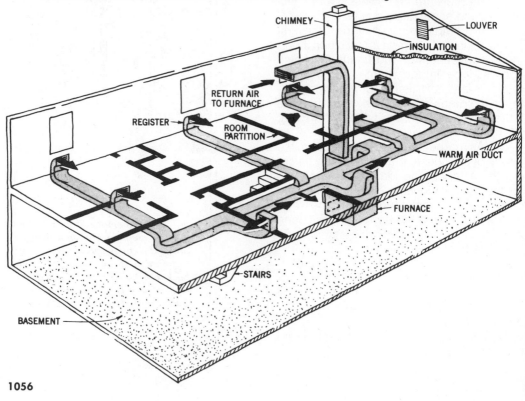

24. At this point the electricians can return to install switches, outlets, and fixtures.

25. The plumbing fixtures can now be installed by the plumbing contractor.

26. One of the last jobs on the interior of the house is to finish the fine wood flooring. Many homes are completely covered by carpeting and require no floor finishing. Carpets are often laid directly over a good plywood or particle board base. However, if hardwood floors are used, floor sanding should be done after the interior painting, to remove any paint drops or spillage. The actual finishing is done as one of the last jobs so that traffic will not raise dust when the floor finish is drying. Hardwood flooring can be purchased prefinished, which greatly simplifies this part of the job.

27. After the wood floors are finished, carpeting is installed.

28. The general contractor is responsible for the final cleanup. A good contractor will make sure that the windows are washed and all waste materials removed from the job site.

Throughout the many stages of building a home, a variety of inspections are done

54-10. *Two-pipe, forced hot water systems have two supply pipes or mains. One supplies the hot water to the room heating units and the other returns the cooled water to the boiler.*

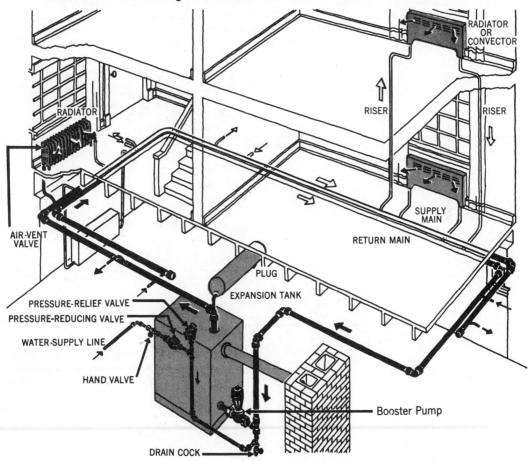

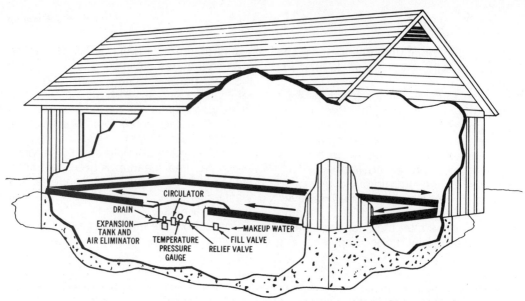

54–11. *Electrically heated hydronic baseboard systems are made in units so that several units may be connected to form a single loop installation. Water is circulated through the entire loop by pump. Each baseboard unit has a separate heating element so that the circulating water can be kept at a uniform temperature around the entire house.*

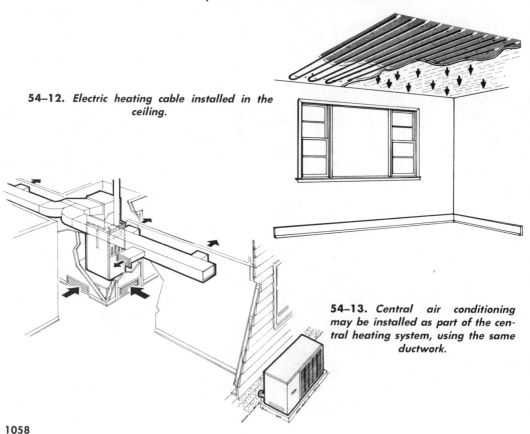

54–12. *Electric heating cable installed in the ceiling.*

54–13. *Central air conditioning may be installed as part of the central heating system, using the same ductwork.*

54–14. *Ripping 2 × 4s into 2 × 2s for construction of the kitchen cabinet soffits.*

54–15. *Installing insulation. Notice that the installer is wearing goggles and gloves.*

54–16. *Installing interior wall finish. The wood panel ceiling is installed first. Notice the rough-in electric wiring in the studs.*

by the local building inspector. Frequently, the contractor must obtain not only a general building permit but also special permits for plumbing, heating, electrical, and roofing work. These building permits specify when inspections must be made before further building can be done. The inspectors check to see that local building codes are being followed. If they are not, the inspector can require the contractor to make the necessary changes before further work can be done. The final inspection is made before a certificate of occupancy is issued.

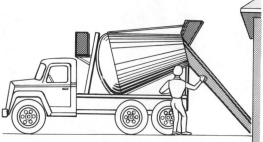

54-17. *Pouring the garage floor.*

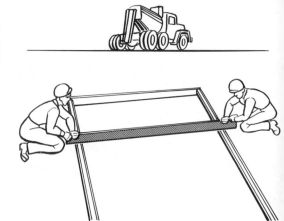

54-19. *Using a straightedge to strike off the concrete for a driveway.*

54-18. *Installing moldings. The finish floor has been laid and the wood paneling has already been installed.*

54-20. *Installing floor tile.*

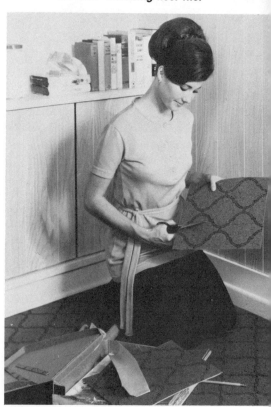

QUESTIONS

1. What are the two types of scheduling needed when building a home?

2. Why should the general contractor subcontract those jobs that require special skills and equipment?

3. List several jobs that are normally subcontracted.

4. What factors control the material deliveries?

5. How are missing or extra materials handled?

6. Why is job scheduling so important?

7. If you were the contractor, at what stage of the construction would you backfill? Why?

8. What kinds of jobs will the carpenters be doing while the mechanical subcontractors are doing their rough-in work?

9. Why must the interior trim work be delayed until the plaster and concrete are thoroughly dry?

10. Who is responsible for the final cleanup?

11. How does the contractor know when certain inspections are to be made by the local building inspectors?

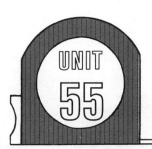

Manufactured Housing and Recreational Vehicles

UNIT 55

For hundreds of years, the only method of building homes was to construct them stick by stick on the site. All of these buildings were part of the *construction* industry. However, the trend today is toward *manufactured* (industrialized) housing. Fig. 55-1. It has been estimated that in the next 20 to 30 years, 60% of all housing will be produced in a factory. This method of building will not necessarily reduce the cost of housing. However, it will require fewer carpenters and other skilled craftsmen. In spite of this trend, carpentry will continue to be the single largest skilled trade in North America.

HISTORY OF INDUSTRIALIZED HOUSING

Industrialization was introduced to home building in the 1950s when prefabricated trusses became common. Builders and lumber dealers could assemble trusses on a simple jig with hammers, nails, and some adhesive. Later, trusses were produced in large quantities in manufacturing plants. The same idea has been utilized in developing prehung doors, prefinished paneling, and many other items that speed the work of building a home on site.

The next step in industrialization was the development of factory-built panels. Fig. 55-2. These panels were used along with prefabricated roof trusses or flat roof panels to speed the building of the house shell. In this way the house could become enclosed in a short period of time and be completed in inclement weather.

The *Unicom Method* of house construction, developed by the National Lumber Manufacturers Association, is another

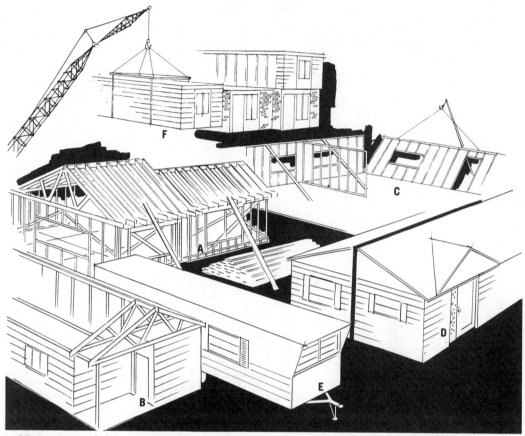

55–1. *Some of the ways to produce housing: A. Custom ("stick-by-stick method"). B. Precut with prefabricated parts. C. Panelized. D. Sectional housing. E. Mobile homes. F. Modular units.*

method of promoting industrialized housing. Fig. 55-3. All dimensions are based on multiples of 4″. The importance of using the modular planning grid as a design control is emphasized. Fig. 55-4.

An important advantage of the Unicom Method is that the wall, floor, and roof elements are not tied to any fixed panel size. A designer need not adhere to a fixed four-foot or larger increment. With this system, a complete house is divided into basic horizontal and vertical elements at regular modular intervals. Fig. 55-5.

Although the elements are shown as sliced planes without thickness, allowance is made for wall thickness and tolerance

variables based on fixed, not imaginary, module lines at the outside faces of the exterior wall studs. Complete exterior walls and partitions may have many overall thickness variables, depending upon whether or not they are load-bearing and which covering materials are used with them. Floor and roof construction elements vary in thickness, depending upon their structural requirements, type of framing, and finishing.

The exterior wall units are separated at natural points between solid partitions and window and door units. Many prefabricated units such as wall panels, trusses, and prehung doors can be used. Fig. 55-6.

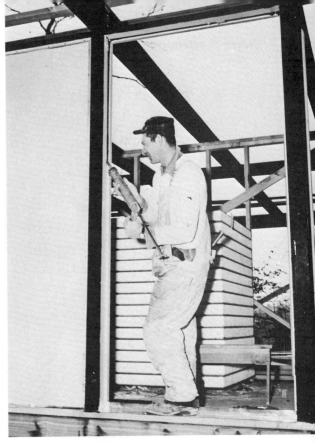

55-2a. *Lightweight, easy-to-handle prefabricated panels are used for rapid enclosure of this building frame.*

55-2b. *Prefabricated roof panels are lowered in place and installed.*

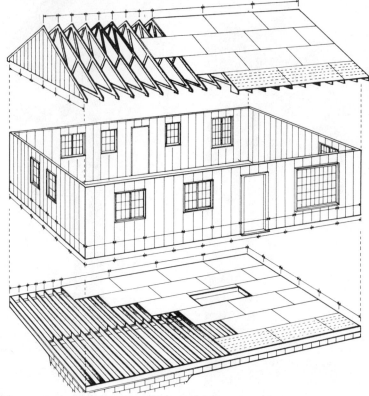

55–3. *The Unicom Method of house construction utilizes lumber framing members in modular sizes to eliminate waste.*

55–4. *Structures can be built with components or by conventional framing using this layout.*

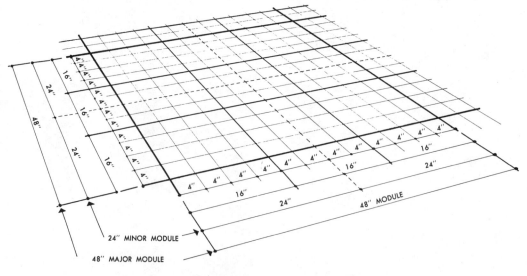

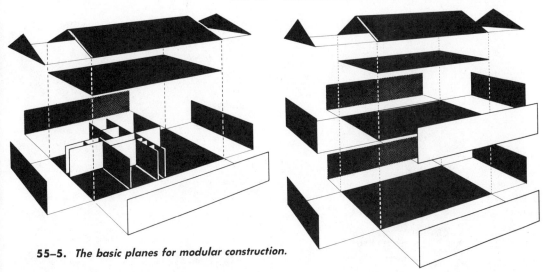

55-5. *The basic planes for modular construction.*

Panels became the heart of the *prefabricated housing* industry in the early 1960s. All parts of the house were precut and prefit at the factory and then moved to the site and assembled. However, prefabricated houses consisted of only the shell. They included none of the interior such as wiring, plumbing, and heating. Since the shell of the house only amounted to 25-35% of its cost, there was little difference in cost between the panelized house and a custom house built "stick by stick." During this entire period, mobile homes began to develop into more and more sophisticated housing. Fig. 55-7.

55-6. *Note how the various units are all standard multiples of 4", such as 32", 48", etc.*

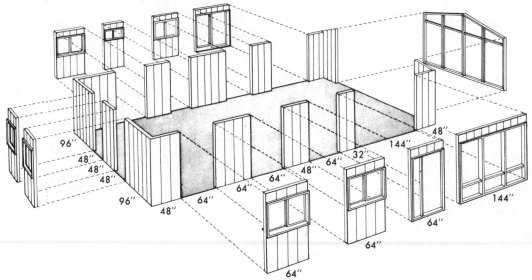

55–7a. *Well-planned and well-designed mobile home parks can be interesting places to live.*

55–7b. *Today's mobile home interiors are very attractive.*

55–8a. *A sectional home being put together on an assembly line.*

Another type of all-factory-built house is the *sectional* house. It is built on an assembly line and is completely finished on the inside. Fig. 55-8. It is moved by sections to the site and then assembled. The sectional house differs from modules in that all the sections are designed for a single home and may not be exactly the same. It is really a complete home built in sections in a factory so that it can be moved to the building site.

The most completely industrialized housing are *modules* (modular units). With the modular boxlike system, complete units of the structure are built, shipped to the site, and assembled with other units. The units are completely finished including rugs, draperies, and even furniture. In one motel built with modules, the beds were made and the towels hung in the bathroom before the motel was assembled. For small houses, a single module may be used; for larger houses, several modules. Fig. 55-9. The same modular technique can be applied to building apartment houses, motels, and commercial buildings.

TYPES OF MANUFACTURED HOUSING

Definitions

In order to understand the different types of manufactured homes and recreational vehicles, it is important to learn the definitions.

Mobile home. A transportable structure, which exceeds either 8 body feet in width or 32 body feet in length, built on a chassis and designed to be used as a dwelling. It may or may not have a permanent foundation when connected to the required utilities.

Double-width mobile home. A mobile home consisting of two sections combined horizontally at the site but retaining their individual chassis for possible future movement.

Expandable mobile home. A mobile home with one or more room sections that

55-8b. *Half of a sectional home being moved along a highway. Like mobile homes, their widths are limited to 12' or 14', depending on state laws.*

55-8c. *Lifting a section of the home in place.*

fold, collapse, or telescope into the principal unit when being transported. The sections can be expanded at the site to provide additional living area.

Modular unit. A factory-built, transportable building unit designed to be used by itself or to be incorporated with similar units at a building site. Modular structures can be used for residential, commercial, educational, or industrial purposes.

Sectional home. A dwelling made of two or more units which are put on a

55–8d. *Certain building parts are added at the site to make the sectional home appear more like a conventional house.*

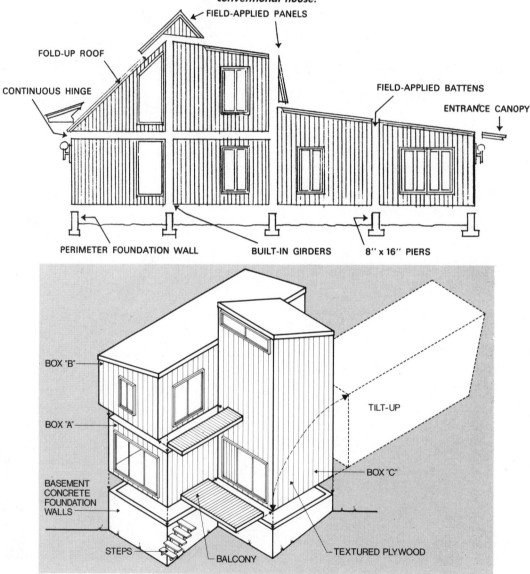

FIELD-APPLIED PANELS

FOLD-UP ROOF

CONTINUOUS HINGE

FIELD-APPLIED BATTENS

ENTRANCE CANOPY

PERIMETER FOUNDATION WALL

BUILT-IN GIRDERS

8″ x 16″ PIERS

BOX "B"

BOX "A"

TILT-UP

BOX "C"

BASEMENT CONCRETE FOUNDATION WALLS

STEPS

BALCONY

TEXTURED PLYWOOD

foundation and joined to make a single house.

Mobile homes, double-wides, sectionals, and modules are transported to their sites by trucks, whose movements are controlled by state highway regulations, or they are shipped on railroad flatcars.

Recreational vehicles. Trailers, campers, and motorized homes. They are designed for moving along highways either as part of the motor vehicle or attached to it by trailer hitch. They contrast with all of the other industrialized housing listed here because they are built for constant movement from one place to another. All other types of manufactured homes are de-

55–9c. *The floor plan for one apartment unit. Note that it is designed with standard width and length modules for easy assembly. Each module is limited to an overall width of 12' for easy transport.*

55–9a. *A single modular unit can be used for one small home.*

55–9b. *This attractive apartment house was assembled from modular units.*

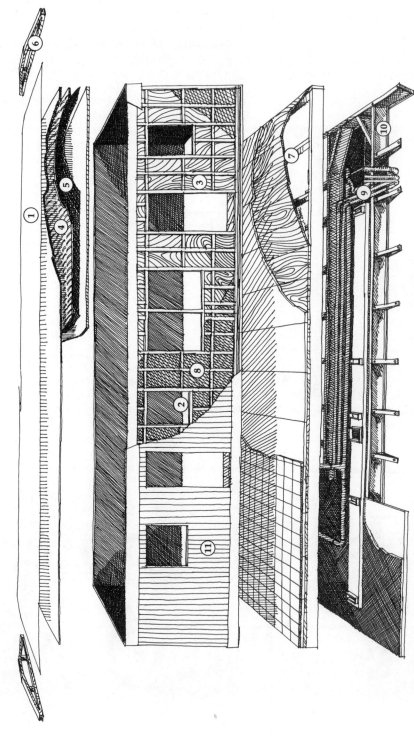

55-10. Construction features of this mobile home include: 1. Galvanized, raised steel roof. 2. Ventilating window in bath. 3. Sturdy 4" sidewall construction. 4. & 5. Double fiber glass insulation with double vapor barrier in roof. 6. Truss-type, built-up roof rafters. 7. Sturdy, reinforced floor joists overlaid with 5/8" plywood or manufactured board, glued and mechanically secured. 8. 1 1/2" high-density fiber glass insulation in sidewalls and floor. 9. Electrical, heating, and plumbing systems meet or exceed specifications of Trailer Coach Association, Mobile Home Manufacturers Association, and American Standards Association. 10. Basement-type frame, with enclosed plumbing system. 11. Prefinished, lifetime aluminum siding.

signed to be moved by heavy truck or by railroad flatcar to a permanent location.

Mobile Homes

The mobile home industry is the most rapidly growing area of industrialized housing. Well over $2½ billion goes into production annually, by approximately 380 different firms. Mobile homes come completely furnished with major appliances, drapes, lamps, carpeting, furniture, and everything else necessary for living. The home is centrally heated by gas, oil, or electric furnace. Mobile homes are available in widths from 8' to 14' and in lengths from 35' to 70'. The width of a mobile home is limited by what each state will permit to pass on the highway. Many states limit this to a 12' width, while other states allow homes 14' wide. By using a double-width mobile home, widths from 16' to 28' are possible.

MANUFACTURE OF MOBILE HOMES

Mobile homes are built on an assembly line. They differ from standard housing in that lighter materials are generally used for framing and all other items such as built-ins and cabinets are on a smaller scale. Typical framing is 2" × 2", rather than the standard 2" × 4". A great deal of plastic and metal is used.

Mobile homes consist of five major parts:
- Frame.
- Floor assembly.
- Wall assembly.
- Roof assembly.
- Interior. Fig. 55-10.

The base, or frame, for mobile homes is a heavy steel chassis equipped with a single, double, or triple axis depending on length. Fig. 55-11. This is moved into place on the assembly line.

Next the floor framing is built. Fig. 55-12. Heating ducts, rough plumbing, and wiring are installed in the floor assembly frame. Fig. 55-13. The frame is insulated and covered with tongue-and-groove plywood

55-11. *Heavy metal frames are used in constructing a mobile home.*

55-12. *Two-by-six (2" × 6") floor joists, 16" on center, are joined and bolted to the frame to make a solid no-sag floor.*

55–13. Plumbing is installed in the floor assembly.

particle board sheets. Figs. 55-14 and 55-15. Another layer of insulation is usually installed over the plywood. When the floor is completed, the necessary heavy mechanical items such as the furnace and hot-water heater are installed. Fig. 55-16.

Next the wall frames are assembled on jigs and lifted into place. Fig. 55-17. The necessary wiring, plumbing, and insulation are installed and the exterior shell is completed. Fig. 55-18. The roof assembly is attached to the frame of the house. Fig. 55-19. The interior is then completed with cabinets, partitions, appliances, furniture,

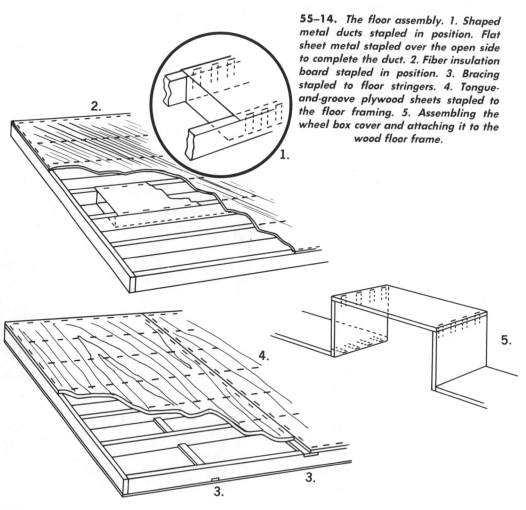

55–14. The floor assembly. 1. Shaped metal ducts stapled in position. Flat sheet metal stapled over the open side to complete the duct. 2. Fiber insulation board stapled in position. 3. Bracing stapled to floor stringers. 4. Tongue-and-groove plywood sheets stapled to the floor framing. 5. Assembling the wheel box cover and attaching it to the wood floor frame.

55-15. The plywood floor is stapled to the floor framing.

55-16. Large items like the hot-water tank are installed before the walls go up.

55-17a. Stapling the wall framings together.

55-17b. A sidewall section before insulation is installed.

55–18a. *Wiring is run through the sidewalls.*

55–18c. *The windows are also stapled into position.*

55–18b. *Sheets of aluminum siding being installed.*

rugs, and all other installed items. Fig. 55-20.

The completed mobile home is moved off the assembly line ready to be trucked to the site. Fig. 55-21. Most mobile homes are moved to the site by heavy-duty trucks and placed on a simple foundation. The necessary mechanical connections are made and the house is ready for use.

Most of the workers in mobile home factories do only one job, such as stapling wall assemblies together, wiring electrical circuits, or installing hot-water heaters. Therefore few highly skilled craftsmen are needed.

Modular Housing

The manufacture of modular housing differs from that of mobile homes. Standard-size materials are used in its construction; so the complete modular unit is at least the same strength as a house built "stick by stick." In fact, with modern industrial techniques the modular unit is stronger, better insulated, and more nearly sound-

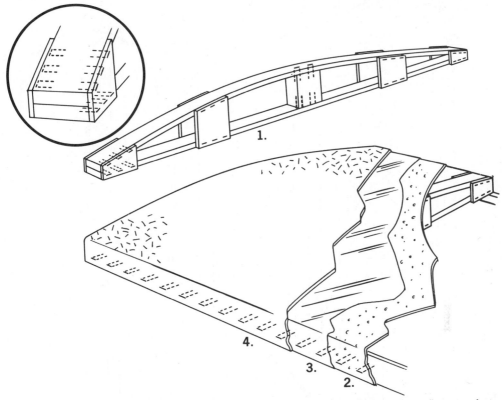

55-19. *Roof assembly: 1. Plywood scab blocks stapled to each side of 1″ × 2″. 2. Insulation sheathing between roof metal and inside ceiling stapled in position. 3. Vapor barrier of film material over roof stapled to wall. 4. Sheet metal placed on roof.*

proof than a house built by the traditional method.

Factories which produce modular housing use a great deal of heavy-duty equipment for cutting, assembling, nailing, stapling, and moving the units along an assembly line. Fig. 55-22. Modules produced on any assembly line are identical in size and shape for a particular model. Several modules can be assembled on site and some details added to give them a wide variety of appearances.

Recreational Vehicles

Trailers, campers, and motorized homes are built a great deal like mobile homes. Fig. 55-23 (Page 1078). However, there is much more plastic and metal used in most of these vehicles. Frequently the entire

55-20a. *Installing electrical outlets on the inside of the mobile home.*

1075

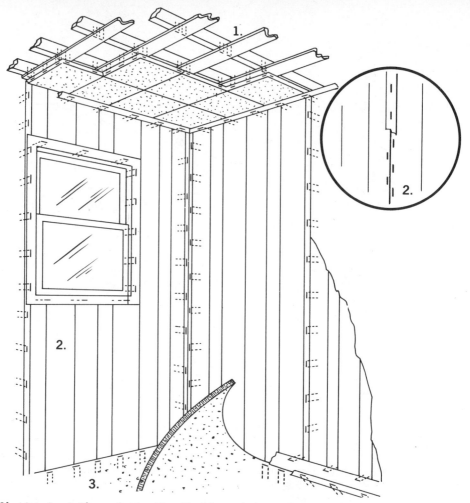

55–20b. *Interior: 1. Plywood strips 4″ to 5″ wide stapled to roof truss. Ceiling tile stapled to plywood strips. 2. Wall paneling stapled to side. 3. Carpeting and/or padding stapled to subfloor.*

55–20c. *Appliances and furniture are moved into the mobile home to complete it.*

55–21. *Moving the finished mobile home out of the factory.*

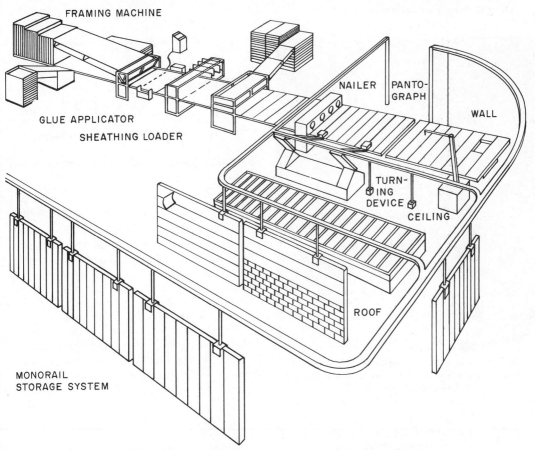

FRAMING MACHINE

GLUE APPLICATOR

SHEATHING LOADER

NAILER PANTO-GRAPH

WALL

TURN-ING DEVICE

CEILING

ROOF

MONORAIL STORAGE SYSTEM

55–22a. *An assembly line section in a modular home factory.*

55–22b. *A section of the assembly line in which wall panels are put together.*

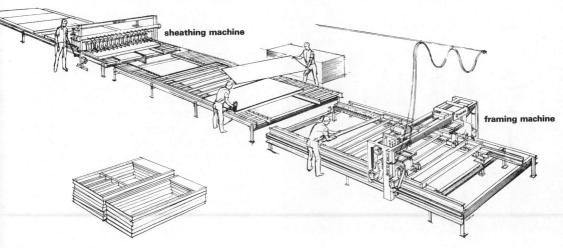

sheathing machine

framing machine

55-22c. *This stapling machine is part of the sheathing machine that is used to fasten the wall panels together.*

55-23a. *Motorized homes are made in many types and styles.*

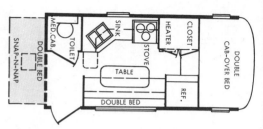

55-23b. *Campers must have a compact design.*

exterior shell of the unit will be of fiber glass construction. Fig. 55-24. Another major difference from other industrialized housing is that the appliances, cabinets, bathroom fixtures, and other items are very small in size and very compact so that everything will fit into the limited space available. Fig. 55-25. Most interior items are also designed to be multi-purpose, such as a davenport by day and a bed by night. Tables that fold into the wall and similar efficiency techniques are also used.

QUESTIONS

1. Approximately what percentage of housing will be produced in factories in the next 20 to 30 years?

2. What part of the house was first prefabricated?

3. What is the basic unit of measurement in the Unicom Method?

4. What is the difference between a sectional house and a modular house?

5. What controls the width of mobile homes? Explain.

6. How do recreational vehicles differ from other industrialized housing?

7. How large is the mobile home industry?

8. How do mobile homes differ from modules in construction?

9. What kind of equipment must a factory have to produce modules?

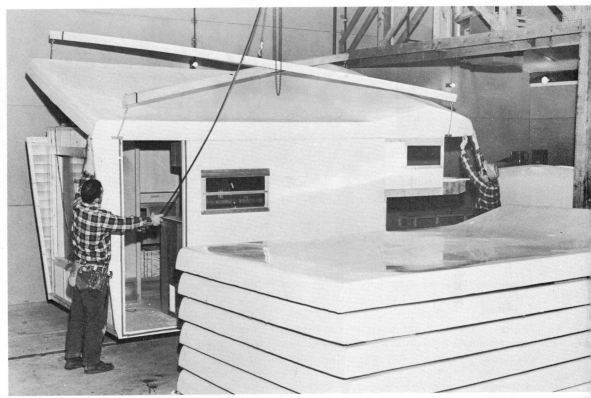

55–24. *This camper has a one-piece fiber glass roof.*

55–25. *Prefabricated plumbing being installed in a camper.*

Building Structural Models

Building a model is an excellent way to convert a drawing into a three-dimensional object that can be used to visualize the overall appearance of the project. There are two basic types of models that are used: the *architectural* and the *structural* model. Figs. 56-1 and 56-2.

The architectural model is designed primarily to show the exterior of a building. It is often constructed of solid materials, thin balsa wood, illustrating paper, and other kinds of simple materials. It is usually built to a scale of $1/4'' = 1'$ for homes, and to a scale of $1/8'' = 1'$ for larger buildings. However, this kind of model is of primary interest to architects and builders who need to determine exterior appearance.

A structural model on the other hand is built to the exact structure of the building but to a smaller scale. Framing members of pine or some other soft wood are cut to scale and assembled exactly the same as a house. Nothing, of course, replaces the experience of actually cutting, fitting, and nailing full size 2 × 4s and other structural members to produce a house. Yet, much can be learned about building construction methods by building a structural model. This is an excellent way to study working drawings, to learn the names and parts, and to get some idea of the way to construct a wood-frame building. It is particularly useful when space does not permit the construction of a full-sized building.

HOUSE PLANS

Any standard set of house plans can be used to build a structural model. The plans can be followed as though they were being used for full-sized construction.

SCALE

The best scale for a structural model is $1^{1}/_{2}'' = 1'$, or one-eighth as large as a full-size house. For example, if the real home measures 24' × 40', the miniature house would be 3' × 5'. Framing members are cut to the *nominal,* or *name,* size. For example, 2 × 4s will measure $1/4'' \times 1/2''$, and a 2 × 6 will measure $1/4'' \times 3/4''$. An architect's scale can be used for most of the measuring and layout.

MATERIALS

A wide variety of materials can be used in constructing the model. These include:

Lumber. Structural members such as the joists, studs, plates, and so forth, should be referred to by their trade name (nominal size) such as 2 × 4s, or 2 × 6s rather than by the scale dimensions.

The structural parts can be cut from basswood, redwood, yellow poplar, or pine of softer variety. Woods such as Douglas fir can be used for larger members such as the foundation walls or footing. The woods used for structural model constructions should be the type that do not split easily. For parts that require a good deal of shaping, balsa wood is also useful. The lumber can be secured from the mill ends or shorts available at a lumberyard, or from waste stock found around any building project.

Plywood and other board materials. Plywood, particle board, or hardboard can be used for the base of the model.

56–1. *Architectural model.*

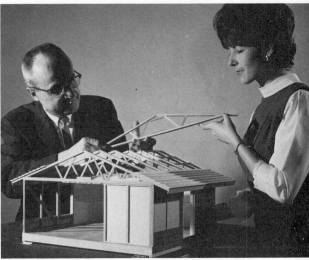

56–2. *Structural model.*

Veneer. Veneer can be used for exterior and interior walls. Heavy illustration board can also be used.

Nails. Nails and other metal fasteners should be used in proportion to the size of the model. The nails for most of the house framing should be 1/2", 19- or 20-gauge wire nails. For spiking together joists, headers, and other large members, use 3/4" No. 18 wire nails. Use 1/2" nails or staples for attaching shingles.

Other materials. A wide variety of common household items and other materials may be needed, including plaster of paris or rigid foam plastic for a foundation, materials like sponges for shrubbery, and sandpaper for shingles.

TOOLS FOR MODEL BUILDING

The tools needed for model building are relatively simple and easy to obtain. They include the following:
- Rafter square.
- Framing square.
- Model builder's knife.
- Pencil.
- Tack hammer.
- Eight-ounce magnetized or upholstery hammer; seven-ounce magnetized hammer to drive nails.
- Long-nosed pliers.
- Hacksaw, 11-point, or dovetail saw to cut scale model materials.
- Chisels for trimming rafters, bases, and other fine work.
- T square, triangle, and architectural scale.
- Sandpaper and masking tape.
- Bench hook, used for holding stock while sawing. Fig. 56-3.

56–3. *Bench hook for holding structural model parts for sawing.*

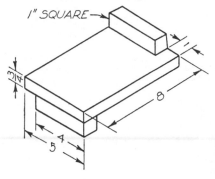

- Brad awl.
- Stapler.
- Miter box for cutting angles for such pieces as rafters and bases, Fig. 56-4.

JIGS

Simple jigs can be made for supporting framing members 16" on center for floor and wall framing. Another method of doing floor and wall framing is to make a scale drawing of each section; namely, wall, front elevation, back elevation, right elevation, and left elevation. Fasten this to a piece of plywood to use as a guide in assembling the section.

A simple jig can also be made for holding the parts for making roof trusses. Fig. 56-5. Use a fast-acting glue to help hold the parts together while nailing.

CONSTRUCTION PRACTICES

Model Base

A model base can be made of 3/4" plywood or particle board that is cut to lot size. Fig. 56-6. A slightly lighter base can be made by utilizing 1/4" plywood with a 1" × 2" framework underneath. Diagonal bracing is necessary to give it rigidity. By using lot size for the base, all the landscaping can be put in to give the house a completed appearance.

Foundation

Carefully lay out on the model base the outline for the foundation of the home. If there is a basement, make sure that the foundation wall is the same height above the model base (grade) as it would be in actual construction. If the house has no basement, a 1/4" piece of plywood can represent the concrete slab.

The foundation walls can be made of rigid plastic foam, molding plaster, or spackling compound. The foundation can also be made of wood scaled to the correct thickness and to the height above grade as indicated in the working drawing. Fig. 56-6. Make the termite shield of thin aluminum or copper foil.

Floor Framing

If the house plans call for a basement, the first step is to install a "steel" beam to support the floor framing. This beam can be made of several layers of thin wood that are glued and nailed together and painted an orange color to resemble the steel I-beam.

Now build the sill plate, the floor joists, and the box sill to complete the floor framing just as it would be done in full-sized construction. Cover the framing with thin veneer to serve as the subfloor. If you wish to expose part of the floor framing, cut a free-form opening in the veneer. Install bridging between the floor joists at the exposed sections.

Wall Framing

Construct each wall section according to a plan. Using a jig on a tabletop will facilitate spacing of studs and locating and framing of window and door openings.

Nail the first top plate solidly to the studding. The second top plate should be nailed to the ceiling joists or roof trusses. Exterior sheathing can cover the crack between the two top plates. This arrangement will permit the removal of the entire roof section so that the interior of the home can be easily seen.

56–4. *Miter box for cutting angles.*
4½" WIDTH & 4½" LENGTH MAKE 45° CUTS

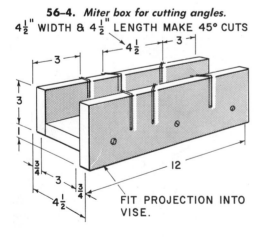

FIT PROJECTION INTO VISE.

Nail the wall sections to the subfloor, one at a time. Make sure that the walls are plumb and that the corners are tight. Now build the interior walls. Fig. 56-7.

Roof Framing

Build the roof by starting with a second top plate, adding the ceiling joists, and then the roof framing members or roof trusses. Fasten these all together to comprise one unit that can be removed when necessary. Fig. 56-8.

Exterior Wall

Complete at least the front of the model by adding the exterior wall, consisting of the sheathing, building paper, and exterior wall. The sheathing can be made of veneer

56-7. Wall framing.

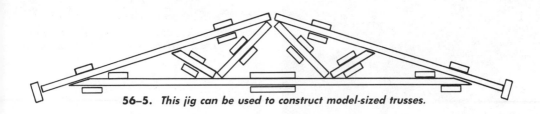

56-5. This jig can be used to construct model-sized trusses.

56-6. Base and foundation of the model.

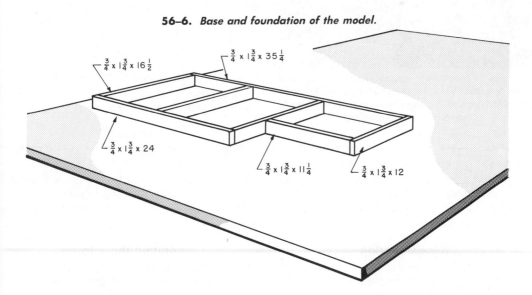

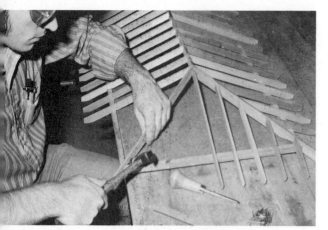

56–8a. *Roof framing.*

56–9. *Full-size mock-up. Standard-sized materials are used, including such mill items as windows.*

56–8b. *A sectional model showing standard construction. Note the use of various materials to simulate actual building materials such as brick, roofing, steel beams, etc.*

or heavy construction paper. The building paper can be plain, thin black paper and the wall material can be made of thin balsa sheets. Wood siding can also be made from thin balsa wood.

Roof Covering

A thin veneer can be used for the roof sheathing. Many other materials can be used for the roof itself. For example, sandpaper might be used or miniature wood shingles can be cut. Only a small portion of the roof should be completed. On a gable or hip roof, for example, most of the front slope should be covered while the roof frame remains exposed.

Chimneys and Other Details

Chimneys and other details of brick or cement can be made of molding plaster.

Millwork

It is extremely difficult to build in miniature the millwork, such as doors and windows. In some cases, clear plastics can be used for windows and the framing can be painted on. Doors can be made of thin plywood.

Completing the Interior

The interior of the home can be completed in small details to show some of the construction materials. For example, dry wall construction can be made by fastening a section of construction board to part of the interior. Several kinds of materials can be used to simulate the finished flooring such as tile, ceramics, and rugs.

LANDSCAPING

The main purpose in constructing the structural model is to gain experience in framing a house and building a roof. Therefore it is not necessary to add every exterior detail unless you wish a model for display purposes. Many kinds of materials can be used for landscaping. Grass areas, for example, can be made from green flocking, and the walls, driveways, and other items with various kinds of masking tape. Trees and shrubs can be made of wire and foam rubber. The creative model maker will find uses for many common household items to complete the model. Fig. 56-1.

FULL-SCALE PARTS OF A BUILDING

To gain experience in working with full-size materials, it may be desirable to construct a full-size section of a house, wall, or roof. Fig. 56-9. Standard house plans can be followed.

QUESTIONS

1. What is the difference between an architectural and a structural model?
2. When building a structural model, why is a ⅛ scale recommended?
3. What are some of the materials that might be used for the foundation wall in a structural model?
4. What can be done to permit the removal of the entire roof section of the structural model?
5. What are some of the benefits gained from building a structural model?

UNIT 57 Energy Conservation

ENERGY USE IN THE HOME

About one-fifth of our total national use of energy is for climate conditioning of houses, apartments, condominiums, mobile homes, and other living units. Energy is needed for heating, air conditioning, humidifying, dehumidifying, and air cleaning. As the cost of energy rises and the supply diminishes, it becomes increasingly important to make all types of living units more energy efficient. Anyone planning to build or remodel a living unit must consider every possible way of reducing energy needs. Also, this must be done without adding too much to the cost of construction. There are energy-saving materials and devices available that are not too expensive and save enough energy to justify their cost.

Energy is used in daily living in a wide variety of ways. Precious energy is consumed in almost everything we do, such as cooking a meal, switching on a light, watching television, taking a shower, or washing clothes. When the weather turns cold or hot, the wind blows, or we have rain, energy is consumed to compensate for these conditions.

Of the many factors that influence internal temperature, the most important is outside weather. If it is cold outside, heat is lost through the living unit in many ways:
• By *exfiltration,* the conduction of heat through the exterior surfaces and leakage of warm air to the outside through cracks in windows, walls, and doors.
• By *infiltration,* the leakage of cold outside air into the house.
• By *radiation,* in which warmth is lost to the sky and the surroundings. (A warm body tends to lose its heat by infra-red radiation.)
• By *heat loss through water drains.* Energy is used to heat water. As the hot water goes down the drain, this energy is lost from the home. Fig. 57–1.

It has been estimated that as much as 40 percent of heat loss is due to exfiltration and infiltration.

Heat is generated and gained in living quarters by appliances, electric lights, and people. Some heat is also gained through solar energy, even in the standard home. In cold weather, the difference between heat loss and heat gain must be made up by some type of heating unit to maintain a comfortable temperature range of 68 to 78° F. (20 to 26°C). In addition,

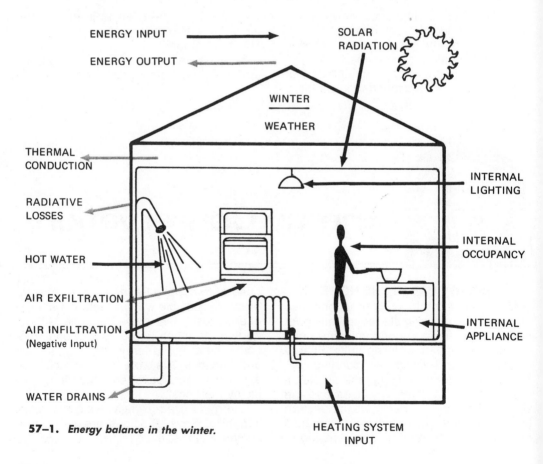

57–1. *Energy balance in the winter.*

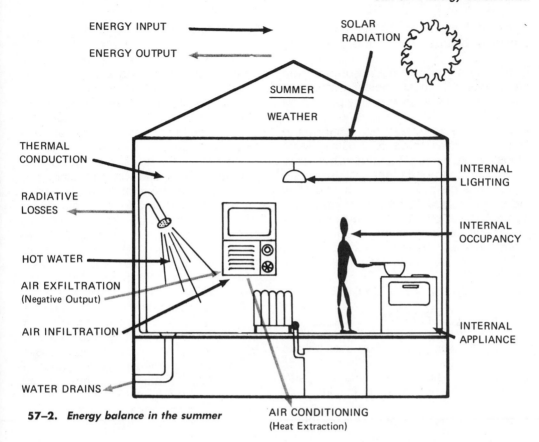

57-2. *Energy balance in the summer*

energy is used to clean the air and to add humidity for comfort.

In warm weather, many of the same factors are at work. When the outside temperature is greater than the indoor temperature, heat tends to flow into the house (infiltration). To maintain comfort, an air conditioner is often used to remove this excess heat. Excess moisture is removed from the air by a dehumidifier. Fig. 57–2.

There are three ways construction can aid in energy conservation:

1. Standard homes can be made more energy efficient with such things as proper insulation, energy-saving sheathing, storm doors and windows, proper calking, and a light color of roofing.

2. Energy-saving (conservation) homes can be designed using different materials and methods to increase energy conservation.

3. Solar energy can be used for heating water and/or for a large part of the space heating and cooling. However, even homes planned for solar energy must also have some kind of supplementary heating and cooling units.

In planning any structure to include energy-saving devices, the following must be given careful consideration:

• Will the device or material save sufficient energy to pay for itself within a ten- to fifteen-year period? At current costs, many solar energy units cannot meet this requirement without government assistance.

• Will the structure be within current construction capabilities and local building codes?

• Will the structure be acceptable from an architectural standpoint? For example, will tiny windows sparsely placed be acceptable?

• Will the design drastically change the lifestyle of the residents? In the years ahead, most living units must be smaller and more space efficient. For example, it may not be possible to have both a living room and a family room. Many current house plans call for a *great room* as an all-purpose living area. Today, fewer homes have separate dining rooms, and bedrooms are smaller.

BUILDING MATERIALS

Plastics

The use of plastics in building construction is increasing rapidly. It is estimated that by the twenty-first century, plastics may be the major building material, surpassing even wood. Much of this increase is due to demand for energy-efficient materials in insulation, sheathing, and siding. Plastic materials offer characteristics and properties different from many natural materials. Some advantages of plastics are:

• Light weight.
• Good electrical insulating properties.
• Good heat insulating properties.
• Resistance to atmospheric corrosion.
• Attractive appearance.
• Lower cost.
• Variety in design and styling.

Plastics may be defined as nonmetallic materials that are capable of being formed or molded with the aid of heat, pressure, chemical reaction, or a combination of these. There are many different types of plastics used in building construction. The following nine types are the most important:

PVC (polyvinyl chloride) is by far the most important plastic material for building construction. It is made in both rigid and flexible form. PVC is used widely for siding, sash, roofing, and interior trim. It can be used for all types of wiring insulation and in home plumbing systems for cold water lines and for sanitation.

ABS (acrylonitrile-butadiene-styrene) is a tough, colorful material that has many uses in building construction for drain-waste-ventilating pipe and fittings. Fig. 57–3. Solid ABS is used for sewer pipe, conduit, and air and water pressure pipe. It can be chrome-plated for such products as faucet handles and soap dispensers. ABS can be formed by heat and pressure (thermoformed) into bath and shower units. There is also an ABS foam-core pipe that is somewhat less expensive than solid ABS pipe and which has wide usage in sewer, conduit, and duct work.

Acrylic is popular for glass replacement since it will not break. Acrylic is thermoplastic, which means that it can be easily formed for such items as skylights, covers for basement windows, and lighting fixtures. Acrylics make excellent sealants and calks that adhere to most construction materials, such as metal, glass, ceramics, plastics, wood, masonry, and concrete.

57–3. *Pipe for drainage, waste, or ventilation can be made of ABS or PVC plastics.*

Rigid plastic foams play an important part in energy conservation. The materials are superior insulators against heat loss and are being used more and more for sheathing. However, plastic foams have no structural strength, and great care must be taken in installing them. The most common kinds of rigid plastic foams are Styrofoam® (polystyrene), urea-formaldehyde, and polyurethane. Another foam used for sheathing is isocyanurate.

High-pressure laminates, such as the decorative laminates described in Unit 46, are important interior design materials. They are commonly used for counter tops, walls, and many other interior surfaces.

Fiber-reinforced plastics are made primarily of spun glass and polyester. These materials are widely used for bathtubs, showers, and vanity units. In fact, fiber-reinforced plastic bathtub and shower units account for a majority of the market, surpassing steel and ceramic units by a wide margin.

Epoxy resins are used in various ways in construction, such as for finishings, or assembling parts, and for adhesives. Epoxies can be used in assembling uints of wood, steel, or concrete.

Acetal plastics have high impact strength and chemical resistance and have been widely used n furniture hardware for such pieces as drawers, swivel components in chairs, and casters. Two types of acetal are now extremely important in the plumbing industry. Acetal homopolymer is replacing copper and brass for plumbing parts. Acetal copolymer is used for plumbing fittings formerly made of brass. Its uses are endless for such products as faucets, faucet underbodies, lavatory basins, valves, fittings, shower heads, faucet sprayers, pipe couplings, and many other items.

Polycarbonate resins and sheets can be used in place of glass to make non-breakable windows. Extruded polycarbonates are used for window and door frame componets.

Adhesives, Sealants, and Calks

Many different types of adhesives are needed in the manufacture of building products, particularly such materials as plywood, hardboard, and particle board. On the job, the builder uses several different kinds, the primary ones being contact adhesives (cement), elastomeric construction adhesives (mastics), and calks. Other types of adhesives are used by tile setters and floor covering specialists.

Contact adhesives, or cements, are used in installing plastic laminates for counter tops and walls. Elastomeric construction adhesives (mastics) have the consistency of putty and are used for fastening plywood, hardboard, and particle board to floor joists, ceiling joists, and studs. Mastics usually come in cylindrical containers that fit into calking guns or in cans. They are also available as ribbons of material that can be pressed into place. Mastics have good gap-filling properties and may be used with or without supplementary nailing or stapling. There are many kinds of mastics, some synthetic and others of natural rubber. Most of the modern mastics are of plastic origin, such as acrylic products.

Many different types of calks and sealants are available. Polysulfide, polyurethane, and silicone types are preferred. Calking may be semi- or slow-drying. Most calking comes in a sealed cartridge for use in a calking gun. Calking is also available in a can for use with a putty knife. Fig. 57-4.

When using a calking gun, hold it at a 45° angle to the surface and squeeze with steady pressure. Keep the rear of the calking gun slightly slanted toward the direction you are moving. Slowly draw it along so that the sealant not only fills the crack, but also overlaps the edges. Move the gun with a pulling, not a pushing, motion. To get a smooth bead, fill a single seam in one stroke.

It is important that the builder select the correct kind of adhesive and/or calk for

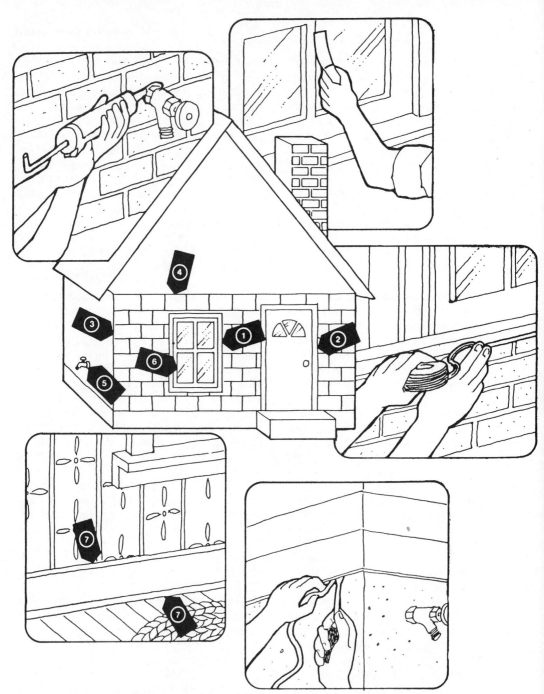

57-4. *Calking should be applied: 1. around the window where the frame meets brick, siding, or sheetrock, 2. along the top and sides of the door where the frame meets brick, siding, or sheetrock, 3. where wall meets wall, 4. where wall meets roof overhang, 5. around water faucets, using tube-type calking and a calking gun, 6. around window panes and frames, using glazing compound where glass meets frames, 7. where baseboard meets wall, with a ribbon-type calk.*

Sheathing Materials

	¾" foil-faced kraft-fiber ply	1" polystyrene (beadboard)	1" Styrofoam	¾" foil-faced urethane	½" plywood	½" fiberboard (standard grade)	½" gypsum
Can it be overlapped to stop air infiltration?	yes	no	no	no	no	no	no
Can it be used without corner bracing?	yes	no	no	no	yes	no	yes
Is it damage resistant, <u>not</u> vulnerable to vandalism, chipping of edges, breakage, or puncture?	yes	no	no	no	yes	no	no
Can it increase R-value of wall beyond traditional ½" wood fiber?	yes	yes	yes	yes	no	no	no
Lightweight, easy to lift?	yes	yes	yes	yes	no	no	no
Cuts easily with a knife?	yes	yes	yes	yes	no	no	no
Provides strong base for siding application?	yes	no	no	no	yes	yes	yes
Low burn rate?	yes	no	no	no	yes	no	yes
Requires no jamb modification for traditional windows or door sizes?	no	no	no	no	yes	yes	yes
Adds structural strength beyond traditional ½" wood fiber?	yes	no	no	no	yes	no	yes
<u>Does not</u> require any special interior wall materials such as drywall for safe installation?	yes	no	no	no	yes	yes	yes

57-5. *Advantages and disadvantages of common sheathing materials.*

57-6. *Styrofoam® (extruded polystyrene) applied horizontally as sheathing.*

each particular purpose and to follow the instructions of the manufacturer. For example, acetate calks are used primarily by builders to replace earlier calking made of oils, artificial rubber, and polyvinyl acetate.

Wall Sheathing

In Unit 26 the four most common types of sheathing—wood, plywood, fiberboard, and gypsum board—were discussed. While these materials are still used, more and more builders are choosing sheathings designed specifically for energy conservation. Each sheathing material has certain advantages and disadvantages that must be considered in making the best selection for each job. Fig. 57-5. Some of the most common of the energy-conserving sheathing materials include the following:

Styrofoam® insulation can be used for both wall sheathing and the exterior of basement walls. Fig. 57-6. Styrofoam® for

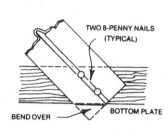

STUD

2, 8D NAILS

57-7. *Sheet metal let-in braces must be used with Styrofoam® sheathing*

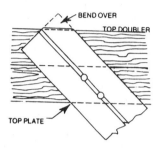

BEND OVER

TOP DOUBLER

TOP PLATE

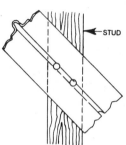

STUD

TWO 8-PENNY NAILS
(TYPICAL)

BEND OVER

BOTTOM PLATE

walls can reduce energy use by as much as 14 percent. Applying the insulation to the outside of the foundation adds another 10 percent savings, giving a total of 24 percent.

Styrofoam® sheathing adds no structural strength to the wall; so sheet metal let-in (wind) braces are needed Fig. 57-7. These heavy T-shaped sheet metal braces are similar to the wood type described on Pages 370 to 373. The sheet metal braces can be installed diagonally across either the outside or the inside of a wall. A shallow, narrow kerf (gain) must be cut into each stud to insert the sheet metal brace. The brace is nailed in place with the cross section of the T nailed to the outside of the studs. Siding must be nailed directly to the studs using long, thin siding nails because the porous Styrofoam® cannot serve as a nailing base.

Foil-faced kraft fiber ply sheathing is made of high-quality long-fibered plies that are pressure-laminated together using a special water-resistant adhesive. Reflective aluminum foil is fastened to each side. Fig. 57-8. The sheathing is a highly effi-

57-8. *Installing foil-faced kraft fiber plies as sheathing. Note that the material is being stapled to the studs.*

cient energy-saving material. It has good structural strength and therefor can be used without corner bracing.

Polystyrene is another type of insulating board. It is made from expanded polystyrene. It has many of the same advantages and disadvantages as Styrofoam®.

Foil-faced urethane insulating board is a panel made of a core of plastic (urethane) covered on both sides by aluminum foil.

STANDARD HOME CONSTRUCTION

Many things can be done to improve the energy efficiency of the standard home. Some of these things can be done at small cost; others are expensive.

● Limit the size of the living unit (in square feet) by redesigning the living quarters. If each area can be made more flexible (have several uses), the original cost of construction can be reduced and energy efficiency improved.

● Use the correct kind of sheathing. Installation of energy-efficient sheathing instead of standard materials provides great fuel savings.

● Add proper insulation in walls, ceilings, and floors and around the perimeter of the building.

● Add proper, well-fitted storm windows and doors and double-glazed windows.

● Install the correct type of weather stripping around all doors and windows.

● Add calking around all windows and doors where the frame meets the siding and around all other possible openings. Calking should be applied around an outside faucet, between the basement and floor framing, the drip cap and siding, the corners formed by siding, where pipes and wires penetrate the ceiling, below an unheated attic or chimney, and where chimney and masonry meet siding.

● Add attic ventilation. Attics with a ceiling vapor barrier must be ventilated with one square foot of vent area for each 300 square feet of ceiling. Attics without a ceiling vapor barrier must be ventilated

with one square foot of vent area for each 150 square feet of ceiling. The vents may be in the roof, soffit, or gable. One of the most efficient venting systems is that installed in the ridge of the roof.

● Insulate duct work. Duct work may be insulated by adding insulating tape around each joint and furnace opening. Also, batt insulation completely covering all exposed duct work can be installed.

ENERGY-SAVING (CONSERVATION) HOME

An energy-saving home was sponsored by the U.S. Department of Housing and Urban Development and built in Little Rock, Arkansas. It has become known as the *Arkansas Plan or Arkansas Home.* Fig. 57-9. Basically, this home involved the redesigning of wall and ceiling construction to allow for 6 inches of insulation in walls and 12 inches in ceilings. The design also called for smaller windows equipped with storm windows, metal exterior doors with insulation cores and magnetic weather stripping, power attic ventilation, humidifier, dehumidifier, and air filtration equipment. Fig. 57-10, 11, and 12. Specifications include the following, most of which are shown on Fig. 57-10:

1. Vapor barriers covering walls, ceilings, and floors.

2. Windows that have storm windows and are calked.

3. Metal exterior doors 1¾" thick with urethane core.

4. Attic space with power roof ventilators and eave vents.

5. Inspection catwalk.

6. Wiring and piping installed to permit correct placement of insulation.

7. Humidifier.

8. Dehumidifier.

9. Air filtering device.

10. Sill and window flashing.

11. Wall studs (2" × 6") spaced 24" on center.

12. Window headers.

13. Proper structural support.

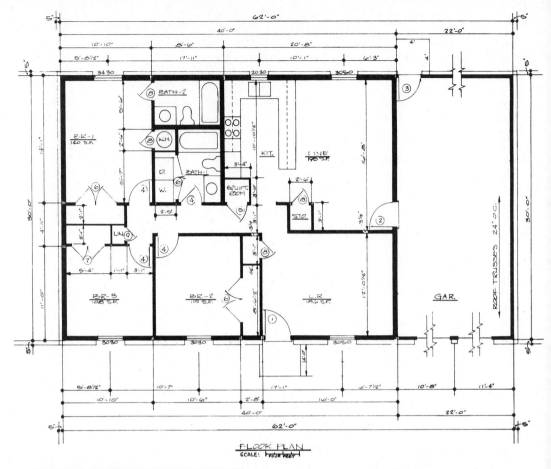

57-9 Floor plan. Most rooms have one small window. The bathrooms have no windows. All plumbing runs are on inside walls. Full insulation, therefore, can be installed in the outside walls.

	DOOR SCHEDULE
1.	3·0 x 6-8 x 1¾" THERMA-TRU
2.	2-8 x 6-8 x 1¾" THERMA-TRU
3.	2-8 x 6-8 x 1¾" H.C. EXT.
4.	2-6 x 6-8 x 1⅜" H.C. INT.
5.	2-6 x 6-8 x 1⅜" H.C. INT. W/LOUVER
6.	2-6 x 6-8 x 1⅜" H.C. INT. PAIR
7.	2·0 x 6-8 x 1⅜" H.C. INT. PAIR
8.	2·0 x 6-8 x 1⅜" H.C. INT.
9.	1-6 x 6-8 x 1⅜" H.C. INT.
10.	9-0 x 7-0 O.H. GAR. DOOR

14. Tie plates and drywall back-up clips.

15. Insulated ducts.

16. Centrally located climate conditioning equipment.

17. Partition walls of 2" × 3" studs.

18. Construction strength that is soundly engineered.

The design of the house is unique in many ways. The home has 1200 square feet of living space. It was designed with windows and doors on the front and rear walls only, with no windows or doors on either the left or the right side of the house. Fig. 57-8. As a result, there is only one small window in each major room of the house. Fig. 57-9. This necessitates the use of air conditioning in almost every location in the country in which the house is built.

DESIGN FEATURES OF THE ARKANSAS ENERGY CONSERVATION HOME

(Illustrative Perspectives)

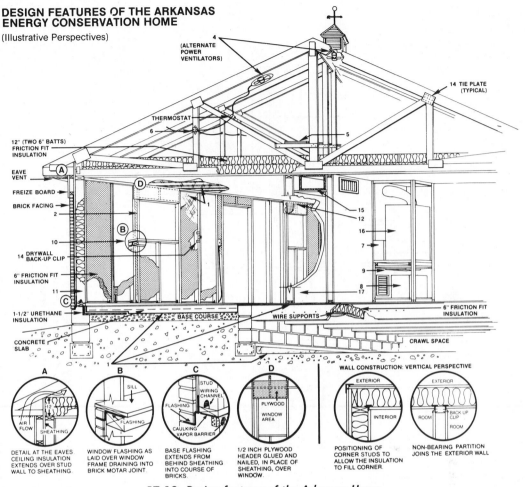

4 (ALTERNATE POWER VENTILATORS)

14 TIE PLATE (TYPICAL)

THERMOSTAT

5

6

12" (TWO 6" BATTS) FRICTION FIT INSULATION

EAVE VENT

FREIZE BOARD

BRICK FACING

2

10

14 DRYWALL BACK-UP CLIP

6" FRICTION FIT INSULATION

11

1-1/2" URETHANE INSULATION

CONCRETE SLAB

15
12

16

7

9

8
17

6" FRICTION FIT INSULATION

BASE COURSE

WIRE SUPPORTS

CRAWL SPACE

A — AIR FLOW / SHEATHING / DETAIL AT THE EAVES CEILING INSULATION EXTENDS OVER STUD WALL TO SHEATHING.

B — SILL / FLASHING / WINDOW FLASHING AS LAID OVER WINDOW FRAME DRAINING INTO BRICK MOTAR JOINT.

C — STUD / WIRING CHANNEL / FLASHING / CAULKING VAPOR BARRIER / BASE FLASHING EXTENDS FROM BEHIND SHEATHING INTO COURSE OF BRICKS.

D — PLYWOOD / WINDOW AREA / 1/2 INCH PLYWOOD HEADER GLUED AND NAILED, IN PLACE OF SHEATHING, OVER WINDOW.

WALL CONSTRUCTION: VERTICAL PERSPECTIVE

EXTERIOR / INTERIOR / POSITIONING OF CORNER STUDS TO ALLOW THE INSULATION TO FILL CORNER.

EXTERIOR / ROOM / BACK UP CLIP / ROOM / NON-BEARING PARTITION JOINS THE EXTERIOR WALL

57-10 *Design features of the Arkansas Home.*

This home reduced the heat loss 66 percent over a typical standard home of exactly the same size (1200 square feet) built to FHA standards. Both homes were built without a basement, but with a crawl space for heating ducts. Conservation in the *energy-saving house* was achieved as follows:

● Windows/doors—32.2 percent. Achieved by using metal doors with urethane cores, magnetic weather stripping, and double-glazed windows. The windows were limited to an area of not more than 8 percent of the floor area.

● Flooring—18.2 percent. Achieved with floor insulation over crawl space and around the slab perimeters.

● Duct loss—18.4 percent. Achieved by using special insulation on the heating and air-conditioning ducts.

● Walls—7.7 percent. Achieved by using 2 × 6 studs on 24" centers, instead of the standard construction 2×4 studs on 16" or 24" centers. Six inches of insulation (R-19) were put into the walls.

● Ceiling—7.5 percent. Achieved by using 12 inches (R-38) of insulation in the ceiling and by adding attic fans.

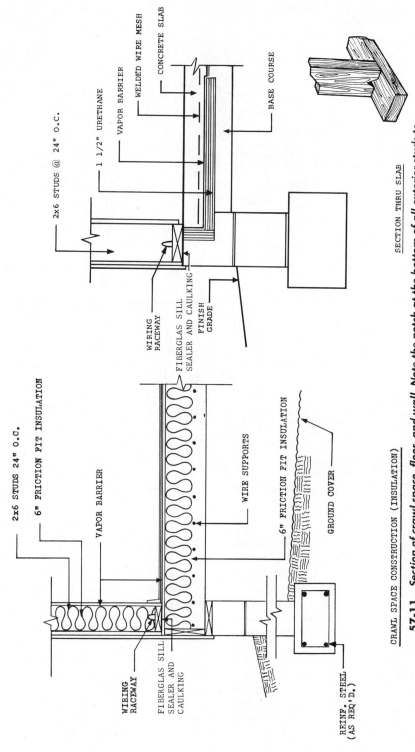

2x6 STUDS @ 24" O.C.

1 1/2" URETHANE

VAPOR BARRIER

WELDED WIRE MESH

CONCRETE SLAB

BASE COURSE

WIRING RACEWAY

FIBERGLAS SILL SEALER AND CAULKING

FINISH GRADE

SECTION THRU SLAB

2x6 STUDS 24" O.C.

6" FRICTION FIT INSULATION

VAPOR BARRIER

WIRE SUPPORTS

6" FRICTION FIT INSULATION

GROUND COVER

WIRING RACEWAY

FIBERGLAS SILL SEALER AND CAULKING

REINF. STEEL (AS REQ'D.)

CRAWL SPACE CONSTRUCTION (INSULATION)

57-11. Section of crawl space, floor, and wall. Note the notch at the bottom of all exterior studs to form a wiring raceway. This eliminates the need to drill holes in the studs and allows full insulation in the walls without any obstruction.

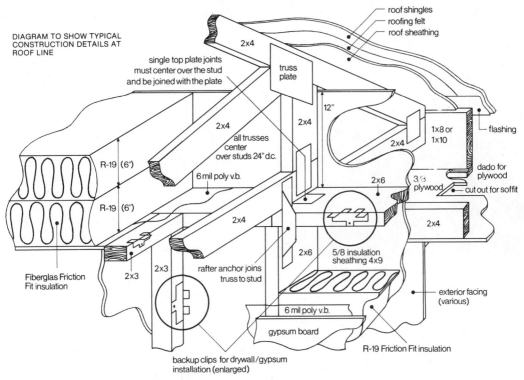

DIAGRAM TO SHOW TYPICAL
CONSTRUCTION DETAILS AT
ROOF LINE

roof shingles
roofing felt
roof sheathing

2x4

truss
plate

single top plate joints
must center over the stud
and be joined with the plate

12"

2x4

1x8 or
1x10

flashing

2x4

2x4

all trusses
center
over studs 24" d.c.

dado for
plywood

R-19 (6")

6 mil poly v.b.

2x6

3/8
plywood

cut out for soffit

R-19 (6")

2x4

2x4

Fiberglas Friction
Fit insulation

2x3

2x3

rafter anchor joins
truss to stud

2x6

5/8 insulation
sheathing 4x9

exterior facing
(various)

6 mil poly v.b.

gypsum board

R-19 Friction Fit insulation

backup clips for drywall/gypsum
installation (enlarged)

57-12. *Typical construction details at the roof line.*

● Infiltration—15 percent. Achieved by using friction fit batts of polyurethane vapor barriers for walls, floors, and ceilings instead of the batts with integral vapor barriers used in standard construction.

SOLAR HOMES

There are two basic types of solar energy systems: active and passive. An active system makes use of large solar collecting panels, usually on the roof of the house. Fig. 57-13. A heat storage unit is located apart from the collector. Heated liquid or air is transferred from the collector to the storage unit and then to the living spaces by mechanical means. A complex system

57-13. *A home equipped with an active solar energy system.*

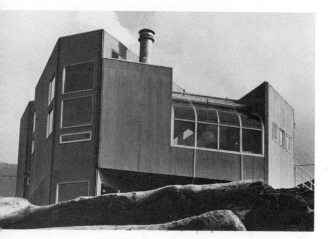

57-14. A passive solar energy home.

collectors installed in or on the roof. The builder must use a roof design that accommodates these panels. Many of the collectors are built into the roof itself. Others are free-standing units on or over a waterproof roof. In a few homes, the solar collectors are designed as a unit separate from the house itself.

The two major types of heat storage are the liquid storage unit and the rock storage unit. Providing the space needed for these units can be a problem for the builder. The builder must also provide a piping system from the collector to the storage unit and on to the heating and air conditioning units.

The homeowner must decide whether an active solar energy system is worth its cost. The combined cost of the equipment, construction, and servicing of the system may be greater than the energy savings over a ten-to-fifteen-year period.

of pumps, fans, pipes, and ducts is needed to transfer the heat. With a passive system, the building itself collects and stores solar heat. Fig. 57-14. Large south-facing windows act as collectors. Thick walls or floors store the heat. Heat is transferred to living spaces through natural, rather than mechanical, means.

The Active System

For the builder of an active solar system, the major problems are posed by the collector and the storage unit. Most active solar energy systems specify flat plate

The Passive System

A passive solar system is less expensive because it has little or no mechanical equipment. Many homeowners, therefore, prefer the passive system.

There are five basic elements in a complete passive system.

● *Collector.* The collector is the large glass or plastic area through which sunlight enters the structure. Fig. 57-15. The

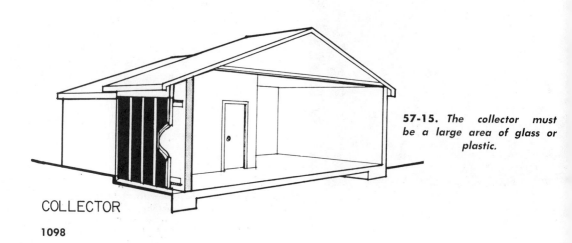

57-15. The collector must be a large area of glass or plastic.

COLLECTOR

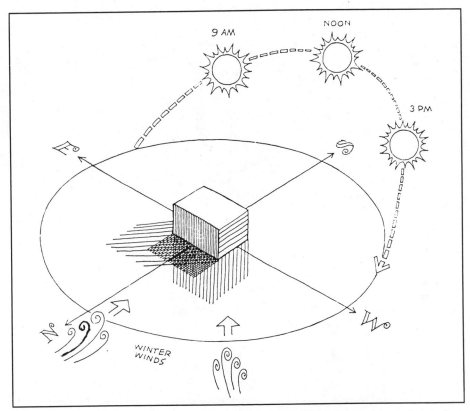

57-16. *To make the best use of sunshine and to protect against cold north winds, solar collectors should face within fifteen degrees of true south.*

collector must face true south, or at least within fifteen degrees either way of south. The collectors must not be shaded by buildings or trees from 9:00 A.M. to 3:00 P.M. each day during the heating season. Fig. 57-16.

• *Absorber.* The sunlight passing through the collector is absorbed by the hard, darkened surface of the storage element. The surface is dark because dark colors absorb more heat.

• *Storage.* The storage element is usually either a thick masonry structure (such as a wall, floor, or room divider) or large containers of water. Both water and masonry store heat effectively. The storage unit is sometimes referred to as a thermal mass.

The absorber and the storage unit are often the same wall or floor. Absorber applies only to the exposed surface. *Storage* refers to the material below or behind the surface. Fig. 57-17.

• *Distribution.* In a passive system, heat circulates from the storage unit to the living spaces by three methods: *conduction, convection,* and *radiation.* Fig. 57-18. *Conduction* refers to heat moving through solid objects. For example, a spoon placed in a cup of hot coffee conducts heat up its handle. *Convection* refers to the motion of hot air or water. Passive solar systems make use of the fact that warm air rises to create currents to heat rooms. *Radiation* is heat moving as a wave similar to light. A

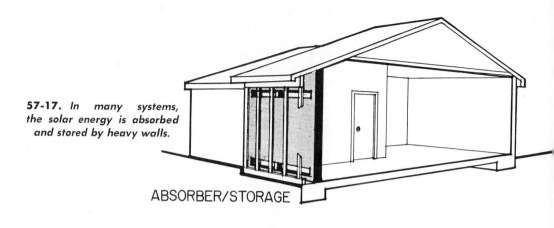

57-17. *In many systems, the solar energy is absorbed and stored by heavy walls.*

ABSORBER/STORAGE

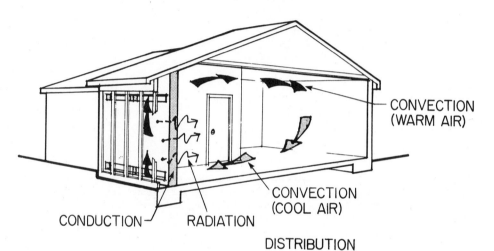

CONVECTION
(WARM AIR)

CONVECTION
(COOL AIR)

CONDUCTION

RADIATION

DISTRIBUTION

57-18. *Most passive solar energy systems use natural means of distributing the heat.*

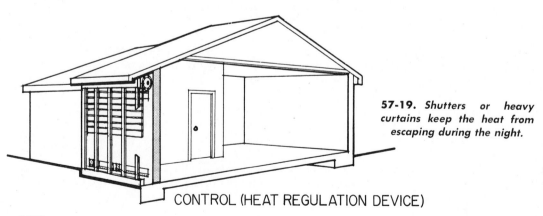

57-19. *Shutters or heavy curtains keep the heat from escaping during the night.*

CONTROL (HEAT REGULATION DEVICE)

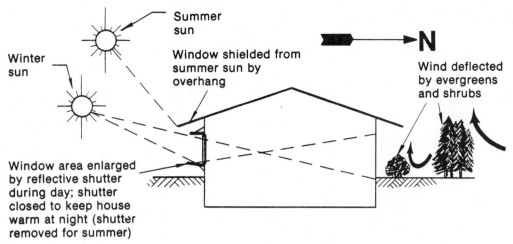

Winter sun

Summer sun

Window shielded from summer sun by overhang

N

Wind deflected by evergreens and shrubs

Window area enlarged by reflective shutter during day; shutter closed to keep house warm at night (shutter removed for summer)

57-20. *Note how the angle of the sun changes from season to season.*

warmed surface inside a home will emit heat (infrared radiation) that will travel towards cooler areas. In addition to these natural means, fans and ducts are sometimes used to distribute heat.

• *Controls.* Controls include the insulating methods used to prevent heat loss back through the collector at night. Fig. 57-19. Curtains and shutters are normally used. Roof overhang or awnings that shade the collector in the summer and prevent overheating are also controls. Fig. 57-20. Sometimes fans, vents, and dampers are part of the control system.

DIRECT GAIN SYSTEMS

There are three types of passive solar energy systems: direct gain, indirect gain, and isolated gain. With a direct gain system, sunlight heats the living spaces directly. Sunlight enters through south-facing window and heats the air in the room. It is also absorbed in masonry walls or floors, which are typically 4″ to 6″ thick. At night, as the air in the room begins to cool, the walls and floor radiate the stored heat into the room. Fig. 57-21.

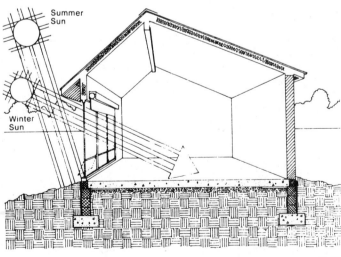

Summer Sun

Winter Sun

57-21. *In the direct gain system, sunlight passes through the window to heat the air in the room and to be absorbed by the walls and floors.*

INDIRECT GAIN SYSTEMS

With an indirect gain system, sunlight does not directly heat the living spaces. Rather, it heats a thermal mass that in turn heats the living spaces. A Trombe wall is a good example of an indirect gain system. Fig. 57-22. It is a solid masonry wall 8″ to 16″ thick on the south side of the house. It is often made of concrete blocks that have been filled with concrete. Glass or plastic glazing is mounted about 4″ in front of the wall's surface. Sunlight passes through the glazing and heats the air between the glazing and the masonry wall. Some of the heat is absorbed by the wall, passes through it by conduction, and radiates into the room on the opposite side of the wall. There are usually also vents in the Trombe wall at the top and the bottom. As the air between the wall and the glazing is heated, it rises and passes through the upper vents into the opposite room. This warm air is replaced by cooler air being drawn through the lower vents from the room. As the air is warmed by the sun, it continues to rise and pass into the room. This convection current is used to heat the living spaces opposite the Trombe wall.

The water wall is a variation of the Trombe wall. Fig. 57-23. In place of a masonry wall, large containers of water are located between the living spaces and the

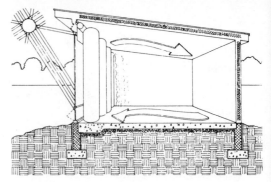

57-23. A water wall serves the same purpose as the Trombe wall in an indirect gain system.

glazing. The water absorbs the sunlight and slowly radiates the heat into the living area.

ISOLATED GAIN SYSTEMS

An isolated system is one built apart from the area being heated. One such system is a solar greenhouse built as part of a house. Solar heat is collected through the greenhouse glazing. It is absorbed and stored by masonry or containers of water. The stored heat is transferred to the house by convection and radiation. Fig. 57-24.

Insulation

A high insulation level is important to a solar energy system. Fig. 57-25. Basements should have two inches of rigid insulation between the wall and the ground, and if possible, three inches above the grade line. If a permanent wood foundation is used, the stud cavities should be well insulated. For slab-on-grade construction, the footings and walls should be insulated along the entire perimeter from two to four feet from the exterior walls. In cold climates, it is a good idea to insulate under the entire slab with rigid insulation.

Walls should be built with 2 × 6 studs, which will increase the amount of insulation that can be installed. No wiring or plumbing runs should go through outside walls. Use wiring raceways at the bottom of the walls. Fig. 57-11.

57-22. The indirect gain system places a heat-storage mass between the windows and the room. This system reduces the problems of excessive glare and too rapid heating that often occur with direct gain systems.

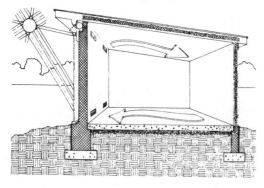

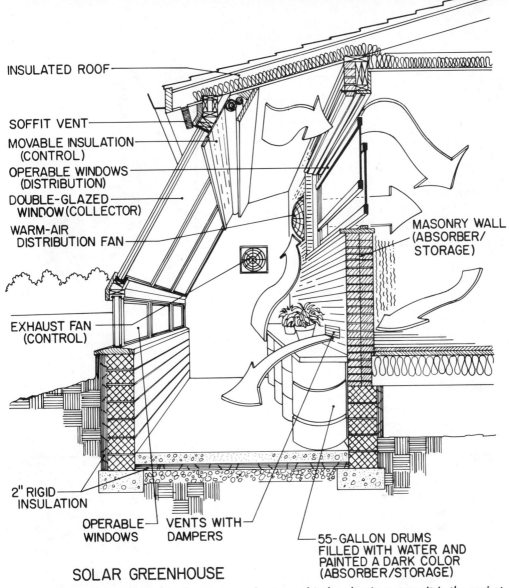

INSULATED ROOF

SOFFIT VENT

MOVABLE INSULATION
(CONTROL)

OPERABLE WINDOWS
(DISTRIBUTION)

DOUBLE-GLAZED
WINDOW (COLLECTOR)

WARM-AIR
DISTRIBUTION FAN

EXHAUST FAN
(CONTROL)

MASONRY WALL
(ABSORBER/
STORAGE)

2" RIGID
INSULATION

OPERABLE
WINDOWS

VENTS WITH
DAMPERS

55-GALLON DRUMS
FILLED WITH WATER AND
PAINTED A DARK COLOR
(ABSORBER/STORAGE)

SOLAR GREENHOUSE

57-24. *The solar greenhouse is the most popular type of isolated gain system. It is the easiest to add to an existing home.*

There should be adequate insulation in the roof and the attic. Batts at least 5 ½" thick should be installed between 2 × 10 rafters. The floor of the attic should also be well insulated. A vapor barrier of 6-mil polyethylene should be carefully overlapped with the plastic sheet used in the wall. In hot, humid climates the roof or attic must be ventilated in the summer.

All exterior doors and windows should be insulated and should fit tightly. Double or triple pane insulated glass should be used. Fig. 57-26. Calking and weatherstripping are extremely important throughout the house.

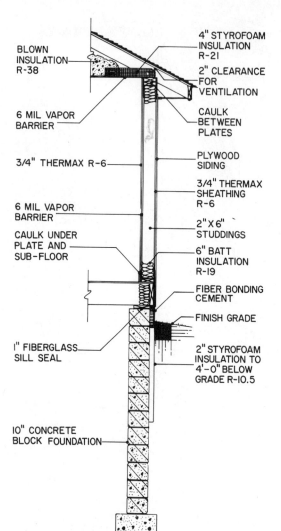

BLOWN
INSULATION
R-38

4" STYROFOAM
INSULATION
R-21

2" CLEARANCE
FOR
VENTILATION

6 MIL VAPOR
BARRIER

CAULK
BETWEEN
PLATES

3/4" THERMAX R-6

PLYWOOD
SIDING

3/4" THERMAX
SHEATHING
R-6

6 MIL VAPOR
BARRIER

2"X6"
STUDDINGS

CAULK UNDER
PLATE AND
SUB-FLOOR

6" BATT
INSULATION
R-19

FIBER BONDING
CEMENT

FINISH GRADE

I" FIBERGLASS
SILL SEAL

2" STYROFOAM
INSULATION TO
4'-0" BELOW
GRADE R-10.5

10" CONCRETE
BLOCK FOUNDATION

57-25. *A typical wall section for a passive solar energy house.*

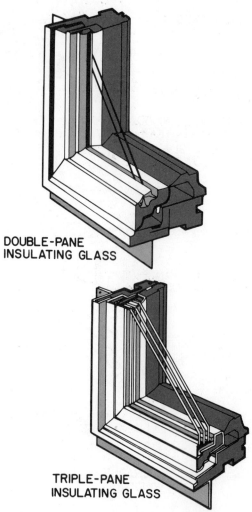

**DOUBLE-PANE
INSULATING GLASS**

**TRIPLE-PANE
INSULATING GLASS**

57-26. *Double- or triple-pane windows are necessary for an effective passive solar energy system.*

QUESTIONS

1. What is the most important factor influencing internal temperature?

2. Name three considerations in planning an energy-efficient house.

3. What type of plastic is primarily used for bathtubs and showers?

4. Name three types of energy-efficient wall sheathing.

5. Name four things that can improve the energy efficiency of a standard house.

6. How does the framing in the Arkansas Home differ from that in a standard house?

7. What are the two main types of solar energy systems?

8. Name the five basic elements in a passive solar energy system.

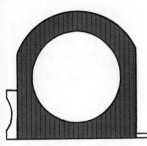

APPENDIX

Interior Finishes

Type of Finish	Preparation of Surface	Sealers and Primers	Application of Finish	Characteristics
General Use (Lumber, Boards, Planks and Paneling)				
Stains —Water Stains	Sand or scrape wood to flat, clean surface; for some hardwoods, fill pores with filler before sanding; if new wood, should already be flat and clean.	One coat of sealer often brushed into softwoods before finishing to reduce depth of penetration between springwood and summerwood; also used to reduce prominence of grain; no primers required for stains and most clear finishes.	Water stains are brushed onto surface; one coat is usually sufficient.	Easy to apply; water can raise wood grain making sanding necessary after stain has dried; fabric dyes in water are typical water stain.
—Spirit Stains			Spirit stains are brushed onto wood rapidly and evenly; one coat is usually sufficient.	Quick drying, so apply evenly; little time to restroke spots; little tendency to raise wood grain.
—Oil Stains			Oil stains are brushed onto wood, all strokes made along the grain; one coat is usually sufficient; sometimes used under varnishes.	Penetrate well; do not raise grain; dry slowly so easy to distribute evenly.
Clear Finishes —Waxes			Two coats of paste wax can be applied over shellac in methanol sealer; surface damage repaired by spot cleaning with mineral spirits and applying more wax.	Low-gloss finish; rarely used because of excellent performance of synthetic varnishes; great refinish problems if wax allowed to penetrate wood
—Synthetic Varnishes (such as polyurethane)			Three coats of synthetic varnish brushed onto surface (two coats if filler used); can be polished with wax but not necessary.	Varying gloss finishes, synthetics give hard, tough finish, resistant to oil, water, and alcohol; dries quickly by reaction with moisture in air.
—Shellacs			Two coats of shellac are brushed onto surface (used as methanol solution); patched in spots using methanol remover and applying more shellac.	Brittle finish; water spots easily; not recommended because synthetic varnishes do better job.
—Boiled Linseed Oil			Two coats of boiled linseed oil are spread evenly onto surface using brush or rag; 24-hour drying period is required between coats.	Seldom used; long drying time required, so surface susceptible to marking.

(Continued on next page)

Type of Finish	Preparation of Surface	Sealers and Primers	Application of Finish	Characteristics
Paints —Alkyd Enamels	Sand or dust wood to ensure good paint adhesion.	Apply alkyd or oil-based primer with good enamel holdout for an enamel undercoat.	Two coats of alkyd-based enamel are brushed or rolled onto surface.	High- or semi-gloss finish; resistant to solvents; good color retention; optimum results using enamel undercoater.
—Latexes		Recommend application of latex primer; not absolutely necessary.	Two coats of latex are brushed or rolled onto surface; may be used without primer but primer recommended.	Full range of glosses available; high- or semi-gloss recommended for kitchens and bathrooms; fast drying; little yellowing; easy clean-up with water; no solvent vapor; spot touch-up without patch effect; reduced fire hazard.
Plywood				
Stains	Hardwood and softwood plywood are usually factory sanded on face to be finished; require no surface preparation other than filling and sanding surface blemishes.	Priming and edge sealing not necessary; sealer will subdue grain contrast if dark stain preferred.	One coat of combined wax and stain is applied; after a few minutes wipe with rag to desired shade, then apply a coat of self-polishing wax and buff surface. Stains can be used alone as described under General Use.	Wax penetrates wood making surface unsuitable for refinishing.
Clear Finishes —Synthetic Varnishes	Fill nail holes with tinted filler, sand smooth, and spot prime.	Apply coat of sealer.	Brush two coats of varnish onto prepared surface; semi-gloss varnish often applied over flat varnish.	
—Blond Finish			Apply one coat of interior white undercoat (thinned so grain shows through) under one coat of flat synthetic varnish.	Easy and inexpensive; offers features of synthetic varnishes.
—Waxes			Wax systems commercially available but not recommended; if used, follow manufacturer's instructions closely.	Not recommended because wax imbedded in wood fibres cannot be removed; this wax will interfere with alternative refinishing.
Paints —Alkyd Enamels	Sand or scrape wood to smooth, clean surface.	Apply one coat of enamel undercoat.	Apply either one coat of undercoater tinted to finish color and one coat of alkyd enamel or two coats of alkyd enamel (for better gloss).	Good washable finish; checking and cracking may become problem.
—Latexes		Apply one coat of latex-based check-retardant primer.	Apply two coats of latex paint.	Primer effectively eliminates cracking and checking on new wood.

Interior Finishes (Continued)

Type of Finish	Preparation of Surface	Sealers and Primers	Application of Finish	Characteristics
			Particle Board	
Stains	Surfaces available smooth or porous depending on type of particle board; porous surfaces must be filled and sanded; some boards available with resin-impregnated fibrous sheet applied in factory.	A primer or sealer may be necessary to isolate additives from the finish; finish a scrap to determine if primer or sealer is needed.	Apply stains as described under General Use.	Interesting decorative effects can be achieved with stains or clear finishes; shape and color contrast of wood particles is emphasized.
Clear Finishes			Apply any clear finish as described under General Use.	
Paints			Apply alkyd or latex paints as described under General Use.	Paint is most common finish for particle board.
			Hardboard	
Paints —Alkyd Enamels —Latexes	Nailheads should be countersunk and puttied, or treated with anti-corrosive primer; surface should be smooth and clean.	Apply one coat of primer unless hardboard has been factory primed.	Apply alkyd enamels or latexes as described under General Use.	Hardboard is usually painted because it has no grain, or natural characteristics to be emphasized.
			Floors (Hardwood and Softwood)	
Pigmented Stains	Sand floor with power sander, first cut across the grain and successive cuts with the grain; after sanding, vacuum off dust; apply finish as soon as possible; some hardwoods, such as oak, require filler before sanding.	Not required unless sealer used alone as finish for industrial occupancies.	One coat of stain (oil stain if synthetic varnish is used) is brushed onto surface.	Stains change color of the wood and emphasize grain; often used under varnish.

(Continued on next page)

Interior Finishes (Continued)

Type of Finish	Preparation of Surface	Sealers and Primers	Application of Finish	Characteristics
Floors (Hardwood and Softwood)				
Clear Finishes				
—Synthetic Varnishes (Polyure-thane)			Two coats of synthetic varnish are brushed onto stained or filled floor; bubbles should be removed by brushing back area lightly.	Varying degrees of gloss; extremely wear-resistant; no wax needed for protection; difficult to spot refinish.
—Floor Sealers			Apply sealer across grain as a spray or with a brush; bluff with steel wool and vacuum off dust; apply second coat of sealer with the grain; sealed floor may be waxed or varnished for greater gloss.	Inexpensive; provides protection against water damage and warping in industrial occupancies.
—Shellac			Three coats of shellac are applied by brush, following wood grain.	Easily repaired; water staining becomes chronic problem, so shellac not recommended.
Paints				
—Alkyd Floor or Deck Enamels		Sealer may be required for softwood floors unless enamel is self-sealing.	Three coats of floor or deck enamel brushed onto surface.	Good wear resistance; wide color range; softwood and hardwood floors are easily painted.

Exterior Finishes

Type of Finish	Preparation of Surface	Sealers and Primers	Application of Finish	Characteristics
General Use (Lumber, Siding and Panel Products)				
Penetrating Finishes —Preservatives			Brush one coat of water-repellent preservative onto surface.	Imparts mildew and decay resistance; some protection from ultraviolet light destruction depending on pigment content; easy to maintain.
			Brush a solution of fungicide onto surface; best method is to add fungicide to finish.	May tend to leach out with rain.
—Stains	Sand or scrape wood to flat, clean surface; if new wood, no preparation required.	None required.	Brush latex or oil-based stain onto surface; working area should be small enough to maintain a wet edge; one or two coats applied according to manufacturer's recommendations.	Easy to apply; attractive finish for even rough surfaces; easy to maintain; choice of semi-transparent or opaque finish.
—Madison Formula			Brush one coat of Madison Formula onto new wood.	Semi-transparent oil-based stain intended for western red cedar; imparts water repellency and mildew and decay resistance; contains wax which leads to refinish problems
—Oil			One coat of oil, applied by brush.	Not recommended; oil remains tacky on surface, collecting insects and dirt.
Surface Finishes —Solvent-Based Paints (Including Alkyds)	Sand or scrape off badly deteriorated finish; if new wood, no preparation is required.	Apply knot sealer or shellac to knots and pitch streaks; apply oil-based primer to new wood.	Painting should be done in dry weather with temperatures above 45° F; paint following sun around house, staying one side behind sun; apply one or two coats (depending on color) by brush.	Alkyd paints overcome blistering and excessive chalking associated with traditional oil-based paints; only white traditional oil-based paint is recommended; alkyd paints are recommended if color is required; alkyd trim paints should be used above masonry; solvent-based paints with other synthetic resins are available.
—Latex Paints		Apply knot sealer or shellac to knots and pitch streaks; apply oil-based primer as undercoat or special latex primer over previous coats of oil-based paints.	Follow application instructions above for solvent-based paints; apply two coats of latex paint by brush over primer.	Easy to apply; adhere well to damp surface; dry rapidly; easy equipment clean-up; chalk-resistant; slower to erode than oil-based paints; thinner surface film, therefore less leveling off of surface irregularities

(Continued on next page)

Exterior Finishes (Continued)

Type of Finish	Preparation of Surface	Sealers and Primers	Application of Finish	Characteristics
General Use (Lumber, Siding and Panel Products)				
Synthetic Varnishes	For refinishing scrape off loose flaking materials, sand area, wash wood surface and stain bleached areas to match the rest of the wood; if new wood, no preparation required.	None required.	Three coats of synthetic varnish are applied by brush to new wood; fewer coats required for refinishing.	Refinishing required about every two years; finish allows penetration of ultraviolet light which degrades film and wood surface; results in darkening of wood color.
Penetrating Finishes				
—Stains			Immersing shingles or shakes in stain (semi-transparent or opaque) is best technique; refinishing done with a brush.	Rough surface of shingles and shakes readily absorbs stain.
—Madison Formula	Same as for General Use.	None required.	Apply as for other pigmented stains.	Madison Formula specifically formulated for western red cedar.*
—Preservatives			Pentachlorophenol (5% to 10% solution) may be added to stain; alternatively, wood may be pressure treated with preservatives.	Preservatives desired but not absolutely necessary for western red cedar.
Surface Finishes				
—Alkyd Paints	None required.	None required.	Two coats of specially formulated shingle and shake alkyd paint applied by brush.	Flat finish; covers all but extreme surface irregularities.
—Latex Paints		Specially formulated sealer required; factory primed shingles and shakes are available.	Two coats of latex paint applied by brush.	Covers all but extreme surface irregularities.

*Pour 1 gallon of mineral spirits into an open-top 5-gallon can. Heat 1 pound of paraffin and 2 ounces of zinc stearate in the top of a double boiler and stir until the mixture is uniform. Pour this into the mineral spirits, stirring vigorously. This should be done outside to avoid the risk of fire. Add ½ gallon of pentachlorophenol concentrate 10:1 and 3 gallons of boiled linseed oil to the cooled solution. Stir in 1 pint of burnt sienna color-in-oil and one pint of raw umber color-in-oil until the mixture is uniform. One gallon covers 400-500 square feet on a smooth surface and 200-250 square feet in a sawn-textured surface.

INDEX